Lecture Notes in Artificial Intelligence   11831

Subseries of Lecture Notes in Computer Science

Series Editors

Randy Goebel
   University of Alberta, Edmonton, Canada
Yuzuru Tanaka
   Hokkaido University, Sapporo, Japan
Wolfgang Wahlster
   DFKI and Saarland University, Saarbrücken, Germany

Founding Editor

Jörg Siekmann
   DFKI and Saarland University, Saarbrücken, Germany

More information about this series at http://www.springer.com/series/1244

Jia-Fei Hong · Yangsen Zhang ·
Pengyuan Liu (Eds.)

Chinese Lexical Semantics

20th Workshop, CLSW 2019
Beijing, China, June 28–30, 2019
Revised Selected Papers

 Springer

Editors
Jia-Fei Hong
National Taiwan Normal University
Taipei, Taiwan

Pengyuan Liu
Beijing Language
and Culture University
Beijing, China

Yangsen Zhang
Beijing Information Science
and Technology University
Beijing, China

ISSN 0302-9743   ISSN 1611-3349   (electronic)
Lecture Notes in Artificial Intelligence
ISBN 978-3-030-38188-2   ISBN 978-3-030-38189-9   (eBook)
https://doi.org/10.1007/978-3-030-38189-9

LNCS Sublibrary: SL7 – Artificial Intelligence

This Springer imprint is published by the registered company Springer Nature Switzerland AG
The registered company address is: Gewerbestrasse 11, 6330 Cham, Switzerland

Preface

The Chinese Lexical Semantics Workshop (CLSW 2019) was the 20th conference in the series established in 2000. CLSW has been held in different Asian cities, including Beijing, Hong Kong, Taipei, Singapore, Xiamen, Hsin Chu, Yantai, Suzhou, Wuhan, Zhengzhou, Macao, Leshan, and Chia-Yi. In 2019, CLSW returned to Beijing, one of its founding sites, to celebrate its 20th anniversary. Over the years, CLSW has become one of the most important venues for scholars to report and discuss the latest progress in Chinese Lexical Semantics and related fields, including theoretical linguistics, applied linguistics, computational linguistics, information processing, and computational lexicography. CLSW has significantly impacted and promoted academic research and application development in the aforementioned fields, and acted as one of the most important meetings in Asia for Chinese Lexical Semantics.

CLSW 2019 was hosted by the Beijing Information Science and Technology University (BISTU), China. This year, 254 papers were submitted to the conference, setting the highest record to date. All submissions went through a double-blind review process by at least two independent reviewers. Of all the submissions, 95 (37%) were accepted as oral papers and 79 (31%) as poster papers. Among accepted papers, selected English papers are included in this Springer LNAI series. They are organized in topical sections covering all major topics of lexical semantics, semantic resources, corpus linguistics, and natural language processing.

On behalf of the Program Committee, we are most grateful to conference chairs: Yong-Sheng Wang (President of BISTU) and Min Zhang (Soochow University), as well as honorary members of the Advisory Committee: Shiwen Yu (Peking University), Cheng Chin-Chuan (University of Illinois), Benjamin Ka-Yin T'sou (City University of Hong Kong), and other members of the Advisory Committee for their guidance in promoting the conference. We sincerely appreciate the invited speakers for their outstanding keynote talks. Also, we would like to acknowledge the chairs of the Organization Committee: Yang-Sen Zhang (BISTU), Shui-Cai Shi (TRS and BISTU), Ning Li (BISTU), and Jiun-Shiung Wu (National Chung Cheng University); the Beijing Information Science and Technology University; and student volunteers for their tremendous contribution to this event.

Our gratitude goes to the Program Chairs: Juan-Zi Li (Tsinghua University), Wei-Dong Zhan (Peking University), and Jia-Fei Hong (National Taiwan Normal University), as well as all the Program Committee members and reviewers for their time and efforts in reviewing all submitted papers. We are pleased that the accepted English papers are published by Springer as part of their *Lecture Notes in Artificial Intelligence* (LNAI) series and are indexed by EI and SCOPUS.

Last but not least, we thank all the authors and attendees for their scientific contribution and participation, which made CLSW 2019 a successful event.

June 2019

Yangsen Zhang

Organization

Program Chairs

Juanzi Li	Tsinghua University, China
Jia-Fei Hong	National Taiwan Normal University, Taiwan
Weidong Zhan	Peking University, China

Organizing Chairs

Yangsen Zhang	Beijing Information Science and Technology University, China
Shuicai Shi	Beijing Information Science and Technology University, TRS, China
Ning Li	Beijing Information Science and Technology University, China
Jiun-Shiung Wu	National Chung Cheng University, Taiwan

Program Committee

Xiaojing Bai	Tsinghua University, China
Baobao Chang	Peking University, China
Ruoyu Chen	Beijing Information Science and Technology University, China
Bo Chen	Hubei College of Arts and Sciences, China
Yinlin Chen	National Cheng Kung University, Taiwan
Minghui Dong	Institute for Infocomm Research, Singapore
Renfeng Duan	National Taitung University, Taiwan
Ruixue Duan	Beijing Information Science and Technology University, China
Hong Gao	Nanyang Technological University, Singapore
Shu-Ping Gong	National Chiayi University, Taiwan
Peirong Guo	National Chiayi University, Taiwan
Shu-Kai Hsieh	National Taiwan University, Taiwan
Donghong Ji	Wuhan University, China
Yuxiang Jia	Zhengzhou University, China
Minghu Jiang	Tsinghua University, China
Yuru Jiang	Beijing Information Science and Technology University, China
Peng Jin	Leshan Normal University, China
Arhens Kathleen	The Hong Kong Polytechnic University, Hong Kong, China
Huei-ling Lai	National Cheng Chi University, Taiwan

Contents

Lexical Semantics

Applications of Natural Language Processing

Lexical Resources

Corpus Linguistics

Lexical Semantics

Spatiality and Its Semantic Consequence of the Quantitative Expression "一CCN" in Mandarin Chinese

Jing Sun[✉] and Yulin Yuan

Department of Chinese Language and Literature,
Peking University, Beijing 100871, China
jsunc@pku.edu.cn

Abstract. In Mandarin Chinese, reduplicative classifiers can be used to express plurality, which is closely related to the spatiality of nouns. This paper takes "一CCN" (short for the construction "One + Classifier + Classifier + Noun") as an example to analyze the spatial presentation of "一CCN" and discusses its semantic consequence. Different from other quantificational expressions such as "Number + Classifier + Noun", "一CCN" does not express a small quantity and "一CC" cannot fall in the scope of quantitative adverb 只 and 才. By analyzing the contexts in which it can or cannot appear, and comparing it with 很多/许多 ("many"), we conclude that the "many" meaning of "一CCN" is not its literal but inferential meaning.

Keywords: Reduplicative classifiers · Spatiality · Plurality · Focus

1 Introduction

In [1], the function of classifiers is to make count nouns enumerable by individualizing and classifying them. According to [2], reduplicated classifiers in Mandarin Chinese reflect people's perspective of observing the objects in the word. [3] proposes two human cognitive modes: synoptic mode and sequential mode. Synoptic mode adopts a stationary distal perspective point with global scope of attention; while sequential mode adopts a moving proximal perspective point with local scope of attention. For example:

(1) a. There are some houses in the valley.

b. There is a house every now and then through the valley.

According to [3], the different grammatical forms such as plural number, agreement, and the determiner some in (1a) reflect the synoptic mode; while in Example

(This study is supported by the project of "Research on the Lexical Semantic Knowledge Representation and Its Computing System within the Framework of Chinese Parataxis Grammar" (Project No. 18JJD740003), and "The Study of International Chinese Language Teaching Resources and Wisdom Education Platform Based on 'Internet +'" (Project No.18ZDA295). We hereby express our sincere thanks.).

© Springer Nature Switzerland AG 2020
J.-F. Hong et al. (Eds.): CLSW 2019, LNAI 11831, pp. 3–13, 2020.
https://doi.org/10.1007/978-3-030-38189-9_1

(1b), the singular case of be verb, the quantified adverbial phrase "every now and then" and the spatial preposition "through" all indicate the view of the sequential pattern adopted by the speaker. We argue that the construction of "Numeral + Classifier + Noun" and the reduplication of classifiers ("一CCN") in Mandarin Chinese can also express the two cognitive modes by which the speaker observes the outside world.

(2) a. 桌　上　放　了　五　本　书。
　　　 desk　up　put　ASP　five　CL　book
　　　 "There are five books on the table."
　　b. 桌　上　放　了　一　本　本　书。
　　　 desk　up　put　ASP　one　CL　CL　book
　　　 "Books are put on the desk one by one."

The number-natural property of bare nouns in Mandarin Chinese is discussed in [4], the bare noun 书 can be singular or plural. However, here 书 in both 五本书 and 一本本书 is plural. Comparing with (2b), (2a) adopts a synoptic mode to observe the books on the table, and the number of the books is bounded and known, while in (2b), a sequential mode is adopted by the speaker, and the total number of the books is unknown and cannot be inferred by the listener. [5] argues that all types of classifiers can be reduplicated, and reduplicative classifiers are true plural markers in Mandarin Chinese.

In [6], the most typical entities occupy space, objects belonging to different categories can presents the characteristic of big/small, high/low, dispersion/reunion. According to [7], spatiality is the essential feature of nouns, [7] takes the ability combing with different classifiers as a criterion texting the degree of spatiality of nouns. Nouns which can co-occur with individual classifiers (such as 个) have the strongest spatiality, while those can only combine with kind classifiers (such as 种, 类) have the weakest spatiality.

Based on the studies above, this article mainly discusses the spatial representation of nominal construction "一CCN", here C stands for classifiers and N for head nouns. We propose that, different from the more frequently used structure "Number + Classifier + Noun", the plurality expressed by "一CCN" is constrained by its spatiality. In order to investigate this issue, we firstly observe the three spatial relationships established by "一CCN" and other noun phrases under the framework of [8], then discuss the semantics of "一CCN" and its co-occurrence with quantificational adverb 只 and 才, and we argue that, though "一CC" expresses "many", comparing with other quantitative expression such as "Number + Classifier + Noun" and 很多/许多 ("many"), the distribution of "一CCN" is more limited, and its inherent spatial characteristics is an important causative factor.

2   Spatiallity of "一CCN"

2.1   "一CC" as an "Dispersion Marker"

In [9], Cantonese classifiers are be regarded as "count markers", and a count marker can mark the countability of nouns on syntactic levels. The classifier 本 in Cantonese or "piece(s)" in English are the examples of count markers given by [9]. Comparatively,

"一CC" can be seen as a "dispersion marker", since no matter the noun phrases are count or mass, their donation are always dispersed in space. For example:

(3) a. 一　张　张　脸
    one   CL   CL   face
    "faces (existing side by side in a certain space)"

   b. 一　滴　滴　水
    one   CL-drop   CL-drop   water
    "many   drops   of   water   (existing   side   by   side   in   a   certain   space,   or
falling from a container drop by drop)"

[10] argues that the extension of all common nouns in Mandarin Chinese are mass, and the mass/count distinction reemerges at some phrasal level: some classifiers make a noun count, while others keep them mass. In (3), the classifier 张 and 滴 can be seen to make the noun 脸 and 水 countable respectively. Both 脸 and 水 are dispersed in space, and a moving proximal perspective is adopted by the speaker.

Besides, duplicative classifiers also have topological variability in referentiality:

(4) a. 一　颗　颗　星星
    one   CL   CL   stars

    "many stars"

   b. 一　颗　颗　药丸
    one   CL   CL   pills
    "many pills"

星星 and 药丸 are dotted-like objects, the duplication of the classifiers 一颗颗 encodes their dispersion in space. However, intuitively speaking, the distance between stars is much longer than pills in the real word. This reflects that the denotation of reduplicated classifiers is topologically variable.

2.2   Three Spatial Relationships

In [8], a spatial representation has two functions: (1) providing a place for an entity; (2) establishing spatial relationships between entities. Entities providing a place can be seen as referents, and entities occupying space as targets. For example, in the sentence 桌子上放着电脑 ("There is a computer on the desk"), here 桌子 ("desk") is the referent and 电脑 ("computer") is the target[1]. However, we do not think the boundary between the two functions of spatial representation is rigid. In fact, the referent's providing a place for an entity has already implicated the establishment of a certain spatial relationship. For "一CCN", it individualizes and disperses the denotation of the nouns, like the common noun 电脑 in the example given above, the donation of "一CCN" also can establish spatial relationships with other larger entities, and they can form a relationship

---

[1] In [3], "targets" and "referents" are called as "primary objects" and "secondary objects" respectively.

of contact, interior and dispersion. In the theory of [8], the three spatial relationships can be marked by different kinds of postpositions (Abbreviated as PostP in the following examples), which are listed in Table 1:

Table 1. Three spatial relationships and their representation

spatial relationship	the space provided by the referents	postpositions
contact	surface space	上
interior	inside space	里/内/中
dispersion	out space	前/后/外/下/东/西/南/北

2.2.1 Contact Relationship

Examples of contact relationship are shown below:

(5) 阳光   在   地毯   上   投下   一   幅   幅   图案。
    sunshine   ZAI carpet PostP.up cast   one CL   CL   pictures
    "Sunshine casts many pictures on the carpet."

(6) 泥土   上   长着   一   个   个   白嫩的鲜蘑菇。
    soil   PostP.up grow   one   CL   CL   fresh white mushrooms
    "There are many fresh white mushrooms growing on the soil."

(7) 他   从   轿   底   取出   了   个   灯笼,
    He   from   sedan chair bottom   take out ASP   CL   lattern,

上面   画   着   一   朵   朵   盛开的梅花。
    surface   draw   ASP   one   CL   CL   blooming Plum Blossoms
    "He took a lantern from the bottom of the sedan chair, and on the surface of the lantern, many blooming plum blossoms were painted."

一幅幅图案, 一个个白嫩的鲜蘑菇 and 一朵朵鲜红的梅花 in the examples are the instances of "一CCN" which can be regarded as the targets, and 地毯, 泥土, 灯笼 are the larger objects providing surfaces for them. In (5) and (6), the postposition 上 attached to the noun 地毯 and 泥土 indicates a contact relationship between referents and targets; however, in (7), there is no postposition added on 灯笼. It is the local word 上面 that indicates the contact relationship. Both the postpositions and the local word cannot be deleted:

(8) *阳光在地毯投下一幅幅图案。

(9) *泥土长着一个个白嫩的鲜蘑菇。

(10) *前面的轿夫，自轿底取出了个灯笼, 画着一朵朵鲜红的梅花。

In (5), 在 and 上 form a frame preposition. Semantically, 上 makes the flatness of the 地毯 more prominent. Besides, it can also be used independently without 在, like (6). The local word 上面 in (7) anaphoricaly refers to 灯笼,and provides 一朵朵鲜红梅花 a surface. Interestingly, the spatial relationship would become obscured if these spatial markers are deleted.

2.2.2   Inside Relationship

Inside relationship indicates the objects denoted by "一CCN" is in the inner space of a larger entity:

(11) 市　　　内　　　有　　一　　座　座　戏院　　和　　一　家　家　餐馆。

city　PostP.inside　have　one　CL CL　theater　and　one CL CL restaurant

"There are many theatres and restaurants in the city."

(12) 指挥舰　　　里，　　　一　　台　台　现代化的　仪器,

warship　PostP.inside　one　CL　CL　modern　　equipments

不断　　　　　　变换　　着　　各种　　数据。

continuously　change　ASP　various　data

"There are many modern equipments changing various data continuously in the warship."

(13) 这　　里　　　　　有　一　个　个　　俄国　马厩,

this　PostP.inside　have　one　CL　CL　Russia　stable,

里面　养　　着　　性子　　刚烈的　骏马

inside　raise　ASP　temperament　fiery　horses

In the examples above, 市, 指挥舰, and 这里 provide inner spaces for the donation of "一CCN". Like the contact relationship shown above, the postpositions 内, 里, and the local word 里面 cannot be deleted either, especially for (12) and (13):

(14) *市有一座座戏院和一家家餐馆。

(15) *指挥舰，一台台现代化的仪器不断变换着各种数据。

(16) 这有一个个俄国马厩，养着性子刚烈的草原骏马。

In (11), 市 and 内 have been lexicalized into a word. 市 cannot be used alone in modern Mandarin Chinese, thus (14) is ungrammatical. Different from市, 指挥舰 in (12) is an independent noun, but 里 cannot be deleted either. In (13), 里 occurs together with the determiner 这. The meaning of the sentence would be slightly different if 里 is deleted. In fact, 这有 in (16) expresses an possessive rather than a spatial relationship. Besides, the local word 里面 in (13) anaphoricaly refers to 一个个俄国马厩. 这里 provides a space for 一个个俄国马厩 which also provides spaces for 性子刚烈的草原骏马. This reflects the multiple inside relationship between objects.

2.2.3   Dispersion Relationship

The donation of "一CCN" can form a dispersion relationship with another objects. For example:

(17) 隔　　　着　　铁路　　的　一　家　家　店铺。
separate   ASP   rail way   DE   one   CL   CL   shops
"Many shops across the railroad."

(18) 庄园　　的　　四周，　　矗立　着　一　所　所　住宅。
manor   DE   surroundings,   stand   ASP   one   CL   CL   residence
"There are many residences around the manor."

(19) 往事历历，　　　　像　一　幅　幅　画　展示　在　她　眼　前。
the past remain fresh,   like   one   CL   CL   picture   show   ZAI   she   eyes   before
"The past events are so fresh that they are like many pictures before her eyes."

In dispersion relationship, targets are located outside the referents. In (17), 一家家店铺 is the target and 铁路 is the referent. In (18), the referent 庄园 is surrounded by the targets 住宅, and 住宅 is coerced to express plural. (19) is an unrealistic sentence, 一幅幅画 is metaphorically referred to 往, 事 and the dispersed pictures are compared to the experiences in the different stages of life. Here 一幅幅画 is the target while 她 is the referent. The referents and the targets given above are all contrapositive in a certain space.

From the analysis above, we can see that there are various ways to express the spatial relationship of contact, inside and dispersion, and that the well-formedness of a sentence containing "一CCN" can be influenced by postpositions or local words, which reflects a strong relationship between spatiality and the plurality expressed by "一CCN".

2.3   The Absence of Reference

In natural languages, referents often do not clearly appear. However, they can be inferred by pragmatic enrichment. For example:

(20) 他们　的　军队　打到　了　长乐街，
they   DE   army   attack   ASP   Changle Street
驻　了　十　天，　留下　了　一　堆　堆　猪毛　和　鸡毛。
station   ASP   ten   day,   leave   ASP   one   CL   CL   pig hair   and   chicken feather
"Their army attacked the Changle Street. After they stayed for more than ten days, many piles of pig hairs and chicken feathers had been left behind."

In the example above, the target 猪毛和鸡毛 appears without an explicit referent, however, it can be deduced with the help of general knowledge:

(21) a. ……, [在大街上] 留下了一堆堆猪毛和鸡毛。 (contact relationship)

"many piles of pig hairs and chicken feathers had been left behind [on the street]."

b.……, [在老百姓的院子里] 留下了一堆堆猪毛和鸡毛。 (inside relationship)

"many piles of pig hairs and chicken feathers had been left behind [in people's yards]."

c. ……, [在长乐街的对面] 留下了一堆堆猪毛和鸡毛。 (dispersion relationship)

"many piles of pig hairs and chicken feathers had been left behind [across the Changle street]."

From the examples above, the referents in the blanket can have different spatial relationships with the target. In the framework of [3], the referents have known properties that can characterize the targets' unknowns. When the targets are perceived, the referents will be more grounded. On the other hand, it can be inferred that the ellipsis of the inferents permits the listener space for meaning constructing. The construction of the spatial relationship above is limited by people's experience and the ability of pragmatic inference. The donation of 一CCN will be more exact through the establishment of the spatial relationship with a larger object, and the informativeness of the utterance can be improved as well.

3 The Co-occurrence of "一CCN" and Quantitative Adverbs "只" & "才"

According to [11], the quantitative adverb 只 ("only") in Mandarin Chinese can help to express "exclusiveness" or "small quantity". For example:

(22) a. 我　只　　学过　　[英语]$_f$。
      I　only　studied　English.
      "I have only studied English."

b. 教室　　　里　　只　　有　　[三四个]$_f$　人。
   classroom PostP. inside only have three or four people
   "There are only three or four people in the classroom."

For (22a), "English is the only language I have studied". For (22b), the quantity (three or four) of the people in the classroom is small in the view of the speaker. According to [12], 只 is sensitive to the focus and can be seen as a "focusing operator". In the sentences above, 英语 and 三四个 are the focuses (marked by []$_f$) and are constrained by 只 respectively. Meanwhile, different focuses may lead to different truth conditions and interpretation of the same sentence.

(23) a. 教室　　　里　　　　只　　有　　三　　张　　　[桌子]f (没有其他东西)。

classroom　PostP. inside　only　have　three　Cl　　desks

"There are only three desks in the classroom (no other things)."

　　b. 教室　　　里　　　　只　　有　　[三　张]f　桌子 (不是四张或五张)。

classroom　PostP.inside　only　have　three　Cl　　desks

"There are only three desks in the classroom (not four or five)."

(24) a. 江　　面　　只　　有　　两　　只　[小　　兵船]f(没有其他类型的船只)。

river　surface　only　have　two　Cl　small　warship

"There are only two small warships on the river (not other kinds of ships)."

　　b. 江　　面　　只　　有　　[两　只]f　小　　兵船 (不是三只或四只)。

river　surface　only　have　two　Cl　small　warship

"There are only two small warships on the river (not three or four)."

For "Number + Classifier + Noun" constructions (三张桌子, 两只小兵船), both "Number + Classifier" and "Noun" can be the focus. When the stress falls on "Number + Classifier" (三张, 两只), 只 expresses a small quantity, and when it falls on the nouns (桌子, 小兵船), 只 expresses exclusiveness. However, this is not the case for "一CCN". "一CC"can never be the focus constrained by 只. 只 are only related to the "N".

(25) a. 墙　　　上　　　只　　有　　一　　个　　个　[小　　黑　　洞]f。

wall　PostP. up　only　have　one　CL　CL　small　black　hole

"There are only many [small black holes]f on the wall."

　　b. *墙　　　上　　　只　　有　　[一　　个　　个]f　小　　黑　　洞。

wall　PostP. up　only　have　one　CL　CL　small　black　hole

"There are only [many]f small black holes on the wall."

(26) a. 屋　　　里　　　只　　有　　一　　排　　排　　　[字幕]f。

room　PostP.inside　only　have　one　CL-row　CL-row　captions

"There are only many rows of [captions]f in the room."

　　b. *屋　　　里　　　只　　有　　[一　　排　　排]f　字幕。

room　PostP.inside　only　have　one　Cl-rows　Cl-rows　captions

For both (25) and (26), the stress can only fall on the nouns (小黑洞, 字幕). Here "只" expresses exclusiveness. The two examples have a common interpretation: "in a certain place, there is no other object except for those in the sight of the speakers". This can be further demonstrated by adding the following sentences:

(27) a. 墙上只有一个个小黑洞，没有窗户。

"There are only small black holes on the wall, no windows."

b. *墙上只有一个个小黑洞，太少了。

"There are only many small black holes on the wall, the quantity is too small."

(28) a. 屋里只有一排排字幕，看不到人。

"There are only captions in the room, no people there."

b. *屋里只有一排排字幕，还缺一些。

"There are only many captions in the room, it's not enough."

An important difference between "Number + Classifier + Noun" and "一CCN" is, the former can be used to express a large or small quantity, which is context-depended, while "一CC" can never be constrained by 只 to express a small quantity.

According to [7] and [13], "一CC" expresses "many", thus it contradicts with the "small quantity" meaning of "只". This can be further tested by its co-occurrence with 才. [14] argues that,the meaning of 才 is identical to 只 in some cases. "才" can express a small quantity, but not exclusiveness, so "一CCN" cannot be bound by "才". For example:

(29) a.

墙	上	才	有	[三	个]$_f$	小	黑	洞,	不	够。
wall	PostP.up	only	have	three	CL	small	black	hole	not	enough

"There are only [three]$_f$ small black holes on the wall, it's not enough."

b. ?

墙	上	才	有	三	个	[小	黑	洞]$_f$,	没有	窗户。
wall	PostP.up	only	have	three	CL	small	black	hole	no	windows

"There are only three [small black holes]$_f$, no windows"

c. *

墙	上	才	有	一	个	个	小	黑	洞
wall	PostP.up	only	have	one	CL	CL	small	black	hole

"There are only many small black holes."

(30) a.

屋	里	才	有	[一	排]$_f$	字幕,	太	少	了。
room	PostP. inside	only	have	one	Cl-row	captions	too	few	ASP

"There are only [one row]$_f$ of captions, it's too few"

b. ?

屋	里	才	有	一	排	[字幕]$_f$,	其他什么也没有。
room	PostP.up	only	have	one	CL-row	captions	nothing else

"There are only one row of [captions]$_f$, nothing else."

c. *

屋	里	才	有	一	排	排	字幕。
room	PostP.up	only	have	one	CL-row	CL-row	captions

"There are only many captions in the room."

In the examples above, the stress can only fall on "Number + Classifier" (三个, 一排), rather than the noun (小黑洞, 字幕) and "一CC" (一个个, 一排排). Here, the "Number + Classifer" is in the scope of 才. For (29), "there are only three small black holes on the wall and the number is too small in the opinion of the speaker", and for (30), "there are only one row of captions in the room, it's not enough". 才 cannot constrain the nouns and "一CC", and this can be seen by 29 (b, c) and 30 (b, c).

According to [15] ,reduplicative classifiers are ambiguous and can express the meaning of "all", "many", "one after another", and "continuous". [7] argues that "many" is the core meaning of reduplicative classifiers. [13] holds a similar opinion, and puts it further. In the theory of [13], although the reduplicated classifiers expresses "many", they have no inherent quantificational force:their quantificational interpretation is context-depended, and is influenced by their syntactic distribution or quantificational adverbs (such as 都). Through the analysis of a large amount of corpus, [13] infers that, the minimum number of objects denoted by "一CCN" is 4. [5] regards reduplicated classifiers as the true plural markers in Mandarin Chinese, and they actually encodes abundant plutality.

In this article, we argue that "many" is not the literal but the inferential meaning of "一CCN". If the objects denoted by nouns are not located side by side in space, or if the speaker intends to judge the quantity rather than observes the individuals one by one, the meaning of "many" cannot be expressed by "一CCN". For example:

(31) a. 他   早上   一口 气   吃   了   五个   馒头。
    he   morning   one   breath   eat   ASP   five   steamed buns
    "He ate five steams buns without break in the morning."

b. *他   早上   一口 气   吃 了   一   个   个   包子。
    he   morning   one   breath   eat ASP   one   CL   CL   steamed buns
    "He ate many steamed buns without break in the morning."

(32) a. 她   父亲   写   过   许多/很多 本   书，   颇有名气。
    she   father   write   ASP   many   CL   book   very famous
    "Her father has written many books and is very famous."

b. *她   父亲   写   过   一   本   本   书，   颇有名气。
    she   father   write   ASP   one   CL   CL   book   very famous
    "Her father has written many books and is very famous."

In the examples above, the quantitative expression 十几个 and 许多/很多 express "many". However, they cannot be replaced by "一CCN". In (31), 馒头 has already been consumed and is beyond the sight of the speaker. (32) is a non-event sentence, 许多/很多 expresses the judgment of the speaker towards the number of books.

4 Conclusive Remarks

This paper attempts to explore the relationship between the plurality and spatiality of a nominal phrase. Reduplication is a frequent way to express plurality not only in Mandarin Chinese, but also in other Asian languages. It should be noted that, there are two forms of reduplicative classifiers in Mandarin Chinese: "CC" (short for "Classifiers + Classifiers) and "一CC". The former has more fixed meanings and distributions, while the latter has no definite quantificational force according to [13]. The usage of "一CCN" is limited by locality of the objects in the real world and the method of information processing adopted by the speakers, which make it different from the plural markers in Indo-European languages.

References

1. Bisang, W.: Classifiers in east and southeast Asian languages: counting and beyond. In: Gvozdanović, J. (ed.) Numeral Types and Changes Worldwide, pp. 113–185. Mouton de Gruyter, Berlin (1999)
2. Li, W.H.: The interaction between classifier reduplication and construction. Chin. Teach. World **3**, 354–362 (2010). (in Chinese)
3. Talmy, L.: Toward a Cognitive Semantics: Concept Structuring Systems, vol. 1. The MIT Press, Cambridge (2000)
4. Doetjes, J.: Count/Mass distinction across languages. In: Maienborn, C., Heusinger, K., Portner, P. (eds.) Semantics: An International Handbook of Natural Language Meaning, vol. 3, pp. 2559–2580. Mouton de Gruyter, Berlin and New York (2012)
5. Zhang, N.: Classifier Structures in Mandarin Chinese. De Gruyter Mouton, Berlin/Boston (2013)
6. Chen, P.: The tripartite-structure of temporal system in modern Chinese. Stud. Chin. Lang. **6**, 401–421 (1988). (in Chinese)
7. Li, Y.M.: The semantics of reduplication. Lang. Teach. World **1**, 10–19 (1996). (in Chinese)
8. Chu, Z.X.: Study on Spatial Phrases in Chinese. Peking University Press, Beijing (2010). (in Chinese)
9. Sybesma, R.: Markers of countability in the nominal domain in Mandarin and Cantonese. Essays Linguist. **35**, 234–245 (2007). (in Chinese)
10. Chierchia, G.: Plurality of mass nouns and the notion of "Semantic Parameter". In: Rothstein, S. (ed.) Events and Grammar, pp. 53–104. Springer Science + Business Media, Dordrecht (1998)
11. Lü, S.X.: 800 Words in Modern Mandarin Chinese. The Commercial Press, Beijing (1999). (in Chinese)
12. Yin, H.H.: The scalar usage and noun usage of the focus-sensitive operator "Zhi" (只). Lang. Learn. Linguist. Stud. **1**, 49–56 (2009). (in Chinese)
13. Xiong, Z.R.: Nominal phrases containing reduplicated classifiers. Lang. Teach. Linguist. Stud. **2**, 44–57 (2018). (in Chinese)
14. Wang, H.: "就" and "才". Chin. Lang. Learn. **12**, 35 (1956). (in Chinese)
15. Song, Y.Z.: Essays on Modern Chinese Grammar. Tianjin People's Publishing Company, Tianjin (1981). (in Chinese)

Difference and Analysis Between the Structures of "Shai(晒) + NP" and "Xiu(秀) + NP"

Cuiting Hu and Yanqiu Shao[✉]

Information Science School, Beijing Language and Culture University,
Beijing, China
1457167081@qq.com, yqshao163@163.com

Abstract. With the popularity of phrases such as "shai(晒) + 幸福(happiness)" and "xiu(秀) + 恩爱(loves)", more and more nouns or noun phrases are coming into the structures of "shai(晒) + NP" and "xiu(秀) + NP". In such a structure, the intuitive and cognitive perception is that " shai(晒) + NP" and " xiu(秀) + NP" express similar semantic connotations. But in the process of observing the corpus, we find that some nouns or noun phrases are unable to replace each other. With regard to this language phenomenon, we take the BCC Corpus of Beijing Language and Culture University as the research corpus, from which the relevant corpus is extracted, and discuss the similarities and differences between the two language structures from two aspects of word formation ability and collocation words by observing the collocation of index lines.

Keywords: shai(晒) + NP · xiu(秀) + NP · Semantic · Syntactic · Pragmatic

1 Introduction

We can see that "shai(晒)" and "xiu(秀)" in "shai(晒) + NP" and "xiu(秀) + NP" are verbs, such as "shai(晒) + 照片(photos)" and "xiu(秀) + 恩爱(loves)", which semantically means to show others some aspects of individual behavior and actions. This semantic concept is spread with the emergence of hot words on the Internet, and has been repeatedly mentioned and used by people. But the sixth edition of the Modern Chinese Dictionary has not yet included this meaning.

The semantics of "shai(晒)" and "xiu(秀)" as examined in this paper has great similarity in origin, and the semantic of "shai(晒)" in this paper comes from the transliteration of English "share", that is, to take out one's acquisition and beloved things to bask in the sun and share with others, which means to show, share or show off. While the semantic of "xiu(秀)" comes from English "show", which has many meanings, such as "make clear", "let somebody see something", "to point to something" and so on. In this sense, the two words are loanwords, but they have similar meanings and connotations, the so-called synonyms. This paper intends to examine the semantic differences between the structures of " shai(晒) + NP" and " xiu(秀) + NP" from the perspective of synonyms.

At present, there are different definitions of synonyms in linguistics circle, and abundant achievements have been made in synonym discrimination. Fu [1, 2] thinks that synonyms mean the same and similar meanings, simply speaking. In fact, it is to

© Springer Nature Switzerland AG 2020
J.-F. Hong et al. (Eds.): CLSW 2019, LNAI 11831, pp. 14–23, 2020.
https://doi.org/10.1007/978-3-030-38189-9_2

divide the constituent elements of word meaning into two parts, that is, it divides the phonetic form of words and the object they refer to into two parts. Jiang [3, 4] also points out that "synonyms have the same meaning in the same semantic field. They have common meanings, and the meanings are similar but different." Ge [5] has raised objections to this view. He believes that this definition of synonyms is not appropriate and should separate words of the same meaning and similarity. Zhou, Qu [6] and Fu [7] systematically summarize the nature and scope of modern Chinese synonyms. Ouyang [8], Pan [9] and An [10] have in-depth research on the analysis of synonyms. Guo [11] puts forward constructive suggestions on the analysis of synonyms in teaching Chinese as a foreign language, while Cui [12] advocates the discrimination of synonyms in Chinese as a foreign language from the perspective of color meaning of vocabulary.

In this paper, we take Fu Huaiqing's definition of synonyms. In this sense, "shai (晒)" and "xiu(秀)" are synonyms, they have similar semantic categories and they can enter the same language structure. Looking at these two structures carefully, we can see that both "shai(晒) + NP" and "xiu(秀) + NP" belong to the verb-object structure. Their structures are basically the same, the meanings of verbs are similar, both of them show some aspects of individuals to others semantically at the same time. We know that different forms, different function, so two structures should also be different in word formation ability and collocation. Referring to the interpretation of synonym analysis methods of Zhang and Liu [13], this paper hopes to explore the differences between the two aspects of semantic differences and syntactic combination.

2　Semantic Comparative Analysis

To observe semantics, we first start with dictionary explanation. The interpretation of these two words in the sixth edition of the Modern Chinese Dictionary is as follows:

"shai(晒)": ① The sun illuminates the object with heat: for example, the scorching sun makes the head dizzy.② Absorbing light and heat in the sun: for example, drying grain. ③ Ignoring and neglecting of metaphor.
"xiu(秀)"[1]: it refers to the heading and flowering of crops; "xiu(秀)"[2]: it is used to describe people, such as beautiful appearance, smart, excellent and excellent people.

From these interpretations, there is not the meaning we want, but considering the evolution of words, combining with the English words "share" and "show" mentioned in the introduction and the dictionary definition of the two words, we can find that the part of speech of "shai(晒)" in the dictionary has always been a verb and has not changed. These meanings can be found in Shuowenjiezi. "shai(晒)" in the ancient prose means drying the collections to prevent mold. Therefore, the meaning of "shai(晒)" in "shai(晒) + NP" is metaphor from the original meaning, which is meant to take things out and put them in the sunshine, so its original meaning is that the sun shines heat on objects called "shai(晒)". For example, "晒照片(share photos)" is to expose photos to people, but the "sunshine" is not the true sense of sunshine but through some social networks or platform media, and this process is like being placed in the sun, so there is a certain correlation between these two meanings.

The "xiu(秀)" is different, in which the original part of speech of the "xiu(秀)" is verb, and the other meanings are adjective or a noun, which is to describe the meaning of beauty and warmth. While "xiu(秀)" is a verb in the "xiu(秀) + NP", the possible speculation is that the part of speech of "xiu(秀)" has changed.

But combined with the pronunciation of "show", it is more reasonable to speculate that the "xiu(秀)" of "xiu(秀) + NP" and the "xiu(秀)" in Chinese are not the same word. "show" of "xiu(秀) + NP" is transliteration words, and the pronunciation is similar to the "xiu(秀)" in Chinese, so its meaning is basically not related to the meaning of "xiu (秀)" in Chinese. Therefore, the conclusion is different from "shai(晒)", the "xiu(秀)" of "xiu(秀) + NP" is transliteration word. The transliteration word is "xiu(秀)" and the "xiu (秀)" in Chinese is a homonym, that is, the pronunciation is similar and the meaning is not related. This is the connection and difference between "shai(晒)" and "xiu(秀)" in Chinese and "shai(晒) + NP" and "xiu(秀) + NP" in language structure.

According to above analysis, it can be seen that the "shai(晒)" and "xiu(秀)" in "shai(晒) + NP" and "xiu(秀) + NP" both have the meaning of showing and displaying in semantics, but there are also differences in semantic connotation: "shai(晒) + NP" emphasizes actions and displays things to others, such as "晒照片(share photos)" and "晒宝宝(share something of his baby)", without emphasizing the nature and attributes of the subject, while "xiu(秀) + NP" tends to focus more on the subject's outstanding advantages and initiative, and show to people, such as "秀身材(show figure)", "身材(show abdominal muscles)" etc., in which "秀腹肌(stature)" and "腹肌(abdominal muscles)" are known as advantages by the public. Therefore, the meaning of "xiu(秀)" comes from "show", which is related to performance. Most of the things that can be performed are worth seeing, emphasizing their advantages, while the use scope of "shai(晒)" is free and there are no restrictions, relatively speaking.

3 Differences in Syntactic Combinations

3.1 Difference from Word Combinations

In synonym discrimination, Fu [14] pointed out that it is appropriate that the description of the meaning of a word combines with the description of its combination ability. Lyons said: "There is an intrinsic relationship between the meaning of words and their distribution" [15]. For a long time, the meaning of words reflects objective things and the objective world. The combination and co-occurrence of words and words embodies the connection between words. Therefore, the examination of combination ability is an important aspect.

In the aspect of combination ability, we firstly searched for two collocation index lines from the BCC corpus, and then extracted the index lines that match the meaning of the two words "shai(晒)" and "xiu(秀)" discussed in this article, and set the threshold of the frequency greater than 5, and got the number of the two words after the statistical quantity so as to measure the formation ability of the two words.

We extracted 2375 "shai(晒) + NP" from the corpus, of which 113 met the requirements, accounting for 4.76%; and 4552 "xiu(秀) + NP", of which 227 were eligible, accounting for 4.99%. After the statistics, it can be seen the number of "xiu

(秀) + NP" is larger, so the formation ability of "xiu(秀)" is maybe stronger. However, because of the difference in cardinal number, the result may not be rigorous enough. Therefore, taking the 2000 index rows randomly, and the proportion of the two words is 5.65% and 11.35%, so it is basically concluded that the formation ability of "xiu (秀) + NP" is stronger, but the specific differences need to observe specific combination of the two words and do further analysis.

In words combination, "shai(晒) + NP" and "xiu(秀) + NP" are both verb-object structures, and most of them are predicates in sentences. However, the "NP" as the object is different. The following are the extractions of the Top30 of the two structures, as shown in Figs. 1, 2:

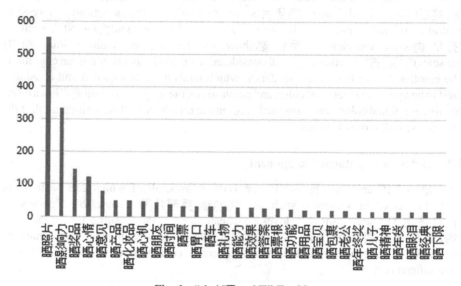

Fig. 1. "shai(晒) + NP" Top30

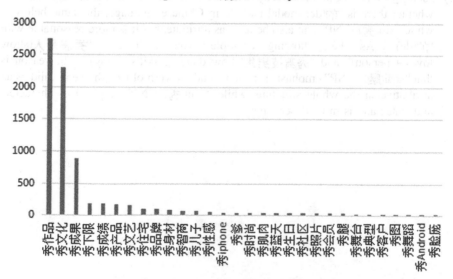

Fig. 2. "xiu(秀) + NP" Top30

As can be seen from the above two figures, the "NP" of "shai(晒) + NP" can be a word that describes things, such as "晒照片(share photos)", "晒奖品(share award)", or can be a noun for people, such as "晒宝宝(share something of his baby)", "晒老公(share something of her husband)", so the use scope of "NP" is relatively free, not too restrictive, structurally more inclusive. The structure of "xiu(秀) + NP" is different. Most of the "NP" are nouns that describes things, such as "秀照片(show photos)", "秀(show abdominal muscles)", "腹肌秀身材(show figure)" etc. Some structures can also be used to describe people. Comparatively speaking, the "NP" of "shai(晒) + NP" is more free and inclusive in the selection range.

In terms of emotional color, "shai(晒) + NP" and "xiu(秀) + NP" can be used both for themselves and for commenting on others. Among them "shai(晒) + NP", such as "晒奖品(share award)" and "晒照片(share photos)", emphases actions, expression neutral emotional color, basically no deviation. And while "xiu(秀) + NP", such as "秀肌肉(show muscles)", "秀甜蜜(show sweetness)", the main words "肌肉(muscles)", "甜蜜(sweetness)" are all considered to be good aspects of human cognition. The emotional color has a clear tendency, which tends to be self-owned and considers itself to be a positive emotional color, and emphasis on the subject, so "xiu(秀)" has a more positive emotional color. But when used to comment on others, both structures are slightly envious, scornful, or ridiculous.

3.2   Difference Syntactic Component

"shai(晒) + NP" and "xiu(秀) + NP" are verb phrases, all of which can be used as predicate center in sentence, such as "我们晒/秀照片(We share/show photos.)", "他们秀恩爱(They show loves.)"; can be the subject, such as "晒/秀恩爱死得快(Showing loves dies quickly.)"; can be independent sentences (especially in language communication); generally cannot be complement, but they have differences:

(1) "shai(晒) + NP" can be directly used as attribute of the noun, regardless of whether there is "的(de: modal particle in Chinese language, the same below)"; while "xiu(秀) + NP" can also be used as attribute, but it is more reasonable with "的(de)". As the following example, you can't say "秀恩爱人(show loves + person)" and "秀舞技男生(show dancing skills + boy)". The reason is that "shai(晒) + NP" emphasizes actions, and the verb-object phrase is valid as an attributive in the whole structure, while " xiu(秀) + NP" emphasizes characters and status and is more descriptive.

1.   晒视频（的）活动

Activity of showing video.

2.   晒年终奖（的）风潮

A wave of showing annual bonus.

3.   秀恩爱 的 人

The person of showing loves.

4.   秀舞技 的 男生

The boy of showing dance skills.

(2) "shai(晒) + NP" can be used as adverbial; "xiu(秀) + NP" is generally not adverbial. The adverbial before the predicate is an objective explanation of the state or trait itself when the actions occurs. From the corpus, relevant examples can be found, such as "晒月亮睡觉(sleeping with the moon)", that is, sleeping in the moonlight. "晒月亮" is the state of "sleeping", and the same thing is the "晒朝霞漫步(strolling in the sunshine)".

5.   晒月亮睡觉

Sleeping with the moon.

6.   晒朝霞漫步

Strolling in the sunshine.

"xiu(秀) + NP" generally does not do adverbial, and no relevant examples are found from the corpus, because "xiu(秀) + NP" is expressive and descriptive. Although it can be used as attribute, it is more reasonable to add "的(de)", which indicates nature and attributes. So it is not appropriate to modify verbs as adverbial.

As can be seen from the above examples and analysis, "shai(晒) + NP" can do more syntactic components, not only can be used as predicate central, subject, but also adverbial and attribute, whose syntactic function is relatively free. And while syntactic component of "xiu(秀) + NP" has certain limitations, and it cannot be directly used as attribute, which is more reasonable to add "的(de)" to be attribute.

3.3 Difference by Collocation

Fu Huaiqing [2] points out that when analyzing multiple pairs of synonyms and comparing in many aspects, the replacement test is the most commonly used method, so we use this method to observe the difference between the two configurations.

Observing the corpus, we can find that the same noun can enter the two structures "shai (晒) + NP" and "xiu(秀) + NP", that is, these collocations can be interchanged, for example:

7.   晒恩爱   秀恩爱

Showing loves.

8.   晒功能   秀功能

Showing functions.

9.   晒业绩   秀业绩

Showing achievements.

Observing these collocations, we find that the noun phrases can represent skillfulness, such as "功能(functions)"; they can be productive, such as "业绩(achievements)", both of which have positive color. There are many similar structures, but there are also many combinations that are not interchangeable (* mark, the same below), for example:

10.   晒包裹   *秀包裹

Showing package.

11.   晒食物   *秀食物

Showing food.

12.   晒账单   *秀账单

Showing bill.

Analysis of these examples shows that the "NP" in "xiu(秀) + NP" often does not combine with the nouns that represent person (not to exclude some special usage), so (1) does not make sense, which is contrary to the semantic features of the two words; "xiu(秀)" emphasizes semantics, while "shai(晒)" emphasizes actions, that is, "take something out and show something to people". The former emphasizes skill or advantage and characteristics. The latter is mostly for people to see, but "包裹(-package)" and "食物(food)" and "账单(bill)" are not dominant or skilled, so they cannot enter "xiu(秀) + NP". In addition, there is a situation in which they can enter the "xiu(秀) + NP", but they cannot enter the "shai(晒) + NP", for example:

13.   秀舞蹈   *晒舞蹈

Showing dance.

14.   秀球技   *晒球技

Showing ball skills.

15.   秀时装   *晒时装

Showing latest fashion.

In the above examples, the former can show that "xiu(秀)" is generally used instead of "shai(晒)". If you analyze it carefully, you will find that the words that can't enter "shai(晒) + NP" but can enter "xiu(秀) + NP" mostly represent things. Words that express skills or dominant features as mentioned above, such as "球技(ball skills)" and "舞蹈(dance)", which indicate acquired skills, highlight differences. So there are some differences between the two structures.

In addition, in the analysis of the collocation, the coordination of the pronunciation and rhythm should also be a place worth considering. In the process of word combination, there is a certain tendency to choose. Although we think that "晒恩爱(share loves)" can make sense, although the meaning can be expressed, it is a bit awkward in terms of language sense. And in the process of actual verbal communication, the language of the public habits tends to choose a structure such as "秀恩爱(show loves)". In addition that the structure is deeply rooted in the development of language, the coordination of speech is also one of the important reasons, which is also where we need to study further in the future.

4 Conclusion

Through semantic comparison and syntactic analysis, we can see that the two structures of "shai(晒) + NP" and "xiu(秀) + NP" have the same meaning under certain circumstances, and the combinations can be interchanged, but they cannot be interchanged in many cases because they have different semantic features: the object of "shai(晒)" can be a word that describes a person, or a word that describes something, while the object of "xiu(秀)" is mostly a word that describes something. "shai(晒)" emphasizes actions and shows things, and "xiu(秀)" not only has the meaning of showing things, but also has some requirements for the attributes of the subject, which often contains the meaning of skill or advantage. So these two constructions have both similarities and differences.

Research and discussion on the popular trend of "shai(晒) + NP" and "xiu(秀) + NP" are conducive to the exchange and development of language and culture, and have a positive impact on the better integration of foreign words into local languages and the standardization of foreign words. Especially in recent years, a variety of new words, popular words and loanwords have emerged endlessly. They play an important role in how to do a good job of blending these words with Chinese, how to protect the purity of Chinese, and how to make Chinese better accept new words and better serve the public. At the same time, in order to guide the application of language, the study of the meaning and usage of modern Chinese words can also better study vocabulary and word meaning and promote the systematic development of language research.

Acknowledgements. Thanks for National Natural Science Foundation of China (61872402), the Humanities and Social Science Project of the Ministry of Education (17YJAZH068), Science Foundation of Beijing Language and Culture University (supported by "the Fundamental Research Funds for the Central Universities") (18ZDJ03).

References

1. Fu, H.: Modern Chinese Lexicology. Peking University Press, Beijing (2004). (符淮青.现代汉语词汇学. 北京大学出版社, 2004). (in Chinese)
2. Fu, H.: Some problems in the study of synonyms. Chin. Lang. Lit. (03) (2000). (符淮青.同义词研究的几个问题. 中国语文, (03) (2000)). (in Chinese)
3. Jiang, S.: Some problems of synonyms and antonyms. J. Peking Univ. (Philos. Soc. Sci. Ed.) (03) (2015). (蒋绍愚.同义词和反义词的几个问题.北京大学学报. (哲学社会科学版, (03) (2015)). (in Chinese)
4. Jiang, S.: Some problems of synonyms and antonyms. J. Peking Univ. (Philos. Soc. Sci. Ed.) (3) (2015). (蒋绍愚.同义词和反义词的几个问题.北京大学学报. (哲学社会科学版), (3) (2015)). (in Chinese)
5. Ge, B.: On synonyms again. Hist. Philos. (01) (2003). (葛本仪.再. 论同义词.文史哲, (01) (2003)). (in Chinese)
6. Zhou, Y, Qu, J.: A Review of the research on the nature and scope of synonyms in modern chinese. Chin. Character Cult. (04) (2016). (周玉琨,曲娟现代汉语同义词性质和范围焦点研究述评. 汉字文化, (04) (2016)). (in Chinese)
7. Fu, N.: A review of the research on the definition of chinese synonyms in sixty years——from the perspective of meaning. J. Hainan Norm. Univ. (Soc. Sci. Ed.), (01) (2014). (付娜.六十年汉语同义词界定研究述评——从意义观的角度. 海南师范大学学报(社会科学版), (01) (2014)). (in Chinese)
8. Ouyang, L., Li, S.: The distribution of synonyms in modern chinese from the perspective of part of speech. J. Guangxi Inst. Educ (02) (2013). (欧阳丽文,李仕春. 从词性的角度看现代汉语同义词的分布规律广西. 教育学院学报, (02) (2013)). (in Chinese)
9. Pan, Q.: Discrimination of several groups of modern chinese synonyms. Young Writ. (05) (2013). (潘茜. 辨析几组现代汉语同义词. 青年文学家, (05) (2013)). (in Chinese)
10. An, C.: The definition and discrimination of synonyms in modern Chinese. J. Liaoning Univ. Eng. Technol. (Soc. Sci. Ed.) (05) (2011). (安春燕.现代汉语同义词的界定与辨析辽宁工程技术大学. 学报(社会科学版, (05) (2011)). (in Chinese)

11. Guo, X.: Analysis and teaching of synonyms in teaching Chinese as a foreign language. Educ. Sci. (01) (2008). (郭雪玫.对外汉语教学中的同义词辨.析与教学.教育科学家, (01) (2008)). (in Chinese)
12. Cui, H.: Analysis of the synonyms of Chinese as a foreign language from the perspective of the color meaning of vocabulary. Chin. J. (19) (2010). (崔海燕.从词汇的色彩意义看对外汉语同义词辨析语文学刊, (19) (2010)). (in Chinese)
13. Zhang, J., Liu, P.: A general method based on the analysis of corpus synonyms. J. PLA Univ. Foreign Lang. (6) (2005). (张继东,刘萍.基于语料库同义词辨析的一般方法. 解放军外国语学院学报, (6) (2005)). (in Chinese)
14. Fu, H.q.: Word meaning and word distribution. Chin. Lang. Learn. (01) (1999). (符淮青.词义和词的分布. 汉语学习, 1999(01)). (in Chinese)
15. Lyons, J.: Introduction to Linguistics. Foreign Language Teaching and Research Press, pp. 486–487 (2000). (Lyons, J. 语言学引论. 外语教学与研究出版社, pp. 486–487 (2000)). (in Chinese)

Relationship Between Discourse Notions and the Lexicon: From the Perspective of Chinese Information Structure

Shun-Hua Fu[✉]

School of Foreign Languages, Xiangtan University,
Xiangtan, Hunan Province, China
fshunhua@163.com

Abstract. This paper investigates the properties of information structure (IS), focusing on the relationship between the discourse notions (i.e. topic and focus) and the lexicon. The feature-based approach suggests that topic and focus are formal features which are numerated from the lexicon, active in the computational system and encoded in syntax. However, this approach has received some criticisms regarding the generation of IS, such as the violation of the inclusiveness condition, the employment of the "look-ahead" technique, the neglect of discourse pragmatic properties of topic/focus and the exclusion of the non-configurational IS. The key issue is that the semantic-syntactic properties of the lexical items in IS are relevant to the lexicon, but the pragmatic properties of IS are determined by factors out of the lexicon. IS should not be the result of pure syntactic computation, but a phenomenon of syntactic-pragmatic interface.

Keywords: Information structure · Topic · Focus · Lexicon · Feature-based approach · Syntactic-pragmatic interface

1 Introduction

Information Structure (IS) describes the way information is formally packaged in a sentence [1]. Topic and focus are the only IS primitives needed to account for all information structure phenomena [2, p. 8]. Pragmatically, topic conveys the old or given information, while focus conveys the new information unmentioned in the context[1]. Syntactically, IS might optionally exhibit word orders divergent from the "ordinary" word order. Take Chinese IS for instance.

---

[1] It should be noted that the given information is old materials including but not limited to topic and the new information is new materials including but not limited to focus [5, p. 683].

© Springer Nature Switzerland AG 2020
J.-F. Hong et al. (Eds.): CLSW 2019, LNAI 11831, pp. 24–36, 2020.
https://doi.org/10.1007/978-3-030-38189-9_3

(1)   Q: 你   喜欢   什么   水果？
      nǐ   xǐhuān   shénme   shuǐguǒ
      You   like   what   fruits
      'What fruits do you like?'
      A: [Topic 水果]，   我   喜欢   苹果。
              shuǐguǒ   wǒ   xǐhuān   píngguǒ
              Fruits   I   like   apples
      'Speaking of fruits, I like apples.'

(2)   Q: 谁   打破了   窗户的   玻璃？
      shéi   dǎpòle   chuānghù de   bōlí
      Who   broke   the window's   glass
      'Who broke the glass of the window?'
      A: 是 [Focus 小王]   打破了   窗户的   玻璃。
      shì   xiǎo wáng   dǎpòle   chuānghù de   bōlí
      Is   Xiao Wang   broke   the window's   glass
      'It is Xiao Wang who broke the glass of the window.'

The determiner phrase (DP) 水果 in (1A), which conveys the given information mentioned in (1Q), can be interpreted as topic, while the DP 小王 in (2A), which conveys the information that (2Q) does not mention, can be interpreted as focus. Both the topic and the focus occupy specific syntactic positions.

The above instances indicate that IS serves as a bridge connecting syntax and discourse pragmatics. Such a property of IS has given rise to heated debates on whether IS notions belong to the syntactic component or the pragmatic component with respect to the generation of IS. This paper aims to answer this question by observing the IS in modern Mandarin Chinese, which is a topic-prominent language [3] and a topic-configurational language [4] [2]. After the introduction, Sect. 2 reviews the definition of the lexicon in generative grammar and then briefly inspects the debates on the relationship between IS and the lexicon in the literature. Section 3 analyzes the problems that a formal feature approach confronts. It paves a way for Sect. 4, which comes to a conclusion that topic/focus features are distinct from case or φ(phi)-features[3] and IS might be a syntactic-pragmatic interface phenomenon.

2   Background

2.1   The Lexicon

In the generative linguistic tradition, a feature-based approach is often used to account for the properties of IS. This approach mainly includes the mainstream generative grammar [6–10], and a lot of relevant studies and the syntactic cartographic approach

---

[2] Xu [4] also points out that some informational foci in modern Mandarin Chinese are configurational.

[3] Φ-feature is a cover term for gender, number, and person.

[11, 12], and other cartographic studies. In the feature-based approach, topic and focus are treated as formal features numerated from the lexicon, similar to other formal features like case or φ-features.

According to the mainstream generative grammar, the lexicon[4] is the base of grammar. As to the content of the lexicon, Chomsky [13, p. 84] has given the following description:

> ...the base of the grammar will contain a lexicon, which is simply an unordered list of all lexical formatives. More precisely, the lexicon is a set of lexical entries, each lexical entry being a pair (D. C), where D is a phonological distinctive feature matrix "spelling" a certain lexical formative and C is a collection of specified syntactic features (a complex symbol).

The lexicon receives little attention at the stage of Standard Theory in the early days of generative grammar. With the development of the lexicon theory, however, more and more items are attributed to the lexicon. Jackendoff [14] has proposed the Extended Lexical Hypothesis, which has been accepted by Chomsky [15], claiming that inflectional affixes like number, gender, case, person and tense exist in the lexicon. In Chomsky's recent work within the framework of the Minimalist Program, the lexicon is given an important place and considered as the locus of language varieties [16]. The lexicon is "a list of exceptions" [6, p. 235]. To put it in another way, apart from differences from the lexicon, there is only one human language in the world [17, p. 50]. Till now, the lexicon includes the categories of entities like nouns, verbs, adjectives, adverbs, and functional categories like tense, complementizers. It also includes idioms, affixes, features, etc.. Topic and focus exist in the lexicon as formal features of Top and Foc[5]. In light of the feature-based approach, notions of IS are closely related to the lexicon.

2.2 The Debate

Based on this lexicon theory, the generation of IS, just like that of wh-structures, begins with Numeration (N) of lexical items from the lexicon. N is defined as follows [6, p. 225]:

> A set of pairs (LI, i), where LI is an item of the lexicon and i is its index, understood to be the number of times that LI is selected.

This leads to the further formulation on IS from Aboh [10, p. 14]:

> A numeration N pre-determines the Information Structure of a linguistic expression.

---

[4] The term *lexicon* in this paper is *the mental lexicon*, a kind of mental dictionary where the lexical information is stored.

[5] In order to avoid terminological chaos, hence, in this paper, respectively the terms of *Topic* and *Focus* in the examples like (1) and (2) refer to the topic constituent and the focus constituent, *Top* and *Foc* describe the formal features of topic and focus, and *TopP* and *FocP* are the abbreviated forms of Topic Phrase and Focus Phrase.

Following this view, the cartographic approach [11, 12] also regards topic and focus as features, emphasizing that the core syntax must involve information-sensitive functional projections (e.g. TopP, FocP). Put differently, topic and focus are not pragmatic notions, but merely syntactic notions active in the computational system.

In contrast to the feature-based approach, there is another approach, which is called the non-feature-based approach, arguing that the generation of IS should not be driven by formal features through the core syntactic operation, because it is a syntactic-pragmatic or syntactic-information interface phenomenon [2, 18–22]. Although topic and focus are inseparable from their syntactic carriers, they are still pragmatic notions. The generation of IS requires the interaction of both syntactic and pragmatic components.

This section has reviewed the two typical approaches to accounting for IS under the generative framework: the feature-based approach and the non-feature-based approach, both of which have totally different viewpoints towards the relationship between the pragmatic notions (i.e. topic and focus) and the lexicon. The next section will discuss the challenges the feature-based approach faces and explain why IS cannot be decided by pure syntactic computation.

3 The Feature-Based Approach and the Generation of Information Structure

3.1 The Inclusiveness Condition and Information Structure

Given that topic and focus are formal features, the derivation of IS should observe the conditions stipulated by the computational system, one of which is the Inclusiveness Condition (IC). According to Chomsky [6, p. 228], a perfect language should meet the IC:

> ...any structure formed by the computation (...) is constituted of elements already present in the lexical items selected for N [the numeration]; no new objects are added in the course of the computation apart from rearrangements of lexical properties (...). Let us assume that this condition holds (virtually) of the computation from N to LF.

The IC requires that no new content is to be added into the computation during the syntactic derivation. Therefore, the extraction of syntactic components and features from the lexicon should be done in one go.

Following the IC and the phase theory in the minimalist program [7, and Chomsky's later works], the derivation of a topic or focus structure begins with selecting the lexical items and related features from the lexicon to form a lexical array (LA). Such items are merged according to the selective restrictions of the predicate to form the v* phrase (v*P)[6]. Then the v*P is merged with the tense (T) to form the tense phrase (TP), which is merged with the complementiser (C) to form the complementiser phrase (CP). Consider the Chinese topic structure in (3).

---

[6] Here v is a light verb, which is null with much the same causative interpretation as the verb *make* [16, p. 339]. Call v with full argument structure $v*$: transitive v or experiencer [8, p. 43].

(3)　a. [$_{TopP}$ 这部　　电影，[$_{TP}$ 我　喜欢]]。
　　　　 zhèbù　diànyǐng,　wǒ xǐhuān
　　　　 This　　movie　　I　like
　　　　 'This movie, I like it.'

　　b. LA= {这部电影, 我, 喜欢, v*, V, T, C}

　　c. [$_{v*P}$ 我 [$_{v*}$ [$_{VP}$ 喜欢 这部电影]]]

　　d. [$_{CP}$ C[$_{TP}$ 我[T[$_{v*P}$ 我[$_{VP}$ 喜欢 这部电影]]]]]

　　e. [$_{CP(TopP)}$ 这部电影 [C(Top) [$_{TP}$ 我[T[$_{v*P}$ 我—[$_{v*}$ Ø+喜欢][$_{VP}$ 喜欢这部电影]]]]]]
　　　　　　 zhèbù diànyǐng,　　　　 wǒ　　　　　　 xǐhuān
　　　　　　 This movie　　　　　　　I　　　　　　　 like
　　 'This movie, I like it.'

In order to derive (3a), it is necessary first to extract the lexical items and features from the lexicon to form the LA in (3b). Then, in terms of the selectional restrictions of the predicate verb 喜欢, the verb merges with DP$_1$ 这部电影 to form the verb phrase (VP) 喜欢这部电影. The light verb v merges with the VP to form the v*' (v* bar). The V 喜欢 is attracted to v*. The v*' merges with the external argument DP$_2$ 我 to generate the v*P 我 Ø+喜欢这部电影, which continues to merge with the tense T to form the TP. In accordance with the Extended Projection Principle (EPP), which requires that clauses must contain a determiner phrase in the subject position [23, p. 10], the DP$_2$ 我 should be raised to [Spec, TP] to form the TP, which later merges with C to form the CP. Due to the EPP feature of C, the DP$_1$ is raised to [Spec, CP]. According to the "Split CP" hypothesis [11], the head C of CP can be split into heads like Force, Top, Foc, Fin, etc., and a sentence can include more than one of such heads, or only one of them. Example (3b) includes a head Top and the DP$_1$ is interpreted as topic.

So far, the whole process seems to go smoothly in (3). However, the problem is that the topic/focus features are different from the features like the case and φ-features - they are features that occur selectively in the syntactic projection. The occurrence of functional heads is related to the discourse pragmatics. Which one is specifically given an interpretation should be determined by the discourse context, not the lexicon. If this is the case, C in (3) should not be pre-specified to be a [Top] feature by the lexicon. It is the discourse context that adds the discourse property of topic to DP$_1$. If the [Top] feature is added during the operation, then the IC will be violated.

Besides the configurational IS like (3a), there are non-configurational ones in Chinese, whose derivation also might violate the IC.

(4) Q: 谁　　喜欢　　这部　　电影？
　　　shéi　xǐhuān　zhè bù　diànyǐng?
　　　Who　likes　　this　　movie
　　　'Who likes this movie?'
　　A: [Focus/Subject 我] 喜欢　　这部　　电影。
　　　　　　　wǒ　xǐhuān　zhè bù　diànyǐng.
　　　　　　　I　like　　this　　movie.
　　'I like this movie.'

In (4A), the DP 我 conveys new information and can be interpreted as topic. The focus 我 and the subject of (4A) coincide. The discourse property of 我 as focus is determined by the discourse context. The generating of the focus 我 can hardly satisfy the requirements of IC, which can be further read as the following two conditions [24, p. 238].

a. *The properties of a non-terminal node are fully recoverable from its daughters.*
b. *The properties of terminal nodes are recoverable from the mapping principles [between the lexical–conceptual system and the grammar].*

These conditions require that the form of a non-terminal node must be recovered by its daughter node, and that the nature of the terminal node can be recovered by the mapping principle between the lexical-concept system and the grammar. In (4A), the DP 我 as focus can be recovered to be the specifier of TP, not the specifier of FocP[7]. Therefore, the interpretation of 我 as focus is not likely to be the result of formal feature driving. In fact, it is achieved by the accent mark of main stress. Stress is not a lexical property of any lexical item that it bears main stress [24, p. 243]. If a focus is decided during the derivation, then the IC is violated.

From the above analysis, it is found that with the feature-based approach the numeration of topic and focus from the lexicon is incompatible with the properties of the topic and focus. It holds true no matter whether the IS is configurational or non-configurational.

3.2 The "Look-Ahead" Technique and Information Structure

In the current minimalist program of generative grammar, a structure is derived by phase [8], so is the IS. Phases, including the CP and the transitive verb phrase (v*P), have propositional properties. A v*P phase, which expresses the basic argument relationship and depends on the lexical semantics of relevant items, is the most basic element of necessity. A CP, whose functional categories are selective, expresses what is related to pragmatics. In order to form a structure, the derivation of the lexical phrase is carried out first, and then comes the CP phase operation.

---

[7] The syntax of a sentence consists of three layers: the VP layer, the inflectional layer and the complementiser layer, [11], which are built in a bottom-up fashion successively. Therefore, the DP 我 in (4A) is recovered to be the specifier of TP on the lower layer, not the specifier of CP on the higher layer.

Phase derivation is performed with a local domain in a cyclic way. As to when to transfer, Chomsky [7, 8] has proposed two versions of Phase Impenetrability Condition (PIC): the former is called Strong PIC (PIC$_1$), and the latter Weak PIC (PIC$_2$), as juxtaposed by Citko [25, pp. 33–34] in (5).

(5)   a. SRTONG PIC/PIC$_1$

The domain of H is not accessible to operations outside HP; only H and its edge are accessible to such operations.

b. WEAK PIC/PIC$_2$

The domain of H is not accessible to operations at ZP; only H and its edge are accessible to such operations.

PIC determines the transfer and the local scope[8]. The scopes of locality are different in the two versions of PIC, as (6) shows [25, p. 34]:

(6)   a.

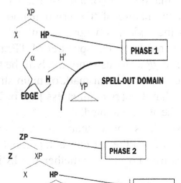

b.

In (6a), by PIC$_1$, X is incapable to check the features with YP because YP has been spelled out before X merges with it. In (6b), by PIC$_2$, X can undergo the feature-checking with YP, because YP is to be spelled out when Z is merged. The functional head in CP splits after the completion of the v*P phase derivation. Due to the selective occurrence of functional heads in CP phases, a technique of "look-ahead" toward the functional heads in CP has to be employed [26]. Consider the Chinese topic structure.

---

[8] The local scope of a phase is the size of the phase.

(7) a. [$_{TopP}$ 吴先生， [$_{TP}$ 我 认识]]
     wú xiānshēng, wǒ rènshi
     Mr. Wu, I know
     'Mr. Wu, I know.'
  b. [$_{CP(TopP)}$ 吴先生， [$_{TP}$ 我 [$_{v*P}$ 我 Ø+[认识] [$_{VP}$ 认识 吴先生]]]]
     wú xiānshēng, wǒ rènshi
     Mr. Wu, I know
     'Mr. Wu, I know.'

The DP 吴先生 is topicalized in (7a) and the process of the topicalization is shown in (7b). According to PIC$_1$, the DP (t$_i$), as an internal argument of the V 认识, is a part of the v*P phase complement and must be transferred after the derivation of the v*P phase. In this case, it is impossible for the DP to act as topic in the upper phase of CP. PIC$_2$ might help to avoid the transfer of the complement elements, which is to act as topic or focus in the upper CP phase, for these elements are moved to the position of [Spec, v*P] before the v*P phase is completed. However, the preserving of the constituent served as topic or focus on the position of [Spec, v*P] is a technique of "look-ahead". As the functional head within the CP domain, "C seems irrelevant to selectional requirements of lexical heads and it is unmotivated for C to be included in the numeration" [20, p. 52]. The functional projection of TopP or FocP is in fact prespecified by employing the "look-ahead" technique, which is incompatible to the discourse properties of IS.

3.3   The Functional Projection and Information Structure

In terms of the cartographic approach [11, 12], all the topic phrases and focus phrases can be represented in syntax. However, it is not always true. For instance, there are two types of IS in Chinese: one with specific configuration and one without specific configuration [4]. The latter cannot be projected in a proper way.

One phenomenon that attracts attention is the subject-topic. The topic and the subject often coincide [12, p. 221]. The topic has the function of transmitting old/given information, and is often located at the beginning of the sentence. Meanwhile, the subject is often placed before the verb and in the sentence-initial position. Therefore, when the subject highlights the old information, the topic coincides with the subject [27]. For instance, in (8 B) [28, p. 139], the DP *John* is a subject-topic.

(8) A: – Tell me about John.
    B: – Well, John is a student from Canada.

This is also often the case in Chinese, which is a "topic-prominent" language [3]. In Chinese, most typical subjects are topics [29, p. 219]. The topic often overlaps with the subject, which has given rise to a long-lasting debate on how to tell whether a constituent is served as topic or subject, or even whether there are notions of subject and topic [30–38].

A subject topic can be verified by applying syntactic tests. For instance, the topic can be tested by 是不是 or by adding topic markers like 呢、啊, etc.

(9) a. [Topic 李四] 打了   王五。
              lǐ sì   dǎle   wáng wǔ.
              Li Si   beat   Wang Wu.
         'Li Si beat Wang Wu.'
     b. 李四   是不是   打了   王五?
          lǐ sì   shì bùshì   dǎle   wáng wǔ?
          Li Si   yes or no   beat   Wang Wu
         'Did Li Si beat Wang Wu?'

The DP 李四 in (9a) is tested to be a topic. As the external argument of the predicate, it is also the subject of the sentence.

If a constituent is a subject topic, it is not easy to decide whether it is projected as the spec(ifier)-TopP or the spec-TP. Take (9a) as an example, whose syntactic projection is shown as follows:

(10)

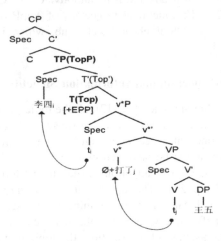

In (10), the DP 李四 is a topicalized subject, as is tested in (9b). If the topic is a feature from the lexicon, it should be projected in the syntax of the sentence. Following the feature-based analysis, there should be a TopP and a TP in (10). TopP and TP are projected on two separate positions in order to meet the requirements of "one feature one head" principle [39] and of the EPP [23]. Thus, we will come to an incorrect conclusion that (9a)/(10) is ungrammatical. The fact, however, is that the coincidence of spec-TP (the subject position) and speci-TopP (the topic position) does not lead to the ungrammaticality of the sentence. This indicates that topic and subject are concepts of different nature. The subject as an argument of the predicate is decided by the lexicon, while the topic conveying old/given information should be decided by factors out of the lexicon.

In addition, a topic constituent might also overlap with an adverbial in Chinese [3, 33, 40, 41]. Adverbial is not a basic constituent of a sentence; a sentence may or may not have adverbials. It follows that a sentence might be projected with or without adjuncts. The adverbial appears on the syntactic projection selectively, which is very similar to the TopP projection. There are two typical solutions: one is to analyze adverbial topics as topics (i.e. [33]), and the other as adjuncts (i.e [40]). Although adverbial and topic are closely related, they have different syntactic and pragmatic properties.

(11)   [Adjunct/Topic 今天]   我   买   红酒。
　　　　　jīntiān   wǒ   mǎi   hóngjiǔ.
　　　　　Today   I   buy   red wine
　　'I'll buy the red wine today.'

If topic is a formal feature, a topic phrase must be projected separately from other constituents to meet the requirement of "one feature one head". The problem this analysis faces is when an adverbial is interpreted as topic, it is projected into the specifier position of a TopP; when it is just a modifier, it is projected into the adjunct position within the CP. The indeterminacy of adverbial topics in syntactic projection suggests that the topic property of the adverbial is not determined by the lexicon.

Furthermore, there is a special type of non-configurational topic—stage topic (also called spatio-temporal topic), which refers to the asserted content at a specific time and space. Stage topics can be divided into two categories: overt stage topics and covert stage topics [42]. The former includes adverbial topics on the sentence initial position, representing time or space. The latter includes the topics completely determined by the context of the discourse [21]. Unlike the subject-topic or adverbial subject, which might coincide with a specific grammatical constituent, covert stage topics cannot even be found syntactically.

(12) Q: 发生了   什么   事情？
　　　　fāshēngle   shénme   shìqíng?
　　　　happened   what   matter
　　　　'What happened?'
　　A: 我 的 腿   受伤   了！
　　　　wǒ de tuǐ shòushāng le!
　　　　My   leg   be injured
　　　　'My leg was injured!'

(12Q) asks what the event that happened was, and the corresponding answer should be related to the event mentioned. The whole sentence of (12A), which is the answer to the question, conveys new information and is interpreted as focus. In terms of the discourse context, (12A) contains a covert stage topic about the event. This topic does not have any syntactic configuration.

This section has discussed several problems the feature-based approach needs to solve. Firstly, the pragmatic properties of topic and focus seem to be at odds with the

Inclusiveness Condition (IC) [6]. Secondly, a "look-ahead" technique is employed during the computation. Thirdly, when interpreting a discourse configurational language like Chinese, a dual method is to be used to deal with the syntactic projection of topic/focus constituents. Such problems show that information notions are different from other formal features; IS has the property of context dependency, which is distinct from the characteristic of the items in the lexicon.

4 Conclusion

This paper has observed two types of IS: one is configurational and the other is non-configurational. From the investigation, it is found that for the former, a feature-driven approach has limited power to explain its generation with regard to the problems mentioned in Sect. 3, while for the latter, it is directly the result of syntactic-pragmatic interaction, which is neglected by the feature-based approach. Although the semantics of the lexical items acting as the information constituents is closely relevant to the lexicon, the pragmatic properties of topic and focus are decided by factors out of the lexicon. The discourse notions might not be formal features existing in the lexicon. Information structure is not the result of pure syntactic computation, but a phenomenon of syntactic-pragmatic interface.

Acknowledgments. This paper is supported by the Key Research Fund Project from Hunan Planning Office for Philosophy and Social Sciences, China (Project NO.: 18ZDB034; Project Title: A Study of the Recursion-Only Hypothesis in Language: from a Biolinguistic Perspective). I thank Professor Liu Li-min from Sichuan University, and two anonymous reviewers, as well as other participants of CLSW 2019 for helpful comments. All errors and misinterpretations remain entirely my responsibility.

References

1. Lambrecht, K.: Information Structure and Sentence Form. Cambridge University Press, Cambridge (1994)
2. Erteschik-Shir, N.: Information Structure: The Syntax-Discourse Interface. Oxford University Press, Oxford (2007)
3. Li, C., Thompson, S.: Subject and topic: a new typology of language. In: Li, C. (ed.) Subject and Topic, pp. 457–489. Academic Press, New York (1976)
4. Xu, L.J.: Whether Chinese is a configurational language. Stud. Chin. Lang. **5**, 400–410 (2002). (in Chinese)
5. Yan, M.Y.: Topicaliazation in the Model of Sentence Formation. In: Liu, P., Su, Q. (eds.) CLSW 2013. LNAI 8229, pp. 682–691. Springer, Heidelberg (2013)
6. Chomsky, N.: The Minimalist Program. MIT Press, Cambridge (1995)
7. Chomsky, N.: Minimalist inquiries: the framework. In: Martin, R., Michaels, D., Uriagereka, J. (eds.) Step by Step: Essays on Minimalist Syntax in Honor of Howard Lasnik, pp. 89–155. MIT Press, Cambridge (2000)
8. Chomsky, N.: Derivation by phase. In: Kenstowicz, M. (ed.) Ken Hale: A Life in Language, pp. 1–52. MIT Press, Cambridge (2001)

9. Chomsky, N.: What Kind of Creatures Are We? Columbia University Press, New York (2016)
10. Aboh, E.O.: Information structuring begins with the numeration. Iberia: Int. J. Theor. Linguist. 2(1), 12–42 (2010)
11. Rizzi, L.: On the fine structure of the left periphery. In: Liliane, H. (ed.) Elements of Grammar, pp. 281–337. Kluwer, Dordrecht (1997)
12. Cinque, G. (ed.): Functional Structure in DP and IP-The Cartography of Syntactic Structures, vol. 1. Oxford University Press, New York (2002)
13. Chomsky, N.: Aspects of the Theory of Syntax. MIT Press, Cambridge (1965)
14. Jackendoff, R.S.: Some rules of semantic interpretation for English, Ph. D. dissertation. MIT (1969)
15. Chomsky, N.: Studies on Semantics in Generative Grammar. Mouton, The Hague (1972)
16. Radford, A.: Minimalist Syntax: Exploring the Structure of English. Cambridge University Press, New York (2004)
17. Smith, N.: Chomsky: Ideas and Ideals. Cambridge University Press, Cambridge (1999)
18. Reinhart, T.: Pragmatics and linguistics: an analysis of sentence topics. Philosophica 27(1), 53–94 (1981)
19. Reinhart, T.: Interface Strategies: Optimal and Costly Computations. MIT Press, Cambridge (2006)
20. Munakata, T.: The division of C-I and the nature of the input, multiple transfer, and phases. In: Grohmann, K.K. (ed.) Interphases: Phase-theoretic Investigations of Linguistic Interfaces, pp. 48–81. Oxford University Press, New York (2006)
21. Erteschik-Shir, N.: The Dynamic of Focus Structure. Cambridge University Press, Cambridge (1997)
22. Horvath, J.: discourse features, syntactic displacement and the status of contrast. Lingua 120(6), 1346–1369 (2010)
23. Chomsky, N.: Some Concepts and Consequences of the Theory of Government and Binding. MIT Press, Cambridge (1982)
24. Szendrői, K.: Focus and the interaction between syntax and pragmatics. Lingua 114(3), 229–254 (2004)
25. Citko, B.: Phase Theory: An Introduction. Cambridge University Press, Cambridge (2014)
26. Bošković, Z.: On the locality and motivation of move and agree: an even more minimal theory. Linguist. Inquiry 38(4), 589–644 (2007)
27. Mallinson, G., Blake, B.J.: Language Typology: Cross-linguistics Studies in Syntax. North-Holland, Amsterdam (1981)
28. Vermeulen, R.: On the position of topics in Japanese. Linguist. Rev. 30(1), 117–159 (2013)
29. Shen, J.X.: Asymmetry and Markedness Theory. Jiangxi Education Publishing House, Nanchang (1999). (in Chinese)
30. Chao, Y.R.: A Grammar of Spoken Chinese. University of California Press, Berkeley (1968)
31. Chafe, W.L.: Givenness, contrastiveness, definiteness, subjects, topics, and point of View. In: Li, C. (ed.) Subject and Topic, pp. 25–56. Academic Press, New York (1976)
32. Huang, C.-T.J.: Logical Relations in Chinese and the Theory of Grammar. Ph.D. dissertation, MIT, Garland Press, New York (1982)
33. Xu, L.J., Langendoen, D.T.: Topic structures in Chinese. Language 61(1), 1–27 (1985)
34. Shi, D. X.: The Nature of Topic Comment Constructions and Topic Chains. Ph.D. dissertation, University of Southern California (1992)
35. Ning, C.Y.: The Overt Syntax of Relativization and Topicalization in Chinese. Ph.D. dissertation, University of California, Irvine (1993)
36. Liu, L.M.: Theoretical reflections on the nature of "subject". Foreign Lang. Res. 179(4), 70–77 (2014). (in Chinese)

37. Liu, D.Q.: Non-topic subjects in Chinese. Stud. Chinese Lang. **41**(2), 259–275 (2016). (in Chinese)
38. Shen, J.X.: Does Chinese have the subject predicate structure? Mod. Foreign Lang. **40**(1), 1–13 (2017). (in Chinese)
39. Pollock, J.-Y.: Verb movement, universal grammar, and the structure of IP. Linguist. Inquiry **20**(3), 365–424 (1989)
40. Chen, G.H., Wang, J.G.: Unmarked non-subject topics in Chinese. Chin. Teach. World **24** (3), 310–324 (2010). (in Chinese)
41. Paul, W.: New Perspectives on Chinese Syntax. Mouton de Gruyter, Berlin (2014)
42. Lahouse, K.: Implicit stage topics: a case study in French. Discourse **1**(1), 1–23 (2007)

A Collostructional Analysis of Ditransitive Constructions in Mandarin

Huichen S. Hsiao[✉] and Lestari Mahastuti

Department of Chinese as a Second Language,
National Taiwan Normal University, Taipei, Taiwan
huichen.hsiao@ntnu.edu.tw, lestarimahastuti@gmail.com

Abstract. By investigating the frequency distribution of 37 verbs in Mandarin ditransitive constructions and adopting a collostructional analysis (cf. Gries & Stefanowitsch [1]), this study aims to clarify the construction meaning of each type of ditransitive construction. The preliminary result shows that two constructions differ in terms of fine-grained aspects, such as the number and completion of transfer events. Based on the corpus findings, this study claims that the transfer meaning expressed by double-object constructions entails only one entire macro-event while the transfer event expressed by prepositional dative constructions highlights and involves more than one event, thus increasing the possibility of the prepositional dative conveying incomplete transfer meaning.

Keywords: Ditransitive construction · Verb meaning · Construction meaning · Collostructional analysis

1 Introduction

The correlation between verbs, constructions, and construction meaning has been one of the research focuses in the past decade in both linguistics and psycholinguistics. The pattern of nouns or noun phrases that follow a verb, known as verb argument structure, remains a complex issue due to some problems often addressed as syntactic alternation. First, a certain verb can occur with multiple argument structures or sentence patterns. Second, some semantically similar verbs can occur in what appear to be competing syntactic patterns; these patterns seem to be semantically synonymous while the others can only participate in one of the patterns [2]. In addition, the theory of Construction Grammar believes that a construction is a form-meaning pairing, which implies that two different constructions cannot have identical construction meaning [3]. In other words, each different form represents a different function of meaning (or discourse).

On the other hand, the development of corpus linguistics research has encouraged more researchers to extend their studies by adopting the occurrence frequency of certain words or constructions in corpora. However, Gries & Stefanowitsch [1] and Gries et al. [4] suggest that only reporting frequencies and percentages alone is often quite unreliable in drawing objective and scientific conclusions. By suggesting an alternative solution to the problems above, Gries & Stefanowitsch [1] have developed a

© Springer Nature Switzerland AG 2020
J.-F. Hong et al. (Eds.): CLSW 2019, LNAI 11831, pp. 37–51, 2020.
https://doi.org/10.1007/978-3-030-38189-9_4

distributional corpus analytical method known as "collostructional analysis," and they have conducted a series of studies to confirm the superiority of that method.

Therefore, by adopting the collostructional analysis method as well as the corpus method in the present study, this study seeks to investigate the frequency distribution of 37 verbs in three types of Mandarin ditransitive constructions (including double-object construction and prepositional dative construction), attempting to answer the following research questions: (1) Which types of verbs have strong preference towards each of the ditransitive constructions as indicated by collostructional strength?; (2) What are the semantic features of each type of Mandarin ditransitive construction?

2 Literature Review

Mandarin ditransitive constructions[1] [5] have three syntactic permutations which include [V NP1 NP2], [V *gei* NP1 NP2], and [V NP2 *gei* NP1]. According to Liu [6], while all of the constructions carry a semantically more-or-less equivalent meaning of *transfer,* they actually differ in terms of range of transfer and argument (indirect object/NP1) role. By analyzing the distribution of Mandarin verbs in the above three different sentence patterns of ditransitive constructions, Liu [6] has summarized that the range of transfer and argument role of each pattern as follows:

(i)   [V NP1 NP2]: The range of transfer involves act of transfer, communicated message, and future having, while the argument role involves recipient, goal, and patient.

(ii) [V *gei* NP1 NP2]: The range of transfer involves act of transfer, manner of transfer, and instrument of transfer, while the argument role involves only recipient.

(iii) [V NP2 *gei* NP1]: The range of transfer involves act of transfer, manner of transfer, instrument of transfer, and precondition of transfer, while the argument role involves only recipient.

Overall, Liu [6] indicates that [V NP1 NP2] and [V *gei* NP1 NP2] are more semantically synonymous, which formulated as [SUBJECT transfer NP2 *gei* NP1]; however, [V NP1 NP2] accepts wider types of argument role compared to [V *gei* NP1 NP2]. Meanwhile, the semantics of [V NP2 *gei* NP1] is slightly different since its range of transfer also involves precondition of transfer. Thus, the semantics of [V NP2 *gei* NP1] is [SUBJECT (act and) transfer NP2 *gei* NP1], which signifies that the transfer process may actually entail its involvement of two events of preconditioning acts and transfer itself. Liu's [6] argument about the synonymy of [V NP1 NP2] and [V *gei* NP1 NP2] is in accordance with Huang & Ahrens's [7] results. They have argued that the *gei* in [V *gei* NP1 NP2] construction is simply an affix of the main verb, thus forming a

---

[1] In this paper, we use the term "ditransitive construction" (in contrast with Liu's [6] term of "dative construction") to sum up all types of Mandarin double-object sentences. In addition, we classify the prototype of double-object ([V NP1 NP2]) and [V *gei* NP1 NP2] as the same category of "double-object construction" while the sentences that consist of preposition are classified into "prepositional dative construction".

complex predicate. This point of view suggests that Mandarin ditransitive constructions should only be categorized into two patterns: (1) Double-object construction (DO), including [V NP1 NP2] and [V *gei* NP1 NP2]; (2) Prepositional dative construction (PD), including only [V NP2 *gei* NP1].

As for the number of events entailed by the constructions, Liu [8] provided detailed insight by classifying constructions into four patterns based on the expression form of the transfer act from a typological viewpoint as shown below:

(1) Double Object Construction, which entails one single event and one single process;

> *Wo song-le*   *ta*   *yi-shu*   *hua.*
> I   gave-PERF   she   one-CL   flower
> 'I gave her a bunch of flowers.'

(2) Prepositional Complement Construction, which entails one single event and two processes;

> *Wo song-le*   *yi-shu*   *hua*   *gei*   *ta.*
> I   gave-PERF   one-CL   flower GEI   her
> 'I gave a bunch of flowers to her.'

(3) Serial Verb Construction, which entails two minor events and one complex event;

> *Wo mai-le*   *yi-shu*   *hua*   *gei (-le)*   *ta.*
> I   bought-PERF   one-CL   flower GEI(-PERF)   her
> 'I bought a bunch of flowers for her.'

(4) Complex Sentence, which entails two events;

> *Wo mai-le*   *yi-shu*   *hua,*   *gei (-le)*   *ta.*
> I   bought-PERF   one-CL   flower, GEI(-PERF)   her
> 'I bought a bunch of flowers, (and) gave (it) to her.'

Based on the two examples of prepositional complement and serial verb construction as explained above, it is evident that Liu [8] claimed the main verb (the first verb/V) as the criteria to determine whether *gei* acts as a verb or as a preposition in a dative construction. In one exactly identical [V NP2 *gei* NP1] pattern, the word *gei* plays a role as a preposition of a prepositional complement construction in which the main verb is *song* 'give'. However, *gei* acts as one of the verbs (and thus can be followed by the perfective aspect *le*) of the serial verb construction in which the main verb is *mai* 'buy'. The difference between those two verbs lies in whether or not the verb itself implies a transfer meaning, hence requiring at least two arguments. *Song* 'give' is a verb that indicates a transfer meaning and requires at least two arguments, while *mai* 'buy' is not and therefore can form a complete grammatical sentence with just one argument (e.g. *wo mai le yi shu hua* 'I bought a bunch of flowers'). The implication of this point of view is that Liu's [8] classification clearly distinguishes between event and process. When *gei* acts as a preposition, the sentence (prepositional complement construction) entails only one macro-event (which consists of two processes); however, when *gei* acts as a verb, the sentence (serial verb construction) entails two sub-events (which forms one complex event).

On the other hand, while comparing two datasets containing the occurrence frequency of 18 Dutch verbs in Dutch ditransitive constructions, Colleman & Bernolet [9] notice a contrasting difference in general preference towards constructions between two entirely identical sets of verbs. The corpus data indicated a strong tendency (64.7%) to occur with double object constructions while the experimental data showed a clear preference (78.2%) towards prepositional dative constructions. However, when the collostructional analysis was applied for further analysis, more consistent results were supported by both datasets. This means that most of the target verbs exhibited the same preference for one construction as opposed to the other, regardless of which dataset those verb frequencies originated from. Therefore, Colleman & Bernolet [9] point out that the collostructional analysis is a more superior method for corpus-based construction analysis, and hence also further proves Gries & Stefanowitsch [1] and Gries et al.'s [4] claims regarding the advantages of collostructional analysis.

To conclude, investigating the interactions between verbs and constructions can be testified and certified by the application of the collostructional analysis to the corpus collection method. Thus, the results are more accurate and have significant impact on verb or construction meaning analysis. Despite that, while collecting verb distribution data in three types of ditransitive constructions, Liu [6] pays attention merely to whether a certain verb occurs in one certain construction or not. As for the frequency of occurrence and the frequency distribution of overall target verbs, Liu [6] does not discuss them in detail. In view of this, the present study attempts to adopt a collostructional analysis to analyze the frequency distribution of various Mandarin verbs in ditransitive constructions so that the semantic features of each construction are clarified.

3 Research Methodology

3.1 Corpus Data Collection

The corpus data in this study were gathered from "Academia Sinica Balanced Corpus of Modern Chinese" (abbr. Sinica Corpus). By using the "Chinese Word Sketch" (CWS) system, the authors searched for the corpus data of verbs that occur with three different sentence patterns of Mandarin ditransitive constructions. The research scope covered 37 Mandarin verbs at first, yet only 34 verbs were included in the analysis. This is because the writers failed to find data for three verbs (*jia* 'pick up food with chopsticks', *ti* 'kick', *baogao* 'report') that matched the target constructions.

The selection of target verbs was based on Liu's [6] categories, from which at least one representative verb of each category was chosen. A list of each category and its chosen representative verb(s) is shown below:

(1) Act of Transfer
   (a) Possession: *song* 'give', *huan* 'return', *jie* 'borrow/lend', *zu* 'rent', *shang* 'award', *pei* 'compensate', *jiao* 'deliver'
   (b) Knowledge/Information: *chuanshou* 'pass on knowledge', *chuanda* 'convey', *jiao* 'teach', *jiaodao* 'advise'
   (c) Provision: *tigong* 'provide', *gongying* 'supply'
   (d) Giving Up Possession: *shu* 'lose', *mai* 'sell'
   (e) Referral: *jieshao* 'introduce', *tuijian* 'recommend'
   (f) Contribution: *juan* 'donate', *xian* 'dedicate'
   (g) Promise: *xu* 'promise'
(2) Manner of Transfer, including *diu* 'throw', *na* 'take', *jia* 'pick up food with chopsticks', *ti* 'kick', *chuan* 'pass on'
(3) Instrument of Transfer, including *ji* 'send', *hui* 'remit'
(4) Precondition of Transfer
   (a) Obtaining: *mai* 'buy', *zhua* 'grab', *ying* 'win'
   (b) Creation: *zuo* 'make', *xie* 'write'
(5) Communicated Message, including *wen* 'ask', *huida* 'answer', *gaosu* 'tell', *baogao* 'report'
(6) Future Having, includes *daying* 'agree'

The search method used involved the two following stages: (1) Inputting "verbs" into CWS system to search for their possible occurrence in the [V NP1 NP2] and [V NP2 *gei* NP1] patterns; (2) Inputting the verb "*gei*" into CWS system to search for its possible occurrence in the [V *gei* NP1 NP2] pattern. Once CWS displayed corresponding search results, the authors then initiated the manual screening process to identify and compile sentences that match the above three patterns for further analysis.

3.2   Collostructional Analysis

Collostructional analysis, also known as distinctive-collexeme analysis, is a corpus-based construction analysis method introduced by Gries & Stefanowitsch [1]. With a certain construction (e.g. ditransitive construction) as a base, this method could identify the behavior of particular semantic slots (e.g. subject, predicate, direct object, indirect object) in the construction in terms of whether the construction attracts or rejects certain words (e.g. measuring the degree of attraction or repulsion of a lemma to a slot in one particular construction). Therefore, to achieve the objectives of the present study, collostructional analysis was adopted to analyze the behavior of predicates in Mandarin ditransitive constructions in terms of whether they showed tendencies to accept or reject certain types of verbs.

The most remarkable advantage of collostructional analysis lies in its ability to identify the tendency of certain words to occur in certain constructions just based on simple data of word frequency distribution. A measure of those tendencies, which is known as collostructional strength or distinctiveness, is basically the significance level (p-value) of one distributional statistics analysis method called Fisher's Exact Test. The lower the significance level (p-value) of words, the stronger the association between words and constructions. Furthermore, when p-value reaches <0.05, the frequency of words in particular constructions is definitely not a chance occurrence. In other words, it can be concluded with a 95% confidence interval that there is a considerable degree of association between the word and its corresponding construction.

4 Research Results

In accordance with Liu's [6] results, ([V NP1 NP2] and [V gei NP1 NP2] are more semantically synonymous) and Huang & Ahrens's [7] argument about the function of gei in Mandarin ditransitive constructions ([V NP1 NP2] and [V gei NP1 NP2] should be classified as one same pattern), the data of verb frequency distribution in this paper are presented in only two categories: (1) Occurrence frequency of verbs in double-object (DO) constructions that consist of [V NP1 NP2] and [V gei NP1 NP2] patterns; (2) Occurrence frequency of verbs in prepositional dative (PD) constructions that include the [V NP2 gei NP1] pattern. The collostructional strength of each target verb in both construction categories is shown in Table 1 below.

Table 1. Collostructional strength of target verbs in DO and PD constructions.

DO Construction (N = 1600)			PD Construction (N = 281)		
Verb	Freq.[a]	Collostructional Strength	Verb	Freq.	Collostructional Strength
tigong 'provide'	814:65	1.78E-18	xie 'write'	1:37	1.12E-30
gaosu 'tell'	230:0	6.07E-18	ji 'send'	6:24	1.62E-15
wen 'ask'	103:0	7.83E-08	mai 'sell'	4:21	1.70E-14
jiao 'teach'	105:1	7.86E-07	mai 'buy'	0:14	2.08E-12
song 'give'	120:48	2.05E-06	jieshao 'introduce'	1:14	2.71E-11
huan 'return'	67:0	2.90E-05	na 'take'	1:13	1.76E-10
jiaodao 'advise'	46:1	6.60E-03	juan 'donate'	2:12	6.90E-09
			chuan 'pass on'	4:6	1.30E-03
			zuo 'make'	0:2	2.20E-02

Neutral[b]: chuanshou 'pass on knowledge', zhua 'grab', jiao 'deliver', diu 'throw', jie 'borrow/lend', shang 'award', chuanda 'convey', ying 'win', huida 'answer', daying 'agree', gongying 'supply', xian 'dedicate', pei 'compensate', shu 'lose', xu 'promise', zu 'rent', tuijian 'recommend', hui 'remit'

[a] Frequency is the ratio of verb occurrence in double-object (DO) constructions to verb occurrence in prepositional dative (PD) constructions. For example, based on corpus data collection, the verb tigong 'provide' occurred 814 times in DO and 65 times in PD; thus the frequency of tigong 'provide' is 814:65.

[b] Neutral verbs are verbs that did not reach any significance level ($p < 0.05$).

4.1   Verbs of Double-Object Construction (DO)

According to Table 1, there are seven verbs that show a tendency to occur with double-object constructions, which based on their meanings could be classified into two categories: (1) Act of Transfer, including *tigong* 'provide', *song* 'give', *huan* 'return'; (2) Knowledge/Information Transfer, including *jiao* 'teach', *jiaodao* 'advise', *gaosu* 'tell', *wen* 'ask'. Among three verbs that belong to the Act of Transfer category (*tigong* 'provide', *song* 'give', *huan* 'return'), only the verb *tigong* 'provide' has a relatively high collostructional strength, which is also the highest compared to the other verbs in double-object constructions. The clearest difference between *tigong* 'provide' and *song* 'give' & *huan* 'return' is the concreteness of the direct object (NP2) because *tigong* 'provide' collocates with abstract objects (e.g. *jihui* 'opportunity', *fuwu* 'service', *xuanze* 'option') more frequently. Meanwhile, among the four verbs of knowledge/information transfer, only one verb, *jiaodao* 'advise', has less collostructional strength than *song* 'give' and *huan* 'return'. This could reasonably lead to a conclusion that, in general, verbs of knowledge/information transfer have strong association with double-object constructions. Moreover, from a direct object point of view, these verbs obviously often collocate with abstract objects of knowledge or information, such as *jishu* 'skill', *fangfa* 'method', *wenti* 'problem', *(yijian) shiqing* '(one) thing', and so on.

Using Liu's [6] categories, the seven verbs above can be classified into two ranges of transfer as below: (1) Act of Transfer, which includes subcategories of possesion (*song* 'give', *huan* 'return'), provision (*tigong* 'provide'), and knowledge/information (*jiao* 'teach', *jiaodao* 'advise'); (2) Communicated Message (*gaosu* 'tell', *wen* 'ask'). Note that in Liu's [6] categories, *jiao* 'teach' and *jiaodao* 'advise' are regarded as members of verbs under the act of transfer category. Hence, they are different from *gaosu* 'tell' and *wen* 'ask' which are classified into another separate category of communicated message. The reason behind this classification is because Liu [6] concludes that communication-related verbs could only appear in the [V NP1 NP2] pattern while knowledge/information (the subcategory of act of transfer) verbs could occur in all three sentence patterns of ditransitive constructions. However, the significance of this separate classification currently becomes irrelevant. This is because the collostructional analysis method has revealed that though knowledge/information related verbs could occur with all three sentence patterns, they demonstrate a significant close association only with double-object constructions. Therefore, in this paper, *jiao* 'teach', *jiaodao* 'advise', *gaosu* 'tell', and *wen* 'ask' are classified into a single category of knowledge/information transfer. The new classification we proposed is clearer, neater, and more straightforward.

Regarding the discussion regarding the superiority of the collostructional analysis method, one thing worth mentioning is Liu's [6] argument that the above three act of transfer verbs (*tigong* 'provide', *song* 'give', *huan* 'return) could appear in all three sentence patterns and therefore serve no significant function related to the differentiation of construction meaning. However, collostructional analysis in the present study has confirmed that act of transfer verbs in fact show a distinctive preference only towards double-object constructions. Thus, collostructional analysis has a quite significant function in differentiating semantic contrasts between various constructions.

4.2   Verbs of Prepositional Dative Construction (PD)

According to Table 1, there are nine verbs that display a strong tendency to occur with prepositional dative (PD) constructions as follows: (1) Precondition of Transfer, which includes *xie* 'write', *zuo* 'make'; (2) Manner/Instrument of Transfer, which includes *chuan* 'pass on', *na* 'take', *ji* 'send'; (3) Contribution Transfer, which includes *juan* 'donate'; (4) Referral Transfer, which includes *jieshao* 'introduce'; (5) Transaction Transfer, which includes *mai* 'sell' and *mai* 'buy'. Compared to verbs of double-object constructions, these verbs clearly have a higher possibility to collocate with concrete direct objects (NP2), such as *xie xin* 'write (a) letter', *zuo bing* 'make cookies', *na yi bei yinliao* 'take a cup (of) drink', *ji shengri ka* 'send birthday card', *juan qian* 'donate money', *jieshao chanpin* 'introduce product(s)', *mai yifu* 'buy clothes', *mai dongxi* 'sell things', and so on.

The difference between verb classification in this paper and in Liu's [6] is reflected in the classification of *mai* 'buy' and *mai* 'sell'. Since Liu [6] concludes that *mai* 'buy' can only occur with the prepositional dative while *mai* 'sell' is found in two sentence patterns of ditransitive construction, the two verbs were classified under two different categories (*mai* 'buy' is under the precondition of transfer, while *mai* 'sell' is classified under the act of transfer). However, according to the results of collostructional analysis, even if *mai* 'sell' can appear in a double-object construction, it has a significantly stronger tendency to occur in a prepositional dative construction. Despite this fact, this study argues that it is unreasonable to classify *mai* 'buy' and *mai* 'sell' under the same category of precondition of transfer, since both verbs have different meanings with the other two precondition of transfer verbs (*xie* 'write' and *zuo* 'make' are verbs of precondition of transfer-creation, while *mai* 'buy' is a verb of precondition of transfer-obtaining).

For the above reasons, the present study classifies *mai* 'buy' and *mai* 'sell' under the transaction transfer category. Another underlying reason is that they share one similar characteristic related to the involvement of a two-way transfer process; that is to say, the completion of the verb action (i.e. buying, or selling) depends on the completion of two-way transfer (explicit level and implicit level) that involves the transfer of exchange object (e.g. money) and transfer of direct object (NP2). Besides, Gu [10] also considered *mai* 'buy' and *mai* 'sell' as meaning-pair verbs, which refers to include groups of words with two-way interactivity; for example, *mai* 'buy' – *mai* 'sell', *shu* 'lose' – *ying* 'win', *shou* 'receive' – *fa* 'send', *juanzeng* 'donate' – *jieshou* 'accept', etc. The following example containing the verb *mai* 'sell' can provide some insight regarding two-way transfer.

(5) *Zhangsan   mai   yi-liang   che   gei   Lisi.*
      Zhangsan   sell   one-CL   car   GEI   Lisi
      "Zhangsan sold a car to Lisi.'

On the explicit level, the above sentence will be understood as "Zhangsan transferred the ownership of one car to Lisi", while on the implicit level, it can be understood as "Lisi transferred the ownership of an equivalent-value object (currency) to Zhangsan." Therefore, this paper claims that *mai* "buy" and *mai* "sell" belong to the same category. In this study, it is referred to as the transaction transfer category.

Moreover, Liu [5] also points out that the manner/instrument of transfer verbs have slightly weaker influence on semantic differentiation since they occur in both double-object and prepositional dative constructions. However, the results of the collostructional analysis have indicated that the manner/instrument of transfer verbs (*chuan* 'pass on', *na* 'take', *ji* 'send') have a clear distinct preference towards the prepositional dative construction. Namely, the results show that these manner/instrument verbs have no significant association with double-object constructions.

4.3   Constructing Meaning of Dative Alternation

This section focuses on the discussion of dative alternation[2] based on the above findings regarding the association of verbs and constructions. Please refer to Table 2 for the detailed list of representative verbs of each construction along with their respective meanings.

According to the comparison of verb meanings between two constructions as displayed in Table 2, the present study concludes that the semantic contrasts between the double-object construction and the prepositional dative construction could be reflected in two aspects: (i) The number of transfer events, that is, whether a construction entails only one transfer event or may also entail two transfer subevents; (ii) The completion of the transfer event, that is, whether the recipient (indirect object/NP1) has received the transferred NP2.

Table 2. Verb meaning in ditransitive constructions.

Double-Object Construction (DO)		Prepositional Dative Construction (PD)	
Verb Meaning	Representative Verbs	Verb Meaning	Representative Verbs
Act of Transfer	*song* 'give' *huan* 'return' *tigong* 'provide'	Precondition of Transfer	*xie* 'write' *zuo* 'make'
Knowledge/Information Transfer	*jiao* 'teach' *jiaodao* 'advise' *gaosu* 'tell' *wen* 'ask'	Manner/Instrument of Transfer	*chuan* 'pass on' *na* 'take' *ji* 'send'
		Contribution Transfer Referral Transfer Transaction Transfer	*juan* 'donate' *jieshao* 'introduce' *mai* 'buy' *mai* 'sell'

---

[2] Based on the definition of ditransitive construction (see footnote 1), the term dative alternation in this paper is used to refer to the alternation between DO (double-object construction (including prototype of double-object [V NP1 NP2] and V-gei complex [V *gei* NP1 NP2]) and PD (prepositional dative construction).

Regarding the number of transfer events, the transfer event expressed by a double-object construction entails only one event because the verb has a transfer meaning by itself. This is shown in two sentences from "Academia Sinica Balanced Corpus of Modern Chinese" in example (6). The verbs *song* 'give' and *jiao* 'teach' each represent only one transfer event of *song dangao* 'give a cake' and *jiao guowen* 'teach Chinese' respectively. On the other hand, since some verbs of precondition of transfer and transaction transfer basically do not imply a transfer meaning, the transfer event expressed by prepositional dative construction may entail two transfer subevents as illustrated in example (7). The phrase *"xie-le yi-feng chang xin gei wo"* actually involves two transfer subevents of *xie yi-feng xin* 'write a letter' and *gei wo yi-feng xin* 'give me a letter'. Moreover, the phrase *"mai-le yi-zhi xiao gou gei ni"* also clearly represents two transfer subevents of *mai yi-zhi gou* 'bought a dog' and *gei ni yi-zhi gou* 'gave you a dog'.

(6) *Furen* **song** *Qiqi* *yi-ge*   *dangao, ganxie*   *shang ci*
Lady gave Qiqi one-CL   cake, thank you   last time
ta-de   *bangmang, Qiqi shen shou gandong.*
her-POS help,   Qiqi deeply touched
'The lady gave Qiqi a cake (to) thank her for her help last time. Qiqi was deeply touched.'

Hu Shi zai Shanghai, xinhui yileng, zai Huatong Gongxue   **jiao**
Hu Shi in Shanghai, disheartened,   at Huatong public school   taught
xiao   *xuesheng*   *Guowen.*
elementary school student   Chinese.
'Hu Shi was in Shanghai (and felt) disheartened. (He) taught Chinese to elementary school students in Huatong public school.'

(7) *Qunian*   *Wu Dayou xiansheng*   *hui-lai*   *jiao-le*
Last year Wu Dayou Mr.   came back-DIRECT   taught-PERF
si-ge   *yue*   *de-shu,*   **xie-***le*   *yi-feng*
four-CLASS   month MOD-book,   wrote-PERF   one-CLASS
chang   *xin*   **gei**   *wo, shuo Taiwan Daxue*   *xuesheng*
long   letter GEI   me, said Taiwan University   student
zhen   *ke'ai.*
very   lovely.
'Last year, Mr. Wu Dayou came back (and) taught courses (for) four months. (He) wrote one long letter for me, saying (that) Taiwan University students (were) very lovely.'

Ni   *xin-li*   *shuo,*   *wo* *li*   *jia*   *yihou, ni mama   jiu*
You letter-LOC said,   I   left home   after, you mama immediately
mai-*le*   *yi-zhi*   *xiao*   *gou*   **gei**   *ni.*
bought-PERF   one-CLASS   little   dog   GEI   you.
'Your letter said (that) after I left home, your mother immediately bought a little dog for you.'

The finding that a prepositional dative construction could actually involve two transfer sub-events is in accordance with Huang & Ahrens's [7] analysis regarding the function of *gei* in ditransitive constructions. They argue that the word *gei* in [V NP2 *gei* NP1] pattern is rather a verb instead of a preposition, which occurs together with the first verb (V), forming a serial verb construction[3]. With this point of view as a basis, the phrase *"xie-le yi-feng chang xin gei wo"* in example (7) could be comprehended as a phrase that contains two serial actions with *xie yi-feng xin* 'write a letter' as the first action (e1) and *gei wo yi-feng xin* 'give me a (previously written) letter' as the second action (e2).

By considering Liu's [8] point of view regarding construction classification based on the number of events and processes implied in a sentence, the double-object construction (e.g.) in both examples each involves only a macro-single event, [*Furen* TRANSFER *dangao* GEI *Qiqi*] and [*Hushi* TRANSFER *Chinese (knowledge)* GEI *xiao xuesheng*], in which its completion requires also only one single process, [*gave a cake*] and [*teach Chinese*]. Meanwhile, the prepositional dative construction (example (7)) contains two minor events ([*Wu Dayou xiansheng xie yi-feng xin*] and [*Wu Dayou xiansheng* TRANSFER *yi-feng xin* GEI *wo*]) and one complex event ([*Wu Dayou xiansheng xie yi-feng xin* GEI *wo*]). Accordingly, another sentence in example (7) also contains two minor events of [*Mama mai yi-zhi gou*] & [*Mama* TRANSFER *yi-zhi gou* GEI *ni*] and one complex event of [*Mama mai yi-zhi gou* GEI *ni*]. In addition, some verbs of prepositional dative construction may frame sentences to be involved with one single event and two processes which Liu [7] classifies under a category of prepositional complement construction. This is shown through the sentences containing the verbs *chuan* 'pass on' and *jieshao* 'introduce' in example (8). One single event refers to [*Ruyu* TRANSFER *zhitiao* GEI *ta*] and [*A Pu* TRANSFER *nüren* GEI *ta*]. Furthermore, [*Ruyu* TRANSFER *zhitiao* GEI *ta*] includes two event processes of [*chuan zhitiao*] and [GEI *zhitiao*] (so that *zhitiao* 'note' is received by the recipient), while [*A Pu* TRANSFER *nüren* GEI *ta*] includes two processes of [*jieshao nüren*] and [GEI *nüren*] (so that *nüren is* recognized by the recipient).

(8) *Xiake hou, Ruyu*   **chuan** le   *yi-zhang*   *zhitiao* **gei**   *ta.*
Class after, Ruyu   pass on-PERF   one-CL   note   GEI   her.
"After class, Ruyu passed on a note to her.'
Houlai,   *A Pu*   **jieshao**   *yi-ge*   *xia yan*   *de-nüren*
Afterwards, A Pu   introduced   one-CL   blind   DE-woman
— *A Tong sao*   *gei*   *ta.*
— A Tong sister   GEI   him.
'Afterwards, A Pu introduced a blind woman, A Tong, to him.'

In terms of the **completion** of the transfer event, the prepositional dative construction has a higher possibility of conveying incomplete transfer meaning than the

---

[3] This paper attempts to identify the construction meaning of two types of double-object construction based on the meanings of their representative verbs. Therefore, whether *gei* in [V NP2 *gei* NP1] is a preposition or a verb will not actually affect the analysis result.

double-object construction. That is, the situation of the recipient (indirect object/NP1) not receiving the direct object NP2 has a higher possibility to occur in a transfer process/event that entails two transfer subevents. For example, the verbs *xie* 'write' and *mai* 'buy' in example (7) obviously convey a completed transfer meaning. Meanwhile, in another context (example (9)), the same verb expresses an incomplete transfer process, since only the preconditioning acts of *xie xin* 'write letter' and *mai hua* 'buy flower' are considered in the event processes (or perhaps completed). A more evident example can also be seen in example (10), in which the speaker is only certain about the fact that he/she had sent the *dongxi* 'thing' but is completely unaware whether the intended recipient had received it. It also proves that [*gei* NP1] in prepositional dative constructions does not necessarily represent one complete process. In addition, the prepositional dative verbs *na* 'take' and *juan* 'donate' in examples (11) and (12) do not necessarily express complete transfers either.

(9) *Zhengzha-le*   hen jiu,   *juede haishi yinggai* **xie**   *zhe-feng*
Struggle-PERF   very long time, feel still   should write this-CL
xin **gei** *ni,* *wo zhenxi women xiangju-de*   *mei yi fen*
letter GEI you, I   cherish our   time together-POS each one minute
'(Having) struggled (for) a very long time, (I) still felt (that I) should write this letter to you; I cherish each minute of our time together.'

You xie shi lai   **mai** *hua* **gei** *nan pengyou de,*   *you xie que*   *shi*
Some   are come   buy flower GEI boyfriend   MOD, some   however are
lai   *gen*   *dianzhang*   *liaotian de*
come   with   store manager   chat   MOD
'Some are coming (to) buy flowers for (their) boyfriend(s), but others are coming (to) chat with store manager.'

(10) *Lingwai, wo* **ji-le**   *yang*   *dongxi* **gei** *ni, shoudao-le*   *ma?*
Besides, I   send-PERF CL   thing   GEI you, receive-PERF   Q
'Besides, I sent you a thing, (have you) received (it)?'

(11) *Zui hou* *Guifu* *jiao*   *ren*   **na**   *zhi*   *bi*   **gei**   *ta,*
Finally   Guifu   order   person take CL   pen   GEI   her,
yao   *ta*   *xie-chulai.*
request   her   write-DIRECT.
'Finally, Guifu asked someone to take a pen for her and requested her to write.'

(12) *Lin jiaoshou geng chang*   *xiang*   *wo baoyuan, lian ta yao* **juan**   *shu*
Lin professor more frequently to   me complain, event he want donate book
gei *tushuguan,*   *guanyuan you shi*   *dou*   *buyao!*
GEI library,   librarian sometimes   all   not want
'Professor Lin complained to me more frequently, even when he wanted (to) donate books to the library, sometimes (the) librarian did not want them at all.'

Furthermore, Lin [11] believes that the ditransitive construction basically is a structure of multiple events composition that involves three events of cause, motion, and possession change. The combination of cause and motion events accomplishes a condition in which the action of the subject causes a transfer of the possesive relationship of the direct object to the indirect object. Since the ditransitive construction conceptually embodies three different events, it is reasonable to assume that there is a possibility of ditransitive construction conveying an incomplete transfer event. That is to say that one or two event(s) among all three events is/are still incomplete.

To sum up, the form, verb meaning, construction meaning, and semantic feature of each Mandarin dative alternation are summarized in Table 3. The meaning of double-object construction is [X transfer Y to Z] with semantic feature of [+ transfer], which means that the double-object construction delivers a meaning of completed transfer. Meanwhile, the meaning of prepositional dative construction consists of two types: (i) Within a condition in which the recipient (NP1) has not yet received the transferred object (NP2), the construction meaning is [X preconditioning act (+ transfer Y to Z)], which means that the process that is in process (or perhaps has been completed) by X is only the preconditioning act; hence, the transfer process probably has not been initiated (example (9)) or has been initiated but not yet received (example (10)); (ii) Within a condition in which the recipient (NP1) receives the transferred object (NP2), the construction meaning is [X (precondition act +) transfer Y to Z], which implies that the prepositional dative construction may entail two events of preconditioning act and transfer (example (7)). This suggests a semantic feature of [± transfer] that indicates a higher possibility of prepositional dative construction to convey an incomplete transfer meaning.

Table 3. Form, verb meaning, construction meaning, and semantic feature of Mandarin dative alternation

	Double-Object Construction (DO)	Prepositional Dative Construction (PD)
Form (Structure)	V (gei) NP1 NP2	V NP2 (gei) NP1
Verb Meaning	1. Act of Transfer	1. Precondition of Transfer
	2. Knowledge/Information Transfer	2. Manner/Instrument of Transfer
		3. Contribution Transfer
		4. Referral Transfer
		5. Transaction Transfer
Construction Meaning	X transfer Y to Z	1. X precondition act (+ transfer Y to Z)
		2. X (precondition act +) transfer Y to Z
Semantic Feature	+ transfer	+/- transfer

5   Conclusions

Mandarin ditransitive constructions can be divided into two categories: (i) Double-Object Construction, which includes the [V NP1 NP2] and [V gei NP1 NP2] patterns; (ii) Prepositional Dative Construction, which includes only the [V NP2 gei NP1]

pattern. By conducting a collostructional analysis on corpus data, the present study demonstrates that most of the verbs indicating a significant close association with double-object constructions are verbs that typically collocate with abstract direct objects, such as *tigong* 'provide', *jiao* 'teach', *jiaodao* 'advise', *gaosu* 'tell', and *wen* 'ask'. In contrast, the verbs that show a significantly strong tendency to occur with prepositional dative constructions are verbs that often collocate with concrete direct objects, such as *xie* 'write', *ji* 'send', *mai* 'buy', *mai* 'sell', *jieshao* 'introduce', *na* 'take', *juan* 'donate', *chuan* 'pass on'. From the lexical verb meaning point of view, verbs of double-object construction could be classified into two categories of act of transfer and knowledge/information transfer. On the other hand, verbs of prepositional dative construction consist of verbs from five different categories that include pre-condition of transfer, manner/instrument of transfer, contribution transfer, referral transfer, and transaction transfer.

Based on the distribution of various verb meanings in two sentence patterns of ditransitive construction, it can be concluded that the construction meaning of double-object construction is [X transfer Y to Z] with the semantic feature of [+ transfer], indicating that the double-object construction conveys a completed transfer meaning (e.g. *Furen* **song** *Qiqi yi-ge dangao, ganxie shang ci ta-de bangmang, Qiqi shen shou gandong.* "The lady gives Qiqi a cake (to) thank (her) for her help last time. Qiqi (is) deeply touched."). Meanwhile, the construction meaning of prepositional dative construction should be separately described in respect of each different context. In a context in which the recipient has not yet received the transferred object, the con-struction meaning of [X precondition act (+ transfer Y to Z)] indicates that the process that is in process (or perhaps has been completed) by X is only a preconditioning act. Accordingly, this implies that the transfer process probably has not been completed yet. In the second context in which the recipient receives the transferred object, the con-struction meaning is [X (precondition act +) transfer Y to Z]. This indicates that the transfer expressed by prepositional dative construction may imply two transfer sube-vents, a preconditioning act and the transfer event itself. Therefore, the semantic feature of prepositional dative construction is [± transfer], which indicates that the construc-tion may convey a complete transfer meaning (*Ni xin-li shuo, wo li jia yihou, ni mama jiu mai-le yi-zhi xiao gou gei ni.* 'Your letter said that after I left home, your mother immediately bought a small dog for you.') as well as an incomplete transfer meaning (*Lingwai, wo ji-le yang dongxi gei ni, shoudao-le ma?* 'Besides, I sent you a thing, have you received it yet?').

Moreover, Liu's [5] study does not make a distinction between the functional role of different verb categories (act of transfer, manner/instrument of transfer) in deter-mining construction meaning. However, the results of this current study are able to further demonstrate the tendency (or preference) of various verb categories towards a specific construction. By employing the collostructional analysis method, our research results indicate that act of transfer verbs only show a significant association with double-object constructions while manner/instrument of transfer verbs display a strong preference only towards prepositional dative constructions. Consequently, the present study is able to differentiate construction meanings based on the meanings of corre-sponding verbs, while at the same time, highlights the interaction between the occur-rence tendency of verbs and construction meaning [12]. The results show that collostructional analysis is not only feasible as well as reliable in drawing objective and scientific conclusions but is also useful in making accurate inferences for corpus construction.

Acknowledgments. This research was supported by a Taiwan Ministry of Science and Technology research grant (MOST 107-2410-H-003-025-MY2). We would like to thank you for the reviewers' comments and Anwei Yu's proofreading. All errors are of course our sole responsibility.

References

1. Gries, St. Th.: Extending collostructional analysis: a corpus-based perspective on 'alternations'. Int. J. Corpus Linguist. **9**(1), 97–129 (2004)
2. Tyler, A.: Cognitive Linguistics and Second Language Learning. Routledge, New York (2012)
3. Goldberg, A.: Constructions: A Construction Grammar Approach to Argument Structure. Chicago University Press, Chicago (1995)
4. Gries, St.Th., Hampe, B., Schönefeld, D.: Converging evidence II: more on the association of verbs and constructions. In: Rice, S., Newman, J. (eds.) Empirical and Experimental Methods in Cognitive/Functional Research, pp. 59–90. CSLI Publications, Stanford (2010)
5. Haspelmath, M.: Ditransitive constructions. Annu. Rev. Linguist. **1**, 19–41 (2015)
6. Liu, F.: Dative constructions in Chinese. Lang. Linguist. **7**(4), 863–904 (2006)
7. Huang, C., Ahrens, K.: The function and category of GEI in Mandarin ditransitive constructions. J. Chin. Linguist. **27**(2), 1–26 (1999)
8. Liu, Danqing: A typological study of giving-type ditransitive patterns in Chinese. Zhongguo Yuwen **5**, 387–397 (2001)
9. Colleman, T., Bernolet, S.: Alternation biases in corpora vs. picture description experiments: DO-biased and PD-biased verbs in the Dutch dative alternation. In: Divjak, D., Gries, St.Th. (eds.) Frequency Effects in Language Representation, pp. 87–126. De Gruyter Mouton, Berlin/Boston (2012)
10. Gu, Mingdi: Internal inheritance research on Chinese ditransitive. Chin. Lang. Learn. **6**, 45–51 (2014)
11. Lin, Y.: Analysis of the double-subject construction's event structure and it's requirement of syntactic representation. Wuhan Univ. Technol. **28**(4), 679–773 (2015)
12. Goldberg, A.E.: Verbs, constructions, and semantics frame. In: Rappaport Hovav, M., Doron, E., Sichel, I. (eds.) Lexical Semantics, Syntax, and Event Structure, pp. 39–58. Oxford University Press, Oxford (2010)

The Semantic Analysis and Representation of "*Hai-NP-Ne*" Construction with NP Quoted from Context

Xiaoyu Cao[✉]

Department of Chinese Language and Literature,
Peking University, Beijing, China
caoxiaoyu619@163.com

Abstract. This paper mainly analyzes the semantics of three types of "*hai-NP-ne*" construction with NP quoted from context, formalizes their semantics with the scale structure, and compares similarities and differences among these three types of construction semantics. It is found that all these three types of "*hai-NP-ne*" are at the lower point of entailing scales. Meanwhile, three different types of entailing scales, including likelihood scale, felicity scale, and truth value scale, are activated by the various types of verbs that are omitted in the "*hai-NP-ne*" construction. As a result, various meanings are derived.

Keywords: *Hai-NP-ne* · Scale structure · Entailment sequence · Semantic analysis

1 Introduction

The construction *"hai-NP-ne"* is commonly used in spoken Chinese. In traditional Chinese grammar, an adverb cannot modify noun phrases, but in this construction, the adverb *"hái"* modifies noun phrases. This phenomenon with grammatical specificity has attracted the attention of several researchers, such as Zong [1, 2], Ding [3, 4], Yang [5], Zheng [6], Hu [7], Wang [8], Wang [9]. They hold the view that *"hai-NP-ne"* construction expresses a broad sense of negation and negative emotions.

Among many instances of *"hai-NP-ne"* construction, some NPs in this construction are quoted components with "non-initiality" [2]. This category of *"hai-NP-ne"* construction are called *"hai-NP-ne"* construction with NP quoted from context (quotative *"hai-NP-ne"* for short).

Zheng [6] and Zong [2] noticed that some NPs in *"hai-NP-ne"* construction are quoted components, but they do not take it as classification basis for further exploration. Wang [9] studied the quotative responsive construction *"Hai + X + (Ne)"* and considered that the construction expresses counter-expectation. But "X" is not limited to NP in his research. And therefore, it is necessary to explore specifically the semantics of *"hai-NP-ne"* construction with NP quoted from context.

The semantics of this construction will involve the relationship with its context. Meanwhile, the semantic relationship between the construction and context has not been fully explored and described. It can be illustrated by the performance of Chinese-English

© Springer Nature Switzerland AG 2020
J.-F. Hong et al. (Eds.): CLSW 2019, LNAI 11831, pp. 52–62, 2020.
https://doi.org/10.1007/978-3-030-38189-9_5

machine translations of "*hai-NP-ne*" construction in the following conversation that is adapted from Shen [10]. S1: Zhè liàng dà kǎchē néng guò nà gè qiáo ma? (S1:这辆大卡车能过那个桥吗？) S2: Lián xiǎo qìchē dōu bù néng guò, hái dà kǎchē ne. (连小汽车都不能过，还大卡车呢。) The second speaker S2's response is translated into "Even cars can't pass. *They're big trucks*." by Baidu Translation and "I can't even have a car, *but also* a big truck." by Google Translation. Neither translation can accurately translate the semantics of "hái dà kǎchē ne" and the semantic relationship between the construction and its context.

In order to interpret "*hai-NP-ne*" construction accurately, we should complete the predicate omitted in the construction. According to the types of omitted verbs, it is divided into three types, Type A that omits general activity verb, Type B that omits speech act verb, and Type C that omits state verbs such as "*shì* (be)" and "*yǒu* (have)".

This paper will mainly analyze the semantics of three types of "*hai-NP-ne*" construction with NP quoted from context, and formalize their semantics with the scale structure, which makes it easier to compare similarities and difference between construction semantics, and helpful for computers to interpret sentences automatically.

2   The Tool for Representing Semantic Structure: Scale Structure

Scale structure can formally represent semantic structure. Fauconnier [11] used a pragmatic scale to explain why the superlative adjectives in English can convey the meaning of universal quantification. Guo [12] used the pragmatic scale of Fauconnier [11] to explain the linguistic phenomena related to entailment in Chinese, such as the "*lián*" sentence. Fillmore et al. [13] interpreted a *let alone* sentence in a scalar model. A scalar model is a set of propositions with a certain structure; that structure can be thought of as a generalization to *n* dimensions of what is known in social psychology as a Guttman scale [13]. Shen [10] used the scalar model proposed by Fillmore et al. [13] to interpret the two sentence patterns with the additive adverb "*hái*" in Chinese. Kennedy et al. [14] analyzed the scale structure of gradable adjectives. Luo [15, 16] used scale structure to analyze the gradable adjectives and nouns in Chinese from the perspective of degree semantics.

Scale has a set of propositions, a dimension (such as height, degree, likelihood), and an ordering relation, as shown in Fig. 1.

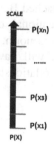

Fig. 1. Scale structure

The scale structure ranks the propositions in a propositional set according to a certain dimension. The scale structure has three crucial parameters: a proposition set, a dimension and an ordering relation.

(1) a proposition set

When there are n comparison items in the utterance, n propositions can be constructed directly. For example, *"hai-NP1-ne, ... NP2..."*, there are two comparison items NP1 and NP2, then two propositions can be constructed according to the context.

When there are not overt comparison items in the utterance, it is necessary to activate and construct a set of related propositions according to presupposition of the utterance, speaker's expectation, the ideal cognitive model (ICM), and the like. For example, "Tā lián shí yuán qián dōu méi yǒu (He doesn't even have ten *yuan*)", we can construct a set of propositions such as "Tā méi yǒu èr shí yuán qián (He doesn't have twenty *yuan*)", "Tā méi yǒu yì bǎi yuán qián (He doesn't have one hundred *yuan*)", and "Tā méi yǒu yì qiān yuán qián (He doesn't have one thousand *yuan*)".

(2) a dimension

In the same structure, different dimensions are activated by different lexical items. Take "dōu NP le" for example, "dōu shí'èr diǎn (twelve o'clock)le" activates a temporal dimension, and "dōu Běijīng le" activates a locational dimension.

In the comparative structure, the comparative items are able to activate a same dimension, otherwise the sentence is ungrammatical. Take "A bǐ B hái adj. (A is even more adj. than B)" as an example, in this sentence "Tā bǐ Yáo Míng hái gāo (He is even taller than Yao Ming)", the comparison items activate the dimension of height. But these two comparative items in this sentence "*Tā bǐ zhuōzi hái rèqíng (*He is even more enthusiastic than a table)" cannot activate a same dimension, so the sentence is not grammatical.

(3) an ordering relation

The ordering relation between propositions depends on the degrees that represent measurement values of propositions on the scale.

Some degrees have no relation to the context with an established order. Take "dōu dà xué shēng (college student)le" for example, "elementary school students", "middle school students", "college students" and "graduate students" are in the academic dimension with a ready-made order.

However, some degrees are related to the context. For example, in this sentence "Hái Shànghǎi ne, dōu Běijīngle" (it reached Beijing, let alone Shanghai), the order of "Shanghai" and "Beijing" in the locational dimension is related to the context knowledge of flights flying from south to north.

Logical negation of a proposition will have the effect of reversing the corresponding scale [11], and the ordering relationship between propositions will change accordingly.

3 Semantic Analysis and Semantic Representation of Quotative *"Hai-NP-Ne"* Construction Type A

（1）S1:王涛过六级了吗？

Wáng Tāo	guò	liù jí	le	ma
Wang Tao	pass	CET-6	ASP	Q

'Has Wang Tao passed CET-6? '

S2:他连英语四级都没过，还六级呢。

Tā	lián	yīngyǔ	sì jí	dōu	méi	guò,	hái	liù jí	ne.
3SG	even	English	Level 4	DOU	NEG	pass	*HAI*	Level 6	*NE*

'He does not even pass CET-4, let alone CET-6. ' (cf. Feng[7])

（2）S1:王涛过四级了吗？

Wáng Tāo	guò	sì jí	le	ma
Wang Tao	pass	CET-4	ASP	Q

'Has Wang Tao passed CET-4? '

S2:还四级呢，他英语六级都过了。

Hái	sì jí	ne,	tā	yīngyǔ	liù jí	dōu	guò	le.
HAI	Level 4	*NE*	3SG	English	Level 6	DOU	pass	ASP

'He has passed CET-6, let alone CET-4.' (self-made)

According to the above examples, Speaker 2 (S2) responds with *"hai-NP1-ne"* construction. In the response, *"hai-NP1-ne"* part quotes NP1 and responds to the above proposition about NP1, and the other part is a proposition about NP2, in which NP1 and NP2 are contrastive components. The order between these two parts can be reversed without affecting the semantics. For the convenience of analysis, this structure is formalized as *"hai-NP1-ne, ... NP2..."*.

3.1 The Semantics of "Hai-NP1-ne, … NP2…" Construction

The part containing NP2 is a proposition with complete form and meaning, which is denoted as P (NP2). If we set the comparison item in P (NP2) to variable X, we will get P(X). P(X) in example (1) denotes Tā méiguò yīngyǔ X (he has not passed English X). P(X) in example (2) denotes Tā guò yīngyǔ X le (he has passed English X).

"Hai-NP1-ne" quotes "NP1" and responds to the above proposition about NP1. It is incomplete in form with an activity verb omitted. Its semantic interpretation needs reconstructing according to the context. Replace X by NP1, and we will get P (NP1) with complete form and semantics. P (NP1) in example (1) denotes Tā méiguò yīngyǔ liù jí (he has not passed CET-6). P (NP1) in example (2) denotes Tā guò yīngyǔ sì jí le (he has passed CET-4).

What is the relationship between P (NP1) and P (NP2)? If these two propositions in example (1) are connected by coordinating connective "hé (and)", this sentence will be interpreted as "Tā yīngyǔ sì jí hé liù jí dōu méiguò (he has not passed CET-4 and CET-6)". Then two facts will be ignored. One is that possibilities of two events are different, and the other is that "Tā méi guò yīngyǔ liù jí (he has not passed CET-6)" can be inferred from "Tā méi guò yīngyǔ sìjí (he has not passed CET-4)", but this sentence "Tā méi guò yīngyǔ liù jí (he has not passed CET-6)" is not necessarily true to life.

According to previous studies, the function of "hái" is to activate a sequence of time, rank or expectation [17], and to serialize entities and propositions within its scope. Deng [18] argues that the adverb "hái" locates the semantic focus towards the lower end of the pragmatic scale, indicating that the proposition that is unlikely to hold is still true. We find that the two propositions P (NP1) and P (NP2) in "hai-NP1-ne, ... NP2..." construction activate the likelihood scale and these two propositions are ordered by the likelihood.

In example (2), P (NP2) denotes Tā guò yīngyǔ liùjí le (he has passed CET-6), and P (NP1) denotes Tā guò yīngyǔ sì jí le (he has passed CET-4). As we know, CET-6 is more difficult than CET-4, so it is less likely to pass CET-6 than CET-4. That means LIKELIHOOD [P (NP2)] < LIKELIHOOD [P (NP1)] . If P (NP1) and P (NP2) are expressed as two points on the likelihood scale, then P (NP1) should be located at the higher point of the scale. In the negative situation, for example (1), P (NP2) denotes "Tā méiguò yīngyǔ sìjí (he has not passed CET-4)", and P (NP1) denotes "Tā méi guò yīngyǔ liù jí (he has not passed CET-6)". Because CET-4 is less difficult than CET-6, the possibility of not passing CET-4 is less than that of not passing CET-6, which can be expressed as LIKELIHOOD [P (NP2)] < LIKELIHOOD [P (NP1)]. According to the analysis of both positive and negative cases, we find that the proposition of "hai-NP1-ne" is more likely to occur.

In example (1), P (NP2) denotes "Tā méi guò yīngyǔ sìjí (he has not passed CET-4)", and P (NP1) denotes "Tā méiguò yīngyǔ liù jí (he has not passed CET-6)". When "Tā méiguò yīngyǔ sìjí (he has not passed CET-4)" is true, it can be inferred that "Tā méiguò yīngyǔ liù jí (he has not passed CET-6)" is true as well. But when "Tā méiguò yīngyǔ liù jí (he has not passed CET-6)" is true, we cannot infer that "Tā méiguò yīngyǔ sìjí (he has not passed CET-4)" is true. That means P (NP2) unilaterally entails P (NP1), which can be expressed as P (NP2) → P (NP1). In example (2), P (NP2) denotes Tā guò yīngyǔ liùjí le (he has passed CET-6), and P (NP1) denotes Tā guò yīngyǔ sì jí le (he has passed CET-4). When we know Tā guò yīngyǔ liùjí le (he has passed CET-6), we can infer that Tā guò yīngyǔ sì jí le (he has passed CET-4), but not vice versa, that is, P (NP2) → P (NP1). According to the analysis of both positive and negative cases, we find that P (NP2) unilaterally entails P (NP1), and the proposition of "hai-NP-ne" is at the lower point of the entailing scale.

To sum up, in the semantic interpretation of "hai-NP1-ne, ... NP2...", there are two propositions P (NP1) and P (NP2), P (NP2) unilaterally entails P (NP1) and the possibility of P (NP1) is higher than that of P (NP2). This semantic interpretation can be formalized as {P (NP2), P (NP1)} and P (NP2) → P (NP1), LIKELIHOOD [P (NP2)] < LIKELIHOOD [P (NP1)].

3.2 The Semantic Representation of "Hai-NP1-ne, ... NP2..." with Scale Structure

How to represent the semantics of "hai-NP1-ne, ... NP2..." with scale structure? "Hai-NP1-ne, ... NP2..." construction activates the likelihood scale. The degree on the scale corresponds to the possibility of the proposition, ranging from 0% to 100%. Along the arrow direction, propositions are more and more likely to occur. Because P (NP1) is more likely to occur than P (NP2), P (NP1) is located higher than P (NP2) on the scale. Since P (NP2) with a small possibility has occurred, we have greater reason to believe that P (NP1) with bigger possibility can occur. P (NP2) unilaterally entails P (NP1) and the entailment direction is the same as that of arrow.

The two propositions in "*hai-NP1-ne, ... NP2...*" construction and the semantic relationship between them are represented on the scale structure, as shown in Fig. 2.

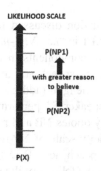

Fig. 2. The scale structure of the semantics of "*hai-NP1-ne, ... NP2...*"

The following are scale structures of the semantics of "*hai-NP1-ne, ... NP2...*" in the specific examples. Figure 3 is the scale structure of the semantics of "Tā lián yīngyǔ sì jí dōu méi guò, hái liù jí ne (he has not passed CET-4,let alone CET-6)" in the example (1). Figure 4 is the scale structure of the semantics of "Hái sì jí ne, tā yīngyǔ liù jí dōu guò le (he has passed CET-6, let alone CET-4)" in the example (2).

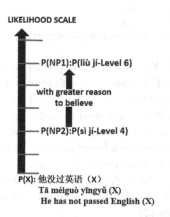

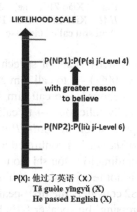

Fig. 3. The scale structure of the semantics of "Tā lián yīngyǔ sì jí dōu méiguò, hái liù jí ne (he has not passed CET-4, let alone CET-6)"

Fig. 4. The scale structure of the semantics of "Hái sì jí ne, tā yīngyǔ liù jí dōu guò le (he has passed CET-6, let alone CET-4)"

In the semantic scale structure of "hai-NP1-ne, … NP2…", the two propositions and the semantic relationship between them are clearly expressed. There is an entailment relationship between these two propositions. The proposition of "*hai-NP-ne*" is at the lower point of the entailing scale.

4 Semantic Analysis and Semantic Representation of Quotative *"Hai-NP-ne"* Construction Type B

In Type A of "*hai-NP-ne*" construction discussed in the previous section, there are contrastive components NP1 and NP1 in "*hai-NP1-ne, … NP2…*", while the other two types do not have overt contrastive components in the context. Type B of quotative "*hai-NP-ne*" construction is related to speech act, and the omitted verb is speech act verb, such as "shuō (say, speak)", "chēnghu (address)", "jiào (call)".

In the conversation, the first speaker S1 performs an NP-related speech act, and then the second speaker S2 directly quotes NP and responds to this NP-related speech act. "*Hai-NP-ne*" activates the felicity scale of NP-related speech act. The degrees on the scale indicate the felicity of speech acts. P (X) denotes how felicitous the NP-related speech act is, and X denotes the felicity of the speech act, ranging from not (0%) to completely (100%). In the direction of the arrow, the degree of felicity is getting bigger and bigger.

(3) S1: 小翟，你也来了？

Xiǎo Zhái, nǐ   yě   lái   le
Xiao Zhai 2SG   also come   ASP
'Xiao Zhai, do you come here too? '

S2: 还小翟呢，都是老头了。

Hái   Xiǎo Zhái ne, dōu   shì   lǎotou   le.
HAI Xiao Zhai *NE* DOU   be   old man ASP
'You still called me Xiao Zhai, but I am already an old man' (CCL corpus)

In the felicity scale of speech act, to call him Xiao Zhai is completely felicitous (felicity is 100%) → to call him Xiao Zhai is very felicitous → to call him Xiao Zhai is fairly felicitous → to call him Xiao Zhai is felicitous → …… → to call him Xiao Zhai is hardly felicitous → to call him Xiao Zhai is not felicitous (felicity is 0%)

The proposition of "*hai-NP-ne*" is at the lower point of the entailing scale. That *"hái Xiao Zhai ne"* is confirmed to locate the proposition at a lower point of the felicity scale, according to "Dōu shì lǎo tou le (I am already an old man)" in example (3). That means it is not felicitous to call him Xiao Zhai. It can be further inferred that the second speaker S2 considers the first speaker S1 should not perform the speech act in this way, thus expressing the speaker S2's blaming emotions [2].

The semantics of "*hai-NP-ne*" construction Type B is represented on the scale structure, as shown in Fig. 5.

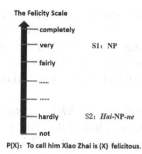

Fig. 5. The felicity scale structure of NP-related speech act in "*hai-NP-ne*" construction

5 Semantic Analysis and Semantic Representation of Quotative "*Hai-NP-Ne*" Construction Type C

The verbs omitted in "*hai-NP-ne*" construction Type C are state verbs such as "shì (be)" and "yǒu (have)". This type of "*hai-NP-ne*" construction quotes NP and responds to the first speaker S1's assertion about NP. It activates a truth-value scale to determine whether it is (or exists) NP or not. P(X) indicates whether it is X or whether there is X, and X refers to the NP in the utterance.

(4) S1:他现在是博士了吧？

Tā	xiànzài	shì	bóshì	le	ba
3SG	now	be	doctor	ASP	Q

'Is he a doctor now?'

S2:还博士呢，他连大学都没毕业。

Hái	bóshì	ne,	tā	lián	dàxué	dōu	méi	bìyè
HAI	doctor	*NE*	3SG	even	college	DOU	NEG	graduate

'He has not graduated from college, let alone being a doctor.' (this example adapted from Zheng [6])

The first speaker S1 asks a question whether he is a doctor now. It activates a truth-value scale to determine whether he is a doctor. The second speaker S2 utters "hái bóshì ne" to respond to S1's question and places his assertion at the lower point of the entailing scale. That is to say, he is not a doctor, which is proved by the context "Tā lián dàxué dōu méi bìyè (He has not graduated from college)" in example (4).

(5) S1：你家可是欠了我六百斤核桃。

Nǐ	jiā	kě	shì	qiàn	le	wǒ	liù bǎi	jīn	hétao
2SG	family	indeed	be	owe	ASP	1SG	six hundred	CL	walnut

'Your family owed me six hundred *jin* of walnuts.'

S2：还核桃呢，地都卖了。

Hái	hétao	ne,	dì	dōu	mài	le.
HAI	walnut	*NE*	land	DOU	sell	ASP

'Walnuts? The land is already sold' (this example cited from Zong [2])

In the presupposition of the first speaker S1, there are walnuts in the second speaker's house. The second speaker S2 responds by "hái hétáo ne", which activates a truth-value scale to determine whether there are walnuts or not, and places his assertion at the lower point of the entailing scale. That means there are not walnuts in his house, and it can be confirmed by the following utterance "Dì dōu mài le (the land is already sold)".

How to represent the semantics of "*hai-NP-ne*" construction Type C with the scale structure? The scale that determines whether it is NP or whether there is NP is binary opposition, yes or no. Yes is represented as T, and no is represented as F. T is at the upper endpoint, and F is at the lower endpoint. Because there are only two values, T and F, the line connecting these two values is dotted on the scale.

The semantics of "*hai-NP-ne*" construction Type C is represented on the scale structure, as shown in Figs. 6 and 7. Specifically, Fig. 6 is a truth-value scale to determine whether it is NP or not, and Fig. 7 is a truth-value scale to determine whether there is NP or not.

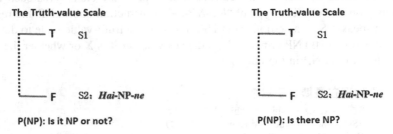

The Truth-value Scale

┌──── T   S1

└──── F   S2: *Hai*-NP-*ne*

P(NP): Is it NP or not?

The Truth-value Scale

┌──── T   S1

└──── F   S2: *Hai*-NP-*ne*

P(NP): Is there NP?

Fig. 6. A truth-value scale to determine whether it is NP or not

Fig. 7. A truth-value scale to determine whether there is NP or not

6   Conclusion

This paper mainly analyzes the semantics of three types of "*hai-NP-ne*" construction with NP quoted from context, and formalizes their semantics with the scale structure. Now the scale structures of three types of "*hai-NP-ne*" construction are summarized into Table 1 in order to compare their similarities and differences.

From Table 1, we can see that these three types of "*hai-NP-ne*" construction are different in scale sequences, dimensions, proposition sets, but consistent in the ordering relation. The proposition of "*hai-NP-ne*" is at the lower point of entailing scales. It means what should not happen in the speaker's view still happens.

This paper uses the scale structure to represent the semantics of three types of "*hai-NP-ne*" construction and semantic relationship between the proposition of "*hai-NP-ne*" and other related propositions. The introduction of a scale structure into semantic analysis is helpful for representing formally the meaning of utterances containing two or more propositions.

Table 1. Scale structures of three types of "*hai-NP-ne*" construction

Three types of "*hai-NP-ne*" construction	Type A	Type B	Type C
The types of omitted verbs in the construction	General activity verbs	Speech act verbs	State verbs
Discourse structure	S1: ......NP1...... S2: Hai NP1 ne, ...NP2 ...	S1: ......NP...... S2: Hai NP ne, ...	S1: ......NP...... S2: Hai NP ne, ...
Dimension	Likelihood of propositions	Felicity of speech act	Truth value of entity state
Degrees	[0%, 100%]	{completely, very, fairly, ..., hardly, no}	{T, F}
Proposition set	{P(NP1), P(NP2)}	{S1's speech act related to NP, *Hai NP ne*}	{S1's assertion about NP, *Hai NP ne*}
Ordering relation	The proposition of "*hai-NP-ne*" is at the lower point of the entailing scale	The proposition of "*hai-NP-ne*" is at the lower point of the entailing scale	The proposition of "*hai-NP-ne*" is at the lower point of the entailing scale
Semantics	P (NP2) with a small possibility has occurred, we have greater reason to believe that P (NP1) with bigger possibility can occur	The NP-related speech act performed by S1 is not felicitous	It negates S1's assertion of NP

"*Hai-NP-ne*" construction can derive different meanings by interacting with three different scales of likelihood scale, felicity scale, and truth-value scale. But the questions as to how to complete the proposition of "*hai-NP-ne*", how to construct the corresponding scale, and how to establish the entailment relationship among utterances, need further research. The answers to these questions will help computers to automatically interpret "*hai-NP-ne*" construction.

Acknowledgments. I would like to express my sincere gratitude to Professor Zhan Weidong, Dr. Huang Sisi, Dr. Kou Xin, Ke Lizhen for their helpful advice and insightful comments. This research is supported by the major project of Humanities & Social Science Fund of Ministry of Education of China (Project No. 15JJD740002).

References

1. Zong, Sh.-Y.: Analysis of "Hai+NP+Ne" and "Bi+N+Hai+N". J. Zhangjiakou Teach. Coll. (Soc. Sci.) **2**, 23–27 (1995). (in Chinese)
2. Zong, Sh.-Y.: The construction "Hai+X+ne" in three conceptual domains: acting, knowing, and uttering. Lang. Teach. Linguist. Stud. **4**, 94–103 (2016). (in Chinese)
3. Ding, L.: Comparison of two inference patterns "Hai+NP+ne" (S1) and "Dao di shi+NP+a" (S2) in two layers of thinking. J. Hanzhong Teach. Coll. (Soc. Sci.) **20**(1), 26–30 (2002). (in Chinese)

4. Ding, L.: Other perspectives on the Chinese sentence structure "hai+NP+ne". J. Shaanxi Univ. Technol. (Soc. Sci.) **25**(3), 46–50 (2007). (in Chinese)
5. Yang, Y.-L.: On the construction of "Hai+NP+ne". Contemp. Rhetoric **6**, 48–50 (2004). (in Chinese)
6. Zheng, J.-M.: On the construction "Hai+NP+Ne". Lang. Teach. Linguist. Stud. **2**, 9–15 (2009). (in Chinese)
7. Hu, F.: An exploration on the sentence pattern of "Hai+NP+Ne". J. Changchun Normal Univ. (Humanit. Soc. Sci.) **30**(6), 91–95 (2011). (in Chinese)
8. Wang, T.-Y.: The semantic analysis model of pair construction "A, hai B ne" based on "(De/Re)-Categorization+". Foreign Lang. Teach. **6**, 16–21 (2015). (in Chinese)
9. Wang, Ch.-W.: The negation focus, pragmatic function and solidification motivation of quotative responsive construction "Hai+X+(Ne)". J. Xinjiang Univ. (Philos. Humanit. Soc. Sci.) **45**(1), 128–134 (2017). (in Chinese)
10. Shen, J.-X.: Two constructions with Hai in Mandarin Chinese. Stud. Chin. Lang. **6**, 483–493 (2001). (in Chinese)
11. Fauconnier, G.: Pragmatic scales and logical structure. Linguist. Inq. **6**(3), 353–375 (1975)
12. Guo, R.: Entailment and negation. Chin. Teach. World **2**, 5–19 (2006). (in Chinese)
13. Fillmore, C.J., Kay, P., O'Connor, M.C.: Regularity and idiomaticity in grammatical constructions: the case of let alone. Language **64**(3), 501–538 (1988)
14. Kennedy, C., McNally, L.: Scale structure, degree modification, and the semantics of gradable predicates. Language **81**(2), 345–381 (2005)
15. Luo, Q.-P.: Gradability, scale structure and classification of simple adjectives in Chinese. Chin. Lang. Learn. **1**, 27–38 (2018). (in Chinese)
16. Luo, Q.-P.: Gradable nouns and the degree interpretations of "Da+NP" constructions. Chin. Lang. Learn. **3**, 43–52 (2016). (in Chinese)
17. Gao, Z.-X.: The basic meaning of adverb "Hai". Chin. Teach. World **2**, 28–34 (2002). (in Chinese)
18. Deng, Ch.-L: A semantic-pragmatic interface study of adverb Hai. Chin. Teach. World **32**(4), 475–482 (2018). (in Chinese)

The Centennial Controversy: How to Classify the Chinese Adverb *Dōu*?

Hua Zhong[(⊠)]

Overseas Education College of Fujian Normal University, Fuzhou, China
jtingshan@163.com

Abstract. The centenary research history of the adverb *dōu* is very controversial. There are severe divergences in academic circles on how to classify the semantic and pragmatic functions of the adverb *dōu*. In the existing literature, there are mainly three versions: the *trichotomy,* the *dichotomy*, and the *univocal. Trichotomy* and *Dichotomy* define the semantic and pragmatic functions of each sub-*dōu* with synonyms in Chinese analytical formulas. This kind of description is simple, convenient, and intuitive but also relatively vague and hard to verify. Moreover, there is a lack of extractions and instructions of the distinctive oppositions for different sub-*dōu*s. *Univocal* takes *totalizing universal* or the *universal/distributive quantification* as the consistent feature of the adverb *dōu*. The verification cannot cover all corpora of *dōu*, besides, it cannot evade the distinctive oppositions between $dōu_a$ and $dōu_b$, which are opposite to the consistent feature. After reviewing various views on the classification of *dōu* in the existing literature, this paper illustrates the distinctive semantic and pragmatic oppositions between $dōu_a$ and $dōu_b$. The first is the quantificational semantic features with objective truth values that $dōu_a$ has, but $dōu_b$ does not. The second is the pragmatic features of a subjective evaluation that $dōu_b$ has, but $dōu_a$ does not. So this paper approves of the *Dichotomy* that the adverbs *dōu* are $dōu_a$ and $dōu_b$.

Keywords: The adverb *dōu* · The semantic and pragmatic classifications of *dōu* · Distinctive oppositions between $dōu_a$ and $dōu_b$

1 Introduction: Making Clear the Starting Point and Direction in the Dispute

In the history of Chinese grammar research for more than a century, the adverb *dōu* has been studied not only for a long time but also with many participants and various research methods. Almost all schools and grammar theories have been applied here in various periods. Since [1] first called *dōu* as *tǒngkuò* (totalizing) adverb, the study on the adverb *dōu* has been going on for nearly a century. Most of the famous scholars of Chinese grammar have studied the related problems of the adverb *dōu*, such as [1–35]. According to the author's incomplete statistics, there are about 813 relevant

© Springer Nature Switzerland AG 2020
J.-F. Hong et al. (Eds.): CLSW 2019, LNAI 11831, pp. 63–73, 2020.
https://doi.org/10.1007/978-3-030-38189-9_6

pieces/works of literature[1] in China and foreign countries, which is very rare in the history of Chinese grammar research.

Especially since the 1980s, the description of the semantic and pragmatic functions of the adverb *dōu* has gradually changed from intuition and roughness to technicality, rationality, and refinement, and has achieved fruitful results. For example, [11, 16] revealed the distribution of *dōu*'s quantification. [12, 17, 27, 28, 30, 32, 36, 37] and others advocate *dōu* as an quantifier on events. [19, 27, 30] described the information structure of *dōu* sentence, and the topicality and definiteness of a plural NP before *dōu*. [21] characterized the semantic and pragmatic conditions of *dōu*'s quantification. [24] tried to explain the semantics of *dōu* by using the focus structure theory and the tripartite structure theory. Also, the semantics and pragmatics of *dōu* were character-ized and interpreted from many new perspectives by [25, 26, 28, 29, 31–35, 37–47]. However, while applauding these achievements, it is also regrettable. Because so far, there are still great controversies over basic issues, such as how to classify the semantic and pragmatic functions of the adverb *dōu* and how to define the semantic and prag-matic functions of sub-*dōu*s. Moreover, there is a growing trend.

In the existing literature, there are mainly *Trichotomy, Dichotomy*, and *Univocal* on how to classify the semantic and pragmatic functions of the adverb *dōu*. Because the semantic and pragmatic functions of adverb *dōu* are incredibly complex, the semantic and pragmatic functions for verification of each sub-*dōu* can be later defined precisely and accurately before a consensus on its classification. However, as far as the classi-fication is concerned, it cannot postpone extracting the distinctive oppositions between different sub-*dōu*s. According to the principle of mutual verification via a form and its meaning, namely the empirical method, any classification that has not extracted dis-tinctive oppositions is still literary, rather than scientific and rational. If you want to group *dōu*s into one category, you need to extract identical and consistent features of the semantic and pragmatic functions from all sub-*dōu*s. Also, they must be so precise and accurate to be verified, and any sub-*dōu* cannot have a feature against to their consistency. Otherwise, they cannot be grouped into one category.

Trichotomy and *Dichotomy* define the semantic and pragmatic functions of each sub-*dōu* by Chinese analytical synonyms, which are concise, convenient, intuitive, and straightforward. However, without the extraction and explanation of the distinctive oppositions between sub-*dōu*s, they are too vague to verify. *Univocal* takes *totalizing universal* or the *universal/distributive quantification* as the consistent feature of sub-*dōu*s. Not only its verification fails to cover all language facts fully but also fails to evade that the distinctive oppositions between *dōu*$_a$(*dōu*$_1$) and *dōu*$_b$(*dōu*$_2$, *dōu*$_3$) opposite to the consistency. Therefore, there is a need for other solutions.

---

[1] As of July 13, 2018, with the keyword of *DOU* or Chinese *fùcí dōu, Lián …dōu, zǒngkuò fùcí dōu, fànwéi fùcí dōu, yǔqì fùcí dōu, shíjiàn liànghuà dōu…* at FULink (Fujian Universities' Digital Library, the Web site: www.fulink.edu.cn) the data has been retrieved. The result was 9,640. Then the repeated and irrelevant documents were manually excluded, and the relevant documents that the author collected in many years were added. So finally, 813 articles/works were obtained. However, this is still incomplete statistics. The list of articles/works is available to interested readers by email and is welcome to be added.

So, after examining all the classifications in the existing literature, this paper tries to explain the distinctive oppositions between $dōu_a$ and $dōu_b$, to clarify the classification of the adverb *dōu*. Also, this paper hopes to eliminate controversies over the classification of the adverb *dōu* and reach a consensus of the *Dichotomy* so that further-in-depth study on the adverb *dōu* can start from the *Dichotomy*.

2 How Are the Semantic and Pragmatic Functions of the Adverb *Dōu* Classified?

2.1 The Dilemma Between the Classification and the Definitions of the Semantic and Pragmatic Functions of the Adverb *Dōu*

All scholars who pay attention to the adverb *dōu* must first think about two closely related issues. Such as, how to classify the semantic and pragmatic functions of the adverb *dōu*, and to define them/it? The solution of these two issues is the fundamental basis for the other related studies of the adverb *dōu*. Ignoring these two issues often leads to the story of *blinds touching elephants with different experiences*.

However, the solution of these two issues has a difficulty that is similar to *eggs first or chickens first*.

If the semantic and pragmatic functions were clearly defined, the classification would be easy. However, the semantic and pragmatic functions of $dōu_a(dōu_1)$ and $dōu_b(dōu_2, dōu_3)$ are too complex to define.

If the classification were precise, it could eliminate tangles between sub-*dōu*s. Also, it is easy to reveal the internal consistency of different contextual usage of the same sub-*dōu* and to define the semantic and pragmatic functions of sub-*dōu*s precisely. However, it is not easy to classify well when the semantic and pragmatic functions are not clearly defined.

So, which of these two issues should be solved first? This paper thinks that the classification only needs to find a few distinctive features, and not to comprehensively and accurately define the semantic and pragmatic functions of different sub-*dōu*s. Easy before difficult, so this paper should first have classified the adverb *dōu*, and then define the semantic and pragmatic functions of different sub-*dōu*s (developed in another article).

Therefore, the following reviews the views of classifying the adverb *dōu* in the existing literature, compare their advantages and disadvantages, explain the distinctive oppositions between $dōu_a$ and *dōu*, and determine the classification of the adverb *dōu*.

2.2 To Review on the Existing Views of Classifying the Adverb *Dōu*

With a dilemma, there is a natural divergence on the classification of the semantic and pragmatic functions of adverb *dōu*. The academic circles of Chinese grammar not only have various views on *dōu*'s classification but also some scholars have different opinions in him/herself different papers, such as [18, 25, 26, 48]. Even a few scholars put forward different opinions in the same article.[2] However, according to the number of sub-*dōu*s, four views of classification can be found in the existing literature, which are *Quartering*, *Trichotomy*, *Dichotomy*, and *Univocal*. The latter three are the main ones.

In the existing literature, the first two works, which advocated *Trichotomy* and described three sub-*dōu*s' semantic and pragmatic functions roundly, are [4] (Hereinafter *Speech*) and [50] (Hereinafter *Explanations*).

Speech (pp. 183–6) holds that the adverb *dōu* has three usages (this is *Trichotomy* mentioned above): I. Expressing a scope or a totality, the totalized components are always before *dōu*. II. *Lián...dōu...*means *shènzhì* (even). III. *Dōu...Le* means *yǐjīng* (already)...*Le* (Asp.). From now on these three usages are called as $dōu_1$, $dōu_2$, and $dōu_3$ respectively.

In *Explanations* (pp. 163–6) the adverb *dōu* is also summarized into three usages: I. To totalize the things mentioned before *dōu*, there is no exception to the following actions. II. *Dōu* indicates an emphatic tone, which can be on different occasions. 1. *Dōu* is used in the subordinate clause of the compound sentences of inference to indicate the tone of emphasis while still playing the role of totalizing the scope, and means *shènzhì* (even) and often matches with *lián* (including). 2. The emphasis is powerful, and also often matches with *lián* (including). 3. *Dōu...le* means *yǐjīng* (already) ...*le* (Asp) and emphasizes that the time is approaching or the situation already exists. III. *Dōu shì* (All are) ... is used to explain the reason, often with a complaining tone. From now on these three usages are called as $dōu_a(dōu_1)$, $dōu_b(dōu_2, dōu_3)$, and $dōu_c$ respectively.

Since then, the roots of the various classification views of the adverb *dōu* seem to be found in *Speech* or/and *Explanations*. For example, in *Modern Chinese Dictionary*, from the trial edition (p235) to the 7th edition (p315), the adverb *dōu* has been interpreted as four usages. I. Expressing totalizing, and the totalized component is before *dōu* except for interrogative sentences. II. *Dōu* collocated with *shì* explains the reason. III. It expresses *shènzhì* (even). IV. It expresses *yǐjīng* (already). This view of *Quartering* seems to be the addition of $dōu_1$, $dōu_2$, and $dōu_3$ in *Speech* and $dōu_c$ in *Explanations*.

$Dōu_c$, which is used to explain the reason, is attributed to $dōu_1$ in [6]. In addition to *Explanations* and *Modern Chinese Dictionary*, which take $dōu_c$ as an independent sub-*dōu*, the academic circle usually classifies it as $dōu_1$. If there were no $dōu_c$, the adverb *dōu* in *Explanations* could only be $dōu_a(dōu_1)$ and $dōu_b(dōu_2, dōu_3)$. Therefore, it is not difficult to understand that the academic circle later developed the viewpoint of *Dichotomy*, such as [22, 25, 49, 51–61].

After further developed the claim that $dōu_b$ in *Explanations* indicates the tone of emphasis while still playing the role of totalizing a scope, there is naturally the view of *Univocal* that *dōu* has only one meaning and usage of totalizing universal or a universal quantification, such as [12–14, 17, 18, 24, 29, 35, 42, 45, 62–67], and so on.

The shared grammar works and documents, represented by [6], describe *dōu* as $dōu_1$, which means totalizing universal; $dōu_2$, which is equivalent of *shènzhì* (even);

---

[2] For example, Paris [49] points out that *dōu* has two meanings. However, what seems to be confusing is that she holds that there is only one *dōu* (in the footnote 1).

and $dōu_3$, which is equivalent to *yǐjīng*(already). They define the semantic and pragmatic functions of each sub-*dōu* with synonyms in Chinese analytical formulas. This kind of description is simple, convenient, and intuitive, but also relatively vague. Moreover, it is impossible to carry out full-coverage verifications that maintain consistency internally and exclusive externally. Moreover, according to different collocations of words (such as *Lián...dōu...*, *Dōu...le*) and different contextual meanings of different sentence patterns, the original $dōu_b$ of the same sub-*dōu* is divided into $dōu_2$ and $dōu_3$. This classification stops at the surface of contextual usage and ignores the internal consistency of $dōu_2$ and $dōu_3$. Although widely used in teaching, these descriptions can only be an expedient measure before more rigorous and accurate definitions are produced, and a consensus is reached.

The diachronic relationship between $dōu_a$ and $dōu_b$ can be linked by totalizing universal or the universal/distributive-quantification. However, *Univocal* takes totalizing universal to define the synchronic relationship between $dōu_a$ and $dōu_b$, ignoring distinctive oppositions between them.[3] This kind of scheme cannot evade distinctive oppositions between $dōu_a$ and $dōu_b$. This kind of scheme cannot evade distinctive oppositions between $dōu_a$ and $dōu_b$. First is the quantificational semantic features with the objective truth value, which $dōu_a$ has but $dōu_b$ does not. Second is the pragmatic features of subjective evaluation, which $dōu_b$ has but $dōu_a$ does not (see below for details). Besides, *Univocal* regards the universal/distributive-quantification on a plural disordered-set or ordered-set as the consistency of $dōu_a$ and $dōu_b$ such as [17, 24, 42, 65, 66]. However, this proposal couldn't evade these problems: First, in a $dōu_a$ sentence, the disordered set may be singular, such as *Nà ge mántou* (that steamed bun). Second, in some sentences, even if there is a plural disordered set, such as in **Tāmen liǎ dōu shì fūqī* (*Both of them are husband and wife), the $dōu_a$ sentence may be untenable. Third, the plural quantification meaning of $dōu_a$ is an objective truth-meaning of a proposition itself, rather than a pragmatic deduction meaning, which cannot be canceled. While $dōu_b$'s universal/distributive-quantification of an ordered-set is not an objective truth-meaning of a proposition itself, which is entirely the result of a plausible inference and can be canceled (see [59–61]). For example:

(1) Lián yéye *dōu* qù le, nǐ hái bú qù
    Including grandpa even go Le, you still not go
    Even Grandpa's gone. You're not going yet.

In this sentence, the universal set implied in the clause *Lián...dōu...* is canceled by the latter clause.

In this way, *Univocal* cannot rule out tangles between originally different categories of $dōu_a$ and $dōu_b$. So that while they advocate *Univocal*, they also advocate a classification similar to *Dichotomy* such as [17, 24, 42, 62, 65–68]. Therefore, *Univocal* is challenged to be univocal.

---

[3] [68] believes that the semantic core of $dōu_1$ includes two parallel semantic components, namely the distributive meaning and a relatively large amount. While the semantic core of $dōu_2$ and $dōu_3$ only contains a relatively large amount. [26] advocates that the condition of *dōu* should be unified to reflect a certain degree of subjective-relative-value that does not need to be quantified objectively. These two theories are too ad hoc and not commented.

Although correctly classified the adverb *dōu*, the existing *Dichotomy* describes *dōu*$_b$ as a modal particle of subjectivity and emphasis (see [22, 25, 51–58]), which is also closer to the natives' intuition. The definitions of semantic and pragmatic functions of *dōu*$_a$ and *dōu*$_b$ in most documents are still not rigorous and accurate enough. So they still unable to be carried out full-coverage verifications that are internally consistent and externally exclusive, and also could not extract the distinctive oppositions between *dōu*$_a$ and *dōu*$_b$. The first is the quantificational semantic feature with objective-truth values, which *dōu*$_a$ has, but *dōu*$_b$ does not. The second is the pragmatic features of subjective evaluation, which *dōu*$_b$ has, but *dōu*$_a$ does not. For example:

(2) a. Tāmen dōu$_a$ shàng dàxué le.
       They all go college LE.
       They all went to college.
    b.* Tā dōu$_a$ shàng dàxué le.
       *He all go college LE.
       *He all went to college.
(3) a. (Lián) Tāmen dōu$_b$ shàng dàxué le.
       (Including) They even go college LE.
       Even they went to college.
    b. (Lián) Tā dōu$_b$ shàng dàxué le.
       (Including) He even go college LE.
       Even he went to college.
(4) Lǎo Lǐ, Lǎo Wáng hé Lǎo Zhāng dōu$_a$ Kuài wǔshí le.
    Lǎo Lǐ, Lǎo Wáng, and Lǎo Zhāng all near fifty LE
    Lǎo Lǐ, Lǎo Wáng, and Lǎo Zhāng are all approaching their fifties.
(5) a. Dōu$_b$ Kuài wǔshí le, yīnggāi xiǎngshòu xiǎngshòu le, búyào zài nàme láolèi le.
       Even near fifty LE, should enjoy enjoy LE, not again so tired LE.
       It's almost 50 years old. I should enjoy it. Don't be so tired.
    b. Dōu$_b$ Kuài wǔshí le, gèng yīnggāi zhuājǐn shíjiān le,
       Even near fifty LE, more should grasp time LE,
       néng yǒuxiào gōngzuò de shíjiān bù duō le.
       able effective work DE time not much LE.
       It's almost 50 years old. More time should be grabbed. There is not much time to work effectively.

The difference between (2) and (3) illustrates the first distinctive opposition. Also, the difference between (4) and (5) illustrates the second distinctive opposition.

Therefore, it is necessary to clarify the distinctive oppositions between *dōu*$_a$ and *dōu*$_b$, to reach a consensus of the *Dichotomy* and more scientifically and rigorously define the semantic and pragmatic functions of *dōu*$_a$ and *dōu*$_b$.

Recently, more and more scholars advocate *Dichotomy*, that divides the adverb *dōu* into *dōu*$_a$ and *dōu*$_b$. In this way, not only can they avoid *Univocal* ignoring the distinctive oppositions between *dōu*$_a$ and *dōu*$_b$ but also *Trichotomy* dividing *dōu*$_b$ into *dōu*$_2$ and *dōu*$_3$ on the surface of contextual usage. Compared with the traditional *Trichotomy* and the more modern *Univocal* (with more technical rationality), the *Dichotomy* is more in line with the natives' intuition and more accessible to be verified after extracting and explaining distinctive oppositions between *dōu*$_a$ and *dōu*$_b$.

3 Conclusion

The centenary study of the adverb *dōu* is full of controversy, among which the most urgent one is the dispute of its classification. There are severe divergences in academic circles on how to classify the semantic and pragmatic functions of the adverb *dōu*. In the existing literature, there are mainly three versions: *Trichotomy, Dichotomy,* and *Univocal.*

In *Trichotomy,* there are three sub-*dōu*s. The first is $dōu_1$, which means totalizing universal; the second is $dōu_2$, which is equivalent of *shènzhì* (even); and the third is $dōu_3$, which is equivalent to *yǐjīng* (already). *Trichotomy* defines the semantic and pragmatic functions of each sub-*dōu* with synonyms in Chinese analytical formulas. This kind of description is concise, convenient, intuitive, and straightforward, but relatively vague. Also, it is impossible to carry out full-coverage verifications that maintain consistency internally and exclusive externally. Moreover, according to different collocations of words (such as *Lián ... dōu ..., Dōu ... le*) and different contextual meanings of different sentence patterns, the original $dōu_b$ in the same sub-*dōu* is divided into $dōu_2$ and $dōu_3$. This classification stops at the surface of contextual usage and ignores the internal consistency of $dōu_2$ and $dōu_3$. Although widely used in teaching, these descriptions can only be an expedient measure before more rigorous and accurate definitions are produced, and a consensus is reached.

The diachronic relationship between $dōu_a$ and $dōu_b$ can be linked by totalizing universal or the universal/distribution-quantification. However, *Univocal* takes totalizing universal to define the synchronic relationship between $dōu_a$ and $dōu_b$, ignoring distinctive oppositions between them. This kind of scheme cannot evade distinctive oppositions between $dōu_a$ and $dōu_b$. $Dōu_a$ has quantification semantic features with the objective truth value, which $dōu_b$ does not; and $dōu_b$ has the pragmatic features of subjective evaluation, which $dōu_a$ does not. Besides, *Univocal* regards a universal/distribution-quantification of a plural disordered-set or ordered-set as the consistency of $dōu_a$ and $dōu_b$. However, this proposal couldn't evade these problems: First, in a $dōu_a$ sentence, the disordered set may be singular, such as *Nà ge mántou* (that steamed bun). Second, in some sentences, even if there is a plural disordered-set, such as in **Tāmen liǎ dōu shì fūqī* (*Both of them are husband and wife), the $dōu_a$ sentence may be untenable. Third, the plural quantification-meaning of $dōu_a$ is the objective truth-meaning of the proposition itself, rather than the pragmatic deduction meaning, which cannot be canceled. However, $dōu_b$'s universal/distribution-quantification of an ordered-set is not the objective truth-meaning of the proposition itself, which entirely results of plausible inference and can be canceled.

Then *Univocal* cannot rule out tangles between originally different categories of $dōu_a$ and $dōu_b$ and fails to fill up gaps in its theory. So it is better to give up *Univocal* and adopt *Dichotomy.*

Although correctly classified the adverb *dōu*, the existing *Dichotomy* describes $dōu_b$ as a modal particle of subjectivity and emphasis, which is also closer to the natives' intuition. In the existing literature, definitions of semantic and pragmatic functions of $dōu_a$ and $dōu_b$ are still not rigorous and accurate enough to extract the distinctive oppositions between them. Also, they are unable to be carried out full-coverage verifications that are internally consistent and externally exclusive. This paper shows that there are two distinctive oppositions between $dōu_a$ and $dōu_b$. So this paper approves of the *Dichotomy* that the adverbs *dōu* are $dōu_a$ and $dōu_b$.

Recently, more and more scholars advocate *Dichotomy* that divides the adverb *dōu* into $dōu_a$ and $dōu_b$. In this way, not only can they avoid *Univocal* ignoring the distinctive oppositions between $dōu_a$ and $dōu_b$ but also *Trichotomy* dividing $dōu_b$ into $dōu_2$ and $dōu_3$ on the surface of contextual usage. Compared with the traditional *Trichotomy* and the more modern *Univocal* with more technical rationality, *Dichotomy* is more in line with the natives' intuition and more accessible to be verified after extracting and explaining distinctive oppositions between $dōu_a$ and $dōu_b$.

Acknowledgment. The work was supported by the National Social Science Foundation of China (Grant No. 15BYY141).

References

1. Li, J.X.: A New Grammar of the Chinese National Language. The Commercial Press, Beijing (1924/ 1992). (In Chinese)
2. Wang, L.: Modern Chinese Grammar. The Commercial Press, Beijing (1943/1985). (In Chinese)
3. Gao, M.K.: On Chinese Grammar. The Commercial Press, Beijing (1948/1986). (In Chinese)
4. Ding, S.: Lectures on Modern Chinese Grammar. The Commercial Press, Beijing (1961/1980). (In Chinese)
5. Zhao, Y.R.: Spoken Chinese Grammar. The Commercial Press, Beijing (1968/1979). (In Chinese)
6. Lü, S.X.: Eight Hundred Words of Modern Chinese. The Commercial Press, Beijing (1980). (In Chinese)
7. Zhu, D.X.: Lectures on Grammar. The Commercial Press, Beijing (1982). (In Chinese)
8. Ma, Z.: On the Location of the Objects totalized by *Dou/Quan*. Chin. Learn. 4, (1983). (In Chinese)
9. Ma, Z.: The Methodology of Research on Function Words in Mandarin Chinese. The Commercial Press, Beijing (2004). (In Chinese)
10. Xing, F.Y.: On the structure of NP + Le. Linguist. Res. **3**, 21–26 (1984). (In Chinese)
11. Lee, H.T.: Studies on quantification in Chinese. Ph.D. Dissertation, University of California, Los Angeles (1986)
12. Huang, S.Z.: Dou as an existential quantifier. In: Proceedings of the 6th North American Conference on Chinese Linguistics, vol. 11, pp. 114–125 (1994)
13. Huang, S.Z.: Quantification and Predication in Mandarin Chinese: A Case Study of Dou. Ph. D. Dissertation, University of Pennsylvania, Philadelphia (1996)
14. Huang, S.Z.: Universal Quantification with Skolemization as Evidence in Chinese and English. The Edwin Mellen Press, New York (2005)
15. Lin, J.W.: Polarity licensing and Wh-phrase quantification in Chinese. Unpublished Ph.D. Dissertation, University of Massachussets, Amherst, MA (1996)
16. Lin, J.W.: Distributivity in Chinese and its implication. Nat. Lang. Semant. **6**, 201–243 (1998)

17. Jiang, Y.: Pragmatic reasoning and syntactic/semantic characterization of *Dōu*. Modern Foreign Lang. **1**, 11–24 (1998). (In Chinese)
18. Dong, X.F.: Definite objects and relevant questions of *Dōu*. Stud. Chin. Lang. **6**, 495–507 (2002). (In Chinese)
19. Dong, X.F.: The position of *dōu* and some related issues. Chin. Teach. World. **1**, 495–507 (2003). (In Chinese)
20. L, J.M.: A Course in the Study of Modern Chinese Grammar. Peking University Press (2004). (In Chinese)
21. Zhang, Y.S.: On the selective restrictions of Chinese adverb *Dōu*. Stud. of the Chin. Lang. **5**, 392–398 (2003). (In Chinese)
22. Zhang, Y.S.: Grammaticalization and subjectivization of adverbs *Dōu*. J. Xuzhou Norm. Univ. **3**, 56–62 (2005). (In Chinese)
23. Liu, D.Q.: Atypical *Lián* sentence as typical construction. Lang. Teach. Linguist. Stud. **4**, 1–12 (2005). (In Chinese)
24. Pan, H.H.: Focus Point and Trisection Structure and Semantic Interpretation of Chinese *Dōu*. Grammar Study and Exploration. vol. 13. The Commercial Press, Beijing (2006). (In Chinese)
25. Xu, L.J.: Similarities and differences of Shanghai Dialect Chai (侪) and Mandarin *Dōu*. Dialects **2**, 97–102 (2007). (In Chinese)
26. Xu, L.J.: Is *Dōu* a universal quantifier? Stud. Chin. Lang. **6**, 498–507 (2016). (In Chinese)
27. Yuan, Y.L.: A new explanation of the semantic function and association direction of *dōu*. Stud. Chin. Lang. 2 (2005). (In Chinese)
28. Yuan, Y.L.: The summative function of *Dōu* and its distributive effect. Contemp. Linguis. **4**, 289–304 (2005). (In Chinese)
29. Yuan, Y.: The Information Structure of the *Lián* construction in Mandarin. Linguist. Sci. **2**, 14–28 (2006). (In Chinese)
30. Yuan, Y.L.: On the implicit negation and NPI licensing of *Dōu*. Stud. Chin. Lang. **2**, 306–320 (2007). (In Chinese)
31. Yuan, Y.L.: On the coordinate and restrictive relation between *Měi* and *Dōu*. J. Sino-Tibetan Lang. **2**, 105–123 (2008). (In Chinese)
32. Yuan, Y.L.: Route of grammar research pushed by question. J. Suzhou Inst. Edu. **2**, 2–11 (2010). (In Chinese)
33. Xiong, Z.R.: *Dōu*'s rightward semantic association. Modern Foreign Languages, 1 (2008). (In Chinese)
34. Xiong, Z.R.: The differences of quantificational objects between *zǒng* and *Dōu*. Stud. Chin. Lang. 3 (2016). (In Chinese)
35. Shen, J.X.: Leftward or rightward? The quantifying of *Dōu*. Stud. Chin. Lang. 1 (2015). (In Chinese)
36. Huang, Z.H.: The quantification adverb *Dōu* in Mandarin Chinese and the focus structure of *Dōu*-sentence. Ph.D. dissertation Peking University (2004). (In Chinese)
37. Huang, Z.H.: A comparative analysis of *dōu* and *zǒng* in event quantification. Stud. Chin. Lang. **3**, 251–264 (2013). (In Chinese)
38. Cheng, L.-S.: On *dōu*-quantification. J. East Asian Linguist. **4**, 197–234 (1995)
39. Wu, J.X.: Syntax and semantics of quantification in Chinese. Ph.D. dissertation, University of Maryland at College Park (1999)
40. Xiang, M.: Plurality, maximality and scalar inferences: a case study of Mandarin *dōu*. J. East Asian Linguist. **17**, 227–245 (2008)
41. Liu, C.F.: On pragmatic number of Modern Chinese. Ph.D. dissertation, Fudan University (2007). (In Chinese)

42. Jiang, J.Z., Haihua, P.: How many *Dōu*s do we need? Stud. Chin. Lang. **1**, 38–50 (2013). (In Chinese)
43. Xiang, Y.M.: Interpreting questions with non-exhaustive answers. Ph.D. dissertation, Harvard University. Cambridge, Massachusetts (2016)
44. Liu, M.M.: Varieties of alternatives: Mandarin focus particles. Linguist. Philos. **1**, 61–95 (2017)
45. Feng, Y.L., Haihua, P.: Revisiting the semantics of *dōu*: From the perspectives of exhaustivity and exclusiveness. Stud. Chin. Lang. **2**, 177–194 (2018). (In Chinese)
46. Wu, H., Hongyin, T.: Expressing (inter)subjectivity with universal quantification: a pragmatic account of Plural NP + *dou* expressions in Mandarin Chinese. J. Pragmatics **128**, 1–21 (2018)
47. Yang, X., Yicheng, W.: A dynamic account of *lian...dou* in Chinese verb doubling cleft construction. Lingua **217**, 24–44 (2019)
48. Dong, X.F.: Quantity and Emphasis. In: On Quantity and Plural Number: A Cross-temporal Survey of Languages in China, pp. 312–328. The Commercial Press, Beijing (2010). (In Chinese)
49. Paris, M.C.: *Lián...Yě/Dōu* in Mandarin Chinese. Linguist. Abroad. **3**, 50–55 (1981)
50. 1955, 1977 L. CL. of Chin. D. of Pek. Un.: Explanations of Functional Words in Modern Chinese. The Commercial Press, Beijing (1982). (In Chinese)
51. Yao, X.B.: Basic usages of *Dōu* in Modern Chinese. J. Jinzhou Normal College, **4** (1984). (In Chinese)
52. Wang, H.: Analysis of grammar meaning of adverb *Dōu*. Chin. Learn. **6**, 55–60 (1999). (In Chinese)
53. Wang, H.: Syntactic, semantic, and pragmatic analysis of adverb Dōu. Jinan University, M.'s Thesis (2000). (In Chinese)
54. Wang, H.: Semantic and pragmatic analysis of modal adverb *Dōu*. J. Jinan Univ. Coll. Lib. Arts. **6**, 41–45 (2001). (In Chinese)
55. Jiang, J.: Evolution and classification of totalizing. Chin. Learn. **4**, 72–76 (2003). (In Chinese)
56. Xiao, S.R.: Item of total quantity and meaning of implicit comparison of *Dōu*. J. Hunan Inst. Hum. Sci. Technol. **6**, 98–104 (2005). (In Chinese)
57. Xu, Y.Z., Yiming, Y.: Subjectivity, objectivity, and pragmatic ambiguity of adverb *Dōu*. Lang. Res. **3**, 24–29 (2005). (In Chinese)
58. Zhang, Y.J.: Semantic recognition strategies and discourse understanding of adverbs *Dōu* in authentic texts. J. Yangzhou Univ. **2** (2007). (In Chinese)
59. Zhong, H.: 都$_b$[*Dou$_b$(Dou$_2$, Dou$_3$)*] as a counter-expectation discourse-marker: on the pragmatic functions of *Dou$_b$* from the perspective of discourse analysis. In: CLSW 2015, LNAI, vol. 9332, 392–407 (2015)
60. Zhong, H.: On the quantification of events in *Dōu$_a$* construction. In: CLSW 2017, LNAI, vol. 10709, 41–60 (2018)
61. Zhong, H.: The conventional implicature of *Dōu$_b$(Dōu$_2$, Dōu$_3$)*: On Semantics of *Dōu$_b$* from the Perspective of Discourse Analysis. In: CLSW 2018, LNAI, vol. 11173, pp. 44–60 (2018)
62. Nakagawa, C.: Contextual analysis and analysis of the tone of Chinese Adverbs *Dōu*. Chin. Transl. Jpn. Res. Pap. Anthol. Mod. Chin. (In Chinese) Translated by Xun Chunsheng, Yusunori Ohkochi, Shi Guangheng. Beijing, Beijing Languages College Press, pp. 309–322 (1993[1985])
63. Lin, S.: Determination principles of range adverbs. J. Shanghai Norm. Univ. **1**, 125–126 (1993). (In Chinese)
64. Ren, H.B.: Semantic function and ambiguity of *Dōu*. J. Zhejiang Univ. **2**, 101–106 (1995). (In Chinese)

65. Jiang, Y.: Scalar Model and Semantic Features of *Dōu*, Conference Articles of The First ISCFF (2006). (In Chinese)
66. Luo, Q.P.: *Dōu*'s Quantification and *Lián* XP *dōu* Construction. Linguist. Theor. Ser. **34**, 243–265 (2006). The Commercial Press, Beijing. (In Chinese)
67. Wu, P., Chou, M.: An Interpretation of *Dōu* from perspective of semantics and pragmatics. Chin. Teach. **1**, 29–41 (2016). (In Chinese)
68. Li, W.S.: On the semantic complexity of *Dōu* in Mandarin: a partially unified account. Chin. Teach. **3**, 319–330 (2013). (In Chinese)

Research on the Hidden 'De' in Basic Noun Compounds Based on the Large-Scale Corpus

Ying Zhang[1], Pengyuan Liu[1(✉)], and Qi Su[2]

[1] School of Information Science, Beijing Language and Culture University,
Beijing, China
zhangyingblcu@163.com, liupengyuan@pku.edu.cn
[2] School of Foreign Languages, Peking University, Beijing, China
sukia@pku.edu.cn

Abstract. Basic noun compounds refer to phrases with nominal functions composed of two nouns. The study on the concealment of "de" in basic noun compounds is helpful to discover the implicit knowledge of noun and noun combinations, and the transformational rules between "N-N phrases" and "N-de-N phrases". Scholars have described the hidden "de" from different perspectives, but no statistical analysis has been carried out from a large-scale corpus. This paper takes newspapers in Dynamic Circulation Corpus (DCC) for ten years as the corpus source and extracts basic noun compounds from it. We find out part without "de" in corpus by searching corpus for verification and after that, summarize types of basic noun compounds from two levels of syntactic structure and semantic relationship. We also analyze the reason for this phenomenon.

Keywords: Basic noun compounds · Structural relationship · Semantic relations · Hidden "de"

1 Introduction

In modern linguistic theories, nouns are one of the most basic word classes in human language. As the largest part of speech in Chinese, nouns occupy a very important position in various aspects of language research and application, and the number of nouns and noun combinations is even more and varied. Basic noun compounds refer to a phrase with noun function consisting of two nouns [1]. Different from compounds, it expresses the most complete information with the least number of words in the form of two noun combinations. The hidden "de" mainly refers to whether the word "de" appears in the basic noun compounds. This study aims to find out the transformational and generative rules between "N-N phrases" and "N-de-N phrases", helping to analyze the deep connection between the two and find out the syntactic structure and semantic relations. In this way, we can excavate the implicit knowledge of combining nouns.

Scholars have made fruitful achievements in describing the concealment of "de" from different perspectives such as whether phrases denote appellations, the conceptual distance between two nouns, or the number of syllables and prosody. Most of these studies are based on example analysis; however, few are based on corpus. This paper extracts high-frequency "N-N phrases" and compares them with the "N-de-N phrases"

© Springer Nature Switzerland AG 2020
J.-F. Hong et al. (Eds.): CLSW 2019, LNAI 11831, pp. 74–83, 2020.
https://doi.org/10.1007/978-3-030-38189-9_7

in a large-scale corpus, expecting to analyze and summarize the occurrence and possible reasons of the word "de" from different levels of syntax and semantic meaning.

2 Research Review

Chen first noticed the looming phenomenon of the word "de" in the "N-N phrase" [2]. From the point of view of whether it denotes the appellation of things, Chen argued that when it is regarded as the appellation of a thing or a kind of thing without specifying its shape and characteristics or the relationship of ownership between two nouns, it is not necessary to use "de", such as "中国人民(Chinese people)". When we use the structure to describe a thing or a kind of thing, we do not regard it as the appellation of the thing, that is to say, when we emphasize the modification and differentiation of modifiers, we use the word "de", such as "铜的墨盒(copper cartridge)". Lv concluded that the meaning of the two noun phrases is specialized, and "de" was not used; if modifiers and central nouns are not often combined, it should be used [3]. Xu interpreted the former as "plate principle", that is, the whole plate represented the appellation, while the latter, because of the "prominence principle", adding "de" is to give prominence to modify partial terms [4, 5].

Some scholars have proposed to judge whether "de" is added between two nouns according to their conceptual distance. Zhang held that the concealment of "de" follows the principle of "distance iconicity" [6], requiring that the formal distance between linguistic components should be parallel to the conceptual distance, while "de" can increase the formal distance between attributive and core nouns. Therefore, the farther the conceptual distance between attributive and core noun is, the easier it is to take "de"; the closer it is, the easier it is not to take "de". Lu put forward the "distance-marker correspondence law" [7]. He believed that "the farther an adjunct is from the core, the more it needs to use explicit markers to express the semantic relationship between it and the core." Zhang proposed the theory of "the tightness of structure" and "the tightness of semantic relations" [8], that is, the tightness of the combination of two nouns to classify whether to add "de". But these statements only apply to some of the hidden phenomena of "de". For example, they can explain the "大瓷的杯子(large porcelain cup)". They are not established because the concepts of "瓷(porcelain)" and "杯子(cup)" are relatively close, but they cannot say that "大(large)" is farther away from the core "杯子(cup)" than "瓷(porcelain)", so it is more necessary to carry an explicit mark "de".

There are also scholars who explain the disappearance of the word "de" from the perspective of syllable number and rhythm. Xu explained that the disappearance of the word "de" to meet the needs of rhythmic harmony, and the balance of context structure also has an influence [4, 5]. Zhang put forward "the adaptation of the number of syllables" [8], that is, the number of syllables in the nominal combination of a certain collocation does not use the word "de". In fact, there are two kinds of structures in the collocation relationship of various syllable numbers, namely "modifier-noun" and "modifier-de-noun". The collocation of syllable numbers and the concealment of the word "de" not simply play an adjusting role.

In addition, some scholars have described and explained it from other angles. Sun, Liu and Wang summarized the rules of the hidden word "de" according to the two nouns used as the syntactic elements [9–11]. Zheng described that the different meanings expressed when "de" exists or not [12]. Qian divided it into two categories: possession and non-possession from the perspective of semantic relations [13], indicating that possession needs to be marked by "de"; otherwise "de" is generally not used. The above scholars mainly describe the concealment of the word "de" in terms of the part of speech and the relationship between the two nouns, but they fail to explain the situation that some phrases in the possessive relationship can be added or not added.

3 Data Preparation

3.1 Data Collection

The Dynamic Circulation Corpus (DCC) [14] collects newspaper data from 2005 to 2015, which includes 14 to 16 newspapers each year, totaling more than three billion words.

3.2 Data Screening

The original corpus was segmented and part-of-speech tagged by using the Jieba word segmentation tool [15], and the phrase extraction was carried out with the "N + N" sequence as the extraction mode (including proper names such as person names, place names and institution names), in this way a total of more than 19 million were obtained. In this paper, the top 20,000 phrases with the highest frequency were studied, and the phrases were manually screened with the illegal structures one by one, mainly in three forms:

There is no direct syntactic connection between nouns in compound noun phrases. This kind of error mainly occurs in the multi-layer nested noun structure. Simply relying on the recognition pattern of "N + N" cannot recognize the direct and indirect connections in the multi-layer nested relations. For example, the complete phrase "*中国国家(China-nation)" should be "中国国家统计局(the National Bureau of statistics)". "国家(nation)" and "统计局(Statistical Bureau)" are direct elements, while "中国(China)" and "统计局(Statistical Bureau)" are indirect components. Thus there is no grammatical relationship between "中国(China)" and "国家(nation)", so they do not meet our requirements.

Word classification errors, such as "中国传统节日 (Chinese traditional festivals)", are mainly due to the coarse participle size of the word segmentation procedure itself, which divides "传统节日(traditional festivals)" into a noun. In fact, it is a phrase consisting of adjectives "传统(traditional)" and the noun "节日(festival)".

Part-of-speech tagging errors, such as "专家研究 (expert research)" and "定位高端(targeted at high-end)", are also the problems of part-of-speech tagging tools. At present, there is no tagging tool with 100% accuracy, so it is inevitable that part-of-speech tagging errors exist, especially for part-of-speech tagging of multi-category words, so this paper uses manual screening.

A total of 10,946 qualified basic noun compounds were obtained through manual screening. According to word "de" can be added between two nouns, compounds can be divided into two categories, that is, "de" can be added and "de" cannot be added. For some noun phrases that cannot be clearly determine whether to add the word "de", we also find out the original sentence in the corpus for judgment. Finally, we get 1,334 basic noun compounds which cannot add the word "de" and 9,612 which can add the word "de".

4 Statistical Analysis

4.1 Basic Information

The basic frequency distribution of 10,946 basic noun compounds is shown in Table 1:

Table 1. Statistics of basic noun compounds

Category	Count
Basic noun compounds	10,946
Noun	4,693
Noun1	2,967
Noun2	2,975
The highest frequency of phrases	192,722
The lowest frequency of phrases	1,142
Average frequency of phrase	3,592

A total of 4,693 nouns are used in the basic noun compounds. Among them, different nouns have different combing abilities. For example, 328 phrases with the word "中国(China)" and two phrases with the word "实体(entity)". Moreover, there are 1,239 nouns in the front and back positions, such as "生活工作(life and work)", "工作生活(work and life)" and "文化艺术(art and culture)", "艺术文化(art and culture)". Some nouns belong to equal juxtaposition, whose positions can be interchanged, and some of them express different meanings after transposition. The number of high-frequency phrases is relatively small, while the number of low-frequency phrases is large, so the average frequency of phrases appears relatively low, which conforms to the law of long-tailed distributions.

4.2 Looming Analysis

In this paper, the frequency distribution of basic noun compounds and the real frequency comparison of NN phrases with "de" in corpus are analyzed and shown in Fig. 1.

Figure 1 shows that the frequency of "N-N phrase" in the corpus is higher, mainly between 1,000 and 8,500, while the frequency of "N-de-N phrase" in the corpus is between 0 and 1,000.

The part of "NN phrases" that cannot be added with "de", however, about 70% of them added the word "de" appear in the corpus at a frequency of 0, which indicates that most phrases that cannot be added with "de" do not exist in the corpus. Although the remaining 30% frequency is not 0, the phrase intercepts part of the sentence which was meaningless. About 8.8% of the phrases that can add the word "de" have a frequency of 0, because the corpus is relatively not large enough, and some phrases with the word "de" are used less.

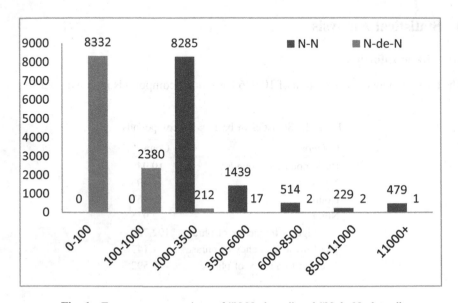

Fig. 1. Frequency comparison of "N-N phrase" and "N-de-N phrase"

For phrases whose frequency is not 0, we use the "NN phrase" and the "N-de-N phrase" frequency to calculate, defined as "word looming rate":

$$Word\ looming\ rate = NN\ frequency / N\ de\ N\ frequency \qquad (1)$$

The lower frequency of the "N-de-N phrase", the lower the usage rate of the phrase in the corpus, the value of "word looming rate" is larger; the higher frequency of the "N-de-N phrase", the higher the usage rate of the phrase in the corpus, the value of "word looming rate" is smaller.

This paper compares two types of basic noun compounds: those that can add the word "de" and those that cannot add the word "de". The distribution of the "word looming rate" of the two phrases is shown in Fig. 2.

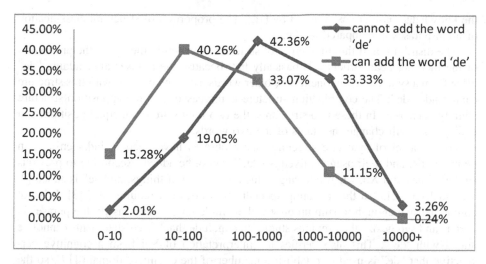

Fig. 2. Comparison of the distribution of the word looming rates for the two types of phrases

According to the statistics, in the basic compound noun phrase that can add the word "de", the mean value of the "word looming rate" is 423, and the average value of the word "de" that cannot be added is 1,645. It can be seen from the formula that the phrase "de" cannot be added, the value of "word looming rate" is large, and it shapes a left bias nearly normal distribution curve, indicating it is low-frequency in the corpus; the phrase that can be added the word "de", the value of "word looming rate" is small and shapes a right bias nearly normal distribution curve, indicating it is high-frequency in the corpus.

Syntactic Structure. There are three types of basic noun compounds that cannot be added with the word "de": modifier-head structure, combination structure and apposition structure. This paper counts the proportion of each type of phrase structure of basic noun compounds that cannot be added with the word "de", and all the phrases in

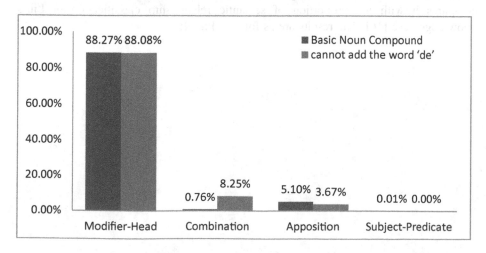

Fig. 3. Contrast of structural types

the knowledge base are established by [16]. The proportion of structural classification of noun phrases is compared as Fig. 3.

The distribution of the three types of structures is roughly the same: the proportion of the modifier-head structure is basically flat, because the phrases after adding "de" often form a syntactic component with other words, so some phrases with this structure cannot add "de"; The combination structure is increased, and the apposition structure slightly decreases. In these two structures, the two nouns are in an equal position, and adding "de" will change the status of the two words.

At the level of syntactic structure, many phrases have become indispensable in people's life, and their usage is fixed, so "de" cannot be added. According to previous studies, Chen and Xu, when referring to the appellation of things, no "de" is added [2, 4], and Lv concluded that meaning specialization does not require "de" [3]. It is not comprehensive enough to sum up only with a single law, so we classify the phrases in corpus and synthesize the previous studies to conclude that "fixed statements cannot be added with 'de'". This also verifies the interpretation of Shi from a cognitive perspective that "de" is used to establish a member of the cognitive domain [17], so the structures that have been specialized or often combined, the membership they represent has been established and does not necessary or need to be identified by "de". According to the different usage, we divide the "fixed statements" into three categories:

Specialization of meaning, such as "保安亭(security booth)", "电子烟(electronic cigarette)", "篝火晚会(bonfire party)".

Names of things, such as "蓓蕾剧院 (Buds Theatre)", "北京西路 (Beijing Xi Road)" and "惠州学院 (Huizhou University)".

Common combinations, such as "法制机构 (legal department)", "房屋管理局 (housing authority)" and "工商机关 (industrial and commercial authority)".

Semantic Relationship. We have labeled basic noun compounds that cannot be added with the word "de" using the semantic relationship classification system in the basic noun compounds knowledge base [1]. Thirteen types of basic noun compounds are summarized as follows: Have, Be, Locate, For, Content, Do, Make, From, Have, Use, And, Like, Cause. This paper also counts the proportion of each type of semantic relationship of the basic noun compounds that cannot be added with the word "de" and compares it with the proportion of semantic relationship classification in Liu's knowledge base [16]. The results are as follows Fig. 4:

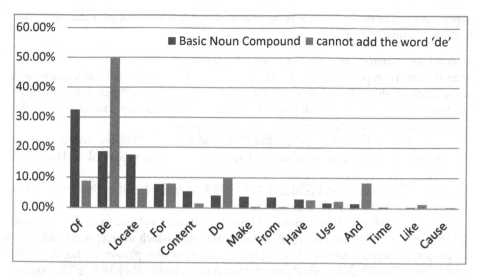

Fig. 4. Comparison of semantic relationship

Among all the noun phrases, there are fourteen kinds of semantic relations. Only the phrases expressing "Time" do not appear in the phrases that cannot be added with the word "de". Among the basic noun compounds that cannot be added the word "de", the category of "Be" accounts for 49.78%, because most of the words in this relation are fixed expressions, and people have formed the habit of not adding "de" in their daily life. In addition, some phrases with "de" usually form an element in the sentence, which cannot be used alone, such as "Have, Of, Location, Do" and so on, while phrases expressing the semantic relationship of "And, Like, For" do not form a phrase after adding "de".

In terms of the semantic relationship, many basic noun compounds are commonly used in people's lives, but when such truncated phrases added with "be" enter sentences, the integrity of semantics will be destroyed or semantic conflicts will occur. Previous studies mainly divide the semantic relationship into two categories from a broader semantic level. For example, Qian divided the relationship into two types: possession and non-possession, and he also believed that possession needs to be marked by "de"; otherwise "de" is generally not used [13]. In this paper, we have a more fine-grained classification of the semantic relations between the two nouns. Because when we label it, we find that it cannot be directly judged by whether it's a relationship of possession. For example, the phrase extracted from the corpus "气象灾害(meteorological disaster)" expresses the possessive relationship, but the phrase "*气象的灾害" does not appear in the corpus, whereas non-possession phrases such as "广州地铁(Guangzhou Metro)" and formed "广州的地铁(Guangzhou Metro)" also appears 206 times in the corpus. Zhang proposed that the law of the hidden "de" depends on the degree of semantic relationship [8]. The distinction between the three semantic relationships of commentary, classification and markup is not high, and the regularity is not strong. Based on the analysis of phrases extracted from corpus, this paper summarizes two reasons why basic noun compounds cannot be added with "de" from the semantic level:

Incomplete semantic representation cannot be added "de"
After the noun compound is added with "de", it usually cannot be used as a component in the sentence alone. It may be that N1 and other words are combined into phrases, or N2 and other words are combined into a part of the structure extracted from the sentence, and the semantic representation is incomplete.

E.g.1: "昨日一场以传递爱心的活动在杭州朝晖实验小学展开。(An activity to convey love was launched in Zhaohui Experimental Primary School in Hangzhou yesterday.)"

E.g.2: "正是这样的生活逻辑在左右着法律的职业逻辑。(It is this logic of life that governs the professional logic of law.)"

In E.g.1, "传递爱心 (transmission of love)" constitutes a phrase to modify the noun "活动(activity)", and "*爱心的活动" belongs to phrase which intercepts N1 and other words and it cannot express the complete meaning of the phrase; in E.g.2, "职业 (professional)" and "逻辑(logic)" constitutes a phrase, and "*法律的职业" belongs to N2 combined with the noun behind it, intercepting the former part of the whole phrase, and cannot express the complete meaning of the phrase correctly.

Non-established semantic representation cannot be added "de"
First is the combined or appositive relationship, two nouns represent the same thing and refer to each other, or the two things belong to the parallel joint relationship of equal status. After adding "de", the phrase makes no sense. For example, "总统奥巴马 (President Barack Obama)": "*总统的奥巴马"; "地震海啸(earthquakes and tsunami)": "*地震的海啸".

Second is the hypernym-hyponym or meaning-containing relationship: N1 and N2, one's meaning covers the other and the phrase after added "de" makes no sense. For example: "东莞地区(Dongguan area)": "*东莞的地区"; "厨余垃圾(Kitchen Waste)": "*厨余的垃圾".

5 Conclusion

This paper extracts 10,946 basic noun compounds from the ten-year newspaper corpus of the Dynamic Circulation Corpus (DCC) which are manually screened. They are summarized from the two levels of syntactic structure and semantic relationship, and the hidden phenomenon of "de" in basic noun compounds is analyzed. Through the verification and analysis of corpus in large-scale corpus, this paper sums up the concealment of "de" in basic noun compounds from two aspects. In terms of syntactic structure, there are three types of phrases: modifier-head, combination and apposition structure. Many phrases have become fixed expressions which cannot be added "de" in people's daily life. In terms of semantic relations, there are thirteen types, such as "Have, Be, And, Like, Use" and so on. When these phrases are added with "de", their corresponding meanings are incomplete or the semantics will be conflicted, so "de" cannot be added.

This paper mainly analyses the situation that "de" cannot be added to basic noun compounds. The deeper relationship between the structural relationship, however, semantic relationship and the concealment of the word "de" remains to be explored, and the influence of the appearance of the word "de" on meaning changes in compound noun also needs to be studied in the future.

Acknowledgments. This study was supported by the Ministry of education of Humanities andSocial Science project (18YJA740030), and supported by Science Foundation of Beijing Language and Culture University (supported by "the Fundamental Research Funds for the Central Universities") (19YJ040003).

References

1. Liu, P., Liu Y.: Semantic relations hierarchy and knowledge base construction of chinese basic noun compounds. In: The 17th China National Conference on Computational Linguistics, Changsha (2018). (in Chinese)
2. Chen, Q.: The study of the word "DE" between modifiers and nouns. Stud. Chin. Lang. (10) (1995). (in Chinese)
3. Shuxiang, L.: Modern Chinese Eight Hundred Words. The Commercial Press, Beijing (1980). (in Chinese)
4. Xu, Y.: The restrictive factors of the hidden "De". Contemp. Rhetoric **02**, 33–34 (2003). (in Chinese)
5. Yangchun, X.: Plate, protrusion and the appearance of the word "De". Lang. Teach. Linguist. Stud. **06**, 76–82 (2011). (in Chinese)
6. Ming, Z.: Cognitive Linguistics and Chinese Noun Phrases. China Social Sciences Press, Beijing (1998). (in Chinese)
7. Lu, B.: "Distance-Marker Correspondence Law" as a Common Language. Stud. Chin. Lang. (01), 3–15 + 95 (2004). (in Chinese)
8. Zhang L.: In the Implied Rules of De in the "Attributive-Noun Structure". J. Hubei Univ. (Philosophy and Social Science). (in Chinese)
9. DeXuan, S.: Auxiliaries and Interjections. Shanghai Education Press, Shanghai (1985). (in Chinese)
10. Yongsheng, L.: On the implication of "De" in the "N_1/P + de + N_2" structure from the sentence level. Contemp. Rhetoric **06**, 51–52 (2004). (in Chinese)
11. Guangquan, W., Yinglu, L.: Rules of using "De" in nominal endocentric phrases. Jilin Univ. J. Soc. Sci. Ed. **46**(2), 115–121 (2006). (in Chinese)
12. Zheng, Y.: The use and non-use of the word "De" after the attributive. Contemp. Rhetoric (01), 9–12 + 45 (2004). (1n Chinese)
13. Shuxin, Q.: The usage of "De" in the structure "N_1 + N_2". J. Nanchang Univ. **06**, 160–163 (2004). (in Chinese)
14. DCC. http://dcc.blcu.edu.cn
15. Jieba. https://pypi.org/project/jieba/
16. Liu, Y.: Establishment of knowledge base of chinese noun compounds and structural and semantic analysis. Master Thesis. Beijing Language and Culture University, Beijing (2018). (in Chinese)
17. Shi, Y.: Cognitive Semantic Basis of Grammar. Jiangxi Education Publishing House, Nanchang (2000). (in Chinese)

Research of Speech Act Verb Interpretations About Dictionaries of Learning Chinese as a Foreign Language from the Perspective of Frame Semantics

Weili Wang[✉]

Chinese Department of Capital Normal University, Beijing, China
wangweili100@cnu.edu.cn

Abstract. Based on the Frame Semantic Theory of Fillmore, C.J., this research summarizes the interpretation modes of 9 speech act verbs with the meta-language interpretations. According to the actual occurrence of the frame elements in the corpora, the original frames are modified, and the standard interpretation models of 9 lexical units are constructed. This research analyzes and compares the interpretations of related meanings in the four dictionaries, and puts forward some suggestions to the verb interpretations of the Dictionaries of Learning Chinese as a Foreign Language from the perspective of frame semantics.

Keywords: Frame Semantic Theory · Speech act verbs · Lexical units · Interpretations of dictionaries of learning Chinese as a foreign language · Suggestions

1 Introduction

Lexical interpretations are the core parts in the lexicographic processes in which the verb interpretation is the most difficult. In addition to the sememe interpretations of the verb itself, it is necessary to clarify the collocation relations of verbs and nouns, which are also the generalization of verbal grammar and meanings. We make a research on verb interpretations in the Dictionaries of Learning Chinese as a Foreign Language. Based on the latest Frame Semantic Theory, we start our research with a small class of verbs, speech act verbs, combing with the meta-language theories to puts forward some suggestions to lexical interpretations for Dictionaries of Learning Chinese as a Foreign Language.

© Springer Nature Switzerland AG 2020
J.-F. Hong et al. (Eds.): CLSW 2019, LNAI 11831, pp. 84–93, 2020.
https://doi.org/10.1007/978-3-030-38189-9_8

2 Frame Presuppositions and Interpretations of Speech Act Verbs

2.1 Lexical Interpretations from the Perspective of Frame Semantics

The frame proposed by Fillmore [1] refers to a "structuralized category system consistent with some motivating contexts." The category of "Activated Context" is very large. It is based on human experience and contains a number of concepts for processing natural languages, such as "schema", "script", "scenario", "ideational scaffolding", "cognitive model", "folk theory". These concepts are structured and stored in human cognition in the form of frames. The vocabularies mentioned in people's conversations continually evoke these frames, and these frames are the bases of human communication unrestrictedly.

The frame semantic interpretations provide a new perspective for the description of semantic cores and relationships of lexical units, which make interpretation content more richer and more practical. However, each lexical unit is described and interpreted by the frames, which does not conform to the concise principles of the language interpretations, so the Frame Semantic Theory only provides some suggestions for lexicography.

2.2 The Establishment of Speech Act Verbs as the Interpretation Objects

Speech act verbs refer to "verbs that a person can use some words to express different things". We think that there are about 109 speech act verbs in Chinese. We will compare these verb interpretations with each other in the *New HSK 5000 Words Grading Dictionary*, the *8000 Chinese Words Dictionary about HSK Chinese Proficiency Test Vocabulary Outline* and the *Teaching and Learning Chinese Dictionary*[1]. And the speech act verbs are required to be included in the four dictionaries with detailed Chinese interpretations at the same time. There are a total of 78 objects meeting these conditions. Since each verb requires a lot of detailed descriptions, 9 verbs are selected in this paper.

Lexical units refers to the meanings of words which can arouse the frames. The lexical unit is not a word but a meaning of the word. In principle, a lexical unit only arouse a frame, but a frame can be aroused by multiple lexical units. A polysemous lexical unit arouses different frames in different scenes. These implication relations of lexical units make these frames stronger, and these relationgs also activate these frames from different perspectives. Since a "frame" is aroused by a meaning of the verbs, what we studied in this paper are verb meanings which are lexical units in the frame semantics. The 9 verbs selected are actually 9 lexical units.

These 9 lexical units are as follows: "announce 发表 (Expression 表达义)" "declare 声明", "admit 承认", "apologize 道歉", "review 检讨 (self-reflection 自我反省义)", "confess 坦白", "reveal揭发", "expose 揭露", "disclose 透露". We arrange these

---

[1] *New HSK5000 Words Grading Dictionary* 《新 HSK5000 词分级词典》, *8000 Chinese Words Dictionary about HSK Chinese Proficiency Test Vocabulary Outline* 《HSK 中国汉语水平考试词汇大纲汉语 8000 词词典》, *Teaching and Learning Chinese Dictionary* 《汉语教与学词典》.

9 speech act verbs into a vocabulary, filter out valuable terms, and then set up description frames in order to classify the meanings of each term (i.e., the lexical units) into each description frame. After determining the lexical units, we use the CCL corpus to search these lexical units and select the items that match the meanings of the lexical units. In this paper, the number of valid items extracted per token is 300.

3 Analyses of Interpretation Modes of Speech Act Verbs and Comparisons of Dictionaries Interpretations

Since interpretations of Dictionaries of Learning Chinese as a Foreign Language usually refer to interpretations of the *Modern Chinese Dictionary*[2], we also refer to the *Modern Chinese Dictionary (seventh edition)* (hereinafter the abbreviation is *Xian-Han*). We compare and analyze the interpretation modes of different dictionaries, the difference between their frame elements selections and their highlighting degrees of the frame elements between Chinese Dictionaries and Dictionaries of Learning Chinese as a Foreign Language. Each Chinese dictionary has its different selections and highlighting degrees of frame elements. So we construct the interpretation modes of these 9 lexical units with the *New HSK5000 Word Grading Dictionary* (hereinafter the abbreviation is *New HSK*), the *8000 Chinese Words Dictionary about HSK Chinese Proficiency Test Vocabulary Outline* (hereinafter the abbreviation is *8000 Words*) and the *Teaching and Learning Chinese Dictionary* (hereinafter the abbreviation is *T and L Chinese*). According to the interpretation modes, the relevant interpretations are compared, and the core frame elements are analyzed. And then we put forward some suggestions about verb interpretations to Dictionaries of Learning Chinese as a Foreign Language from the perspective of frame semantics.

3.1 Frame Interpretations of the Lexical Units "Announce 发表" and "Declare 声明"

For the lexical unit "announce 发表", the three Dictionaries of Learning Chinese as a Foreign Language highlight the "specific statement content" (nr) and the "listening object" (dx) elements, without "statement manner" (mf) and "statement person" (per) elements. We believe that the interpretation "to the collective or society" implies the "openness" element of "statement manner" (mf), which is highlighted. The "statement person" (per) has no distinguishing features and can be highlighted by the paraphrasing contexts. In terms of the levels of element description details, the three dictionaries have made a description of distinguishing features for the "specific statement content" (nr) elements. The *T and L Chinese* describes the features in the most detailed way (Table 1).

The descriptions of the frame elements in three dictionaries are consistent with their syntactic implementation characteristics in the corpus. And there are no differences in the selection of frame elements between learning dictionaries and Chinese dictionaries.

---

[2] *Modern Chinese Dictionary* 《现代汉语词典》.

Table 1. Frames of "Announce 发表 "and" Declare 声明"

lexical unit	Modern Chinese Dictionary (The abbreviation is XianHan)	New HSK5000 Words Grading Dictionary (The abbreviation is New HSK)	8000 Chinese Words Dictionary (The abbreviation is 8000 Words)	Teaching and Learning Chinese Dictionary (The abbreviation is T and L Chinese)
announce 发表 Fābiǎo	Express opinions to the collective or society. 向集体或社会表达意见。 Xiàng jítǐ huò shèhuì biǎodá yìjiàn	Express opinions to the collective or society. 向集体或社会表达意见。 Xiàng jítǐ huò shèhuì biǎodá yìjiàn	Express (opinions) to the community or the collective. 向社会或集体表达(意见)。 Xiàng shèhuì huò jítǐ biǎodá (yìjiàn)	Express (views, opinions, attitudes, etc.) to the society or a group. 向社会或某个集体表达(看法, 观点, 态度等)。 Xiàng shèhuì huò mǒu gè jítǐ biǎodá (kànfǎ, guāndiǎn, tàidù děng)
declare 声明 Shēngmíng	To express an attitude or state the truth publicly. 公开表明态度或说明真相。 Gōngkāi biǎoshì tàidù huò shuōmíng zhēnxiàng	To express an attitude or state the truth publicly. 公开表明态度或说明真相。 Gōngkāi biǎoshì tàidù huò shuōmíng zhēnxiàng	To express an attitude or state the truth publicly. 公开表示态度或说明真相。 Gōngkāi biǎoshì tàidù huò shuōmíng zhēnxiàng	(The government, unit or individual) publicly express attitudes or state the truth. (政府, 单位或个人) 公开表示态度或说明真实情况。 (Zhèngfǔ, dānwèi huò gèrén) gōngkāi biǎoshì tàidù huò shuōmíng zhēnshí qíngkuàng

The *T and L Chinese* gives more detailed descriptions for the elements of "the specific statement contents". The three Dictionaries of Learning Chinese as a Foreign Language do not reduce the difficulty levels of interpretation languages.

For the lexical unit "declare 声明", the *New HSK* and the *8000 Words* only highlight the "specific statement content" (nr) and "statement manner" (mf) elements, while the *T and L Chinese* highlights three core framework elements. In terms of the description details of the frame elements, the three dictionaries all have detailed descriptions for the protruding frame elements, and the descriptions of frame elements in the three dictionaries conform to the characteristics of their syntactic implementation in corpus. The four dictionaries highlight the "specific statement content" (nr) and the "statement manner" (mf) elements. The *T and L Chinese* adds and details the "statement person" (per) element. And the "truth" has been changed to "real situation", which reduces the difficulty levels of interpretations (Table 2).

Table 2. Frames of "Admit 承认", "Apologize 道歉", "Review 自我检讨" and "Confess 坦白"

lexical unit	Modern Chinese Dictionary (The abbreviation is XianHan)	New HSK5000 Words Grading Dictionary (The abbreviation is New HSK)	8000 Chinese Words Dictionary (The abbreviation is 8000 Words)	Teaching and Learning Chinese Dictionary (The abbreviation is T and L Chinese)
admit 承认 Chéngrèn	Express affirmation, consent, and approval. 表示肯定，同意，认可。 Biǎoshì kěndìng, tóngyì, rènkě	Express affirmation, consent, and approval. 表示肯定，同意，认可。 Biǎoshì kěndìng, tóngyì, rènkě	Express affirmation, consent, and approval. 表示肯定，同意，认可。 Biǎoshì kěndìng, tóngyì, rènkě	Affirmation of the existence or nature of the facts. 对事实的存在或性质表示肯定。 Duì shìshí de cúnzài huò xìngzhì biǎoshì kěndìng
apologize 道歉 Dàoqiàn	Apologize, specifically acknowledging the mistake. 表示歉意，特指认错。 Biǎoshì qiànyì, tè zhǐ rèncuò	Apologize, specifically acknowledging the mistake. 表示歉意，特指认错。 Biǎoshì qiànyì, tè zhǐ rèncuò	Apologize, specifically acknowledging the mistake. 表示歉意，特指认错。 Biǎoshì qiànyì, tè zhǐ rèncuò	Apologize (to the other party). (向对方)表示歉意。 (Xiàng duìfāng) biǎoshì qiànyì
review 检讨 Jiǎntǎo	To identify shortcomings or mistakes and make self-criticism. 找出缺点或错误,并做自我批评。 Zhǎo chū quēdiǎn huò cuòwù, bìng zuò zìwǒ pīpíng	To identify and criticize one's shortcomings or mistakes. 找出并批评自己的缺点或错误。 Zhǎo chū bìng pīpíng zìjǐ de quēdiǎn huò cuòwù	To identify shortcomings or mistakes and make self-criticism. 找出缺点或错误, 并做自我批评。 Zhǎo chū quēdiǎn huò cuòwù, bìng zuò zìwǒ pīpíng	To analyze, recognize, and criticize work, ideological shortcomings, mistakes, their nature, and causes. 对工作, 思想上的缺点, 误及其性质, 原因进行分析, 认识, 做自我批评。 Duì gōngzuò, sīxiǎng shàng de quēdiǎn, cuòwù jí qí xìngzhì, yuányīn jìn háng fēnxī, rènshì, zuò zìwǒ pīpíng
confess 坦白 Tǎnbái	Truthfully tell one's mistakes or crimes. 如实说出(自己的错误或罪行)。 Rúshí shuō chū (zìjǐ de cuòwù huò zuìxíng)	Tell the whole mistakes, crimes, etc. 把错误、罪行等全说出来。 Bǎ cuòwù, zuìxíng děng quán shuō chūlái	Truthfully tell one's mistakes or crimes. 如实说出(自己的错误或罪行)。 Rúshí shuō chū (zìjǐ de cuòwù huò zuìxíng)	Speak out unreservedly (one's mistake or crime). 毫无保留地说出(自己的错误或罪行)。 Háo wú bǎoliú dì shuō chū (zìjǐ de cuòwù huò zuìxíng)

3.2 Frame Interpretations of the Lexical Units "Admit 承认", "Apologize 道歉", "Review 检讨" and "Confess 坦白"

The *New HSK* and the *8000 Words* are not detailed enough to explain the lexical unit "admit 承认". The *T and L Chinese* highlights elements of the "concrete contents of admission" (nrr), which is described in great detail, but the distinguishing features are short of "approved views". The elements of "concrete contents of admission" (nrr) are generally found in the object positions in the forms of noun phrases, while the *T and L Chinese* uses the forms of preposition phrases and puts the elements in the adverbial positions. The *XianHan*, the *New HSK* and the *8000 Words* are all interpreted by words, which makes it difficult to analyze the frame elements. The *T and L Chinese* has kept the core word "affirmation" and refines the "concrete contents of admission" (nrr) elements, which is more detailed and accurate interpretations. However, the difficulty levels in interpretation languages are not reduced (Table 3).

For the lexical unit "apologize 道歉", the *New HSK* and the *8000 Words* only highlight the "apology content" (nrd) elements, ignoring the "faulty person" (perd) elements and "the object of receiving apology" (dxd) elements. The *T and L Chinese* should highlight the two elements of "the object of receiving apology" (dxd) and "apology content" (nrd). The element "faulty person" (perd) should be characterized by distinction and should be interpreted. In terms of the detail levels of interpretations, the three dictionaries do not describe the features of the "fault" specifically. So we believe that the interpretations of the three Dictionaries of Learning Chinese as a Foreign Language are not perfect. In the syntactic implementation of the frame elements, the interpretations "to the other party" of the *T and L Chinese* conform to the distribution characteristics of the frame elements. All in all, the four dictionaries highlight the "apology content" (nrd), but the descriptions are not detailed. The *T and L Chinese* adds the "the object of receiving apology" (dxd) elements, and the interpretation languages do not reduce the difficulty levels.

For the lexical unit "review 检讨", the *New HSK*, the *8000 Words* and the *T and L Chinese* all highlight the "statement content" (nrc) elements, but ignore the "faulty person" (perd) elements and "listening object" (dx) elements. The distinguishable feature of the "listening object" (dx) elements can not be omitted, and the "faulty person" (perd) elements should be interpreted. In terms of the detail levels of described interpretations, the three dictionaries have detailed descriptions for the "statement content" (nrc) elements.

In terms of the syntactic implementation modes of the frame elements, the "concrete content of the statement" (nrc) elements often appear in the object positions as noun phrases. And the *New HSK* and the *8000 Words* are more consistent with this principle, and the *T and L Chinese* does not match it. The four dictionaries select and highlight the "concrete content of the statement" (nrc) elements, but the three Dictionaries of Learning Chinese as a Foreign Language do not reduce the difficulty levels of the interpretation languages.

For the lexical unit "confess 坦白", the three dictionaries highlight the "concrete content of the statement" (nrc) elements, but they do not highlight the "faulty person" (perd) elements. In terms of the details level of interpretations, the three dictionaries

Table 3. Frames of "reveal 揭发", "expose 揭露" and "disclose 透露"

lexical unit	*Modern Chinese Dictionary* (The abbreviation is *XianHan*)	*New HSK5000 Words Grading Dictionary* (The abbreviation is *New HSK*)	*8000 Chinese Words Dictionary* (The abbreviation is *8000 Words*)	*Teaching and Learning Chinese Dictionary* (The abbreviation is *T and L Chinese*)
reveal 揭发 Jiēfā	To reveal (bad guys and bad things). 揭露(坏人坏事)。 Jiēlòu (huàirén huàishì).	Expose (bad or bad) to the authorities or the public. 向有关部门或群众揭露(坏人坏事)。 Xiàng yǒuguān bùmén huò qúnzhòng jiēlù (huàirén huàishì)	Show the bad things of the bad guys that have not been discovered. 使没有发现的坏人坏事显露出来。 Shǐ méiyǒu fāxiàn de huàirén huàishì xiǎnlù chūlái	(No interpretation) (无解释)
expose 揭露 Jiēlù	Make hidden things visible. 使隐蔽的事物显露。 Shǐ yǐnbì de shìwù xiǎnlù	Make hidden things visible. 使隐蔽的事物显露。 Shǐ yǐnbì de shìwù xiǎnlù	Make hidden things come out. 使隐蔽的事物显露出来。 Shǐ yǐnbì de shìwù xiǎnlù chūlái	Make hidden things (mostly negatives) exposed. 使被掩盖的事物(多指消极的)暴露出来。 Shǐ bèi yǎngài de shìwù (duō zhǐ xiāojí de) bàolù chūlái
disclose 透露 Tòulòu	Give way. 泄露。 Xièlòu	To tell (a secret or information) intentionally. 有意告诉别人(秘密或消息)。 Yǒuyì gàosù biérén (mìmì huò xiāoxī)	Telling secretly. 暗地里告诉[a]。 Àndìlǐ gàosù	Speak out intentionally or unintentionally; show. 有意或无意地说出; 表现出。 Yǒuyì huò wúyì dì shuō chū; biǎoxiàn chū

[a]There is no explanation for the word "disclosure" in the *8000 Words*. The meaning of the word in the table is the meaning of the Chinese character "透" in the phrase "透露话" and "透露消息".

have detailed descriptions of the distinguishing features. The *8000 Words* and the *T and L Chinese* also highlight the non-core frame elements "mode" (m).

In terms of the syntactic implementation modes of the frame elements, the "concrete content of the statement" (nrc) elements often appear in the object positions as noun phrases. The *8000 Words* and the *T and L Chinese* are more consistent with the distribution rules of the frame elements in the language surface level. There is no difference in the selections and interpretations of the frame elements between the

Dictionaries of Learning Chinese as a Foreign Language and the Chinese Dictionaries, and the difficulty levels of the interpretation languages are not reduced (Table 2).

3.3   Frame Interpretation of the Lexical Unit "Reveal 揭发", "Expose 揭露" and "Disclose 透露"

For the lexical unit "reveal 揭发", the *New HSK* highlights the two frame elements of "exposure content" (nrj) and "receiving message object" (dxj). The *8000 Words* only highlights the "exposure content" (nrj) elements, and both dictionaries ignore the frame elements of "exposure person" (perj). We think that "exposure person" (perj) has the "informed feature" which should be interpreted. Both dictionaries use complicated words ("expose 揭露" and "show 显露") to explain the frame elements, which is not appropriate. In terms of the difficulty degrees of the interpretation languages, both dictionaries have made a description for the distinct feature ("bad guys"), which is feasible. In terms of the syntactic implementation modes of the frame elements, the *New HSK* conforms to the distribution rules of the framework elements. On the other hand, the *XianHan* as the Chinese Dictionaries uses the modes of interpretating words by other words, so it is difficult to decompose the frame elements. Then we can conclude that the interpretations of the three Dictionaries of Learning Chinese as a Foreign Language are more simpler and more detailed than the *XianHan*.

For the lexical unit "expose 揭露", the three Dictionaries of Learning Chinese as a Foreign Language highlight the "exposure content" (nrl) elements, ignoring the "exposure person" (perj) elements and "receiving message object" (dxl) elements. And the two elements have their "distinct features" which should be described, and the "hidden features" should be also emphasized in detail. The *T and L Chinese* also emphasizes the "negative features". In terms of the syntactic implementation modes of the frame elements, the "exposure contents" (nrl) generally appear in the object position in the form of noun phrases, and the contents should be interpreted by "speaking (undisclosed, negative things)". The interpretations of the three dictionaries are similar to those of the *XianHan*. The *T and L Chinese* describes the "exposure content" (nrl) elements in more detail.

For the lexical unit "disclose 透露", the *New HSK* highlights the two frame elements of "disclosed content" (nrt) and "receiving message object" (dx), ignoring the "exposure person" (perj) elements. The *8000 Words* and the *T and L Chinese* do not highlight the core frame elements. The three dictionaries highlight the "exposure modes" (m) which are the non-core frame elements, but the *New HSK* and the *T and L Chinese* have contradictory interpretations. In terms of detailed degrees of interpretations, the *New HSK* describes the distinguishing feature of "secret or message" which is the "disclosed content" (nrt) elements. In terms of the syntactic implementation modes of the frame elements, the *New HSK* conforms to the distribution rules of frame elements. The *XianHan* as the Chinese Dictionaries uses the modes of interpretating words by other words, the other three Dictionaries of Learning Chinese as a Foreign Language are more detailed and more simpler than the *XianHan*.

4 Suggestions on the Interpretations of Dictionaries of Learning Chinese as a Foreign Language

The perfect interpretations must be based on real corpora. we put forward some suggestions to the Dictionaries of Learning Chinese as a Foreign Language from the perspective of Frame Semantic Theory by the analysis and research of real corpora:

The first one is describing the verb meanings based on "real situations". The Frame Semantic Theory advocates the real contexts in which the verbs are triggered and describes such cotexts in different frames. "Situation" is a concept put forward from the perspective of human experience and cognition. And "situations" are more vivid and close to reality, and are more conducive to non-native speakers learning a second language.

The second one is describing as many frame elements as possible and studying their corresponding interpretation languages. The frame elements are the constituent elements of frames, and frame elements are participants of "situations". The more comprehensively frame elements are described, the more accurate verbs interpretations are made. We Compare the interpretations of the three Dictionaries of Learning Chinese as a Foreign Language with each other, it is obvious that the interpretations of the Dictionaries of Learning Chinese as a Foreign Language are not comprehensive. The interpretations of many dictionaries directly copies the *XianHan*, which is not helpful to the foreign learners. There is almost no difference between the Dictionaries of Learning Chinese as a Foreign Language and the Chinese Dictionaries in the selections and descriptions of the frame elements. Whether it is in line with the learning rules for foreign learners remains to be explored. The frame elements can be used to analyze the interpretation contents of the dictionaries, but they can not be directly used in dictionary interpretations because the frame elements are refined and are very broad concepts which must be described and be elaborated in a paraphrasing language acceptable to foreign learners. Therefore, it is necessary to study the interpretation language of the frame elements.

The third one is describing the frame elements as comprehensively as possible. The frame elements mentioned above are relatively broad and need to be defined and standardized. According to the analysis of dictionary interpretations in this paper, the frame elements of many lexical uints have not been described in detail. To describe the frame elements in detail, we must focus on real corpora and make a statistical analysis for the frame elements in real corpora to summarize the features of frame elements.

The fourth one is starting with the syntactic representation of the frame elements to study the meta-syntax of the interpretations. When analyzing the interpretations of Dictionaries of Learning Chinese as a Foreign Language, we clearly find that the jaggedness of the meta-syntax. We think that Frame Semantic Theory is a very good solution for this problem. Statistics is carried out through the syntactic implementation of the frame elements, based on the syntactic structures with the largest frequency, which is very considerable and scientific.

The fifth one is that the quality and quantity of the frame elements are used as a measure of dictionary interpretations. The interpretations of Dictionaries of Learning Chinese as a Foreign Language is very subjective. It is feasible to introduce the

Framework Semantic Theory into the field of dictionary interpretations. Whether dictionary interpretations containing the core frame elements and how many core frame elements contained are all useful to dictionary interpretations.

Acknowledgments. This research is supported by the National Social Science Fund Project (17BYY211), Youth Project of Humanities and Social Sciences Research of the Ministry of Education (14YJC740087), National Language Committee Project (ZDI135-79) and National Social Science Fund Major Project (14@ZH036), Science and Technology Innovation Service Capacity Building Project (0251-8530-5000/183).

Reference

1. Fillmore, C.J.: Frame semantics. In: The Linguistic Society of Korea (ed.) Linguistics in the Morning Calm, vol. 117. Hanshin Publishing Company, Seoul (1982)

A Study on the Expressions of Modal Particles of the Suggestion Function in Spoken Chinese

A Case Study on "ba", "ma", and "bei"

Xie Jingyi[✉]

Fujian Normal University, Fuzhou 350007, FJ, China
xjybnu@qq.com

Abstract. This paper focuses on the analysis of how modal particles "ba", "ma", and "bei" can help express the suggestion function in spoken Chinese. It mainly describes the characteristics of expressions of the suggestion function and the contextual variables which affect the selection of linguistic forms. Through video transcription, this study summarizes three typical means of expressions of the modal particles "ba", "ma", and "bei" of the suggestion function from 604,585 words of a spoken Chinese corpus. The study analyzes the contextual variables involved in the expressions of discourse function, making the context tagging available, so as to facilitate future research on the expressions of other functions.

Keywords: Suggestion function · Contextual variables · Modal particles

1 Introduction

1.1 Topic Background

In verbal communication, "function" means some special ways of expressions taken by speakers to achieve a certain purpose or to perform certain speech acts, that is, what acts the speakers intend to accomplish by using language, such as greetings, thanks, farewell, congratulations, etc. [1] The suggestion function is one of the most important functions of oral communication and often expressed by modal particles "ba", "ma", and "bei" in daily spoken language [2]. However, the functions of modal particles "ba", "ma", and "bei" are not unique, and the functions of suggestion are not only realized by these modal particles. There are complex corresponding relations between them. For example, on analyzing the corpus of this paper containing modal particles "ba", "ma", and "bei", 357 cases realize the suggestion functions, accounting 31.05% among the entire corpus with the suggestion functions; however, these 357 cases account for only 12.05% of all the corpus containing modal particles. That is to say, the expressions of the suggestion function with modal particles have their characteristics and only can be realized in specific contexts. It is necessary to pay attention to both expressions of modal particles of the suggestion function and corresponding contextual factors [3].

© Springer Nature Switzerland AG 2020
J.-F. Hong et al. (Eds.): CLSW 2019, LNAI 11831, pp. 94–101, 2020.
https://doi.org/10.1007/978-3-030-38189-9_9

Therefore, this study is organized into three parts as follows:

1. What are the characteristics of distribution for the modal particles "ba", "ma", and "bei" in the expressions of the suggestion function?
2. What are the expressive characteristics of modal particles "ba", "ma", and "bei" in the expressions of the suggestion function? What contextual variables affect the selection of modal particles?
3. What typical means of expressions can be summed up when the modal particles "ba", "ma", and "bei" are used to express the suggestion function?

1.2 Theoretical Foundation and Research Methods

A given meaning can be realized in different ways in the grammar through metaphorical modes of expression. Metaphorical modes of expression are characteristic of all adult discourse [4]. Every act of meaning has a context, an environment within which it is performed and interpreted. The linguistic form selection depends on the contextual variables [5]. For example, the social role relationship determines the interpersonal function of the discourse, and the interpersonal function will influence the speaker's selection of linguistic forms, which are with various mood and modality [6].

These contextual variables are referred to as "field", "tenor" and "mode" in [7]. The environment, or social context, of language is structured as a field of significant social action, a tenor of role relationships and a mode of symbolic organization. However, not all environmental factors can be used as contextual variables affecting the form selection. All possible contextual factors can be determined as contextual variables if they promote the relationship with specific pragmatics. If it does not influence pragmatics, it will not be qualified to act as a context variable [8].

The corpus in this thesis is collected by transcription of six reality TV programs: If You Are the One, Go Fighting, Food Road-map, Sisters over Flowers, Back to High School and Dad Came Back in 2015. There are 604,585 words in the video transcription script. The entire corpus is analyzed by turn-taking. In corpus segmentation, a turn is the basic unit of research and turn-taking should be seen as the symbol of segmentation. The corpus segmentations which express the suggestion functions are extracted by 30 Chinese native speakers via a 30-min text. Correlation analysis is conducted on linguistic forms and contextual factors to locate contextual variables. Typical means of expression are described by the function, Linguistic forms, and contextual variables.

2 Review of Previous Studies

This study explores the expressions of the suggestion function of modal particles in spoken Chinese, including the linguistic forms of expressions of the suggestion function, the semantics and contextual factors. It mainly involves the study of linguistic forms (especially Chinese), context, and functional expressions.

2.1   Study of Linguistic Forms and Functions

Meaning is the core of the language expression of humankind [9]. The linguistic form is the material carrier through which language communication can be realized [10]. In this process, linguistic forms can realize conceptual and interpersonal meanings and constitute various discourses to express various functions [11].

In previous studies, the classification and definition of linguistic functions are quite different. For example, three metafunctions and seven basic functions in language are proposed in [9]. From the perspective of discourse, a study summed up eight language functions [12]. From the perspective of teaching, the communicative functions of language were classified into six categories [13].

In the study of spoken Chinese, since the 1980s, many types of research have been conducting on the constraints of linguistic functions to linguistic forms, as well as the expressions of linguistic functions by linguistic forms [14]. For example, there is a survey result imposing that the language function has a selective effect on the modal particles at the end of imperative sentences [15]. Modal particles such as "ha" have the dual function of expressing both modality and meaning [16]. From the perspective of interactive linguistics, a study revealed the interactive function of the variation of modal particles in Beijing discourse [17].

2.2   Study of Context and Functional Expressions

Determining the selection of linguistic form in context requires a series of conditions [5]. Since the 1960s, the study of context began to pay attention to the classification of contextual variables and the relationship between contextual variables and semantic structures. Contextual factors were classified into three categories (scope of discourse, tone of discourse, and mode of discourse), six categories (speaker, receiver, information, context, code, and contactor), and more categories by different researchers [10, 18]. In addition, linguists also observed dozens of contextual factors from their respective research perspectives, such as location, scope, field, topic, scene, sequence of action, participants, purpose of action, formality of discourse, discourse medium, intention of discourse, and structure of discourse, etc. [5, 19–21]

In this study, we use the above theories for reference to study the contextual factors and extract the contextual variables that affect linguistic forms selection in daily conversations.

3   Distribution and Contextual Variables of Modal Particles "ba", "ma", and "bei" in Expressing the Suggestion Function

When expressing the suggestion function, the modal particles "ba", "ma", and "bei" need to be appended to other forms to realize the function. These expressions are also constrained by context.

3.1 Distribution of Modal Particles "ba", "ma", and "bei" in Expressing the Suggestion Function

Modal Particles and Other Forms Work Together to Express the Suggestion Function. In imperative sentences, modal particles are often added after clauses or complex sentences in an imperative tone. In these imperative sentences, the mood is relatively euphemistic [22]. The commonly used modal particle is "ba". The subjects could be the plural form of the first-person pronouns such as "zanmen" or "women" [23]. This is also the most common use of modal particles in expressing the suggestion function. There are also some fixed forms which are often used to express the suggestion function, such as "háishì......ba" , "yàobu......ba" ,"gǎnjǐn......ba",etc. For example, in Sisters over Flowers 20150329, Li Zhiting's suggestion "Háishì jì yīxià ba. (Let's count it down.)" used the fixed form "háishì......ba".

The characteristics of these expressions are closely related to the components before "ba". For example, "gǎnjǐn......ba" presents a kind of urgent suggestion; "háishì...... ba" presents a kind of deliberative suggestion, because "háishì......" usually expresses the suggestion after comparison, consideration or choice.

Modal Particles Enable the Form that Seldom Expresses the Suggestion Function to Realize the Expression of the Suggestion Function. The modal particles "ba", "ma", and "bei" can be used to express the suggestion function in some forms that do not express the suggestion function in ordinary circumstances. For example, in the Go Fighting, 20150621, a passerby put forward the suggestion of "100kuài ba." in response to Huang Bo's questions, considering that he denied the proposal of "singing a song for 50 cents". Here, the "ba" is directly attached to the quantitative phrase "100kuài", which seldom expresses the suggestion function.

Modal Particles Are Used as Discourse Markers to Lead the Expressions of the Suggestion Function. Before the exact sentence expressing the suggestion function, sometimes there are some specific forms to guide the function. The specific forms often include "ba", such as "zhèyang ba", "lái ba", etc. In Dad Came Back 20150606, Du Jiang found his son crying over losing a game, so he suggested Jia Nailiang should change the rule of the game. Then, Dujiang said "Yàobu zhèyang ba", followed by the exact suggestion expressions "Wǒmen gǎigai guīzé. (Let's revise the game rule.)".

But what determines a speaker whether to use a modal particle and which modal particle will be selected? This requires further elaboration.

3.2 Contextual Variables of the Expressions of the Suggestion Function

Not all the contextual factors can affect the selection of the linguistic forms [24]. Therefore, this study addressed a correlation analysis between contextual factors and the selection of linguistic forms. These contextual factors were obtained after two rounds of selection. First, 26 contextual factors were summed up from the former linguists' research results. In the second round of selection, 30 Chinese native speakers were told to select context factors that affected the selection of linguistic forms in the expressions of the suggestion function. Then, linguistic forms and contextual factors were tagged and calculated. Finally, the result of the correlation showed that only six contextual factors were significantly related to the selection of linguistic forms. Detailed contextual variables were summarized in Table 1.

Table 1. Contextual variables of modal particles expressing the suggestion function

Category	Contextual variables
Clues in context	An unsolved predicament
	A behavior to be guided
	Ideas or actions need to be improved or corrected
	An inquiry needs to be answered
Speaker's attitude	Prediction of the recipient's acceptance of the suggestion
	Intensity of suggestion

Furthermore, this paper uses the above contextual variables to describe the typical means of expressions of the suggestion function.

4 Expressive Characteristics of Modal Particles in the Expressions of the Suggestion Function

4.1 Features of the Suggestion Function Expressed by Modal Particles "ba", "ma", and "bei"

The corpus statistics show that, when expressing the suggestion function, speakers would tend to use modal particles "ba", "ma", and "bei" when they identify a higher level of acceptance from the recipient, reflecting a lower intensity of suggestion than expressions with other linguistic forms.

Although the modal particles "ba", "ma", and "bei" often appear in groups, the three modal particles also have their features in the sentences when expressing the suggestion function:

Feature 1. When speakers express the suggestion function with "ma", they take a dimmer expectation of the suggestion acceptance. The corpus statistics show that 70.7% of the expressions with "ba" and 80% of the expressions with "bei" are used when speakers predict others will accept their suggestions. While only 42.1% of the expressions in the same context are with "ma".

Feature 2. When speakers have a strong intention to achieve confirmative response to their suggestions, they are more likely to use "ma". The corpus statistics show that 57.14% of the expressions with "ma" have a higher level of suggestion intensity. By contrast, only 20% of expressions in the same context are with "bei" and 25.49% are with "ba".

Feature 3. When speakers note there is an unsolved predicament or a behavior to be guided in context, they are more likely to offer a suggestion with the modal particle "bei". When speakers note there are ideas or actions to be improved or corrected, they are more likely to offer a suggestion with the modal particle "ma". When the suggestion is put forward to answer an inquiry, they are more likely to use "ba" in the expressions.

4.2 Typical Means of Expressions When the Modal Particles "ba", "ma", and "bei" Are Used to Express the Suggestion Function

According to the above description, we can sum up three typical means of expressions when the modal particles "ba", "ma", and "bei" are used to express the suggestion function.

First, Modal particle "ba" is more often used in the context with these features:

(1) There is an inquiry for a suggestion.
(2) The speaker desires little of others' acceptance.
(3) The speaker identifies that others are probable to take his/her suggestion.

For example, in the Dad Came Back 20150530, Zheng Jun asked Du Jiang: "What can I do now?" Du Jiang gave him a suggestion by "Nǐ jiù kànzhe nòng ba. (Help yourself.)". In this conversation, Zheng Jun asked Dujiang for advice. There was a great chance that Du Jiang's suggestion would be accepted by Zheng Jun, while Du Jiang had little intention that Zheng Jun must accept the suggestion.

Secondly, Modal particle "ma" is more often used in the context with these features:

(1) There are ideas or actions to be improved or corrected.
(2) The speaker strongly desires others' acceptance.
(3) The speaker identifies that others are unlikely to accept his/her suggestion.

For example, in Sisters over Flowers 20150322, Wang Lin was dissatisfied because Li Zhiting turned down her proposal to buy drinks and tried to put forward suggestions to change his mind. She first adopted "......, xíng ma? (Is it possible to...?)" to propose the first suggestion. After being denied, she tried more specific suggestions by "Ná, Ná jǐgè kōng bēi ma! (Just get some empty cups!)". In this context, Wang Lin wanted to change Li Zhiting's mind and she really wanted him to listen to her. But actually, she knew that the suggestion might get denied again.

Third, Modal particle "bei" is more often used in the context with these features:

(1) There is an unsolved predicament, or a behavior needs to be guided.
(2) The speaker desires little of others' acceptance.
(3) The speaker identifies that others are probable to take his/her suggestion.

For example, in Dad Came Back 20150523, Jagger was trying to learn to ride a bicycle, but he wasn't able to balance it. At this time, Zheng Jun, his father, put forward the suggestion of "huázhe bei (Just keep sliding!)". In this context, Zheng jun wanted to teach his son. He knew it's not easy, so he was not urgent. Jagger was in trouble and he was likely to try his dad's suggestion.

5 Conclusion

Modal particles are often used to express the suggestion function. This study describes the characteristics of distribution, the expressive characteristics, the contextual variables, and the typical means of the expressions of the modal particles of the suggestion function in spoken Chinese.

When expressing the suggestion function, the modal particles "ba", "ma", and "bei" can be appended to other forms to realize the function expression. They can even enable the form that seldom expresses the suggestion function to realize the expression of the suggestion function. In daily conversations, it can also be used as discourse markers to lead expressions of the suggestion function.

Context clues and speakers' attitudes determined whether speakers choose modal particles to express the suggestion function. When expressing the suggestion function, speakers who use modal particles "ba", "ma", and "bei" have a higher expectation of accepting the suggestion and a lower intensity of suggestion than expressions with other linguistic forms. The modal particles "ba", "ma", and "bei" also have their respective typical means of expressions.

This research is helpful to deepen the study of the relationship of the language function, linguistic forms, and context in the real spoken environment. It is also useful to improve the teaching system related to the category of suggestion in the teaching of spoken Chinese as a second language.

Acknowledgments. This study is sponsored by Fujian Normal University Ph.D. Graduate Research Project: "Research on International Chinese Education Model Based on Mobile Devices".

References

1. Halliday, M.A.K.: Explorations in the Functions of Language. Edward Arnold, London (1973)
2. Jianzhou, Y.: Functional Outline for the Primary Stage of Teaching Chinese as a Foreign Language. Beijing Language and Culture University Press, Beijing (1999)
3. Zhiping, Z., Lili, J. Siyu, M.: Commentary on teaching materials of Chinese as a foreign language in 1998–2008. J. Beijing Norm Univ. (Social Science Edition) 5 (2008)
4. Halliday, M.A.K.: Computing meanings: some reflections on past experience and present prospects. In: 1995 The Second Conference of the Pacific Association of Computational Linguistics. University of Queensland, Brisbane (1995)
5. Gregory, M., Carroll, S.: Language and Situation: Language Varieties and Their Social Contexts. Routledge and Kegan Paul, London (1978)
6. Halliday, M.A.K., Hasan, R.: Cohesion in English. Longman, London (1976)
7. Halliday, M.A.K.: Text as a semantic choice in social contexts. In: van Dijk, A., Petöfi, J.S. (eds.) Grammars and Descriptions. De Gruyter, Berlin (1977)
8. Wang, J.-H., Zhou, M.-Q., Zhou, A.-P.: Context Study of Modern Chinese. Zhejiang University Press, Hangzhou (2002)
9. Halliday, M.A.K.: Learning How to Mean: Explorations in the Development of Language. Edward Arnold, London (1975)

10. Halliday, M.A.K.: Categories of the theory of grammar. Word **17**(3), 242–292 (1961)
11. Zhu, W.-J.: Translations of Sociolinguistics. Peking University Press, Beijing (1985)
12. Searle, J.R.: Speech Acts. Cambridge University Press, Cambridge (1969)
13. Wilkins, D.A.: Notional Syllabuses. Oxford University Press, Oxford (1983)
14. Zhang, W.-X., Han, C.: Persuasion scene and empathy function of personal pronoun "I/We": based on discourse analysis of three TV plays. Lang. Teach. Res. **11** (2011)
15. Yang, Q.H., Min, Z.: Selective study of modal words at the end Beijing: of imperative sentences in modern Chinese. J. Shanghai Normal Univ. (Philosophy and Social Sciences Edition) (2005)
16. Liang, C.X.: The modality meaning and function of the inter subjective marker "Ha (哈) in Mandarin Chinese. Lang. Teach. Res. 4, 39–45 (2011)
17. Mei, F.: International functions of the sentence final practical variables in Beijing Mandarin: a case study onya, na and la. Lang. Teach. Res. **2**, 75–98 (2016)
18. Lyons, J.: Semantics. Cambridge University Press, Cambridge (1977)
19. Firth, J.R.: Papers in Linguistics, 1934–1951. Oxford University Press, London (1957)
20. Van Dijk, T.A.: Cognitive situation models in discourse production. In: Forgas, J.P. (ed.) Language and Social Situations. Springer- Verlag, New York (1985)
21. Cook, G.: The Discourse of Advertising. Routledge, London, New York (1992)
22. Yulin, Y.: A Study of Imperative Sentences in Modern Chinese. Peking University Press, Beijing (1993)
23. Lv, S.-X.: 800 Words in Modern Chinese. Commercial Press, Beijing (1980)
24. Zhang, B.-J., Fang, M.S.A.: Study of Chinese Functional Grammar. Business Press, Beijing (2014)

A Study of the Characteristics of ABB-Type Adjectives in Shaoxing Dialect

Bihua Wang[1,2], Yueming Du[1,2], and Lijiao Yang[1,2(✉)]

[1] Institute of Chinese Information Processing,
Beijing Normal University, Beijing, China
282843696@qq.com, ddddym@yeah.net,
yanglijiao@bnu.edu.cn
[2] Ultra Power-BNU Joint Laboratory for Artificial Intelligence, Beijing, China

Abstract. Shaoxing lies in the north of Zhejiang Province. The dialect of Shaoxing has an ancient pedigree and deep cultural connotation. There are abundant ABB adjectives in Shaoxing dialect, which are different from those in Mandarin. The previous scholars either compared this kind of adjectives with other forms of reduplicated words in the Wu dialect, or made a simple and closed description of its grammatical function and ways of word formation. There was no exploration from semantic aspects. This paper explains ABB adjectives in Shaoxing dialect from the aspects of phonetics, grammar and semantics, and uses ABB adjectives in Mandarin as a reference for comparative analysis to have a more complete description of the characteristics of ABB adjectives in Shaoxing dialect.

Keywords: ABB-type adjective · Shaoxing dialect · Comparative analysis

1 Introduction

The ABB-type adjective is a kind of reduplicated Chinese words, which has been paid attention to quite a few scholars. There was abundant research on ABB-type adjectives from diversified angles. Some scholars have turned their attention to abundant ABB-type reduplicated words in the Wu dialect. Ruan classified the ABB-type reduplicated words in Ningbo dialect according to part of A's speech, explored the relevance of word formation between different forms of reduplications [1]. Zhu et al. collected the ABB-type reduplications in Xiaoshan dialect and pointed out the uneven age distribution of speakers [2]. Lou paid attention to the word-formation and grammatical functions of the AAB-type and ABB-type adjectives in Shaoxing dialect, and made a simple comparison of their semantic degrees [3]. Although these works are good precedents for research of special reduplicated words in the Wu dialect, they generally focus on summing up the grammatical functions and internal composition of ABB-type reduplicated words in a simple and closed way, lacking semantic exploration.

Shaoxing dialect is a representative one of the Wu dialect. It is included in the Linshao cluster of Taihu zone according to Language Atlas of China [4]. There are abundant ABB-type adjectives in it. This paper will take Mandarin as a reference, use examples to describe and analyze the ABB-type adjectives in Shaoxing dialect from three aspects: tone change, grammar and semantics. The samples of the ABB adjectives

© Springer Nature Switzerland AG 2020
J.-F. Hong et al. (Eds.): CLSW 2019, LNAI 11831, pp. 102–109, 2020.
https://doi.org/10.1007/978-3-030-38189-9_10

of Shaoxing dialect in this paper mainly come from Yue Proverbs [5], other literature studies on Shaoxing dialect and daily collection of the author, while the samples of the ABB adjectives of Mandarin come from Eight Hundred Modern Chinese Words [6].

2 Tone Change

Shaoxing dialect, like most other dialects of the Wu dialect, retains the Ru tone whose typical manifestation is the glottal stop terminal [-?]. Four tones (平上去入) are devided into yin and yang respectively giving the Shaoxing dialect eight tones in total. The following table shows the tone pitch[1]:

In the actual speech, the tones of some words will change due to the pronunciation of words before or after them. These changes are called tone sandhi. The tone sandhi of the ABB adjectives in Shaoxing dialect observes the following rules:

1. If the tone of A is Shang or Ru, it retains its original tone, otherwise its tone will become Shang.
2. If the tone of BB is Ping, Shang or Qu, its tone will change as below:

    (1) Yin-Ping + Yin-Ping → Yin-Shang + Yin-Ping
        such as Wen Tuntun(温吞吞), Shi Kengkeng(实坑坑).
    (2) Yang-Ping + Yang-Ping → Yang-Qu + Yang-Ping
        such as Nuo Chouchou(懦绸绸), Bai Tutu(白涂涂)
    (3) Yin-Shang + Yin-Shang → Yin-Shang + Yin-Ping
        such as Hua Duoduo(花朵朵), Liang Shanshan(亮闪闪)
    (4) Yang-Shang + Yang-Shang → Yang-Shang + Yang-Ping
        such as Shi Niuniu(湿扭扭), Kong Laolao(空佬佬)
    (5) Yin-Qu + Yin-Qu → Yin-Qu + Yin-Ping
        such as Ai Pangpang(矮胖胖), Ning Diaodiao(宁吊吊)
    (6) Yang-Qu + Yang-Qu → Yang-Qu + Yang-Ping
        such as Ruan Fanfan(软泛泛), Xie Tata(懈沓沓)

3. If the tone of BB is Ru, the tone of the former B will not change, while the latter B will be weakened or the ending of the entering-tone terminals will fall off.

According to Table 1, we find that the tone of Ping in Shaoxing dialect is a falling tune, the tone of Shang is a rising tone, and the tone of Ru is a flat regulation. Generally speaking, unlike the stable Yin-Ping tone of BB in Mandarin, the pitch of the ABB-type adjectives in Shaoxing dialect shows a rising and then falling trend, which makes the word full of rhythms and read fluently.

---

[1] The tones of Shaoxing dialect proposed by Zhang [7] is adopted in this paper.

Table 1. Tone pitchs of tones of Shaoxing dialect.

	Ping	Shang	Qu	Ru
Yin	52	35	33	5
Yang	31	13	22	3

3 Grammatical Features

3.1 Internal Word-Formation

There are two ways to form ABB adjectives in Shaoxing dialect:

1. A+BB: this type of adjectives can be divided into two categories according to the relationship between A and BB.
   (1) Affixation: A assumes the meaning of the whole ABB-type adjective, while the meaning of BB is bleached and BB can only be used as affixes, for example:

   Qian Sese (千色色)   Kong Laolao (空佬佬)   Suan Jiji (酸叽叽)
   Gan Baba (干巴巴)   Huang Xixi (慌兮兮)   Da Zhouzhou (大粥粥)

   (2) Mid-supplement structure: A assumes the main meaning of the whole ABB-type adjective while BB is a dynamic supplementary explanation of A, for example:

   Cui Songsong (脆松松)   Shou Jingjing (瘦精精)
   Yuan Gungun (圆滚滚)   Qi Zhanzhan (齐盏盏)

2. AB + B: ABB is the extension of AB. Part A and B assume the semantics of the whole word together. This type of adjectives can be divided into three categories according to the relationship between A and B:

   (1) Coordinate structure: A and B have the same grammatical status and express the same or similar meanings, such as:
   Leng Qingqing(冷瀮瀮): "瀮" means cold. The whole word means gloomy and cold.
   Ai Zizi(矮呰呰): "呰" means emaciated. The whole word means short and thin.
   Gan Kaokai(干滂滂): "滂" means the water has dried up. The whole word means dry.
   (2) Modifier modified mode: A is the modifier of B, and B overlaps, for example:
   Yi Xiexie(一歇歇): The whole word means a short period of time.
   Tong Genggeng(同庚庚): The whole word means the same age.
   Nian Dongdong(黏涷涷): The whole word means viscous things.
   Tang Huohuo(烫火火): The whole word means the temperature is too high.

(3) The relationship between A and B can also be subject predicate mode, such as Lu Didi(卤涕涕), "涕" means water seeps slowly. There are special cases where A and B form a two alliterated word, such as "依稀稀", but these cases are relatively rare in Shaoxing dialect and not separately divided into a category.

The type A + BB also shows flexibility and openness in word formation. The same A can be attached with different BBs, for example:

Tian(甜): Tian Mimi(甜蜜蜜)　Tian Zizi(甜滋滋)　　　Tian Sisi(甜丝丝)
Hong(红): Hong Xuexue(红血血)　　Hong Tongtong(红通通)
Huang: Huang Jiaojiao(黄焦焦) Huang Songsong(黄松松)
Conversely, the same BB can follow different As, such as:
Haha(哈哈): Huang Haha(黄哈哈)　　Jiu Haha(旧哈哈)
Dudu(督督): Shou Dudu (寿督督)　　Rou Dudu (肉督督)　Yang Dudu (洋督督)
Honghong(烘烘): Xiang Honghong(香烘烘)　　Chou Honghong(臭烘烘)

3. It is worth noting that some ABB-type adjectives in the Wu dialect have variant AAB, which is not the characteristic of ABB-type adjectives in Mandarin, for example:

Xiang Penpen(香喷喷) –　Penpen Xiang(喷喷香)
Baixuexue(白雪雪)　　–　Xuexue Bai(雪雪白)
Shou Jingjing(瘦精精) –　Jingjing Shou(精精瘦)
Tang Huohuo(烫火火) –　Huohuo Tang(火火烫)

It can be found that the adjectives with AAB variants mainly come from the mid-supplement structure of A+BB-type and the modifier modified mode of AB+B-type. The semantic core of these words is A. BB part of these words has not lost semantics completely and is a supplementary explanation of A. The semantics core of these words remains unchanged after transforming into the AAB type while the form chage can cause some changes to semantic of the whole word, which will be further explained later. In this way, the words whose AB+B-type modifier modified mode produces extended meaning can be regarded as special mid-supplement structure of A+BB-type.

3.2　Grammatical Function

The grammatical function of ABB-type adjectives in Shaoxing dialect is almost as same as that of ABB-type adjectives in Mandarin, which is shown in the following four aspects:

1. Serve as an attributive, for example:
依话依有木佬佬匀事体。(He said he had many things to do.)
焦焦黄匀脸色。(A waxy complexion.)
香喷喷匀荷饭。(A delicious meal)

When ABB-type adjectives serve as attributives to modify nouns, the auxiliary word "ge"(旮) should be attached, which is consistent with the principle that multi-syllable adjectives should add auxiliary word "de"(的) when serve as attribute in Mandarin. There are, of course, special cases, such as "一歇歇时光" (short time), can choose not to use auxiliary words.

2. Some ABB-type adjectives can be used as adverbials directly. In this case, they should be attached by auxiliary words "ge"(旮), such as:

我是硬邦邦旮熬牢旮。(It was hard for me to bear it.)

侬要笑呵呵旮跟我话么才对呀。(You have to talk to me with a smile.)

伊老早醉醺醺旮走哉。(He had already left in drunk)

3. The ABB-type adjectives generally do not act as predicates independently, otherwise, they should be followed by auxiliary words "ge"(旮), such as:

侬旮宁慌兮兮旮。(You're too scared.)

伊哩空佬佬弄。(He has a lot of space there.)

侬旮甜滋滋旮,蛮好吃。(It tastes sweet and delicious.)

4. There is usually an auxiliary word "le"(了) in front of ABB-type adjectives when they act as a complement.

地板被侬弄了湁獲獲哉。(You've soiled the floor)

勿好端端读书，书么扯了破纷纷旮(You tore the book to pieces and didn't work hard.)

The reduplicated adjectives in Mandarin can not be modified by degree adverbs since they have degree implication themselves. Conversely the ABB-type adjectives in Shaoxing dialect can still be modified by degree adverbs indicating low degree, such as "Zheng Dang"(真当) and "You Dian"(有点), but not by adverbs indicating high degree, such as "Man"(蛮) and "Tai"(太). For example:

侬旮宁么真当慌兮兮旮。(You're too scared.)

? 侬旮宁么太慌兮兮旮。

伊哩空有点佬佬弄。(He has a lot of space there.)

? 伊哩空蛮佬佬弄。

4 Semantic Features

The semantic features of the ABB-type adjectives in Shaoxing dialect are mainly reflected in degrees and emotional colors. The following paragraphs will focus on these two aspects.

4.1 Semantic Degree

As a typical form of adjective reduplication, ABB-type adjectives are generally considered to have an effect on deepening semantics, such as:

Wu Cucu(乌簇簇)   Yuan Gungun(圆滚滚)   Gan Baba(干巴巴)
Liang Qinqin(凉沁沁) Bai Xuexue(白雪雪)

The variant AAB of some ABB-type adjectives in the Wu dialect often deepens the degree of semantics. Taking "Bai Xuexue"(白雪雪) as an example, if an ABB adjective has the variant of AAB-type, the degree ranking of the ABB-type, the variant AAB-type and the part A is as follows.

Xuexue Bai(雪雪白)>Bai Xuexue(白雪雪)>Bai(白)

On the other hand, the ABB-type adjectives in Shaoxing dialect also weaken the degree of semantics, for example:

Tang Huohuo(烫火火) Suan Jiji(酸叽叽)   Lao Qieqie(老茄茄)
Da Zhouzhou(大粥粥) Hone Tongtong(红通通)

The third chapter of this paper mentions that the same A can bring different BBs. In terms of semantics, different BBs give A different semantic degrees, for example:

Huang Jiaojiao(黄焦焦)>Huang Hah(黄哈哈)>Huang Songson(黄松松)
Tian Mimi(甜蜜蜜)>Tian Zizi(甜滋滋)>Tian Sisi(甜丝丝)
The semantic degree of A is weakened from left to right.

4.2 Emotional Color

The lexical meaning is generally divided into the rational meaning and the emotional meaning, and reduplicated words are considered to have obvious emotional color. This paper collects 413 ABB-type adjectives from Eight Hundred Modern Chinese Words and Shaoxing dialect. Five subjects whose mother tongue is the Wu dialect were asked

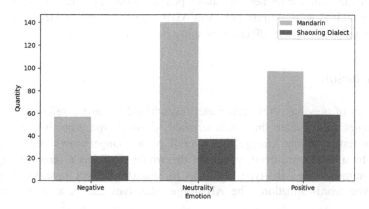

Fig. 1. The emotion distribution of ABB-type adjectives in Mandarin and Shaoxing dialect.

to mark the emotion tendencies of these words intuitively. Considering that the intuition of subjects might be different, the controversial sample was marked twice, the emotion of the sample was the most voted for emotion. With this method, the emotion distribution of the ABB-type adjectives in Mandarin and Shaoxing dialect is shown Fig. 1.

The positive samples of ABB-type adjectives in Shaoxing dialect are like "Nuo Chouchou"(懦绸绸), "Xiang Penpen"(香喷喷), neutral samples are like "Hua Jie-jie"(滑结结), "Cui Songsong"(脆松松), negative samples are like "Ai Zizi"(矮呰呰), "Bei Chichi"(背迟迟). According to the analysis results, ABB-type adjectives with positive emotion are fewer than ABB-type adjectives with other emotions, they overlap with ABB-type adjectives in Mandarin and the semantics of BB are not completely blurred. Unlike the fact that neutral ABB-type adjectives have numerical superiority in Mandarin, negative ABB-type adjectives account for the largest proportion in Shaoxing

Table 2. The influence of A, BB in Mandarin and dialects on the emotion tendency of the whole word.

Variable	F score
Mandarin A	6.02035977
Mandarin BB	3.27209197
Shaoxing dialect A	0.05227057
Shaoxing dialect BB	0.16556674

dialect. The BB part of these negative ABB-type adjectives are often highly blurred, and most of them do not have AAB variants.

With a small amount of annotated data, we take the emotion of words as labels and take A and BB as variables for regression analysis. The results are shown in Table 2.

According to the F score, BB in ABB-type adjectives of Shaoxing dialect have greater influence on the emotion tendency of the whole word than A in the ABB-type adjectives. For example, "Ga"(嘎) in "Ga Naonao"(嘎獿獿) and "Feng"(尌) in "Feng Naonao"(尌獿獿) are derogatory terms themselves, "Naonao"(獿獿) is often attached to these words to reinforce negative feelings. A more obvious example is that part A of the words "Bai Xuexue"(白雪雪) and "Bai TuTu"(白涂涂) are the same, but part BB of them are different, the former word has a positive tendency, while the latter has only a neutral meaning. As a contrast, part A in ABB-type adjectives of Mandarin has more significant influence on the affective meaning of the whole word.

5   Conclusion

On the basis of previous work, this paper takes Mandarin as the reference object and use examples to illustrate the characteristics of ABB-type adjectives in Shaoxing dialect. In terms of tone change, the overall pitch change shows an upward trend followed by a downward trend, which let the whole word has a sense of rhythm. In terms of grammar, the ABB-type adjectives in Shaoxing dialect have A+BB-type and A B+B-type word-formation. The ABB-type adjectives can be used as attributive, predicate, some are used as adverbial and complement in sentences. They can also be modified by degree adverbs indicating low degree. Semantically, the ABB-type adjectives in Shaoxing dialect can reduce or enhance the degree of semantics. They tend to express negative meanings as a whole, BB has a greater impact on the emotional tendency of the whole word.

The ABB-type adjectives often reflect an intuition that comes from the human senses. In Shaoxing dialect, this kind of reduplication adjective not only shows dynamic beauty and musicality in use, but also embodies the flexible way of thinking and language expression habits of the Yue.

Acknowledgments. This work is supported by the Fundamental Research Funds for the Central Universities, National Language Committee Research Program of China (No. ZDI135-42) and National Social Science Fund of China (No. 18CYY029).

References

1. Ruan, G.-J.: Reduplication of adjectives in Ningbo dialect. Yangtze River Acad. **04**, 102–109 (2008). (in Chinese)
2. Zhu, L.-Q., Huang, Y.: Overlapping of three-syllable adjectives in Xiaoshan dialect. Xue Yuan. **10**, 18–19 (2014). (in Chinese)
3. Lou, J.-T.: Study on AAB-type and ABB-type Adjectives in Shaoxing Dialect. Overseas English. (22), 180 (2016). (in Chinese)
4. Li, R.: Language Atlas of China. Longman, Hongkong (2012). (in Chinese)
5. Fan, Y.: Yue Proverbs. The Commercial Press, Beijing (1999). (in Chinese)
6. Lv, S.-X.: Eight Hundred Modern Chinese Words. The Commercial Press, Beijing (1999). (in Chinese)
7. Zhang, J.-S.: The Correlations between Initial Consonant and Tone in the Wu dialect. Contemporary Linguistics. (02), 121–128 + 189 (2006). (in Chinese)
8. Hou, Y.-L.: Notes on Yue Proverbs. People' s Publishing Company, Beijing (2006). (in Chinese)
9. Wu, Z.-H.: A Study on Shaoxing Dialect from the Perspective of Wuyue Culture. Zhejiang University Press, Hangzhou (2007). (in Chinese)
10. Huang, B.-R., Liao, X.-D.: Modern Chinese. Higher Education Press, Beijing (2007). (in Chinese)
11. Shao, J.-M.: A comprehensive study on ABB-type adjectives. Chin. Teach. World **01**, 19–26 (1990). (in Chinese)
12. Lu, Z.-Q.: Historical development of adjective repetition patterns. J. Hubei Univ. Philos. Soc. Sci. **06**, 30–35 (2000). (in Chinese)
13. Meng, S.-J.: ABB-type state adjectives in Zhuji Dialect. J. Suzhou Univ. Sci. Technol. (03), 27–30 + 20 (1987). (in Chinese)
14. Li, J.-R.: A Cognitive Study of Descriptive Adjectives in Mandarin Chinese. Shanghai Normal University (2004). (in Chinese)
15. Zhang, L.-L.: An Overview of The ABB Adjective Patten. Shanxi Normal University (2001). (in Chinese)

The Reclassification of Chinese Nominal Measure Words Based on Definition Mode

Wang Enxu[1(✉)] and Yuan Yulin[2]

[1] International School of Education and Exchange, Jinan University,
No. 336 Nanxinzhuang West Road, Shizhong District, Jinan 250022, China
wangbush000@126.com
[2] Department of Chinese Language and Literature, Peking University,
No. 5 Yiheyuan Road, Haidian District, Beijing 100871, China

Abstract. The classification of nominal measure words is a long-standing problem in Chinese syntactics circle. This paper tries to solve it based on dictionary definition. Through analyzing the definition mode, we find there are some rules hidden in the meaning of nominal measure words, which can be used to reclassify these words into the following categories. When the definition mode is "yongyu(used before)…xing (shape)…", they belong to the subcategory of highlighting shape of things, such as "tiao (strip)"; when the definition mode is "yongyu (used before) fen (divide)…", they fall into the subcategory of highlighting constitutive aspect of things, such as "ban (piece)"; when the definition mode is "yongyu (used before)… cheng (become)…", they fit into the subcategory of highlighting agentive aspect of things, such as "dui (pair)", etc. …. Compared with the previous classifications, this paper unifies the classification criteria, and enhances the universality and exclusiveness of classification.

Keywords: Nominal measure words · Definition mode · Reclassification · Qualia structure

1 Introduction

The classification of nominal measure words is a long-standing problem in Chinese syntactics circle. For the classification results, scholars unanimously agree that Chinese nominal measure words include at least three subcategories, which are units of weights and measures, individual measure words and collective measure words. Apart from these subcategories, however, they haven't reached an agreement on how many other subcategories exist in nominal measure words. Some scholars think there are two other subcategories [1], some believe there are three [2], while others hold the opinion that there are four or five [3–5]. For further details, see Table 1:

© Springer Nature Switzerland AG 2020
J.-F. Hong et al. (Eds.): CLSW 2019, LNAI 11831, pp. 110–122, 2020.
https://doi.org/10.1007/978-3-030-38189-9_11

Table 1. Comparative table of classification of nominal measure words in Modern Chinese (highlighting the main differences with shadows)

Reference [5]	Reference [2]	Reference [3, 4]	Reference [1]		Reference [6]	
weights and measures	weights and measures	weights and measures	Special-ized	weights and measures	Special-ized	weights and measures
individual	individual	individual		individual		individual
collective	collective	collective		collective		
	nonquantitative					kind
partial		partial				
temporary	temporary	temporary	borrowed	from noun	borrowed	from noun
quasi	quasi	quasi		from verb		from verb
container		container				
among V-O						

The first reason for the inconsistency in classification of nominal measure words is the characteristics (such as unclear boundaries between subcategories, unclear rules of collocation, etc.); the second reason is the classification criteria. Although many criteria (such as morphological, syntactic, semantic, etc.) have been used, there is no consensus on which should be the major criterion and which should be the supplement criterion. It is the inconsistency of classification criteria that has caused the various divergence in classification of nominal measure words [6]. In order to solve such problem, this paper tries to reclassify Chinese nominal measure words based on dictionary definition.

2 The Reclassification of Nominal Measure Words[1]

There are some rules hidden in the semantic construction of words. Revealing and utilizing these rules can not only reduce the difficulty of dictionary definition and explain the meaning of a group of words with one single mode [11–13], but also help to guide the classification research of words; moreover, it will make the subcategories of words more scientific and systematic. Taking nominal measure words as an example, they have a close relationship with nouns from the perspective of dictionary definition. According to whether the nominal measure words are related to things that nouns refer to, this paper divides them into two subcategories: one is unrelated to things, and the other related to things.

2.1 The Nominal Measure Words Unrelated to Things

Such nominal measure words are widely distributed in world languages, which have common semantic constructions and definition modes. Generally speaking, their semantic constructions include three parts: semantic type, conversion relationship with

---

[1] **Data sources: (1) Except for some nominal measure words added by authors in this paper, most of them are cited from the reference [7], total 179; (2) The definitions of nominal measure words mainly refer to references [8–10].**

their hyponyms and aliases, and **the definition mode is as follows**: "…fading/fei-fading (legal/illegal) jiliang danwei (metrological unit), 1 X dengyu (equal to) …, he (convert into) … (jiancheng (abbr.)/yeshuo (also called) …)" (X represents definiendum, similarly hereinafter). For example:

(1) 【qianmi (kilometer)】 changdu fading jiliang danwei, 1 qianmi dengyu 1000 mi, he 2 shili, ye shuo gongli (It is a metrological unit of length. One kilometer equals to 1000 m, which converts into 2 Chinese miles, also called "gongli").

　　【pingfangmi (square metre)】 mianji fading jiliang danwei, 1 pingfangmi dengyu 10000 pingfang-limi, he 9 pingfang-shichi. Jiancheng pingmi huo pingfang (It is a metrological unit of area. One square metre equals to 10000 cm^2, which converts into 9 Chinese square feet, abbreviated as "pingmi" or "pingfang").

　　【xiaoshi (hour)】 shijian fading jiliang danwei, 1 xiaoshi dengyu 60 fen, shi 1 tian de 1/24 (It is a metrological unit of time. One hour equals to 60 min, which converts into 1/24th of a day).

The similar nominal measure words include "du (degree), fenzhong (minute), qianke/gongjin (kilogram), gongli (kilometer), ka (card), ke (gram), ke (quarter), li (Chinese mile), liang/shiliang (Chinese ounce), mi (meter), nian/sui (year), tian/ri (day), shi (hour), jin/shijin (Chinese pound), chi (Chinese inch), cun (Chinese foot), dun (ton), hao/haomi (millimeter), limi (centimeter), miao (second), mu (Chinese acre), sheng (litre), zhang (three meters), bang/yingbang (pound), qing/gongqing (hectare), xun (ten days), lifang/lifangmi (cubic meter), qianwa (kilowatt)", etc.

If there is no hyponym in a nominal measure word, it is not necessary to explain the conversion relationship between it and its hyponym. But sometimes we need to explain its source. Under this circumstance, **the definition mode of nominal measure words is** "…fading/fei-fading (legal/illegal) jiliang danwei (metrological unit), (X laiyuan (source of X)). (jiancheng (abbr.)/yeshuo (also called) …)". For example, "wa/wate (watt), fu/fute (volt), mali (horsepower), jiao/jiaoer (joule)" and so on.

(2) 【wate(watt)】 gonglü fading jiliang danwei, 1 miao-zhong zuo 1 jiaoer de gong, gonglü shi 1 wate. Wei jinian yingguo famingjia wate er mingming. Jiancheng wa (It is a metrological unit of electrical power. The power of 1 W equals to doing 1 J of work per second. It is named in memory of the British inventor Watt, abbr. "wa").

　　【fute(volt)】 dianya he diandongshi fading jiliang danwei, 1 anpei dianliu tongguo dianzu shi 1 oumu de daoxian shi, daoxian liangduan de dianya wei 1 fute. Wei jinian yidali wulixuejia fute er mingming. jiancheng fu (It is a metrological unit of electric voltage and electromotive force. When one-ampere current passes through of 1 Ω, the voltage at both ends of the conductor is 1 V. It is named in memory of Italian physicist Volt, abbr. "fu").

2.2   The Nominal Measure Words Related to Things

According to whether or not the nominal measure words are related to things that nouns refer to, this paper divides them into two subcategories. And details are as follows.

2.2.1 Nominal Measure Words Related to the Container of Things

This kind of nominal measure words is widely distributed in Modern Chinese, English, Spanish, Japanese, Korean, Vietnamese, etc., **whose definition mode is** "yongyu (used before) X zhuang/cheng/yang/zai(load/fill/rear/carry) de…". For example:

(3) 【wan (bowl)】 yongyu wan cheng de shiwu, ru "fan, cai, jiu, shui, yao, mian, tang" deng (used before food filled in bowls, such as rice, dishes, wines, water, traditional Chinese medicine, noodles, soup, etc.).[2]

【ping (bottle)】 yongyu ping zhuang de dongxi, ru "yao, shui, cu, yinliao, xianhua" deng (used before things filled in bottles, such as traditional Chinese medicine, water, vinegar, beverage, flowers, etc.).

【juan (corral)】 yongyu juan yang de shengchu, ru "zhu, yang" deng (used before domestic animals reared in corral, such as sheep, pigs, cows, etc.).

Other similar nominal measure words include "bei (cup), cang (warehouse), dai (bag), fang (room), gang (tank), guan (jar), gui (cabinet), guo (pot), he (box), jiao (cellar), lan/kuang/lou (basket), pen (basin), tan (bombonne), tong (barrel), tong (tube), wo (nest), xiang (bin), zhong (wine cup)" and so on.

In addition, some nouns of implements, places and body parts can also be temporarily borrowed as nominal measure words. In this case, **their definition mode is** "yongyu (used before) X shang (upon) chuan (wear)/bai (display)/jinxing (go on) de…". Only a few nominal measure words like this, such as "zhuo (table), chuang (bed), kang (Chinese heated bed), tang (hall), tai (stage), shen (body), xi (mat)". For example:

(4) 【shen (body)】 yongyu shen shang chuan de yifu(used before clothes worn on bodies).

【zhuo (table)】 yongyu zhuo shang bai de dongxi, ru "fancai, jiuxi, haixian, majiang" deng (used before things put on tables, such as meal, dinner, seafood, mahjong, etc.).

2.2.2 Nominal Measure Words Related to Things Themselves

According to the different aspect related to things, this paper further divides these nominal measure words into two subcategories: one related to abstract aspect of things, such as "zhong (kind), ji (grade)"; the other related to specific aspect of things, such as "tiao (strip), jie (section)". The details are as follows:

2.2.2.1 Related to Abstract Aspects of Things, or Highlighting Abstract Aspects of Things

Similar to words in Part 2.2.1, this kind of nominal measure words is also widely distributed in Chinese, English, Spanish, Japanese, Korean, Vietnamese and so on, and **their definition mode is** "yongyu (used before) … de zhonglei (kind) huo (or) dengji (grade), cezhong yu (focus on)…". For example:

---

[2] The following kinds of nominal measure words can only be used before certain nouns. Therefore, it is necessary to annotate their collocation objects when explaining them.

(5) 【zhong (kind)】 yongyu ren huo shiwu de zhonglei, cezhong yu yiban leibei de butong, ru "ren, yanse, youqi, cailiao" deng (used before kinds of people or things, focusing on the differences in general categories, such as people, color, paint, materials, etc.).

【yang (style)】 yongyu shiwu de zhonglei, cezhong yu yangshi de butong, ru "cai, huose, shuiguo, dianxin, yifu" deng (used before kinds of things, focusing on the differences in styles, such as dishes, vegetables, goods, fruits, dessert, clothes, etc.).

Other similar nominal measure words include "men (door), lu (road), liu (rate), lei (type), deng/ji (grade), pin (level)" and so on.

2.2.2.2 Related to Specific Aspects of Things, or Highlighting Specific Aspects of Things

Unlike words in Part 2.2.1, this kind of nominal measure words is mainly distributed in Sino-Tibetan, South Asian, South Island and American Indian languages [14]. According to the differences of highlighting specific aspect of things, this paper divides them into the following types:

Type I: Highlighting the Shape of Things

Many studies notice that nominal measure words can highlight the shape (including appearance, color, size, etc.) of things [15–19]. Type I is the most typical one. When highlighting shape of things, **the definition mode** of nominal measure words is "yongyu (used before)... xing/zhuang (shape)...". For example:

(6) 【tiao (stripe)】 yongyu changtiao xing de dongxi, ru "shanchuan, ling, gou, jie-dao, xianglian, tiegui, chuan, yu, she, gou, niu, haohan, sangzi, bianzi, gebo, weiba, qiang, biandan, changzhuo, huanggua, madai, shengzi, maojin, lingdai, kuzi, qunzi, chuangdan, kouzi, caihong" deng (used before long stripe-shaped things, such as mountains, rivers, collars, ditches, streets, necklaces, rails, boats, fishes, snakes, dogs, cattle, brave men, voices, braids, arms, tails, guns, poles, long tables, cucumbers, sacks, ropes, towels, ties, trousers, skirts, sheets, slits, rainbows, etc.).

【li (granule)】 yongyu keli zhuang de dongxi, ru "mi, liangshi, zhenzhu, shazi, niukou, yaowan, laoshu-shi" deng (used before granular-shaped things, such as rice, grains, pearls, sands, buttons, pills, rat droppings, etc.).

【zhi (branch)】 yongyu xichang de gan-zhuang-wu, ru "bi, qiang, jian, xiang, lazhu, dengguan, dizi, yashua" deng (used before long slender rod-shaped things, such as pens, guns, arrows, incenses, candles, lamp tubes, flutes, toothbrushes, etc.).

Other similar nominal measure words include "ke (pellet), dao (road), di (drop), gen (root), guan (tube), jiao (angle), ke/mei/zhu (straw), kuai (dice), lun (wheel/round), lü (strand), mian (aspect), pian (slice), ting (upright), tou (head), tuan (daigh)" and so on.

Type II: Highlighting the Constitutive Aspect[3] of Things

References [3–5, 20] have noticed the nominal measure words can highlight the constitutive aspect of things. From the perspective of dictionary definition of nominal measure words, there are two kinds of prominence, which correspond to two different definition modes.

One highlights a part included in the whole thing. Such as "ba (handle), bi (pen), ding (top), zhi (branch), gan (pole), jie (session), kou (mouth), ming (fame), jia (shelf), zhi (legged)" and so on. **Their definition mode is** "yongyu (used before) … dai/you (with/have)…". For example:

(7) 【ba (handle)】 yongyu you ba huo leisi you ba de dongxi, ru "liandao, baojian, yüsan, shanzi, yaoshi, shuzi" deng (used before things with handles or similar handles, such as sickles, swords, umbrellas, fans, keys, combs, etc.).

【bi (pen)】 yongyu you jilu de qiankuan huo yewu, ru "qian, zhang, shengyi, zhaiwu, lixi, sunshi, kuikong" deng (used before recorded money or business, such as moneys, accounts, businesses, debts, interests, losses, deficits, etc.).

【ding (top)】 yongyu mouxiu dai ding de dongxi, ru "caomao, maozi, huajiao" deng (used before some things with a top, such as straw hats, hats, sedan chairs, etc.).

The other highlights a separate part from the whole thing or event, such as "bu (step), ceng (layer), duan (segment), ji (episode), ji (compilation), jian (room), jian (part), jü (round), sheng (sound), ye (page), xiang (item), ban (petal), bei (generation)" and so on. **The definition mode is** "yongyu (used before) … fen (separate/split)…". For example:

(8) 【bu (step)】 yongyu fen bu zou de xingwei, ru "lu, qi" deng (It is used before step-by- step behavior, such as roads, chesses, etc.).

【ban (petal)】 yongyu tianran huo renwei fen ban de dongxi, ru "hua, youzi, dasuan, xiguan, pingguo" deng (used before natural or artificial partitional things, such as flowers, grapefruits, garlics, watermelons, apples, etc.).

The nominal measure words mentioned above mainly highlight part; there is a "part-whole" relationship between the part and the whole thing. The number of such nominal measure words is the largest, accounting for the majority of type II. In addition, there are still some nominal measure words that do not highlight the part but attributes of things, and there is an "attribute-thing" relationship with things, such as "jia (family), zun (dignified)" and so on. When defining these nominal measure words, definers often use **the definition mode of** "yongyu (used before) (jüyou mouzhong shuxing (having certain attribute)) de…". For example:

---

[3] According to Qualia Structure theory, the so-called constitutive aspect of things refers to what a thing is constituted by, what the thing can constitute, and whether there is a "part-whole' relationship with other things [13, 19].

(9) 【jia (family)】 yongyu xiangdui duli de shengchan huo shenghuo danwei, ru "shangdian, gongchang, yiyuan, ren, baoshe" deng (used before relatively independent social organizations or living groups, such as shops, factories, hospitals, human beings, newspaper presses, etc.).

　　　【zun (dignified)】 yongyu mouxie zungui huo weiyan de shiwu, ru "foxiang, pusa, luohan, paiwei, tieta" deng (used before some noble or dignified things, such as Buddhas, Bodhisattvas, arhats, memorial tablets, iron towers, etc.).

Type III: Highlighting the Agentive Aspect of Things[4]

In terms of dictionary definitions, it is generally collective nominal measure words that can highlight agentive aspects of things, such as "cu (cluster), shuang (couple), hang (line), pai (row), pi (batch), dui/fu (pair), zu (team), dan (dan), dui (pile), fen/tao (suit), kun (bundle), lie (column), qun (group), xi (mat), chuan (string), peng (fluffy), tan (stall)" and so on. Only a few individual nominal measure words, such as "ben (bound-book), juan (volume), pan (plate), juan (roll), ji (dose)" and so on, can highlight agentive aspect of things. When highlighting the agentive aspects of things, **the definition mode** is unified, that is, "yongyu (used before) ...cheng (form/be)...". For example:

(10) 【dui (pair)】 yongyu cheng dui de ren huo shiwu (duo bu duichen), ru "fuqi, shuangbaotai, shizi, ernü" deng (used before people or things that are in pairs (usual asymmetric), such as couples, twins, lions, children, Chinese ducks, etc.).

　　　【ben (book-bound)】 yongyu zhuangding cheng ce de shuji bu ce, ru "shu, zhang, huace, xiaoshuo, zazhi, biji, caidan" deng (used before bound books for reading, note-taking or accounts-keeping, such as books, accounts, picture albums, stories, notes, magazine, menus, etc.).

　　　【pan (plate)】 yongyu chanrao cheng panxing de dongxi, ru "cidai, tiesi, she, wenxiang" deng (used before the winding plate-shaped things, such as tapes, iron wires, snakes, mosquito-repellent incenses, etc.).

Type IV: Highlighting Other Aspects of Things

As mentioned above, most nominal measure words highlight the shape, constitutive or agentive aspects of things, yet there are a few exceptions. Among them, some highlight the telic aspect of things, whose **definition mode** is "yongyu (used before)... keyi/neng (can)...", such as "zhang (open), fa (shoot)"; some highlight the handling aspect of things, whose **definition mode** is "yongyu (used before) (chuzhi (handle)) ...", such as "feng (seal), peng (hold in both cupped-hands)"; while others highlight two aspects of things at the same time, such as "ba (hold), gua (hang)". For example:

(11) 【zhang (open)】 yongyu keyi kai-he de dongxi, ru "zui, gong, yuwang" deng (used before things that can be opened and closed, such as mouths, bows, fishing nets, etc.). (highlighting telic aspect)

---

[4] According to Qualia Structure theory, the so-called agentive aspect of things focuses on how things are formed, by a natural way, an artificial way, or through a special causal relationship [13, 19].

【feng (seal)】 yongyu fengzhe de dongxi, ru "xin, qingshu, mijian, xiang" deng (used before the sealing things, such as letters, love letters, confidential letters, incenses, etc.). (highlighting handling aspect)

【ba (hold)】 yongyu keyi yong shou zhuazhu de xi-tiao-zhuang dongxi, ru "shuzhi, toufa" deng (used before the thin strip-shaped things that can be grasped by hand, such as branches, hairs, etc.). (highlighting shape and handing aspects)

3   Deficiencies of Previous Classifications

Ideal classifications of nominal measure words should be internally universal and externally exclusive. However, as a matter of fact, previous classifications have not achieved this goal.

If the classification is internally universal, members of the same subcategory should share the same properties, such as the same syntactic distribution, same collocation objects, same semantic structure and same definition mode. As far as nominal measure words are concerned, in addition to same syntactic distribution (they all appear in "number + nominal measure words + noun" structure), the collocation objects, semantic structure and definition mode are all different. Taking "ke (pellet), zhi (legged), zhang (open)" as the examples, they are all individual nominal measure words, but the collocation objects are different, the semantic structures and definition modes are also different.

(12)   【ke (pellet)】 highlighting the shape aspect, and the definition mode is "yongyu (used before) …xing (shape) …".

　　　　【zhi (legged)】 highlighting the constitutive aspect, and the definition mode is "yongyu (used before) …you (have/with) …".

　　　　【zhang (open)】 highlighting the telic aspect, and the definition mode is "yongyu (used before) keyi (can) …".

If the classification is externally exclusive, members of different subcategories are often mutually exclusive. That is to say, they should have different syntactic distributions, different collocation objects and different definition modes. As far as nominal measure words are concerned, different subcategories usually overlap rather than exclude each other. As reference [5:268] points out, individual nominal measure words are opposite to collective nominal measure words in semantics, "but such opposition is only a point of view, because a group of individuals can be parts of a larger collection at the same time". Reference [6:296] also mentions that "there is no absolute boundary between the borrowed and specialized nominal measure words. From the diachronic perspective, most of the specialized nominal measure words are derived from borrowed nouns or verbs". The views mentioned above indicate that the boundaries between individual and collective nominal measure words, specialized and borrowed nominal measure words are not clear, and the overlap between subcategories is common. In other words, non-exclusiveness is a common feature of previous classifications of nominal measure words.

4 Advantages of Reclassification in This Paper

Compared with previous classifications, the reclassification in this paper has the following two advantages.

4.1 Advantages in Classification Standards

Previous classification criteria are not unified, it is mainly because they rely too much on the subjective language experiences and neglect the verification of language forms. In order to avoid the recurrence of these problem, this paper abandons the previous multiple classification criteria which rely on language sense, and replaces it with a single classification criterion based on dictionary definition.

The reason for this is the meaning and usage of a word are interdependent. On the one hand, the meaning of a word depends on its usage, and words appearing in the same context often have the same meaning [21:157]. If we intend to understand the meaning of a word accurately, the best way is to observe what context it appears in and what words it collocates with [22]. On the other hand, the usage of a word depends on its meaning. If we know the meaning of a word, then we can predict what context it may appear in and what word it may collocate with. Taking "kuai (dice), pian (slice), li (granule)" as an example. According to the definition mode "yongyu (used before) ... zhuang (shape) de ...", we can know that "zhuang (shape)" is the core semantic component of these words; according to the above conclusion, we can predict a word's usage from its meaning, and then we know that "zhuang(shape)" is the core collection component of these words, often appearing in the context of these words or collaborating with these words. In fact, from our survey through Baidu News, we find that "kuai (dice), pian (slice), li (granule)" do often appear together with "zhuang (shape)". For example:

(13) **kuai (dice)**:   kuaizhuang wu (dice-shaped things) kuaizhuang jirou (dice-shaped muscle) kuaizhuang zhifang (dice-shaped fat), ba luobo/rou/tudou/xihongshi/nailao/huotui qie cheng kuai-zhuang (to cut radishs/meat/potatoes/tomatoes/cheeses/hams into dices);

pian (slice):   pianzhuang mianmo (slice-shaped mask) pianzhuang yaowu (slice-shaped medicine), pianzhuang shimo (slice-shaped graphite), ba shanyao/jiang/baicai/rou/nailao/doufu/huanggua qie cheng pianzhuang (to cut yams/gingers/meat/cheeses/tofus/cucumbers into slices);

li **(granule)**:   keli-zhuang wu(granule-shaped things) keli-zhuang shiwu (granule-shaped foods), keli-zhuang dongxi/mifen/zhenzhu-mian/yüzi/nitu/putaogan deng (granule-shaped things/rice flour/pearl cottons/fish seeds/soils/raisins, etc.)

Nouns that often collocate with "kuaizhuang (dice-shaped), pianzhuang (slice-shaped), lizhuang (granule-shaped)" also often collocate with nominal measure words "kuai (dice), pian (slice), li (granule)". For example, the "jirou (muscle), zhifang (fat)" and other dice-shaped things often collocate with "kuai (dice)"; and the "mianmo (mask), zhongyao (Chinese medicine), shimo (graphite)" and other slice-shaped things often collocate with "pian (slice)". In this way, the usage feature of nominal measure words can be inferred from their semantic structure and definition characteristics, which can help to achieve the mutual verification of meanings and usages of nominal measure words.

Like "kuai (dice), pian (slice), li (granule)", the usage feature of other nominal measure words (such as "wan (bowl), ceng (layer), qun (group)") can also be inferred from their semantic structure and definition characteristics. The inference method is the same as that of "kuai (dice), pian (slice), li (granule)", and thus it will not be illustrated here in detail.

4.2   Advantages in Classification Results

This paper improves the inherent universality of classification by division. According to the observation of this paper, words with less internal universality are mainly individual, partial and collective nominal measure words. Moreover, whether their definition modes are internally universal or not has nothing to do with whether they are individual, partial or collective nominal measure words, but is closely related to whether they can highlight the same aspect of things. If so, their definition mode will be the same; if not, the definition mode will be different.

In the light of highlighting different aspects of things, this paper further divides nominal measure words into five subcategories: highlighting the shape aspect, highlighting the constitutive aspect, highlighting the agentive aspect, highlighting the telic aspect and highlighting the handling aspect. After reclassification, members of the same subcategory share the similar syntactic distribution, semantic structure and common definition mode, which basically achieves the unification of pragmatic functions and definition modes of the same subcategory of nominal measure words.

Moreover, this paper enhances the exclusiveness of classification by combination. According to Part 3, the boundaries of individual, collective, partial, specialized and borrowed nominal measure words are not clear, and the overlap of subcategories is common. In order to solve these problems, this paper adopts a method of combination. Based on definition mode, this paper makes three types of combination: (1) Since there is no substantial difference in the syntactic distribution, pragmatic function and definition mode between some specialized and borrowed nominal measure words, this paper combinates them together; (2) this paper combinates some partial and individual nominal measure words together, and forms a type of highlighting the constitutive aspect; (3) this paper combinates some individual and collective nominal measure words with the same pragmatic function and definition mode, and forms a types of highlighting the agentive or handling aspect.

Besides enhancing the internal universality and external exclusiveness of classification, the reclassification in this paper can also contribute to the reduction of classifying difficulty of nominal measure words. Take "jia (shelf)" as an example. It would be troublesome to classify jia (shelf) according to previous methods, because "jia (shelf)" seems to fall into different subcategories, namely the borrowed category (given it is derived from the noun "gujia (skeleton), zhijia (trestle)"), the specialized category (given the dictionary marks it as a measure word), the individual category (given it can collocate with individual nouns plane, ladder, bridge, etc.) or the collective category (given it can collocate with collective nouns carriage, grapes, cucumbers, etc.). Employing the method in this paper, however, the classification of it will be easier. According to the definition "yongyu mouxie you zhizhu huo gujia de wuti (used before some objects with pillars or skeletons)", we can infer that the definition mode of "jia (shelf)" is "yongyu (used before)...you (have/with) ..." Therefore, it should be classified as the subcategory of highlighting constitutive aspect of things.

5 Conclusion

The classification of nominal measure words is not a strictly syntactic problem but a semantic problem. Moreover, the previous classifications of nominal measure words are mostly based on semantics rather than syntax. But the classification problem of "lack of internal universality and external exclusiveness" is still unsolved for the absence of suitable theory or method. It is not until the emergence of Qualia Structure theory that we find a practical solution.

Under the guidance of Qualia Structure theory [13, 19, 23, 24], this paper reclassifies nominal measure words based on the definition mode (see Sect. 2.2.2), and the result of reclassification ensures the internal universality and external exclusiveness to the greatest extent (see Table 2).

Table 2. Reclassification of Nominal Measure Words (Total 179, X represents definiendum)

Reclassification in this paper				Definition mode	Amount	Example	Previous classification
Unrelated to nouns				... fading/fei-fading (legal/illegal) jiliang danwei (metrological unit), 1 X dengyu (equal to) ..., he (convert into) ... (jiancheng (abbr.)/yeshuo (also called) ...)	47	Mi (meter), sheng (litre)	Weights and measures
Related to nouns	Related to the container of things			yongyu (used before) X zhuang/cheng/yang/zai (load/fill/rear/carry) de ...	34	ping (bottle), bei (cup)	container
	Related to things themselves	To abstract aspects		yongyu (used before) ... de zhonglei (kind) huo (or) dengji (grade), cezhong yu (focus on) ...	10	zhong (kind), ji (grade)	kind
		To specific aspects	The shape aspect	yongyu (used before) ... xing/zhuang (shape)...	21	tiao (stripe), kuai (dice)	individual
			The constitutive aspect	yongyu (used before) ... dai/you (with/have) ... yongyu (used before) ... fen (separate/split) ...	37	gan (pole), bu (step), ji (episode)	Individual or partial
			The agentive aspect	yongyu (used before) ...cheng (form)...	27	dui (pair), juan (roll)	individual or collective
			The handling aspect	yongyu (used before) (chuzhi (handle)) ...	3	feng (seal), peng (hold in cupped hands)	individual or collective
			The telic aspect	yongyu (used before) ... keyi/neng (can)...	3	zhang (open), fa (shoot)	individual

The classification method in this paper is mainly applicable to Chinese. From the cross-linguistic point of view, the classification problem of nominal measure words not only exists in Chinese, but also in Korean, Japanese, Thai, Vietnamese and so on [see references 25–28]. Whether this method can be used to these languages or not remains to be tested.

Acknowledgments. This paper is supported by the National Social Science Fund (19BYY030), the Youth Fund of the Ministry of Education (18YJC740095, 18YJC740058), the National Basic Research Program of China (2014CB340502), the Project of the National Language Commission (YB135-45). The anonymous reviewers of CLSW2019 put forward many valuable comments. Here, we express our sincere thanks for them!

References

1. Huang, B., Liao, X. (eds.): Modern Chinese, 5th edn. Higher Education Press, Beijing (2011). (In Chinese)
2. Zhu, D.: Lectures on Grammar. Commercial Press, Beijing (1982). (In Chinese)
3. Lü, S.h. (ed.) Modern Chinese 800 Words. Commercial Press, Beijing (1984)
4. Ji, X. (ed.): Lü Shuxiang's Anthology. Northeast Normal University Press, Changchun (2002). (In Chinese)
5. Zhao Y.: A grammar of spoken Chinese. Commercial Press, Beijing (1968/2010) (In Chinese)
6. Zhang, B.: New Modern Chinese, 2nd edn. Fudan University Press, Shanghai (2018). (In Chinese)
7. Examination Center of the Office of the National Chinese Proficiency Examination Commission: Syllabus of Chinese words and characters (Revised Version). Economic Science Publishing House, Beijing (2001) (In Chinese)
8. Li, X. (ed.): Dictionary of Modern Standard Chinese, 3rd edn. Foreign Language Teaching and Research Press & Language and Culture Press, Beijing (2014). (In Chinese)
9. Dictionary editorial office of Linguistic Research Institute of CASS (eds.): Modern Chinese Dictionary, 7th edn. Commercial Press, Beijing (2016). (In Chinese)
10. Zhang, B. (ed.): Collocation Tables of Nominal Measure Words and Nouns (Appendix 1 of Modern Chinese Functional Words Dictionary). Commercial Press, Beijing (2013). (In Chinese)
11. Chen, Z.h., Yuan, Y.: Semantic representation and automatic reasoning of Chinese kinship. Stud. Chinese Lang. **1**, 44–56 (2010). (In Chinese)
12. Tan J.: Talking about the three basic principles of dictionary definition: take the definition revision of Modern Chinese Dictionary, 6th edn. as an example. Lexicographical Studies, vol. 2, pp. 20–25 (2015). (In Chinese)
13. Wang, E., Yuan, Y.: The Qualia role distribution in word meaning and its influence on the word interpretation. J. Foreign Lang. **2**, 21–31 (2018). (In Chinese)
14. Gil, D.: Numeral Classifiers. In: Dryer, M.S., Haspelmath, M. (eds.): The World Atlas of Language Structures Online. Max Planck Institute for Evolutionary Anthropology, Leipzig (2013). http://wals.info/chapter/55, Accessed on 30 Dec 2018
15. Chen, W.: Unit and Unit Words in Modern Chinese. Shanghai People's Publishing House, Shanghai (1973)
16. Chi, Ch. (ed.): Complete Works of Chen Wangdao, vol. 2. Zhejiang University Press, Hangzhou (2011). (In Chinese)

17. Yuan, Y.: The iconicity of Chinese nominal measure words. Chinese Lang. Learn. **6**, 33–36 (1981). (In Chinese)
18. Shi, Y.: The cognitive foundations of shape-based classifiers in Modern Chinese. Lang. Teach. Res. **1**, 34–41 (2001). (In Chinese)
19. Pustejovsky, J.: The Generative Lexicon. MIT Press, Cambridge (1995)
20. Hui, H.: A Study on Chinese Nominal Measure Words. Southwest Jiaotong University Press, Chengdu (2011). (In Chinese)
21. Harris, Z.: Distributional structure. Word **10**, 146–162 (1954)
22. Firth, J.: A synopsis of linguistic theory, 1930–55. In F. Palmer (ed.) Selected Papers of J. R. Firth 1952-59, pp. 168-205. Indiana University Press, Bloomington (1957/1968)
23. Yuan, Y.: Description system and application case of Chinese noun Qualia Structure. Contemporary linguistics **1**, 31–48 (2014). (In Chinese)
24. Song, Z.: The Generative Lexicon Theory and Event Coercion in Chinese. Peking University Press, Beijing (2015). (In Chinese)
25. Zhao, B.: Chinese-Japanese Comparative Grammar. Jiangsu Education Press, Nanjing (1999)
26. Qi, T.: A comparative study of nominal measure words between China and Korea. Master Dissertation of Shandong University, Jinan (2009). (In Chinese)
27. Yuan, S.h., Long, W.: A comparative study of nominal measure words in Chinese and Thai —starting from the errors of Thai students. J. Yunnan Normal Univ. (Teach. Res. Ed. Chinese Foreign Lang.) **1**, 75–79 (2006). (In Chinese)
28. Li, Zh.: A cross-linguistic study of nominal measure words. Ph.D. Dissertation of Peking University, Beijing (2011). (In Chinese)

A Brief Analysis of Semantic Interactions Between Loanwords and Native Words in the Tang Dynasty

Yuchen Zhu[1] and Renfen Hu[2(✉)]

[1] School of Chinese Language and Literature, Beijing Normal University,
Beijing, China
zhuyuchen81@qq.com
[2] Institute of Chinese Information Processing, Beijing Normal University,
Beijing, China
irishere@mail.bnu.edu.cn

Abstract. Loanwords are words borrowed from another language. This paper conducts case studies of Chinese loanwords in the Tang Dynasty, including both transliterated and liberally translated words, e.g. "Kurung slave", "lion", "camel bird" and "nail aromatic". With synchronic and diachronic analysis, the study finds that the incorporation of loanwords not only brings in new words, but also triggers semantic interactions between loanwords and native words, resulting in the misconception of both loanwords and native words.

Keywords: Loanword · Semantic interaction · Misconception · The Tang Dynasty

1 Introduction

A loanword, in a broad sense, refers to a word that at some point in the history of a language entered its lexicon as a result of borrowing (or transfer, or copying) [1]. After loanwords enter another language, they can trigger a series of lexical changes, e.g., the birth of new words, the disappearance of old words, and the change of word meaning. Previous studies on Chinese loanwords have summarised several types of "word-borrowing" (Luo 2017) [2], and discussed many specific examples by tracing their lexical origins from the western regions of China, Buddhism and the Western world (Wang 2015) [3]. However, those studies have not examined the relation between loanwords and native words in the Sinicisation process. Therefore, the influences of loanwords on existing vocabulary system are still worthy of further investigation.

In the Tang Dynasty, the flourishing communication between China and foreign countries greatly influenced the society and culture of ancient China, as well as the development of Chinese vocabulary. Therefore, it is an ideal period for studying the relations between loanwords and native words in Chinese. To retrieve the loanword vocabulary in the Tang dynasty, this study refers to Edward (2016)'s work on Tang exotic, which includes 1,493 exotic nouns [4]. After removing the names of people, places, books, and word groups, 700 loanwords remain for further analysis. This study

J.-F. Hong et al. (Eds.): CLSW 2019, LNAI 11831, pp. 123–129, 2020.
https://doi.org/10.1007/978-3-030-38189-9_12

shows that the loanwords firstly join the existing vocabulary system as new words, and then they may gradually interact with the existing words, resulting in misconceptions. Case studies will be illustrated in the following sections.

2 Semantic Evolution Introduced by Transliterated Loanwords

Chinese loanwords are mostly introduced via transliteration. For example, in the Tang Dynasty, the dance performed near Tashkent was called "*zhe-zhi*" (柘枝, *Chaj* in Farsi), the bodhi tree was called "*bo-luo*" (波罗, *pippala* in Sanskirt), the lead oxide was called "*mi-tuo-seng*" (密陀僧, *mirdāsang* in Farsi), and the date palm tree was called "*gu-mang*" (鹘莽, **gurmang* in Farsi). Usually, transliterated loanwords do not influence the usages and meanings of existing words. However, after analysing the transliterated Chinese loanwords in the Tang Dynasty, this study finds that some specific loanwords have semantic interactions with native Chinese words.

2.1 Kurung Slave and Kunlun Mountain

The prosperous world trade in the Tang Dynasty imported South Asian slaves to China. Those slaves, says the *Old Book of Tang* [5], "…all have curly hair and black bodies, and are collectively called 'Kurung' (昆仑)." According to Edward (2016) [4], "Kurung" means that they were "Malays" in the broadest sense who came from Kurung Bnam (an ancient name for Cambodia). Takeda [6] proposes that the name of "Kurung" came from Gunong Api (a volcano) in Malay. According to existing studies, "Kurung" is a transliterated loanword. In addition to "Kunrung", "*Zānjī*" (僧祇), "*Turmi*" (突弥), "*Kurdang*" (骨堂) are also used to refer to these slaves in Tang's literature.

Among these transliterated names, "Kurung" was most widely used because its pronunciation was most familiar to people in the Tang Dynasty. In the *Classic of Mountains and Seas*, "Kunlun" (昆仑) is a western holy mountain where the Queen Mother of the West resides, and it is also the source of the Yellow River. Because of the similar pronunciation, people also used the same characters as "Kunlun"(昆仑) to record "Kurung" (昆仑) in Chinese. These two words became homonyms with different meanings, which led to the misconception of the two words. In the process of semantic evolution, "Kurung" derives the mythological semantics of "Kunlun", and "Kunlun" obtains the geographical semantics of "Kurung".

There are many superpower descriptions of Kurung slaves in Tang legends, such as long-term diving, flying, and immortality. According to Kang [7], Peoples in the Tang Dynasty perceived Kurung as a Buddhist nation and believed the Kurung slaves were followers of Buddhism. Thus, their religious belief made the Tang people imagine that these Kurung slaves have superpowers. However, rather than being consistent with Buddhism, superpowers such as flying and immortality are more consistent with the legends about the Queen Mother of the West. A story in the *Huai-nan Tzu* [8] says, "Yi asked the Queen Mother of the West for an immortality drug. His wife stole it and flew to the moon". Thus, it seems that the Tang people integrated their legends of Kunlun

Mountain into their cognition of Kurung slaves. In this way, the culture behind the native word influenced the meaning of the homonymous loanword.

On the other hand, Youngae [9] suggests that the Tang people could not distinguish between the Kunlun in the west of China and the Kurung in Southeast Asia. A piece of evidence is shown in the silver case in Fig. 1. It depicts the Kunlun Kingdom with elephants, which comes from Southeast Asia. Figure 2 shows another example. It is a map drawn by Luo (罗洪先) in the Ming Dynasty. On this map, the source of the Yellow River, a cucurbit-shaped lake on the map, is not located in the western regions of China, as was believed in ancient times, but is closer to Southeast Asia. Possibly, the Chinese people's geographical concept of Kunlun has changed since the Tang Dynasty. It becomes clear that "Kunlun", previously a geological noun, experienced semantic deviation under the influence of the loanword "Kurung".

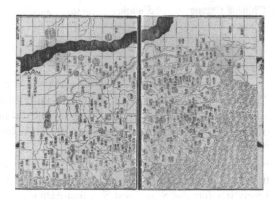

Fig. 1. Six-panelled silver case with the names of the seven countries of the Protectorate General

Fig. 2. *Enlarged Terrestrial Atlas* by Luo

2.2   Lion: From Animal to Kingdom

In the Manjusri Buddha paintings of the Tun-Huang frescoes, Kurung slaves often appear together with lions as a particular Buddhist art theme. Youngae [9] analyses this theme in detail, holding that Kurung slaves mainly served as lion tamers. Similar to Kurung slaves, lions were also exotic in ancient China, and the loanword for this animal underwent a complex semantic evolution.

Lions were called "*suangi*" (狻猊) when imported into China in the Han Dynasty. "*suangi*" is a transliterated word from ancient India. After the seventh century, lions were imported from Persia, and their Chinese name turned to be "*shi-zi*" (/ʂʐ̩ tsi/) since the Tocharian pronunciation for lion is *śiśäk* [10].

In the Buddhist art of the Tang Dynasty, the lions often appeared together with Kurung slaves, so peoples believed that lions also originated from Southeast Asia. That is why the Tang people called Ceylon "Country of Lions". However, Ceylon was not home of lions, but was home of gems. Arabic people called it Jazīrat al-Yakūt (Isle of

Gems), whose direct transliterated name was "*Sinhala*" (僧伽罗). By comparing the word frequencies of "*Sinhala*" (僧伽罗国) and "Country of Lions" (师子国) in the corpus of Chinese Classic Ancient Books and CBETA, one can easily see a significant difference between the two words:

Table 1. Word frequencies of *Sinhala* (僧伽罗国) and Country of Lions (师子国) in CBETA (http://cbetaonline.dila.edu.tw) and Corpus of Chinese Classic Ancient Books (http://dh.ersjk.com)

		Han-Wei and the Six Dynasties	Sui-Tang and the Five Dynasties	Song Dynasty
CBETA	*Sinhala*	0	25	3
	Country of Lions	70	121	54
Chinese Classic Ancient Books	*Sinhala*	0	22	3
	Country of Lions	29	121	80

As shown in Table 1, Ceylon received much attention in the Tang Dynasty. "*Sinhala*" and "Country of Lions" coexisted for a short period, and "Country of Lions" was used more widely later. The Tang people did not translate the toponym on the basis of Ceylon being perceived as a place of gems. Instead, they probably re-transliterated "*Sinhala*" into "*shi-zi*" (/ʂʅ tsi/, lion).

The result of Ceylon being named "Country of Lions" was to make people believe that Ceylon was home of lions. As the *New Book of Tang* [11] describes, "Country of Lions inhabits the Southwest Sea that extends over 2,000 *li*, with Leng-Jia Mountain and abounding in rare jewels… People there can tame lions. So, the state is named 'Country of Lions'". This phenomenon is similar to that of "Kurung". In both cases, the native word causes the meaning shift of the homonymous loanword.

3   Semantic Evolution Introduced by Liberally Translated Loanwords

In addition to transliterated loanwords, many foreign words came into China by liberal translation in the Tang Dynasty. In this section, we will discuss the semantic interaction between the native words and the liberally translated loanwords.

3.1   How Did the Ostrich Become a Monster?

The ostrich was also imported into China in the Han Dynasty. According to the *Book of the Later Han* [12], "Tazi is adjacent to the Western Sea. It is the homeland of lions and huge birds". "The neck, body, and hoof of the huge bird are similar to those of a camel. It is eight to nine *chi* (0.33 m) tall. Its wingspan can extend over one *zhang* (3.33 m)." Thus, people used the term "camel bird" (驼鸟, *tuo-niao*) to refer to the ostrich at that time.

Camel bird is a liberal translation from *ushtur murgh* in Farsi (Edward 2016) [4] to highlight the bird's camel-like hoof. The famous poet Li (李白) wrote in his *Autumn Estuary Song*, "Tapestried camel-bird at Autumn Estuary. As rare among men as up in Heaven." Since the ostrich was so rare, people in later dynasties could hardly understand the bird's translated name. Figure 3 shows an image of ostrich portrayed by people in the Qing Dynasty: a camel with a pair of wings, which is a complete monster. Since the camel bird was not a common animal in ancient China, people misinterpreted it with its morphemes. In human cognition, the image of the camel bird changed from a bird with a camel-like hoof to a monster, i.e., a camel with wings in Fig. 3.

Fig. 3. An ostrich as depicted in the *Collection of Ancient and Contemporary Books*

3.2 Nail Aromatic and Chicken Tongue Aromatic

When a liberally translated loanword encounters a similar native word, it may result in misconceptions of both words, such as "nail aromatic" (丁香, *ding-xiang*) and "chicken tongue aromatic" (鸡舌香, *ji-she-xiang*).

The use of the chicken tongue aromatic can be dated to the Han Dynasty. The Ancient Chinese name for "chicken tongue aromatic" refers to the appearance of its flower buds. The *Ritual of Officials of Han* records that officers were required to have chicken tongue aromatic in their mouths when they made reports to the emperor. However, when the Tang people imported this plant from Indonesia, trouble arose with its new name.

The English word for chicken tongue aromatic is "clove". In Latin, it is "*clavis*" derived from "*clou*" (nail) in Old French (Edward 2016) [4]. It was so named because the herb looks like a nail. In China, after imported from Indonesia in the Tang Dynasty, "chicken tongue aromatic" was also renamed "nail aromatic" for the same reason with "clove". However, "nail aromatic" was used to refer to lilac in earlier times. Thus, these two words became homonyms denoting two different things, and soon made people confused about the meaning of the word. Although Scholars in the Song Dynasty, such as Shen (沈括), pointed out "chicken tongue aromatic" (clove) and "nail aromatic" (lilac) were different species. Peoples in the Ming Dynasty still believed that both words referred to the same flower, as *Compendium of Materia Medica* says, the pistillate flower named "chicken tongue aromatic" and the staminate flower named "nail aromatic".

In short, Chinese people initially used "chicken tongue aromatic" to refer to clove and "nail aromatic" to lilac. After the "nail aromatic" was used as a loanword, it can denote both clove and lilac. Meanwhile, when referring to clove, people used "chicken tongue aromatic" and "nail aromatic" to distinguish between different genders. Thus, the loanword "Nail aromatic", which referred to both lilac and clove, led to the misconception of those plants.

4 Conclusion

Loanwords are not only new words to a vocabulary system, but they also have semantic interactions with existing words. This paper investigates four loanwords in the Tang Dynasty: "Kurung slave", "lion", "camel bird" and "nail aromatic". "Kurung slave" and "lion" are transliterated, and "camel bird" and "nail aromatic" are liberally translated.

When a loanword is introduced by transliteration, if its orthography is similar to or the same as an existing native word, misconceptions will occur between these two words. Taking "Kunlun" as an example, people attached the myth of Kunlun Mountain to Kurung slaves, and as a result, the geographical position of Kunlun Mountain in human cognition was moved southward since Kurung slaves came from south Asia. A similar case is "shi-zi" (lion). The Chinese did not insist on using the transliteration of the word Sinhala but transliterated it into "shi-zi-guo" (country of lions). After that, Chinese people also believed that Ceylon is a country abounds with lions because of its name.

For loanwords translated liberally, misconceptions can also be easily observed. "Camel bird" is a word that was not related to any native word, and its meaning was misinterpreted as "a camel with wings" when the animal became rare in daily life. By contrast, "nail aromatic" was used for both a native word and a loanword. This homonym phenomenon also led to the misunderstanding of word meanings in ancient Chinese.

This preliminary study on Chinese loanwords reveals the complicated relationship between loanwords and native words. In the future, historical linguistic data and methods can be used to deepen the study of loanwords in medieval Chinese vocabulary.

Acknowledgements. This paper is supported by the Fundamental Research Funds for the Central Universities.

References

1. Martin, H., Uri, T.: Loanwords in the World's Languages: A Comparative Handbook. De Gruyter Mouton, Berlin (2009)
2. Luo, C.-P.: Language and Culture. Peking University Press, Beijing (2017). (in Chinese)
3. Wang, L.: History of Chinese Language. Chunghwa Book Company, Beijing (2015). (in Chinese)
4. Edward, S.: The Golden Peaches of Samarkand: A Study of Tang Exotics. University of California Press, Berkeley (2016)
5. Liu, X.: Old Book of Tang. Chunghwa Book Company, Beijing (1975). (in Chinese)

6. Takeda, M.: Construct Another Universe: Traditional Chinese Thinking of Space-Time. Chunghwa Book Company, Beijing (2017). (in Chinese)
7. Kang, H.: Kunlun and Kunlun slaves as Buddhists in the eyes of the Tang Chinese. KEMANUSIAAN 1(22), 27–52 (2015)
8. Liu, A.: Huai-nan Tzu. Chunghwa Book Company, Beijing (1998). (in Chinese)
9. Youngae, L.: The "Lion and Kunlun Slave" image: a motif of Buddhist art found in Unified Silla funerary sculpture. Sungkyun J. East Asian Stud. 2(18), 153–178 (2018)
10. Berthold, L.: The Si-hia language, a study in Indo-Chinese philology. Toung Paos 2(17), 1–126 (1916)
11. Ouyang, X., Song, C.: New Book of Tang. Chunghwa Book Company, Beijing (1975). (in Chinese)
12. Fan, Y.: Book of the Later Han. Chunghwa Book Company, Beijing (1975). (in Chinese)

On the Verb *Zao* in Ha-Fu Northeastern Mandarin

Bing Shao[1(✉)] and Minglong Wei[2]

[1] Chinese Language and Culture College,
Beijing Normal University, Beijing, China
`shaobing1789@163.com`
[2] School of Foreign Languages, Peking University, Beijing, China

Abstract. As a multi-ethnic area, Northeast China has a long history and abundant culture. Northeastern Mandarin is a symbol of the unique multicultural phenomena in the Northeastern Mandarin area, which plays an irreplaceable role in the communication network in this region. Based on the dialect corpus collected from local citizens, this article conducts a research on the frequently-used universal verb *zao* (造) in Ha-Fu Northeastern Mandarin from the perspectives of semantics, grammar and pragmatics, with a view to some references for further researches on the lexicology, typology, and computational linguistics concerning Northeastern Mandarin.

Keywords: Northeastern Mandarin · Ha-Fu dialect · Universal verb · *Zao*

1 Introduction

Northeastern Mandarin is a subgroup of Mandarin which is mostly spoken in Northeast China. The *Language Atlas of China* [1] divides it into three subgroups: Ji-Shen (Jilin and Shenyang), Ha-Fu (Harbin and Fuxin) and Hei-Song (Heilongjiang and Songhuajiang), among which the Ha-Fu subgroup is spoken in 68 counties of 5 provinces. There have been substantial studies on the lexicon in Northeastern Mandarin. For example, Chang [2] expounds the usages of various degree adverbs; Chen [3] investigates the etymology and usage of Northeastern Mandarin words borrowed from the Manchu language; Fu [4] describes the lexicon in terms of usage, attitudes, regional varieties among others.

There are some "universal verbs" in Northeastern Mandarin which are frequently used and have different meanings under different conditions, such as *zheng* (整) and *zao* (造). Zhang [5] summarizes the meaning and usage of *zheng* and suggests that it might enter Standard Mandarin in the future. Lu [6] does a systematic research on *zheng* from many aspects. And Yang and Tong [7] compare the semantics and pragmatics of *zheng* and *zao*.

In summary, previous studies on universal verbs in Northeastern Mandarin are lacking, and most of them prefer *zheng*. This article, therefore, focuses on *zao* as another important universal verb which is often used in Ha-Fu Northeastern Mandarin based on the description of the speech by speakers of this dialect. Systematic analysis

© Springer Nature Switzerland AG 2020
J.-F. Hong et al. (Eds.): CLSW 2019, LNAI 11831, pp. 130–138, 2020.
https://doi.org/10.1007/978-3-030-38189-9_13

and summary are carried forward in this article from semantics, grammar and pragmatics, in hopes of some insights for further research on *zao* and other similar phenomena.

2 Semantics of *Zao*

According to the author's investigation and field research, it is found that the universal verb *zao* in Ha-Fu Northeastern Mandarin is very expressive and frequently used in vernacular language. This verb is so rich in meaning that it can even replace many action verbs.

2.1 "To Eat, To Drink"

Zao is often used in place of *chi* (吃, to eat) or *he* (喝, to drink), particularly between family members or close friends, which manifests the vivid precision and humor with intimate persons, as in (1) and (2).

(1) 他现在咋这么能吃呢？一锅饭全让他造了！
ta xianzai za zheme neng chi ne?   yi guo fan quan rang ta zao   le!
he now why so able eat Modal one pan rice all let he *zao* ASP
'Why can he eat so much now? The whole pan of rice was eaten up by him!'

(2) 他们几个昨天晚上造没了三箱啤酒。
tamen ji ge zuotian wanshang zao mei   le san xiang pijiu.
they several CL yesterday night   *zao* gone ASP three case beer
'They drank up three cases of beer last night.'

2.2 "To Waste, To Ruin"

Zao can mean "to waste, to ruin", which refers to the harm to entities, to persons and to virtual things like money.

(3) 听说他弟弟是个败家子儿，不到一周就把父母辛辛苦苦攒的钱造光了。
tingshuo ta   didi   shi ge baijiazir, budao yi zhou jiu ba
fumu xinxinkuku zan de qian   zao guang le.
hear   he younger brother be CL spendthrift, within one week just BA
parents   hard   earn DE money *zao* all   ASP
'It's said that his younger brother is a spendthrift, who spent up all his parents' hard-earned money in one week.'

When used in this sense, *zao* can also combine with some free morphemes to express meanings similar to *zao* alone, such as *zaohuo* (造祸, to *zao* misfortune), *zaohai* (造害, to *zao* harm), *zaojian* (造践, to *zao* spoiling).

2.3 Equivalent to Other Universal Verbs

Sometimes *zao* is interchangeable with other universal verbs such as *zheng* (整), *nong* (弄), *gao* (搞), *gan* (干). It tends to refer to negative changes imposed on the object due to external causes, which involves deeper effect from the action and stronger emotional function than other universal verbs.

(4) 本来挺好一计划，你一插手全给造黄了。
benlai ting hao yi jihua, ni yi chashou quan gei zao huang le.
original very good one plan you once take part all Voice *zao* fail ASP
'It was a very good plan, but your taking part made a mess of it.'

The *zao* in (4) shows an exemplary usage of *zao*, whose occurrence can be replaced by other universal verbs *zheng, nong, gao, gan* in Northeastern Mandarin.

2.4 Other Meanings

2.4.1 Commendatorily "To Be Capable (Of)"

(5) 这小子现在造得厉害，手底下管着百八十号人呢。
zhe xiaozi xianzai zao de lihai, shou dixia guan zhe bai bashi hao ren ne.
this guy now *zao* AUX great hand under rule ASP hundred eighty CL person INTERJ
'This guy is capable and doing great now. He is supervising almost one hundred people.'

2.4.2 "To Go, To Leave For"

(6) 我说这几天咋都没见着你，造北京去了？
wo shuo zhe ji tian za dou mei jianzhao ni, zao Beijing qu le?
I say this several day why all not see you *zao* Beijing go ASP

'No wonder I haven't seen you these days at all. Went to Beijing?[1]'

2.4.3 "To Shock or To Be Shocked"

(7) 他一听这话，当时造一脑袋冷汗。
ta yi ting zhe hua, dangshi zao yi naodai leng han.
he once listen this word at that moment *zao* one head cold sweat
'Once he heard this, cold sweat (of fear) broke out over his head instantly.'

---

[1] Since the instances are from oral communication, translations are also done in a relatively informal way.

2.4.4   (Price, Age, Etc.) "To Reach Certain Degree"

(8) 食堂这学期炒菜造到八块钱一盘了。

shitang   zhe xueqi chao   cai zao dao ba   kuai qian   yi pan le.
dining hall this term stir-fry dish *zao* to eight yuan money one plate ASP
'The stir-fried dishes in the dining hall rose to 8 yuan per plate this term.'

In the examples above, *Zao* has revealed both the specificity as a content verb and the integrity and generality as a universal verb, which manifests its abundance in meaning and flexibility in usage.

3   Grammatical Functions of *Zao*

Taxonomically, *zao* can be classified into action verbs. Thus, it can achieve various grammatical functions in collocation with other syntactic constituents. This section will give a detailed grammatical analysis of the verb by virtue of the corpus.

3.1   Can Be Pre-modified by Modal Verbs, Adverbs, and Descriptive Adjectives

As an action verb, *zao* can be preceded by modal verbs, as in *neng zao* (能造, can "zao"), and *gai zao* (该造, should "zao"), and by adverbs like *bai zao* (白造, to "zao" in vain) and *bu zao* (不造, not to "zao"). Adjectives in Mandarin can also modify verb directly or with a *de* (地) [8], as in *kuai zao* (快造, to "zao" quick(ly)) and *hutude zao* (糊涂地造, to "zao" muddled(ly)).

3.2   Can Take an Object

When used as a transitive verb, *zao* can take various kinds of objects as listed below.

3.2.1   Resultative Object

The resultative object of *zao* is usually an endocentric noun phrase composed of a classifier phrase and a noun phrase as the head.

(9) 告诉你干活加点小心，大衣咋又造了个大口子？

gaosu ni gan huo jia dian xiaoxin, dayi za   you zao   le ge da kouzi?
tell   you do work add a little care   coat why again *zao* ASP CL big rip
'I told you to take care while working. Why did you get the coat ripped again?'

Bare nouns usually cannot function as the object of *zao*. For example, *zao yishen han* ("zao" a body of sweat) is acceptable whereas **zao han* ("zao" sweat) is not.

3.2.2 Locative Object

(10) 你这速度挺厉害啊，没几天从东北造上海去了？

ni zhe sudu ting lihai a,   mei ji   tian cong dongbei   zao Shanghai
qu le?
you this speed so great INTERJ. not several day from Northeast *zao* Shanghai
go ASP
'You had such a high speed that you reached Shanghai from Northeast within
a few days?'

The locative object of *zao*, as a rule, cannot be a bare noun either. If it is to take a bare noun, the noun must be part of a locative or directional complement, which are to be introduced in Sect. 3.3.

3.2.3 Numeral Object

The numeral object of *zao* is more often a numeral phrase than a single word, like (11):

(11) 小外甥走路时不小心摔了个屁股蹲儿，手里的饮料造了一地。

xiao waisheng zoulu shi bu xiaoxin shuai le   ge pigudunr, shou li de
yinliao zao le   yi di.
little nephew   walk time not careful fall ASP CL to the ass hand in DE
drink *zao* ASP one floor
'Little nephew fell carelessly to the ground on his ass, and dropped the drink
in his hand to the floor.'

3.2.4 Target Object

(12) 这大胖小子感冒好了胃口特好，一顿就造了两大碗饭。

zhe da pang xiaozi ganmao hao   le   weikou te hao, yi dun jiu
zao le liang da wan fan.
This big fat   boy   cold recover ASP appetite very good one meal just
zao ASP two big bowl rice
'This big fat boy had a very good appetite after recovering from the cold and
ate two big bowls of rice in just one meal.'

3.3 Can Take a Complement

Zao may also precede a verb, an adjective or another component as a complement. There are five types of complement listed below, among which a resultative complement is the most common case. The action of *zao*, therefore, is inherently already finished in most cases.

3.3.1 Resultative Complement

(13) 你也太败家了，一个月工资几天就让你造没了。

ni ye tai baijia   le, yi ge yue gongzi ji   tian jiu rang ni
zao mei le.

you MOD too spendthrift MOD one CL month salary several day just let you
zao gone ASP

'You are so spendthrift that all the monthly salary was spent by you in just a
few days.'

3.3.2 Stative Complement

(14) 新买的衣服就让你造这么埋汰。

xin mai de yifu jiu rang ni zao zheme maitai.

new buy DE clothes just let you *zao* so   dirty

'The newly bought clothes were made so dirty by you.'

3.3.3 Locative Complement

(15) 国产货现在行了啊，都造到国外去了。

guochan huo xianzai xing le a,   dou zao dao guowai qu le.

national goods now   capable ASP INTERJ MOD *zao* to abroad go ASP

'National goods are so good that they have spread abroad.'

3.3.4 Directional Complement

(16) 我大姐开车特别猛，车刚一发动就像闪电一样造出去好远。

wo da jie kai che tebie meng, che gang yi   fadong jiu xiang
shandian yiyang zao chuqu hao yuan.

I eldest sister drive car very violent car just once start (a car) just like
thunder same *zao* out so far

'My eldest sister drives cars very violently. The car rushed out like a thunder
once started.'

3.3.5 Numeral Complement

(17) 小田丁架给人起外号，一个人的外号就造四五个。

Xiaotian dingjia gei ren qi   waihao, yi ge ren de waihao
jiu zao si wu ge.

Xiaotian always give person name nickname one CL person POS nickname
just *zao* four five CL

'Xiaotian always gives other people nicknames. The number of his nickname
for one even reaches four or five.'

3.4   Can Take an Aspectual Marker

Zao can take an aspectual marker such as *zhe*, *le* and *guo* to indicate the relavant time of the action.

(18) 今天是小张的生日，妈妈给他煮的长寿面他正造着呢。

jintian shi Xiaozhang de   shengri, mama gei ta zhu de chang shou mian ta

zheng zao zhe ne.

today be Xiaozhang POS birthday mother for he boil DE long   life noodle he

AUX *zao* ASP INTERJ

'Today is Xiaozhang's birthday. The long-life noodles his mother cooked for him, he is eating.'

But it should be noted that when *zao* means "to eat" (Subsect. 2.1), it does not normally directly co-occurs with *zhe* (progressive), but accompanies other function words such as *zheng* (progressive auxiliary, literally "right") and *ne* (interjection which often indicates progression), as is shown in (14).

3.5   Cannot Be Reduplicated or Used in "VP-Neg-VP" Questions

Zao cannot be reduplicated or used in "VP-neg-VP" questions like other universal verbs *nong* and *gan*:

(19) a1. 别着急，我再弄弄。

bie   zhaoji, wo zai   nong nong.

don't hurry   I   again *nong nong*

'Don't hurry. I'll handle it again.'

a2. *别着急，我再造造。

*bie   zhaoji, wo zai   zao zao

*don't hurry   I   again *zao zao*

b1. 还干不干了？

hai gan bu gan le ?

still *gan* not *gan* ASP

'Will you do it any more or not?'

b2. *还造不造了？

*hai zao bu zao le ?

*still *zao* not *zao* ASP

4   Pragmatic Features of *Zao*

4.1   Emotional Features

Zao adds to some emotional insights in Ha-Fu Northeastern Mandarin which are absent in Standard Mandarin. While it is a neutral verb in nature, certain emotions may get involved as it co-occurs with some words, which is often part of the addresser's subjective appraisal.

(20) 小张连干一天活都不累，这身体真抗造！

    Xiaozhang   lian   gan yi   tian huo dou   bu   lei, zhe shenti zhen kang zao!

    Xiaozhang continue do one day work even not tired this body really resist *zao*

    'Xiaozhang doesn't get tired even after working all day. His body is so tough!'

The emotional features of *zao* are endowed by the context. *Zao* in (20) is emotionally commendatory, implying the admiration and praise for Xiaozhang's quality.

4.2 Colloquial Features

The usage of *zao* is a concise and vivid one, which is of apparent colloquial features. Such features can cause an enthusiastic or humorous effect in communication.

(21) 都多少年不碰篮球了，造两下子？

    dou   duoshao nian bu peng   lanqiu   le,   zao liang xiazi?

    MOD how many year not touch basketball ASP *zao* two AUX

    'You haven't touched a basketball for so many years. Why not play a little?'

The *zao* used here is a so colloquial that the intimacy and informality between the speaker and the hearer is easy to discover. This is what the neutral counterpart "to play" *wan* (玩) and *da* (打) cannot express.

4.3 Regional Features

Developed in a multi-ethnic and multi-cultural area, Northeastern Mandarin combines the features of various dialects of Chinese and languages of minority nationalities, shaping a humorous, generous and powerful image.

(22) 咱都是哥们儿就别这么见外，吃的喝的可劲儿造！

    zan dou shi gemenr jiu bie zheme jian wai, chi de   he   de   kejinr zao!

    we all be brother then don't this   see out eat DE drink DE full power *zao*

    'Since we are all brothers, don't treat yourself as an outsider. Just have whatever you want!'

5 Conclusion

This article has comprehensively analyzed the Ha-Fu Northeastern Mandarin from the three panels of semantics, grammar and pragmatics based on relevant studies and actual speech. *Zao* is found to be a distinctive sign of Ha-Fu Northeastern Mandarin from three perspectives: firstly, it has rich and complicated meanings; secondly, it has flexible and various syntactic structures; thirdly, it has highly characteristic emotional, colloquial and regional features.

Such dialectal words as *zao* reflect the linguistic and cultural features of the area where the dialect is spoken. Studying Northeastern Mandarin may help provide evidence for the study of Chinese. It is wished that this article could add a block in the further study not only by and of the dialect itself, but also from typological and computational linguistic perspectives.

References

1. Institute of Linguistics, Chinese Academy of Social Sciences; The Institute of Ethnology and Anthropology, Chinese Academy of Social Sciences; Language Information Sciences Research Centre, City University of Hong Kong. Language Atlas of China, 2nd edn. The Commercial Press, Beijing (2012)
2. Chang, C.: shilun dongbei fangyan chengdu fuci [On degree adverbs in Northeastern dialect]. J. Qiqihar Univ. (Philos. Soc. Sci. Edn.) (3), 115–121 (1983)
3. Chen, B.: manyu zai dongbei fangyan zhong de yiliu [Remnant of Manchu in Northeastern dialect]. Heilongjiang Natl. Ser. (4), 109–113 (1994)
4. Fu, N.: Dongbei fangyan ciyu shiyong he yuyan taidu de daiji ji diqu chayi yanjiu - yi jilinsheng songyuanshi weili [On the intergenerational and regional differences in the use and language attitudes of Northeastern dialect - a case study in Songyuan, Jilin]. Master's dissertation, Minzu University of China (2010, unpublished)
5. Zhang, W.: fangyan dongci "zheng" jianshuo [Brief comments on the dialectal verb *zheng*]. J. Zhaowuda Mongolian Teach. Coll. (Soc. Sci.) **20**(1), 36–39 (1999)
6. Lu, Y.: The Research of the Universal Verb "Zheng" of Northeast Dialect. Master's dissertation, Shanghai Normal University (2012, unpublished)
7. Yang, C., Tong, X.: On two universal verbs of "Zheng" and "Zao" in dialect of Northeast China. J. Dalian Univ. **34**(2), 42–46 (2013)
8. Zhu, D.: yufa jiangyi [Lecture notes on Grammar]. The Commercial Press, Beijing (1982)

Linguistic Synaesthesia of Mandarin Sensory Adjectives: Corpus-Based and Experimental Approaches

Qingqing Zhao[1(✉)], Yunfei Long[2], and Chu-Ren Huang[3]

[1] Institute of Linguistics, Chinese Academy of Social Sciences, Beijing, China
zhaoqingqing0611@163.com
[2] School of Medicine, University of Nottingham, Nottingham, UK
Yunfei.Long@nottingham.ac.uk
[3] Department of Chinese and Bilingual Studies,
The Hong Kong Polytechnic University, Hong Kong, China
churen.huang@polyu.edu.hk

Abstract. This study examines linguistic synaesthesia based on both the corpus distribution and the modality rating of Mandarin synaesthetic adjectives. We find that the tendencies attested through the corpus-based and the experimental approaches are compatible, including: (1) the modality exclusivity is negatively correlated with the usage of Mandarin sensory adjectives in linguistic synaesthesia; and (2) the ratings on sensory modalities of Mandarin synaesthetic adjectives are consistent with the synaesthetic directionality of these adjectives. The paper thus argues for the cognitive reality of linguistic synaesthesia, which can be evidenced by both the language production in the corpus and the language processing in the behavior experiment.

Keywords: Linguistic synaesthesia · Corpus-based · Modality exclusivity · Mandarin

1 Introduction

Linguistic synaesthesia is a cross-linguistic language usage of lexical items in one sensory domain to conceptualize or describe objects, events, or properties in another sensory domain [1–5]. Take linguistic synaesthesia in Mandarin Chinese for example. The lexical item 味 *wei4* "taste" originally representing a gustatory perception can be used to denote the odor in the olfactory domain, as in the word 氣味 *qi4-wei4* "odor"; the word 聞 *wen2* "to hear" etymologically conceptualizing the auditory action (as indicated by the radical 耳 *er3* "ear") can be employed to denote smelling, as in the Mandarin expression 聞一聞這個花香不香 *wen2-yi1-wen2 zhe4-ge4 hua1 xiang1-bu4-xiang1* "to smell and judge whether the flower is fragrant or not"; and the Mandarin tactile adjective 冷 *leng3* "cold" can be utilized to describe the visual perception, as in the sentence 色調很冷 *se4-diao4 hen3 leng3* "The color is cold" [6].

There are basically two approaches adopted to examine linguistic synaesthesia in the literature. One is the corpus-based approach, through which transfer directionalities of sensory items among different sensory modalities in linguistic synaesthesia are

© Springer Nature Switzerland AG 2020
J.-F. Hong et al. (Eds.): CLSW 2019, LNAI 11831, pp. 139–146, 2020.
https://doi.org/10.1007/978-3-030-38189-9_14

generalized on the basis of the distributions of the items in the corpus. Studies employing the corpus-based approach are such as [4, 5] and [7–9]. The other approach is experimental. Research based on experiments (mostly behavior experiments) mainly focuses on the directionality constraints of linguistic synaesthesia in the language processing. More specifically, synaesthetic expressions following transfer directionalities are found to be more easily processed than the expressions violating the directionalities [3, 10–14]. However, few studies have combined the two different methods to further investigate linguistic synaesthesia. This paper aims to show that the tendencies found through the two approaches are in fact compatible based on the corpus and experimental data of Mandarin synaesthetic adjectives.

2 Methods

[8] utilized a well-built linguistic synaesthesia identification procedure to extract Mandarin sensory adjectives with synaesthetic usages in the Sinica corpus [15]. In [8], she identified 260 Mandarin synaesthetic adjectives, of which 199 adjectives are composed of morphemes with the etymological origin in the same sensory domain (e.g., 甜 tian2 "sweet" and 甜美 tian2-mei3 "tasty") and 61 adjectives are compounded by morphemes originally from different sensory modalities (e.g., 苦澀 ku3-se4 "bitter-rough [bitter]"). This study employs the corpus data of linguistic synaesthesia from the distributions of Mandarin synaesthetic adjectives in the Sinica corpus collected by [8].

[16] followed [17, 18] to ask native speakers to rate the extent (from 0 to 5) to which Mandarin sensory adjectives can be perceived via each of the five sensory modalities (including touch, taste, smell, vision, and hearing). In [16], the modality ratings of 171 Mandarin adjectives consisting of one single sensory concept (e.g., 冷 leng3 "cold" and 朗朗 lang3-lang3 "bright") and 61 Mandarin compound adjectives comprising morphemes from different sensory domains (e.g., 濃厚 nong2-hou4 "of intense taste-thick [dense]") were collected. The experimental data of linguistic synaesthesia this study relies on is from [16]'s modality ratings on Mandarin sensory adjectives.

Based on the two data sources, we extract Mandarin adjectives occurring in both [8] and [16], thus obtaining 72 Mandarin adjectives involving one sensory concept and 61 Mandarin adjectives compounded by different sensory concepts. As the compound adjectives would show different and more complex patterns in linguistic synaesthesia (see [8, 9]), we only focus on the 72 adjectives containing one single sensory concept in this paper and leave the compound adjectives for future research.

As shown in Table 1, tactile and visual adjectives are the top two among the extracted 72 Mandarin adjectives, with 32 adjectives (e.g., 脆 cui4 "crisp" and 暖 nuan3 "warm") and 26 adjectives (e.g., 暗 an4 "dark" and 薄 bao2 "thin") respectively. In contrast, adjectives originally from auditory and olfactory domains are the two least, with only one (i.e., 吵 chao3 "loud") and two adjective types (i.e., 臭 chou4 "smelly" and 香 xiang1 "fragrant") respectively. The gustatory adjectives are in the middle on the rank of the numbers of adjective types among the extracted adjectives, with 11 adjectives such as 淡 dan4 "of mild taste" and 酸 suan1 "sour". The number rank of adjective types is generally in line with the one of the whole Mandarin synaesthetic adjectives attested by [8]. The two differences between this study and [8] are both on the relative order: tactile adjectives are the most and auditory adjectives the least in this study, while visual adjectives are the most and olfactory adjectives the least in [8]. Therefore, the tendency of the extracted 72 Mandarin synaesthetic adjectives employed by this study would be representative of the whole synaesthetic adjectives of Mandarin Chinese.

Table 1. Mandarin adjectives from five sensory modalities

Sensory domain	Number of adjectives	Example
Touch	32	脆 cui4 "crisp"
Vision	26	暗 an4 "dark"
Taste	11	酸 suan1 "sour"
Smell	2	臭 chou4 "smelly"
Hearing	1	吵 chao3 "loud"

In the following, we will demonstrate that: (1) the modality exclusivity scores of Mandarin adjectives calculated based on the sensory ratings show a significantly negative correlation with the usages of these adjectives in linguistic synaesthesia; and (2) the sensory ratings of Mandarin synaesthetic adjectives are consistent with the transfer directionalities of these adjectives in linguistic synaesthesia.

3 Modality Exclusivity of Mandarin Synaesthetic Adjectives

Following [16–18] normalized the mean ratings for 171 Mandarin single-concept adjectives on each of the five sensory modalities. In addition, the study measured the modality exclusivity for these adjectives as the range of the ratings divided by the sum, where the exclusivity scores thus range from 0 to 1:0 meaning an entirely multimodal, and 1 meaning an entirely unimodal. Based on the exclusivity scores of the 171 Mandarin adjectives, we compare the extracted 72 Mandarin adjectives which have been attested to involve synaesthetic usages (see [8]) and the remaining 99 adjectives which have not been reported with synaesthetic uses. As demonstrated in Table 2, the adjectives with synaesthetic usages have a less exclusivity score on the average than the adjectives without synaesthetic usages. Moreover, the synaesthetic adjectives show a lower standard deviation.

We conduct an ANOVA test on the exclusivity for synaesthetic and non-synaesthetic adjectives. The result shows that the difference is significant ($p < 0.05$). Thus, it can be concluded that the modality exclusivity is negatively correlated with the synaesthetic usage of Mandarin sensory adjectives. In other words, if the adjectives are more multimodal (i.e., with lower exclusivity scores), they are more likely to be used for linguistic synaesthesia.

Table 2. Comparisons between Mandarin adjectives with and without synaesthetic uses

Type of adjectives	Count of adjectives	Sum of exclusivity	Mean of exclusivity	Standard deviation	Confidence interval
Synaesthetic adjectives	72	30.757	**0.427**	**0.165**	±0.038
Non-synaesthetic adjectives	99	60.855	**0.615**	**0.208**	±0.041

Another interesting pattern can also be observed with respect to the mean modality exclusivity scores of adjectives originally from different sensory domains. That is, adjectives originally from touch and vision show the two least modality exclusivity scores (with 0.366 and 0.450 respectively), while adjectives from hearing and smell have the two most modality exclusivity scores (with 0.841 and 0.525 respectively). The rank of the mean modality exclusivity for adjectives in five sensory domains, i.e., Touch (0.366) < Vision (0.450) < Taste (0.497) < Smell (0.525) < Hearing (0.841), is consistent with the reverse rank of the number of the adjectives in each modality, i.e., Touch (32) > Vision (26) > Taste (11) > Smell (2) > Hearing (1). Therefore, the mean modality exclusivity scores of adjectives from different sensory modalities would also suggest a negative correlation between the modality exclusivity and linguistic synaesthesia.

4   Directionality and Ratings of Mandarin Synaesthetic Adjectives

Table 3 presents three pieces of numeral information for five modalities based on the collected 72 Mandarin synaesthetic adjectives: (1) the first numeral is the synaesthetic transferability (i.e., the percentage of the adjective types in a sensory domain used for a specific modality in all the synaesthetic adjectives in the sensory domain, see [5]); (2) the second is the count of the synaesthetic tokens showing a specific transfer; and (3) the third number in the second line in each cell is the mean rating scores for adjectives in their non-original sensory domains. Take the tactile adjectives with synaesthetic transfers to taste for example. There are five tactile adjectives found with synaesthetic uses for taste among all the 32 tactile adjectives, thus the mapping from touch to taste with the synaesthetic transferability of 15.6% (5/32), where 121 synaesthetic examples are attested to show the transfer. Besides, the 32 tactile adjectives receive the mean rating score of 2.190 in the gustatory domain based on [16]'s exclusivity data.

We follow [8] to generalize the directionality of linguistic synaesthesia, based on both the lexical type and the lexical token (i.e., the two numerals in the first line in each cell in Table 3). Therefore, the synaesthetic transfers from touch to smell, from touch to hearing, and from taste to hearing show the absolute directionality, as the respective transfers with a reverse direction are unattested (cf. Table 3 for the numeral 0 for the transfers from smell to touch, from hearing to touch, and from hearing to taste). The synaesthetic transfers from touch to vision, from taste to smell, and from taste to vision demonstrate the tendencies-based directionality, whose synaesthetic transferabilities and token examples are both larger than the respective transfers with a reverse direction, i.e., from vision to touch, from smell to taste, and from vision to taste. Although the synaesthetic transferability of the mapping from vision to smell (26.9%) is smaller than the one from smell to vision (50%), the number of the tokens of the former mapping is five times larger than that of the latter one (i.e., 52 vs. 10). In addition, as there are much more visual adjectives than olfactory adjectives used in linguistic synaesthesia (i.e., 26 vs. 2, see Sect. 2), the tendencies-based directionality could also be figured out for the transfer from vision to smell for the collected Mandarin synaesthetic adjectives. Similarly, the synaesthetic transfer from vision to hearing is also tendencies-based, whose transferability is close to that from hearing to vision, while the token examples of the former transfer are much more frequent than that of the latter one (i.e., 1849 vs. 1). Different from the transfers elaborated above, the number of the transfer tokens from touch to taste are more than that from taste to touch (i.e., 121 vs. 99), while the synaesthetic transferability for the former mapping is less than that of the latter one (i.e., 15.6% vs. 36.4%), both of which, however, are close. Therefore, we assume a bi-directionality for the synaesthetic transfers between touch and taste for the collected Mandarin synaesthetic adjectives. As the synaesthetic transfers from smell to hearing and from hearing to smell for the adjectives are both unattested, there is no synaesthetic directionality between these two sensory modalities. The transfer directionality of the collected 72 Mandarin synaesthetic adjectives is in line with the hierarchy of the whole Mandarin synaesthetic adjectives generalized by [8].

Table 3. The synaesthetic transferability, number of synaesthetic tokens, and rating score of Mandarin synaesthetic adjectives in five modalities

Original domain	Target domain				
	Touch	Taste	Smell	Vision	Hearing
Touch	–	15.6%|121 2.190	25%|17 0.853	87.5%|1563 2.653	53.1%|605 1.531
Taste	36.4%|99 0.523	–	72.7%|94 2.867	63.6%|662 1.151	54.5%|87 0.457
Smell	0|0 0.071	50%|22 3.808	–	50%|10 0.531	0|0 0.051
Vision	30.8%|205 2.757	23.1%|49 1.124	26.9%|52 0.617	–	92.3%|1849 1.899
Hearing	0|0 0.071	0|0 0	0|0 0.010	100%|1 0.860	–

Figure 1 summarizes the transfer hierarchy for the collected 72 Mandarin synaesthetic adjectives, where the mean rating scores for adjectives in their non-original sensory domains are also included. Specifically, the bold numbers are the rating scores of adjectives in the modalities consistent with the transfer directionality. For example, with respect to the mapping from touch to smell, 0.853 (in bold) means the score that tactile adjectives receive on smell, while 0.071 is the olfactory adjectives rated on touch. A closer look at the transfer directionality and the rating scores could suggest an intriguing correspondence between them. ANOVA tests can show significant differences (p < 0.05) between tactile adjectives rated on taste and gustatory adjectives rated on touch (2.190 vs. 0.523) and between tactile adjectives rated on smell and olfactory adjectives rated on touch (0.853 vs. 0.071). Thus, it can be concluded that tactile adjectives are more preferred to be utilized and perceived in smell, rather than vice versa. Although the corpus data shows similar probabilities of synaesthetic transfers between touch and taste, the experimental data of ratings could add new evidence to reveal the preferred directionality from touch to taste in the collected Mandarin synaesthetic adjectives.

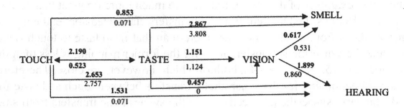

Fig. 1. The synaesthetic hierarchy with the transferred mean rating score

In addition, though the ANOVA test cannot work for the auditory domain with only one adjective, it could still be observed that hearing is more difficult to be rated on other sensory domains than to be perceived through other domains, just analogous to the pattern that hearing is less preferred to serve as the source domain than as the target domain in linguistic synaesthesia. To be more specific, the only one auditory adjective receives a less rating score on touch than tactile adjectives rated on hearing (i.e., 0.071 vs. 1.531), which tendency also appears in the relationship between hearing and taste and between hearing and vision (i.e., 0 vs. 0.457 and 0.860 vs. 1.899 respectively).

Although significant differences have not been shown by the ANOVA test for rating scores between touch and vision, between taste and smell, between taste and vision, and between smell and vision. However, two of them still exhibit a correspondence with the synaesthetic directionality, namely, gustatory adjectives with a higher rating score on vision than visual adjectives rated on taste (i.e., 1.151 vs. 1.124) and visual adjectives rated higher in smell than olfactory adjectives rated on vision (i.e., 0.617 vs. 0.531). It should be noted that the other two pairs, i.e., between touch and vision and between taste and smell, without a strict correspondence to the synaesthetic directionality, are both involving behaviorally and neutrally-integrated modalities, as argued by Winter (2016) (cf. Figure 1 for all the rating scores over 2.5 in the range from 0 to 5). This might override the directionality effect of linguistic synaesthesia. It can therefore be summarized that the transferred ratings of Mandarin synaesthetic adjectives are generally consistent with the transfer directionality of linguistic synaesthesia of these adjectives, which are generalized based on the distributions in the corpus.

5 Conclusion

This study employs both the corpus distributions and the experimental ratings of Mandarin synaesthetic adjectives to investigate linguistic synaesthesia. We have found that the tendencies attested based on the two different kinds of data are compatible, including: (1) the usage of linguistic synaesthesia shows a negative correlation with the modality exclusivity for Mandarin sensory adjectives; (2) the transfer directionality generalized based on corpus distributions are consistent with the sensory ratings of synaesthetic adjectives in their non-original sensory domains. These findings could reveal the cognitive reality of linguistic synaesthesia in both language production and perception. Furthermore, our findings could indicate the conceptual nature of linguistic synaesthesia, but not just a kind of linguistic expression.

Acknowledgments. We would like to thank The National Social Science Fund of China for the financial support on the project (No. 19CYY006).

References

1. Ullmann, S.: The Principles of Semantics. Basil Blackwell, Oxford (1957)
2. Williams, J.M.: Synaesthetic adjectives: a possible law of semantic change. Language **52**(2), 461–478 (1976)
3. Shen, Y.: Cognitive constraints on poetic figures. Cogn. Linguist. **8**(1), 33–71 (1997)
4. Strik Lievers, F.: Synaesthesia: a corpus-based study of cross-modal directionality. Funct. Lang. **22**(1), 69–95 (2015)
5. Zhao, Q., Huang, C.-R., Long, Y.: Synaesthesia in Chinese: a corpus-based study on gustatory adjectives in Mandarin. Linguistics **56**(5), 1167–1194 (2018)
6. Zhao, Q., Xiong, J., Huang, C.-R.: Linguistic synaesthesia, metaphor, and cognition: the systematicity and significance of linguistic synaesthesia in Chinese [通感、隱喻與認知——通感現象在漢語中的系統性表現與語言學價值]. Zhuoguoyuwen [中國語文] **2**, 240–253 (2019)
7. Zhao, Q., Huang, C.-R.: A corpus-based study on synaesthetic adjectives in modern Chinese. In: Lu, Q., Gao, H. (eds.) Chinese Lexical Semantics. LNCS (LNAI), vol. 9332, pp. 535–542. Springer, Cham (2015). https://doi.org/10.1007/978-3-319-27194-1_54
8. Zhao, Q.: Synaesthesia, metaphor, and cognition: a corpus-based study on synaesthetic adjectives in Mandarin Chinese. Doctor of Philosophy, The Hong Kong Polytechnic University, Hong Kong (2018)
9. Zhao, Q. (趙青青), Huang, C.-R. (黃居仁): Mapping models and underlying mechanisms of synaesthetic metaphors in Mandarin [現代漢語通感隱喻的映射模型與制約機制]. Language Teaching and Linguistic Studies [語言教學與研究] **1**, 44–55 (2018)
10. Shen, Y., Cohen, M.: How come silence is sweet but sweetness is not silent: a cognitive account of directionality in poetic synaesthesia. Lang. Lit. **7**(2), 123–140 (1998)
11. Shen, Y., Eisenman, R.: Heard melodies are sweet, but those unheard are sweeter: synaesthetic metaphors and cognition. Lang. Lit. **17**(2), 107–121 (2008)
12. Shen, Y., Gil, D.: Sweet fragrances from Indonesia: a universal principle governing directionality in synaesthetic metaphors. In: Auracher, J., Peer, W. (eds.) New Beginnings in Literary Studies, pp. 49–71. Cambridge Scholars Publishing, Newcastle (2008)

13. Shen, Y., Gadir, O.: How to interpret the music of caressing: target and source assignment in synaesthetic genitive constructions. J. Pragmat. **41**(2), 357–371 (2009)
14. Werning, M., Fleischhauer, J., Beseoglu, H.: The cognitive accessibility of synaesthetic metaphors. In: Proceedings of the 28th Annual Conference of the Cognitive Science Society, pp. 2365–2370 (2006)
15. Chen, K.-J., Huang, C.-R., Chang, L.-P., Hsu, H.-L.: Sinica corpus: design methodology for balanced corpora. In: Proceedings of the 11th Pacific Asia Conference on Language, Information and Computation (PACLIC 11), pp. 167–176 (1996)
16. Chen, I., Zhao, Q., Long, Y., Lu, Q., Huang, C.-R.: Mandarin Chinese modality exclusivity norms. PLoS ONE **14**(2), e0211336 (2019)
17. Lynott, D., Connell, L.: Modality exclusivity norms for 423 object properties. Behav. Res. Methods **41**(2), 558–564 (2009)
18. Lynott, D., Connell, L.: Modality exclusivity norms for 400 nouns: the relationship between perceptual experience and surface word form. Behav. Res. Methods **45**(2), 516–526 (2013)

The Negation Marker *mei* in Northeastern Mandarin

Minglong Wei[✉]

School of Foreign Languages, Peking University, Beijing, China
weiminglong@pku.edu.cn

Abstract. Negation in Mandarin is a field that has been substantially studied, but researches on the negation of dialects are relatively lacking. This article conducts an analysis into the negation marker *mei* in Northeastern Mandarin and finds three different types of *mei*, including mei_1, a verbal negator complementary to *bu*, mei_2, an unaccusative intransitive verb which is characteristic in this dialect but has rarely been mentioned in previous researches, and mei_3, a telic and static aspectual negator which involves a tone sandhi pattern. The tone sandhi of mei_3 in Northeastern Mandarin has something to do with the 4-tone-transformation in Mandarin and can be extended to other negators and even some numerals.

Keywords: Northeastern Mandarin · Negation marker · *mei* · *bu* · Tone sandhi

1 Introduction

There are two negation markers which can be used complementarily in Mandarin according to the negated predicate and the aspect: *mei* (没) and *bu* (不) [1]. For example:

(1) a. wo **mei** you qian.
          *wo **bu** you qian.
          I   not have money
          'I don't have money.'
  b. wo **bu** chi zaofan.
          *wo **mei** chi zaofan. (means 'I haven't eaten breakfast.')
          I   not eat breakfast
          'I don't eat breakfast.'
  c. wo zuotian **mei** chi zaofan.
          *wo zuotian **bu** chi zaofan.
          I   yesterday not eat breakfast
          'I didn't eat breakfast yesterday.'
  d. wo mingtian **bu** chi zaofan.
          *wo mingtian **mei** chi zaofan.
          I   tomorrow not eat breakfast
          'I will not eat breakfast tomorrow.'

J.-F. Hong et al. (Eds.): CLSW 2019, LNAI 11831, pp. 147–155, 2020.
https://doi.org/10.1007/978-3-030-38189-9_15

Bu has a tone sandhi pattern whereby it changes from Tone 4 to Tone 2 when it precedes a Tone 4 character, such as **bù shì* (to be not) to *bú shì*. *Mei* has different patterns, which remains Tone 2 in Beijing Mandarin, but has the same pattern as *bu* in Northeastern Mandarin except for *méi yǒu* (not to have), as is shown below:

Table 1. Tonic patterns of *bu* and *mei* in Northeastern Mandarin.

Followed by	*bu*	*mei*
Tone 1: chī (吃, eat)	Tone 4: bù chī	Tone 4: mèi chī
Tone 2: qí(骑, ride)	Tone 4: bù qí	Tone 4: mèi qí
Tone 3: gǎi (改, revise); yǒu (有, have)	Tone 4: bù gǎi; *bu you	Tone 4: mèi gǎi; Tone 2: méi yǒu
Tone 4: qù (去, go)	Tone 2: bú qù	Tone 2: méi qù
Tone 0 (neutral)	Tone 4: bù de 不的 (no, I) won't	Tone 4: mèi you 没有 (no, I) haven't

2 *Mei*₁ as a Verbal Negator of *You*

The basic interpretation of Northeastern Mandarin *mei* is a common usage of Mandarin, which serves as a negation marker of the verb *you*. *You* has possessive (2a) and existential (2b) meanings, both of which can be negated by *mei*. When *mei* is used in this way, *you* may be optionally deleted anywhere in the sentence except in final positions [2].

(2)   a1. wo **mei** (you) yaoshi.
        I   not have key
        'I don't have a key.'
      a2. yaoshi wo **mei** you.
        *yaoshi wo mei.
        key   I   not have
        'I don't have the key.'
      b1. zher **mei**   (you)   ren.
        here not there be person
        'There is nobody here.'
      b2. zher yige ren dou **mei** you.
        *zher yige ren dou mei.
        here one person all not have
        'There isn't anybody here.'

Some researchers [3, 4] tend to treat this *mei* as a verb which means "not to have". Because of the tight connection between *mei* and *you*, however, we might well take this *mei* as a prefix specially on *you* [5]. When it occurs alone in non-final positions, a morphophonological rule can be applied to delete the following *you*.

Mei and *bu* are complementary when used to negate predicates in declarative sentences in the unmarked aspect. *Mei* negates possessive and existential *you*, whereas *bu* negates all the other predicates, be it a verb, an adjective, or even a noun. When noun predicates are to be negated, as in *Laowang Shanghairen* (Laowang (is) a Shanghaier), the copula verb *shi* (是, "to be") is added for negation. Generally, the negation pattern can be summarized as follows:

> (3) Negation pattern of unmarked predicates:
> a. verbs other than *you*:
> wo **bu** chi rou.
> I   not eat meat
> 'I don't eat meat.'
> b1. possesive *you*:
> wo **mei** you qian.
> I   not have money
> 'I don't have money.'
> b2. existential *you*:
> zher **mei** you   ren.
> here not there be person
> 'There is nobody here.'
> c. adjectives:
> ta **bu** gao.
> he not tall
> 'He isn't tall.'
> d. nouns:
> Laowang **bu shi** Shanghai ren.
> Laowang not be Shanghai person
> 'Laowang isn't a Shanghaier.'

This type of *mei*, which we call *mei*$_1$ hereafter, always takes Tone 2. It corresponds to the only circumstance where *bu* and *mei* do not contrast to each other tonally in Northeastern Mandarin, because *you* cannot be negated by *bu* as *mei*$_1$ can, which is marked with the asterisk in Table 1.

3 *Mei*$_2$ as an Intransitive Verb

In Northeastern Mandarin, *mei* can be specially used as an intransitive verb, meaning "to be gone, to disappear". This *mei*, which is typical in Northeastern Mandarin[1] but hardly mentioned in previous researches, we will call *mei*$_2$. Though it always takes Tone 2 like *mei*$_1$, *mei*$_2$ is a real verb and can serve as a predicate further negatable as in (4a).

---

[1] *Mei*$_2$ is not included as an independent entry in dictionaries of Standard Mandarin, and speakers of other Mandarin dialects such as Beijing Mandarin do not have this usage but would use a specific verb such as *diu* (丢, "to lose") instead.

(4) a. qianbao **mei**   le.
      wallet be gone Aspect
      'The wallet has been gone.'
    b. qianbao mei[2] **mei**.
      wallet   not **be gone**
    'The wallet hasn't been gone.'

Levin and Rappaport [6] divide intransitive verbs into unergative verbs and unaccusative verbs. Some unaccusative verbs like *kai* "to open" often have a transitive counterpart, together forming an "ergative pair", as in (5a).

(5) a1. men **kai** le.
       door **open** Aspect
       'The door has **opened**.'
    a2. wo kai men le.
       I **open** door Aspect
       'I have **opened** the door.'
    b1. qian   mei   le.
       money be gone Aspect
       'The money has been gone/disappeared.'
    b2. wo mei (you) qian   le.
       I not (have) money Aspect
       'I don't have money any more.'

Mei$_1$ and *mei*$_2$ seem to form such an ergative pair superficially in (5b), but as discussed in Sect. 2, the *mei*$_1$ in (5b2) is not a real verb, but just a prefix which cannot even be negated. To compare it with a real verb *kai*, we may negate the examples in (5) to get (6):

(6) a1. men mei **kai**.
       door not **open**
       'The door hasn't **opened**.'
    a2. wo mei kai men.
       I   not open door
       'I haven't opened the door.'
    b1. qian   mei   mei.
       money not be gone
       'The money hasn't been gone/disappeared.'
    b2. *wo mei mei qian.
       I   not mei$_1$ money
       '(intended: I haven't lost the money).'

---

[2] This mei used before a verb is an aspectual negator to be discussed in detail in Sect. 4.

Mei_1 and mei_2 are, therefore, a "pseudo-ergative pair". But mei_2 is still an unaccusative verb of existence and appearance like "to disappear" in that its single argument bears the Theme, instead of the Agent role, as Levin and Rappaport [6] define.

4 *Mei₃* as an Aspectual Negator

Having discussed mei_1 as a verbal negator on *you* and mei_2 as an intransitive verb, we will move on to the third usage of *mei* – mei_3 as an aspectual negator for verbs. Unlike mei_1 and mei_2, both of which always take Tone 2, mei_3 sometimes presents phonological distinctions between Beijing Mandarin and Northeastern Mandarin as in (7a).

> (7) a. wo **méi** chī zaofan. (Beijing, Tone 2)
>   wo **mèi** chī zaofan. (Northeastern, Tone 4)
>   I   not eat breakfast
>   'I haven't eaten breakfast.'
>   b. wo **méi** dài yaoshi. (Beijing, Tone 2)
>   wo **méi** dài yaoshi. (Northeastern, Tone 2)
>   I   not bring key
>   'I haven't brought the key.'

4.1 Aspectual Features of *Mei₃*

Mei_3 and *bu* have different aspectual selections [7]. While mei_3 aspectually selects an event as its complement, *bu* selects a stative situation that requires no input of energy in order to obtain that situation. In other words, mei_3 negates the existence or realization of an event. Li [8] summarizes the aspectual features of *bu* and mei_3:

Table 2. The aspectual features of *bu* and mei_3.

	Telic	Static	Resultative	Progressive
bu	–	–	–	o
mei_3	+	+	o	o

Corresponding to Table 2, the compatibility of *bu* and mei_3 with the major aspectual morphemes as Smith and Erbaugh [9] describe, is summarized in Table 3, in which "+" and "–" stand for the possibility of coexistence, and "neg" means that Ø becomes *bu*, and *le* becomes mei_3 in negations, which is shown in (8).

Table 3. The compatibility between negators and aspectual morphemes.

	le 了	*guo* 过	*zai* 在	*zhe* 著	Ø (zero)
bu	–	–	+	–	–, neg
mei_3	–, neg	+	+	+	–

(8)   a1. wo mingtian chi zaofan.
          I   tomorrow eat breakfast
          'I (**will**) **eat** breakfast tomorrow.'
      a2. wo mingtian **bu** chi zaofan.
          I   tomorrow not eat breakfast
          'I **won't eat** breakfast tomorrow.'
      b1. wo zuotian chi   **le**   zaofan.
          I yesterday eat **Aspect** breakfast
          'I **ate** breakfast yesterday.'
      b2. wo zuotian **mei₃** chi zaofan.
          I yesterday **not** eat breakfast
          'I **didn't eat** breakfast yesterday.'

To make the table more precise, two ostensible "counterexamples" should also be taken into account. In *wo bu qu le* (I won't/wouldn't like to go, literally 'I not go *le*'), although *le* and *bu* co-occur, *le* functions as a modal marker rather than an aspectual marker, thus not incompatible with the table. *Mei₂*, as an intransitive verb, can be given aspectual features by *le* as well, hence the possibility of the co-occurrence of *mei le* in Northeastern Mandarin; but this *mei* is actually not an aspectual negator and thus not a counterexample either.

4.2   A Typological and Diachronic View

So far we have discussed three types of *mei* in Northeastern Mandarin: *mei₁* as the negator of the lexical verb *you*, *mei₂* as an intransitive verb, and *mei₃* as an aspectual negator.

From a typological perspective, it can be observed that *mei₁* bears resemblance to the negation of a special usage of the English verb "to have" like an auxiliary verb, and *mei₃* to "to have" as a perfective aspect marker.

(9)   a.   wo mei (you) qianbao.
           I   not have wallet
           'I have not a wallet.'
      b.   wo mei (you) chi zaofan   ne.
           I   not have eat breakfast Modal.
           'I have not eaten breakfast yet.'

Shi and Li [10] explain this resemblance as a universal phenomenon in human language that possessive verbs may develop into perfective markers. They also introduce the three diachronic steps in the grammaticalization of *mei*:

8th century–13th century: the original meaning of "sinking, burying" (which is still used now, pronounced as *mò*) changed to "lacking, not having";

14th century: *mei* gradually lost its status as an independent verb;

Since 15th century: *mei* gradually began to be used as a negator.

Zhu [11] discovers the correspondence between *mei₁* and *wu* (无), *mei₃* and *wei* (未) in Ancient Chinese. As for *mei₂*, it is a further derivation from *mei₁*. Park [12] attributes the diachronic change from *wu* and *wei* to *mei* in modern Mandarin to the loss of the onset [m] in *wu* and *wei*, which disabled them to function as negators of the original type.

4.3 The Tone Sandhi of *Mei₃*

Unlike *mei₁* and *mei₂*, which maintain Tone 2, *mei₃* in Northeastern Mandarin has a characteristic tone sandhi, which takes Tone 2 before a Tone 4 and Tone 4 before all other tones. The examples in (2) cover all the situations: *mèi chī, mèi qí, mèi gǎi, méi qù* and *mèi you*.

Zhang [13] regards this tone sandhi as a form of derivation from the 4-tone-transformation (四声别义), a word-formation method inherited from ancient Chinese, which changes the word class or meaning of a word by giving it a Tone 4. This type of derivation exists to a larger extent in Northeastern Mandarin than in Beijing Mandarin. For example, *zai* (在, "at"), which always takes Tone 4 in Beijing Mandarin, has two variants in Northeastern Mandarin: locational *zǎi₁* in Tone 3 and progressive *zài₂* in Tone 4. Interestingly, the two *zai*s correspond to *mei₃* in two tones due to the tone sandhi of *mei₃*.

> (10) a. wo nei shihou **méi zài₂** ti   zuqiu.
>   I   that time   not at (Pro) play soccer
>   'I wasn't playing soccer at that time.'
>   b. wo zuotian **mèi zǎi₁**   jia.
>   I yesterday not at (Loc) home.
>   'I wasn't at home yesterday.'

Zhang [14] posited a phonological rule in Beijing Mandarin which changes the tones of *yī* (一, "one"), *qi* (七, "seven"), *ba* (八, "eight") and *bu* to Tone 2 from other tones when occurring before Tone 4. Although this rule hardly applies to *qi* and *ba* in contemporary Beijing Mandarin anymore, it fits perfectly into Northeastern Mandarin. What's more, even *san* (三, "three"), *bie* (别, imperative negator), and *mei₃* also follow this phonological rule, enabling it to cover all the numerals in Tone 1 and negators. We have the expanded tone sandhi pattern, therefore, as is shown in (11):

(11) Tone 2 Sandhi in Northeastern Mandarin:
      For a character X which is a numeral in Tone 1 (*yi, san, qi, ba*) or a negator (*mei₃, bu, bie*), if it precedes a character in Tone 4, X takes Tone 2.

The three different *mei*s and the tone sandhi pattern enable Northeastern Mandarin to disambiguate some structures ambiguous or absent in other dialects. A typical example is the well-known "*mei mei mei*" [3, 15], which comprises three same characters "没没没". It is a form of the interrogative construction "VP-neg-VP" [16], which means "VP or not?". The first and third *mei*s are both verbal *mei₂* in Tone 2, and the second is mei_3 negating the latter mei_2. The tone of mei_3 changes to Tone 4 as it precedes a Tone 2 character. So "*mei mei mei*" can be interpreted as "*méi₂ mèi₃ méi₂*", meaning "(is …) lost or not?".

5 Conclusion

This paper has discussed the lexical semantic interpretations of the negation marker *mei* in Northeastern Mandarin and found three different types of *mei*. Mei_1 is the negative prefix of possessive and existential *you*, and serves complementarily to *bu* as a negator. Mei_2, which is peculiar in Northeastern Mandarin, is an unaccusative intransitive verb which means "to be gone, to disappear". Mei_3 is an aspectual negator inherently telic and stative, which is the negation of the aspectual marker *le*.

Both mei_1 and mei_2 take Tone 2 invariably, but mei_3 has a tone sandhi pattern whereby it takes Tone 2 when followed by Tone 4, and Tone 4 elsewhere. The tone sandhi pattern of taking a Tone 2 before Tone 4 can also be extended not only to other negators *bu* and *bie*, but even to the numerals which take Tone 1 inherently, i.e. *yi* (一, "one"), *san* (三, "three"), *qi* (七, "seven") and *ba* (八, "eight").

References

1. Li, C.N., Thompson, S.A.: Mandarin Chinese - A Reference Functional Grammar. University of California Press, Berkeley (1981)
2. Chao, Y.-R.: A Grammar of Spoken Chinese. The Commercial Press, Beijing (2011)
3. Yin, S.: Dongbei guanhua de foudingci [The Negative Words in Northeastern Mandarin]. In: Wang, G. (ed.) Hanyu fangyan yufa yanjiu – dier jie guoji hanyu fangyan yufa xueshu taolunhui lunwenji [Research on Chinese Dialects - Proceedings of the 2nd International Chinese Dialect Grammar Workshop], pp. 399–417. Central China Normal University Press, Wuhan (2007)
4. Jing, Y.: Qianxi dongbei fangyan zhong de "mei" [On "mei" in Northeastern dialect]. Modern Commun. 2, 28–29 (2015)
5. Ernst, T.: Negation in Mandarin Chinese. Nat. Lang. Linguist. Theory 13, 665–707 (1995)
6. Levin, B., Rappaport, M.: Nonevent -er nominals: a probe into argument structure. Linguistics 26, 1067–1083 (1988)
7. Lin, J.-W.: Aspectual selection and negation in Mandarin Chinese. Linguistics 41, 425–459 (2003)
8. Li, M.: Negation in Chinese. Shanghai Foreign Language Education Press, Shanghai (2007)
9. Smith, C.S., Erbaugh, M.S.: Temporal interpretation in Mandarin Chinese. Linguistics 43, 713–756 (2005)
10. Shi, Y., Li, C.N.: A History of Grammaticalization in Chinese - Motivations and Mechanisms of Evolution of Chinese Morpho-Syntax. Peking University Press, Beijing (2001)
11. Zhu, D.: Yufa jiangyi [Lecture Notes on Grammar]. The Commercial Press, Beijing (1982)

12. Park, J.K.: Cong leixingxue de jiaodu kan hanyu foudingci "mei" de chansheng yu fazhan [On the Emergence and Development of the Chinese Negator "mei"]. In: Department of Chinese Language and Literature, Fudan University (ed.) 2014 nian "yuyan de miaoxie yu jieshi" guoji xueshu yantaohui lunwenji [Proceedings of 2014 International Workshop on the Description and Interpretation of Language], pp. 81–84 (2014)

13. Zhang, F.: Qiantan dongbei fangyan zhong de sishengbieyi xianxiang [On the 4-tone-transformation in Northeastern dialect]. Songliao J. **1**, 83–85 (1989)

14. Zhang, Z.-S.: Tone and tone sandhi in Chinese. Doctoral dissertation, The Ohio State University (1988, unpublished)

15. Li, S.: Dongbei fangyan zhong de "méi mèi méi" chuyi [On "méi mèi méi" in Northeastern dialect]. J. Mudanjiang Normal Univ. **4**, 90, 93 (2013)

16. Zhu, D.-X.: A preliminary survey of the dialectal distribution of the interrogative sentence patterns V-neg-VO and VO-neg-V in Chinese. J. Chin. Linguist. **18**, 209–230 (1990)

Resource Construction and Distribution Analysis of Internal Structure of Modern Chinese Double-Syllable Verb

Guirong Wang[1,2], Gaoqi Rao[1(✉)], and Endong Xun[1,2(✉)]

[1] Research Institute of International Chinese Language Education,
Beijing Language and Culture University, Beijing 100083, China
guirongwang@126.com, edxun@126.com,
raogaoqi@blcu.edu.cn
[2] Institute for Language Intelligence, Beijing Language and Culture University,
Beijing 100083, China

Abstract. In this paper, we analyze the necessity of the construction of internal structure resources of verbs from the perspectives of linguistics and NLP application. We also introduce the method and process of the internal structure annotation of double-syllable verbs in Modern Chinese Dictionary (7th Edition). A total of 9697 double-syllable verbs are annotated. From the results of the annotation, it is found that there are 61.19% of words with a character inside the word as the center of the verb. Among them, the words with verb-object structure is the most (64.5%), followed by the adverbial-head structure (27.77%). This paper can provide basic resources for syntactic and semantic analysis so as to realize the unified analysis of lexicon and syntax.

Keywords: Verb internal structure annotation · Syntactic and semantic analysis · Resource construction · Distribution analysis

1 Introduction

In the traditional syntactic and semantic analysis methods, scholars paid more attention to the differences between lexical and syntactic. Among these methods, the lexical analysis was first performed, followed by syntactic analysis. The results of lexical analysis only contain word boundary without word structure. As the boundaries between words and morphemes, words and phrases in Chinese are relatively vague, in practice, there is no consistent criterion for word segmentation in corpus construction and system evaluation. Moreover, different applications have different requirements for word granularity, which requires complex post-processing to obtain the appropriate granularity for various applications. In order to solve the above problems encountered in the industry, in this paper, we label the internal structure of verbs from the perspective of consistency of lexicon and syntactic, hope to break the boundary between lexicon and syntax, and truly unify lexical analysis and syntactic analysis.

© Springer Nature Switzerland AG 2020
J.-F. Hong et al. (Eds.): CLSW 2019, LNAI 11831, pp. 156–164, 2020.
https://doi.org/10.1007/978-3-030-38189-9_16

2   The Necessity of Structure Annotation

In this section, we will discuss the necessity of the internal structure annotation of verbs from three perspectives, which takes into account two major aspects of linguistic theory and NLP application, so that our work will be useful in both theoretical research and language application, rather than in either of them.

2.1   The Ambiguity of Word Boundary

A definition of words generally accepted by the linguistic community is that "words are the smallest language units that can be used independently [1]". However, the understanding of "minimum" and "be used independently" varies from person to person, which is subjective and difficult to define in practice. The Modern Chinese Dictionary, a widely used authoritative reference book, also contains such problems. Among them, the words "挨批(ai2pi1, be criticized)" and "挨整(ai2zheng3, be attacked)" are included. The structure of the words "挨骂(ai2ma4, be scolded)" and "挨饿(ai2e4, suffer from hunger)" are exactly the same as the previous two words, but they are not included in the dictionary. Besides, different word segmentation models give different word boundaries, which takes an important impact on tasks results with high requirements for word boundary consistency. If words and phrases are no longer identified, they are all treated as language units with a certain structure. There will be no problems with the unclear boundary between words and phrases, words and morphemes, and the above problems will also be solved.

2.2   Separable Words

Separable word is a typical linguistic phenomenon between lexical and syntactic in Chinese. Theoretically, it is difficult to define whether it belongs to lexical or syntactic level. For example, "吃饭(chi1fan4, to eat), 结婚(jie2hun1, to marriage), 睡觉 (shui4jiao4, to sleep)", they can be used as a word in communication, such as "他正在 吃饭(ta1zheng4zai4chi1fan4, He is eating)". We can also insert other complex syntactic components in the middle and use the two characters of a separable word separately as two words. For instance, "吃完饭(chi1wan2fan4, finished the meal)". The traditional word segmentation result was "吃(chi1)/完(wan2)/饭(fan4)", which would divide "吃饭(chi1fan4, to eat)" into two words. It is impossible to show that "吃饭 (chi1fan4, to eat)" is a holistic concept. When counting the word frequency of "吃饭 (chi1fan4, to eat)", the words "吃(chi1, eat)" and "饭(fan4, rice)" were used as two single words to count the word frequency in this sentence, rather than a word as "吃饭 (chi1fan4, to eat)". We present the structure of Fig. 1 (using the annotation model of CTB) in this paper. This structure can not only reflect the integrity of the concept of "吃 饭(chi1fan4, to eat)", but also make it more convenient to count the frequency of words. It is only necessary to count the frequency of the language units of the same level. At the same time, the analysis of two usages of separable words will be unified. The separable words and the general words will be no longer different, they can be analyzed in the same way in the syntactic and semantic analysis, which will reduce the complexity of syntactic and semantic analysis.

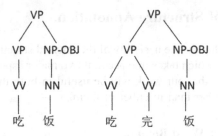

Fig. 1. Annotation results for different usages of separable words

2.3 Requirements for Different Applications

Different NLP applications have different requirements for word granularity. It would be hard for any specific word segmentation criteria to meet a variety of practical needs at the same time. For example, machine translation usually prefers fine-grained words, while for information extraction, coarser-grained word segmentation is more appropriate. Still taking the verb "吃饭(chi1fan4, to eat)" as an example, since there is no word directly corresponding to "吃饭(chi1fan4, to eat)" in English, it is necessary to translate "吃(chi1, eat)" and "饭(fan4, rice)" separately, which requires dividing "吃饭 (chi1fan4, to eat)" into two words in word segmentation. In the case of information extraction, it is more inclined to output "吃饭(chi1fan4, to eat)" as a whole to better reflect the integrity of information, which requires dividing "吃饭(chi1fan4, to eat)" into one word. Thus, different applications require different word segmentation criteria.

Many scholars have noticed the disconnection between word segmentation results and application requirements, and explored ways to automatically convert between different word segmentation standards. However, this method lacks the support of linguistic theory, which is difficult to be accurate in practice, and would increase the complexity of the syntactic and semantic analysis system. In fact, if the analysis results in Fig. 1 can be given, any NLP applications could extract the language units of the required granularity from it. In this way, there is no need to distinguish between morphemes, words, and phrases, and the post-processing of word segmentation results will be avoid.

3   State of Art

From previous studies, it has been found that there are many resources for lexicon or syntactic. The former is the Modern Chinese Grammar Information Dictionary [2] developed by the Institute of Computational Linguistics of Peking University. It is an electronic dictionary developed for the computer to realize Chinese analysis and generation. The dictionary described grammatical properties as detailed as possible for each word that was included and performed detailed formal descriptions. It has played an important role for many years in research and engineering practice and provided great convenience for Chinese analysis.

The latter, such as the CTB [3], has grown to version 8. 0 since the summer of 1998. Based on the phrase structure grammar, it annotated phrase structure, phrase function and empty element. On the basis of CTB, the University of Pennsylvania has completed the construction of the Chinese Proposition Library 1. 0 of the predicate argument structure and the construction of the Chinese Discourse Tree Bank with the textual connection. It has a direct impetus to the development of applications such as machine translation, information retrieval, information extraction, and question-answering system.

At present, the existing resource promoted the development of academic research and engineering practice to a certain extent, but the lexicon and syntax are still independent to each other. Zhao first proposed a character dependency tree bank [4]. He proposed a character-level dependency representation, used for the basic structure representation and analysis of Chinese. The annotation was based on the absolute part-of-speech(POS) of the word and the POS tag of each character inside, which is determined by the annotator as well. With a modifier head view, all Chinese words can be easily annotated in an iterative analysis and finally a dependency tree structure will be built for the given word. He pointed out that word dependence could be perfectly integrated into the existing standard dependency parser, and it is superior to the traditional pipelined joint learning strategy in system accuracy. However, he only annotated words extract from CTB, and only gave the direction of the dependency relationship, regardless of the type of dependency.

4   The Process of Structure Annotation

4.1   The Annotated Objects

In this paper, we annotated the internal structure of two-syllable verbs in the Modern Chinese Dictionary (7th Edition), the following factors were considered when selecting the annotation objects:

(1) From the perspective of linguistics, Chinese is a language with predicates as its core. In 1942, Mr. Lv proposed that "the center of the sentence is a verb [5]" in the *"Chinese Grammar Summary"*. For example, when someone only says "我弟弟 (*wo3di4di, my brother*)", "他女朋友(*ta1nv3peng2you, his girlfriend*)", "一束花 (*yi1shu4hua1, a bunch of flowers*)", the listener does not understand the meaning of the speaker. If the verb "送(*song4, send*)" is used to associate these three language fragments together to form a sentence, "我弟弟送了他女朋友一束花 (my brother sent his girlfriend a bunch of flowers)". Listeners can easily understand, because that verb "送(*song4, send*)" is the center of the sentence. When we hear "送(*song4, send*)", the verbal frame associated with "送(*song4, send*)" appears in mind, as shown in Fig. 2. The listener only needs to put the corresponding words into the corresponding slots according to the predicate framework. Therefore, we chose verbs as the objects of annotation.

(2) From the perspective of syntactic analysis, firstly, whether dependent grammar or phrase structure grammar, people always focus on the core predicate of the sentence. And then analyze it layer by layer from the core component to the edge

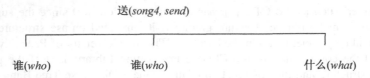

Fig. 2. The verbal frame of "送(song4, send)"

component, from the inner component near the core to the outer component far from the core. The inner layer component here refers to the modifier and complement of the core predicate. Whether it belongs to a structure of multi-layer or a single layer, is related to the core predicate. It is an explanation of the modality, degree and result expressed by the core predicate of a sentence, which often affects the expression of the core meaning of a sentence. Therefore, this part needs detailed analysis in the syntactic analysis. The outer component refers to the subject and object, which are related to the core predicate as a whole. So we took Lv's view, and regarded the subject and the object as a whole analysis unit, without internal detailed analysis.

(3) There are 16,882 two-syllable verbs in the Modern Chinese Dictionary, including some spoken words, dialect words, classical written Chinese, and words that are not used frequently. The usages of spoken words and dialect words are special. The analysis of internal structure is not conducive to use. The way of word formation in classical written Chinese is different from that in modern Chinese. It is more difficult to annotate them. Therefore, these words are not the object of structure annotation.

As Zipf's Law reveals, in the natural language corpus, the number of occurrences of a word is inversely proportional to its ranking in the frequency table. Therefore, after eliminating special words, we counted the frequency of other verbs in the BCC corpus of Beijing Language and Culture University, and selected words with a frequency of more than 500 as the annotation objects, a total of 9697 (Table 1).

Table 1. Distribution of subcategories in modern Chinese two-syllable verbs

Total number	Spoken words	Dialect words	Classical written Chinese	Word frequency below 500	Number of annotated words
16882	150	277	1127	5631	9697

4.2 The Guidelines of Annotation

To mark the internal structure of modern Chinese verbs, it is necessary to figure out the patterns of word generation in linguistics. It is generally believed that both simple words and synthetic words exist in Chinese. There is no internal structure in simple words, and it express meaning as a whole. There are three ways to form synthetic words, which are compound, overlap and addition. The compound word formation is similar to phrase, including combined structure, modifier-head structure (including attributive-head structure and adverbial-head structure), predicate-complement structure, verb-object structure and subject-predicate structure. The annotation objects in this paper are two-syllable verbs, but the phrase of attributive-head structure is generally nominal, it is not the type that needs to be annotated. In addition, in order to carry out the viewpoint of predicate core and improve the adaptability between resources and applications, we labeled the single word and the words of combined structure as concrete words. In summary, the structure types that would be labeled were defined as subject-predicate, verb-object, adverbial-head, predicate-complement, addition and concrete. The definition and example of each type are shown in Table 2.

Table 2. The types and definitions of verb structure

Type	Definition	Example
Subject-predicate	The former character is the agent of the latter character	地震, di4zhen4, earthquake
	The former character is the object of the latter character	肢解, zhi1jie3, dismember
Verb-object	The former character is the object of the latter character	上班, shang4ban1, go to work
	The latter character is the agent of the former character	变形, bian4xing2, out of shape
	The latter character is the result of the former character	打包, da3bao1, to pack
Adverbial-head	The latter character is modified by the former character, and the central word is the latter	不可, bu4ke3, not allowed
Predicate-complement	The former character is supplemented by the latter character, and the central word is the former	看见, kan4jian4, to see
Concrete	Combined structure, the central word is the word itself	开始, kai1shi3, begin
	Lianmian words	唠叨, lao2dao, be garrulous
Addition	Words derived from affixes	处于, chu3yu2, be in

In addition, a verb with multiple internal structures will also be encountered in the annotation. If a word is only tagged as a verb, it is marked according to the structure type mentioned above, if the verb is a categorical word, the internal structure is marked only when it is used as a verb. For example, "表情, biao3qing2, to express", it is tagged as a verb and a noun. When it is used as a noun, it means "the feelings expressed on a person's face". It is an attributive-head structure. When it is used as a verb, it means "to express one's thoughts and feelings from a change in face or posture", it is a verb-object structure. When labeling the word "表情, *biao3qing2, to express*", only the verb-object structure is marked.

4.3   Results Analysis

After the annotation of verb structure completed, we made a statistics and distribution analysis of the verbs of each type of structure (Fig. 3).

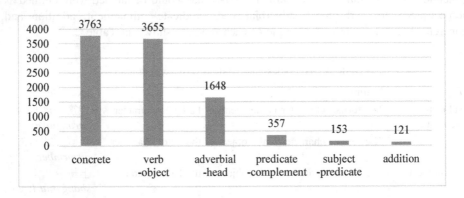

Fig. 3. Statistics on the number of verbs in each structure type

As can be seen from the Fig. 3, the internal structure type of the verb is consisted mainly by the phrase structure type, and there are a few types such as subject-predicate, verb-object, adverbial-head and predicate-complement. The number of verbs except concrete words is 5934, accounting for 61.19%, which has exceeded half of the total number of annotations. They all take a character inside the disyllabic verb as the center, which is more consistent with the phrase structure and more suitable for syntactic analysis. In fact, a verb is a concrete structure does not mean that there is no structure, but that the central word of the verb is itself, not a character in the word. In the strict sense of linguistics, there is a certain internal structure in these words. From this point of view, the internal structure exists in almost all words, and the word structure and phrase structure can form a continuum.

From the Fig. 4, it can be found that in addition to the verbs with concrete structure, the verbs with verb-object structure and adverbial-head structure account for 61.59% and 27.77% respectively. At the syntactic level, the most common components in sentences are predicate, object and adverbial. This also indicates that lexical analysis can be synchronized with syntactic analysis. At the phrase level, the phrases with adverbial-head and predicate-complement are the easiest to insert other components. As a matter of fact, these two structures account for a large proportion of separable words, which explains the inevitability of usage of separation and conjunction. Therefore, the annotation of the internal structure of the verb links the same structure of the word and the phrase, which weakens the concept of separable words to a certain extent, and solves the difficult problem of separable words in the syntactic and semantic analysis. At the same time, when learning the usage of the separable word, we can also analogize the phrase of the same structure to improve the learning effect.

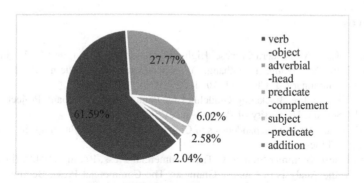

Fig. 4. Statistics on the proportion of verbs in each structure type

5   Conclusion

At present, we have initially completed the construction of the internal structure resources of the high-frequency two-syllable verbs in the Modern Chinese Dictionary, which can provide basic data for the training model in syntactic and semantic analysis so as to unify the analysis of lexicon and syntax. At the same time, it also provides some convenience for Chinese learners and reduces the difficulty of learning the separable words. After reaching a certain data scale, the annotation resources of this paper will also be opened to the academic and industrial circles free of charge for research and development.

From the perspective of resource construction, the resources can be further improved in the following aspects. First, annotating the internal structure of the low-frequency two-syllable verbs in the Modern Chinese Dictionary; secondly, extending from the two-syllables to the three-syllables and the four-syllables, trying to cover all verbs in the Modern Chinese Dictionary; finally, combing with the existing corpora or dictionaries, such as CTB, HowNet, etc., to cover more verbs used in NLP.

From the perspective of resource application, in the aspect of linguistics research, we can excavate the rules of compound verb formation on the basis of this resource. Based on the internal structure information of compound words, we can analyze the relationship between the part of speech of words and the part of speech of the internal components of words, and the relationship between the semantic classes of components of words and the semantic classes of words. In addition, this resource can be used to set up a model of lexical syntactic unified analysis. Based on the model, a syntactic parser with unified analysis of lexical and syntactic is developed to promote the development of syntactic and semantic analysis.

Acknowledgments. This paper is supported by the National Social Science Fund (16AYY007), the Funds of Beijing Advanced Innovation Center for Language Resource (TYR17001), the funds of Major Projects of Key Research Bases of Humanities and Social Sciences of the Ministry of Education (16JJD740004).

References

1. Huang, B., Liao, X.: Modern Chinese. Higher Education Press, Beijing (2007). (in Chinese)
2. Yu, S., Zhu, X., Wang, H., Zhang, Y.: Specification of modern Chinese grammar information dictionary. J. Chin. Inf. **10**, 1–22 (1996). (in Chinese)
3. Xue, N., Xia, F.: The Bracketing Guidelines for Penn Chinese Treebank Project. Technical report IRCS00-08, University of Pennsylvania (2000)
4. Zhao, H.: Character level dependencies in Chinese: usefulness and learning. In: EACL 2009 (2009). (in Chinese)
5. Lv, S.: Chinese Grammar Summery. The Commercial Press, Beijing (1942). (in Chinese)
6. Lv, S.: On the Analysis of Chinese Grammar. The Commercial Press, Beijing (1979). (in Chinese)
7. Lu, B.: Core Derivation Syntax. Shanghai Education Press, Shanghai (1993). (in Chinese)
8. Lu, C.: The Parataxis Network of Chinese Grammar. The Commercial Press, Beijing (2001). (in Chinese)
9. Yang, M.: A Study on Compound Word Formation in Modern Chinese. Nanjing Normal University, Nanjing (2006). (in Chinese)
10. Ye, F., Xu, T., Wang, H., et al.: Outline of Linguistics. Peking University Press, Beijing (2010). (in Chinese)
11. Fang, Y.: An Analysis of the Internal Structure of Words. Suzhou University, Suzhou (2014). (in Chinese)
12. Xun, E., Rao, G., Xiao, X., Zang, J.: Development of BCC corpus in the context of big data. Corpus Linguist. (2016). (in Chinese)
13. Modern Chinese Dictionary. The Commercial Press, Dictionary Editing Room, Institute of language Studies, Chinese Academy of Social Sciences (2016). (in Chinese)
14. Su, B.: The implication pluralism and Cognitive principles of the structure of Chinese compound words. Acad. Res. 162–165 (2016). (in Chinese)
15. Zhang, L.: A brief introduction to Tesnier's structural syntactic basis. Foreign Linguist. (1985). (in Chinese)
16. Song, L.: A brief comment on the main points of Tesnier's structural syntactic basis. Cult. Educ. Mater. 30–34 (2017). (in Chinese)

Gradability, Subjectivity and the Semantics of the Adjectival *zhen* 'real' and *jia* 'fake' in Mandarin

Fan Liu[1] and Qiongpeng Luo[2(✉)]

[1] School of Liberal Arts, Anhui Normal University, Wuhu 241003, China
fanliu_ling@foxmail.com
[2] School of Liberal Arts, Nanjing University, Nanjing 210023, China
qpluo@nju.edu.cn

Abstract. In this paper, we provide an empirical description and a theoretical analysis of the adjectival *zhen* 'real' and *jia* 'fake' in Mandarin Chinese. The two adjectives manifest resistance to degree modifiers, and thus have been traditionally treated as non-gradable adjectives. Empirical evidence, however, shows that they can actually fuse both degree intensification and expressive meanings together. Based on their semantic behaviors, we follow recent advances in multidimensional semantics to propose that *zhen* and *jia* are mixed items with bi-dimensional semantics, i.e., the judge of truth-value as the descriptive meaning, and the degree of similarity/deviation between the facts and the subjective expectations as the expressive meaning.

Keywords: Adjectives · Gradability · Subjectivity · Multidimensional semantics · Mandarin Chinese

1 Introduction

This study provides a semantic account for the adjectival *zhen* 'real' and *jia* 'fake' in Mandarin Chinese. In general, the intersection rule is applicable to most of the [A+NP] constructions, as shown in (1). However, *zhen* and *jia* in (2) seem to be counterexamples to this rule.

(1) $[\![\text{A NP}]\!] = [\![\text{A}]\!] \cap [\![\text{NP}]\!]$
$[\![\text{hei zhenzhu}]\!] = [\![\text{hei}]\!] \cap [\![\text{zhenzhu}]\!] = \lambda x.\mathbf{black}(x) \wedge \mathbf{pearl}(x)$
(2) a. $[\![\text{zhen junzi}]\!] = \lambda x.\mathbf{real}(x) \wedge \mathbf{gentleman}(x)$
$\qquad = \lambda x.\ \mathbf{gentleman}(x) \wedge \mathbf{gentleman}(x) = 1$ [Redundancy]
   b. $[\![\text{jia zhengju}]\!] = \lambda x.\mathbf{fake}(x) \wedge \mathbf{proof}(x)$
$\qquad = \lambda x.\ \mathbf{proof}(x) \wedge \mathbf{proof}(x) = 0$ [Contradiction]

In addition, based on their original logical meanings that denote truth-values, *zhen* and *jia* have been treated as non-gradable adjectives in many analyses ([1, 2], a. o.); while some other works [3, 4] have pointed out that *zhen* and *jia* can actually give rise

© Springer Nature Switzerland AG 2020
J.-F. Hong et al. (Eds.): CLSW 2019, LNAI 11831, pp. 165–172, 2020.
https://doi.org/10.1007/978-3-030-38189-9_17

to the gradable readings. The examples are provided in (3). So far, there has been no consensus regarding whether the adjectival *zhen* and *jia* are gradable or not.

(3) a. 烟台缴获 4 万元很真的假饮料。

Yantai jiaohuo si wan   yuan hen zhen de jia yinliao.
Yantai capture 4 ten.thousand yuan very real DE fake drink
'Yantai officers have captured the fake drinks that look very real, valued RMB 40,000.'

b. 这个数据可能不准确，但其他数据更假。

Zhege shuju keneng bu zhunque, dan qita shuju geng jia.
this data probably not accurate, but other data more fake
'These data are probably not accurate, but other data look more fake.'

In this paper, we present a novel solution to the semantic puzzles of the adjectival *zhen* and *jia* in Mandarin. The goals of the study are twofold. First, we report some fresh observations to demonstrate that the gradable uses of *zhen* and *jia*, differing from their logical uses, can pass the tests of faultless disagreement and judge dependence, and thus show the subjectivity. To our knowledge, this (non-)subjective patterns of *zhen* and *jia* have received little attention in the theoretical literature. Second, we develop a multidimensional analysis of *zhen* and *jia*. Unlike the traditional truth-conditional semantics, the multidimensional semantics assumes that meanings operate on both the at-issue dimension and the CI (conventional implicature) dimension. In line with this approach, *zhen* and *jia* are analyzed as mixed items to capture both of their logical and subjective meanings.

2 Previous Analyses

As mentioned above, the lexical semantics of *zhen* and *jia* has been a long debated issue in the literate (cf. [1–6], a. o.). Roughly speaking, the existing accounts can be divided into two frameworks, i.e., descriptive linguistics and degree semantics. The former was argued by [1, 3, 5, 6]; whereas the latter was proposed by [2, 4]. In this section, we provide a brief review of these accounts.

In the descriptive approach, the gradability of certain items is diagnosed by their co-occurrences with degree modifiers. [1, 5, 6] claim that *zhen* and *jia* cannot be further modified by any intensifiers or degree adverbs, and therefore should be treated as non-gradable adjectives. However, [3] has adopted a corpus survey, which clearly confirms that *zhen* and *jia* can co-occur with some typical degree modifiers in Mandarin, like *hen* 'very' and *geng* 'more'. The examples are shown above in (3). In short, this descriptive approach fails to clarify the gradability of *zhen* and *jia*, as it largely depends on the degree modifiers as the only diagnosis, whereas the empirical distribution is actually complicated.

Unlike the descriptive account, the works based on degree semantics, following [7], assume that gradable adjectives, which denote sets of ordered degrees on particular property dimensions, can usually appear in comparatives. Under this framework, [2] identifies *zhen* and *jia* as non-gradable adjectives, although [2] has noticed that *zhen*/*jia* is felicitous in Mandarin comparatives, as illustrated in (4) and (3b).

(4) 梦想比现实更真。

   Menxiang bi   xianshi geng   zhen.
   dream   COMP reality more   ture
   'A dream is more true than reality.'

To explain these counter-examples, [2] proposes that *zhen* and *jia* are used in their non-literal senses. Their gradable uses do not represent their primary meanings, so they can still be categorized as non-gradable adjectives. This analysis, however, is not convincing enough, as [2] has not provided persuasive arguments to explain why the gradable uses of *zhen* and *jia* are not idiosyncratic but productive.

Contrary to [2, 4], on the grounds that *zhen* and *jia* are found in the environments where typical gradable adjectives occur, concludes that they are essentially gradable adjectives. [4] further claims that they are associated with an upper-closed and a bottom-closed scale respectively, Although [4] adopts a formal treatment of *zhen* and *jia*, he has not paid enough attention to their subjective uses.

To summarize, it remains to be explored whether the two adjectives are gradable or not. Moreover, the semantics of *zhen* and *jia* have not yet received formal analyses except [4], and the combination of [A+NP] also requires explanation. In the following sections, we will present some fresh empirical generalizations and then propose our novel solution.

3   Some Empirical Observations

In this section, we argue that *zhen* and *jia* in gradable uses have some additional subjective flavor, while this flavor is gone in their logical uses. The subjectivity can be tested by two methods, namely, Faultless Disagreement and Judge Dependence.

3.1   Faultless Disagreement

One of the usual tests to identify subjective adjectives is faultless disagreement ([8, 9], a. o.), as illustrated below:

(5) a. Speaker A: The chili is tastier than the soup!
       Speaker B: No, the soup is tastier!   **(faultless)**
    b. Speaker A: Anna is taller than Zoe.
       Speaker B: No, Zoe is the taller of the two!   **(factual only)**

In (5a), the statements of both speaker A and speaker B can be true at the same time. In line with this effect, the gradable uses of *zhen* and *jia* show the faultless, i.e., subjective disagreement:

(6) Speaker A: 我可能{喝了假酒，读了假大学}！
　　　　　　Wo keneng {hele jiajiu, dule jia daxue}!
　　　　　　'I probably {have drunk counterfeit wine, registered in fake
　　　　　　colleges...}!'
　Speaker B: 不，你没有。你{喝的是真酒，读的是真大学}！
　　　　　　Bu, ni meiyou. Ni {he de shi zhen jiu, du de shi zhen daxue...}!
　　　　　　'No, you haven't. What {you have drunk is real wine, registered in
　　　　　　is real colleges...}!'   (**<u>faultless</u>**)

By contrast, the logical uses of *zhen* and *jia* are factual only, as shown in (7), which means either speaker A or speaker B is right in a certain possible world, i.e., their statements cannot be true at the same time. Hence, the logical uses are factual only, parallel to the typical case (5b).

(7) Speaker A: 这个命题是真的。
　　　　　　Zhege mingti shi zhen de.
　　　　　　'This proposition is ture.'
　Speaker B: 不，这个命题是假的。
　　　　　　Bu, zhege mingti shi jia de.
　　　　　　'No, this proposition is false.'   (**<u>factual only</u>**)

3.2 Judge Dependence

Another widely accepted diagnosis of subjectivity is whether the implicated judge can be introduced by a PP (cf. [8, 10], a. o.). The predicates of personal taste allow an overt opinion-holder, i.e., the 'judge-PP'; while the non-subjective ones cannot. See the contrast below:

(8) a. The book was interesting <u>to/for me</u>.
　　b. ?? Anna is intelligent <u>to/for me</u>.

The gradable uses of *zhen* and *jia*, once again, in line with the subjective predicates (8a), allow the implicated judge, as illustrated in (9). By contrast, the logical uses resist the judge-PPs as (10), showing that they are not subjective.

(9) a. 让假发在别人看来更真。
　　　Rang   jiafa   zai   bieren kanlai   geng zhen.
　　　make   wig   at   others see.from more real
　　　'Make the wig look more authentic for others.'
　　b. 这些数据对读者来说很假。
　　　Zhexie shuju   dui   duzhe   laishuo   hen   jia.
　　　these data   to   readers   from.say   very fake
　　　'These data are very fake to the readers.'

(10) a. ?? 他的陈述对法官来说很真。

  ?? Ta de   chenshu dui faguan laishuo   hen zhen.
   he DE   statement for judge   from.say very real
   'His statement looks very real to the judge.'

 b. ?? 这个命题对教授来说更假。

  ?? Zhe ge mingti   dui jiaoshou   laishuo   geng jia.
   this CL proposition for professor from.say more fake
   'This proposition is more fake to the professor.'

3.3   Interim Summary

To recap, *zhen* and *jia* have some mixed properties. On the one hand, they pattern with non-subjective items in their logical uses, as shown in the examples above, where the predicates that they are involved resist the implicated judge argument; on the other, *zhen* and *jia* pattern with evaluative/subjective adjectives in their gradable uses, where the faultless disagreements are natural, and the overt judge-PP are well accepted. In the next section, we will develop a formal account for such mixed properties in a multi-dimensional semantics framework.

4   A Multidimensional Account

The mixed properties of *zhen* and *jia* are liable to two possible analyses: (a) to claim that they are homophonous, i.e., the different uses are attributed to the logical *zhen1/jia1* vs. the gradable *zhen2/jia2*, and (b) to propose a unified account in a novel framework. The homophonous approach, however, fails to predict the various readings of [A+NP] combinations. As shown by the examples in (11), *jia jiu* 'fake/counterfeit wine' can either belong to the wine or not, whereas *jia lingzi* 'extra collar' is essentially a collar, and *zhen qiang* 'real gun' is a gun as well. The homophonous account cannot predict these readings correctly, since it assumes all *zhen/jia* are homophonous, and thus have two readings, i.e., the logical meaning and the gradable one, which is obviously not the case.

(11) a. 假酒 jia jiu 'fake wine[-wine]; counterfeit wine[+wine]'
  b. 假领子 jia lingzi 'extra collar [+collar]'
  c. 真枪 zhen qiang 'real gun [+gun]'

In this section we provide a unified account in the framework of multidimensional semantics. While the idea that certain expressions specifically convey attitude, emotions and evaluations on the part of the speaker has been around for several decades, only in recent years expressive meanings have received a significant amount of attention in formal semantics ([11–13], a. o.). [11] provides the first attempt to formalize this class of meaning, outlining several properties that distinguish it from truth-conditional ones. Most recently, [14, 15] have shown how the subjective/expressive meanings of adverbs in Mandarin can receive a formal semantic treatment as well.

The most fundamental assumption of multidimensional semantics is that meanings operate on different dimensions. An utterance may express both an at-issue (truth-conditional) content in the descriptive dimension and a conventional implicature in the expressive dimension. Informally, the expressive meaning is like "double assertion", or some side comment by the speaker. [11] introduces a new semantic type for CI (expressive content) in the semantic system. The semantic types organize the semantic lexicon, and they index the denotation domains. The semantic types in L_{CI} are defined as below:

(13) a. e^a and t^a are basic at-issue types.
　　b. e^c and t^c are basic CI types.
　　c. If τ and σ are at-issue types, then $\langle\tau, \sigma\rangle$ is an at-issue type.
　　d. If τ is an at-issue type and σ is a CI type, then $\langle\tau, \sigma\rangle$ is a CI type.
　　e. The full set of types is the union of the at-issue and CI types.

The at-issue content is marked by the superscript "a", and the expressive content by the superscript "c". [11] proposes the following CI application rule for semantic composition when an expressive item (such as *damn*) combines with an item that only has descriptive meaning (such as *Republicans*), as illustrated below:

(14) CI application

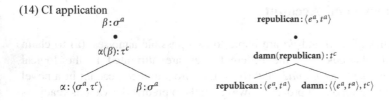

(14) serves as the standard rule of functional application. It states that if α is a term of type $\langle\tau, \sigma\rangle$, and β is a term of type σ then $\alpha(\beta)$ is a term of type τ. The bullet "•" is a metalogical symbol to separate the at-issue content from the CI content. This provides a straightforward account of the distributional pattern of *zhen* and *jia*, especially their logical and subjective uses.

Following this line of thought, we assume that *zhen* and *jia* have bi-dimensional meanings as well. In the descriptive dimension, *zhen* and *jia* maintain their original logical functions, i.e., to judge the truth-value of propositions, and thus display non-gradable uses. In the expressive dimension, *zhen* and *jia* convey the speaker's judge towards the degree of similarity/deviation between the facts and the subjective expectations. To formalize the subjective expectations, we incorporated [16] 's PRO-TOTYPE function into their lexical semantics. This has enabled us to capture the expressive content of *zhen* and *jia*, as the prototypes for each individual's perspective can be variable. Hence, the semantic expressions are shown below:

(15) ⟦zhen⟧= λPλx.P(x) • λPλx. **similarity** (x, **prototype**(P)) ≥ !s
    (x has the property of P, and the similarity between x and the prototype of P has
    exceeds a certain standard !s)
(16) ⟦jia⟧= λPλx.¬P(x) • λPλx. **deviation** (x, **prototype**(P)) ≥ !s
    (x does not have the property of P, and the deviation between x and the prototype
    of P has exceeds a certain standard !s)

Unlike the exiting analyses, this account treats the subjective content as a part of *zhen* and *jia*'s lexical semantics. This semantics contains a degree component which measures the similarity/deviation between the individual x and the prototype of P, thus correctly predicting that *zhen* and *jia* are gradable. Moreover, the semantic puzzle of [*zhen/jia*+NP] is solved, since *zhen junzi* 'real gentleman' is more than gentleman, but also conveys the subjective attitude that the speaker evaluates this individual is very close to her/his certain standard of being a gentleman. Therefore, the meaning of *zhen* is not redundant according to this analysis. And this semantics works well for *jia* similarly, check the formulas:

(17) ⟦zhen junzi⟧=λx. **gentleman** (x)
           • λx. **similarity** (x, **prototype(gentleman)**) ≥ !s
(18) ⟦jia zhengju⟧=λx. ¬**evidence** (x)
           • λx. **deviation** (x, **prototype(evidence)**) ≥ !s

5   Conclusion

To conclude, in this paper we discussed the semantic behaviors of the adjectival *zhen* and *jia* in Mandarin Chinese. The two adjectives have been traditionally treated as non-gradable and cannot be modified by degree words (cf. [1, 2, 5, 6], a. o.). However, there are ample data showing that both *zhen* and *jia* can present in gradable context. Here we provided some fresh empirical evidence to suggest that the two adjectives display both gradable and non-gradable uses. The mixed properties of *zhen* and *jia* can be attributed to their bi-dimensional semantics, i.e., the judge of truth-value as the descriptive meaning, and the degree of similarity/deviation between the facts and the subjective expectations as the expressive meaning. Hence, it is the descriptive dimension that makes *zhen* and *jia* act as non-gradable adjectives in their pure logical uses; while the expressive dimension is responsible for their gradable distribution and semantic behaviors like judge-dependence and subjectivity.

Acknowledgments. This work is supported by the National Social Science Fund of China (NSSFC) (grant number: 16BYY006) to Qiongpeng Luo. We thank the audience at CLSW 2019 for helpful comments and feedback. All errors remain our own.

References

1. Zhu, D.: On Mandarin adjectives. Stud. Lang. Linguist. **1**, 1–37 (1956). (Xiandai hanyu xingrongci yanjiu, Yuyan Yanjiu) (in Chinese)
2. Lin, J., Peck, J.: Classification of Mandarin Chinese simple adjectives: a scale-based analysis of their quantitative denotations. Lang. Linguist. **17**(6), 827–855 (2016). https://doi.org/10. 1177/1606822X16649845
3. Nie, Z., Tian, X.: Can't zhen and jia really be modified by the degree words? Stud. Lang. Linguist. **2**, 6–11 (2014). ("Zhen" "jia" zhende buneng bei chengdu fuci xiushi ma, Yuyan Yanjiu) (in Chinese)
4. Luo, Q.: Degree, scale and the semantics of zhen and jia. Stud. Lang. Linguist. **2**, 94–100 (2016). (Chengdu, liangji yu xingrongci "zhen" he "jia" de yuyi. Yuyan Yanjiu) (in Chinese)
5. Lü, S.: On the distribution of monosyllabic adjective. Stud. Chin. Lang. **2** (1966). (Danyin xingrongci yongfa yanjiu, Zhongguo Yuwen) (in Chinese)
6. Chao, Y.-R.: A Grammar of Spoken Chinese. University of California Press, Berkeley (1968)
7. Kennedy, C., McNally, L.: Scale structure, degree modification and the semantics of gradable predicates. Language **81**(2), 345–381 (2005). https://doi.org/10.1353/lan.2005. 0071
8. Lasersohn, P.: Context dependence, disagreement, and predicates of personal taste. Linguist. Philos. **28**, 643–686 (2005). https://doi.org/10.1007/s10988-005-0596-x
9. Kennedy, C.: Two sources of subjectivity: qualitative assessment and dimensional uncertainty. Inquiry: Interdisc. J. Philos. **56**, 258–277 (2013). https://doi.org/10.1080/ 0020174X.2013.784483
10. Bylinina, L.: The grammar of standards: judge-dependence, purpose-relativity, and comparison classes in degree constructions. Doctoral dissertation, Utrecht University (2014)
11. Potts, C.: The Logic of Conventional Implicature. Oxford University Press, Oxford (2005)
12. McCready, E.: Varieties of conventional implicature. Semant. Pragmatics **3**(8), 1–57 (2010). https://doi.org/10.3765/sp.3.8
13. Gutzmann, D.: Use-Conditional Meaning. Oxford University Press, Oxford (2015)
14. Luo, Q., Liu, F.: The expressive content of the ad-adjectival *tai* 'too' in Mandarin Chinese: evidence from large online corpora. In: Hong, J.-F., Su, Q., Wu, J.-S. (eds.) CLSW 2018. LNCS (LNAI), vol. 11173, pp. 311–320. Springer, Cham (2018). https://doi.org/10.1007/ 978-3-030-04015-4_26
15. Xie, Z., Luo, Q.: Degree intensifiers as expressives in Mandarin Chinese. Lang. Linguist. **20** (2), 256–281 (2019). https://doi.org/10.1075/lali.00033.xie
16. Kamp, H., Partee, B.: Prototype theory and compositionality. Cognition **57**, 129–191 (1995)

The Repetition of Chinese Onomatopoeia

Mengzhen Xu[1,2(✉)]

[1] University of Chinese Academy of Social Sciences, Beijing, China
xmzget@163.com
[2] College of Chinese Language and Literature, Central China Normal
University, Wuhan, China

Abstract. The boundary between the reduplication and the repetition of Chinese onomatopoeic forms is relatively vague, which brings some problems to Chinese information processing and Chinese teaching. Based on language practice, we focus on the repetition of modern Chinese onomatopoeia, and consider it a rhetorical device to emphasize repetitive sounds or rhythms. The resulting language forms should be regarded as onomatopoeic phrases, in which pauses can be inserted freely. We summarize some typical features of the repetition of onomatopoeia, and briefly introduce relevant punctuation marks, which can be used to express pauses. In addition, we agree that there is a subtle relationship between the reduplication and the repetition of onomatopoeia in syntax, semanteme and other aspects. In essence, they are both the imitation of repetitive sounds.

Keywords: Chinese onomatopoeia · Repetition · Reduplication ·
Lexicalization

1 Introduction: About Chinese Onomatopoeic Forms

People imitate natural sounds by voice, and gradually create a kind of systematic language forms, which we call onomatopoeic forms, including onomatopoeia and onomatopoeic phrases. Due to the iconicity between their phonetic forms and the objective sounds to be imitated, there was a sharp division of opinions in earlier studies (such as [1–3] et al.) on whether onomatopoeia could be defined as linguistic symbols. Thanks to the exploration and research of many experts, onomatopoeia has been widely recognized as a special part of speech in Chinese. However, compared with other parts of speech, the current attention to onomatopoeia in academia is still insufficient, and some problems in practical application have not been well solved for the time being. For example, in order to maximize the imitative effect, people tend not to use only one onomatopoeic syllable when they want to describe specific sounds in Chinese, but to repeat it multiple times. For the convenience of description, here we use a capital English letter to represent an onomatopoeic syllable. So is an onomatopoeic form like AAA one word or three words? And how to distinguish the repetition from the reduplication of onomatopoeia? This comes down to the division of onomatopoeic forms into different levels of linguistic units.

© Springer Nature Switzerland AG 2020
J.-F. Hong et al. (Eds.): CLSW 2019, LNAI 11831, pp. 173–179, 2020.
https://doi.org/10.1007/978-3-030-38189-9_18

The boundary between onomatopoeia and onomatopoeic phrases has always been vague, and those dividing standards formulated by different experts are not completely consistent. In addition to the above AAA, there are abundant onomatopoeic forms composed of duplicate syllables in practice, such as AAAA, AAAAA and AAAAAA etc. Most existing studies categorized them as words. For example, Geng [4] called AAA "Legato Word". Shao [5] named AA, AAA and AAAA "Duple Word", "Triple Word", and "Quadruple Word" respectively, and considered that "Quintuple Word" and "Sextuple Word" should also exist objectively. While Xing [6] noticed the awkward question of segmentation of these "conjunctive repetition", namely it was difficult to distinguish exactly how many words constituted such forms.

It is generally believed that onomatopoeic forms with inserted pauses (expressed as punctuation marks in writing) should be integrally regarded as phrases. Take Zhang's view [7] as an example: "The segmentation of onomatopoeia should be based on pauses instead of the quantity of syllables, which mark rhythm and manifest as dots or dashes." Ma [8] also regarded successive onomatopoeic morphemes as an integral word when there were no pauses among them, otherwise as a phrase. While Wang [9] and Du et al. [10] argued that some onomatopoeic forms separated by dashes, which only imitated the extension or change of sounds, actually reflected "a whole sound" and should be regarded as an onomatopoeia.

When using onomatopoeic forms, Chinese native speakers only vaguely realize that they are imitating the objective sounds, but do not pay any more attention to the above problems deliberately. And we can also observe that onomatopoeic phrases and compound onomatopoeia have roughly syntactic functions in common (refer to [8]), so that their difference in structure even can be ignored when constructing a sentence with them. For now, therefore, people tend to regard all onomatopoeic forms as generalized "onomatopoeia". However, it means there are no clear dividing standards as a reference for non-native speakers, and even for computer programs. Taking automatic word segmentation and part-of-speech tagging in the field of Chinese information processing as an example, there are two common problems in the segmentation of onomatopoeic forms: First, the establishment of reference words. Second, the syntagmatic relation judgment after segmentation. Onomatopoeic forms are too abundant and various to be recorded exhaustively. Once there are any problems with their segmentation and annotation, it will not only fail to reflect their structural features faithfully, but also affect the accuracy of subsequent syntactic or semantic analysis. The above situation means that it is necessary to further refine the existing theoretical system. Hence, in order to provide more suggestions for relevant problems, we focus on the repetition of modern Chinese onomatopoeia and try summarizing typical features based on language practice.

2 Repetition of Onomatopoeia

Repetition is a rhetorical device with the function of "highlighting contents and emphasizing emotions" [11]. For example:

(1) *Ta xiang yongyou henduo qian, bingqie jueding xuyuan shi buting de shuo "QIAN QIAN QIAN".*

He wanted to have a lot of money, and decided to make a wish by keeping saying "money money money".

(2) *Wo xiangwang hunyin, bingqie jueding xuyuan shi shuo "JIEHUN JIEHUN JIEHUN".*

I long for marriage and decide to make a wish by saying "marriage marriage marriage".

The repetition of words can enhance the expression of subjective attitudes, but is not necessary from the perspective of syntax or semanteme. Its absence (E.g. Change the expression in sentence (1) from "keeping saying 'QIAN QIAN QIAN'" to "keeping saying 'QIAN'". Or change "JIEHUN JIEHUN JIEHUN" in sentence (2)) does not affect the integrity of a syntactic structure or semantic expression.

We can also test onomatopoeic forms with the same principle. Namely, the repetition of onomatopoeia is not necessary in context and can be replaced with other briefer forms, while the reduplication as a word-building device is indispensable and irreplaceable to a specific onomatopoeia. Compare the following sentences:

(3) *Mama qichuang hou, DONG DONG zouxia loudi.*

Mother got up and drummed down the stairs.

(4) *Tom you DONG DONG DONG daole sanlou, xiang zhao yishan chuanghu.*

Tom thumped his way to the third floor, looking for a window.

(5) *Ta DONG DONG DONG DONG paoshang louti, bing meiyou qu ziji de bangongshi.*

She thumped up the stairs, instead of going to her office.

In the above sentences, "DONG DONG", "DONG DONG DONG" and "DONG DONG DONG DONG" are all used to imitate the footsteps on the stairs, and they all serve as adverbials without attached auxiliary word in the sentences. In other words, They are in basically the same contexts. By comparing them, we can see that two repetitive syllables are enough to imitate the successive repetition of the sound. The onomatopoeic forms in sentences (4) and (5) can also be replaced with "DONG DONG". Disyllable AA like "DONG DONG" meet the requirements of the contexts, and at the same time, it can express the meaning of repetitive sounds or rhythms which is not available to monosyllable A like "DONG". So we can regard the structure of AA as reduplication. More repetitive syllables like AAA (e.g. "DONG DONG DONG") or AAAA (e.g. "DONG DONG DONG DONG") are not necessary for the contexts, but only play a rhetorical role of further emphasizing rhythms, which is better regarded as repetition.

The features of the repetition of modern Chinese onomatopoeia can be grasped from four aspects in detail.

2.1 Optionality

Although the repetition of onomatopoeia also means "repetition" in semanteme, it is not necessary to repeat the same syllable more than twice. We can use less onomatopoeic syllables for synonymous substitution. Repetition is just a convenient option when the context calls for emphasis on repetitive sounds or rhythms.

2.2 Wholeness

The repetitive onomatopoeic syllables are often used as a whole in sentences, and their syntactic functions are similar to that of compound onomatopoeia (composed of more than two onomatopoeic morphemes). They can serve as independent elements, adverbials, predicates, objects and attributes in sentences, but not as complements or subjects.

2.3 Recursiveness

Due to its wholeness, in practice, the repetition of onomatopoeia can still be integrally used as a basic onomatopoeic unit to continue a larger scale repetition so as to imitate more complex periodic sounds as precisely as possible.

2.4 Insertability

The repetition of onomatopoeia is a specific kind of phrase with a relatively loose internal structure. Pauses can be inserted among the same onomatopoeic syllables and actually shown as the use of punctuation marks (such as commas, dashes, exclamation marks and so on) in writing, without obvious influence on the realization of the original syntactic function or semantic expression, which is generally not allowed for other parts of speech in Chinese.

3   About Inserted Pauses

The most obvious difference between reduplication and repetition is that pauses can be inserted freely among several onomatopoeic syllables of the latter.

Pauses are not a device of constructing onomatopoeic phrases, but rather separated markers of adjacent onomatopoeia. We know that sound is caused by the vibration of a specific sound source, and stops as the vibration stops. When the sound source vibrates again, it produces a new sound. That is to say, we can separate several relatively independent sounds assisted by pauses. Therefore, pauses, represented in writing by various punctuation marks inserted into onomatopoeic forms, can be regarded as effective markers of distinguishing onomatopoeic phrases from onomatopoeia.

Depending on the rhythms, pauses can exist not only among different sounds, but also among the successive same sounds. The essence of rhythms is a kind of time structure, "derived from, but not limited to periodicity", and governed by such factors as "frequency, amplitude, duration and pause time" [12]. Wu [13] also summarized the rhythm of language as "the combination of speed, pitch, length, intensity and timbre of our voice, which alternate with each other and cycle regularly in a certain period of time". Therefore, it contributes to the vivid imitation of rhythmic characteristics by combining pauses of different properties with the repetition of onomatopoeia, which not only greatly improves the imitating ability and iconicity of onomatopoeic forms, but also enhances the sense of rhythm of sentences.

As a written expression of various pauses, punctuation marks in Chinese are divided into dots and marks. The common dots inserted into onomatopoeic forms are caesura signs, commas and exclamation marks. And the marks mainly include quotation marks, dashes and ellipses.

3.1   Common Dots: Caesura Signs, Commas, and Exclamation Marks

As the most basic kind of pauses, caesura signs ("、") are often inserted among repetitive onomatopoeia to indicate the internal intervals of repeated sounds. While the pauses indicated by commas ("，") are slightly longer than that by caesura signs. The insertion of commas in the onomatopoeic forms can lead to a distinct separation of rhythms. Depending on the duration of pauses, caesura signs and commas can also be nested to represent intervals of varying length, for example:

(6) *BA, BA, BA, BA, BA, BA, sanqiang yige xiao jianxie, Tom lianxu dale liuqiang.*
     Bam, bam, bam. Bam, bam, bam. Every three shots followed by a little interval, Tom successively fired six.

According to CNS *General Rules for Punctuation*, exclamation marks ("!") are used after onomatopoeia to indicate the shortness or suddenness of sounds. And the sequence of the three kinds of dots in terms of pause duration is: Exclamation marks > Commas > Caesura signs [14]. In fact, exclamation marks can be put not only at the end of a sentence, but also into an onomatopoeic phrase. In addition to pauses, they can emphasize the short, strong, loud, crisp end of partial sounds. For instance:

(7) *Nanhai shijin jiaoban miantuan, yizhiyu penzi fachu "KUANG! KUANG!" de shengxiang.*
     The boy stirred the dough so hard that the basin clanged.

3.2   Common Marks: Quotation Marks, Dashes, and Ellipses

Dashes ("——") are one of the most commonly used punctuation marks in onomatopoeic forms. It indicates that the pause is simultaneously accompanied by the prolongation of the ending, which corresponds to the length of sounds. For example, in (8), the monosyllabic onomatopoeia "DANG" and dashes are used alternately and repeatedly to imitate the bell with an even rhythm and periodic repetition.

(8) *Wuli you yizhi lao zuozhong. DANG——DANG——DANG——DANG—— DANG——DANG——DANG——Qidian le.*
     There is an old clock in the room. Clang——Clang——Clang——Clang—— Clang——Clang——Clang——It's seven o'clock.

The ellipses ("……") relevant to onomatopoeic forms are mainly used in three ways: First, similar to dashed, they can indicate the extension or continuation of sounds; Second, they help to imitate intermittent sounds; Third and the most common one, they can represent the omission of repetitive sounds. In addition, ellipses are often in combination with dots like caesura signs and commas, such as (9).

(9) *PA, PA, PA... ...Shuipao yigege polie.*
     Crack, crack, crack...The bubbles burst one by one.

Quotation marksz (" " or ' ') are optional when using onomatopoeic forms in sentences. As to the repetition of onomatopoeia, some people tend to use quotation marks outside the basic onomatopoeic unit for clearer boundary of multiple sounds. For example:

(10) *Miaozhen "DI DA" "DI DA" de zouzhe.*
      The second hand was ticking away.

Although pauses in repetitive forms play an auxiliary role in the representation of rhythms, they are not necessary sentence constituents and will not affect semantic expression after deletion. For instance, "PA, PA, PA" in (9) can also be recorded as "PA PA PA". However, the insertability for pauses is an important difference between the repetition and the reduplication of onomatopoeia. The latter like "HA HA" (laughter) should not be expressed as "HA, HA" or "HA! HA!" in Chinese.

4 Conclusion

We focus on the repetition of onomatopoeia and argue that it is a rhetorical device to emphasize repetitive sounds or rhythms. The resulting language forms should be regarded as onomatopoeic phrases, which differ from the reduplication of ono-matopoeia in structure. For instance, pauses can be inserted freely among the ono-matopoeic syllables of the former. Pauses of different properties are shown in writing as caesura signs, commas, exclamation marks, quotation marks, dashes and ellipses etc. respectively, which play different auxiliary roles in onomatopoeic effect and are con-ducive to the construction of various rhythms.

In practical application, the repetition of onomatopoeia is not so easy to distinguish and some forms even cannot be adequately interpreted in terms of rhetoric. Tak-ing AAA and ABAB as examples, due to frequent use, they form certain semantic features and show a relatively obvious trend of lexicalization. Moreover, there is a subtle relationship between the reduplication and the repetition of onomatopoeia, which should not be distinctly separated only according to the present construction. The reduplication of onomatopoeia like AA can be regarded as "secondary reduplication" derived from the repetition of speech [15]. This repetitive feature is still evident today. For example, AA (reduplication) and "A——A" or "A! A!" (repetition) have some-thing in common semantically. In fact, the boundary between the reduplication and the repetition of Chinese onomatopoeic forms is not so clear-cut. Xu [16] called the former "grammatical repetition" and the latter "rhetorical repetition", and supposed that they were in "a continuum of repetitive iconicity". Therefore, whether an onomatopoeic form is a reduplication or a repetition structurally, it is essentially the imitation of repetitive sounds. It needs more diachronic language materials to verify whether there is a progress of lexicalization from "repetition" to "reduplication" in onomatopoeic forms.

Acknowledgments. This work is part of the project of *Onomatopoeic Forms in Mandarin*, supported by the Fundamental Research Funds for the Central Universities (Innovation Project 2018CXZZ112).

References

1. Wen, L.: Questions on linguistic symbols. J. Stud. Chin. Lang. **2**, 52–62 (1991)
2. Geng, E.L.: The question of signs in relation to onomatopoeia. J. Stud. Chin. Lang. **3**, 186–189 (1994)
3. Liu, D.W.: The semiotic nature of onomatopoeia. J. Linguist. Res. **2**, 11–16 (1996)
4. Geng, E.L.: The functions of onomatopoeia. J. Chin. Lang. Learn. **8**, 52–53 (1980)
5. Shao, J.M.: A preliminary study on onomatopoeia. J. Lang. Teach. Linguist. Stud. **4**, 57–66 (1981)
6. Xing, F.Y.: Inner consistency of onomatopes. J. Stud. Chin. Lang. **5**, 417–429 (2004)
7. Zhang, B.: A brief discussion on onomatopoeia. J. Hebei Normal Univ. (Philos. Soc. Sci. Ed.) **3**, 83–90 (1982)
8. Ma, Q.Z.: Problems on Chinese Semanteme and Grammar. Beijing Language and Culture University Press, Beijing (1998)
9. Wang, J.H.: Research on Chinese Language and Culture (Series 4). Tianjin People's Publishing House, Tianjin (1994)
10. Du, L.R., Shao, W.L.: The basic characteristics and causes of onomatopoeia. J. Linguist. Res. **4**, 25–28 (2003)
11. Xing, F.Y., Wang, G.S.: Modern Chinese. Central China Normal University Press, Wuhan (2011)
12. Tang, L.: The Introduction to Music Physics. University of Science and Technology of China Press, Hefei (1991)
13. Wu, J.M.: What is the rhythm of language. J. Lang. Plan. **5**, 13–15 (1991)
14. AQSIQ, SAC: General Rules for Punctuation. Standards Press of China, Beijing (2011)
15. Liu, D.Q.: Primary and secondary reduplications: multiple diachronic sources of reduplicative forms. J. Dialect. **1**, 1–11 (2012)
16. Xu, M.F.: Grammatical repetition and rhetorical repetition. J. Rhetoric Learn. **2**, 1–10 (2009)

A Degree-Based Analysis
of 'V+A+*le2*' Construction

Mengjie Zhang[1(✉)], Wenhua Duan[2], and Yunqing Lin[3]

[1] School of Foreign Languages, Beihang University, Beijing, China
by1612102@buaa.edu.cn
[2] School of Humanities, Zhejiang University, Hangzhou, China
11504034@zju.edu.cn
[3] English Department, Beijing International Studies University, Beijing, China
ylin@bisu.edu.cn

Abstract. This paper examines both the result-realization and the result-deviation interpretations of 'V+A+*le2*' construction under the framework of degree semantics. It is argued that the 'V+A+*le2*' construction is a comparative construction and that different interpretations arise because different standards of comparison are adopted. Result-realization interpretations are seen in all instances of the construction while result-deviation interpretations can be found only in instances with open-scale adjectives. The syntactic structure determines whether the 'A' encodes a measure function or the 'V+A' encodes a measure of change function in the 'V+A+*le2*' comparative construction.

Keywords: 'V+A+*le2*' comparative construction · Measure function · Measure of change function

1 Introduction

This paper explores the semantics of 'V+A+*le2*' construction in Mandarin Chinese. The construction, taking the form of 'NP+ verb+ adjective+ *le2*', normally expresses the meaning that the denotation of the adjective is the result (or is to modify the result) of the action denoted by the verb, as exhibited in 1 and 2.

> 1　这　个　坑　挖　深　了。
> *Zhe ge　keng　wa　shen　le2*
> This CL　pit　dig　deep　-LE
> 'This pit has been deepened.'

> 2　这　件　衣服　晒　干　了。
> *Zhe jian　yifu　shai gan le2*
> This CL　cloth　bask dry-LE
> 'This cloth has dried.'

Both 1 and 2 have result-realization interpretations: 1 means that the pit has been deepened as a result of the activity of digging, and 2 means that the piece of cloth changed form not being dry to being dry as a result of the activity of basking. It seems

J.-F. Hong et al. (Eds.): CLSW 2019, LNAI 11831, pp. 180–188, 2020.
https://doi.org/10.1007/978-3-030-38189-9_19

that similar syntactic structures would give rise to similar meanings of 1 and 2. Some result-deviation interpretation, however, can be found in 1 rather than in 2. The pit in 1 was not only deepened via the digging, it might also be deepened to some depth that exceeds the interlocutors' expectation. Thus, the result brought about by the action of digging deviates from the presupposed result in 1. In contrast, the result of drying cloth in 2 is always in accordance with the expectation of the interlocutors, i.e., the cloth dried. Previous studies mainly concentrate on deriving result-deviation interpretations and finding out semantic differences between adjectives in instances of different interpretations. However, whether all instances of 'V+A+le2' construction have the result-realization interpretations and how this kind of interpretation is derived are rarely accounted for. This paper aims to give a united analysis of how the result-realization and result-deviation interpretation of 'V+A+le2' construction arise under the framework of degree semantics.

2 Literature Review

Lu [1] points out that the result-derivation meaning is the only possible interpretation for instances in which the action denoted by the verb has no effect on the event participant or the influence on the participant is abnormal, as shown in 3 and 4:

3　这　　双　鞋子　买　大　了。
　　Zhe Shuang xiezi　mai　da　le
　　This　CL shoes　buy　big -LE
　　'The shoes bought is too big.'

4　这　口　　井　挖　浅　了。
　　Zhe kou　jing　wa　qian　le
　　This CL　　well　dig　shallow -LE
　　'The well was dug to the depth that is shallower than is expected.'

In general, the size of the shoes cannot be influenced by the action denoted by *mai* 'to buy'. Hence, Lu [1] believes that *da* 'big' is to modify the size of the shoes, leading to the result-deviation interpretation. Likewise, the pitch cannot become *qian* 'shallow' with the action of digging going on. Therefore, 4 has an interpretation that the depth of the pitch is shallower than the speaker's expectation. However, result-realization interpretations are not impossible for 3 and 4 when some conditions are added, as shown in 5 and 6.

5 上   一   双   鞋子 买 大 了， 这 次   按   你 的 要求，
   Shang yi Shuang xiezi mai da le, zhe ci an ni de yaoqiu
   previous one CL shoes buy big -CL this time follow your requirement
   买 小 了。
   mai xiao le
   buy small -LE
   'The pair of shoes bought last time is too big. Following your instruction, shoes bought this time are smaller than the previous ones.'

6 上 口   井   挖 深 了， 这 次 挖   浅 了，   你 过来 看看
   shang kou jing wa shen -le, zhe ci wa qian -le, ni guolai kankan
   last CL well dig deep -LE this CL dig shallow-LE you come see
   行 不 行。
   xing bu xing
   ok not ok

'The last well is too deep. This currently dug one is shallower than that. Please check whether it meets your requirement or not.'

It is clear that *maixiao* 'to buy to be small' in 5 indicates that the shoes bought this time are smaller than the ones bought previously and that they are as small as what the hearer requires. Similarly, *waqian* 'to dig to be shallow' in 6 means that the pitch is shallower than the previous one. But it is unclear whether its depth is in accordance with the interlocutors' expectation or not. Hence, any explanation [1, 2] that attributes the result-realization interpretation to the match between interlocutors' expectation and the value of the property measured by the adjective goes wrong.

Apart from that, syntactic accounts holding such a stance need some revision as well. Lu [1] and Ma and Lu [2] point out that the sentence structure of 'V+A+*le2*' construction is [[V A] *le2*] if the result-realization interpretation is derived while it is [[V] A *le2*] if the construction receives a result-deviation interpretation. Shen and Peng [3] explains the account further by arguing that the result-realization interpretation is obtained when the result denoted by the adjective is directly caused by the action. In contrast, for the instances of result-deviation interpretations, adjectives modify the existing results of the actions. If the analysis is on the right track, the syntactic structure of 5, having the result-realization interpretation, would be [[*mai da*] *le*], which indicates that *da* 'big' is the direct result brought about by the action of *mai* 'to buy'. This cannot be true because the size of the shoes cannot be changed as a result of buying. Therefore, neither the semantic analyses nor the syntactic explanation explains the 'V+A+*le2*' construction correctly. Taking all of these considerations into account, this paper argues that all instance of 'V+A+*le2*' construction can have a result-realization interpretation and that both the result-realization and the result-deviation interpretations can have the [[V] A *le2*] structure and the [[V A] *le2*] structure.

It should be noted that Shen and Peng [3] is insightful in pointing out that different scales of adjectives influence the interpretation of 'V+A+*le2*' construction as well. They assume that result-deviation interpretations arise only in instance where the adjectives use open-scales. This paper agrees to this point of view and will utilize formal tools to make it clearer.

3 Theoretical Background

Following Kennedy and McNally [4] and Kennedy [5], it is assumed that adjectives encode measure functions. A proposition can be true only if another value or degree is introduced to compare with the degree or value measured by the given adjective. The truth of plain adjectives depends on comparisons between degrees to which they manifest in relevant scales and degrees of standards of comparison in contexts, as shown in Eq. (1)

$$pos = \lambda g \epsilon D_{<e,d>} \lambda t \lambda x. g(x)(t) \gg stand(g). \tag{1}$$

(1) says that the measure function g encoded by the adjective takes an entity variable x and a time variable t and returns a degree d. The degree is compared with the standard of comparison stand(g), from which the truth value of the positive form is derived.

The standard of comparison is influenced by the nature of the scale associated with the adjective [5, 6]. Based on the Principle of Interpretive Economy [5], the standard of comparison corresponds to the maximal or minimal point in the scale if the adjective is associated with an upper or lower closed scale. In contrast, if the adjective use open scales, the standard of comparison is contextually dependent. It can be the expectation of participants or certain comparison class.

7   Interpretative Economy
       'Maximize the contribution of the conventional meanings of the elements of a
       sentence to the computation of its truth conditions.'

Furthermore, the measure of change function m_Δ [6] is put forward, when this line of thought is combined with event semantics, to measure the changes an object undergoes as a result of participating in an event, as shown in Eq. (2).
For any measure function m,

$$m_\Delta = \lambda x \lambda em^{\uparrow}_{m(x)(init(e))}(x)(fin(e)). \tag{2}$$

The measure of change function m_Δ measures the difference between the degrees to which an entity variable x manifests when it is at the initiation of the event (init(e)) and when it is at the end of the event (fin(e)). According to Kennedy and Levin [6], the truth of a proposition with an element encoding measure of change function is shown in Eq. (3).

$$pos(m_\Delta) = \lambda x \lambda em^{\uparrow}_{m(x)(init(e))}(x)(fin(e)) \gg stand(m_\Delta). \tag{3}$$

A proposition is true iff the application of the measure of change function to x returns non-zero value[1] relative to the scale measured by that function. In other words,

---

[1] Measure of change function are derived from the difference function that is always associated with the minimum standard, i.e., zero value. See the discussion in Kennedy and McNally [4].

it is greater than the stand (m_Δ). So, the default value of stand (m_Δ) is always zero if no other context information is provided.

4 Data Analysis

The semantics of 'V+A+*le2*' construction is captured based on the assumption that the construction is a comparative construction and that different interpretations arise because different standards of comparison are adopted. Moreover, the syntactic structure plays a role in determining whether 'A' encodes a measure function or 'V+A' encodes a measure of change function. The interpretation of an instance having [[V] A *le2*] syntactic structure depends on the degree to which 'A' measures in the gradable property and the actual value of the standard of comparison, while that of an instance having [[V A] *le2*] syntactic structure is determined by the value differences measured by measure of change function denoted by 'V+A' and the standard of comparison.

4.1 The Derivation of Result-Realization Interpretation

Result-Realization Interpretation with Respect to Closed-Scale Adjectives. If the adjective in 'V+A+le2' construction is associated with a closed-scale, only result-realization interpretation is allowed no matter what the syntactic structure is. Let us firstly examine instances of [[V A] *le2*] structure. An example is shown in 8.

8   衣服   洗 净 了。
yifu   xi jing le
cloth wash clean –LE
'The cloth has been washed clean.'

Since the result denoted by *jing* 'clean' is directly brought about the action of *xi* 'to wash', the syntactic structure of 8 is [[*xi jing*] *le*]. Here, *xijing* 'to wash to be clean' encodes a measure of change function shown in Eq. (4).

8 is true iff $\text{clean}_\Delta = \text{clean} \uparrow_{\text{clean (cloth) (init (washing event))}} (\text{cloth}) (\text{fin (wahsing event)})$
$\geq \text{stand (clean}_\Delta)$.

$$(4)$$

The measure of change function denoted by *xijing* 'to wash to be clean', in the light of Eq. (3), measures the difference between the degree to which the cloth is clean at the beginning of washing event and the degree to which the cloth is clean at the end of the event. Because the clean scale has a maximal element, the degree to which the cloth is clean at the end of the event is the maximal point of the scale. The difference between the two degrees at the initiation and the end of the event is definitely greater than the stand (clean$_\Delta$) that equals to zero. Therefore, the result-realization interpretation is derived.

As for 9 with the syntactic structure of [[*hua*] *wan le*], the result of the drawing event is the appearance of the line and it is further modified by *wan* 'bent'. Thus, a measure function rather than a measure of change function is denoted by *wan* 'bent'. The truth value of 9 is given in Eq. (5).

9   线　画　　弯　了。
    xian hua   *wan* *le*
    line draw   bent -LE
    'The line drew is bent.'

$$9 \text{ is true iff bent (line) (t)} \geqslant \text{minimal value in bent scale} \tag{5}$$

The measure function denoted by *wan* 'bent' measures the degree to which the line manifests at the current speech time in the scale of bentness. Because *wan* 'bent' is associated with a lower closed-scale, Eq. (5) says that the line drawn has a non-zero degree of bentness. That is, the standard of comparison is the minimal zero point in the bentness scale. The degree to which the line manifests in the scale is greater than the minimal zero point. Hence, result-realization interpretation is obtained.

Result-Realization Interpretation with Respect to Open-Scale Adjectives. In the line with instances where adjectives are associated with closed-scales, instances of both sentence structures can have result-realization interpretations when the adjectives are associated with open-scales. For example, of [[V A] *le2*] structure, 10 means that the road is wider than it was before the widening event.

10   马路　(按上　　级　　　　要求)　　　拓宽　了。
    malu　*(an*　　*shangji　yaoqiu)*　　　*tuokuan　le*
    road following authority requirement　widen　-LE2
    'As is required by the authorities, the road has been widened.'

The direct result of the widening event is that the road becomes wider under the circumstance of 10. Given that the syntactic structure is [[*tuo kuan*] *le*], *tuokuan* 'to widen' encodes a measure of change function that measures the difference between degrees to which the road manifested in the scale of width at the initiation and the end of the widening event. This difference value is greater than the comparative standard, as shown in Eq. (6).

$$10 \text{ is true iff wide}_\Delta = \text{ wide} \uparrow_{\text{wide (road) (init (wideningevent))}} \text{ (road) (fin (widening event))} \geqslant \text{stand (wide}_\Delta). \tag{6}$$

Since difference function is always associated with a minimal value, stand (wide$_\Delta$) is of zero value. Therefore, the result that the road had become wider is realized as long as the road underwent some change in the width. This gives rise to the result-realization interpretation of 10.

Furthermore, result-realization interpretation can also be found in instances of the [[V] A le2] structure. For example:

11   画儿   这   回   才   挂   高   了。
    huaer   *zhe*   *hui*   *cai*   *gua*   *gao*   *le2*
    painting this   CL only   hang   high   -LE
    'Only in this time is the painting hung high enough.'

Because *gao* 'high' is to modify the position of the painting resulting from the action of *gua* 'to hang', the syntactic structure [*gua* [*gao le2*]] licenses *gao* 'high' to encode a measure function that measures the height of the painting in current event of hanging. The truth condition of 11 is shown as follows.

$$11 \text{ is true iff high (painting) (this time)} \geqslant \text{high (painting) (previous time)}. \qquad (7)$$

Since *gao* 'high' is an open-scale adjective, the standard of comparison is the degree provided by the context of 11, namely, the height of the painting in a previous hanging event. Equation (7) says that the height of the picture in the current hanging event is higher than that of the previous hanging event. Result-realization interpretation is obtained.

4.2   The Derivation of Result-Deviation Interpretation

The result-deviation interpretation arises due to the incompatibility between the interlocutors' expectation and the degree measured by the measure function or the degree difference measured by the measure of change function. According to the Interpretative Economy Principle in 7, the comparison standard uses the maximal or minimal point in the relevant scale. Interlocutors' expectations cannot not be involved in instances whose adjectives are associated with closed-scales. Thus, result-deviation interpretations can only be found in instances of 'V+A+*le2*' construction with open-scale adjectives.

Let us firstly examine instances of [[V A] *le2*] structure. Sentence 12 means that the change in width the road undergoes is greater than the interlocutors' expected change. The truth condition is shown in Eq. (8).

12   马路   拓宽   了。
    malu   *tuokuan*   *-le2*
    road   widen   -LE
    'The road has been widened.'

$$12 \text{ is true iff wide}_\Delta = \text{wide} \uparrow_{\text{wide (road) (init (widening event))}} \text{(road) (fin (widening event))}$$
$$\geqslant \text{expectation (wide}_\Delta).$$

$$(8)$$

Given that the change of degree in the width of the road is greater than the interlocutors' expectation, result-deviation interpretation is available.

For an instance of the [[V] A *le2*] structure, result-deviation interpretation can arise as well if the degree 'A' measure is greater than the interlocutors' expected degree. A revised version of 11 is shown in 13.

13 画儿   挂 高 了。
    huaer   *gua gao le2*
    painting hang high -LE
    'The painting is hung to a position that is too high.'

Similar to 12, *gao* 'high' encodes a measure function that measures the height of the painting. The composition of result-deviation interpretation of 13 is shown in Eq. (9).

$$13 \text{ is true iff high (painting) (reality)} \geqslant \text{high (painting) (expectation)} . \qquad (9)$$

Equation (9) says that the actual position of the painting is higher than that of interlocutors' expected position, resulting in a result-deviation interpretation of 13.

Table 1. Interpretations of 'V+A+le2' construction

	Result-realization			Result-deviation		
	Syntactic structure	Encoded function	Comparison standard	Syntactic structure	Encoded function	Comparison standard
Closed-scale	[[V A] le2]	Measure of change function	Minimum value of degree difference = 0	×	×	×
	[[V] A le2]	Measure function	Minimal or maximal degree of the scale	×	×	×
Open-scale	[[V A] le2]	Measure of change function	Degree difference designated by context	[[V A] le2]	Measure of change function	Degree difference designated by expectation
	[[V] A le2]	Measure function	Degree designated by context	[[V] A le2]	Measure function	Degree designated by expectation

5   Conclusion

This paper examines how both the result-realization and result-deviation interpretations of 'V+A+le2' construction arise under the framework of degree semantics. Interactions among scales associated with adjectives, syntactic structures, standard of comparisons and different interpretations are shown in the Table 1.

As seen in the table, the 'V+A+le2' construction is a comparative construction. Its different interpretations depend on choices of standards of comparison which are further influenced by the syntactic structure of the construction and the scale associated with the adjective. The sentence structure determines the kind of function denoted by the construction while the scale associated with 'A' determines whether the construction is allowed to have a result-deviation interpretation or not.

References

1. Lu, J.: The semantics of 'VA'-*le2* predicate-complement structure. Chin. Lang. Learn. **1**, 1–6 (1990)
2. Ma, Z., Lu, J.: A survey of adjectives functioning as resultative compound. Chin. Lang. Learn. **1**, 14–18 (1997)
3. Shen, Y., Peng, G.: The syntactic and semantic analysis of the excessive resultative "VA le" in Chinese. Chin. Lang. Learn. **5**, 3–10 (2010)
4. Kennedy, C., McNally, L.: Scale structure and the semantic typology of gradable predicates. Language **2**, 345–381 (2005)
5. Kennedy, C.: Vagueness and grammar: the semantics of relative and absolute gradable predicates. Linguist. Philos. **1**, 1–45 (2007)
6. Kennedy, C., Levin, B.: Measure of change: the adjective core of degree achievements. In: McNally, L., Kennedy, C. (eds.) Adjectives and Adverbs: Syntax, Semantics and Discourse, pp. 156–183. Oxford University Press (2008)

Semantic Distinction and Representation of the Chinese Ingestion Verb *Chī*

Meichun Liu[1] and Mingyu Wan[2,3,4(✉)]

[1] Department of Linguistics and Translation, City University of Hong Kong,
Kowloon Tong, Hong Kong
meichliu@cityu.edu.hk
[2] Department of Chinese Bilingual Studies, The Hong Kong Polytechnic
University, Hung Hom, Hong Kong, China
mywan4-c@my.cityu.edu.hk
[3] MOE, Key Laboratory of Computational Linguistics, Peking University,
Beijing, China
[4] School of Foreign Languages, Peking University, Beijing, China

Abstract. Research on the Chinese high-frequency verb *chī* 'eat' is manifold with quite diverse observations by various analytical proposals. Representative works include the five-element semantic chain [1], the emergent argument structure hypothesis [2], and the MARVS-based semantic accounts [3–6]. However, little consensus has been reached on the polysemy of *chī* and its semantic-to-syntactic properties. In this paper, a comprehensive study of *chī* with in-depth lexical semantic analysis is conducted by adopting a corpus-driven, frame-based constructional approach. It proposes that *chī* can be viewed as having 'one frame, three profiles and seven constructional meanings' under the assumption that semantic distinctions can be made only if there are sufficient collo-constructional evidence. This study also demonstrates how the polysemy of *chī* can be understood by a two-dimensional analytical model to account for its semantic extensions based on the interaction of spatial and eventive readings.

Keywords: Ingestion verb *chī* · Chinese verbal semantics · Verbal polysemy · Meaning representation and categorization · Frame-based constructional approach

1 Introduction

Ingesting events (e.g., *eat* and *drink*) pertain to the basic human need to take in substance that the human body requires to survive. They also provide a familiar and salient channel for humans to conceptualize the external world [7]. Encoding the universal need, verbs of eating have been studied by a large number of researchers who aim at characterizing the different arrays of usages of the verbs cross-linguistically and cross-culturally. For instance, [8] introduces the range of linguistic behaviors associated with 'eat' and 'drink' verbs in English and depict the two processes as two distinct metaphorical bases for cognitive semantic extensions. It provides a full semantic characterization of the predicates as they reflect the experiential realities of eating and drinking, including literal and generic uses, figurative uses, morphosyntactic variations,

© Springer Nature Switzerland AG 2020
J.-F. Hong et al. (Eds.): CLSW 2019, LNAI 11831, pp. 189–200, 2020.
https://doi.org/10.1007/978-3-030-38189-9_20

and grammaticalized extensions of the predicates. [9] discusses the cross-lingual differences and similarities of eating verbs by focusing on languages of Papua New Guinea and Australia. An interesting observation is that some morphemes, such as k_ϑ - in Manambu or Kalam ñb, cover a diverse range of meanings encompassing 'eat', 'drink', 'suck', 'breast-feed', and 'smoke', while 'eat' remains as its central or prototypical sense. It is further suggested that all languages display some sort of a continuum in terms of the specificity of their verbal predicates. In the domain of ingestion, it amounts to recognizing a continuum of highly differentiated predicates in terms of manners of eating and types of food to encode salient experiential distinctions in the domain.

With regard to Chinese, there are numerous previous studies on ingestion verbs and they vary in theoretical frameworks and analytical approaches. This study focuses on the most frequently used verb chī 吃 'eat', which ranked 124th among 18813 verbs in Chinese Treebank. Existing works all agree that the verb chī is highly polysemous, demonstrating a high degree of semantic multiplicity, ambiguity and flexibility in its meaning representation and categorization. For illustration, the earliest Chinese dictionary *Shuowen Jiezi* includes **nine** basic meanings[1], the Chinese Wordnet and [4] enumerate **28** meaning distinctions[2], while [10] distinguishes and defines **five** basic meanings for chī:

(1) Put food into the mouth to chew and swallow it:
e.g., *chī miàntiáo* 吃面条 'eat noodles',
(2) Rely on other people or sources to live:
e.g., *chī fùmǔ* 吃父母 'live on parents':
(3) Destroy (used in military and chess games):
e.g., *chī le yī-gè jūn* 吃了一个军 'defeat an army':
(4) Absorb: e.g., *chīshuǐ* 吃水 'absorb water'
(5) Suffer (from): e.g., *chī yī-quán* 吃一拳 'get a punch'.

To explain the multi-facet definitions of the verb, [3, 11] provide a cognitive account of the semantic variation by means of metaphor and metonymy, and [12–15] concern the different argument structures and subcategorization frames. [12, 13] suggest that the different argument realizations cause the different interpretations of chī. Specifically, [14] indicates that the varied semantic roles of 'chī+N' are triggered by the verb, while [15] argues that the different meanings result from the unconventional involvement of the post-verbal noun as a non-mandatory semantic role, that is, a pseudo-patient, such as *chī shítáng* 吃食堂 'eat in canteen' (Location), *chī kuàizi* 吃筷子 'eat with chopsticks' (Instrument), *chī fùmǔ* 吃父母 'live on parents' (Source), etc. [16] suggests that there are fast and slow variables in the verb-object structure where fast variables (the verb) usually adapt to slow variables (the object) to reach a stable status. Similarly, [17] and [15] adopt the generative model and propose the notions of 'light verb incorporation' and 'light noun incorporation'. In addition, [18] explores the semantic prominence of the object role of 'eat' by testing the N400

---

[1] http://www.zdic.net/z/16/js/5403.htm.

[2] http://lope.linguistics.ntu.edu.tw/cwn/query/.

amplitude in an ERP experiment, and proposes the following prominence ranking: Patient > Means > Source > Locative > Instrument. Despite the substantial attention the verb has attracted, the studies of *chī* have not reached much consensus in terms of lexical semantic distinction, categorization and representation. Several key issues still await further investigations as shown in the following research questions.

1.1 Research Questions

The diverged proposals and approaches to the verb *chī* in existing works call for a re-examination of the verbal semantics of *chī*, as well as the cognitive semantic mechanisms behind its semantic-to-syntactic behavior. In line with the basic tenants of the prototype theory in cognitive linguistics [19], a number of essential questions are hereby raised:

First, what is the core frame or most prototypical meaning of the verb *chī*? In what ways and by what motivations can the core meaning give rise to other extended meanings? What are the correlations between the core and extended meanings?

Second, in terms of syntax-semantics interface of verbal behaviors, a lexical meaning is usually associated with some syntactically distinct patterns according to the form-meaning mapping principle. What are then the salient syntactic-semantic associations found in the verbal behavior of *chī*? What are the collo-constructional evidences for its meaning distinctions?

Third, since '*chī*+N' is the most frequently and productively used construction that renders ambiguous or polysemous readings, what is a well-motivated account that can explain the interrelations between the various interpretations of '*chī*+N'?

Fourth, is there any superiority of the various representations of *chī* showing its different semantic prominence in use?

2 Lexical Semantics of *Chī* in Previous Works

In addition to the above briefly reviewed studies, the following three representative works on polysemy are highly relevant to the current study, which are concerned with different issues and offer distinct answers.

2.1 Semantic Chain Theory

The 'semantic chain theory' was first proposed by [20] that the multiple meanings of a polysemous word are formed through a semantic chain, in which meaning A relates to meaning B with some shared properties, and C is extended from B with another set of shared properties, and so on and forth: A → B→C → …. With this approach, [1] proposed a five-element semantic chain of the verb *chī* to represent and account for its different meanings, as shown below:

(a) **Eat for survival** →
   e.g., *chī shípǐn* 吃食品 'eat food', *chī shítáng* 吃食堂 'dine at canteen', *chī xiǎozào* 吃小灶 'treated with special care'

(b) **Import/get/in-take** →

e.g., *chīshuǐ* 吃水 'absorb water', *chī huíkòu* 吃回扣 'get rebate', *chīxiāng* 吃香 'have the advantage', *chī tòu jīngshén* 吃透精神 'understand the case'

(c) **Suffer (from)** →

e.g., *chīkuī* 吃亏 'suffer losses', *chīguānsī* 吃官司 'get sued', *chīkǔ* 吃苦 'endure hardships', *chīqiāngzǐ* 耻笑 'being hostile', *chī tā chǐxiào* 吃枪子吃他 'be mocked by him'

(d) **Consume** →

e.g., *chīlì* 吃力 'consume energy', *chī xià tóuzī* 吃下投资 'consume the funding'

e) **Destroy** →

e.g., *chī diào dírén yī-gè shī* 吃掉敌人一个师 'defeat the enemy'.

However, such meaning categorization may lead to some problems. For example, syntactically similar usages with metaphorical extensions may fall into different meaning categories. Thus, *chī diào dírén yī-gè shī* is syntactically homogenous with *chī diào tājiāyī-guōfàn*, but they are differentiated into two sense categories. Moreover, *chī xià tóuzī* differs syntactically and semantically from *chīlì* (e.g., *hěn chīlì* vs. **hěn chī xià tóuzī*), but they are categorized into the same sense. And idiomatic expressions such as *chī xiǎozào*, have lost the original meaning pertaining to the eating event. Lastly, it is unclear how the five senses should be ordered in the semantic chain.

2.2 The Emergent Argument Structure Hypothesis

The emergent argument structure hypothesis was proposed in [2], which maintained that a verb's argument structure is not fixed, but open, dynamic, and adaptive to the change of language use. With this hypothesis, [2] attempted to explain the dynamic properties of *chī* in different contexts, including its diachronic evolvement of de-subcategorization, de-nominalization, de-intransitivity and de-stereotype. It distinguished the argument structures of *chī* in two broad classes via the notion of 'argument extension path'. The typical class of *chī* contains the core arguments of agent and patient, while the non-typical class contains an open set of extended argument structures, such as instrument, place, etc. Based on 'argument extension path' theory, patient is regarded as the core argument, and place is the secondary argument. Other relevant arguments are extended from the core to the peripherals, forming a path as 'Patient → Place → Instrument → Manner…' The work in [2] provides a dynamic and historical perspective to the polysemous relations of the verb *chī*, but it is unclear how the extensions in the semantic path are postulated and to what extent the path can be exhaustive in reaching an endpoint. In addition, there is no account of the semantic interactions between the dynamic arguments, and the data used in [2] was from *Beijing Profiles* in 1980s, which is quite outdated compared to current usages.

2.3 The MARVS-Based Accounts

The Module-Attribute Representation of Verbal Semantics (MARVS) was proposed by [21] to analyze the lexical semantics of verbs. MARVS consists of two types of modules, namely, role modules and event modules, and two sets of internal attributes, role internal attributes and event-internal attributes, which are meant to delimit the eventive and semantic factors of verbal polysemy.

Using the MARVS theory, [3] studied the meaning representations and categorization of *chī* and divided it into six senses by introducing six pairs of role modules, including <Agent, Patient>, <Agent, Place>, <Agent, Instrument>, <Agent, Source>, <Experiencer, Theme>, and <Location, Locutum>. Among the six pairs, the last two indicate novel semantic relations for the use of *chī*. For example, in *tā xǐhuān chī tián* 她喜歡吃甜 'she likes sweets', *tā* is considered to be an experiencer, not an agent, and *chī* a theme, such as a taste, flavor, or cuisine. As for *dàmǐ chīshuǐ* 大米吃水, [3] proposes that *dàmǐ* 'rice' is the location and *shuǐ* 'water' is the locatum, indicating the spatial relation of rice containing water.

The MARVS-based analysis of *chī* is theoretically insightful with many interesting linguistic tests showing the various internal attributes in relation to semantic roles and event modules. However, it is argued that <Experiencer, Theme> interpretation is triggered by the emotion verb *xǐhuān*, instead of the verb *chī*. And the <Location, Locutum> role module seems to only indicate a spatial relation of *dàmǐ* and *shuǐ*, but fails to infer the dynamic change of absorbing or consuming between the two components.

2.4 Significance of This Work: A Framed-Based Constructional Approach

Given that there are still potential problems in the above-mentioned approaches, the present work attempts to provide a theoretically sound, Chinese-specific and cognitively motivated analysis of the verbal semantics of *chī*, which is not only meant to differentiate senses but also to explain the interrelations of the different but related senses with cognitively plausible mechanisms. This paper adopts the **frame-based constructional approach (FC)** proposed in [22–24] by incorporating the theoretical tenets of frame semantics [25] and construction grammar [26] to identify the semantically motivated syntactic variations [27]. Moreover, the analysis is corpus-based by looking at naturally-occurring data in *Sinica balanced corpus* (10 million tokens) and the *Chinese Gigaword corpus* (1 billion tokens). The syntactic distributions of the verbs are extracted automatically from the most prominent collocational forms of the verb *chī* with Stanford CoreNLP toolkit and the semantic analysis is conducted with manual validation based on the FC approach. The data are used to explore and define the frame-verb relation by identifying collo-constructional features distinctively associated with different verb frames or classes. Verbs and constructions, both viewed as meaning-bearing units, go hand-in-hand in defining distinct semantic frames realized with specific argument realization patterns characteristic of a given background frame. Thus, the current approach is able to represent semantic frames with clear formal criteria of constructional associations.

3 Syntactic and Semantic Properties of *Chī*

3.1   Syntactic Distribution

Verbs display distinct syntactic behavior, which provides evidence for their semantic membership. The salient syntactic features and distributions of *chī* are extracted in order to find sufficient syntactic evidence for categorizing the verb meanings of *chī*, as presented in Table 1 below.

Table 1. The overall syntactic distribution of the verb *chī*.

Grammatical roles		Syntactic patterns and examples	%
Nominalized		*jīngguó xiānshēng duì chī bìng bù jiǎngjiū* 经国先生对[吃]ₙ并不讲究 'Mr. Jingguo is not keen on food'	10.2
Pre-noun modifier		*shā zhù chīhēfēng, bànnián jiēzhī 600 wàn yuán* 刹住[吃喝]ₐdj [风]ₙ，半年节支 600 万元 'Banquet remediation saves 6 million in half year'	0.5
Intransitive		a. **S-V:** *wǒ-men bùdébù cáitài rǒngyuán, dàn bùnéng bù chī* [我们]s 不得不裁汰冗员，但不能不[吃]ᵥ 'We had to dismiss employers, but cannot afford to not eat' b. **S-V-C :** *yǒu zhè-zhǒng máobìng de hái-zǐ tōngcháng chī hē guòliàng* [有这种毛病的孩子]s 通常[吃喝]ᵥ [过量]c 'Children with this disease usually eat excessively' c. **V :** *kuài chī ba* 快[吃]ᵥ吧! 'Finish the food quickly!' d. **S-V-P-N:** *wǒ-men chī zài xiānggǎng* [我们]s[吃]ᵥ[在]ₚ[香港]ₙ 'We eat at HK'	12.4 (6.8+4.3 +1.2+0.1)
Transitive	Without post-verbal noun	1. **O-V-C:** *huìyì fàn běnlái jiù chī bù wán* [会议饭]ₒ本来就[吃]ᵥ[不完]c 'Meeting gathering goes and comes' 2. **passive :** *gùyǒuzīchǎn bèi chī kōng le* [国有固定资产]ₒ[被]∗Bₑᵢ[吃]ᵥ[空了]c 'State-owned assets have been consumed' 3. **Ba construction:** *zhuómùniǎo bǎ chóngzi chī le* [啄木鸟]s[把]∗Bₐ[虫子]ₒ[吃]ᵥ 了 'The woodpecker ate the worm' 4. **dative :** *wǒ-men gěi tā-men yú chī* [我们]ₙ₁[给]∗Gₑᵢ[他们]ₙ₂[鱼]ₙ₃[吃]ᵥ 'We feed them fish' 5. **S-*CoV-O-V:** *Zéng Bǎoyí shīmián, zhīhǎo zhǎo ānmiányào lái chī* [曾宝仪]s 失眠，只好[找]∗CoV[安眠药]ₒ[吃]ᵥ 'Zeng Baoyi suffers from insomnia, and she needs to take pills' 6. **O-*CoV-V:** *fàncài gòu chī* [饭菜]ₒ[够]∗CoV[吃]ᵥ 'Foods are fairly enough'	34.9 (5.5+8.3 +10.5+3.1 +2.6+4.9)
	With post-verbal noun	**V+N:** *chīfàn* 吃饭 'dine', *chī màidāngláo* 吃麦当劳 'dine at Macdonald', *chī* *cáizhèng bǔtiē* 吃财政补贴 'live on financial subsidy', *chī yī mèngùnzi* 吃 一闷棍子 'Be beaten', *chīkuī* 吃亏 'suffer loses', *chī lǎoběn* 吃老本 'live off one's past gains', *chīkǎ* 吃卡 'bank card stuck in ATM', *chī láofàn* 吃牢饭 'to do prison time', *chīàn* 吃案 'conceal the case' and so on.	42.0

S: subject, V: verb, O: object, C: complement, N: noun, P: preposition, Adj: adjective, Adv: adverb, *Ba: *ba* marker, *Bei: *bei* marker, *Gei: dative marker, *CoV: light verb marker.

In view of its syntactic distribution, *chī* is mainly used as a verb (intransitive: 12.4% and transitive: 76.9%), or occasionally nominalized (10.2%), or as a pre-modifier of a head noun (0.5%). This paper focuses on the verbal usage.

In its verbal uses, *chī* can be intransitive or transitive. The intransitive *chī* tends to refer to the stereotypical eating event without specifying the object for eating. The transitive *chī* may occur with or without a post-verbal noun. When used without a post-verbal noun, it is normally marked by other syntactic devices (1–6), such as object inversion with verbal complement, serial verb construction, or object fronting in the *GEI, BA,* and *BEI* constructions, indicating a higher degree of transitivity and agentivity. While allowing certain syntactic variations, these marked transitive uses tend to express the prototypical sense of *chī* without too much ambiguity. In contrast, the default unmarked transitive pattern '*chī*+N' is semantically more complicated and poses a number of research issues as discussed in many related studies. The section below will further demonstrate how a frame-based constructional approach can help identify its syntactic and semantic characteristics.

3.2   The Meaning Categorization and Representations

Verb meanings are anchored in semantic frames which describe distinct proto-events. Hence, the key issue is to determine which frame or what proto-event is encoded in the verb *chī* that allows further sense extensions, and whether the extended senses interact with other different frames. Following the "one frame, one sense" principle, it is proposed that the various uses of '*chī*+N', can be viewed as deriving from '**one frame, three profiles, and seven constructional variations**'. This principled account allows this work to classify, explain and represent its syntactic and semantic variations, as shown in Table 2.

Following the '**One semantic frame, one core meaning**' principle in the frame-based constructional approach, the most salient meaning of *chī* involves the basic event of 'taking something in (to consume)', i.e., the act of ingesting, which gives rise to all possible extensions from the semantic intersection of *chī*. In the event of in-taking, there must be an intaker and intaked in a directed relation: In-taker (X) ← In-taked (Y). The action of intaking may vary in terms of agentivity and volitionality. There are thus '**three potential profiles**' along the event schema that further distinguish the uses of *chī* by encoding different degrees of volitional agentivity: (1) profiling the proto-typical, highly volitional and agentive physical act of ingestion, involving an animate Ingestor, as in *chī dōngxi* 吃東西 'eat something'; (2) profiling the effect of digesting as a result of consumption in a dependency relation (the intaker demands and consumes the intaked), which triggers the interpretation of an consumable support or demand, as in *chī-lì* 吃力 'effort-consuming'; (3) profiling the change of state on the consumed intake as it suffers from a decrease of volume, mass or number, thus referring to negative or undesirable experiences of suffering, misfortune or unwillingness, such as *chī guānsī* 吃官司 'suffer a lawsuit'. In view of the varied semantic elements associated with the surface form *chī*+NP, the three profiled semantic cores may render seven different constructional associations with '**seven constructional meanings' (CMs)** that are non-compositional in nature. Each of the seven CMs demonstrates a distinct form-meaning mapping unit with its own collocational preferences. For instance, Constructional Meanings 1–4 (CMs 1.1–1.4) express the metaphorical or metonymic transfers of the prototypical meaning of Ingestion, and their varied constructional meanings can be differentiated by different collocational features as shown in Table 2.

Table 2. The semantic distribution of the verb *chī*

Semantic classes			Example	Collocational features	%
Core	Basic	Extended			
0. In-take {A←B}	1. Ingest {A←B}ₐ𝒸ₜ	1.1 Typical: 'X eats Y' Ingestor + Ingestibles	*chī tāngyuán* 吃汤圆 'eat sweet dumpling'	**Features in Table 1**	79.9
		1.2 CM1: 'X lives on Y' Ingestor + Source	*chī fùmǔ* 吃父母 'live on parents'	NO classifier/determiner/quantifier referring to 维生 *'to make a living'*	4.5
		1.3 CM2: 'X dines at Y' Ingestor + Place	*chī shítáng* 吃食堂 'dine at canteen'	+ **Quantifier:** *chī yī-cì/sān-tiān shítáng* 吃一次/三天食堂 'Dine at canteen once/for three days'	3.1
		1.4 CM3: 'X favors X' Ingestor + Attribute	*chī tián* 吃甜 'keen on sweet food'	+*hěn*-**verbs of liking** *tā hěn xǐhuān chī là* 她很喜欢吃辣 'She likes spicy food very much'	3.0
		1.5 CM4: 'X eats by means of Y' Ingestor +Instrument	*chī kuàizi* 吃筷子 'eat with chopsticks'	NO Classifier/quantifier **Symmetrical couplets:** *dàrén chī kuàizi, xiǎohái chī sháozi* 大人吃筷子, 小孩吃勺子 'Grown-ups use chopsticks; children use spoons'	0.1
	2. Digest-Consume {A←B}𝒸ₕₐₙg𝑒	2.1 CM5: 'X consumes Y'	*chīlì* 吃力 'consuming strength'	+**Degree** *hěn* 很 '**very**': *dài xiǎohái hěn chīlì* 带小孩很吃力 'Baby-sitting is exhausting.' **V-***tòu* 透 '**thoroughly**': *shàngbān ràng wǒ chī tòu jìn le* 上班让我吃透劲了 'I am exhausted after working'	1.1
		2.2 CM6: 'X absorbs Y'	*chīshuǐ* 吃水 'absorb water'	+ *hěn* 很 '**very**': *dàmǐ hěn chīshuǐ* 大米很吃水 'Rice absorbs water well'	4.4
	3. Suffer {A←B}ᵤₙd𝑒ᵣg𝑜	CM7: 'X suffer from Y'	*chī yì mèn gùn* 吃一闷棍 'hit by a stick'	+**Adv:** *bùxìng* 不幸 '**unfortunate**': *tā bùxìng chī guānsī le* 他不幸吃官司了 'Unfortunately, he was sued.	3.9

The above proposed analytical scheme may help answer the research questions raised earlier, with regard to meaning representation and categorization of the polysemy of *chī*. As for semantic prominence of the various senses, it is found that corpus-based distributions may provide some preliminary ranking. According to simple frequency counts, the different uses may be plotted along a continuum of constructional associations: 'Typical:Ingestible > CM1:Source > CM6:Consume > CM7:Suffer > CM2:Place > CM3:Attribute > CM5:consume → digest → destroy > CM4:Instrument'.

In addition to the answers to the several research questions, further consideration is given for analyzing the possible inter-relations between the proposed semantic sub-classes with a cognitively motivated two-dimensional analytical model which combines the spatial and eventive dimensions, as elaborated in the following section.

3.3   The Interrelations: Two-Dimensional Analytical Model

In this section, the possible interrelations between the above-proposed senses are examined by using the Two-dimensional Analytical Model, which highlights the spatial dimension and eventive dimension. The two dimensions work together to clarify the different spatial-physical relations of the frame elements of *chī*, as well as to reveal the temporal sequences between the profiled eventive processes.

Spatial Dimension (horizontal axis): The spatial dimension provides a schematic viewpoint in understanding the spatial-physical relations in the different sub-classes of *chī*. Participant A in Fig. 1 below, as highlighted in the orange circle, typically represents the agent-consumer of *chī* in each sub-class, indicating the spatial directionality of the in-taking event. By labeling the different semantic roles of the agent-consumer, the three basic profiles of *chī* can be semantically distinguished. On the other end, Participant B represents the ingested or consumed entity/substance as the 'moved object' in the event, which can also be further distinguished to different semantic roles with different interpretations. The various types of 'moved object', as highlighted in blue circles in Fig. 1, serve as an indicator of semantic features in rendering the different constructional meanings of *chī*. The dashed boundaries help to segment the different semantic components in representing the three basic profiles of *chī*, and the arrows indicate the directed path of the in-taking motion (A ← B). By representing the common semantic components and the default path of the event, it highlights the core meaning encoded in the verb *chī* as a representative lemma in the In-taking Frame (0: In-take), which serves as a common ground and intersection of all the distinguished meanings of *chī*.

Eventive Dimension (vertical axis): The eventive dimension provides a temporal-progressive viewpoint in understanding the possible causal relations of the in-taking processes of *chī*. As shown in Fig. 1 below, the three basic profiled meanings of *chī* correspond to the three common processes in a dining event, forming a coherent series of the sub-stages of the eating experience: [starting point: Ingesting] → [process: Digesting-absorbing] → [result: Suffering]. As for the result of ingesting, one might argue on the negative polarity in the use of *chī*. In Chinese, only a few cases are found with a positive implication in the use of *chī*, such as in *chīxiāng chīxiāng* 吃香 'have the advantage of'. Indeed, a greater proportion of the eating terms in Chinese are negative in reflecting the unpleasant feelings with bad dining experiences, such as *chī huài dùzǐ* 吃坏肚子 'get the run due to bad foods', *chībùxiāo* 吃不消 'unable to digest', etc.

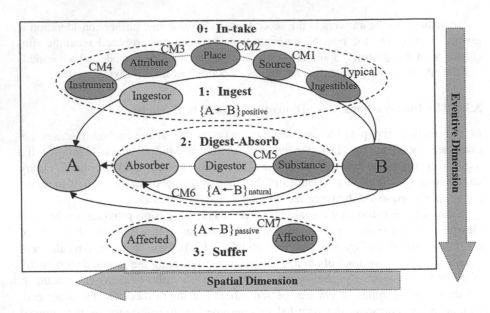

Fig. 1. The two-dimensional semantic model of *chī*

3.4   Idiomization, Grammaticalization and Ambiguity

As a high frequency word in daily usage, the Chinese verb *chī* demonstrates a high flexibility in its lexicalization patterns. As a monosyllabic form in Chinese, it allows an open set of non-typical V-N collocations, and hence encompasses a great number of extended uses, as shown in the above-proposed analytical account. In addition to such meaning extensions, the syntactic openness of the sequence '*chī*+X' leads to proliferated exceptional uses, including idiomatic expressions (e.g., *chī bìméngēng* 吃闭门羹 'be brushed off'), argument alternation with category shift (e.g., *chī tā biànliǎn* 吃她变脸 'suffer her bad emotion'), dual readings or ambiguity (e.g., *chī yādàn* 吃鸭蛋 'eat duck egg/perform badly in exam'), and contextually-triggered implication (e.g., *lǎobǎn xiàng yào chī le tā* 老板像要吃了她 'The boss seems to hit on her') and so on. These uses require considerations beyond the scope of lexico-grammatical properties and taking into account the involved discourse-pragmatic factors. Despite of this, it always complies with the general cognitive semantic principles of sense extension. The key issue is then how to identify the core event schema for a verb form and derive the extended sense with a well-motivated path of cognitive transfers for sense extension and association. In sum, the syntactic and semantic behavior of *chī* may vary on the surface, but there must be a semantic core that serves to link the semantic extensions from the prototypical sense associated with the event of in-taking, absorbing, and consumption.

4 Conclusion

This paper presents a case study of lexical semantic inquiry to distinguish related senses and tease out the interrelations between the senses. It investigates the polysemous uses of the high-frequency verb *chī* in Chinese with a corpus-driven and frame-based constructional approach. Addressing the several key issues in the existing literature in terms of meaning categorization and classification of *chī*, it proposes that the meanings of *chī* are linked via the 'one core frame, three basic profiles, and seven constructional meanings' scheme, which is able to account for the syntactic and semantic variations of the verb. The interrelationship between the involved frame elements (semantic roles) and the proposed sub-classes can be schematically plotted along the 'two-dimensional analytical model'. Along the spatial axis, the different roles of the subject-experiencer of *chī* can vary from Ingestor, to Absorber-consumer, to the Affected, which correlate with the three basic profiles of *chī*, indicating different degrees of volition of the actor in conducting the act of in-taking. As a complementary dimension to the spatial dimension, the eventive dimension helps to provide a developmental viewpoint in understanding the possible causal relations of the processes of *chī*, forming a coherent series of the eating experience: [starting point: Ingesting] → [process: Digesting-Absorbing] → [result: Suffering]. The various types of 'intaked objects' also serve as the semantic motivation for aligning the surface V-N form with seven constructional meanings, rendering an array of semantically diverse but syntactically similar uses. The varied semantic associations of the seemingly identical form (*chī*+X) give rise to seven sub-constructions, each of which is a unique form-meaning mapping unit to highlight a unique semantic relation. Future work will verify the effectiveness of the proposed approach and analytical model for detecting the semantic relations of other major Ingesting verbs in Chinese and ultimately conduct typological studies for cross-lingual comparisons with regard to the lexicalization patterns of ingestion verbs and related verbs in general.

Acknowledgments. This work is partially supported by the internal grant of SRG 2019 at City University of Hong Kong and the PDF project no. 4-ZZKE at the Hong Kong Polytechnic University. Special thanks go to the reviewer, Prof. Chu-Ren Huang and Dr. Jiafei Hong who gave valuable comments on the earlier manuscript of this work.

References

1. Wang, Y.: A corpus-based comparative study on the 'V-N' structure of eating verbs in Chinese and English. Foreign Lang. **28**(2), 1–6 (2007). (in Chinese)
2. Tao, H.: The dynamic properties of the argument structure of the verb chī. Lang. Stud. **40**(3), 21–38 (2000). (in Chinese)
3. Wang, S., Huang, C.-R.: Sense representation in MARVS: a case study on the polysemy of chī. Int. J. Comput. Process. Lang. **23**(03), 285–306 (2011)
4. Hong, J.: Verb Sense Discovery in Mandarin Chinese-A Corpus Based Knowledge-Intensive Approach. Springer, Heidelberg (2015)
5. Hong, J., Huang, C.-R., Ahrens, K.: Event selection and coercion of two verbs of ingestion: a MARVS perspective. Int. J. Comput. Process. Orient. Lang. **21**(1), 29–40 (2008)

6. Hong, J., Ahrens, K., Huang, C.-R.: Event structure of transitive verb: a MARVS perspective. Int. J. Comput. Process. Orient. Lang. **24**(1), 37–50 (2012)
7. Newman, J.: The Linguistics of Eating and Drinking. John Benjamins Publishing, Amsterdam (2009)
8. Newman, J.: Eating and drinking as sources of metaphor in English. Cuadernos de filología inglesa (Spec. Vol. Cogn. Linguist.) **6**(2), 213–231 (1997)
9. Aikhenvald, A.-Y.: 'Eating', 'drinking' and 'smoking': a generic verb and its semantics in Manambu. In: Newman, J. (ed.) The Linguistics of Eating and Drinking. John Benjamins Publishing, Amsterdam (2009)
10. Meng, Z.: Verbal Usage Lexicon. Shanghai Lexicography Publisher, Shanghai (1987). [in Chinese]
11. Wang, Z.: The cognitive study on '*chi shitang*'. Lang. Teach. Res. **2012**(2), 58–64 (2002). (in Chinese)
12. Shen, Y.: Valency Grammar of Modern Chinese. In: Zheng, D. (ed.) Peking University Publisher, Beijing (1995). (in Chinese)
13. Yu, L.: Verb valency analysis to *chi* and its related verbs in modern Chinese. Modern Commun. **2018**(7), 85–86 (2018). (in Chinese)
14. Wu, X.: The syntactic and pragmatic selection of the object by the verb. J. Hangzhou Normal Univ. **1996**(4), 73–81 (1996). (in Chinese)
15. Zhang, J., Yu, L.: The non-typical argument structure of the verb Chi – a case study on *chi shitang*. Modern Chin. **2014**(12), 48–50 (2014). (in Chinese)
16. Chu, Z.: The verb+object phrase and its 'obey principle'. World Chin. Educ. **43**(3), 43–49 (1996). (in Chinese)
17. Feng, S.: "*Xie Maobi*" and rhythm inducing verb incorporation. Lang. Teach. Res. **2000**(1), 25–31 (2000). (in Chinese)
18. Liu, Y., Xu, X., Panther, K.-U.: An ERP approach to thematic hierarchies regarding grammatical objects of the Chinese verb *Chi* (eat). Lang. Sci. **2013**(40), 36–44 (2013)
19. Lakoff, G.: Cognitive models and prototype theory. In: Neisser, U. (ed.) Concepts and Conceptual Development, pp. 63–100. Cambridge University Press, Cambridge (1987)
20. Taylor, J.: Linguistic Categorization. Oxford University Press, New York (1989)
21. Huang, C.-R., Ahrens, K., Chang, L.-L., Chen, K.-J., Liu, M.-C., Tsai, M.C.: The module-attribute representation of verbal semantics: from semantic to argument structure. Comput. Linguist. Chin. Lang. Process. **5**(1), 19–46 (2000)
22. Lui, M.-C., Wan, M.: Chinese verbs and its categorization: construction and application of Mandarin VerbNet. Lexicogr. Stud. **2019**(2), 42–60 (2019)
23. Wan, M., Liu, M.-C.: Supervised word sense disambiguation with frame-based constructional features: a pilot study of fán 'to annoy/be annoying/be annoyed'. Int. J. Knowl. Lang. Process. **9**(2), 33–46 (2018)
24. Liu, M.-C.: A frame-based morpho-constructional approach to verbal semantics. In: Kit, C.-Y., Liu, M.-C. (eds.) Empirical and Corpus Linguistic Frontiers. China Social Sciences Press, Beijing (2018)
25. Fillmore, C.J.: Frame semantics. In: Linguistics in the Morning Calm. Hanshin Publishing Co, Seoul (1982)
26. Goldberg, A.E.: Constructions: A Construction Grammar Approach to Argument Structure. University of Chicago Press, Chicago (1995)
27. Levin, B.: English Verb Classes and Alternations: A Preliminary Investigation. University of Chicago Press, Chicago (1993)

From Repetition to Continuation: Construction meaning of Mandarin AXAY Four-Character Idioms

Chiarung Lu[1](✉) [iD], I-Ni Tsai[2] [iD], I-Wen Su[1], and Te-Hsin Liu[2]

[1] Graduate Institute of Linguistics, National Taiwan University,
Taipei 10617, Taiwan
{chiarung, iwensu}@ntu.edu.tw
[2] Graduate Program of Teaching Chinese as a Second Language,
National Taiwan University, Taipei 10617, Taiwan
{initsai, tehsinliu}@ntu.edu.tw

Abstract. Adopting the theoretical framework of Construction Grammar, the present paper aims to examine the internal structures, constructional meanings, and syntactic categories of Mandarin parallel idiomatic prefabs A-X-A-Y. The analysis once again confirms the importance of semantic integration between lexical and constructional meanings. Although X and Y seem to dictate the syntactic category of a four-character idiomatic expression, the syntactic category of the whole idiom is indeed adjustable in contextualized real-world language use. This prominent feature, i.e., the flexibility in terms of syntactic behavior, results from the markedness of the four-character skeleton, namely, a grammatical construction. This syntactic flexibility is then argued to be the essential property of Chinese four-character idioms.

Keywords: Chinese four-character idioms · Construction · AXAY · Lexical semantics · QIE

1 Introduction

Four-character idiomatic expressions, aka Quadrisyllabic Idiomatic Expression (QIE), in Mandarin Chinese have long been an intriguing subject in linguistics. For example, Cheng Xiangqing has dealt with their syllabic structures and considered them as an extension of disyllabic structures [1]. Other researchers have pointed out some characteristics of QIEs such as the abundance of allusions, and structural simplicity [2, 3]. Besides, many studies have paid attention to the complicatedness of their internal structures [4, 5]. This study also deals with the reanalysis of the internal structure of QIEs targeting on a specific group featured as the AXAY structure.

This paper is organized as follows. Section 1 is the introduction. Section 2 reviews key papers on idioms. Section 3 addresses data and methodology, spelling out data sources, and selection criteria. Section 4 presents our data analysis of AXAY QIEs based on their forms and meanings. Section 5 discusses the underlying mechanisms and their syntactic roles in a sentence. The concluding remarks are found in the last section.

© Springer Nature Switzerland AG 2020
J.-F. Hong et al. (Eds.): CLSW 2019, LNAI 11831, pp. 201–210, 2020.
https://doi.org/10.1007/978-3-030-38189-9_21

2   Literature Review

A detailed review on the compositionality of Chinese QIEs can be found in our previous papers [6, 7]. In this section, we shall inquire briefly into the literature on the internal structures of Chinese idioms and construction grammar.

2.1   The Internal Structures of Chinese QIEs

Hu [8] and scholars such as Liu and Xing [9] categorized Chinese QIEs into two types: parallel and non-parallel. The former can be illustrated by *tian-fan-di-fu* 天翻地覆 'sky-turn-earth-overturn; heaven and earth turning upside down' where *tian-fan* and *di-fu* share a similar syntactic structure, namely, parallel and symmetric. The latter can be realized by *xiong-you-cheng-zhu* 胸有成竹 'chest-have-mature-bamboo', meaning 'to have a well-considered plan', which contains a relatively complete syntactic structure (i.e., subject, predicate and object). Tsou [5] conducted a series of research on four-character idioms and proposed a special term to catalog these idioms, namely, QIE (Quadrisyllabic Idiomatic Expression). A QIE consists of four syllables and boasts parallel and repetition as its two salient features in terms of structure. According to Tsou [5], QIEs with a parallel structure account for 35% of all QIEs: e.g., *long-fei-feng-wu* 龍飛鳳舞 'dragon-fly-phoenix-dance; lively and vigorous flourishes in calligraphy' and *feng-qi-yun-yong* 風起雲湧 'wind-rise-cloud-scud; rolling on with full force'. QIEs with a modification structure account for 21.5%: e.g., *da-qian-shi-jie* 大千世界 'big-thousand-world; the boundless universe'. Other QIEs with a subject-predicate structure make up 17.5%: e.g., *ni-niu-ru-hai* 泥牛入海 'mud-ox-enter-sea; like a clay ox entering the sea, gone without return'. Lastly, QIEs with a predicate-object structure make up 15%: e.g., *yi-jue-ci-xiong* 一決雌雄 'one-decide-female-male; fight a decisive battle'. Details of the percentages are provided in (1).

(1)   a.   ABCD = AB+CD = coordination (35%)
     b.   ABCD = AB+CD = subordination (21.5%)
     c.   ABCD = A+B+CD/AB+C+D = SV (17.5%)
     d.   ABCD = AB+CD = VP sequence (15%)
     e.   ABCD = ABC+D/AB+CD = NP (11%)

Moreover, Tsou provided another analysis of the proportion of QIE types based on a Chinese synchronic corpus (LIVAC), which contains more than 300 thousand QIE entries [10]. He found that among those four characters, the largest group (7.8%) (see Table 1) is the one whose first and third characters (in postions A and C) have a semantic correspondence. The runner-up group (6.0%) is the one whose second and fourth characters (in postions B and D) have a semantic correspondence. Other groups are listed in Table 1. Among them, if the character in position A resembles that in position C or B resembles D, then the lexico-semantic relationships between them could be synonymy, hypernymy, hyponymy, antonymy, broad contextual or situational relations, or collocation or syntagmatic relations.

Table 1. The percentages of types of QIE internal structures [10]

(Characters of QIE) A B C D	
Structural type	%
A __ C __	7.8
__ B __ D	6.0
A B __ __	2.7
__ __ C D	2.1

Furthermore, according to the statistical result based on *The Modern Dictionary of Idioms* published in 1994, Liu and Xing [9] reported that out of 12,703 idiom entries, 39.2% belongs with the symmetric structure, which features symmetricity as the salient characteristic of Chinese QIEs. As for the syntactic categories that QIEs represent in a sentence, Tsou pointed out that they can be used as nouns, adjectives, and verbs [10]. However, Tsou has not dwelt much on the interaction between the internal structure of a QIE and its part of speech as a whole. Based on these previous studies, this paper then sets its particular focus on the "A_C_ type" (alternately, AXAY) aiming to further explore such an interaction. It should be shortly noted here that the compositionality of QIEs is an essential issue. However, as this study is a follow-up study, the discussion of compositionality is not provided here. Please refer to a detailed review on the compositionality of Chinese QIEs in previous papers [6, 7].

2.2 Construction Grammar and Idioms

Fillmore et al. argued that the proliferation and productivity of idioms lie in basic structural conventions, resulting from the multi-operations of syntactic, semantic, and pragmatic phases [11]. This structural convention or regularity has ever since been regarded as a kind of constructional meaning. Goldberg defined "construction" as a meaningful pair of form and meaning, which is an essential element for constituting linguistic structures [12]. Chinese QIEs have salient structural features and, by the same token, may contain some sort of constructional meanings. Bybee and Croft proposed several tiers of construction according to differing degrees of conventionality and substitutability [13–15]. In this notion, the most abstract one should lie in the syntactic configuration such as the ditransitive construction [V NP$_1$ NP$_2$], where a sense of *giving* is generated. The most concrete tier contains abundant prefabs, a type of construction in Bybee's definition, which consist of specific lexemes. Thus, a prefab is a chunk, ready-made for retrieval from memory at any time.

Lu et al. further analyzed the constructional meanings of Chinese QIE *yi-X-#-Y* 'one-X-#-Y', where # represents numbers [6]. They found that *yi-X-#-Y* indeed consists of several construction subgroups such as *yi-X-yi-Y* 'one-X-one-Y', *yi-X-er-Y* 'one-X-two-Y', and *yi-X-qian-Y* 'one-X-thousand-Y'. These subgroups constitute a constructional network. The *yi-X-yi-Y* construction has more constructional meanings compared with other constructions, resulting in its high productivity in actual use. In *yi-X-er-Y*, *er* 'two' is the double of one, therefore, featuring *double* as one of its constructional meanings. Besides, *er* happens to be both a cardinal number (i.e., two) and an ordinal number (i.e., second) in Chinese, so it suggests a sense of either doubled or repeated emphasis in this construction. For instance, *yi-gan-er-jing* 一乾二淨 'one-clean-two-clean; extremely clean' contains a disyllabic word *ganjing* 'clean' as its semantic root and a repeated emphasis as its constructional meaning. Similarly, *yi-X-qian-Y* 'one-X-thousand-Y' has a constructional sense of contrast due to the contrastive nature between the numbers: one and thousand. Besides, the number *one* is polysemous in nature leading to the constructional polysemy of *yi-X-yi-Y*.

Another follow-up study conducted by Lu et al. explored the constructional meanings of six groups of QIEs (i.e., *da-X-da-Y* 'big-X-big-Y', *da-X-xiao-Y* 'big-X-small-Y', *xiao-X-xiao-Y* 'small-X-small-Y', *xiao-X-da-Y* 'small-X-big-Y', *da-X-wu-Y* 'big-X-no-Y', and *wu-X-wu-Y* 'no-X-no-Y') [7]. Their findings not only confirmed the generality of constructional meanings for researching Chinese QIEs, but also identified the crucial contributions of lexical semantics to QIE research. That is to say, a well-interpreted QIE must result from an integrated understanding of both the lexical side (e.g., sense relations) and the structural side (e.g., constructional meanings). A semantic map of the six constructional meanings was further presented to catalog their senses and distances of meanings. Although the above-mentioned studies can be illuminating, further investigations are required to validate its generality.

3   Data and Methodology

This study starts with a collection of QIE examples to be examined. The materials used herein are collected from representative online dictionaries of idioms: Dictionary of Chinese Idioms website[1], compiled by National Academy for Educational Research, Taiwan, and Online Dictionary of Idioms[2], China. Firstly, we searched for and confirmed the meanings of all QIEs that come in the form of AXAY. Since these constructions contain specific lexical items, they are then treated as prefabs in nature [13]. Next, we manually sorted through these prefabs by examining the senses of A in these QIEs. Furthermore, we inquired into the sense relations (i.e., hypernymy, synonymy, antonymy) and the semantic fields between X and Y. Their sense relations were later studied with instances obtained from aforementioned online sources. Lastly, we examined the gestalt meanings of composite AXAY QIE series.

4   Data Analysis

This section presents the analysis of the internal structures of AXAY QIEs—829 AXAY QIEs are found among a total of 39,635 QIEs. Table 2 introduces the top 60 types (571 tokens) of AXAY QIEs. To facilitate readability, we use the question mark (?) to represent X and Y.

---

[1] Dictionary of Chinese Idioms website (http://dict.idioms.moe.edu.tw/cydic/index.htm).

[2] Online Dictionary of Idioms website (http://cy.5156edu.com; http://cy.hwxnet.com/).

Table 2. The top 60 types of AXAY (=A?A?)

A?A?	Tokens	A?A?	Tokens	A?A?	Tokens	A?A?	Tokens	A?A?	Tokens
一?一?	94	相?相?	11	先?先?	5	快?快?	4	杖?杖?	3
不?不?	84	若?若?	9	再?再?	5	所?所?	4	美?美?	3
無?無?	63	難?難?	9	好?好?	5	活?活?	4	捏?捏?	3
如?如?	38	可?可?	8	作?作?	5	盡?盡?	4	能?能?	3
有?有?	31	小?小?	7	見?見?	5	獨?獨?	4	假?假?	3
大?大?	24	同?同?	7	使?使?	5	古?古?	3	閑?閑?	3
自?自?	24	做?做?	7	宜?宜?	5	打?打?	3	誠?誠?	3
半?半?	19	非?非?	6	順?順?	5	各?各?	3	屢?屢?	3
多?多?	15	爲?爲?	6	實?實?	5	克?克?	3	疑?疑?	3
三?三?	14	載?載?	6	十?十?	4	妖?妖?	3	隨?隨?	3
百?百?	12	說?說?	6	必?必?	4	希?希?	3	縮?縮?	3
沒?沒?	12	人?人?	5	全?全?	4	弄?弄?	3	七?七?	2

In the construction of AXAY, the syntactic category of A is also crucial. Table 3 presents the details of the parts of speech of A. Table 4 shows the top 60 types of XY.

Table 3. The details of the syntactic categories of A in A?A?

POS of A	Tokens	POS of A	Tokens
Verb	69	Localizer	3
Adverb	34	Demonstrative	1
Adjective	33	Pronoun	1
Noun	22	Interrogative	1
Numeral	13	Conjunction	1
Auxiliary	10		

The most frequently used XY in the construction of AXAY is *tounao* 頭腦 'brain'. If it collocates with *ben* 笨 'dumb', then the QIE will become *ben-tou-ben-nao* 笨頭笨腦 'stupid-head-dumb-brain; stupid'. More examples will be given in the next section. This study finds that if XY constitutes a noun such as *tounao* 'brain', then A is usually an adjective or a verb. In Table 2, the five most frequently used A?A?-types are *yi?yi?* 一?一? 'one-?-one-?', *bu?bu?* 不?不? 'not-?-not-?', *wu?wu?* 無?無? 'no-?-no-?', *ru? ru?* 如?如? 'like-?-like-?', and *you?you?* 有?有? 'have-?-have-?'. As we have investigated in our previous paper, the high-frequency QIE group *yi-X-yi-Y* 'one-X-one-Y' may result from the polysemy of *yi* 'one' in Mandarin Chinese [6]. Since *bu* and *wu* are negators, they often occur with various predicates, resulting in their high-frequency of usage. It is worthy noting that these three QIEs (i.e., *bu-X-bu-Y*, *wu-X-wu-Y*, and *you-X-you-Y*) are conventional prefabs that usually contain a hidden subjunctive mood. They can usually be interpreted as "if ∼, then ∼." For instance, *wu-quan-wu-yong* 無拳無勇 'no fists, no braveness' can be a short form of "*if you do not have fists, then you are not brave*" [7]. If the lexical sense between X and Y is antonymy, then *wu-X-wu-Y* can be interpreted as 'none of them'. For instance, *wu-hui-wu-yu* 無悔無譽 'no-fault-no-

Table 4. The top 60 types of XY in AXAY

XY	Tokens	XY	Tokens	XY	Tokens	XY	Tokens	XY	Tokens
頭腦 *tounao* 'brain'	15	神鬼 *shengui* 'god-ghost'	4	鬼神 *guishen* 'ghost-god'	3	德心 *dexin* 'virtue-heart'	2	茶飯 *chafan* 'tea-rice'	2
手腳 *shoujiao* 'hand-foot'	13	始終 *shizhong* 'start-end'	4	情緒 *qingxu* 'emotion'	3	時地 *shidi* 'time-place'	2	地天 *ditian* 'earth-heaven'	2
模樣 *moyang* 'appearance'	6	聲氣 *shengqi* 'sound-air'	4	是非 *shifei* 'right-wrong'	3	起坐 *qizuo* 'get.up-sit.down'	2	偏黨 *piandang* 'preference-party'	2
頭尾 *touwei* 'head-tail'	6	情理 *qingli* 'emotion-sanity'	4	眉眼 *mei'yan* 'brow-eye'	3	日夜 *ri'ye* 'day-night'	2	偏倚 *pian'yi* 'preference-dependence'	2
言語 *yan'yu* 'language'	6	時刻 *shike* 'hour-moment'	3	手足 *shouzu* 'hand-foot'	3	人鬼 *rengui* 'man-ghost'	2	即離 *jili* 'reach-leave'	2
嘴舌 *zuishe* 'mouth-tongue'	5	文武 *wenwu* 'letter-arm'	3	來往 *laiwang* 'come-go'	3	好歹 *haodai* 'good-bad'	2	伶俐 *lingli* 'smart-clever'	2
心意 *xin'yi* 'heart-thought'	5	明暗 *ming'an* 'bright-dark'	3	心力 *xinli* 'heart-power'	2	賊贓 *zeizang* 'stolen goods'	2	三四 *sansi* 'three-four'	2
長短 *changduan* 'long-short'	5	山海 *shanhai* 'mountain-sea'	3	吹擂 *chuilei* 'blow-boast'	2	隱現 *yinxian* 'hide-show'	2	法天 *fatian* 'law-sky'	2
天地 *tiandi* 'heave-earth'	5	仁義 *ren'yi* 'benevolence-righteousness'	3	起落 *qiluo* 'rise-fall'	2	癡聾 *chilong* 'dumb-deaf'	2	裡氣 *liqi* 'in-air'	2
頭臉 *toulian* 'head-face'	5	進退 *jintui* 'go.ahead-step.back'	3	入出 *ruchu* 'enter-exit'	2	知覺 *zhijue* 'know-feel'	2	切磋 *qiecuo* 'cut-improve'	2
人事 *renshi* 'people-affair'	5	買賣 *maimai* 'buy-sell'	3	裡外 *liwai* 'in-out'	2	國民 *guomin* 'nation-people'	2	生死 *shengsi* 'life-dead'	2
千萬 *qianwan* 'thousand-10 thousand'	4	死生 *sisheng* 'die-live'	3	心德 *xinde* 'heart-virtue'	2	間界 *jianjie* 'margin-border'	2	死活 *sihuo* 'die-live'	2

honor' does not mean "all of the faults and honors." In fact, it means none of the faults and honors that one can receive. Therefore, it is a QIE referring to a mediocre man. As shown in these examples, the lack of conjunction or subjunctive markers can be taken as one of the characteristics of Chinese QIEs, that is, the structural simplicity with interpretable meanings.

5 Discussion

As shown in Table 4, the most frequently used XY in the prefabs of AXAY is *tounao*, namely, brain. The examples are *tu-tou-tu-nao* 土頭土腦 'muggy brain; acting like a country bumpkin'; *mei-tou-mei-nao* 沒頭沒腦 'no-head-no-brain; without rhyme or reason'; *que-tou-que-nao* 怯頭怯腦 'scare-head-scare-brain; nervous and clumsy'; *hun-tou-hun-nao* 昏頭昏腦 'dizzy-head-dizzy-brain; muddleheaded'; *hu-tou-hu-nao* 虎頭虎腦 'tiger-head-tiger-brain; looking dignified and strong'; *jue-tou-jue-nao* 倔頭倔腦 'stubborn-head-stubborn-brain; blunt of manner and gruff of speech'; *gui-tou-gui-nao* 鬼頭鬼腦 'ghost-head-ghost-brain; evil and shrewd'; *tan-tou-tan-nao* 探頭探腦 'search-head-search-brain; crane one's neck to peer'; *ben-tou-ben-nao* 笨頭笨腦 'stupid-head-dumb-brain; stupid'; *wu-tou-wu-nao* 無頭無腦 'no-head-no-brain; completely without clue'; *shun-tou-shun-nao* 順頭順腦 'smooth-head-smooth-brain; be obedient'; *zei-tou-zei-nao* 賊頭賊腦 'thief-head-thief-brain; behaving stealthily like a thief'; *pi-tou-pi-nao* 劈頭劈腦 'chop-head-chop-brain; straight on the head'; *zhong-tou-zhong-nao* 撞頭撞腦 'hit-head-hit-brain; being rigid; stiff'; *suo-tou-suo-nao* 縮頭縮腦 'shrink-head-shrink-brain; recoil in fear' and so on. The study probes into the interaction between lexical senses and constructional meanings based on the above QIE group. In this series, XY means brain (i.e., an NP) so that AA can be some modifiers such as adjectives or verbs to modify this NP. If AA is an adjective, then it can be used as in (2)–(5). If AA is a verb, it can be used as in (6) *suo-tou-suo-nao*, or in (9) *tan-tou-tan-nao*. Note that the reduplicated form has a constructional meaning depicting a continuous action or state.

(2) 隔壁那個「吝嗇鬼」變鬼變怪、鬼計多端、鬼頭鬼腦 (*gui-tou-gui-nao*)鬼主意。另一個「膨風鬼」最會吹牛 (Sinica corpus).
    'The next is a penny-pincher **full of devilish tricks and cunning ideas**. The other one is a fast talker, boasting all the time.'

(3) 在翻跟斗時, 可以說是從從容容、不慌不忙,也可以說是笨頭笨腦 (*ben-tou-ben-nao*)、憨手拙腳。(Sinica corpus)
    'When doing a somersault, some do it like a cakewalk, while others are just very **clumsy**.'

(4) 現在她走起路來, 已經變得有點賊頭賊腦 (*zei-tou-zei-nao*), 東張西望的, 老疑心迎面走來的都是要對她下手的歹徒 (Sinica corpus)
    'Now when walking on the street, she becomes **overcareful**, always looking around as if everyone coming her way were a bad guy who intends to hurt her.'

(5) 他覺得來的這個土頭土腦(*tu-tou-tu-nao*)的人, 說話氣挺粗。(馮志《敵後武工隊》第四章)
    'He feels that the man who came to him is a **country bumpkin** who speaks in an unpolished way.'

(6) 小偷縮頭縮腦 (*suo-tou-suo-nao*)地溜走了(兩岸萌典)
    'The thief **recoiled in fear** and disappeared skillfully.'

(7) 看那怯頭怯腦 (*que-tou-que-nao*)的神情, 倒很像個莊稼漢子。(《鷹爪王》)
    'He looks **nervous and clumsy**, rather characteristic of a countryman.'

(8) 接連來了幾張訂單, 把他忙得昏頭昏腦 (*hun-tou-hun-nao*) (兩岸萌典)
    'The orders were placed one after another, keeping him busy and **dizzy**.'

(9)   過不久牠(蟋蟀)會受不了淹水的威脅,偷偷的爬到了洞口,探頭探腦 (*tan-tou-tan-nao*)的查看洞外的風吹草動。(Sinica corpus)

'After a while, it (the cricket) could not stand the threat of flooding so it sneakingly crawled to the top of the hole, **craning its neck to detect** any movements outside the hole.'

Besides, if the AA is a negator like *mei* 沒 'no', then *mei-tou-mei-nao* means 'without a clue'.

(10)   貪官愣愣忘了詞, 觀眾忘了戲, 阿小跑出茶館。阿小沒頭沒腦 (***mei-tou-mei-nao***) 的跑, 兩條長辮飛濺身後, 是黑色的淚痕。(Sinica corpus)

'The corrupt official on the stage blanked out forgetting his lines. The audience forget that they were watching the play. A-xiao ran out of the teahouse. A-xiao kept running **without any clue**, her two braids tossing into the air like two lines of black tear.'

(11)   隔壁家的小狗, 沒頭沒腦 (***mei-tou-mei-nao***) 的對我狂吠, 嚇我一大跳。(優學網)

'The dog next door barked at me **without any clue**, frightening me suddenly.'

Although Tsou pointed out that QIEs can behave as an NP, AP or VP, he had not yet dwelt further on the interaction between the internal structure and the part of speech of a QIE [10]. Our findings show that there is usually an incoherency between a QIE's internal structure and syntactic category in a given context. For instance, *zei-tou-zei-nao* 賊頭賊腦 'thief-head-thief-brain' can be taken as an NP compound based on its internal structure. However, in real usage, it could function as an adjective or an adverb. We thus propose that the four-character structure as a whole (i.e., an abstract construction) may impose some extra meaning on the QIE itself so that it can change its syntactic category to go with the context. Figure 1 shows this interaction between the internal structure and the overall syntactic category of the QIE, exemplified by *ben-tou-ben-nao* 笨頭笨腦 'stupid-head-dumb-brain; stupid'. This indeed constitutes a two-fold question regarding the analysis of a Chinese QIE. The first level concerns the transparency of the internal structure of a QIE. If the compositionality of a QIE is clear, then its transparency is high and its meaning is entirely predictable. The second level concerns pragmatic aspects of a sentence where the QIE is used. In this case, the QIE is taken as a chunk representing a unified meaning. It then can change its syntactic category to fit the context if necessary, as shown in Fig. 1. The operation could be seen as a sort of type coercion in Pustejovsky's sense [16].

One of the anonymous reviewers has kindly reminded us that Chinese nouns can be used as attributes without any overt syntactic marking, and if adding the adverbial suffix *de* 地, such nouns can be converted to adverbs naturally. Lu Shuxiang has also pointed out that there are three ways of producing adverbial phrases out of noun phrases, if they are (a) temporal nouns; (b) noun phrases that depict time, location, or locative phrase; (c) nouns if the adverbial suffix *de* 地 is added [17]. Since the internal structures of Chinese QIEs are abundantly varied, the syntactic categories of their composites do not necessarily have to be nouns. However, once a QIE is used in a sentence, its part of speech can be easily realized according to its context or slot in the sentence. In other words, a QIE could be *quasi-nominal* at first glance, but then it

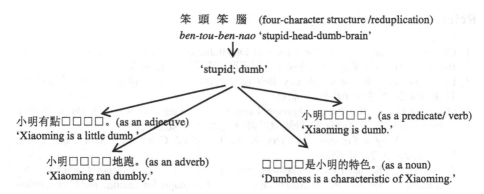

Fig. 1. The interaction between the internal structure and the syntactic category of *ben-tou-ben-nao*

adjusts its part of speech to match the context. Therefore, we can state that the slot in the sentence can "coerce" an emergent part of speech onto the QIE in question. In this sense, the constructional meaning of the QIE is thus apparent as a chunk.

6 Concluding Remarks

Our findings have confirmed the generality of constructional meanings for researching Chinese QIEs as shown in AXAY QIEs. Also confirmed to the QIE research are the crucial contributions of lexical semantics. A well-interpreted QIE results from an integrated understanding of the lexical side (e.g., sense relations), structural side (e.g., constructional meanings), and cultural background knowledge. If a QIE displays a parallel structure, its construction could bring an abstract sense of emphasis to itself, as one of the effects of reduplication. As a case study, we find that the syntactic category of a AXAY QIE usually operates upon the organization of its internal structure. If X or Y is a noun like tounao 'brain', it is highly possible that the *A-tou-A-nao* is a noun, too. It is also possible that *A-tou-A-nao* QIE turns to behave like an adjective or even an adverb such as *ben-tou-ben-nao* and *zei-tou-zei-nao*. To fully understand and use a Chinese QIE, one must take into account not only a semantic analysis of the QIE, but also an adequate understanding of its context. This duality of part of speech found in a QIE may be the most fascinating and mysterious characteristic of Chinese QIEs awaiting future investigations.

Acknowledgments. This paper is partially supported by research grant from the Ministry of Science and Technology, Taiwan (Project numbers: 105-2420-H-002-009-MY2 and 108-2410-H-002-059). The authors would like to express their gratitude for this support.

210   C. Lu et al.

References

1. Cheng, X.: Studies on Chinese Historical Disyllabic compounds. The Commercial Press, Beijing (2003). (程湘清. 漢語史書複音詞研究. 商務印書館, 北京)
2. Fu, Z.: Modern Chinese Vocabulary. Beijing University Press, Beijing (2003). (符淮青. 現代漢語詞彙. 北京大學出版社, 北京)
3. Ge, B.: Chinese Morphology, Foreign Language Teaching and Research Press, Beijing (2006). (葛本儀. 漢語詞彙學. 外語教學與研究出版社, 北京)
4. Su, I.-W.: Why a Construction—That is the Question! Concentric: Stud. English Lit. Linguist. **28**(2), 27–42 (2002)
5. Tsou, B.K.: Idiomaticity and classical traditions in some East Asian Languages. In: Proceedings of the 26th Pacific Asia Conference on Language, Information and Computation (PACLIC 26), pp. 39–55 (2012)
6. Lu, C., Liu, T.-H., Su, I.-W., Tsai, I.-N.: The polysemy of the four-character idiomatic construction 'yi-X-#-Y' in Mandarin Chinese. In: 18th Chinese Lexical Semantics Workshop (CLSW 2017), Sichuan, China, 18–20 May 2017. (呂佳蓉, 劉德馨, 蘇以文, 蔡宜妮: 論漢語四字格「一X#Y」的構式網絡)
7. Lu, C., Tsai, I.-N., Su, I.-W., Liu, T.-H.: Internal structures and constructional meanings: 'Da-X-da-Y' and its related constructions in mandarin Chinese. In: Hong, J.-F., Su, Q., Wu, J.-S. (eds.) CLSW 2018. LNCS (LNAI), vol. 11173, pp. 91–106. Springer, Cham (2018). https://doi.org/10.1007/978-3-030-04015-4_8
8. Hu, Y.: Modern Chinese. Xinwenfeng Publisher, Taipei (1992). (胡裕樹. 現代漢語. 新文豐, 台北)
9. Liu, Z., Xing, M.: The semantic symmetrical features of four-character idioms in Chinese and their effects on cognition, Chin. Teach. World, **1**, 77–81 (2000). (劉振前, 邢梅萍. 漢語四字格成語語義結構的對稱性和認知, 世界漢語教學)
10. Tsou, B.K.: Cantonese quadra-syllabic idiomatic expressions: their conventions and innovations. In: Lee, H. (ed.) Papers on "Quadra-syllabic Idiomatic Expressionsin Languages around South China", pp. 139–158. The Chinese University of Hong Kong, Hong Kong (2016). (粵語四字格慣用語: 傳承與創新, 李行德編. 百越語言研究: 中國南方語言四音節慣用語論文集. 香港中文大學, 香港)
11. Fillmore, C.J., Kay, P., O'Connor, M.C.: Regularity and idiomaticity in grammatical constructions. The case of let alone. Language **64**(3), 501–538 (1988)
12. Goldberg, A.: Constructions: A Construction Grammar Approach to Argument Structure. University of Chicago Press, Chicago (1995)
13. Bybee, J.L.: From usage to grammar: the mind's response to repetition. Language **82**(4), 711–733 (2006)
14. Bybee, J.L.: Language, Usage and Cognition. Cambridge University Press, New York (2010)
15. Croft, W.: Radical Construction Grammar. Syntactic Theory in Typological Perspective. Oxford University Press, Oxford (2001)
16. Pustejovsky, J.: The Generative Lexicon. MIT Press, Cambridge (1995)
17. Lu, S.: Eight hundred words in modern Chinese (revised edition). Shang Wu Publisher, Beijing (1999). (呂叔湘. 現代汉语八百词 (增订本). 商务印书馆, 北京)

A Study on Classification of Monosyllabic and Disyllabic Onomatopoeias Based on the Relation Between the Form and Meaning

Bo Xu$^{(\boxtimes)}$ and Zezhi Zheng

Xiamen University, Xiamen 361005, Fujian, China
xubo_xmu@126.com

Abstract. There is a certain connection between the form and meaning of the onomatopoeia, so its classification may include both formal and semantic criterion. This paper mainly studies on monosyllabic and disyllabic onomatopoeias in the *Modern Chinese Dictionary* (6th edition), summarizes three semantic description perspectives as object of sound production, sound features and action features from the definitions of onomatopoeias in dictionary, and initially classifies them as simple onomatopoeias, compound featured onomatopoeias (with sound feature and action feature) and sound featured onomatopoeias. Combined with the formal criteria, this paper further classifies monosyllabic and disyllabic onomatopoeias as A type simple onomatopoeias, A type sound featured onomatopoeias, A type compound featured onomatopoeias, AA type sound featured onomatopoeias, AA type compound featured onomatopoeias, AB type simple onomatopoeias, AB type sound featured onomatopoeias, AB type compound featured onomatopoeias. Based on the classification, this paper also discusses the semantic, structural characteristics and relations between form and meaning of each type.

Keywords: Onomatopoeia · Classification · Relation between the form and meaning · Modern Chinese · Definition

1 Introduction

As suggested by Meng Cong, the phonetic form of onomatopoeia has a certain objective basis, for this type of words were created by direct imitation of people's own vocal organs towards external sounds. Having a certain connection between its form and meaning rather than being established by usage, it is the feature of onomatopoeia that distinguishes it from most words in modern Chinese vocabulary.

The current classification of onomatopoeia covers three main perspectives (form, meaning and syntactic function), and the specific criteria include the internal structure [2], phonetic structure, sound source [1], sound feature [3], adjectival and verbal [4], *etc.*

The existing studies of onomatopoeia usually classified it from a single perspective of above criteria. However, regarding the correlation between the form and meaning, we insist that the classification of onomatopoeia could combine both form and

© Springer Nature Switzerland AG 2020
J.-F. Hong et al. (Eds.): CLSW 2019, LNAI 11831, pp. 211–219, 2020.
https://doi.org/10.1007/978-3-030-38189-9_22

meaning. According to the statistics provided by Chu Taisong, the monosyllabic and disyllabic onomatopoeias in *Modern Chinese Dictionary* (6th Edition) account for more than 93% of the whole onomatopoeias, whether or not including the heteronyms [5]. Considering the representativeness of onomatopoeia, this study selected the monosyllabic and disyllabic ones in the dictionary (excluding the obvious dialectal or literary words, and words with part of speech transition in use) as the object, to analyze and discuss the classification.

2   Semantic Description Perspectives and Classification

What is the basis of imitating sounds of modern Chinese onomatopoeia? What are the specific perspectives in describing sounds? Both questions could be answered by analyzing the definitions of onomatopoeic words in *Modern Chinese Dictionary*.

Yuan Mingjun concluded the definition modes of onomatopoeic words in *Modern Chinese Dictionary* to eight types [6], from which we summarized the three main perspectives of onomatopoeia in depicting different sounds: **object of sound production** (where is the sound from), **sound feature** (what does the sound like) and **action feature** (how does the sound produce).

2.1   Semantic Description Perspectives of the Onomatopoeia

For onomatopoeia with a highlight on objects that produce sounds, such as the doorbell sound "*ding-dong*" (叮咚), dogs' barking "*wang*" (汪) and sheep's bleating "mie" (咩), it is featured as simple, highly recognizable, easy to imitate, and has a single and stable connection with the object that produces it. Though easy to imitate and recognize, this type of onomatopoeia enjoys a small percentage among all the onomatopoeic words.

Limited to people's auditory perception, vocal organs and phonological systems of different languages, the sound which onomatopoeia imitates usually not merely corresponds to a certain object that produces sound. For instance, "*hong-long*" (轰隆) is used to describe the sounds of thunder, oversize vehicles, the roar of engines, the crash of buildings, *etc.*, that is, this syllable could imitate a type of sounds with the "loud, heavy and muffled" features. Here the description of sound is not a direct imitation towards external sounds but a re-construction based on people's cognition in sound features. The recognizable sound features include the pitch, velocity, continuity and pause, volume and intensity, and so on. To be specific, "*keng*" (铿) is used to describe the resonant sound, "*zhi*" (吱) used for sharp sound. This type of onomatopoeia is of most significant multi-functionality, for each word corresponds to various objects that produce sounds; however, it is impossible to determine how sounds produce from this type of imitation.

For sounds that are difficult to imitate directly, neither categorizing sound features could make the imitation accurately, it is necessary to introduce action features of sounds for categorization to meet the requirement of accurate sound imitation. For instance, the sounds of throwing an iron ball and a discus (of same weight) from the same height, share the same features (short and clear), but are distinguished as a hit on a point and a hit on a plane in terms of the action features. Thus, they are imitated as "*da*"

(嗒) and "*pa*" (啪) by different initial consonants. This type emphasizes action features, but the imitations of sounds represent both action and sound features, such as the sound of knocking wood "*bang*" (梆), the initial consonant shows that the sound comes from collision, and the vowel reveals the bright sound feature. The multifunctional attribute of this type is weaker than that of last type, but its meaning is more explicit.

2.2 Semantic Classification of the Onomatopoeia

Above analysis differentiates the onomatopoeia as three types. This semantic classification presents a clear hierarchy, that it firstly distinguishes the onomatopoeia as simple ones and featured ones, then divides the featured onomatopoeia into compound featured ones (with sound and action feature) and sound featured ones (See Fig. 1).

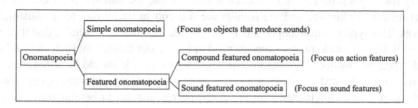

Fig. 1. Semantic classification of onomatopoeia

For the simple onomatopoeia, there are three main characteristics: (1) the words directly correspond to the objects that produce sounds; (2) the sound source is hard to determine, in other word, the action that make the sound is unknown; (3) the sound feature is difficult to summarize. The essential attribute of simple onomatopoeic words is imitation, that the relation between phonetic form and meaning of this type is direct and objective. Therefore, simple onomatopoeic words embody the universality of languages, such as the imitation of cats' sound ("喵/miao" in Chinese and "miaow" in English) and dogs' bark ("汪汪/wang wang" in Chinese and "ワンワン/uan uan" in Japanese) are roughly the same. Unlike simple onomatopoeia, the essential attribute of featured onomatopoeia is categorization, that the relation between phonetic form and meaning is established through the process of categorization, thus this type of words is usually of strong subjectivity and tends to reflect national features.

3 Classification of Monosyllabic Onomatopoeia

As referred in previous study, the formal classification mainly focuses on numbers and constitution of the syllables. Combining the above-mentioned semantic classification, the monosyllabic onomatopoeia could be divided into A type simple onomatopoeia, A type sound featured onomatopoeia and A type compound featured onomatopoeia.

3.1 A Type Simple Onomatopoeia

This type of onomatopoeia mainly includes animals' sounds with distinctive features (see e.g. (1)) or human's sounds produced for physiological reasons (see e.g. (2)):

e.g. (1) 叽/ji (sound of chicken or birds)　　喵/miao (miaow/sound of cats)
　　　咩/mie (bleat/sound of sheep)　　　　哞/mou (moo/sound of cows)
　　　咕/gu (coo/sound of hens or turtledoves)　喔/wo (crow/sound of roosters)
　　　汪/wang (barking/ sound of dogs)　　　喳②/zha (charp/ birds' sound)
e.g. (2) 喀/ka (sound of vomiting or coughing)　哇/wa (sound of vomiting)
　　　哕/yue (sound of vomiting, that come out of one's mouth)

It could be concluded from above examples that A type simple onomatopoeia usually has explicit objects that produce sounds, while the sounds' production mechanism remains unknown, and the sounds are difficult to categorize for its distinctive features. This type of onomatopoeia is most flexible in syntactic position, that it can be used as independent element, or repeats randomly without limitation, such as "wo" (喔) could repeat as "wo-wo" "wo-wo-wo" or "wo-wo-wo-wo". It should be noted that, this "wo-wo" is a non-word element but a successive use of two same onomatopoeic words, which should be distinguished with the AA type onomatopoeia hereinafter.

3.2 A Type Sound Featured Onomatopoeia

This type of words use a single syllable to imitate the sound features, that usually represented by adjective morphemes. There is only a small amount of sound featured onomatopoeic words in *Modern Chinese Dictionary*. (See below examples)

e.g. (3) 轰/hong (loud sound of thunder or blasting)　哄①/hong (roars of laughter or big noise)

　　哄②/hong (sound made by many people)　　訇/hong (loud sound)
e.g. (4) 嘻/xi- 嚯/huo -哈/ha (sound of laughter)

In the third group of examples, the four words choose the same syllable to imitate the "loud/big/heavy" sound. The feature "loud/big/heavy" reflects people's common perception upon the sounds of thunder, blasting, roars of laughter and so on. Based on this common understanding, people relate this type of sound with the syllable hong [xuŋ], that has the similar sound feature of loud, big and heavy.

For the three words in the fourth group, the dictionary gives the same definition without providing any adjective morpheme as explanations of their sound features. However, they have distinctive features in fact, that "xi" uses to describe a slight and subtle chuckle, "huo" uses for a heavy laugh in a low voice, and "ha" for a bright and loud laugh, here the "slight" "low" and "loud" are the distinctive sound features. These words are all used for describing laughter, but the different sound features endow them with the value of existence as well as forming a categorization in expressing laughter.

3.3   A Type Compound Featured Onomatopoeia

This type of words, which presents a comprehensive understanding of both sound and action features within a single syllable, accounts for a large percentage among the monosyllabic onomatopoeias. Usually the initial consonant and head vowel show the action features of the sound while the essential vowel and tail vowel reveal the sound features.

> e.g. (5) 叭/ba (sound of gunshots)   啪/pa (sound of firing, clapping hands or collision)
>
> e.g. (6) 嗖/sou (sound of passing through very quickly)
>
> 唰/shua (sound of scraping along something very quickly)
>
> 哗/hua (sound of collision or water flow)

In the fifth group of examples, both the two initial consonant b[p] and p[p^h] are plosive, which show iconicity with some actions of colliding, cracking, spraying, exploding and so on. They are also bilabial in terms of articulation place, that the contact of labial surface is propitious to imitate the action of colliding between plane objects. Since the lips are in the outside position among all the vocal organs, they are especially appropriate for imitating the spraying and shooting sound. Meanwhile, the same low vowel a[A] is suitable for imitating the loud and brief sound, for its articulation characteristic.

In the sixth group, the three initial consonants are dental fricative s[s], blade-palatal fricative sh[ʂ] and velar fricative h[x] respectively, which are all applied for imitating the sound of quick and smooth action, for the articulation of fricative allows airflow going through easily and smoothly. Among the three fricatives, [x] has the minimum obstruction of airflow so that it is especially good for imitating the blowing (of air) or flowing (of liquid) sound. As for the vowels, rounded and compound vowels ou[ou] of "*sou*" presents the continuous action feature as well as the unhindered sound feature (without obstruction), while the head vowel u[u] of "*shua*" "*hua*" depicts the smooth action feature, and the essential vowel a[A] still uses to imitate the loud and brief sound.

Both the initial consonants and vowels of this type have certain corresponding sound types: the obstructing feature of consonants provides conditions for them to imitate the action feature, while the different tongue positions of vowels are appropriate to describe sound duration, pitch and volume. The correspondences mainly based on the iconicity between the sequence in syllables and the process of sounds' production.

4   Classification of Disyllabic Onomatopoeia

With the introduction of the semantic classification, the disyllabic onomatopoeia could be classified into five types which the AA structure has the function of indicating continuity and visualizing. Indicating continuity refers that the successive and repeated syllables were used to imitate the continuous sound feature. While visualizing function means the AA structure highlighted and visualized a certain sound feature by its reduplicated form, for “the reduplicated form has distinct image color to express the senses of sight, hearing, taste and touch through listeners’ thinking in image” [7].

4.1 AA Type Sound Featured Onomatopoeia

The reduplicated form of AA type sound featured onomatopoeia could imitate the continuous sound as well as strengthening the vividness of sound features. Usually an adjective morpheme, which is the concentrated reflection of people’s perception upon the sounds, could be extracted (concluded) from the definition of each word in this type.

e.g. (7) 溹溹/suo suo (rapid and coherent sound) 喃喃/nan nan (sound of whispering)
咝咝/ si si (slight and continuous sound)
e.g. (8) 索索/suo suo (slight sound) 隆隆/long long (sound of violent vibration)

All the sounds in seventh group have continuous feature. Meanwhile, the “rapid” meaning of “*suo suo*”, “low (voice)” meaning of “*nan nan*”, “slight/subtle” meaning of “*si si*” are presented by their different consonants and vowels, while the “slight” and “violent” adjectival sound features could be concluded from the definitions of “*suo suo*” and “*long long*”, and these sound features are strengthened by the AA structure as well.

4.2 AA Type Compound Featured Onomatopoeia

Besides the structural meaning of imitating continuous sounds or strengthening the vividness of sounds, AA type compound featured onomatopoeia also provides a comprehensive understanding of action and sound features within its every single syllable.

e.g. (9) 辘辘/lu lu (sound of wheel rolling)
e.g. (10) 铿铿keng keng (sound of mental clashing)

In the ninth example, the AA structure imitates the continuous feature of wheel rolling. Meanwhile, the lateral sound l[l] of “*lu lu*”, with the air almost directly flowing from the lung, is also used to imitate the rolling feature, and the rounded vowel u[u] is used to correspond the relatively smooth and stable sound features. For “*keng keng*” in tenth example, it uses the velar plosive k[k^h] to imitate the clashing feature, for the obstructive position of articulating the consonant is at the back. While the vowel eng [eŋ], also with a back articulation position, is used to imitate the heavy sound feature.

4.3 AB Type Simple Onomatopoeia

This type of onomatopoeia mainly refers to distinctive sounds of animals (see e.g. (11)) or human’s sounds produced for physiological reasons (see e.g. (12)), but accounts for a very small number among the onomatopoeic words.

e.g. (11) 布谷/bu gu (sound of cuckoo)

e.g. (12) 阿嚏/a ti (sound of sneezing)

Since this type of onomatopoeia imitates the external sounds directly, its disyllabic structure has no extra structural meaning, thus the structure itself is inseparable.

4.4 AB Type Sound Featured Onomatopoeia

The structure of AB type sound featured onomatopoeia has an additional meaning of complementation, that the two syllables A and B are usually share the same initial consonant but different vowels, or with same vowels but different initial consonants. Even in the case that neither consonants nor vowels are same, there is antagonistic relation or approximate relation between the initial consonant and the vowels.

e.g. (13) 咿呀②/yi ya (sound of children's babble)

e.g. (14) 轰隆/ hong long (sound of thunder, blasting or machines)

e.g. (15) 呼啦/ hu la (sound of sudden collapse, or rapid gathering and dispersal of the crowd)

In e.g. (13), the zero-initialed syllables "*yi ya*" have two distinctively different vowels [i] and [iA]. In e.g. (14), "*hong long*" has the same vowels, but the articulation positions and manners of the two initial consonants distinctively differentiate with each other. For "*hu la*" in e.g. (15), neither the initial consonants nor the vowels are the same. However, there is an approximate relationship between the fricative [x] and lateral [l], for the two consonants are both sounds without obvious obstruction; and the high vowel [u] and low vowel [A] have great difference in tongue position, which form an antagonistic relationship. Here the sameness and similarities present the similarity of the sounds (for imitating a type of sound), while the differences and contrarieties strengthen the generality of onomatopoeia (to determine the range of a sound type).

4.5 AB Type Compound Featured Onomatopoeia

The structural meaning of AB type compound featured onomatopoeia could be divided into two types: to highlight the action state and to highlight the description.

Highlight the Action State

For AB type compound featured onomatopoeic words that represent action states, their first syllable A shows the action feature while second syllable B describes sound feature. In most cases, each word of this type has a corresponding A type compound featured onomatopoeia (with same meaning but weaker in dynamic characteristic).

e.g. (16) 但是他刚飞了进去，方孔上的小盖板就吧嗒 (*ba da*)一声盖上了，这样巴塔基就被关在里面了。   （当代/翻译作品/文学/尼尔斯骑鹅旅行记）

e.g. (17) 我立刻把车钥匙插进去，锁打开了，这嗒 (*da*) 的一声算是为我平反恢复名誉。③   （当代/报刊/作家文摘/1997/1997B）

Both the "*ba da*" and "*da*" in above examples describe the sound of slight impact, and the difference is that the transition between the two syllables "*ba da*" provides a better presentation of the sound source's dynamic characteristic. The initial consonant b [p] of the first syllable imitates the impact of plane objects (sound source/action feature), and d[t] of the second syllable shows the plosive sound feature; furthermore, the spatial sequence of the two syllables maps directly to the time sequence of sound source and sound. In contrast, limited to its monosyllabic form, "*da*" concentrates the attributes of sound source and sound into one initial consonant d[t], as well as combining the complex sound feature into one vowel a[A]; therefore, this monosyllabic onomatopoeia cannot imitate the time sequence of sound source – sound, thus its dynamic characteristic is evidently weaker than that of "*ba da*".

Highlight the Description
The number of this type is very small. For this type of compound featured onomatopoeia, neither initial consonants nor vowels of the two syllables are the same, both the syllables A and B refer to description upon the features of sound source, and there is a time sequence of A and B (See below examples).

e.g. (18) 咯吱/ge zhi (sound of pushing bamboo and wood, usually used as reduplicated form)

扑哧/pu chi (sound of laughter, or squeezing water or gas)

As for "*ge zhi*", the first syllable ge[kə] imitates the crashing between two objects, while the second syllable zhi[tʂʅ] imitates the continuous action of pushing. For "*pu chi*", the first syllable pu[pʰu] imitates the action of gas eruption, while the second syllable chi[tʂʰʅ] imitates the continuous squeezing of water and gas. The two syllables corporate with each other to express the action feature of sounds completely, and the time sequence of the imitated action (crash and then push, erupt and then squeeze continuously) is in accordance with the spatial sequence of the corresponding syllables.

5   Conclusion

The authors summarize the three semantic description perspectives – objects, that produce sounds, sound features and action features, and discuss the characteristics of onomatopoeia combing above-mentioned perspectives. Based on these semantic standards, the onomatopoeia is primarily classified as simple onomatopoeia (imitates sounds directly) and featured onomatopoeia (refers to re-construction based on people's cognition upon external sounds), and among which the featured onomatopoeia is distinguished as compound featured onomatopoeia (presents both sound and action features) and sound featured onomatopoeia (presents sound features only). By introducing the formal standard, the authors further classify monosyllabic and disyllabic onomatopoeia into eight types, and illustrate the form-meaning relationship of each type.

Compared to previous single-perspective studies on onomatopoeia's classification, this paper observes the modern Chinese onomatopoeia from the point of the form-meaning relation, establishes a semantic classifying framework tentatively, and combines the formal and semantic standards for further classification and discussion, to enrich the research view on onomatopoeia's classification. However, since this paper only discusses about the monosyllabic and disyllabic onomatopoeia, this new classification don't cover all the formal types of the onomatopoeia, which needs further research in the future. Meanwhile, the authors find that parts of onomatopoeic words' definitions in the dictionary don't highlight or at least illustrate their semantic features when selecting examples for each type, and for some polysemous onomatopoeic words, the dictionary don't separate the senses but mixes the definition. Aiming at these existing problems, a set of normal and rational definition modes of onomatopoeia in consideration of each types' characteristics should be discussed and concluded in future study.

References

1. Meng, C.: The onomatopoeia in Beijing Dialect. In: Magazine Office of Studies of the Chinese Language (ed.) Research and Exploration on Grammar I, pp. 120–156. Peking University Press, Beijing (1983)
2. Shao, J.M.: A tentative study of the onomatopoeia. Lang. Teach. Linguist. Stud. **4**, 57–66 (1981)
3. Zhang, J.: Speeches on Vocabulary Teaching. Hubei People's Publishing House, Wuhan (1957)
4. Ha, S.: Modern Chinese onomatopoeia. J. Inner Mongolia Normal Univ. (Philos. Soc. Sci.) **5**, 21–25 (2002)
5. Chu, T.S.: Phonetic principles and exceptions of mandarin onomatopoeia. J. An Hui Normal Univ. (Humanit. Soc. Sci.) **1**, 107–112 (2012)
6. Yuan, M.J.: The onomatopoeic words in Modern Chinese Dictionary. In: 6th Chinese Lexical Semantic Workshop, Xiamen, p. 6 (2005)
7. Shi, Q.: A study on reduplicated form of Chinese adjectives. J. Wuhan Univ. Technol. **4**, 618–622 (2005)

Semantic Features and Internal Differences of Ergative Verbs

Fan Jie[(⊠)]

School of Literature and Journalism and Communication,
Xihua University, Chengdu, China
fanjie_219@qq.com

Abstract. The ergative phenomenon in Chinese involves many hot-debated issues in the study of Chinese syntax and semantics, which was explored from various perspectives in previous researches. In this paper, a total number of 123 ergative verbs are sorted out in terms of their syntactic representation, semantic types and semantic features, and result in three findings. Firstly, the common semantic feature of ergative verbs is the meaning of change. Secondly, there are obvious internal differences in transitivity, causativity and volition of ergative verbs: unary ergative verbs indicate spontaneous and uncontrollable changes in events, with low transitivity and obvious non-volitional tendency; binary ergative verbs have higher transitivity and obvious causative tendency. Thirdly, the two relevant structures, 'S+V+N' and 'N+V', represent different stages before and after the change. The former structure represents the origin or motive force of the change, while the latter represents the state after the change. A temporal sequence and logical causality exist between them.

Keywords: Ergative verbs · Semantic features · Transitivity · Causative · Volitional

1   Introduction

Ergativity, originally a feature of linguistic type, refers to the phenomenon that the object of a transitive verb and the subject of an intransitive verb are identical. The subject of transitive verb is different from the ergative, which is called 'absolutive'. The ergative-absolutive language is regarded as a different language type, which is opposed to the nominative-accusative language. The [1] initiates the study of ergative verbs in Chinese by discussing the two types of verbs as *sheng* (胜) and *bai* (败). Generally, the Chinese ergative phenomenon refers to the fact that the unique argument of some intransitive verbs can appear both at the subject and object positions, for example (1).

This research is supported by the General Program of National Natural Science Foundation of China (NSFC): Research on the Mechanism and Model of the Interactive Development between Multiple Industry Forms of Digital Creative Products (Grant No. 71874142).

J.-F. Hong et al. (Eds.): CLSW 2019, LNAI 11831, pp. 220–233, 2020.
https://doi.org/10.1007/978-3-030-38189-9_23

(1)a. *Da   bai   kedui.* (大败客队)——*Kedui   da   bai.* (客队大败)
　　 Great defeat visiting team   ——Visiting team   great defeat
　　 (They) defeated the visiting team badly. ——The visiting team failed badly.

 b. *Lai   keren le.* (来客人了)——*Keren   lai le.* (客人来了)
　　 Come   a guest   ——   The guest   came
　　 Here comes a guest. —— The guest came.

 c. *Kai   yuanmen le.* (开院门了)——*Yuanmen   kai le.* (院门开了)
　　 Open gate of yard   ——Gate of yard   opened
　　 (They) opened the gate of yard.—— The gate of courtyard is open.

The verbs in the two sentences in Example (1) are same, and the noun, as the unique argument of the verb, is in different syntactic positions of subject and object in the two sentences. Although there is no case marker in Chinese, the phenomenon above does exist. It is evident that the verbs in the two sentences are different from each other in terms of event structures, perspectives and foci.

Previous research has made a thorough study on ergative phenomenon in Chinese from the perspective of the relationship between the role of the thesis and the syntactic components in [1–4]. There are abundant discussions on the assignment of themes and the sentence generation, and many discoveries on transitivity, passive, existential sentences, verb-result construction and causative transformation are also insightful, such as [5–10].

As far as ergative phenomenon is concerned, one of the questions researchers have been trying to explain is why ergative verbs can enter two different structures of subject-object transposition. This paper tries to analyze this issue from the perspective of verb semantics. By observing the common semantic characteristics of this kind of verbs, the semantic basis of this syntactic phenomenon is revealed. However, at present, the study of ergative verbs in Chinese has been mostly focused on typical verbs and typical examples. The sorting of ergative verbs based on the common verb lexicon is in scarcity. Thus, the first task of this paper is to search and collect the ergative verbs in Chinese comprehensively. On the basis of observing the syntax and semantics of a considerable number of ergative verbs, it is possible to identify the common semantic basis of ergative phenomenon. This paper aims to select out verbs from *the Dictionary of the Usage of Chinese Verbs,* and investigate their syntactic and semantic features.

2   Semantic Categories of Ergative Verbs

The ergative phenomenon originally needs to be judged by the form of cases, but there is no case marker in Chinese. The so-called ergative phenomenon is only 'a kind of comparison' in a sense in [1]. Therefore, the judgment of Chinese ergative verbs is a kind of semantic ergative rather than a syntactic ergative. The key point of Chinese ergative verbs lies in the phenomenon of the unique argument subject-object transposition. If the position is the most important marker of Chinese syntactic components, the subject-object trans-position is the most dominant basis for judging Chinese ergative verbs. This paper identifies ergative verbs by referring to the criteria of [9].

Firstly, the verb with same meaning (or closely related meaning) can enter the two structures of '(S+) V+N' and 'N+V'.

Secondly, the verb denotes the state of the noun N in the structure 'N+V'.

On the basis of the two points above, this paper collects 123 ergative verbs based on *The Dictionary of Chinese Verb Usage*, and roughly sorts them according to semantic types. The previous researcher have identified some semantic types of ergative verbs, such as [11], who classified ergative verbs into four categories, including 'presenting', 'feeling', 'meteorological' verbs, existential sentence verbs, and 'breaking' verbs. The [4] suggests that ergative verbs include 'directed' verbs, some state verbs. Generally, previous researchers have pointed out some important aspects of the semantic categories of ergative verbs, but there is still some work needs to be done. The 123 ergative verbs collected in this paper can be roughly divided into the following six semantic types.

A. *zui* type: *zui* 'drunk', *e* 'hungry', *lei* 'tire', *xing* 'wake up', *fan* 'trouble', *chou* 'worry', *ji* 'worry', *gandong* 'touch', *dong* 'move', *liulu* 'reveal', *zhendong* 'shock';

B. *dao* type: *dao* 'pour', *dao* 'reach', *xia* 'descend', *diao* 'drop', *jiang* 'descend', *sheng* 'ascend', *tui* 'fall', *jihe* 'gather', *liu* 'flow', *lai* 'come', *zou* 'go', *liu* 'remain', *tang* 'flow', *liutang* 'flow';

C. *jia* type: *jia* 'increase', *jian* 'decrease', *jianshao* 'decrease', *kuoda* 'amplify', *suoxiao* 'reduce', *jiangdi* 'reduce', *zengjia* 'increase', *zengzhang* 'increase', *zengda* 'enlarge', *yanchang* 'extend', *jilei* 'accumulate';

D. *cheng* type: *cheng* 'succeed', *chengli* 'establish', *chuban* 'publish', *miqie* 'close', *diandao* 'overthrow', *jiefang* 'liberate';

E. *fan* type: *dao* 'collapse', *fan* 'reverse', *chen* 'sink', *zhuan* 'turn', *kai* 'open', *guan* 'close', *bi* 'close', *he* 'close', *li* 'erect', *gun* 'roll', *zhendong* 'shock', *yaohuang* 'rock', *kao* 'depend on', *yi* 'depend on', *zuo* 'sit', *tang* 'lie', *pao* 'run';

F. *bai* type: *bai* 'defeat', *diao* 'fall', *diu* 'lose', *po* 'break', *lan* 'rot', *chuan* 'penetrate', *fei* 'scrap', *tou* 'penetrate', *kua* 'collapse', *ta(1)* 'collapse', *ta(2)* 'sink', *duan(1)* 'cut off', *duan(2)* 'break off', *jin* 'tighten', *song* 'loosen', *juan* 'roll up', *wan* 'roll up', *suo* 'draw back', *si* 'die', *mohu* 'obscure', *mie(1)* 'turn off', *mie(2)* 'extin-guish', *hui* 'destroy', *huimie* 'destroy', *fenlie* 'split', *duanjue* 'cut off', *gai* 'change', *bian* 'change', *gaibian* 'change', *gaijin* 'improve', *gaishan* 'improve', *gailiang* 'improve', *tong(1)* 'poke', *tong(2)* 'lead to', *rong* 'melt', *hua* 'melt', *ronghua* 'melt', *jiesan* 'dismiss', *tongyi* 'unify', *fangsong* 'loosen', *xiaohua(1)* 'digest', *xiaohua(2)* 'resolve', *xiaoshi* 'disappear', *suo* 'lock up', *qi* 'appear', *mao* 'appear', *pen* 'spray', *shen* 'leak', *lou* 'leak', *jizhong* 'gather', *zhang* 'grow', *zoulou* 'divulge'.

The criterion for judging ergative verbs is whether they can enter two structures '(S+) V+N' and 'N+V'. However, the syntactic function of these ergative verbs is not entirely consistent. Their expressions in the *ba*-construction, *bei*-construction, exis-tential sentence, 'NP1+NP2+V' structure and topic subject sentences are different. Take the *ba*-construction and *bei*-construction as examples.

(2) N+V   (S+) V+N   ——   *N+*bei*+V   *(S+) +*ba*+N+V

a. *Tao lan le* (桃烂了)—*Lan le jige tao* (烂了几个)—**Tao bei lan le* (桃被烂了)
—* *Ba tao lan le* (把桃烂了)

Peach rot — rot some peach   ——   * Peach *bei* rot—* *Ba* peach rot
The peach is rotten.

b. *Shu dao le* (树倒了)—*Dao le yike shu* (倒了一棵树)—* *Shu bei dao le* (树被倒
了)—* *Ba shu dao le* (把树倒了)

Tree fall down—Fall down tree—— *Tree *bei* fall down—*Ba shu fall down
The tree fell down.

(3) N+V   (S+) V+N   ——   N+*bei*+V   (S+) +*ba*+N+V

a. *Shuzhi yaohuang* (树枝摇晃)— *yaohuang shushi* (摇晃树枝) — *Shuzhi bei
yaohuang* (树枝被摇晃)— *ba shuzhi yaohuang* (把树枝摇晃)

Branch   sway   —   sway   branch——Branch *bei* sway—*ba* branch sway
Branches sway — Shake the branches—— Branches were shaken— Shake
the branches.

b. *Guanzhong gandong le* (观众感动了)—*Gandong le guanzhong* (感动了观众)
—*Guanzhong bei gandong le* (观众被感动了)—*Ba guanzhong gandong* (把观众感
动了)

Audience   moved — moved   Audience——Audience bei moved—Ba
Audience moved

The audience was moved. —Moved the audience.——The audience was
moved. —Moved the audience.

The verbs in Example (2) are incompatible with the *ba*-construction and *bei*-construction, while the verbs in Example (3) are compatible with them. The two groups of verbs are not parallel in syntactic performance, which reflects that there are two subcategories in ergative verbs. According to the arguments of verbs, ergative verbs are divided into two groups: unary ergative verbs and binary ergative verbs.

The verbs in Example (2) are unary ergative verbs, which have only one noun in their argument structure. That is to say, in the two structures of 'S+V+N' and 'N+V', the noun N can appear in both the subject and object positions. Unary ergative verbs are as follows.

Dong 'move', *dao* 'reach', *xia* 'descend', *lai* 'come', *zou* 'go', *liu* 'flow', *tang* 'flow', *liutang* 'flow', *zhang* 'grow', *dao* 'collapse', *lan* 'rot', *chuan* 'penetrate', *fei* 'scrap', *tou* 'penetrate', *cheng* 'succeed', *diao* 'lose', *diu* 'lose', *kua* 'collapse', *ta(1)* 'collapse', *ta(2)* 'sink', *si* 'die', *bian* 'change', *tong(2)* 'lead to', *zengda* 'enlarge', *zengzhang* 'increase', *xing* 'wake up', *kao* 'depend on', *yi* 'depend on', *zuo* 'sit', *tang* 'lie', *pao* 'run', *xiaoshi* 'disappear'.

The binary ergative verb, represented by example (3), has two arguments, which are compatible with the *ba*-construction and *bei*-construction. Binary ergative verbs are shown in the follows.

E 'hungry', *lei* 'tire', *fan* 'upset', *chou* 'worry', *ji* 'worry', *gandong* 'touch', *liulu* 'reveal', *zhendong* 'shock', *dao* 'pour', *jiang* 'descend', *sheng* 'ascend', *tui* 'fall', *jihe* 'gather', *liu* 'remain', *jia* 'increase', *jian* 'decrease', *jianshao* 'decrease', *kuoda* 'amplify', *suoxiao* 'reduce', *jiangdi* 'reduce', *zengjia* 'increase', *yanchang* 'extend', *jilei* 'accumulate', *chengli* 'establish', *chuban* 'publish', *miqie* 'close', *diandao* 'overthrow', *jiefang* 'liberate', *zhuan* 'turn', *kai* 'open', *guan* 'close', *bi* 'close', *he* 'close', *li* 'erect', *gun* 'roll', *zhendong* 'shock', *diu* 'lose', *po* 'break', *fei* 'scrap', *duan(1)* 'cut off', *jin* 'tighten', *song* 'loosen', *juan* 'roll up', *suo* 'draw back', *mohu* 'obscure', *mie (1)* 'turn off', *mie(2)* 'extinguish', *hui* 'destroy', *huimie* 'destroy', *fenlie* 'split', *duanjue* 'cut off', *gai* 'change', *gaibian* 'change', *gaijin* 'improve', *gaishan* 'improve', *gailiang* 'improve', *tong(1)* 'poke', *rong* 'melt', *hua* 'melt', *ronghua* 'melt', *jiesan* 'dismiss', *tongyi* 'unify', *fangsong* 'loosen', *xiaohua(1)* 'digest', *xiaohua(2)* 'resolve', *suo* 'lock up', *qi* 'appear', *pen* 'spray', *shen* 'leak', *jizhong* 'gather', *zoulou* 'divulge'.

The two groups of verbs have different syntactic functions, including selection of structure and subject, volitional, causality and transitivity. In the following sections, the analyses of the syntactic and semantic characteristics of these two types of verbs are presented.

3 Semantic Commonalities and Differences of Ergative Verbs

3.1 Change and State

When the ergative verb enters into the two structures, it reflects two different sentence meanings. The structure '(S+)V +N' denotes the change event, while the structure 'N +V' denotes the state after the change. The change event is the precondition of the state event. A temporal and logical causal relationship exists between the two events, as shown in the following examples.

(4) *Zhangjia chengle zhuang qinshi.* (张家成了桩亲事) — *Qinshi   chengle.* (亲事成了)

      The Zhangs achieve a marriage — A marriage achieve

      The Zhangs brought the marriage to fruition.—The marriage was completed.

      Qiangjiao zhang le ke cao. (墙角长了棵草) — *Cao zhang chulai le.* (草长出来了)

      Corner of wall   grow   a   grass — Grass grow up

      In the corner of the wall, a grass grows. — The grass grew up.

(5) *Zhe shi chousi ta le.* (这事愁死他了) — *Ta chousi le.* (他愁死了)

      This matter upset him — He anxious.

      This matter upset him very much. — He is very upset.

      Ta suo hui le shou. (他缩回了手) — *Shou suo huiqu le.* (手缩回去了)

      He   retract   hand — Hand   retract

      He retracted his hand. — Hands retracted.

The structure 'S+V+N' on the left represents the change event, and the structure 'N+V' on the right represents the state event. In Example (4), the structure 'S+V+N' represents the starting point or scope of the change event without causative meaning, in which the verbs are unary ergative verbs. While in Example (5), the structure 'S+V+N' indicates the starting point or motive force of a change event, and the verbs are binary ergative verbs with the meaning of causation. The structures of 'N+V' in the two groups indicated the changed state.

In other words, the two groups of sentences represent different stages of events' changes. S indicates the starting point of a change event, and V denotes the change. 'N+V' denotes the state after the change. The [6] deduces that the 'S+V+N' is the source structure, and the 'N+V' is derived structure. The [12] points out that the structure 'S+V+N' is unmarked, while 'N+V' is marked. From the perspective of the internal time and logical relationship of events, it is consistent with the above conclusions.

The two basic structures of ergative verbs express the whole process of things' changes, which requires that the verb semantics must be consistent with it and have the characteristic of the change. These features can be highlighted by some syntactic means. Therefore, ergative phenomenon puts forward requirements and restrictions on verb semantics.

There are three different types of events, representing three different stages of the change of things. They are process, transition and state in the theory of event structure of [13]. 'Transition' means that things are in two heterogeneous states before and after the change, namely, E1 and -E2. 'State' means that things are in a homogeneous and unchanged situation. 'Process' refers to the progressive stage of action.

Ergative verbs show a distinct changing feature in their semantics. From the perspective of event type, most of them can be regarded as 'transition' verbs. Not being able to co-occur with 'zhengzai(正在)' is the important judgment basis of 'transition' verbs (instantaneous verbs). Some psychological verbs have been regarded as 'state' verbs, such as 'chou (worry), fan (upset), ji (worry), gandong (touch)'. But in fact, these verbs are obviously different from typical state verbs. Typical state verbs, such as 'ai (love), hen (hate), danyuo (worry)' and so on, cannot enter the structure 'N+V' because they have no causativity in structure 'S+V+N'. However, the verbs as 'chou (worry), fan (upset), ji (worry), gandong (touch)' can do. Examples are shown below.

(6) a. *Xiao Zhang chousi Xiao Wang le.* (小张愁死小王了) — *Xiao Wang chou si le* (小王愁死了).

Xiao Zhang worry much Xiao Wang — Xiao Wang worry much
Xiao Zhang worries Xiao Wang very much. — Xiao Wang very worries.

b. *Xiao Zhang aisi Xiaowangle.* (小张爱死小王了) — * *Xiao Wang aisi le.* (*小王爱死了)

Xiao Zhang love much Xiao Wang   —* Xiao Wang love much
Xiao Zhang loves Xiao Wang very much.——* Xiao Wang loves very much.

It can be seen that the verbs such as '*chou* (worry), *fan* (upset), *ji* (worry)' are used to express the psychological change, while the verbs such as '*ai* (love), *hen* (hate), *danyuo* (worry)' are used to express the psychological states. Only verbs with the changing meaning can enter the ergative two structures. The common semantic features of ergative verbs are the 'change' in orientation, quantity, relationship, attribute, state and posture of persons or things.

3.2   Difference in Transitivity

Unary ergative verbs have only one argument, which is the subject N in 'N+V'. However, in some sentences, there is a different nominal component, which is the subject S in the structure 'S+V+N'. In some cases, S can be a noun referring to a person or a place, and in other cases, S can refer to both, as shown in Examples (7–9).

(7) *Zhudui   da bai   kedui.* (主队大败客队)
Home team utterly defeat visiting team
The home team defeated the visiting team utterly.

Ta zoulou le xiaoxi. (他走漏了消息)
He divulge information
He spilled the beans.

Ta   dong le   zhenqing. (他动了真情)
He   move   true feelings
He moved his true feelings.

(8) *Taishang   zuozhe   zhuxituan.* (台上坐着主席团)
On the stage   sitting   presidium
The presidium are sitting on the stage.

Wuli   laile   keren. (屋里来了客人)
In the room   come   guest
Guest arrived in the room.

Qiangjiao   kaozhe   yiba   chutou. (墙角靠着一把锄头)
The corner of wall   lean on   a   hoe
A hoe was propped against the corner of the wall.

(9) *Xiao Zhang diule qianbao.* (小张丢了钱包) / *Bangongshili diule dongxi.* (办公室丢了东西)
Xiao Zhang lose wallet/   In the office   lose things
Xiao Zhang lost the wallet. / There's something missing in the office.

Wang Mian sile fuqin. (王冕死了父亲) / *Cunli sile liangtou yang.* (村里死了两头羊)

Wang Mian die father/In the village die two sheep
Wang Mian's father died. / Two sheep died in the village.

Ah Q paole laopo. (阿 Q 跑了老婆) / *Jianyuli paole lia fanren.* (监狱里跑了俩 犯人)

Ah Q run away wife / In prison   run away   two prisoners
Ah Q's wife ran away. / Two prisoners ran away from the prison.

Examples from (7–9) show the difference in subject selection of unary ergative verbs, from which unary ergative verbs are divided into two subcategories. The subject S of the first kind of verb can only refer to people, as such as the Example (7). The subject S of the second kind of verb can also refer to person or place, such as Examples (8–9). Examples of verbs are as follows.

① *bai* 'defeat', *dong* 'move', *zoulou* 'divulge', and so on.
② *dao* 'reach', *xia* 'descend', *liu* 'flow', *lai* 'come', *zou* 'go', *liu* 'flow', *tang* 'flow', *liutang* 'flow', *zhang* 'grow', *dao* 'collapse', *lan* 'rot', *chuan* 'penetrate', *fei* 'scrap', *tou* 'penetrate', *cheng* 'succeed', *diao* 'lose', *diu* 'lose', *kua* 'collapse', *ta(1)* 'collapse', *ta(2)* 'sink', *si* 'die', *bian* 'change', *tong(2)* 'lead to', *zengda* 'enlarge', *zengzhang* 'increase', *xing* 'wake up', *kao* 'depend on', *yi* 'depend on', *zuo* 'sit', *tang* 'lie', *pao* 'run', *xiaoshi* 'disappear'.

The first group of verbs can basically enter the '-*shi* (causative)' sentence. On the contrary, the second group of verbs can hardly enter the '-*shi* (causative)' sentence, whether the subject refers to a person or a place. Examples are shown below.

(10) *Zhudui dabai kedui.* (主队大败客队) / *Zhudui shi kedui dabai.* (主队使客队 大败)

Home team utterly defeats visiting team / Home team make visiting team fail
The home team defeated the visiting team utterly. / The home team made the visiting team failed utterly.

Ta zoulou le xiaoxi. (他走漏了消息) / *Ta shi xiaoxi zoulou.* (他使消息走漏)
He divulge information / He make information divulge
He spilled the beans. / He made information divulged.

(11) *Wuli laile keren.* (屋里来了客人) / **Wuli shi keren laile.* (屋里使客人来了)
In the room come guest / * In the room make guest come
Guest arrived in the room. /

Wang Mian sile fuqin. (王冕死了父亲) / **Wang Mian shi fuqin sile.* (王冕使父亲死了)
Wang Mian   die   father/ * Wang Mian make father die
Wang Mian's father died. /

Ah Q paole laopo. (阿 Q 跑了老婆) /*Ah Q shi laopo paole.* (阿 Q 使老婆跑了)

Ah Q run away wife / *Ah Q make wife run away
Ah Q's wife ran away. /

As shown in Example (10), although the verb is a unary verb, the subject has a certain degree of causativity to the verb. There is a relatively direct semantic relationship between the verb and the subject-object nouns, and the transitivity is relatively higher than others. It is also the part that is closer to the binary verbs within the unary verbs.

As shown in Example (11), firstly, the most prominent syntactic feature of the verbs is that they can generally enter existential sentences. Secondly, these verbs can enter the 'N1+N2+V' sentence and the topic-subject sentence when the subject refers to a person; these verbs can also enter the topic-subject sentence when the subject refers to the place, while the first group of verbs cannot. See the examples below.

(12) *Wang Mian sile fuqin.* (王冕死了父亲) / *Wang Mian de fuqin sile.* (王冕的父亲死了) / *Wang Mian fuqin sile.* (王冕父亲死了)
Wang Mian die   father/ Wang Mian' father   die/Wang Mian   father die
Wang Mian's father died. / Wang Mian' father died. / Wang Mian, his father died.

Ah Q paole laopo. (阿 Q 跑了老婆) / *Ah Q de laopo paole.* (阿 Q 的老婆跑了) / *Ah Q laopo paole.* (阿 Q 老婆跑了)
Ah Q   run away wife / Ah Q's wife run away / Ah Q wife run away
Ah Q's wife ran away. / Ah Q's wife ran away. / Ah Q, his wife ran away.

(13) *Zhudui dabai kedui.* (主队大败客队) /*Zhudui de kedui dabai.* (*主队的客队大败) / **Zhudui kedui dabai.* (*主队客队大败)
Home team utterly defeats visiting team / *Home team's visiting team fail utterly. / * Home team visiting team fail utterly.
The home team defeated the visiting team utterly.

Ta zoulou le xiaoxi. (他走漏了消息) / **Ta de xiaoxi zoulou.* (*他的消息走漏了) /* *Ta xiaoxi zoulou.* (*他消息走漏了)
He divulge information/*His information divulge /* He information divulge
He spilled the beans. /

Generally, the transitivity of the second group of verbs is relatively low. The transitivity of these verbs involves some classical syntactic problems such as 'Wang Mian died his father'. The Previous studies have made extensive and in-depth studies on this issue from the perspectives of argument, valence, generation, and construction suppression. The number of arguments of verbs is not consistent with the number of nouns on the surface of the sentence. This phenomenon does not only occur in Chinese, but widely exists in many other languages. It is a syntactic approach to explain how the non-argument 'Wang Mian' enters the surface and becomes the subject of 'die' from the verb point of view. It is a semantic approach to observe the relationship between 'father' (the unique argument of 'die') and 'Wang Mian' from the noun point of view. 'Wang Mian' appears on the surface of syntax, not because of the verb 'die', but because of the qualia structure of the noun 'father'.

In Generative Lexicon Theory, qualia structure is a theoretical system of the semantic knowledge of nouns in [14], which is also the core of various generative attributes. The [15] expands nominal roles from four to ten, including the possession relationship of nouns in 'constitutive role' and the location information of nouns in 'formal role'. As shown in Example (9) above, the subject S in the structure 'S+V+N' refers to the owner or the location of object N. The information is contained in the qualia structure of the noun N, which is the semantic basis for its syntactic function. In other words, the reason why 'Wang Mian' can be the subject of the verb '*si* (die)' is that its semantic information comes from 'father' rather than 'die'. 'There is no syntactic and semantic choice relationship between verbs (die) and subjects (Wang Mian)', said in [16]. Therefore, it is necessary to explain the semantic source of subject 'Wang Mian' on the perspective of qualia structure.

The transitivity of ergative verbs has been controversial in the academic circles, and the key point of the questions is also focused on the second group of unary verbs (*si* 'die', *pao* 'run', *diu* 'lose', etc.). A binary ergative verb has two arguments, and its transitivity is the highest within the ergative verb, followed by the first group of unary ergative verbs (*cheng* 'succeed', *bai* 'defeat', *dong* 'move', *zoulou* 'divulge', etc.). Both subject and object in this group have relatively direct semantic relations with verbs. The lowest transitivity in ergative verbs is the second group of unary verbs (*si* 'die', *pao* 'run', *diu* 'lose', etc.), which have only one argument. Its subject has no direct semantic connection with verbs, but comes from the qualia structure of object nouns. Therefore, its transitivity is the lowest among ergative verbs, but it is still higher than ordinary intransitive verbs, such as '*kesou* (cough)', '*suxing* (awake)', which have been classified as quasi-transitive verbs by [6]. There are differences in transitivity among ergative verbs, which straddle the boundary between transitivity and intransitivity. The binary ergative verbs and the first group of unary verbs can be regarded as transitive verbs, and the second group of unary verbs is intransitive verbs, as shown in the following Fig. 1.

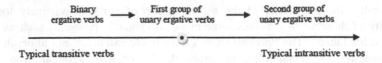

Fig. 1. Transitivity of ergative verbs.

3.3   Difference in Causativity

The syntactic function of binary ergative verbs is related to the characteristics of argument structure. There are two arguments for binary ergative verbs. But in the structures of 'N+V' and 'S+V+N', the argument structures of the verbs are different. Firstly, the nouns appearing in the subject and object position in the 'S+V+N' structure are the external argument and the internal argument respectively. As shown in a set of sentences on the left side in Example (14), because there are both internal and external arguments and no vacancy on the syntactic surface, it is unlikely to enter existential sentences. Secondly, what appears in the subject position of 'N+V' is the external argument of the verb, as shown in a set of sentences on the right side in Example (14).

> (14) *Ta e le haizi liang dun.* (他饿了孩子两顿) / *Haizi e le.* (孩子饿了)
> He starve children two meals   /Children hungry
> He starved the children twice. / The children are hungry.
>
> *Tamen zai sheng qi.* (他们在升旗) /*Qizi sheng qilai le.* (旗子升起来了)
> They   raising   a flag   / Flag   rise   up
> They are raising the flag. / The flag rose.
>
> *Ta diandao shishi.* (他颠倒事实) / *Shishi dinanao le.* (事实颠倒了)
> He reverse truth   / Truth   reverse
> He reversed the truth. / The truth is reversed.
>
> *Qinguo miewang liuguo.* (秦国灭亡六国)/*Liuguo miewang le.* (六国灭亡了)
> Qin Kingdom destroyed six kingdoms / Six kingdoms perish
> Qin Kingdom destroyed the six kingdoms. / The six kingdoms perished.

In the structure of 'S+V+N', the object N is the internal argument of the verb, which is the basis of binary. In the structure of 'N+V', the subject N is the external argument of the verb. This inconsistency in argument structure is the basis of ergative phenomenon. The basic reason lies in that the two structures represent two different types of events. 'S+V+N' structure is a causative event while 'N+V' structure is a state event. Therefore, 'S+V+N' is a necessary condition for causativity of binary ergative verbs. Almost all the "S+V+N" structures of binary ergative verbs can be regarded as causative events, because they can enter the *ba*-construction and *shi*-construction.

Only those unary verbs with higher transitivity have causative properties, such as '*cheng* (succeed), *bai* (defeat), *dong* (move), *zoulou* (divulge)', as shown in the Example (7). Therefore, for the whole ergative verbs, the causativity also shows a gradual change. According to the degree of causativity from the highest to the lowest, the ergative verbs thus span on a continuum of causativity, as shown in Fig. 2.

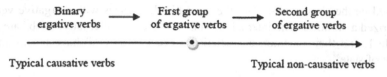

Fig. 2. Causativity of ergative verbs

3.4 Difference in Volition

Volition and non-volition are important semantic features of verbs. Inspired by the study of Tibetan grammar, the [17] divides modern Chinese verbs into volitional verbs and non-volitional verbs to distinguish whether the action is controllable. Within ergative verbs, unary ergative verbs tend to be non-volitional. The [16] and [17] point out that being able to enter imperative sentences is an important basis for judging volitional verbs. According to this criterion, most unary ergative verbs are non-volitional verbs, and only a few are volitional verbs, as shown in follows.

A. *dong* 'move', *dao* 'reach', *liu* 'flow', *tang* 'flow', *liutang* 'flow', *zhang* 'grow', *lan* 'rot', *chuan* 'penetrate', *fei* 'scrap', *tou* 'penetrate', *cheng* 'succeed', *diao* 'lose', *diu* 'lose', *kua* 'collapse', *ta(1)* 'collapse', *ta(2)* 'sink', *si* 'die', *bian* 'change', *tong(2)* 'lead to', *zengda* 'enlarge', *zengzhang* 'increase', *xing* 'wake up', *xiaoshi* 'disappear', and so on.

B. *xia* 'descend', *lai* 'come', *zou* 'go', *dao* 'collapse', *kao* 'depend on', *yi* 'depend on', *zuo* 'sit', *tang* 'lie', *pao* 'run', and so on.

The verbs in group A are non-volitional verbs, which are the majority in unary ergative verbs and cannot enter imperative sentences. Two types of core arguments can be identified for these verbs. The first type indicates inanimate things, reflecting a spontaneous and uncontrollable change of things. As the sentence of Example (2) shows, such as 'peaches rot', 'bridges collapse', 'sweat flow', these phenomena cannot occur independently and controllably. The second type indicates the physical or mental feelings of human beings, such as 'drunk', 'wake up (2)', which are not controlled by the subject.

The verbs in group B are volitional verbs. The core argument of these verbs is the animate noun that can act spontaneously. The verbs indicate the postures of the animate nouns, without affecting other nominal components.

Some studies believe that the unique argument of unary ergative verbs is the agent of the verb. But according to the above analysis of volition and non-volition, only the arguments of the second group can be regarded as the agent. There is no agent component in the conceptual structure of the first group of non-volitional verbs, and the noun N itself cannot dominate these changes. Therefore, it seems to be more appropriate to regard it as the theme of the verbs.

Ergative verbs have obvious tendency of non-volition, which coincides with the strong tendency to enter existential sentences. Because of the non-volition of unarguable ergative verbs, the unique argument can be an object in the structure of 'S+V +N'. Generally, the non-volitional verbs have low transitivity. A close internal relationship exists between non-volition and low transitivity.

Based on the above discussion, the semantic differences within ergative verbs are summarized as follows. The internal semantic differences of ergative verbs are shown in Table 1, and their common characteristics are that they all have the meaning of 'change of state'.

Table 1. The internal differences of semantic characteristics of ergative verbs.

Type of verbs	Transitivity	Volition	Causativity	Examples
Binary ergative verbs	Transitive	Non-volitional	Causative	*E\guan\manzu\chengli*
1 group of unary verbs	Transitive	Volitional	Causative	*Cheng\bai\dong\zoulou*
2 group of unary verbs	Intransitive	Non-volitional	Non-causative	*Si\pao\zhang\xiaoshi*

4 Conclusion

Through the collation and analysis of ergative verbs, the understanding of the ergative verbs in semantics and syntax is further deepened. In semantics, ergative verbs have both commonality and internal differences. On the one hand, the common semantics of ergative verbs is the meaning of change, which indicates the change of things in quantity, trend, attitude, relationship and character. On the other hand, the semantic differences of ergative verbs lie in two points. Firstly, the unary ergative verbs have lower transitivity and obvious tendency of non-volition, indicating the spontaneous and uncontrollable change of things. Secondly, the binary ergative verbs have higher transitivity and obvious tendency of causativity.

In syntax, the two semantically related structures of 'S+V+N' and 'N+V' are the key structures of ergative phenomenon, which indicate the different stages before and after the change. The former represents the origin or motive force of changing, and the latter represents the state of change. Furthermore, there are temporal succession and logical causality between the two structures. Through analyzing the transitivity, causativity and volition of ergative verbs, this paper explains the semantic origin of the structure 'S+V+N' from the perspective of nominal structure, and points out that the subject S (Wang Mian) is based on the qualia structure of the object N (father), and it has no direct semantic relationship with the verb *die*.

In general, ergative phenomenon involves numerous classical issues in the Chinese syntax and semantics studies, including the misplacement of argument assignment, existential sentences, verb-result construction, and the typological significance of Chinese ergative verbs. The exploration of all these issues is closely related to the detailed and in-depth analysis of verb semantics. The work done in this paper may shed some light in the exploration of ergative verbs.

References

1. Lv, S.X.: Discuss the verbs *"Sheng"* and *"Bai"*. Stud. Chin. Lang. **1**, 1–5 (1987)
2. Huang, C.T.J.: Two types of transitive verbs and two types of intransitive verbs in Chinese. In: Collection of the Second World Seminar on Chinese Language Teaching, vol. 1, pp. 39–59. The World Chinese Press, Taipei (1990)
3. Gu, Y.: Generative grammar and some characteristics of verbs in lexicon. Linguist. Abroad **3**, 1–16 (1996)
4. Yang, S.Y.: The unaccusative phenomenon-a study on the relationship between syntax and semantics. Contemp. Linguist. **1**, 30–43 (1999)
5. Xu, J.: Break his four cups and the principle of binding. Stud. Chin. Lang. **3**, 185–191 (1999)
6. Xu, J.: 'Transitivity' and four classes of relevant verbs. Stud. Lang. Linguist. **3**, 1–11 (2001)
7. Teng, S.W.: Ergativization and Chinese passives. Stud. Lang. Linguist. **4**, 291–301 (2004)
8. Zeng, L.Y., Yang, X.W.: On the choice of subject system from the ergative perspective. Chin. Linguist. **4**, 22–30 (2005)
9. Jiang, S.Y.: Ergative verbs in old Chinese. Stud. Hist. Linguist. **11**, 1–28 (2018)
10. Liu, F.X.: Causative alternation in mandarin Chinese. Contemp. Linguist. **3**, 317–333 (2018)
11. Zhou, X.P.: Aspects of Chinese Syntax: Ergativity and Phrase Structure. University of Illinois, Urbana (1990)
12. Shi, Y.Z.: Cognitive Semantic Basis of Grammar. Jiangxi Education Publishing House, Nanchang (2000)
13. Pustejovsky, J.: The syntax of event structure. Cognition **41**, 47–81 (1991)
14. Pustejovsky, J.: The Generation Lexicon. MIT Press, Cambridge (1995)
15. Yuan, Y.L.: On a descriptive system of qulia structure of Chinese nouns and its application in parsing complex Chinese grammatical phenomena. Contemp. Linguist. **1**, 31–48 (2014)
16. Yuan, Y.L.: Cognitive study on univalent nouns. Stud. Chin. Lang. **4**, 241–253 (1994)
17. Ma, Q.Z.: Conscious and unconscious verbs. J. Linguist. Soc. China **3**, 160–191 (1988)
18. Meng, C., Zheng, H.D., Meng, Q.H., Cai, W.L.: Chinese Verb Usage Dictionary. The Commercial Press, Beijing (2012)

Reduplicated Kind Classifier *zhǒngzhǒng* in Mandarin Chinese and the Associated Plurality Type

Hua-Hung Yuan[✉]

School of Foreign Languages, Xiangtan University, Xiangtan, China
yuan.huahung@gmail.com

Abstract. The reduplicated kind classifier, *zhǒngzhǒng*, is related to a nominal plurality. *Zhǒngzhǒng* has to co-occur with abstract nouns only. This paper argues that without any kind-referring interpretation, *zhǒngzhǒng* groups entities into an approximative taxonomic category. This differentiates from another reduplicated kind classifier, *yī-zhǒngzhǒng*, which individualizes entities at the kind level and makes a taxonomic category. *Zhǒngzhǒng*-NP denotes a set of entities which can be inferred as sum, group atoms and individual atoms, which are terms to cover the distinction between distributive and collective interpretations of plural NPs. Some types of predicates will be served to attest the denotation of *zhǒngzhǒng*-NP and describe its representation of plurality.

Keywords: Classifiers · Kind classifiers · Plurality · Distributivity · Collectivity · Set of atoms · Mandarin Chinese

1 Introduction

It has been discussed that in Mandarin Chinese, nominal classifiers (Cl[1]) can be reduplicated and form a unit, like *ClCl*, *yī-ClCl* and *yī-Cl-yī-Cl* which express a nominal plurality in [1–8] among others. This paper focuses on the reduplicative form of one of kind classifiers, *zhǒngzhǒng*, especially its plural denotation.

Cls are used to count nouns when they appear between a numeral and noun (1). Individual Cls (Cl_{IND}), as in (1b) individualize the entities denoted by the nouns. Kind Cls (Cl_K), like in (1a) refer to a subset of individuals to which the noun *pig* applies, a type of entities (Cf. [1, 5, 9–12] among others).

(1) a. *sān-zhǒng zhū*   b. *sān-zhī zhū*
    three-kind pig   three-Cl pig
    'Three kinds of pigs.'   'Three pigs.'

---

[1] Abbreviations: Acc: Accomplished aspect *le* (了); Cl: Classifiers; Cl_{IND}: Individual Classifiers; Cl_K: Kind Classifiers; de: relator *de* (的); Dur: Durative aspect *zhe* (着); neg: Negation marker *bù* (不) and *méi* (没).

I thank the Department of Education of Hunan Province, China for supporting this research (grant KZ13032).

© Springer Nature Switzerland AG 2020
J.-F. Hong et al. (Eds.): CLSW 2019, LNAI 11831, pp. 234–242, 2020.
https://doi.org/10.1007/978-3-030-38189-9_24

[2] indicates that ClCl concerns two types of nominal plurality, plural individuals denoted by ClCl composed of individual Cl (2) ([13] also) and a group with unspecified atomic quantity denoted by the reduplicated Cl_K (3).

(2)   *Tā gègè xuéshēng dōu rèn-dé*
      He ClCl student   all   know
      'He knows all the students (individually).' ([2] p. 6 (13) and (15))

(3)   *fēnxī-shàng   yùjiàn   zhǒngzhǒng kùnnán*
      analysis-on   encounter kind-kind   difficulty

[5] claims that ClCl forms induce abundant plural readings and give rise to non-distributive or non exhaustive readings, as illustrate (4)–(5), shown by the author. This is different from some previous discussions that ClCls denote a distributive interpretation, for instance [11].

(4)   *huòjià-shàng bǎi-mǎn-le   zhǒngzhǒng guàntóu*
      shelf-on   put-full-Acc kind-kind   can
      'On the shelf are many kinds of cans.'

(5)   (*yī-)*zhǒngzhǒng zácǎo   zhǎng-mǎn-le   yuànzǐ*
      one-kind-kind   weed   grow-full-Acc yard
      'Many kinds of weed have grown in the whole yard.'

However, in my opinion, the incompatibility of *zhǒngzhǒng* with a concrete object like weed or can in (4)–(5) makes the sentences unacceptable. Also, *zhǒngzhǒng*-NP can be distributive when it co-occurs with a non-collective adverbial *zhú-gè*, as (6) illustrates. [5] ignores the fact that *yī-zhǒngzhǒng*-NP and *zhǒngzhǒng*-NP focus on different expressions: the former lists types of difficulties while the latter presents various difficulties, as (7) illustrates.

(6)   *Zhǒngzhǒng kùnnán   zhú-gè   jiějué-le*
      kind-kind   difficulty one-after-another solve-Acc
      'Every kind of difficulties are solved one after another.'

(7)   *Yǎnī yùjiàn-le   (yī-)zhǒngzhǒng-de kùnnán*
      YN encounter-Acc one-kind-kind-de   difficulty
      For *yī-zhǒngzhǒng*: 'Yani encountered various kinds of difficulties.'
      For *zhǒngzhǒng*: 'Yani encountered various difficulties.'

It is obvious that Cl_KCl_K and *yī*-Cl_KCl_K are different and *zhǒngzhǒng*-NP is associated with a different type of plurality, which has not been explored in depth yet. Thus, in this paper, it is expected to answer two questions: (i) What differentiates between the two reduplicated forms of Cl_K? (ii) How does *zhǒngzhǒng* pluralize nominal? What kind of plurality is associated with *zhǒngzhǒng*?

The paper is organized in two sections: semantic properties of *zhǒngzhǒng* in Sect. 2 and the plural denotations of *zhǒngzhǒng*-NP in Sect. 3.

2   Semantic Properties of Cl_KCl_K

In this section, I will discuss some semantic properties of *zhǒngzhǒng*.

Two Forms of Cl_K. Firstly, $Cl_{IND}s$ select NPs as individual entities (1b) and Cl_Ks select NPs as kind denoting (1a). Cls are needed to quantify with the numerals over NPs, as the contrast with (8) illustrates. The demonstratives can co-occur with the numeral-NP and Cls individualize entities at the individual level and the kind level respectively in (9). The Cl_K refers the NP to a sub-kind of species, pig.

(8) *sān-Ø   zhū   (9) zhè-zhī /-zhǒng zhū
     three-Ø pig   this-Cl   kind   pig
                              'This pig / this kind of pig'

For $Cl_{IND}s$, when the numeral yī co-occurs, two reduplicative forms are available, one is yī-$Cl_{IND}Cl_{IND}$ and the other is yī-Cl_{IND}-yī-Cl_{IND} while Cl_Ks can form yī-Cl_KCl_K only, not yī-Cl_K-yī-Cl_K when yī co-occurs. Individual ClCl and kind ClCl select different types of nouns. $Cl_{IND}Cl_{IND}$ selects concrete nouns, as in (10) while zhǒngzhǒng selects abstract ones, such as yīnsù 'factor', wēixiǎn 'danger', jìxiàng 'sign', kùnnán 'difficulty', as (11) illustrates (Cf. [1]).

(10) {zhīzhī /*zhǒngzhǒng}{zhū /dòngwù }
      ClCl   kind-kind   pig animal
      For zhīzhī: 'Each pig'

      For zhǒngzhǒng ; impossible reading
(11) *gègè /^{OK}zhǒngzhǒng yīnsù / wēixiǎn / jìxiàng / kùnnán
      ClCl   kind-kind   factor   danger   sign   difficulty
      'Factors/dangers/signs/difficulties'

Different from zhǒngzhǒng, yī-zhǒngzhǒng which is similar to yī-$Cl_{IND}Cl_{IND}$ is able to select concrete nouns, like pig, animal (12).

(12) Yī-zhīzhī /yī-zhǒngzhǒng zhū /dòngwù guān   zài dòngwùyuán-lǐ
      One-ClCl one-kind-kind pig animal   enclose at   zoo-inside
      '{Many / Many kinds of} pigs/animals are enclosed in the zoo.'

It is noticed that the relator de can follow zhǒngzhǒng, not $Cl_{IND}Cl_{IND}$ (13)–(14), which shows zhǒngzhǒng can be a head of a relative clause and qualifies NP.

(13) Zhǒngzhǒng-de kùnnán   (14)   *Zhīzhī-de zhū
      kind-kind-de   difficulty   ClCl-de   pig
      'Various difficulties'

Two Types of Individualization at Kind Level. The above observation leads to assume that the two forms of reduplicated Cl_Ks presents distinct denotations based on the individualization of Cl_K at the kind level. When a concrete noun is selected by yī-zhǒngzhǒng, it denotes sub-kinds of that noun, which are its hyponyms (15). It is referred to a taxonomic usage (Cf. [14, 15]). Notice that zhǒngzhǒng cannot select concrete nouns (See [16] also).

(15) *(Yī-)zhŏngzhŏng shuĭguŏ* : *píngguŏ, xiāngjiāo, căoméi,   xīguā*
　　 one-kind-kind fruit   apple   banana   strawberry water melon
For *yī-zhŏngzhŏng*: 'Different kinds of fruits: apples, bananas, strawberries,
　　　　　　　　　　water melons.'
For *zhŏngzhŏng*: impossible reading.

In the same manner, when an abstract noun co-occurs with *yī-zhŏngzhŏng*, as in
(16a), it denotes a set of sub-kinds (16c) which are viewed as hyponyms of that noun.

(16)　a. *(yī-)zhŏngzhŏng kùnnán*　b. *Quē-zījīn, shēngbìng, tài duō jìngzhēngzhě*
　　　　 one-kind-kind   difficulty　　 lack fund sick   too many competitor
　　　　 'Various difficulties'　　　　　'lack of fund, sickness and too many competitors'
　　　c. *kùnnán* 1,　　*kùnnán* 2,　　*kùnnán* 3
　　　　 difficulty 1　　difficulty 2　　difficulty 3
　　　　 'difficulty 1, difficulty 2 and difficulty 3'

Yī-zhŏngzhŏng and *zhŏngzhŏng* realize two distinct types of individualization of
kinds. *Yī-zhŏngzhŏng* pluralizes over (sub-)kinds whereas *zhŏngzhŏng*-NP denotes a
set of entities which are aggregated by the same quality, as in (16b), instead of entities
viewed as sub-kinds of the NP. In my view, *zhŏngzhŏng* is a marker of approximative
category (scc [14]) and it groups the NPs on the same level of generalization.

In fact, *zhŏngzhŏng* performing a qualifying function (13) induces a collective
reading over NPs. The Cl$_K$ *zhŏng* individualizes entities denoted by a NP at the kind
level and gives rise to a kind-referring interpretation of a NP, especially when the NP
co-occurs with a kind predication or an individual-level predicate, as illustrate (17a)
(Cf. [17, 18]). NPs that *zhŏng* applies to are not compatible with an episodic sentence,
as (17b) show.

(17) *Zhè-zhŏng xiăo huā* /*wèntí zài zhè-lĭ* a *hĕn chángjiàn*/ b*yĭjīng diāoxiè-/chŭlĭ-le*
　　 this-kind little flower problem at herc very common   already wilt   handle-Acc
　　 'The kind of little flowers/ problems is very common here.'

However, when Cl$_K$ *zhŏng* is iterated to form *zhŏngzhŏng*, Cl$_K$Cl$_K$ shifts to a
different function: it rejects concrete nouns (10) and selects only abstract ones (11).
Zhŏngzhŏng does not indicate a (sub-)kind denotation of the abstract noun since it is
able to co-occur with not only a kind predicate *cháng jiàn* 'be common' (18a) but also
an episodic sentence, encoded by the accomplished aspect *le* (18b).

(18)　*Zhè-zhŏngzhŏng wèntí　zài zhè-lĭ* a *hĕn cháng jiàn* /b *yĭjīng   chŭlĭ-le*
　　　 this-kind-kind   problem at here   very common   already handle-Acc
　　　 'These problems are very common here.''These problems have been handled here.'

Distributivity or Collectivity. Firstly, it is claimed that ClCls convey the universality
(Cf. [19]). However, *zhŏngzhŏng*-NP refuses to be modified by *jīhū* 'almost' (19),
which attests the exhaustivity of the universal denotation of NP.

(19) jīhū ^OK*měi-gè* /*zhǒngzhǒng nántí   dōu   lìng rén   tóutòng
      almost every-Cl   kind-kind difficulty all   let   people headache
      'Almost every difficult problem is troubling.'

A universal quantifier like *měi* is perfectly compatible with *jīhū*. Thus, it is shown that non-exhaustive, *zhǒngzhǒng* does not operate universal quantification. Secondly, as [5] suggests, *zhǒngzhǒng* denotes a collective reading, as *suǒyǒu-de* in (20) and when *dōu*, an adverb of quantification co-occurs, the distributive interpretation is triggered ([5] p. 122–124).

(20) *Suǒyǒu-de* /*zhǒngzhǒng nántí   (dōu)   lìng rén   tóutòng*
      all   kind-kind   difficulty all   let people headache
      'All the /various problems are troubling.'

[5] argues that *zhǒngzhǒng* can co-occur with a collective predicate due to its collective reading. For her, the co-occurrence of *dōu* is supposed to trigger a distributive reading of *zhǒngzhǒng*-NP. However, contrary to her analysis, a collective predicate and *dōu* can co-occur with *zhǒngzhǒng*-NP, as in (21), and it still denotes a collective reading.

(21) *Suǒyǒu-de* /*zhǒngzhǒng dǎoméi   shì   (dōu) còu-dào   yīkuài*
      All   kind-kind   unlucky matter all   gather-arrive together
      'All the /various unlucky matters all gathered.'

Therefore, it is not sufficient to consider only the collective and distributive interpretations of *zhǒngzhǒng*. I suggest that *zhǒngzhǒng*-NP can be interpreted as collective due to the nature of Cl_KCl_K. *Zhǒngzhǒng* is a marker of approximative taxonomic categorization, which groups non kind-referring entities into a set.

To sum up, *yī-zhǒngzhǒng* selecting concrete nouns denotes sub-kinds of the NP while *zhǒngzhǒng* taking abstract nouns only regroups entities into an approximative category and the set of entities is yielded to a collective interpretation.

3 Plurality Expressed by *Zhǒngzhǒng*

The collective interpretation may not be enough to describe all the properties of *zhǒngzhǒng*. In this section, by means of its co-occurrence with different types of predicates, I will explain how *zhǒngzhǒng* induces a plural denotation and account for the co-occurrence of *zhǒngzhǒng* and *dōu*.

Firstly, I will define the terms to describe the plural denotations of a NP. I adopt the terms used by [20], a set of atoms to include the three possible denotations of a plural NP, *the boys*, as in (22) that traditionally are distributive, collective and group.

(22) The boys carried the piano upstairs.

Since 'what is lexically specified is what kinds of atoms they take in their extension' ([20] p. 155), some predicates take only individual atoms (distributive), group atoms (collective), individual atoms and group atoms both or sums (predicated with plural predication). [21, 22] analyze the denotations of plural NP subjects with different types of predicates, predicates over atoms, predicates over group, mixed predicate, existential predicates and total predicates. I will adopt these types of predicates in order to show the possible denotations of *zhǒngzhǒng*-NP.

(i) **Predicates over atoms** select a plural NP which denotes individual atoms. The predicate *hěn kěxiào* 'be funny' in (23) attributes the property to individual entities so as to trigger the denotation of individual atoms of *zhǒngzhǒng*-NP. (ii) **Predicates over group** must apply to group atoms. In (24), *zhǒngzhǒng*-NP cannot induce a group atoms denotation because of its incompatibility with the predicate *hěn duō* 'be numerous'.

(23) *Zhǒngzhǒng lǐyóu (dōu) hěn kěxiào*   (24)   **Zhǒngzhǒng yuányīn hěn duō*
    kind-kind   reason all   very funny   kind-kind   reason very many
    Without *dōu*:'Various reasons are funny.'
    With *dōu*: 'Various reasons are individually funny.'

(iii) **Total predicates** have to apply to group atoms. The predicate *còu-zài-yīkuài* 'to gather' co-occurs with *zhǒngzhǒng*-NP because it can induce a denotation of group atoms, as in (25a). The division into sub-groups is allowed since *méiyǒu* negates the predicate whose subject is the group atoms, as (25b) illustrates.

(25) *Zhǒngzhǒng dǎoméi shì* a *(dōu) còu-zào yīkuài* / b *méi-yǒu (dōu) còu-zào yīkuài*
    kind-kind   unlucky matter all gather-arrive together neg-have all gather-arrive together
    For a without *dōu* : 'Various unlucky matters gathered together.'
    For b without *dōu*: 'Various unlucky matters did not gather together.'
    For a with *dōu* : 'Various unlucky matters all gather together.'
    For b with *dōu*: 'All the /various unlucky matters all gathered.'

(iv) **Mixed predicates** can apply to a plural NP subject denoting individual atoms and group atoms both. In (26), the predicate *zàochéng* 'to make' can operate over a part of individual atoms or the entire group atoms that *zhǒngzhǒng*-NP denotes. The co-occurrence with the adverb *xiānhòu* 'one after another' shows the denotation of individual atoms inferred by *zhǒngzhǒng*-NP.

(26) *Zhǒngzhǒng dǎoméi shì*   *(dōu)(xiān hòu)*   *zàochéng tā-de shībài*
    kind-kind unlucky matter all one-after-another make   his   failure
    Without *dōu*:'Various unlucky matters made his failure (one after another).'
    With *dōu*: 'Various unlucky matters all made his failure.'

(v) **Existential predicates**, like *bù-yīyàng* 'be different' apply to group atoms. The group atoms allow the predicate *bù-yīyàng* to compare the NP subject with a referent, introduced by *hé* 'with', *tā xiǎngxiàng-de* 'what he thinks', as (27b) illustrates. The usage is referred to one of two interpretations of *be different*, which is the dependent interpretation due to the presence of a referent of comparison.

(27) *Zhǒngzhǒng lǐyóu* a *(dōu)* **bú-yīyàng* /b *hé   tā xiǎngxiàng de   bú-yīyàng*
kind-kind reason   all   different   with he imagine   de   different
For a without *dōu*: (Intended: 'Various reasons are different from each other.')
For b without *dōu*:'Various reasons are different from what he thought.'
For a with *dōu* : (Intended: 'Various reasons are different from each other.')
For b with *dōu*: 'Various reasons are individually different from what he thought.'

With the other interpretation, the independent interpretation, the predicate *bù-yīyang* in (27a) needs a NP subject inferred as individual atoms, but this is not possible when *zhǒngzhǒng*-NP co-occurs.

Until now, I have shown the denotations of *zhǒngzhǒng*-NP: group atoms and individual atoms according to its co-occurrence with different types of predicates suggested by [21, 22]. Nevertheless, how can the co-occurrence of *zhǒngzhǒng* with *dōu* (23), (25)–(27) be explained? I don't involve the debates about *dōu*-quantification in this paper. In order to treat *zhǒngzhǒng*, I consider one of the representations of *dōu* illustrated below in which it operates a summation over predicates (Cf. [23]). *Dōu* pluralizes rightward over VP because it has to be placed after the additive adverb *yě* (28).

(28) *Nà-xiē   háizǐ dōu chī-le   fàn, (*dōu) yě   dōu xiě-le   zuòyè*
those   child   all eat-Acc meal   all   too   all write-Acc homework
'The kids all ate and also all finished their homework.'

[20] distinguishes two modes of predication: singular predication and plural predication. The former applies a predicate to single units, individuals or groups while the latter applies a plural predicate distributively to a sum of singular individuals, as if the singular predicate attributed to the individual parts of that sum. Thus, *dōu*-VP can be viewed as a plural predication over *zhǒngzhǒng*-NP. This can account for (25a) where *dōu-còu-zài-yīkuài* applies to a sum and each singular predicate *còu-zài-yīkuài* applies to parts of individuals in the sum. The negation *méiyǒu* over the plural predicate *dōu-còu-zài-yīkuài*, as in (25b) indicates some of singular predicates are blocked to apply to a corresponding group atoms in the sum, which results in inducing sub group atoms in the *zhǒngzhǒng*-NP, as (29) shows.

(29) *Zhǐ yǒu   yīxiē   dǎoméi shì   còu-zài   yīkuài*
only have   some unlucky matter gather-at   together
'Only some (of the set of) unlucky matters gathered.'

For existential predicates, *dōu-bù-yīyàng* with the dependent interpretation in (27b) attributes to the sum denoted by *zhǒngzhǒng*-NP that a single predicate applies to parts of individuals. *Dōu-bù-yīyàng* of the independent interpretation in (27a) cannot apply to *zhǒngzhǒng*-NP because it is not allowed to induce individual atoms denotation which is necessary for the application of a singular predicate.

It is easier to show the plural predication over *zhǒngzhǒng*-NP, as in (23) because the plural predicate, *dōu-hěn kěxiào* selects a sum involving individuals and each singular predicate applies to an individual.

The plural predicate *dōu-zàochéng* 'to make' in (26) applies to the sum and each singular predicate is attributed to individuals so that *zhǒngzhǒng*-NP induces a distributive interpretation. The singular predication, as in (26) allows the application to individual atoms or group atoms. The plural predicate has to be attributed distributively (individually) to individual atoms and blocks the group atoms denotation. The denotations of *zhǒngzhǒng*-NP vary according to its co-occurrence with the types of predicates. *Zhǒngzhǒng* sorts entities seen as the same quality into the set denoted by the NP, which is an approximative category on one hand, and it individualizes members in the set on the other hand. The plurality type associated with *zhǒngzhǒng* is expressed with a set of atoms viewed as sum, group atoms and individual atoms.

To summarize, according to its co-occurrence of different types of predicates, *zhǒngzhǒng*-NP can denote sum, group atoms and individual atoms. The plural predication of [20] can explain the co-occurrence of *zhǒngzhǒng*-NP with *dōu* and different types of predicates.

4 Conclusion

This research has discussed the plurality type expressed by *zhǒngzhǒng*. Compatible with abstract nouns only, *zhǒngzhǒng* sorts entities into an approximative taxonomic category denoted by the NP it selects. Different from *zhǒngzhǒng*, the other form of reduplicated Cl$_K$, *yī-zhǒngzhǒng* pluralizes over (sub-)kinds, which contributes to a taxonomic categorization. *Zhǒngzhǒng*-NP denotes a set of atoms which can be viewed as sum, group atoms, and individual atoms according to the types of predicates that the NP co-occurs.

References

1. Paris, M.-C.: Problèmes de Syntaxe et de Sémantique en Linguistique Chinoise. Institut des Hautes Etudes Chinoises. Collège de France, Paris (1981)
2. Paris, M.-C.: Un Aperçu de la Réduplication Nominale et Verbale en Mandarin. Faits de langue **29**, 63–76 (2007)
3. Yang, K.-R.: The similarity between the meaning and structural difference between sentences with classifier reduplication and *mei*. Contemp. Res. Mod. Chin. **6**, 1–8 (2004). (in Chinese)
4. Yang, X.-M.: "Gege", "Meige" he "Yige(yi)ge" de Yu Fa Yu Yi Fenxi [The syntactic and semantic analysis of *gege, meige* and *yi-ge(yi)ge*]. Chin. Lang. Learn. **4**, 26–31 (2002). (in Chinese)
5. Zhang, N.N.: Classifier Structures in Mandarin Chinese. De Gruyter Mouton (2013)
6. Guo, Y.-Y.: The Study of Classifier Reduplication in Mandarin Chinese in the Framework of Generative Grammar. MA thesis, Beijing University (2015). (in Chinese)
7. Sui, N., Hu, J.-H.: The syntax of classifier reduplication. Stud. Chin. Lang. **2017**(1), 22–41 (2017). (in Chinese)

8. Xiong, Z.-R.: Noun phrases containing reduplicated classifiers. Lang. Teach. Linguist. Stud. **190**(2), 44–57 (2018). (in Chinese)

9. Chao, Y.-R.: A Grammar of Spoken Chinese. University of California Press, Berkeley (1968)

10. Huang, C.-R., Ahrens, K.: Individuals, kinds and events-classifier coercion of nouns. Lang. Sci. **25**, 353–373 (2003)

11. Li, C., Thompson, S.: Mandarin Chinese: A Functional Reference Grammar. University of California Press, Berkeley (1981)

12. Cheng, L.L.S., Sybesma, R.: Yi-wan tang, Yi-ge Tang: classifiers and Massifiers. Tsing-Hua J. Chin. Stud. **28**(3), 385–412 (1998). New Series

13. Hsieh, M.-L.: The Internal Structure of Noun Phrases in Chinese. Crane Publishing Co., Taipei (2008)

14. Mihatsch, W.: The construction of vagueness: "sort of" expressions in romance languages. In: Radden, G., Koepcke, K.-M., Berg, T., Siemund, P. (eds.) Aspects of Meaning Constructing Meaning: From Concepts to Utterances, pp. 225–245. John Benjamins, Amsterdam/Philadelphia (2007)

15. Gerhard-Krait, F., Vassiliadou, H.: Lectures Taxinomique et Floue Appliquées aux Noms: Quelques Réflexions. Travaux de linguistique **2014**(2) 69, 57–75 (2014)

16. Yuan, H.-H.: Quelques Aspects de la Quantification en Chinois Mandarin: Pluralité et Distributivité, PhD dissertation. Université Paris-Diderot (2011)

17. Krifka, M., et al.: Genericity: an introduction. In: Carlson, G.N., Pelletier, F.J. (eds.) The Generic Book, pp. 1–124. University of Chicago Press, Chicago (1995)

18. Li, J.: Predicate type and the semantics of bare nouns in Chinese. In: Xu, L. (ed.) The Referential Properties of Chinese Noun Phrases. Ecole des Hautes Etudes en Sciences Sociales, Paris, pp. 61–84 (1997)

19. Lu, J.-M.: Universality and Other. Zhong Guo Yu Wen 3 (1986). (in Chinese)

20. Landman, F.: Events and Plurality: The Jerusalem Lectures. Studies in Linguistics and Philosophy, vol. 76. Kluwer, Dordrecht (2000)

21. Corblin, F.: Des Prédicats Non Quantifiables: les Prédicats Holistes. Langages **169**, 34–57 (2008)

22. Corblin, F.: Definites and Floating Quantifiers (tous/all), Linguista Sum. Mélanges offerts à Marc Dominicy. L'Harmattan, Paris, pp. 265–278 (2008)

23. Huang, S.-Z.: Universal Quantification with Skolemization as Evidenced in Chinese and English. Edwin Mellen Press, Lewiston (2005)

A Study on the Semantic Construal of 'NP *yào* VP' Structure from the Perspective of Grounding

Limei Yang[(✉)]

School of Foreign Languages, Henan University, Kaifeng, China
abc_ylm@126.com

Abstract. In the framework of grounding theory in cognitive grammar, this paper takes *yào* as a grounding element and "*yào* VP" a grounded construction. The conceptualization process of "NP *yào* VP" structure can be described as a process in which the conceptualizer inputs volitional force, deontic force or speculative force into the process profiled by VP and makes the semantic profile of VP an intentional process. Grounding by *yào* has effects on the situation type of VP.

Keywords: *yào* · Modality · Grounding · Agentivity

1 Introduction

As a typical representative of modern Chinese modal verb, modal verb *yào* presents rich semantic features. Please observe the following sentences:

(1) 我要看看它的来历!
   Wǒ yào kàn kàn tā de láilì!
   'I will look at its origins!'
(2) 要把基点放在自力更生的基础上。
   Yào bǎ jīdiǎn fàng zài zìlìgēngshēng de jīchǔ shàng.
   'Lay the foundation on the basis of self-reliance.'
(3) 你要犯大错误的。
   Nǐ yào fàn dà cuòwù de.
   'You will make big mistakes.'

The phrase '看看它的来历' *kàn kàn tā de láilì* 'look at its origin' in sentence (1) is an intentional process; in sentence (2), '把基点放在自力更生的基础上' *bǎ jīdiǎn fàng zài zìlìgēngshēng de jīchǔ shàng* 'put the base point on the basis of self-reliance' expresses an action ordered by the speaker; in sentence (3), '犯大错误' *fàn dà cuòwù* 'make big mistakes' denotes a process that was predicted. The meaning representations lead us to ask the following questions: Why can the combination of modal verb *yào* and

This paper is supported by Henan Province Social and Science Fund Project (2017CYY025) "A Contrastive Study of English and Chinese Modal Verb from the Perspective of Cognitive Grammar".

© Springer Nature Switzerland AG 2020
J.-F. Hong et al. (Eds.): CLSW 2019, LNAI 11831, pp. 243–248, 2020.
https://doi.org/10.1007/978-3-030-38189-9_25

verb express intentional, obligatory and predictive process? What role does *yào* play in the meaning construal of the sentence?

Grounding theory developed by Langacker [1–3] represents new developments in cognitive research. Scholars [4–6] have used this theory to interpret the meaning construction of certain Chinese linguistic structures. This paper intends to discover the semantic construction mechanism of "*yào*+VP" and the semantic contribution of *yào* to the subsequent verb.

2 Semantic and Syntactic Features of "NP *yào* VP"

The combination of modal *yào* and VP form a construction "NP *yào* VP". Syntactically this structure consists of three parts: subject NP, modal verb *yào* and VP. Palmer [7] classifies modality into dynamic modality, deontic modality and epistemic modality. Dynamic modality is related to the ability or willingness of the subject; deontic modality serves to encode the speaker's commitment to the necessity/permissibility of an action [8]. Epistemic modality deals with speaker's evaluation/judgment of, degree of confidence in, or belief of the knowledge upon which a proposition is based. According to this system, *yào* can express all the three meanings. Dynamic modal *yào* indicates that NP is willing to implement the action characterized by VP ('我要学德语' *Wǒ yào xué déyǔ*. 'I want to learn German'.); deontic modal *yào* means that the speaker believes NP has the obligation to carry out the action denoted by VP ('你要好好学习'*Nǐ yào hǎo hǎo xuéxí*. 'You must study hard'.); epistemic modality *yào* means the process expressed by VP has the possibility to occur ('天要下雨' *Tiān yào xià yǔ*. 'It is going to rain'.).

3 The Cognitive Basis of the Semantic Construction of "NP *yào* VP"

3.1 Grounding Theory

According to grounding theory, simple nouns or verbs are abstractions and generalizations of things or processes. Only when a thing or process is anchored in a specific position in space and time can it be instantiated as a specific action or thing. In sentence "He can speak French", modal verb *can* acts as a grounding element and forms a grounded structure "can speak French". The grounding of modal verb *can* helps to anchor the process "speak French" in irreality.

In order to describe the role of the abstract concepts such as tense and modality in our cognition of the world, Langacker [1] proposed an idealized cognitive model of modal verbs, namely dynamic evolutionary model (Fig. 1).

In the diagram above, what is represented by the solid cylinder is "reality", which includes all the events that has occurred and are happening as perceived by the conceptualizer. "Present reality" refers to the reality at the time of conceptualization, which is represented by the solid ellipse of the cylinder. The structure of the present reality determines the development trend of future. The trend is conceptualized by the

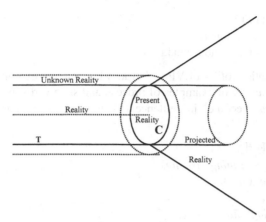

Fig. 1. Dynamic Evolutionary Model (DEM)

conceptualizer in the form of force, which is named as "evolutionary momentum". The conceptualizer's understanding of reality is constantly ongoing. It is moving forward under the impetus of dynamic evolutionary momentum, and then evolves into future reality. Modal verbs represent modal force acting on the known reality. Driven by modal forces, reality evolves into future realities along this path. From the perspective of grounding theory, the role of modal verbs is to locate the process profiled by the verb in the above unreal areas. According to the degree of the evolutionary force and the time zone in which the process is located, Langacker [1] distinguishes three realities: unknown reality, potential reality, and projected reality.

3.2   *Yào* as a Grounding Element

By examining "*yào* VP" in the framework of grounding theory, we can find that *yào* act as a grounding element, and transform the type process into a grounded instance process. The conceptualization process of "*yào* VP" can be regarded as a cognitive process in which the subject inputs modal force into the process profiled by VP and conceptualizes it as an unreal process.

4   Analysis of Semantic Construction Process of "NP *yào* VP"

4.1   Grounding Function of Dynamic Modal *yào*

We believe that dynamic modal *yào* expresses a strong subjective will. Grounding of it makes the conceptualizer conceptualize the verb as a process with stronger agentivity. The strengthening effect of dynamic modal *yào* on the verb can be verified by the following language facts.

Firstly, When *yào* behaves as a dynamic modal, of all the 118 sentences we collected, 97.46% VPs are action verbs, such as '雇保姆' *gù bǎomǔ* 'hire a housemaid' in sentence (4), which means the semantic profiles of VP denote actions with strong agentivity.

(4) 你是不是要雇保姆?
   Nǐ shì bùshìyào gù bǎomǔ?
   'Do you want to hire a housemaid?'

Secondly, semantic profiles of VP and "*yào* VP" structure exhibit great differences. There are a small number of examples of process and state verbs in the structure, such as '成为' *chéngwéi* 'become' in sentence (5) and '姓' *xìng* 'has the surname' in sentence (6).

(5) 他要成为一名医生。
   Tā yào chéngwéi yī míng yīshēng.
   'He wants to be a doctor.'
(6) 他提出要姓高。
   Tā tíchū yào xìng gāo.
   'He asks to have the surname 'Gao'.'

In sentence (5), '成为' *chéngwéi* 'become' is a typical process verb, which is not subject to human control. After the grounding by modal *yào*, the modal force makes the conceptualizer conceptualize the process as an action with strong agentivity. In the same way, in sentence (6), '姓高' *xìng gāo* 'have the surname 'Gao'' means a constant or stative social attribute of the subject. Grounding by *yào* makes its semantic profile a process that can be controlled.

Thirdly, the semantic profiles of VP when it co-occurs with other verb and adverb are different from the semantic profile of VP occurring alone. First, dynamic *yào* tend to co-occur with adverbs such as '偏偏' *piānpiān* in sentence (7) and it seldom co-occurs with adverbs such as '略微' *lüèwēi* and '几乎' *jīhū*.

(7) 彭总真怪,为什么偏偏要提出这个问题呢?
   Péng zǒng zhēn guài, wèishénme piānpiān yào tíchū zhège wèntí ne?
   'Peng is really strange, why does he want to ask this question?'
(8) *我略微/几乎要吃饭。
   Wǒ lüèwēi/jīhū yào chī fàn.
   *'I slightly/almost want to eat.'

Third, when *yào* behaves as a dynamic modal, VP can appear in VV structure and "V *yī* V" structure and form "NP *yào* VV" and "*yào* V *yī* V". VV structure and "V *yī* V" structure mark delimitivity, as is shown below:

(9) 我要看看融化的小白桦。
   Wǒ yào kàn kàn rónghuà de xiǎo báihuà.
   'I want to see the melting white birch.'
(10) 我要听一听你意见。
   wǒ yào tīng yī tīng nǐde yìjiàn.
   'I will listen to your opinion.'

Lakoff [9] argues that "more of form is more of content". More of the verb stands for more of the action. The linguistic fact indicates that the semantic profiles of "VV" and "V *yī* V" structure correspond. The semantic profiles of the verbs in these structures are processes with stronger agentivity.

4.2   Grounding Function of Deontic Modal *yào*

Like dynamic modal *yào*, grounding by deontic *yào* strengthens the agentivity of VP. This can be proved by the following language phenomena.

First, When *yào* behaves as a deontic modal, 79.18% VPs are action verb such as '听妈妈的话' *tīng māmā dehuà* 'listen to your mother' in sentence (11), which means that the semantic profiles of VP denote actions with strong agentivity.

(11)   要好好听妈妈的话。

      Yào hǎohǎo tīng māmā dehuà.

      'You should listen to your mother.'

Second, when *yào* behaves as a deontic modal, a small number of verbs and state verbs appear in "*yào* VP" structure, as shown in sentences (12) and (13):

(12)   大学要成为独立的办学实体。

      Dàxué yào chéngwéi dúlì de bànxué shítǐ.

      'University must have a legal person status.'

(13)   培土要细心。

      Péitǔ yào xìxīn.

      'When cultivating soil, one must be careful.'

In sentence (12), '成为' *chéngwéi* 'become' is a typical process verb. Grounding by *yào* makes the subject participate in the process represented by VP and the semantic profile of verb is a process with greater agentivity.

In (13), '细心' *xìxīn* 'be careful' denotes a constant mental state of the subject. After the grounding by deontic modal *yào*, the state adjective is conceptualized as an active process.

Third, adverbs appearing in front of deontic modality are generally adverbs that strengthen speaker's tone, such as '越' *yuè* in sentence (14).

(14)   我们越要注意做好思想工作。

      Wǒmen yuè yào zhùyì zuò hǎo sīxiǎng gōngzuò.

      'The more we must pay attention to ideology.'

This example shows that the agentivity of the verb VP after *yào* is enhanced. The grounding of deontic *yào* makes the semantic profile of VP process with stronger agentivity.

4.3   Grounding Function of Epistemic *Yào*

Epistemic modal *yào* expresses speaker's judgment of the possibility of the process profiled by VP. On one hand, *yào* affirms the action represented by VP, and on the other hand, it weakens this affirmation. Contrary to root modality, epistemic modal *yào* prompts the conceptualizer to conceptualize the verb as process with weaker agentivity. This can be proved by observing the following linguistic facts.

First, epistemic modal *yào* tends to co-exist with adverbs and modal verbs that express weakened tone, as is shown in sentences (15) and (16):

(15) 我有点要吐了。
Wǒ yǒudiǎn yào tǔle.
'I have to vomit a little.'

(16) 我可能要引火烧身的。
Wǒ kě néng yào yǐnhuǒshāoshēn de.
'I am likely to draw fire against myself.'

Words like '有点' *yǒudiǎn* 'kind of' '可能' *kěnéng* 'may' express weakened tones. The agentivity of VP is obviously weakened.

Secondly, verbs that lies in front of the epistemic modal *yào* generally indicate uncertainty, such as '怕是' (sentence 17) *pà shì* 'fear', as well as verbs that indicate the source of the speaker's information, such as '据说' (sentence 18) *jùshuō* 'it is said'.

(17) 你们怕是要过得很艰难了。
Nǐmen pà shì yàoguò dé hěn jiānnánle.
'I am afraid that you may have a hard time.'

(18) 据说要走上正轨了。
Jùshuō yào zǒu shàng zhèngguǐle.
'It is said that we are on the right track.'

5 Conclusion

In the framework of grounding theory, *yào* is a grounding element that inputs volitional force, deontic force and speculative force into the profiled process. Driven by these forces, processes profiled by the sentences are conceptualized as unreal. Grounding by root modal *yào* makes the conceptualizer regard the process profiled by sentence as a process with stronger agentivity, while grounding by epistemic modal *yào* seems to make it a process with weaker agentivity.

References

1. Langacker, R.W.: Foundations of Cognitive Grammar: Vol. II. Stanford University Press, Stanford (1991)
2. Langacker, R.W.: Deixis and subjectivity. In: Brisard, F. (ed.) Grounding: The Epistemic Footing of Deixis and Reference. Mouton de Gruyter, Berlin/New York (2002)
3. Langacker, R.W.: Cognitive Grammar: A Basic Introduction. Oxford University Press, New York (2008)
4. Niu, B.Y.: A study of the grounding function of modal verb may in English. J. Foreign Lang. **3**, 12–22 (2017). (in Chinese)
5. Shen, S.: The study of content restriction of mandarin *yǒu* measure construction. In: Wu, Y., Hong, J.F., Su, Q. (eds.) Chinese Lexical Semantics. LNCS (LNAI), vol. 10709, pp. 198–209. Springer, Cham (2018). https://doi.org/10.1007/978-3-319-73573-3_17
6. Yang, L.M.: A Study of Chinese Modal Verb *Yào*: a Grounding Perspective. Henan University, Kaifeng (2015). (in Chinese)
7. Palmer, F.R.: Mood and Modality, 2nd edn. Cambridge University, Cambridge (2001)
8. Lyons, J.: Semantics. Cambridge University Press, Cambridge (1977)
9. Lakoff, G., Johnson, M.: Metaphors We Live By. University of Chicago Press, Chicago (1980)

Research Into the Additional Meanings of the Words of Shaanxi-Gansu-Ningxia Border Region Consultative Council—Taking *the Shaanxi-Gansu-Ningxia Border Region Consultative Council Literature* as a Corpus

Yao Zhang(✉)

School of Chinese Ethnic Minority Languages and Literatures,
Minzu University of China, Beijing 100081, China
raitozy@163.com

Abstract. The words used in the Shaanxi-Gansu-Ningxia border region consultative council are different from the words used in the region during different periods, which contain additional meanings. Taking the Shaanxi-Gansu-Ningxia border region's consultative council literature as a corpus, this research aims to research the emotional meanings, writing style meanings thought processes, and denotations of the words used in general official documents words during this period; and also intends to explain the inspiration behind using the interrogative pronoun in topic words, and consider the notion of humanistic care. Hopefully, this research presents a relevant point of reference for present document writing.

Keywords: Shaanxi-Gansu-Ningxia border region · General official documents · Words · Additional meanings

1 Introduction

The additional meaning is the meaning of the subjective attitude and evaluation of people other than the rational meaning of the words. It is a semantic content attached to the rational meaning, which usually reflects the meaning of a certain tendency, charm, style and other aspects of the language units. In the words of the consultative council in the Shaanxi-Gansu-Ningxia Border Region, it includes rich additional meanings. Researching the official documents words of this period has important significance for the writing of modern official documents.

The Shaanxi-Gansu-Ningxia Border Region Consultative Council Literature includes the first council in January 1939, the second council in November 1941, the second meeting of the second council in November 1944 and the first meeting of the third council in April 1946, which contents the programs, reports, resolutions, proposals, regulations, etc. These documents are important historical materials for studying the situation of the anti-Japanese base areas during the Anti-Japanese War.

© Springer Nature Switzerland AG 2020
J.-F. Hong et al. (Eds.): CLSW 2019, LNAI 11831, pp. 249–254, 2020.
https://doi.org/10.1007/978-3-030-38189-9_26

In this corpus, there are 18 kinds of official documents, such as reports, resolutions and programs issued by the four border councils, the details are as follows (Table 1):

Table 1. The general situation of the general documents involved in *the Shaanxi-Gansu-Ningxia Border Region Consultative Council Literature*

The kind of documents	Quantity	The kind of documents	Quantity
Report	8	Resolution	7
Program	2	Proposal	398
Regulation	18	Bulletin	1
Telegram	10	Conference summary	4
Declaration	1	Speech	2
Conference conclusion	6	Roster	7
Scheme	2	Principle	1
Address the meeting	1	Election result	6
Election briefing	1	Statistical table	1

The Shaanxi-Gansu-Ningxia Border Region is a model anti-Japanese base in the country and had played a huge role in promoting the national anti-Japanese democracy movement. The official documents of the border region recorded important achievements in the construction of the border region in the eight years of the Anti-Japanese War. It is an important material for studying the party history and party building of the Communist Party of China. The border official documents are worthy of study and contain a wealth of linguistic values. By studying the documents of the Shaanxi-Gansu-Ningxia Border Region, we can see the basic situation of the use of official documents words in the Shaanxi-Gansu-Ningxia Border Region, and it can reflect the complexity of the use of words in this period. Therefore, this article illustrates the additional meaning of the words research on the contents covered by the Shaanxi-Gansu-Ningxia Border Region.

2   The Emotional Significance of General Official Words in the Shaanxi-Gansu-Ningxia Border Region

2.1   Words Usage Profile of Emotional Significance

When using official language, the appropriate use of emotional words can clearly express the writer's views and attitudes, and the audience, in the process of accepting information, subtly recognizes what they say, thus indirectly shaping the role of public opinion.

According to the phrase "emotional meanings", the general official words of Shaanxi-Gansu-Ningxia Border Region are divided into two main categories; namely, commendatory and derogatory terms.

2.2   Use of Emotional Words

2.2.1   General Usage of Emotional Appraise Terms

Among these emotional words, some words directly indicate their derogatory, such as enthusiasm, positive and narrowness. There are also some special words that are emotionally meaningful by virtue of derogatory affixes. For example, the reason why *you qu*(友区), *you jun*(友军) and *you dang*(友党) express the intimate meaning is because of the affix "*you*(友)". "*you*(友)" stands for closeness and ambiguity, "*qu*(区)" "*jun*(军)" and "*dang*(党)" does not have emotional significance; when paired together with the prefix "*you*(友)", they add extra emotional meaning.

2.2.2   Special Usage of Emotional Significance Words

2.2.2.1 Neutral Words Are Used as Commendatory or Derogatory Terms in Context

In the Shaanxi-Gansu-Ningxia Border Region's official language, some words do not have any additional emotional meaning, but are artificially endowed with emotional valence when used in certain situations or by different users. For example, "*Wang Jingwei zhi liu*(汪精卫之流)" means "*Wang Jingwei, the villain*". "*Wang Jingwei*(汪精卫)" is a person's name and is devoid of any innate, derogatory meaning. "*Liu*(流)" refers to a class of people, and also carries no inherently derogatory meaning. However, when Wang Jingwei became the leader of the Japanese government in China. The anti-Japanese contingent in China deeply loathed him so much so that "*Wang Jingwei zhi liu* (汪精卫之流)" became a phrase that carried a distinctly abhorrent connotation.

2.2.2.2 Commendatory or Derogatory Terms Changes According to Context

In the use of official words, there are some words that originally carry commendatory or derogatory connotations, but the emotional significance of these words alter when they are uttered by different speakers. For example, "Request the government to pay attention to the education and health case of the *xiaogui*(小鬼). Review opinion: It is proposed to change the word *xiaogui*(小鬼) to *xiaoqinwu*(小勤务). The original case passed and submitted to the government for reference" [1, 144]. But examining the sentence, it is evident that the word "*xiaogui*(小鬼)" refers to the orderly. "*Xiaogui*(小鬼)" originally refers to a nickname signifying contempt, but in this instance, it stands orderly here, which expresses intimate meaning. In effect, the emotional significance of the term "*xiaogui*(小鬼)" changed due to different references.

2.3   The Actual Effect of Using Emotional Words

In official documents, writers can express their emotional tendencies to readers through the use of words. In the general document of the Shaanxi-Gansu-Ningxia Border Region, such use is everywhere. This is exemplified in the following phrase: "The government of the border area government leads the loyalty of the whole country to support the National Government and the Generalissimo, and supports the cooperation between the Kuomintang and the Communist Party... Strive to establish close and mutual assistance with the governments and the armed forces in the neighboring regions…" [1, 37]. Emotive language is also deployed in the following statement: "The border government will be able to successfully complete the above tasks, … in order to cooperate with the national friendly military, the friendly district to defeat the Sundial finally, and establish an independent and happy republic country" [1, 40]. The following words "loyal support, support, close mutual assistance relationship, friendly military, friendly district" clearly demonstrate that the relationship between the Kuomintang and the Communist Party of China is friendly and cooperative this time.

3   Words Stylistic Sense of the Shaanxi-Gansu-Ningxia Border Region Official Documents

3.1   The Spoken Style of the General Official Words in the Shaanxi-Gansu-Ningxia Border Region

The language used in official documents in the Shaanxi-Gansu-Ningxia Border Region, can be divided into written words and spoken words. Written and spoken words are a classification of mandarin words that focus on different applicable domains. Spoken words are words used in daily oral communication with a popular, playful style; written words, by contrast, are words that are used in situations requiring formal communication that are often elegant and, solemn colors [2, 100]. In formal document writing, written language is required, so it will not be discussed here. It is worth noting that there are a large number of words with colloquial styles in the general official documents of the Shaanxi-Gansu-Ningxia Border Region. The use of these words transforms the original dullness of the official document content into something lively and vivid.

3.2   The Regional Style of the General Official Words in the Shaanxi-Gansu-Ningxia Border Region

The border Region regime is located in the transition zone of the Shaanxi-Gansu-Ningxia provinces; most of the landforms on the Loess Plateau are situated in a dry climate. The people plant crops for a living while living in cool caves. In this environment, the unique regional culture of northern Shaanxi people was formed. The words that hides the regional culture are also faintly visible in the general official documents of the Shaanxi-Gansu-Ningxia Border Region.

These contain regional style words which are mostly comprised of nouns and verbs. In terms of political construction, there are titles such as *baozhang*(保长), *jiangzhang*(甲长), *fangzhang*(坊长) that are reflected in terms of local powers name; in terms of production and life, there are words such as *yaodong*(窑洞), *yin yang*(阴阳), *playing hemp rope* (打麻绳), *sugar beets* (糖萝卜), *maomaojiang*(毛毛匠), *zhonglan* (种蓝) and other production -related words; in terms of public culture, terms such as *yangge*(秧歌), *shehuo*(社火), *changben*(唱本), *piyingxi*(皮影戏), *daoqingbanzi*(道情班子), *jiaxi*(家戏), *muke*(木刻) express cultural and other regional characteristics.

3.3   The Literary Style of the General Official Words in the Shaanxi-Gansu-Ningxia Border Region

Literary and artistic words refer to those words that have certain aesthetic qualities and are suitable for the needs of literary and artistic communication fields, and are rarely used in other genres [4]. The use of literary style words can increase aesthetic enjoyment and truly convey an emotional experience to others. For example, Zhu De observes that, "the Shaanxi-Gansu-Ningxia people, in under the new leadership of the council, must be more united. The demobilization work was completed quickly, peace-building developed rapidly, and the border area was built into a brighter lighthouse, a stronger fortress for a peaceful and democratic new China, and together with the people of the whole country to smash the conspiracy of the reactionaries and promote the cause of democracy and construction throughout the country" [1, 278]. Zhu De made continuous use of literary and artistic words such as peace, democracy, light and strength to describe the demobilization work in the border areas and the construction of the border area. It is vivid and full of beauty, and it is highly appropriate to express the construction of the border area.

4   The Associative Meaning of the General Official Words in the Shaanxi-Gansu-Ningxia Border Region

4.1   A Thing That Is Associated with What Is to Be Expressed

In the use of official words, the metaphor is often used, and a vivid image is used to refer to other ideas. For this process to work, the nature of the two must be the same. For example, "But because of the importance of the Shaanxi-Gansu-Ningxia Border Region, it has not only been continually tainted by traitors and sectarians, but has also become a thorn in the side of some diehards" [1, 6]. The expression "thorn in the side" refers to a problematic person or situation that is an ongoing source of irritation or concern; in this instance, the Shaanxi-Gansu-Ningxia border area is likened to an embedded thorn of side. The image vividly reflects the bad intentions of the hostile saboteurs on the new red regime.

4.2   A Local Action of a Thing Is Associated with What Is to Be Expressed

In the second session of Vice-President Li Dingming's report on "The Direction of Culture and Education", he discussed the educational role of the blackboard newspaper on the people. He gave an example in which "Two mothers-in-law fights, and can't compromise. Some people say, 'if you fight again, I want you to climb the blackboard', and the two mother-in-law compromise and quietly walk away" [1, 228]. Of course, people cannot literally "climb the blackboard". Here, the phrase "climb the blackboard" is used to refer to the idea of two individuals' quarrels being recorded on the black-board, while the masses are privy to the conflict. The local action of "climb the blackboard" instead of the thing of using the blackboard and newspaper to criticize and educate people publicly. Instead of using basic or simplistic language, it is easy to be expressive and humorous even in an official report.

5   Conclusion

Through the collation and induction of the corpus, it can be found that in the process of selecting and using the official documents words in the Shaanxi-Gansu-Ningxia Border Region, people often instill their own subjective thoughts and emotions into it, which causes the additional meaning of the official documents words in Shaanxi-Gansu-Ningxia Border Region. By studying the additional meaning behind the official documents words, we can see the realistic tendency of the people's emotions, and even reflect the social, political, economic and cultural life at that time.

References

1. Editor-in-Chief of the Institute of Modern History of the Chinese Academy of Social Sciences. Documentary Collection of the Shaanxi-Gansu-Ningxia Border Region. Intellectual Property Publishing House, Beijing (2013)
2. Cao, Yu.: Research on Modern Chinese Vocabulary (Revised). Jinan University Press, Guangzhou (2010)
3. Bai, J.: Official Document Reform in Shaanxi-Gansu-Ningxia Border Region. Friends Secretary (5) (1985)
4. Zhang, L.: The definition of literary style and literary forms. J. Yantai Univ. (Phil. Soc. Sci.) (1) (2005)

From Lexical Semantics to Traditional Ecological Knowledge: On Precipitation, Condensation and Suspension Expressions in Chinese

Chu-Ren Huang[1,2(✉)] ⓘ and Sicong Dong[3] ⓘ

[1] Department of Chinese and Bilingual Studies, The Hong Kong Polytechnic University, Hong Kong, China
churen.huang@polyu.edu.hk
[2] Center for Chinese Linguistics, Peking University, Beijing, China
[3] The Hong Kong Polytechnic University - Peking University Research Centre on Chinese Linguistics, Beijing, China
szechungtung@gmail.com

Abstract. Precipitation, condensation and suspension are different meteorological events involving water in different forms. They are conceptualised and conventionalised with various verbal constructions in Sinitic languages. In this paper, we analyse data from three Mandarin varieties and 229 Sinitic languages, as well as materials from Old Chinese, to support the claim that there is an underlying shared conceptualisation scheme to account for all the variations, and that traditional ecological knowledge (TEK) can be extracted based on the directionality expressed by these linguistic constructions and PoS of weather words. Specifically, we found that across all Mandarin varieties and Sinitic languages, the weather verbs for precipitation (e.g., rain, snow and hail) typically represent downward movement and the weather phenomena words can typically act as verbs in Old Chinese. On the other hand, although the weather verbs for condensation (e.g., dew and frost) also tend to represent downward movement but the weather nouns typically do not have verbal usage in Old Chinese. Lastly, the weather verbs for suspension (e.g., fog and mist) are directionally uncertain and cannot function as a verb in Old Chinese either. The radical 雨 shared by Chinese characters denoting these phenomena provided the conceptual ground for morpho-semantic and grammatical behaviours based on Hantology. Our findings not only have important implications for linguistic ontology and lexical semantics, but also lend support to the emerging area of language-based reconstruction of TEK.

Keywords: Traditional ecological knowledge (TEK) · Chinese · Sinitic languages · Weather · Precipitation · Condensation · Suspension

This work was partially funded by the Hong Kong Polytechnic University (Project 4-ZZHK) and the Hong Kong Polytechnic University - Peking University Research Centre on Chinese Linguistics.

J.-F. Hong et al. (Eds.): CLSW 2019, LNAI 11831, pp. 255–264, 2020.
https://doi.org/10.1007/978-3-030-38189-9_27

1 Introduction

1.1 Weather and Language

We talk about weather when we do not know what to talk about, or when we first meet. Weather events are among the most observed and most discussed natural phenomena, and meteorology is one of the few scientific disciplines regularly discussed in daily conversation and mass media. Weather plays a central role not only in language use but also in language typology [1–4, etc.] as languages typically have special constructions for weather words. The current study is conducted in this general context on the issue of the relation between language and ecological knowledge. In particular, we would like to capture the underlying principles for linguistic conceptualisation of weather expressions and to compare this conventional knowledge to current meteorology.

Our study covers the expressions of weather phenomena involving atmospheric water. They are commonly classified into three types: Precipitation, Condensation and Suspension. First, rain, snow, hail, sleet and graupel, all condensed water vapour that falls under gravity, are the main forms of precipitation. Second, dew and frost, formed by the cooling water vapour condensing and becoming liquid or solid form on some surfaces, are called condensation. Lastly, although fog and mist are also formed by cooling of water vapour, yet since they have rather wide spatial distributions and their water droplets are very small and suspending in the air, hence they are classified as suspension.

1.2 Same Radical, Same Perception?

In this study, we focus on six weather phenomena belonging to the aforementioned three categories in the hydrologic cycle [5]. In particular, we look into the way how Mandarin Chinese and other Sinitic languages (traditionally called Chinese dialects) express the formation and/or appearance of such phenomena. They are: 雨 *yǔ* 'rain', 雪 *xuě* 'snow' and 雹 *báo* 'hail', which are precipitation; 露 *lù* 'dew' and 霜 *shuāng* 'frost', which are condensation; and 霧 *wù* 'fog', which is suspension.

Note that all the six characters referring to the weather phenomena have one component in common: they all contain the radical 雨 *yǔ* 'rain'. It is known that over 80% of Chinese characters are formed by the form-and-sound (形聲 *xíngshēng*) principle in terms of internal composition and contain a radical to represent their semantic classes [6]. Among the six characters for weather phenomena, 雨 *yǔ*, lacking a phonetic part and denoting rain itself, functions as a radical part in the other five characters. Thus, the five characters should, in principle, belong to the semantic category same with or related to 雨 *yǔ* 'rain'. However, as previously introduced, such six phenomena belong to three meteorological categories and have different physical manifestations concerning forms, spatial distributions, directionality, etc., but the shared radical 雨 *yǔ* 'rain' does not seem to provide such taxonomic information. Additionally, the annotations to the six characters in *Shuo Wen Jie Zi* 'Explaining Graphs and Analysing Characters', one of the oldest preserved Chinese dictionaries, as listed below, do not show clear distinctions among the three meteorological categories. So, the question we want to address is: does their sharing of the same radical suggest that such meteorological classification is not part of the linguistic knowledge?

雨 *yǔ* 'rain': 水從雲下也 'water falling from clouds'
雪 *xuě* 'snow': 凝雨, 說物者 'condensed rain, which pleases things'
雹 *báo* 'hail': 雨冰也 'to rain ice'
露 *lù* 'dew': 潤澤也 'to moisturise'
霜 *shuāng* 'frost': 喪也, 成物者 'to perish and preserve; substance to make farming accomplished'
霚(霧) *wù* 'fog': 地气發, 天不應 'ground emits vapour while the sky does not respond'

1.3 Hantology and TEK

Radicals in Chinese characters do represent semantic classifications, but not in a straightforward manner. In fact, a complex conceptual system, i.e., the linguistic ontology of Chinese, is conventionalised by character radicals [7].

An ontology is an explicit specification of a conceptualisation [8]. According to Huang [9], ontology studies the system for knowledge representation in terms of basic concepts and how these concepts are organised in terms of relations. Being a logophonetic writing system, Chinese characters are different from Latin alphabet that people may know the meanings even if they do not know the exact pronunciations. This entails the Chinese orthography a robust conceptual system that has endured over 3,000 years of sound change and adopted by diverse speech communities of Sinitic languages and even neighbouring languages such as Japanese, with more than one billion users at present. Hence, an ontology based on the relations of Chinese characters and their meaning clusters, the Hantology (*hanzi* ontology), is proposed by Chou [10] and Chou and Huang [6], and it has been argued recently that Hantology has its psychological reality among Chinese people [11]. The meanings of the characters are also mapped to Suggested Upper Merged Ontology (SUMO) [12], which makes it possible to share information with other ontologies. Our current study will show that such conceptual system is underlying various linguistic behaviours related to weather phenomena.

Another important implication of the current line of inquiry is the discovery of traditional ecological knowledge (TEK). Huntington [13], for instance, showed that TEK could complement scientific data and methods. Nazarea [14] claimed that with the loss of language, we also lost our irreplaceable accumulated knowledge about the ecology and biodiversity stored in that language. Gagnon and Berteaux [15], among others, demonstrated the significant benefits of combining TEK with ecological sciences. Vossen et al. [16], on the other hand, showed that environmental and ecological knowledge from different languages could be effectively integrated based on ontology and Wordnet. Huang et al. [17, 18] extracted traditional knowledge of ecology in China based on classical texts and ontology. Recently, Huang et al. [19] showed that, with a linked data approach, ecological information can be recovered with very small linguistic data. Our study can complete the above by tackling the issue of how to extract useful ecological knowledge when the linguistic expressions represent such knowledge in an oblique way, or even seem to be infelicitous scientifically. Furthermore, a crossregional comparison of different Mandarin varieties can also shed light on how ecological knowledge is lost due to language variation.

2 Methodology

Our major purpose is to find out whether the classification of precipitation, condensation and suspension in meteorology are linguistically well represented in Mandarin Chinese and other Sinitic languages, based on six weather phenomena whose characters share the radical 雨. From the previous related research such as Dong [20] and Ren [21], we inferred that the directionality of these phenomena and their parts of speech (PoS) in Old Chinese might provide very important starting points for the discussion. Consequently, we used corpora and dictionaries to acquire data concerning the directionality of these weather phenomena.

We consulted three major corpora for the usage in Mandarin Chinese: BCC Corpus [22] and CCL Corpus [23] for Mainland Mandarin, and Sinica Corpus 4.0 [24] for Taiwan Mandarin. 12 compounds were searched in these corpora: '降/下/起/上 + 露/霜/霧'. Among them, 降 *jiàng* 'to fall' and 下 *xià* 'to fall' are verbs indicating downward movement, while verbs 起 *qǐ* 'to rise' and 上 *shàng* 'to rise' indicate upward movement. Rain, snow and hail are not included here because they consistently co-occur with verbs of downward meanings in Mandarin.

For a further comparison among different Mandarin varieties, we also obtained the collocations of dew, frost and fog using Tagged Chinese Gigaword 2.0 [25] via Chinese Word Sketch (CWS) [26]. Three sub-corpora consisting of Mandarin data from Taiwan, Mainland China and Singapore, respectively, were utilised.

As for data of Sinitic languages, dictionaries of Li [27], Xu and Miyata [28], Tao [29] and Zhang and Mo [30] were used. The expressions concerning the appearance of rain, snow, hail, dew, frost and fog were collected, and were classified into four types in terms of directions: upward, downward, both and no obvious vertical direction.

3 Results

The 12 forms ('降/下/起/上 + 露/霜/霧') are distributed in BCC, CCL and Sinica corpora as shown in Table 1 (accessed 8 and 13 Dec 2018 and 21 Jan 2019, respectively).

In the three sub-corpora of Tagged Chinese Gigaword 2.0 (accessed 22 Jan 2019), 露 *lù* or 露水 *lùshuǐ* 'dew' do not have collocated verbs expressing formation or appearance of the phenomenon, while such collocated verbs for frost and fog are found and shown in Table 2.

We have investigated 229 Sinitic languages or dialects. Compounds or phrases meaning the appearance of the six phenomena were collected and grouped according to the directions they indicate. The percentages of directions concerning each phenomenon are shown in Table 3.

Table 1. Distributions in corpora

	BCC	CCL	Sinica
降露 *jiànglù*	7	2	0
下露 *xiàlù*	7	4	0
起露 *qǐlù*	1	0	0
上露 *shànglù*	0	0	0
降霜 *jiàngshuāng*	66	13	0
下霜 *xiàshuāng*	119	26	1
起霜 *qǐshuāng*	14	0	0
上霜 *shàngshuāng*	16	2	0
降霧 *jiàngwù*	11	8	0
下霧 *xiàwù*	174	10	0
起霧 *qǐwù*	352	22	4
上霧 *shàngwù*	5	0	0

Table 2. Collocations in Chinese Gigaword 2.0

霧 *wù* 'fog'	Taiwan	突起 'to rise suddenly', 出現 'to appear', 形成 'to form', 飄 'to drift', 起 'to rise'
	Mainland China	出現 'to appear', 形成 'to form', 降 'to fall', 突起 'to rise suddenly', 起 'to rise', 漸起 'to rise gradually', 驟起 'to rise abruptly'
	Singapore	形成 'to form', 發生 'to occur'
霜 *shuāng* 'frost'	Taiwan	結成 'to condense to', 結 'to condense', 凍成 'to freeze to', 出現 'to appear', 凝結成 'to condense to'
	Mainland China	凍成 'to freeze to', 出現 'to appear', 結 'to condense'

Table 3. Directional distributions in Sinitic languages (%)

	Down	Up	Both	None
Snow	100	0	0	0
Rain	98.3	0	0	1.7
Hail	97.0	0	0	3.0
Fog	51.7	24.7	13.5	10.1
Dew	50.0	15.2	4.5	30.3
Frost	45.5	0	2.0	52.5

4   Discussion

4.1   Different Types of Water Have Different Linguistic Representations

Based on the data we investigated, the three types of water-related phenomena, i.e., precipitation, condensation and suspension, are shown to behave differently in morpho-semantics and grammar. We will illustrate it from modern Sinitic languages and Old Chinese respectively.

First, in Mandarin Chinese, the three weather types exhibit three different patterns in terms of directionality. For the precipitation category, as has been mentioned, rain, snow and hail are consistently expressed with downward movement. The condensation dew and frost, as shown in Table 1, can be expressed both as moving downwards and upwards, but the downward meaning has bigger proportions. Being suspension, fog can also be said to move downwards and upwards, but more cases were found to indicate the upward meaning. In addition, the usage in Taiwan Mandarin from Sinica Corpus can provide interesting support. Due to its corpus size, only five results were obtained, whereas they display clear-cut tendencies: fog can only be 'rising', and frost can only be 'falling'. This distinguishes suspension from condensation in an extreme way.

Secondly, statistical results of Sinitic languages demonstrate a similar picture with Mandarin, but with different details for fog-related expressions, as illustrated in Table 3. In these languages, fog tend to be expressed as 'falling'. It seems that the data cannot tell suspension and condensation apart. However, fog has the biggest propor-tions of 'up' and 'both' among all the six phenomena, which can serve as evidence that fog is most likely to be described as moving upwards by Chinese people, although it is more often said to move downwards. This makes fog standing out in these phenomena with seeming uncertainty about directions.

Thirdly, in Old Chinese, the three types are also linguistically separate from each other, but two sets of criteria are needed for the distinction: directionality and PoS. Based on Ren's [21] research, 雨 *yǔ* 'rain', 雪 *xuě* 'snow' and 雹 *báo* 'hail' can function as nouns, as well as weather verbs in Old Chinese, meaning to rain, to snow and to hail, respectively, while 露 *lù* 'dew', 霜 *shuāng* 'frost' and 雾 *wù* 'fog' have almost no verbal usage. This shows that precipitation and non-precipitation can be set apart by PoS. Moreover, 露 *lù* 'dew' and 霜 *shuāng* 'frost' need verbs with downward meanings to take them as objects to express their appearance, thus indicating the same directional meaning with rain, snow and hail. 雾 *wù* 'fog', on the other hand, show directional uncertainty again. In the ancient documents we investigated, no expression was found describing the occurrence of fog. However, the annotations to two variant Chinese characters meaning fog in *Shuo Wen Jie Zi* present even contradictory views on fog's directions: 霿 is 地气發, 天不應 'ground emits vapour while the sky does not respond', and 霚 is 天气下, 地不應 'sky's vapour falls while the ground does not respond'. Therefore, the three types differ in the way that precipitation is [+V, +Down], condensation is [−V, +Down], while suspension is [−V, ±Down].

4.2   An Account from Hantology

The radical 雨 shared by these characters can account for the above differences based on Hantology. In Hantology, the radical 雨 represents more than one concept, hence is linked to ontology nodes 'water' and 'weather process' in SUMO [6]. We can view the ontological class of 'water' as weather products, and 'weather process' as weather events. Specifically, 露 *lù* 'dew', 霜 *shuāng* 'frost' and 霧 *wù* 'fog' are linked to 'water', and 雨 *yǔ* 'rain', 雪 *xuě* 'snow' and 雹 *báo* 'hail' to both. Such linking conforms to the characteristics of these phenomena and can provide explanations to some scientific infelicity.

Precipitation phenomena rain, snow and hail display conspicuous movement during their process, and have observable products as raindrops, snowflakes and hailstones. Hence, they can be perceived in both ways as dynamic events and static substance. This is why they have verbal and nominal usages in Old Chinese and are linked to both ontological nodes. Dew, frost and fog, on the other hand, do not typically involve obvious movement, thus lack enough activity to function as verbs and can be only perceived as products.

As for directionality, dew, frost and fog exhibit a rather 'extraordinary' fashion. We know from meteorology that they do not move downwards when forming [5]. However, they are said to 'fall' according to the linguistic data. We claim that the 'water' node they share with precipitation phenomena may reflect that they are conceptualised as same as precipitation. With regard to fog, since it distributes in a much larger vertical range, hence unlike dew and frost near ground, its status is possible to be interpreted almost equally as 'fallen' or 'risen'.

4.3   Language Variation and Knowledge Loss

Ecological knowledge will be lost not only to the loss of language, but possibly also to the variation of language. Mandarin is originated in northern China, and has developed various varieties for sociocultural reasons. Therefore, how the knowledge is kept and lost in different varieties is an intriguing topic. With the data collected in Tables 1 and 2, we could take a brief look into this issue.

Several findings are in order here: first, frost cannot find collocated verbs of movement or appearance in Singaporean Mandarin; second, no verbs with directional meanings are found in Singaporean Mandarin; third, verbs with downward meanings are found solely in Mainland Mandarin to describe condensation and suspension phenomena. More details are surely needed for a full-scale account. Nevertheless, an explanation can be given to one thing at present. Frost is rarely seen in the tropical areas, so Singaporean Mandarin might have very limited representations of such phenomenon, and the related knowledge is lost or partly lost accordingly.

5 Conclusions

We claim in this paper that the three types of weather phenomena in hydrologic cycle, i.e., precipitation (rain, snow and hail), condensation (dew and frost) and suspension (fog), are represented differently in Chinese. Data from Mandarin varieties, other Sinitic languages and Old Chinese all suggest that such distinctions in our weather knowledge have linguistic representations in terms of directionality and PoS. Based on Hantology, the radical 雨 shared by the Chinese characters denoting these phenomena is argued to account for the morphosemantic and grammatical behaviours.

Sensory modalities of these weather phenomena can provide another perspective. Based on the collocations found in Chinese Gigaword 2.0 (accessed 25 Dec 2018), rain and hail involve visual, tactile and auditory senses; snow, dew, frost and fog involve visual and tactile senses. According to Strik Lievers and Winter [31] and Zhong et al. [32], sound concepts are more prone to being expressed as verbs in both English and Mandarin. It is shown in our data that 雨 *yǔ* 'rain' and 雹 *báo* 'hail', which contain auditory sense, can function as verbs in Old Chinese. This can serve as evidence from the early stage of one language for the relation between sensory modalities and lexical categories.

Last, but not least, our study shows that the morphosemantic and grammatical generalisations of these weather words in Chinese in fact do represent our knowledge of meteorology with felicity when their linguistic behaviours are given the appropriate ontological interpretation. This offers strong support for the effectiveness of an ontology-driven approach to the interpretation and integration of TEK with scientific knowledge. Besides, our preliminary cross-regional comparison among Mandarin varieties also shows that ecological knowledge may be lost due to language variation but could contribute to the typological studies of weather and language.

References

1. Eriksen, P., Kittilä, S., Kolehmainen, L.: Linguistics of weather: cross-linguistic patterns of meteorological expressions. Stud. Lang. **34**(3), 565–601 (2010). https://doi.org/10.1075/sl.34.3.03eri
2. Eriksen, P., Kittilä, S., Kolehmainen, L.: Weather and language. Lang. Linguist. Compass **6**(6), 383–402 (2012). https://doi.org/10.1002/lnc3.341
3. Van Hoey, T.: Does the thunder roll? Mandarin Chinese meteorological expressions and their iconicity. Cognit. Semant. **4**(2), 230–259 (2018). https://doi.org/10.1163/23526416-00402003
4. Dong, S.: A study on meteorological words in Chinese. Postdoctoral report, Peking University (2019). https://doi.org/10.13140/rg.2.2.22648.29445. (in Chinese)
5. Ahrens, D.: Essentials of Meteorology: An Invitation to the Atmosphere, 6th edn. Brooks/Cole, Cengage Learning, Belmont (2012)
6. Chou, Y.-M., Huang, C.-R.: Hantology: conceptual system discovery based on orthographic convention. In: Huang, C.-R., et al. (eds.) Ontology and the Lexicon: A Natural Language Processing Perspective, pp. 122–143. Cambridge University Press, New York (2010). https://doi.org/10.1017/cbo9780511676536.009

7. Huang, C.-R., Yang, Y.-J., Chen, S.-Y.: Radicals as ontologies: concept derivation and knowledge representation of four-hoofed mammals as semantic symbols. In: Cao, G., et al. (eds.) Breaking Down the Barriers: Interdisciplinary Studies in Chinese Linguistics and Beyond, pp. 1117–1133. Institute of Linguistics, Academia Sinica, Taipei (2013)
8. Gruber, T.: Toward principles for the design of ontologies used for knowledge sharing. Int. J. Hum. Comput. Stud. **43**(5–6), 907–928 (1995). https://doi.org/10.1006/ijhc.1995.1081
9. Huang, C.-R.: Notes on Chinese grammar and ontology: the endurant/perdurant dichotomy and Mandarin D-M compounds. Lingua Sinica **1**(1) (2015) https://doi.org/10.1186/s40655-015-0004-6
10. Chou, Y.-M.: Hantology: a Chinese character-based knowledge framework and its applications. Doctoral dissertation, National Taiwan University (2005)
11. Yang, Y., Huang, C.-R., Dong, S., Chen, S.: Semantic transparency of radicals in Chinese characters: an ontological perspective. In: Proceedings of the 32nd Pacific Asia Conference on Language, Information and Computation, pp. 788–797 (2018)
12. Niles, I., Pease, A.: Towards a standard upper ontology. In: Welty, C., Smith, B. (eds.) Proceedings of the 2nd International Conference on Formal Ontology in Information Systems (FOIS 2001), pp. 2–9 (2001)
13. Huntington, H.: Using traditional ecological knowledge in science: methods and applications. Ecol. Appl. **10**(5), 1270–1274 (2000). https://doi.org/10.2307/2641282
14. Nazarea, V.: Cultural Memory and Biodiversity. University of Arizona Press, Tucson (2006)
15. Gagnon, C., Berteaux, D.: Integrating traditional ecological knowledge and ecological science: a question of scale. Ecol. Soc. **14**(2), 19 (2009). https://doi.org/10.5751/es-02923-140219
16. Vossen, P., et al.: KYOTO: a system for mining, structuring, and distributing knowledge across languages and cultures. In: Proceedings of the 6th Language Resources and Evaluation Conference (LREC 2008) (2008)
17. Huang, C.-R., Lo, F.-j., Chang, R.-Y., Chang, S.: Reconstructing the ontology of the tang dynasty: a pilot study of the Shakespearean-garden approach. In: Proceedings of OntoLex 2004: Ontologies and Lexical Resources in Distributed Environments (2004)
18. Huang, C.-R., Hsieh, S.-K., Chang, R.-Y., Luo, F.-j.: From classical poetry to modern ontology: bridging the knowledge divide with a linked data approach. In: Huang, C.-R. (ed.) Digital Humanties: Bridging the Divide. Springer (to appear)
19. Huang, C.-R., Hsieh, S.-K., Prévot, L., Hsiao, P.-Y., Chang, H.: Linking basic lexicon to shared ontology for endangered languages: a linked data approach toward Formosan languages. J. Chin. Linguist. **46**(2), 227–268 (2018). https://doi.org/10.1353/jcl.2018.0009
20. Dong, S.: Directions of fog. Yuyan Wenzi Zhoubao, 2, 1 Aug 2018. https://doi.org/10.13140/rg.2.2.15796.81289. (in Chinese)
21. Ren, H.: Noun-verb conversion in old Chinese: from the perspective of lexical semantic analysis. Doctoral dissertation, Peking University (2018). https://doi.org/10.13140/rg.2.2.16642.22724. (in Chinese)
22. Xun, E., Rao, G., Xiao, X., Zang, J.: The construction of the BCC corpus in the age of big data. Corpus Linguist. **3**(1), 93–109 (2016). http://bcc.blcu.edu.cn/. (in Chinese)
23. Zhan, W., Guo, R., Chen, Y.: The CCL Corpus of Chinese texts: 700 million Chinese characters, the 11th Century B.C. - present (2003) http://ccl.pku.edu.cn:8080/ccl_corpus
24. Chen, K.-J., Huang, C.-R., Chang, L.-P., Hsu, H.-L.: SINICA CORPUS: design methodology for balanced corpora. In: Park, B.-S., Kim, J.-B. (eds.) Proceedings of the 11th Pacific Asia Conference on Language, Information and Computation, pp. 167–176 (1996). http://asbc.iis.sinica.edu.tw/index_range.htm
25. Huang, C-R.: Tagged Chinese Gigaword Version 2.0 (LDC2009T14). Linguistic Data Consortium, Philadelphia (2009). https://catalog.ldc.upenn.edu/LDC2009T14

26. Huang, C.-R., et al.: Chinese Sketch Engine and the extraction of grammatical collocations. In: Proceedings of the Fourth SIGHAN Workshop on Chinese Language Processing, pp. 48–55 (2005). http://cwk.cbs.polyu.edu.hk/index.html, http://wordsketch.ling.sinica.edu.tw/

27. Li, R., (ed.): Great Dictionary of Modern Chinese Dialects. Jiangsu Education Publishing House, Nanjing (1993–2003). (42 Volumes) (in Chinese)

28. Xu, B., Miyata, I. (eds.): A Comprehensive Dictionary of Chinese Dialects. Zhonghua Book Company, Beijing (1999). (in Chinese)

29. Tao, G. (ed.): Dictionary of Nantong Dialect. Jiangsu People's Publishing Ltd., Nanjing (2007). (in Chinese)

30. Zhang, W., Mo, C.: Dictionary of Lanzhou Dialect. China Social Sciences Press, Beijing (2009). (in Chinese)

31. Strik Lievers, F., Winter, B.: Sensory language across lexical categories. Lingua **204**, 45–61 (2018). https://doi.org/10.1016/j.lingua.2017.11.002

32. Zhong, Y., Dong, S., Huang, C.-R.: Lexical differentiation in Mandarin Chinese sensory lexicon. Presented at the 10th International Conference of the European Association of Chinese Linguistics (EACL-10), Milan (2018)

A Diachronic Study of Structure Patterns of Ditransitive Verbs

Guoyan Lyu[1(✉)] and Yanmei Gao[2]

[1] School of Foreign Studies, Beijing Information Science and Technology University, Beijing 100192, China
guoyankulta@163.com
[2] School of Foreign Languages, Peking University, Beijing 100871, China
ymgao2013@126.com

Abstract. This paper mainly studies English dative alternation from a diachronic point of view. As a verb of caused-possession, the typical ditransitive verb *give* is contrasted with *send* and *tell* in respect of their different structural biases diachronically (12th–19th centuries). We built dynamic, syn-diachronic structural models, and made contrastive observations on structural biases of the verbs. The verb *give* shows a strong tendency to occur in the double object construction, *send* has a structural bias towards the prepositional object construction and *tell* shows preference for the clausal complement diachronically, which suggests the determining role of lexical semantics of the verb in its structural preference and the meaning of syntactic structures in which the verb occurs.

Keywords: Dative alternation · Ditransitive verbs · Diachronic · Structural biases · Lexical semantics

1 Introduction

There is a very large literature on the dative alternation as illustrated in the following sentences[1]:

(1) a. Mayor John Hughes gave Mr. Rees the boot. (DOC)
    b. He gave the photograph of the graveyard to Jacques. (POC).

The dative alternation as shown in (1) consists of a ditransitive verb (e.g., *give*), an Agent, a Recipient and a Theme argument. The alternation features two opposite Object orders. Sentence (1a) is a prototypical example of the double object construction (DOC) in which the indirect object (IO) precedes the direct object (DO). Sentence (1b) is a prepositional object construction (POC) with the opposite order of Objects: DO-to-IO.

With regard to the alternating structures in (1), an important issue existing in literature is whether the two structures in the alternation express the same meaning. In relation to that, another question is on the interaction between the lexical semantics of the verb and its syntactic behaviors in the alternation.

---

[1] Example sentences in (1) are extracted from British National Corpus (BNC).

© Springer Nature Switzerland AG 2020
J.-F. Hong et al. (Eds.): CLSW 2019, LNAI 11831, pp. 265–274, 2020.
https://doi.org/10.1007/978-3-030-38189-9_28

2 Dative Alternation

In China, early attempts concerning the structural meanings in the alternation have been made by Zhu to study different types of alternating double object and prepositional structures in connection with the semantic features of the verb of general purpose of give, "gei", in Chinese [1]. Zhu defined the semantics involved in the dative alternation as follows: (a) there are Giver and Recipient; (b) there exists the Thing that is given and received; (c) the Giver intentionally makes the Thing move from the Giver to the Recipient [1]. In terms of the structural meanings, Peyaube considers that the DOC is the original structure expressing the meaning of giving, and other ditransitive constructions are somehow derived from this structure [2]. Similarly, Zhang employed the term "ditransitive construction" and pointed out the meaning of the construction of V-N1-N2 as "intended giving type of transferring" of the Theme from the Agent to the Recipient [3]. On the other hand, Zhang considered the structural meaning of the double object construction to be determined by the verbs of taking; therefore, the meta-structural meaning of the construction is taking instead of giving [4]. He also made the claim that the DOC in modern Chinese is derived from the POC rather than from the same structure in ancient Chinese.

There are similar debates on English dative alternation especially on the type illustrated in (1) [5–15]. Among these synchronic studies, there are two major classes of analysis for this alternation. One assumes that both variants are associated with the same meaning, with this meaning allowing two argument realization options. The other assumes that the variants are associated with different but related meanings, with each meaning giving rise to a distinct argument realization pattern.

The currently dominant approach is the polysemy approach, which assumes a non-derivational relation between the variants: each is associated with its own meaning, and each gives rise to its own realization of arguments [15, 16]. That is to say, the DOC in English expresses caused-possession ('Agent causes Recipient to have Theme') and the POC expresses caused-motion ('Agent causes Theme to move to Recipient').

However, in monosemy approaches, people consider the two variants to share one single meaning. For example, Rappaport Hovav and Levin argue that English verbs that denote transfer of possession, such as *give*, only denote the sense of caused-possession (CP), even in the *to*-dative structure [8]. It is only in *send*-type verbs that different constructions are associated with different meanings; that is, the caused-motion (CM) event is realized by the POC and the caused-possession by the DOC. This idea bears its early root in Levin's verb classification where verbs are classified into different types based on their syn-semantic behaviors in different syntactic structures [17].

3 Ditransitive Verbs

Jackendoff distinguishes the *give*-type verbs from *throw*-type verbs [18]. He provides an analysis for the alternation with verbs like *give* and *sell*, the meaning of which inherently involves caused-possession (CP), and a distinct analysis for the alternation with verbs like *throw* and *send* (*throw*-type verbs). In this sense, *give* inherently takes three arguments, and it semantically involves a Theme moving along a possessional path from a source to a Goal – the Recipient. Therefore, with *give*-type verbs, the two variants in the dative alternation have basically the same meaning. On the other hand, the *throw*-type verbs, denoting causation of motion (CM), such as *throw, send,* are considered to select two arguments: An Agent and a Theme, allowing the dative alternation with a Goal, and requiring no Recipient.

Chinese scholars also made efforts in classification of verbs in the dative alternation. For example, Ma and Li classified ditransitive verbs and distinguished features of the double object structures in Chinese. Ma grouped the verbs "gei", "song (send)" and "gaosu (tell)" in the first type which includes all verbs of giving [19]. Li classified "gei" and "song" in the same category of verbs denoting outward motions as opposed to inward motions (e.g., "na [take]" or "tou [steal]") [20]. Finer-grained investigations of the lexical features of ditransitive verbs can be found in historical studies of Chinese ditransitive constructions, such as in Shi and later in Xu [21, 22]. They focused on the *give*-type of verbs in different periods of ancient Chinese. Yet, their attention was paid more to the discrete verb classification and exemplification of structural instances, while much less attention to the continuous developmental trends of verb meanings and structural choices.

Zhang in his comprehensive dialectical and typological study, suggests that the evolution of the ditransitive verb *give* may have typological significance and provide valid explanations on the syn-semantic relations between lexical semantics and constructional meanings [23]. To some extent, this motivates us to delve into the diachronic variations and changes in the structure patterns of the ditransitive verbs in English so as to reveal the lexical semantics of verbs and their syntactic behaviors in the dative alternation across different languages.

In our study, we will examine three verbs, *give, send* and *tell*. We assume that the similarity of the POC with the verb *send* and the POC with *give* is only a surface similarity. Since the verbs *send* and *give* lexicalize contrasting inherent meanings, caused-motion and caused-possession, respectively, they should be associated with the respective constructions.

4   Contrastive Diachronic Structural Patterns

We propose to make a diachronic exploration, from Old English (OE), Middle English (ME) and to Modern English (ModE) in an attempt to find out the relation between the meaning of the two variants and the historical evolution of ditransitive verbs along with their syntactic behaviors. The historical manuscripts which form the basis of our diachronic investigation consist of 15 diachronic chronicles or epochs, including six from OE (870–end of 12th century), three from ME (early 13th century – end of 15th century, two from early ModE (16th century-17th century) and three from ModE (18th- 20th centuries). The PDE instances are mainly extracted from the British National Corpus (BNC).

We employed a data-based, and data-driven research. The major type and token frequencies were calculated based on a certain lexical item or a type of structure. With regard to the relative token frequencies, they were calculated by close observation with manual extraction. Based on our research of literature, such diachronic study on the dative alternation has hardly been done before.

We will build diachronic models to reconstruct the diachronic variations in the argument structure types of ditransitive verbs from earlier language stages to the modern time focusing on three verbs, *give*, *tell* and *send*. These verbs are selected for distinct reasons. First, the verb *give* is considered in Zhang to be a general-purpose verb of giving, a representative type of ditransitive verb [23]. According to Kittilä's Universal 1, "If a language has only one ditransitive trivalent verb (on the basis of any feature of formal transitivity), then that verb is *give* [12]." It is also considered in Bresnan and Nikitina to be the prototypical ditransitive verb which has the highest frequency in the parsed SWITCHBOARD corpus, constituting 42% of all alternating dative verbs [24]. Therefore, the verb give is chosen as a representative of the most typical ditransitive verb.

The verb *send* is found in literature to be one of the most frequently used verbs denoting motion-goal relation, which could be a viable candidate in comparison with *give*. In other words, an investigation on their patterns of argument alternation could reveal two types of argument alternation biases if the semantics of verb determines the semantics of the structures in which they occur as stated by Rappaport Hovav and Levin [8]. The verb *tell*, is considered (by Levin) to be a verb denoting transferring of messages [17]. The verb *tell* is found to be the most frequent verb of communication that occurs in dative constructions in the parsed SWITCHBOARD corpus (in Bresnan and Nikitina) [24]. Though it does not express transfer of possession in the literal sense, it is considered to fit into the hypothesis that they are mentally represented as metaphoric extensions denoting the notion of caused-possession (Gropen et al.) [25].

4.1 Different Structural Trends

To construct diachronic models for the evolutions of particular lexical terms and argument structure types, we collected different kinds of token frequencies of lexical items and token frequencies of argument structure types in which *give*, *send* and *tell* occur in the diachronic materials. Our diachronic study of the structural biases of the three verbs indicates that *give*, *send* and *tell* behave differently in terms of the DOC and the POC. In comparing the structural biases of the three verbs, *give*, *tell* and *send*, we first consider the different structural biases of the three verbs in terms of the DOC. In Fig. 1 below, the IO includes both types of full NP and pronouns, so does the DO. The data in Fig. 1 indicates the general comparative distributions of the DOC in which the verbs *give*, *tell*, and *send* occur.

As shown in Fig. 1, *give*, as a verb of causative possession, diachronically has the highest relative frequency in the DOC, the V-IO-DO structure. Its average relative token frequency in the course of seven centuries is around 44%. The verb *tell*, a verb of transfer of a message, has a 12% frequency ratio and the verb *send*, a verb of caused-motion, has about 10% in the DOC. This pattern of distribution suggests that if we take the DOC to be a caused-possession structure denoting successful transfer or reception, then among the three verbs, *give* is the most typical one to occur in this structure. On average, the verb *tell* seems to have a bit higher ratio of token frequency than *send*, yet not so prominent in the long term, especially after the 15th century as the token frequency ratio of the DOC starts to be relatively stabilized.

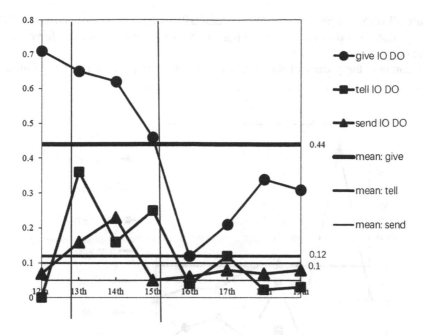

Fig. 1. *Give, Tell* and *Send* in DOC (12–19th centuries).

It is also shown in Fig. 1 that in the era between the 13th century and the 15th century, there has been a notable change for the structural patterns of the three verbs. The structural distributions of the DOC for both *send* and *tell* experience some increase. Then from 16th century until the 19th century, this ratio for both verbs falls back to relatively the same level as that early in the 12th century. This means that the changes in the development of English, especially the change in the case system also has some influence on the structural biases of *send* and *tell*. Yet once the readjustment is stabilized, the two verbs retain their usual structural patterns as those in OE.

However, it is a totally different case for the verb *give*. Between the 13th and 16th centuries, the verb *give* as indicated in Fig. 1, shows a significant modifying trend in its structural biases. The distribution ratio of the DOC has been falling all the way from over 75% (in the 12th century) to 12% (in the 16th century). It suggests that the change in the English language, especially that in the case system and the increased use of prepositions exerts a profound impact on the structural bias of *give* in the manner of dragging down the ratio of the DOC prominently, while at the same time sparing space for the distribution of the newly emerging POC. Since the 17th century until the 19th century, the ratio of the DOC has been elevated to a higher level, yet still much lower than that in the 12th century. This could mean that once the English language has gradually completed its significant changes in ME, the POC has become an alternative structure for the verb *give*, and thus shares the distributional space with the DOC in a correlative way.

If we compare the three verbs in the structural pattern of the DOC, we can see that either CM or CP verb appears in the DOC, the CP-CM distinction becomes irrelevant because all verbs appearing in the DOC necessarily entail a change of possession. This suggests that the constructional meaning of the DOC exhibits strong force on the lexical semantics of the verbs.

In contrast, the pattern of the POC exhibits different patterns as in the following Fig. 2.

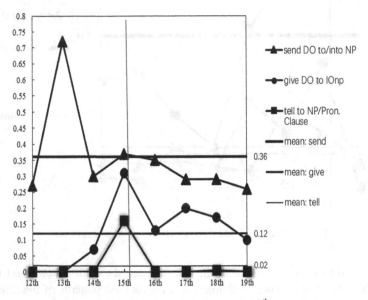

Fig. 2. *Give, Tell* and *Send* in POC (12–19th centuries).

With regard to the structure type POC, V-DO-to-IO, Fig. 2 indicates that the relative frequency of the POC (among all the constructions in which the verb occurs) for the verb *send* (36% on average) has always been at a higher level than that of *give* (12%) and *tell* (2%) altogether. This means that to some extent, the verb *send* is more likely to be seen in the POC diachronically, which is a typical structure denoting a motion towards a Goal than both *give* and *tell*. This corresponds to the observation in Rappaport Hovav and Levin that the verb *send* is one of the verbs of change of location or caused-motion [8]. In terms of semantic roles, aside from the Theme argument, the verb *send* takes a Goal while *give* takes a Recipient argument.

In the case of *send*, the only significant change shown in Fig. 2 in the 13th century is that it seems to have experienced a salient increase of the distribution of the POC. Other than that, there seems to be no noticeable modification in the distributional pattern of the POC during the time from 12th to the 19th centuries. This means that as a CM verb, *send* has always been different from *give*. In OE when *give* has no instance in the POC, *send* used to show preference for it, and in ME until ModE, *send* has always shown to have the same preference for the caused-motion structure POC. Even during the time when the case system changes significantly in English, the structural bias of *send* shows to be influenced by the change in a very limited manner.

As shown in Fig. 2, the most significant and relatively stable trend of the verb *tell* is: it has never been a verb co-occurring with the preposition *to* (except for the 15% ratio in the 15th century when the use of prepositions surged to an unreasonable level). It suggests that it hardly has the sense fitting in the construction of caused-motion which is expressed usually in the POC. In other words, the POC shows to exert less force on the verb *tell*, which may also mean that the lexical meaning of the verb has a strong countering force against the constructional meaning of POC. There is no hint of meaning related to 'cause to go to a goal' in the root meaning of the verb *tell*. The verb *tell* per se has the least sense of motion towards a Goal and hence has the least token frequency in the structure of V-DO-to-IO. Even during the 15th century when the use of prepositions drastically prevails, the verb *tell* still has the lowest rate (15%) of occurring in prepositional constructions among the three verbs. It most frequently occurs with complement clauses headed by that or wh-word. This implies that it is not a verb denoting caused-motion and its core meaning is incompatible with the sense of the POC which conveys the meaning of caused-motion.

As already shown in Figs. 1 and 2 also indicates a significant trend in the distribution of the POC during the era between the 13th century and the 15th centuries. The verbs *give* and *tell* both show to have significant rising distribution level of the POC, which means that the increasing use of prepositions in the transitional period from OE to ME causes some turbulence in the structural biases of the two verbs. However, since the 16th century, the verb *tell* has shown to have almost no instance in the POC, which suggests that the root meaning of *tell* and its structural preference for Tell-IO-Clause have relatively been the same as that in OE. In contrast, *give* shows to spare a portion of structural space for the POC, making it an alternative structure as a substitution for its long-lost structure Give-DO-IO in OE when the case system existed. In other words, the verb *give* forces the POC to convey the meaning of cause-possession which used to be expressed by cased-marked DO-IO in OE. And this is largely due to the erosion of the case system in OE.

4.2 Findings

Taking into consideration the statistics in both Figs. 1 and 2, for the structural bias of *give*, the ratio of DOC vs. POC on average is (11:3), *send* (2:7), and *tell* (17:2). Therefore, in terms of their preference for the DOC, the lexical biases toward the DOC from stronger to lower is: *tell* > *give* > *send*; *send* > *give* > *tell* toward the POC.

As shown in the relative frequencies, the probability of finding *give* in the DOC is almost four times of finding it in the POC based on historical records in the present study. This result echoes what is found in the synchronic data in Wasow and Arnald where *give* shows the lexical preference for the DOC over the POC to be roughly 4:1 [26]. And the verb *tell* shows the preference for the DOC even higher, eight times of its chance in POC. In order words, as the prototypical dative verb of communication, *tell*, has a strong bias toward the dative DOC construction, which matches the findings in Bresnan and Nikitina that over 99% of all dative uses of *tell* in the parsed SWITCHBOARD corpus occur in the dative NP construction [24].

The opposite holds true for the verb *send*, which shows a significant preference for the POC. This result bears much resemblance to what is found in synchronic studies, such as Mukherjee's and Ruppenhofer's in which the ratio of the DOC as opposed to the POC instantiations for *send* was found to be significantly lower [27, 28]. If we take the relative frequency occurring in the POC as a criterion for a caused-motion verb, then the verb *send* would be the one of the most typical caused-motion verbs while *tell,* one of the least typical ones.

4.3   Discussion

Since we found that the Theme and Recipient are the core arguments of the verb *give,* then the meaning of caused-possession is realized structurally in the DOC in OE and later in both the DOC and the POC in ME and ModE. And the Theme and Goal are the core arguments of *send,* and the meaning of caused-motion is mostly realized structurally in the POC. In the case of *tell,* it has always been a verb of message communication which involves a metaphorical meaning of caused-possession in its root meaning, and so it sub-categorizes a Recipient and a Theme. Yet the Theme argument is realized mostly by a clause instead of an NP argument. In a sense, we could say that the verb *tell* has always been a verb of caused-possession, the meaning of which determines its definite structural choice for the Recipient-Theme structure, which is shown in our data as having an 80% of chance to occur in the DOC than in the POC.

In general, the frequency differences in our diachronic study indicate that the verb *give* has a stronger tendency to occur in the DOC. However, the relative distributions of the DOC and the POC tend to be relatively even after the 15–16th century, suggesting that both constructions tend to be equally compatible with the core meaning of the verb *give,* denoting successful transference or reception. This means that for the verb *give,* both constructions convey the meaning of caused-possession and they tend to be evenly distributed among all the alternative constructions of the verb *give,* which reflects the even distribution of the two orderings of IO-DO and DO-IO found in the chronicles in OE. This suggests that the alternation constructions where *give* appears does have the same meaning, which has to do with the historical changes in the language history. However, *tell* exhibits strong countering force against the constructional meaning of POC, yet it exhibits preference for the DOC.

5   Conclusion

By constructing diachronic models of structural biases of ditransitive verbs, we have found that even though some ditransitive verbs belong to the same category and may all have the potential to occur in the dative alternation, yet they show different biases towards syntactic structures.

Our study indicates that fine-grained diachronic studies on the individual lexical biases of verbs towards alternating structures are necessary to reveal the hidden syn-semantic interaction of the verbs and their constructions. In the future, a more elaborated study would be conducted on the comparative analysis of Chinese and English ditransitive verbs and their structural biases. And we will also consider the informational properties and prominence of the NP objects in the dative alternation, which also influence structural biases of ditransitive verbs.

References

1. Zhu, D.X.: Grammatical problems related to the verb "gei". Dialect **02**, 81–87 (1979)
2. Peyraube, A.: Double-object constructions—Historical development from Han times to the Tang dynasty. Stud. Chin. Lang. **3**, 204–216 (1986)
3. Zhang, B.J.: Ditransitive constructions in modern Chinese. Stud. Chin. Lang. **3**, 175–184 (1999)
4. Zhang, G.X.: The Grammaticalization of Chinese Double Object Constructions and the semantics of "Yuan" sentences. In: Jie, X. (ed.) Typological Perspectives in Chinese Studies, pp. 345–373. Beijing Language and Culture University Press, Beijing (2005)
5. Goldberg, A.E.: Constructions: a construction grammar approach to argument structure. Chicago University Press, Chicago (1995)
6. Goldberg, A.E.: Constructions at Work. Oxford University Press, Oxford (2006)
7. Rappaport Hovav, M., Levin, B.: Building verb meanings. In: Butt, M., Geuder, W. (eds.) The Projection of Arguments, pp. 97–134. CSLI (1998)
8. Rappaport Hovav, M., Levin, B.: The English dative alternation: the case for verb sensitivity. J. Linguist. **44**(01), 129–167 (2008)
9. Haspelmath, M.: Explaining the Ditransitive Person-Role Constraint: a usage-based approach. Constructions **2**, 1–71 (2004)
10. Haspelmath, M.: Argument marking in ditransitive alignment types. Linguist. Discovery **3**(1), 1–21 (2005)
11. Haspelmath, M.: Ditransitive constructions: the verb 'give'. In: Haspelmath, M., et al. (eds.) The World Atlas of Language Structures, pp. 426–429. Oxford University Press, Oxford (2005)
12. Kittilä, S.: The anomaly of the verb 'give' explained by its high (formal and semantic) transitivity. Linguistics **44**, 569–612 (2006)
13. Malchukov, A., Haspelmath, M., Comrie, B.: Ditransitive construction: a typological overview. In: Malchukov, A., Haspelmath, M., Comrie, B. (eds.) Studies in Ditransitive Constructions: A Comparative Handbook, pp. 1–64. De Gruyter, Berlin (2010)
14. Bresnan, J., Hay, J.: Gradient Grammar: an effect of Animacy on the syntax of give in New Zealand and American English. Lingua **118**(2), 245–259 (2008)
15. Bresnan, J., Ford, M.: Predicting syntax: processing dative constructions in American and Australian varieties of English. Language **86**(1), 186–213 (2010)
16. Pinker, S.: Learnability and Cognition: The Acquisition of Argument Structure. MIT Press, Cambridge (1989)
17. Levin, B.: English Verb Classes and Alternations. University of Chicago Press, Chicago (1993)
18. Jackendoff, R.: Semantic Structures. MIT Press, Cambridge (1990)
19. Ma, Q.Z.: Constructions with double objects in Modern Chinese. Essays Linguist. **10**, 166–196 (1983)
20. Li, L.D.: Verbal objects and structural objects. Lang. Teach. Res. **03**, 103–123 (1984)
21. Shi, B.: Study of the Double Object Construction in Ancient Chinese (Doctoral Dissertation). Anhui University (2003)
22. Xu, H.: A study on the give-type verbs in ancient Chinese. Unpublished Ph.D. dissertation. Northeast Normal University (2018)
23. Zhang, M.: Revisiting the alignment typology of ditransitive constructions in Chinese dialects. Bull. Chin. Linguist. **4**(2), 87–270 (2011)
24. Bresnan, J., Nikitina, T.: On the Gradience of the Dative Alternation. Stanford University, Stanford (2003)

25. Gropen, J., Pinker, S., Hollander, M.: The learnability and acquisition of the Dative Alternation in English. Language **65**(2), 203–257 (1989)
26. Wasow, T., Arnold, J.: Post-verbal constituent ordering in English. In: Rohdenburg, G., Mondorf, B. (eds.) Determinants of Grammatical Variation in English, pp. 119–154. Mouton de Gruyter, Berlin & New York (2003)
27. Mukherjee, J.: English Ditransitive verbs. Aspects of Theory, Description and a Usage-based Model. Rodopi, Amsterdam/New York (2005)
28. Ruppenhofer, J.K.: The Interaction of Valence and Information Structure. Unpublished Ph. D. dissertation. University of California, Berkeley (2004)

Investigation on the Lexicalization Process and Causes of "Guzhi"

Hong Jin[1(✉)] and Yingjie Dong[2]

[1] Chinese Language and Culture College, Beijing Normal University, Beijing, China
jinhongbnu@163.com
[2] College of Arts, Suchoow University, Suzhou, China
dongyj_ringle@163.com

Abstract. The lexicalization process of "guzhi" underwent a transformation from a verbal cross-layer structure to an adjective. In this process, there are several points worthy of attention: first, the object was omitted after "zhi" and the structure center was inclined to "gu"; second, the high frequency of "guzhi + VP" made it possible for reinterpretation; and third, the VPs in the structure were often disyllabic phrases, which further strengthened the role of foot. All these factors worked together, causing the verb-adverbial structure "gu + zhi" gradually faded out and the adjective "guzhi" appeared in large numbers during the Ming and Qing Dynasties. Basically in the Qing Dynasty, the lexicalization process was finally completed.

Keywords: Guzhi (固执) · Lexicalization · Cross-layer structure · Rhythm · Part of speech

1 Introduction

Many scholars have paid attention to lexicalization of Chinese. From the perspective of cognitive linguistics, lexicalization could be understood as a process. During this process, categories, concepts or meanings were relatively fixed in the form of words. [1] Some scholars believed that lexicalization is a diachronic change from a non-word form to a word [2]. Scholars of semantic typology also explored lexicalization in an attempt to analyze "whether or not lexicalization models have a world-wide consensus" [3]. In recent years, case studies of lexicalization have become increasingly common, such as the study of "Buguo (不过)" [4], the discrimination between "Kanqilai (看起来)" and "Kanshangqu(看上去)" [5], and the research about lexicalization and subjectivity of "Huitou (回头)" [6]. Previous studies had provided good examples for case analysis, but most of the research objects were adverbs or structural phrases. Studies about adjectives' lexicalization are rare.

Our research focuses on the word "guzhi (固执)". We try to figure out the word's lexicalization process from diachronic perspective. To the best of our knowledge, we are the first one to study this word's lexicalization process. The Modern Chinese Corpus of CCL shows that "guzhi" occurs 2048 times, while in the Ancient Chinese Corpus, "gu+zhi" are used 556 times. As a common word, its development has its own

© Springer Nature Switzerland AG 2020
J.-F. Hong et al. (Eds.): CLSW 2019, LNAI 11831, pp. 275–283, 2020.
https://doi.org/10.1007/978-3-030-38189-9_29

uniqueness, but also manifests universal laws. Thus, it is quite meaningful for us to make the lexicalization process clear. Also our work can enrich the study of adjective's lexicalization.

2 The Phenomenon of Early Continuous Use of "Gu" and "Zhi"

In modern Chinese, "guzhi" is a compound word. Contemporary Chinese Dictionary shows "guzhi" is an adjective, which is interpreted as "whose temperament or attitude are stubborn; who are inflexible". Retrieving the CCL corpus and tracing the origin of the word, we find that the continuous use of "gu" and "zhi" appeared in the Warring States Period at the latest:

(1) Natural sincere people can be achieved without reluctance and thinking. Such people are naturally conformed to the principles of heaven. We call them Saints. In order to be sincere, one must choose good goals and pursue them persistently. (From the Doctrine of the Mean)
    诚者，不勉而中不思而得: 从容中道，圣人也。诚之者，择善而固执之者也。(战国《中庸》)
(2) Since the state has different decrees for four seasons and people resolutely carry out the monarch's commands, the affairs to be done seasonally should be arranged in a proper way ahead. At the same time, some decrees should be taken as necessary supplements. (From Guanzi)
    国有四时，固执王事，四守有所，三政执铺。(战国《管子》)

In these two sentences, "guzhi" is directly followed by objects. Here "zhi" is not an adjective because of its strong verbal character. Specifically, the structure "gu+zhi" above means "to do something resolutely". It is a typical verb-adverbial structure and quite different from the adjective "guzhi" in modern Chinese. Therefore, before clarifying the lexicalization process of "guzhi", it is necessary to talk about meanings of "gu" and "zhi" respectively.

The definition of "gu" in Shuowen Jiezi is "be surrounded on all sides". A scholar in the Qing Dynasty commented, "No omissions... The word 'gu' can be used to describe firm things." "The character form of 'gu' just looks like a strong barrier on all sides," one explained, "One sentence in Qince said like this 'there are Yaohan in the east(东有崤函之固)'. It is to say that thanks to Yaohan, which helps make the place a perfect location to repel enemies and protect oneself. " The phrase "prince of Qi named Gu, also called Zicheng(齐公子固, 字子城) [7]" in Zuozhuan is also a proof. Because it was surrounded on all sides and nothing could break in, the meaning of "gu" then extended to "firm". [4] From this we can see that the original meaning of "gu" refers specifically to "firmness of cities' walls". Later it expanded. Every solid thing can be described with "gu". Meanwhile, "one thing's original state" is called "gu" in Chinese. The real state is most stable and most difficult to be broken, that is, the most solid state. When it turns to maintain the stability of man-made state, one must be consistent. In this respect, the word "gu" can be used as "resolutely". The CCL corpus shows that the earliest meaning of "resolutely" roughly appeared in the Spring and Autumn Period.

For example, there is a sentence in Guoyu, "The emperor of Jin lost the hearts of the people. He broke his promise not to give you the city. He killed Ke Li and hated the people around him. People were not satisfied with him resolutely."(晋君大失其众, 背君赂, 杀里克, 而忌忿者, 众固不说) It was the usage like this that laid the foundation for the combination of "gu" and "zhi".

In Shuowen Jiezi, the interpretation of "zhi" is "to catch criminals". The oracle bone characters of "zhi" writes like this . The left part looks like a torture implement while the right like a person who kneels and stretches out his hands. The whole form just as a person in handcuffs. That is to say, "arrest" or "capture". For example, "the marquis of Jin caught the earl of Cao and sent him to Jingshi" (晋侯执曹伯归于京师). Later, the word's meaning extended to something like "take" or "hold". The object of action could be specific, such as utensils, axes, indigo, ritual utensils, etc. It could also be abstract, such as speech, command, ambition, etc.

Therefore, "gu" and "zhi" did not belong to the same level at first, but cross-layer. The so-called "cross-layer structure" refers to "the combination of two adjacent components not at the same syntactic level but in the surface form" [8]. Take "guzhi" as an example, "zhi" is a verb, which must be followed by a noun component as an object. The verb "zhi" and the subsequent noun component form a verb-object phrase, which is modified by the adverbial "gu". For example, the first two examples should be divided into parts in this way:

pursue them persistently [[固]执之]
resolutely carry out the monarch's commands [固[执王事]]

The meaning of the structure [gu [zhi + noun/nominal phrase]] can be roughly understood as "resolutely practicing something". Of course, the specific situation should be understood in the context.

3 The Formation of the Word "Guzhi"

As mentioned earlier, the continuous use of "gu" and "zhi" appeared in the Warring States Period. After the Han Dynasty, the continuous use of "gu" and "zhi" began to increase. The examples of each dynasty are as follows:

(3) The emperor sent Zhonglangjiang (an official position) named Wu Su to the Huns, but the Huns detained Wu and refused to let him return. Wu insisted on holding the Fujie(符节) of the Han Dynasty and never surrendered. (From the Book Pre-Han Dynasty)
遣中郎将苏武至匈奴, 匈奴留武不得归。武固执汉节, 不肯降。(东汉《前汉纪》)

(4) Just like Ning Guan, who insisted on his original opinion and never wanted to be an official. He took Ya Hong, Fu Chao and You Xu as his examples. (From Records of the Three Kingdoms)
若宁固执匪石, 守志箕山, 追迹洪崖, 参踪巢、许。(晋《三国志》)

(5) Dou was angry and deeply blamed Ling Han, who insisted on his own views. However when the incident finally happened, it was just like Han said. (From the History of Latter Han)

窦太后怒, 以切责棱, 棱固执其义。及事发, 果如所言。(南北朝《后汉书》)

(6) Lie Xi was always ambitious, threatening his men to violate my edicts and invade my army. (From Complete Prose Works of the Tang)

希烈又固执凶图, 驱胁将士, 违我诏命, 犯我军兵。(唐《加恩被擒将士诏》)

(7) I dare not insist on my shallow ideas for my own interests. Just because it is a national event, we must be cautious and give more consideration. (From History of the Early Tang Dynasty)

臣非敢固执愚见, 欲求已长, 伏以国之大典, 不可不慎。(五代《旧唐书》)

(8) Pei Zhao saw Yi Jia again and debated with him face to face. However, Jia always insisted on his own words, so Zhao could not clarify for himself. (From Extensive Records Compiled in the Taiping Year)

(赵裴)复见贾奕, 因与辨对。奕固执之, 无以自明。(北宋《太平广记》)

The above 6 cases include the structure of "gu [zhi + N]". Among them, only the "zhi" in Ex (3) has practical action meaning, that is, "holding sth with hands". Nouns follow by "guzhi" are abstract mostly, such as "foolish opinions", "speech", "justice", etc. Since then, the context formed by such objects has gradually become the common context of "guzhi".

It is worth noting that there was object ellipsis after "zhi" at this time. For example:

(9) When Gun Cao came to the court in early days, he violated bans of Jingdu. In the first year of Qinglong (233), the official in charge criticized Cao to the Ming Emperor. Thus the emperor issued an imperial edict saying, "The king of Zhongshan has always been respectful and prudent. This time he comes here by chance. We'd better judge the matter by the rules of evaluating the relatives of the emperor." The official insisted on punishing him. The emperor had no choice but to reduce Cao's feudal territory, including two counties and 750 households. (From Records of the Three Kingdoms)

初, 衮来朝, 犯京都禁。青龙元年, 有司奏衮。诏曰: "王素敬慎, 邂逅至此, 其以议亲之典议之。"有司固执。诏削县二, 户七百五十。(晋《三国志》)

(10) At that time, the country was in trouble at home and abroad. People who were in mourning often did not adhere to the end, except Hong Wang. (From Book of Song)

时内外多难, 在丧者皆不终其衰, 惟弘固执得免。(南北朝《宋书》)

(11) Stop building palaces. Cut down taxes and the amount of forced labor. Be Sympathy with the poor and help them. As a result, our country's reputation will spread all over the world. At the same time, we should advocate integrity and drive enemies out of our country without killing them. Dismiss officials who have no merit but are highly paid. The above regulations should be issued continuously in order and must be carried out consistently without violating them. (From the History of the Han Dynasty)

止诸缮治宫室，阙更减赋，尽休力役，存恤振救困乏之人以弭远方，厉崇忠直，放退残贼，无使素餐之吏久尸厚禄，以次贯行，固执无违。(东汉《汉书》)

In example (9)–(11), according to the contexts, the omissions after "zhi" can be added. As for Ex (11), Shigu Yan explained that "Guan" meant "continuously" and the whole sentence meant "what the monarch said should be practiced sequentially. Never against the monarch's will." From this point of view, the pronouns "zhi(之)" omitted after the verb "Xing(行)", "Zhi(执)" and "Wei(违)" can be supplemented according to the contexts. Another example is "Yousi(an official position) insisted on doing something (有司固执)", the earlier part has already explained that Yousi reported Cao's violation of the prohibition to his superiors, thus here the object of "zhi(执)" is the content of Yousi's report, but it is omitted. The object in Ex (10) can also be added clearly. The above examples show that "zhi" was still a verb at that time. The continuous use of "gu" and "zhi" was an adverbial middle structure, rather than a word.

Since the Northern and Southern Dynasties, "guzhi" began to be closely followed by verbal components:

(12) At first, Daoji planned to take Dinu Bo and Xian Liang from Wucheng as captains of the army, while Qian Fei firmly disagreed. (From the History of the Song Dynasty)
初，道济以五城人帛氏奴，梁显为参军督护，费谦固执不与。(南北朝《宋书》)

(13) Liu Xiubin then told his brother Liu Wenwei, "The outcome is predictable. We are bound to lose the battle. You would better sign the surrender as soon as possible." Wenwei Liu was silent and refused his advice resolutely. At last, he still did not specify the terms of the document. (From the History of the Wei Dynasty)
於是告兄子闻慰曰："事势可知，汝早作降书" 闻慰沉疑，固执不作，遂差本契。(南北朝《魏书》)

(14) Yang Hu said, "In the past, Zhang Liang did not accept the reward from Han Gaozu. Gaozu respected his ambition and withdrew the reward. As for me, Juping was given to me by the previous emperor. How dare I abandon it for other generous rewards? It will only attract more slander." Therefore, Yang refused to accept the reward and promotion firmly. The monarch finally agreed. (From the History of the Jin Dynasty)
祜让曰："昔张良请受留万户，汉祖不夺其志。臣受钜平于先帝，敢辱重爵，以速官谤!" 固执不拜，帝许之。(唐《晋书》)

(15) Zhuangzong changed Li Congshen's name to Li Jihuang and regarded him as his son. One day Zhuangzong asked Congshen to go back to Li Siyuan, Congshen was firmly unwilling to go. Instead, he would rather die to show his loyalty. (From Notes on the History of the Five Dynasties)
庄宗改其名为继璟，以为己子，命再往，从审固执不行，愿死於御前以明丹赤。(北宋《五代史记注》)

Most of the above examples include the structure "gu + zhi + bu + V" (固执不V). In this structure, "bu" is closely combined with the verbs behind it, which leads to the natural combination of "guzhi" in rhythm. [9] Here, "gu" and "zhi" may be regarded as an adhesion group. "The division of an adhesion group will change rhythmic boundaries and cause changes of pause positions." [11] The pause here is "guzhi | bu V". Meanwhile, according to the contexts, it is difficult to determine the noun components after "guzhi" clearly. Hence "guzhi" here can be understood as a verb-adverbial structure or an adjective. If it is the former case, then after adding the omitted objects, "guzhi +N" and the verbal components behind constitute verb-sequence constructions. Such as "gu zhi bu yu (固执不与)", it means "stick to your own opinion and disapprove others". The object after "zhi" can be added as "your own opinion (固执己见不与)". If it is the latter case, then "guzhi" and the following verbal components directly constitute a verb-adverbial structure. "Gu zhi bu yu" can be understood as "stubborn disapproval". We believe that for a long time, the structure of "gu + zhi + bu + V" is influenced by the frequent occurrence of the term "stubborn opinions". It is difficult to completely divide two ends of a continuum. We are not sure which case each sentence belongs to. However, it is exactly this process that provides possibility for the transformation of "guzhi" into an adjective. Until the Ming and Qing Dynasties, the verb-adverbial structure "gu + zhi" gradually faded out. It began to combine into one word and be modified by adverbs, which led to the prominence of its adjective nature.

Basically, by the Qing Dynasty, "guzhi" had changed from a verbal cross-layer structure to an adjective and solidified into a double-syllable compound word.

(16) The Miss said, "I wish I could come, just for the fear that my parents would not agree." The nun answered, "If you are determined to go, your mother will hardly insist. If your mother agrees, your father has no reason to disagree."
小姐道:"我巴不得来, 只怕爹妈不肯。"尼姑道: "若是小姐坚意要去, 奶奶也难固执。奶奶若肯时, 不怕太尉不容。"(明《喻世明言》)

(17) The prime minister was worried about your father's evasion, so he did a little trick to promote the marriage. Why do you insist on not agreeing?
丞相犹恐尊翁推托, 故略施小计, 成此姻缘。小姐何苦固执?(明《封神演义》)

(18) Mrs. Wang smiled and said, "You are too stubborn. Move in quickly. There is no need to alienate relatives for something unimportant."
王夫人凤姐都笑着:"你太固执了。正经再搬进来为是, 休为没要紧的事反疏远了亲戚。"(清《红楼梦》)

(19) The marquis of Lu was extremely disappointed to see that the duke of Song was too stubborn to do it.
鲁侯见宋公十分固执, 怏怏而罢。(清《东周列国志》)

(20) But Kun Huang was a stubborn man. He could not ask his sister-in-law for money when he was poor, so he went to Hangzhou.
无奈黄昆是一个固执人, 他能受穷也不去向嫂嫂要钱去, 故此才奔杭州。(清《三侠剑》)

(21) However, she is so stubborn. Can you persuade her to understand?
然而她那固执的性格, 可以劝说明明白白么?(清《八仙得道》)

Ex (18)–(21) are sentences selected from documents of the Qing Dynasty. Here "guzhi" has obvious grammatical features of adjectives. In Ex (18) and Ex (19), the degree adverb "Tai(太)" and "Shifen(十分)" modify "guzhi", and there are no objects. While in Ex (20) and Ex (21), "guzhi" is used as an attributive to modify nouns directly. Therefore, we believe that in the Qing Dynasty, "guzhi" was lexicalized.

Of course, although the use of "guzhi" as a verb-adverbial structure gradually decreased in the process of its lexicalization, some of them were retained. This kind of usage could still be seen sporadically in some documents of the Qing Dynasty.

(22) You are here today as a distinguished guest. We have arranged everything. The frontier town will be restored soon. Why do you stick to checking on your own as if nobody here can make it?
今客卿到来,诸事益备,不久边城自复,何得固执已往,而轻视下国无材?(清《海国春秋》)

(23) If you stick to your point of view all the time, you will surely be killed. If so, I am willing to die instead of you.
你若固执一己之见, 必欲处斩, 老程愿代一死。(清《说唐全传》)

(24) Although the cemetery has been bought, it is still up to the elders to decide whether to move. Brother, you should not always stick to your point of view.
买后迁与不迁, 仍然由堂上大人作主, 弟弟不必固执己见 。(清《曾国藩家书》)

Among these sentences, "gu zhi ji jian (固执己见)" which means "stubborn self's opinion" is condensed into an idiom. What's more, it has been retained so far. Although its rhythm is divided into two groups of disyllable, the usage of "guzhi" as a verb-adverbial structure can still be seen here. In other words, the hierarchy here should be divided as [gu [zhi ji jian]].

4 Lexicalization Motivation of "Guzhi"

The ellipsis of objects after "guzhi" was an important step in its lexicalization. As for this phenomenon, we believe it was mainly influenced by the principle of economy in the process of language development. As mentioned earlier, objects after "guzhi" were often something like suggestions and opinions. People's suggestions and opinions were often lengthy in content. It was difficult for a single syllable verb to control. Namely, it was easy to cause structural imbalance. In most cases, suggestions and opinions held by someone had already appeared in preceding contents, so there was no need to repeat again. Besides it was very common that abstract nouns such as "speech(言)", "justice (义)", "viewpoint(见)" or pronouns "zhi(之)" were used to refer to preceding texts. Sometimes objects were omitted directly. These three situations were all caused by economic principles.

The omission of objects after "guzhi" created possibility for its lexicalization. First, ellipsis of objects weakened the action of "zhi". As mentioned earlier, "guzhi" was not originally at one syntactic level. "Zhi" and its subsequent nominal components constituted a verb-object structure "zhi + N". "Gu" only modified this verb-object structure. This showed that the verb "zhi" was a strong morpheme, in other words, the structure center. However, with the increasing number of omissions, the action meaning of "zhi" inevitably weakened. The semantic focus of "guzhi" began to shift to the adjective "gu", which later helped change the nature of the whole structure to an adjective. This could also be supported by the phonetic form of the word. According to the Great Dictionary of Modern Chinese, "guzhi" is pronounced as "gù·zhi". "Zhi" is pronounced softly, losing its original tone, also shortening in length and weakening in intensity. It fully illustrates the weakening of its meaning. "Gu" then becomes the strong morphemes.

Secondly, omitting objects led to the high frequency use of the structure "guzhi + VP". It created a fixed context for its lexicalization. In this regard, some one had a similar explanation, "The high frequency of one syntactic construction determines the possibility of its constituents' lexicalization. The form of the syntactic construction determines the choice of constituents, and the overall meaning determines the meaning of its constituents after lexicalization." [10]. Strictly, "gu + VP" here could not be called a structure, but in a broad sense, it was no harm to be included.

Thirdly, the high frequency use of the structure "guzhi + VP" provided possibility for reinterpretation[1]. Owing to the omission of objects, "guzhi + VP" was similar to a verb-adverbial structure on the surface. It should be analyzed as a conjunctive predicate structure at first. Not until the object after "zhi" was omitted, the relationship between its direct components had changed, although the hierarchy had kept the same. That was to say, "guzhi" could be analyzed as an adverbial modifier, which further weakened its action.

Fourthly, ellipsis of objects further strengthened the restrictive role of rhythm. The basic rhythm pattern of Chinese is two syllables constitute one foot [11]. The corpus shows that VPs in the structure of "stubborn + VP" are mostly disyllabic, such as "not to follow (不从)", "seek to go (求去)", "not to allow (不与)", etc. The ellipsis of objects contributed to formations of four-character structures. One four-character structure can be divided into two disyllables. "Guzhi" was naturally classified as a prosodic unit, while the action phrase behind it was classified as another prosodic unit. As a result, during the development of disyllabic, "guzhi" merged further and gradually condensed into a word.

To sum up, "guzhi" as a cross-layer structure was verbal at first, but with the frequent omission of objects, two results arose. First, the grammatical nature of "guzhi" began to transform into adjective; and second, "guzhi" gradually solidified into a word in a prosodic unit.

5   Additional Discussion

It is worth mentioning the change of semantic color in the lexicalization process of "guzhi". In the original corpus, the structure of "gu + zhi" means "resolutely practicing something or insisting on some idea", which is a kind of commendatory color, such as "adhere to justice". But in modern Chinese, "guzhi" is completely derogatory, which means "too old-fashioned, not flexible". There are inherent reasons for this change.

---

[1] Reanalysis and reinterpretation are different. Reanalysis refers to changes of sentence hierarchies within one structure, while reinterpretation refers to changes of direct components' relationships. Sentence hierarchies keep the same.

"Gu, surrounded on all sides… Everything that is solid can be expressed by 'gu' " [12]. It is a positive aspect to derive the meaning of "solid" from "encirclement". But it is also possible to derive the meaning of "being blocked". Because of being blocked, ones do not know how to change. In addition, this change of semantic color also hints that "guzhi" of commendatory color must be replaced by another more commonly used word. Our preliminary guess is "jianchi(坚持)". In modern Chinese, "jianchi" is more commonly used. This substitution is completely equivalent literally: "firm" corresponds to "firm", while "hold" corresponds to "persistent". Due to space limitation, we intend to discuss this issue in another draft.

References

1. Wang, Y.: An Introduction to Cognitive Grammar. Shanghai Foreign Language Education Press, Shanghai (2005)
2. Dong, X.-F.: Syntactic evolution and lexicalization of Chinese. Chin. Lang. **332**(5), 399–409 (2009)
3. Li, Z.: Introduction to Semantic Typology. Guangdong World Book Publishing Co., Ltd, Guangzhou (2016)
4. Shen J.-X.: Remarks on Buguo (不过). J. Tsinghua Univ. (Philos. Soc. Sci. Ed.) **19**(5), 30–36, 42 (2004)
5. Zhang, Y.S.: Kànqilɑi (看起来) and kànshɑngqu (看上去). Chin. Teach. World **77**(3), 5–16 (2006)
6. Li, Z.J.: On the lexicalization and subjectivity of "huitou". Linguist. Sci. **5**(4), 24–28 (2006)
7. Li, Y. (ed.): Ancient Language Qulin. Shanghai Education Publishing House, Shanghai (1999)
8. Dong, X.F.: Lexicalization: Derivation and Development of Chinese Disyllabic Words. Sichuan Ethnic Publishing House, Chengdu (2002)
9. Editorial Board of Research on Chinese as a Foreign Language: Research on Chinese as a Foreign Language (No. 12). Commercial Press, Beijing (2015)
10. Dong, X.-F.: Phenomena and laws of Lexicalization and Grammaticalization of Chinese. Xuelin Press, Shanghai (2017)
11. Zhu, S.-P.: Four-character structures of Chinese. Beijing Language and Culture University Press, Beijing (2015)
12. Xu, S., Duan, Y.-C.: Commentaries on Shuowen Jiezi. Shanghai Ancient Books Publishing House, Shanghai (1988)

When "Natural Nouns" Surface as Verbs in Old Chinese: A Lexical Semantic Exploration

He Ren[✉] [iD]

Department of Chinese Language and Literature, Peking University,
Beijing, China
helenrenhe@outlook.com

Abstract. This paper reports a study of "natural nouns" that are used as verbs in Old Chinese, focusing on the lexical semantic analysis of natural nouns based on Generative Lexicon Theory. The detailed investigation of 39 natural nouns reveals that each of them encodes information of events as various types of Conventionalized Attributes (CAs) in their qualia structures, and is verbalized by activating or exploiting a particular type of CA and realizing this CA as the core meaning of the denominal verb. The discussion shows that the relative salience of a particular CA in the qualia structure largely determines the CA's probability of triggering N-V conversion. Moreover, there is a clear tendency in CA exploitation: CAs encoding events with human participants are much more likely to be exploited by N-V conversion than CAs encoding events without human participants. This tendency could be accounted for by people's anthropocentric view of the world.

Keywords: N-V conversion · Natural nouns · Conventionalized Attributes · Anthropocentrism · Old Chinese

1 Introduction

In Old Chinese, "Noun-Verb" conversion is a relatively active linguistic phenomenon, which has been widely concerned by scholars over the years. According to previous studies [1–4, etc.], the majority of denominal verbs in Old Chinese are those derived from nouns denoting artefacts, people, locations, directions or abstract things, while nouns denoting animals (except for human beings), plants and other natural objects— we call them "natural nouns" in this article—can hardly surface as verbs. This is an important observation as it touches the core mechanism of N-V conversion. Since these studies, however, are conducted not in an exhaustive way, and few cases of natural nouns are examined in them, there is need to make a thorough study on natural nouns in Old Chinese in order to reach an impartial conclusion.

To perform an exhaustive study, we refer to 古辞辨 *Guci Bian* (A Comprehensive Thesaurus of Classical Chinese Words) [5] and 左传 *Zuozhuan*, and make a list of 1,615 common nouns in Old Chinese. After a full investigation into these nouns and their denominal verbs, it is found in our research that 39 out of 415 natural nouns can

© Springer Nature Switzerland AG 2020
J.-F. Hong et al. (Eds.): CLSW 2019, LNAI 11831, pp. 284–291, 2020.
https://doi.org/10.1007/978-3-030-38189-9_30

surface as verbs. Although this proportion (9%) is lower than average (30%)[1], such kind of natural nouns does exist and appears to occupy a fair proportion worth our attention. So questions arise: Why can these 39 natural nouns surface as verbs? Where do the meanings of these denominal verbs come from? Are there any mechanisms or laws behind their verbalization? If so, what are they?

2 Method

Generative Lexicon Theory (GL) uses "Qualia Structure" to represent various kinds of properties of referents of lexical items [6]. Qualia Structure specifies four essential aspects of a word's meaning: CONSTITUTIVE, FORMAL, TELIC (encoding the purpose and function of the referent) and AGENTIVE (encoding the origin of the referent). Based on the analysis of qualia structures, GL divides nouns into three types: Natural Types, Artifactual Types and Complex Types [7, 8]. The qualia structure of a noun belonging to Natural Types only carries values for two roles—the Formal role and the Constitutive role, while that of a noun belonging to Artifactual Types carries values for all four qualia roles.

Ren's [4] research on nouns denoting physical objects in Old Chinese shows that, the high frequency of the verbalization of nouns denoting artefacts results from the fact that these nouns belong to Artifactual Types and encode salient Telic roles or Agentive roles, which express activities or events and are easy to be exploited by N-V conversion. For instance, the Telic role for the noun 权 *quán* 'steelyard', represented as [TEL = 'to weigh with'], is activated during N-V conversion and is realized as the core meaning of its denominal verb 权 *quán* 'to weigh with a steelyard'. (In most cases, the concept of a parent noun is also incorporated into the meaning of its denominal verb.)

Obviously, a natural noun usually belongs to Natural Types and does not carry a value for a Telic role or an Agentive role. Nevertheless, its qualia structure can encode events in a different way. Properties and events conventionally associated with natural entities are identified as "Conventionalized Attributes" (CAs) in GL. CAs are not strictly part of qualia roles, but are also encoded in qualia structures and are available when nouns surface as verbs.[2] Drawing on the views of Yuan [9] and Song [10], we divide CAs into seven types:

a. Natural-Telic (N-TEL)[3]: the natural function of the referent, e.g. 目 *mù* 'eye': [N-TEL = 视 *shì* 'to see with'].

---

[1] About 30% of all 1,615 nouns can surface as verbs.

[2] They can also be exploited during N-V conversion and be realized as the core meaning of the denominal verb.

[3] The Telic role expresses an inherent function of a referent, and the function is what people do with the referent, e.g. 镜 *jìng* 'mirror': [TEL = 'to look in']. A function encoded as "Natural-Telic" is also an inherent function, but is not associated with people's intention, e.g. 心 *xīn* 'heart': [N-TEL = 'to pump blood']. On the contrary, a function encoded as "Additional-Telic" is associated with people's intention, but is not the inherent function of the referent, e.g. 水 *shuǐ* 'water' [A-TEL = {'to flood with'…}]. The Agentive role, "Natural-Agentive" and "Additional-Agentive" are distinguished in the same way as the Telic role, "Natural-Telic" and "Additional-Telic".

b. Additional-Telic (A-TEL): what we usually do with the referent, e.g. 水 *shuǐ* 'water': [A-TEL = {淹 *yān* 'to flood with', 浸 *jìn* 'to soak with', 饮 *yǐn* 'to drink'}].

c. Natural-Agentive (N-AGE): how the referent naturally comes into being, e.g. 冰 *bīng* 'ice': [N-AGE = 凝 *níng* 'to freeze'].

d. Additional-Agentive (A-AGE): how the referent comes into being with human intervention, e.g. 蚕 *cán* 'silkworm': [A-AGE = 养殖 *yǎngzhí* 'to raise'].

e. Conventionalized-Action (CA-ACT): the conventionalized actions of the referent, e.g. 犬 *quǎn* 'dog': [CA-ACT = 吠 *fèi* 'to bark'].

f. Conventionalized-Handle (CA-HAN): how we usually handle with the referent, e.g. 茸 *cǎo* 'grass': [CA-HAN = 割 *gē* 'to cut'].

g. Conventionalized-Evaluation (CA-EVA): how we usually evaluate the referent, e.g. 花 *huā* 'flower': [CA-EVA = 美 *měi* 'beautiful'].

Under the above framework of Qualia Structure, we examine the meaning of each of the 39 natural nouns, exploring which CA is exploited by N-V conversion, and why one CA is more likely to be exploited than another or others (when two or more CAs are available for a noun).

3 Results

Distribution of 39 natural nouns which can surface as verbs in three subcategories is shown in Table 1.

Table 1. Distribution of natural nouns that surface as verbs in three subcategories

	Nouns that surface as verbs	Corresponding denominal verbs	Total number of nouns	Percentage of nouns that surface as verbs
Natural objects	18	23	164	11
Plants	14	22	135	10.4
Animals	7	8	116	6

According to our investigation, six types of CAs can be exploited by the verbalization of nouns denoting natural objects. Table 2 shows the occurrences of activation of these 6 types of CAs.

Among all six types, the A-TEL type shows the highest frequency, which means that this type is most likely to be exploited by N-V conversion of nouns denoting natural objects. For instance, the noun 涂 *tú* 'clay' encodes an Additional-Telic CA, represented as [A-TEL = 'to smear with']. By activating this CA, the noun 涂 *tú* 'clay' converts into a verb 涂 *tú* which means "to smear…with clay".

(1) 彻小屋, 涂大屋, 陈畚、挶。

chè__xiǎo__wū__tú__dà__wū__chén__běn__jū

remove__small__house__smear with clay__big__house__set out__box for holding soil__box for carrying soil

'(People) removed small houses, smeared walls of big houses with clay, and set out tools for holding and carrying soil.' (Duke Xiang 9, in *Zuozhuan*)

Table 2. CAs activated by the verbalization of nouns denoting natural objects

CA type	Occurrence(s)	CA type	Occurrence(s)
N-TEL	1	CA-ACT	1
A-TEL	10	CA-HAN	2
N-AGE	4	Others[a]	1
A-AGE	2	Total[b]	21

[a]Those CAs that cannot be classified into the seven types above (in Sect. 2) go to the category "others".
[b]The total number of denominal verbs derived from "nouns denoting natural objects" is 23. But the total number in Table 2 is 21. Why does this discrepancy exist? Here is the reason. Two denominal verbs ($\pm_1$B and 玉) do not exploit any CA, but rather exploit the conceptual frame of light verbs. Since light verbs do not belong to CAs, they are not included in the table. The instances in Tables 3 and 4 are dealt with in the same way.

Four types of CAs can be exploited by the verbalization of nouns denoting plants. Table 3 shows the occurrences of activation of these 4 types of CAs.

Table 3. CAs activated by the verbalization of nouns denoting plants

CA type	Occurrence(s)
A-TEL	7
N-AGE	6
CA-ACT	2
CA-HAN	6
Total	21

The A-TEL type still shows the highest frequency. For instance, the Additional-Telic CA of the noun 麦 *mài* 'wheat'—represented as [A-TEL = 'to eat']—is activated during N-V conversion, giving rise to the denominal verb 麦 *mài* 'to eat wheat'. Besides, the N-AGE type and the CA-HAN type have a strong tendency to trigger N-V conversion as well. For instance, the Natural-Agentive CA of the noun 花 *huā* 'flower'—'to bloom'— is activated, giving rise to the denominal verb 花 *huā* '(some plant) to bloom'. The Conventionalized-Handle CA of the noun 草 *cǎo* 'grass'—'to cut'—is activated, giving rise to the denominal verb 草 *cǎo* 'to cut the grass'.

(2) 昔我往矣, 黍稷方华。
xī__wǒ__wǎng__yǐ__shǔ-jì__fāng__huā
in the past__I__leave__SFP__millet__PROG__bloom
'When I left, the millet was blooming.' (Chuche, Xiaoya, in *Shijing*)

(3) 未发秋政, 则民弗敢茸也。
wèi__fā__qiū__zhèng__zé__mín__fú__gǎn__cǎo__yě
IMPERF.NEG__issue__autumn__decree__CONJ__people__NEG__dare__cut
the grass__SFP
'When the autumn decree hasn't been issued yet, people dare not cut the grass.'
(Jitong, in *Liji*)

Four types of CAs can be exploited by the verbalization of nouns denoting animals. Table 4 shows the occurrences of activation of these 4 types of CAs.

Table 4. CAs activated by the verbalization of nouns denoting animals

CA type	Occurrence(s)
A-TEL	2
A-AGE	1
CA-ACT	1
CA-HAN	1
Others	1
Total	6

Let's take the noun 蚕 *cán* 'silkworm' for instance. It encodes an Additional-Agentive CA, which is represented as [A-AGE = 'to raise']. By activating this CA, the noun converts into a denominal verb which means "to raise silkworms".

(4) 后妃率九嫔蚕于郊。
hòu-fēi__shuài__jiǔ__pín__cán__yú__jiāo
queen__lead__nine__concubine__raise silkworm__in__suburb
'The queen led nine concubines to raise silkworms in the suburbs.' (Shangnong, in *Lüshichunqiu*)

4   Discussion

4.1   The Relative Salience of CAs and Their Probability of Triggering N-V Conversion

The qualia structures of most natural nouns encode more than one type of CAs. Then, for a particular noun, which CA is the most likely to trigger N-V conversion? Based on a close examination of the 39 natural nouns, we argue that, it is the relative salience of a particular CA in the qualia structure of a particular noun that largely determines the CA's probability of triggering N-V conversion. CAs describe various kinds of conceptual knowledge of a referent that we human beings have in our minds. Which part of the conceptual knowledge is the most salient depends on two aspects: (1) the objective attributes of the referent; (2) the preferences of people within a given culture.

For instance, the nouns 垩 è 'white earth' and 泥 ní 'mud' both refer to some kind of earth and both can convert into verbs. However, the denominal verb 垩 è means "to paint... (usually walls) with white earth", whereas the denominal verb 泥 ní means "(someone) to be mired or trapped like stuck in the mud". Since 垩 è 'white earth' is a certain kind of white earth which is usually used for painting walls, its function which expresses exactly an event is among the most salient properties, as well as its colour. Therefore, the CA encoding this functional property (A-TEL) shows the highest degree of salience in this noun's qualia structure and is easy to be activated in its N-V conversion. As for the noun 泥 ní 'mud', in most cases, its referent is "mud formed by natural causes such as rain, water logging and so on" [5], and the most salient property is 黏稠 niánchóu 'sticky', which is easily associated with the concept "to cause someone or something to get stuck in the mud". Therefore, the event "to get stuck in the mud" is the most salient among all the events associated with the noun 泥 ní 'mud', and accordingly the CA encoding this event is the very one to be exploited during this noun's N-V conversion. Clearly, though 垩 è 'white earth' and 泥 ní 'mud' have similar meanings, their referents have distinct objective attributes, which gives rise to the fact that different CAs stand out in their qualia structures respectively.

Here is another example. Objectively speaking, 桑 sāng 'mulberry leaf' is just one kind of leaf and 蚕 cán 'silkworm' is just one kind of worm. However, for the ancient Chinese people, 桑蚕 sāngcán 'to pick mulberry leaves and to raise silkworms' is one of the most important production activities. Therefore, in the minds of people within this culture, among all the events associated with "mulberry leaf", the event of 采摘 cǎizhāi 'to pick' is the most prominent, so the CA which encodes this event shows the highest degree of salience in the qualia structure of the noun 桑 sāng 'mulberry leaf' (represented as [CA-HAN = 'to pick']), and is exploited in this noun's verbalization—the meaning of its denominal verb is "to pick mulberry leaves". Likewise, in people's minds at that time, the event of 养殖 yǎngzhí 'to raise' is most closely related to the noun 蚕 cán 'silkworm', so the CA which encodes this event shows the highest degree of salience in the qualia structure, and is exploited in this noun's verbalization—the meaning of its denominal verb is "to raise silkworms". This example shows that preferences of people within a given culture also determine the relative salience of CAs to some extent.

4.2 An Anthropocentric Account

We can see a clear tendency in the verbalization of natural nouns: CAs that encode events with human participants are much more likely to be exploited than CAs that encode events without human participants.

According to our statistics, among the 48 denominal verbs[4], meanings of 36 verbs come from CAs encoding events with human participants, accounting for 75% of the total. The majority of CAs involving human beings belong to the following three types: A-TEL, A-AGE and CA-HAN. For instance, 垩 è 'white earth' [A-TEL = 'to paint with'], 水 shuǐ 'water' [A-TEL = {'to flood with', 'to soak with', 'to drink'}], 火 huǒ

---

[4] The sum of the total numbers from Tables 2, 3 and 4 is 48.

'fire' [A-TEL = 'to burn with'], 麦 mài 'wheat' [A-TEL = 'to eat']; 蚕 cán 'silkworm' [A-AGE = 'to raise']; 鱼 yú 'fish' [CA-HAN = 'to catch'], 桑 sāng 'mulberry leaf' [CA-HAN = 'to pick'].

On the contrary, only 12 denominal verbs come from CAs encoding "pure natural events" (events without human participants), with a proportion of 25%. The majority of these CAs belong to the N-AGE type, e.g. 冰 bīng 'ice' [N-AGE = 'to freeze'], 花 huā 'flower' [N-AGE = 'to bloom']; only one belongs to the N-TEL type, namely 露 lù 'dew' [N-TEL = 'to moisten'].

Intuitively, those pure natural events are more closely related to the referents of natural nouns, since they are inherent, natural properties of natural objects, plants or animals. Then, why are CAs encoding this kind of events rarely activated by N-V conversion? Why are CAs encoding human-involved events (not as inherent as natural events) easily activated? We argue that it results from "our anthropocentric view of the world" [11]. Generally, people get to know about external objects through their interaction with them, which means that compared to pure natural events, events associating external objects with people themselves have a relatively higher degree of salience in people's minds. Therefore, CAs encoding human-involved events are much more likely to be exploited during N-V conversion.

5   Conclusions

In this article, we have illustrated how natural nouns convert into verbs in Old Chinese. Even though natural nouns belong to Natural Types and don't carry values for the Telic Role and the Agentive Role (the two typical qualia roles encoding information of events) in GL theory, they are able to encode information of events in their qualia structures as well, in the form of Conventionalized Attributes. CAs can be divided into seven types: Natural-Telic, Natural-Agentive, Additional-Telic, Additional-Agentive, Conventionalized-Action, Conventionalized-Handle and Conventionalized-Evaluation. During the N-V conversion of a noun, a certain CA belonging to a certain type will be activated and realized as the core meaning of the denominal verb.

When a particular noun has more than one type of CAs, the relative salience of a particular CA in the noun's qualia structure largely determines this CA's probability of triggering N-V conversion. The CA with the highest degree of salience is most likely to trigger N-V conversion. The relative salience of CAs depends on the objective attributes of the referent (of the noun) and preferences of people within a given culture.

We also find that there is an important law of CA exploitation: compared to CAs encoding events without human participants, CAs encoding events with human participants are much more likely to be exploited and realized as core meanings of denominal verbs. This law could be accounted for by a cognitive principle of salience: conceptual knowledge closely related to human beings has a higher degree of salience in people's minds than that irrelevant to human beings.

References

1. Wang, K.-Z.: Word-Class Shifts in Classical Chinese. Hunan People's Publishing House, Changsha (1989). (in Chinese)
2. Zhang, J.-W.: A reexamination of verbal uses of nouns in classical Chinese. Res. Ancient Chin. Lang. **3**, 7–11 (1999). (in Chinese)
3. Zhang, W.-G.: Noun-Verb Conversion in Classical Chinese and Its Development. Zhonghua Book Company, Beijing (2005). (in Chinese)
4. Ren, H.: A primary study of "Noun-Verb" conversion in Pre-Qin Chinese: with the example of nouns denoting physical objects. Essays Linguist. **50**, 312–341 (2014). (in Chinese)
5. Wang, F.-Y.: A Comprehensive Thesaurus of Classical Chinese Words. Zhonghua Book Company, Beijing (2011[1993]). (in Chinese)
6. Pustejovsky, J.: Generative Lexicon. The MIT Press, Cambridge (1995)
7. Pustejovsky, J.: Type construction and the logic of concepts. In: Bouillon, P., et al. (eds.) The language of Word Meaning, pp. 91–123. Cambridge University Press, Cambridge (2001)
8. Pustejovsky, J.: Type theory and lexical decomposition. In: Pustejovsky, J., et al. (eds.) Advances in Generative Lexicon Theory, pp. 9–38. Springer, NYC (2013). https://doi.org/10.1007/978-94-007-5189-7_2
9. Yuan, Y.-L.: A study of chinese semantic knowledge based on the theory of generative lexicon and argument structure. J. Chin. Inf. Process. **27**(6), 23–30 (2013). (in Chinese)
10. Song, Z.-Y.: The role of telic features in the lexical meaning and word formation of nouns: from the perspective of language values and linguistic values. Stud. Chin. Lang. **1**, 44–57 (2016). (in Chinese)
11. Radden, G., Kövecses, Z.: Towards a theory of metonymy. In: Panther, K.-U., et al. (eds.) Metonymy in Language and Thought. John Benjamins Publishing Company, Amsterdam (1999)

The Classification of Korean Verbs and Its Application in TCFL

Aiping Tu$^{(\boxtimes)}$ and Duo Qian

International College of Shenyang Normal University, Shenyang 110034, China
tuaiping81@163.com

Abstract. Korean Verbs can be classified according to meanings and usages, verbal characteristics covering four dimensions (action involvement in other items, action influence on other agents, action initiativity) and verbal independency. Generally, all of these are not considered in the classification of Chinese verbs. On the basis of the classification of Korean verbs, we studied the verbs listed in the Outline of HSK (level 1-6) Vocabulary, and discussed its application in teaching Chinese as a foreign language (TCFL).

Keywords: Classification of verbs · Application · TCFL

1 Introduction

The classification of Verbs in other languages often provide references and a new perspective for the classification of Chinese Verbs. For example, according to the concept *valence* referenced in chemistry by French linguist Tesniére [1], Zhu explained the ambiguity phenomenon of '*de (的)*'-constructions [2], which solved many grammar difficulties in Chinese study. Based on introductions of volitional and non-volitional verbs in Tibetan language [3], Ma classified Chinese verbs into controllable verbs and non-controllable verbs according to the controllability of behavior [4], which became a basic classification methods in Chinese verb studies. Perlmutter classified one-valence verbs into unergative and unaccusative verbs [5]. Based on this hypothesis, Xu classified Chinese verbs into four classes: intransitive, quasi-transitive, mono-transitive, and di-transitive verbs [6]. Huang pointed out that Chinese verbs, including one-valence verbs, two-valence verbs and three-valence verbs, can be classified into unergative verbs and unaccusative verbs, depending on the theta role of the only core argument [7].

Korean Verbs can be classified according to different standards, including meanings and usages, verbal characteristics covering four dimensions (action involvement in other items, action influence on other agents, action initiativity) and verbal independency [8]. According to the Classification of Korean verbs, we studied the verbs listed in the Outline of HSK (level 1-6) vocabulary (hereinafter referred to as HSK 1-6), and discussed its application in Teaching Chinese as a Foreign Language (TCFL).

© Springer Nature Switzerland AG 2020
J.-F. Hong et al. (Eds.): CLSW 2019, LNAI 11831, pp. 292–299, 2020.
https://doi.org/10.1007/978-3-030-38189-9_31

2 Categories According to Meanings and Usages of Verbs

As a distinctive verb category in Korea language according to the meanings and usages, symmetric verb indicates one event involved two interactive or correlative agents, e.g. *mannata* (만나다) 'to meet', *kielhonhata (결혼하다)* 'to marry' etc. The most important character of symmetric verbs is that they express the meaning of mutual relationship even without using the relevant word *solo (서 로)* 'mutually'. Consider (1).

(1) Nanen keua mannasta. (나는 그와 만났다)
  I he and meet SF
  I met him. (We met each other.)

As is shown above, subjects of symmetric verbs in Korean are commonly collateral agents. When symmetric verbs refer to the action which is sent out by only one of the agents, the topic and subject in this event is the one who has initiative willing. Consider (2).

(2) Nanen kelel mannasta. (나는 그를 만났다.)
  I him meet SF
  I met him.

Symmetric verbs are common in Chinese. Like symmetric verbs in Korean, Chinese symmetric verbs always have two symmetric agents. Sentence structures include three different cases. Firstly, the subject of the symmetric verb is plural in nature, such as *women (我们)* 'we', *women lia (我们俩)* 'we both'. Secondly, the subject of the symmetric verb is a complex structure involving a conjunction such as *he (和)* 'and', *gen (跟)* 'and' or *yu (与)* 'and', entailing a plural agent. Thirdly, one single agent serves as both topic and subject of the sentence when action is initiated by only one of the agents. The other agent can follow the subject with the help of preposition *he (和)* 'with', *gen (跟)* 'with' or *yu (与)* 'with', and they can be modified or restricted by adverbs and modal verbs in front. Consider below.

(3)
  a. Women (lia) jianmian le. *[我们(俩)见面了。]*
    We (two) meet SF
    We met each other.
  b. Wo he ta jianmian le. (我和他见面了。)
    I and him meet SF
    I met him.
  c. Wo bu xiang he ta jianmian le. (我不想和他见面了。)
    I not want with him meet SF
    I do not want to meet him.

There is a phenomenon needed to pay attention. In different languages, the main subjects of symmetric verbs may not be the same. For example, in Korean and English, the subjects of *mannata (만나다)* 'to meet' could be both parts, multiple parts or single part. But in Chinese, the subjects of *jianmian (见 面)* 'meet' are definitely both parts, while the subjects of *jian (见)* 'meet' could be both parts, multiple parts or single part. This difference results in misuse of symmetric verbs in TCFL. Here are some sentences misused by non-native Chinese learners.

(4)

    a. *Wo jianmian ni. (*我见面你。*)
       I meet you
    b. *wo jiehun ni. (*我结婚你。*)
       I merry you

Based on overall investigation of the verbs in HSK 1-6, Chinese symmetric verbs can be classified into three types according to the number of action agents. In the first category, the subjects of the symmetric verbs are restricted to the both part of the agents. There are only a few verbs in HSK 1-6, such as *jiehun (结婚)* 'marry', *woshou (握手)* 'handshake' *duili (对立)* 'opposite'. In the second category, the subjects of the symmetric verbs are not only the both parts but also multiple parts. *Bi (比)* 'compare', *duihua (对话)* 'dialog', *bianlun (辩论)* 'discuss' are typical in this type of symmetric verbs. In the third category, the subjects of symmetric verbs can be both parts, multiple parts or single part. For example, the verb *renshi (认 识)* 'know' can be knowing each other, and can also be knowing from only one part as the case *I knew him, but he did not know me*. *Lianxi (联系)* 'relate', *gaobie (告别)* 'farewell' and *goujie (勾结)* 'collude' are also typical in this category of symmetric verbs.

3 Categories According to Verbal Characteristics

Korean verbs can be categorized according to different verbal characteristics, such as, action involvement in other items, action influence on other agents and action initiativity.

3.1 Categories According to Action Involvement in Other Items

According to action involvement in other items, Korean verbs can be divided into intransitive verbs, transitive verbs and dual-purpose verbs. Intransitive verbs express the action not involving other items. For example, the verb *nalta (날다)* 'to fly' is only related to the agent, not involved in other items, and not followed by any objects. The similar verbs includes *usta (웃다)* 'to laugh', s*ota (서다)* 'to stand', *tuita (뛰다)* 'to run' etc.

Transitive verbs express the action involved in other items. For example, the verb *mokta (먹다)* 'to eat' is not only related to the agent but also patient, followed by objects. The similar verbs includes *masita (마시다)* 'to drink', *sata (사다)* 'to buy', *pota (보다)* 'to see' etc.

Dual-purpose verbs integrate usages of intransitive verbs and transitive verbs. For example, the verb *momjuta (멈추다)* 'to stop' in the phrase *the car stops* only involves the subject, not involved other things, can not be followed by any objects, while in the phrase *someone stops the car* involves not only the agent but also other things, followed with object *car*. The similar verbs includes *umjikita (움직이다)* 'to move', *hesanhata (해산하다)* 'to disperse', *jaktonghata (작동하다)* 'to operate' etc.

3.2 Categories According to Action Influence on Other Agents

According to action influence on other agents, Korean verbs can be divided into active verbs and causative verbs. Active verbs express the actions which are acted by the agents. Causative verbs express the actions caused by the agents. Active verbs and causative verbs are a pair of relative concepts. The difference is that active verbs have no infixes in the mid of the verb forms, but causative verbs have causative infixes such as *- 이, - 히, - 리, - 기, - 우* etc. For example, the active verb *mokta (먹다)* 'to eat' means only the agent eats, while the causative verb *mokita (먹이다)* 'to let someone eat' means to let or force someone else to eat. Such similar pairs of verbs include *ipta (입다)* 'to wear' and *iphita (입히다)* 'to let someone wear', *usta (웃다)* 'to laugh' and *uskita (웃기다)* 'to let someone laugh', *keta (깨다)* 'to wake' and *keuta (깨우다)* 'to wake up someone' etc.

3.3 Categories According to Action Initiativity

According to action initiativity, Korean verbs can be divided into initiative verbs and passive verbs. Initiative verbs express the actions which are initiated by the agents. Passive words express the actions which are forced to subjects. Initiative verbs and passive words are a pair of relative concepts. The difference is that initiative verbs have no infixes in the mid of the verb forms, but passive verbs have passive infixes such as *- 이, - 히, 리, - 기* etc. For example, the initiative verb *pota (보다)* 'to see' means nobody but the agent sends out the action, while the passive verb *poita (보이다)* 'to be seen' means somebody or something be seen by the agent. Such similar pairs of verbs include *mokta (먹다)* 'to eat' and *mokhita (먹히다)* 'to be eaten', *ielta (열다)* 'to open' and *iellita (열리다)* 'to be opened', *anta (안다)* 'to embrace' and *ankita (안기다)* 'to be embraced' etc.

3.4 Dual-Purpose Verbs in Chinese

Chinese verbs include intransitive verbs, transitive verbs and dual-purpose verbs expressing both intransitive and transitive meaning either. Although Chinese use the prepositions *shi (使)* 'cause' or *bei (被)* 'be caused' instead of causative or passive infixes, there are dual-purpose verbs expressing both initiative and causative meanings and dual-purpose verbs expressing both initiative and passive meanings. Consider below.

(5)

    a. Men kai le. *(门开了。)*
        Door open SF
        The door opened.
    b. Ta kai men le. *(他开门了。)*
        He open door SF
        He opened the door.

(6)

    a. Zhe ge mimi gongkai le. *(这个秘密公开了。)*
    This CL secret disclosed SF
    The secret was disclosed.

    b. Ta gongkai le zhe ge mimi *(他公开了这个秘密。)*
    He disclose SF this CL secret
    He made the secret public.

The verb *kai (开)* 'open' in sentences (5a) can be used as intransitive, initiative and passive verbs to express the patient *men (门)* 'door' reach to the state of *kai (开)* 'opening' after experiencing the action *kai (开)* 'open'. It can also be used as transitive, initiative and passive verbs to express an action *kai (开)* 'open' which causes the patient to the state of *kai (开)* 'opening'.

In these examples above, (5a) and (6a) are unmarked passive sentences, which can express passive meaning without using the mark *bei (被)* 'be caused'. These sentences could be changed into *bei (被)*-sentences if *bei (被)* 'be caused' is added. See below.

(7)

    a. Men bei dakai le. *(门被打开了。)*
    Door be open SF
    The door is opened.

    b. Zhe ge mimi bei gong kai le. *(这个秘密被公开了。)*
    This CL secret be make public SF
    The secret was made public.

(5b) and (6b) are unmarked causative sentences, expressing causative meaning without using marks *shi (使)* 'cause' and *rang (让)* 'let'. And all these sentences could express causative meaning since the whole event includes four elements, which are agent, patient, causative force and result [9]. Agent makes patient produce result under the influence of causative force. These sentences could be changed into marked pivotal sentences if *shi (使)* 'cause' or *rang (让)* 'let' is added. See below.

(8)

    a. Zhangsan shi men dakai le. *(张三使门打开了。)*
    Zhangsan cause door open SF
    Zhangsan cause the door opened.

    b. Ta rang zhe ge mimi gong kai le. *(她使这个秘密公开了。)*
    He let this CL secret make public SF
    This secret was made public by him.

According to the standards above, after investigating all the verbs in HSK 1-6, we found that dual-purpose verbs expressing both intransitive and transitive meanings, dual-purpose verbs expressing both initiative and causative meanings and dual-purpose verbs expressing both initiative and passive meanings are totally coincide. We name all types of verbs as Dual-purpose verbs. The semantic feature of dual-purpose verbs is that the agent causes the patient to change. In the syntactic form, the patient can also be proposed at the position of subject, which expresses the patient experienced some kind of changes either.

4 Categories According to Verbal Independency

According to verbal independency, Korean verbs can be divided into independent verbs and auxiliary verbs. Independent verbs refer to the verbs which can be used independently, and auxiliary verbs are verbs that add functional or grammatical meaning to the main predicates. Auxiliary verbs losing or weakening original concrete meanings, can not act as sentence components, but can be used only in the conjunction as suffixes after other independent verbs, assisting narrative function of independent verbs. According to meanings, auxiliary verbs in Korean can be classified into several types, including terminative auxiliary verbs, result-preserving auxiliary verbs, desire auxiliary verbs, directional auxiliary verbs, subsidiary auxiliary verbs, subsidiary auxiliary verbs, progressive auxiliary verbs, repetitive auxiliary verbs and negative auxiliary verbs etc.

Similarly, Auxiliary verbs in Chinese have lost or reduced their original meaning, can not act as a main sentence component independently. These verbs can serve as sentence components only if combining with other independent verbs, assisting narrative function of independent verbs. Auxiliary verbs in Chinese can express the meanings such as behavior termination, result maintaining, existence, desire, direction and service etc.

Behavior-termination auxiliary verbs *wan (完)* 'deplete' and *diao (掉)* 'disappear' indicate that the object of action has been depleted, as from something to nothing, or the action has already been finished. Consider below.

(9)

    a. Fenbi yong wan le. (粉笔用完了。)
       Chalk use depleted SF
       The chalks have run out.
    b. Zi ca diao le. (字擦掉了。)
       Characters wipe disappear SF
       The characters have been wiped off.

Result-maintaining auxiliary verb *wan (完)* 'finish' indicate that the relevant status continued or maintained after actions completed. It can collocate with other main verbs, which generally cause some results, and objects expressing results are also acceptable after the verb combination. Consider (10).

(10) Maoyi zhi wan le. (毛衣织完了。)
      Sweater knit finish SF
      The sweater is finished.

Existence auxiliary verb *you (有)* 'have' indicate that the relevant status still exists after actions being completed. It can collocate with other verbs. Consider (11).

(11) Qiang shang gua you liang fu hua. (墙上挂有两幅画。)
      Wall on hang have two CL picture
      There are two pictures on the wall.

Desire auxiliary verbs indicate that the speaker or subject hopes, supposes, or speculates actions or statuses expressed by main verbs have happened or appeared. Typical desire auxiliary verbs, such as *xiang (想)* 'want' and *yao (要)* 'want' can collocate with other verbs. Consider (12).

(12) Wo jintian xiang/yao qu xuexiao. (我今天想/要去学校。)
      I today want go school
      I want to go to school today.

Direction auxiliary verbs indicate that people or things change their locations by actions. The typical direction auxiliary verbs, such as *lai (来)* 'come', *qu (去)* 'go', *shang (上)* 'up', *xia (下)* 'down', *jin (进)* 'enter', *chuqu (出去)* 'go out', huilai (回来) 'come back', *guolai (过来)* 'come over', *guoqu (过去)* 'go over', can collocate with other verbs. Consider below.

(13)
   a. Ta zou jin le bangongshi. (*他走进了办公室。*)
      He walk enter SF office
      He entered the office.
   b. Ta zou chuqu le. (*他走出去了。*)
      He walk out SF
      He walked out.

Service auxiliary verbs indicate that actions or behaviors are completed for someone's profit. The typical service auxiliary verbs, such as *bang/bangzhu (帮/帮助)* 'help', and *jiao (教)* 'teach' can collocate with other verbs. Consider below.

(14)
   a. Wo bang mama zuo zaocan. (*我帮妈妈做早餐。*)
      I help mother cook breakfast
      I help my mother cook the breakfast.
   b. Laoshi jiao women xuexi yufa. (*老师教我们学习语法。*)
      Teacher teach us study grammar
      The teacher teach us grammar.

There is not a consistent one-to-one match between Korean auxiliary verbs and Chinese auxiliary verbs. As illustrated above, some auxiliary verbs in Korean corresponds to Chinese verbs, while others correspond to words of other classifications or sentence components. For example, in Korean, progressive auxiliary verbs correspond to Chinese time adverbs, such as *zheng/zai/zhengzai (正/在/正在/)* 'be v. -ing', and repetitive auxiliary verbs correspond to Chinese time adverbs, such as *hai (还)* 'still', *yizhi (一直)* 'continuously', and negative auxiliary verbs correspond to Chinese time adverbs, such as *bu(不)* 'no'. All of these differences above need to be highlighted in TCFL.

5   Conclusion

According to meanings, symmetric verbs are the significant difference in Korean verb classification compared with the Chinese counterpart. In Korean, verbs are classified into active and causative verbs according to action influence on other agents, and initiative & passive verbs according to action initiativity, whereas there are no corresponding word but preposition phrases, such as *shi (使)* 'cause' and *bei (被)* 'be caused' in Chinese. However, there are dual-purpose verbs expressing active & causative and initiative & passive. In addition, Korean verbs can be classified according to verbal independency, which was not considered in Chinese verb classification.

The classification of verbs in Korean enlightened us in some way. Firstly, symmetric verbs suggest that these words share common features and help understand component properties in relevant sentence patterns. Secondly, dual-purpose verbs help us understand meanings and relevance of pairs of concepts and combine unmarked causative sentences, unmarked passive sentences and patient prepositional sentences for investigation. Thirdly, auxiliary verbs, as they are named, indicate their features and contribute to illustration in language teaching.

Cross-linguistic comparison plays an important role in understanding the characteristics of a language, especially the foreign language. Most importantly, it can offer insightful references in TCFL.

Acknowledgments. This work has been supported by the Major Projects of Chinese National Social Science Foundation (11&ZD189). We are indebted to many linguistics researchers including Xiaona Wang, Jun Xia, Xinyi Chen, Yanan Gao, Jun Yan, Caiyun Piao, Minzhi An etc. for the discussions in the writing of this paper.

References

1. Tesniére, L.: Element de Syntaxe Structurale. Klincksieck, Paris (1957)
2. Zhu, D.: The DE structure and the judgment sentence. Chin. Lang., 1–2, 23–27, 104–109 (1978)
3. Jin, P.: The formula of the verb in Lhasa Tibetan and its expression. Natl. Lang. **3**, 9–18 (1983)
4. Ma, Q.: Autonomous and non-autonomous verbs. Chin. Lang. J. **3**, 157–180 (1988)
5. Perlmutter, D.: Impersonal passives and the unaccusative hypothesis. Proc. Berkeley Linguist. Soc. **4**, 157–189 (1978)
6. Xu, J.: 'Transitivity' and four classes of relevant verbs. Lang. Res. **3**, 1–11 (2001)
7. Huang, C.-T.J.: Thematic structures of verbs in Chinese and their syntactic projections. Lang. Sci. **4**, 3–21 (2007)
8. Gao, Y., Gu, P.: The Grammar of Our Language. Gateway Hall, Seoul (2018)
9. Zhou, H.: A Study on Cause Category in Modern Chinese. East China Normal University, Shanghai (2004)

The Independence of Monosyllabic Words

Jinzhu Zhang[✉]

Information Engineering University, Kunshan, China
95339041@qq.com

Abstract. Based on the characteristics of Chinese characters, Chinese characters is the intersection of speech and grammar in Chinese with stable speech performance. One Chinese character is a syllable with a tone as a sign. Moreover, Chinese characters have a tenacious meaning, which makes the meaning of words not easy to lose. In other words, monosyllabic words can always maintain strong independence. But between different word classes, its independence also shows differences. In this paper we take the component of antonymous compounds as the entry point, and find that monosyllabic nouns have the strongest independence, followed by verbs, and adjectives have the weakest independence.

Keywords: Independence · Component · Antonymous compound · Function migration

1 Introduction

According to Haiman's definition, iconicity motivation of independence means that the grammatical separateness of a clause corresponds to the conceptual independence of the proposition expressed by that clause, which can also be called separateness motivation [1]. This mechanism actually refers to the individuation of the linguistic form and the individuation of the concept. An independent word means an independent entity, but the agglutinative morpheme cannot be an independent entity.

Haiman and Mithun studied the "noun incorporation" and found that a completely independent noun is more conceptually independent than a nominal component that has been integrated into a larger word [1, 2]. In fact, this is not only reflected in noun morphemes, but also in verb and adjective morphemes. But between different word classes, the independence also shows differences.

2 Grammatical Attributes and Structural Relationships of Antonymous Compounds and Components

In this paper, antonymous compounds are used as the starting point to explore the strength of independence between different monosyllabic word classes. In Chinese, there is a kind of coordinate compounds composed of two opposite monosyllabic words, which is called Antonymous Compounds, such as 利害 [*lihai*] (*benefits and harms*), 大小 [*daxiao*] (*size*), 动静 [*dongjing*] (*movement*) etc., which is a closed class.

© Springer Nature Switzerland AG 2020
J.-F. Hong et al. (Eds.): CLSW 2019, LNAI 11831, pp. 300–306, 2020.
https://doi.org/10.1007/978-3-030-38189-9_32

The reason why antonymous compounds are chosen is that its components are two monosyllabic words with strong independence. Some antonymous compounds are still in the process of lexicalization, and they are more analytical.

In this article, we collected 232 antonymous compounds from the Modern Chinese Dictionary. The use cases of antonymous compounds, if not specified, are from the Modern Chinese Corpus of Peking University Chinese Linguistics Research Center (CCL).

The grammatical attributes and structural relationships of antonymous compounds and its components are shown in the following Table 1:

Table 1. Distribution of grammatical attributes of antonymous compounds and its components.

Components		Compounds				
		Noun	Verb	Adjective	Pronoun	Adverb
N+N	79	77	1	3		5
V+V	88	25	66	1		2
A+A	61	54	2	2	1	10
A+N	1	1				
A+V	1		1			
V+A	1	1				
P+P	1				1	
Total	232	158	70	6	2	17

We can see from the above table that the grammatical attributes of components are mainly distributed in three major categories: verbs, nouns, and adjectives. Among them, the number of V+V antonymous compounds is the largest, accounting for 37.93%; the second is N+N, accounting for 34.05%; the A+A is the least, accounting for 26.29%. As for part of speech of the whole word, the nominal antonymous compounds account for the most, representing 68.1%; the verbal antonymous compounds account for 30.17%; the adverbial antonymous compounds account for 7.33%; the adjectival antonymous compounds are the least, which is 2.59%.

3   The Binomial Semantic Features of N+N Antonymous Compounds

N+N antonymous compounds are composed of nominal morphemes, which are nouns, such as 前后 [qianhou] (before and after), 功过 [gongguo] (credit and negligence) etc. The utterance function of nouns is reference. Nouns' semantic connotation is essential and stable. The nominal components maintain relative independence and individuality at the level of the whole word, and the semantic features of the binomial are significant.

3.1 Semantic Level

The components are not easy to merge, and the degree of conceptual integration is very low. 86.42% of the N+N antonymous compounds can be derived from the components, or the whole word meaning is separated from the components' meaning. The internal semantic structure of the N+N antonymous compounds is additive or general meaning.

3.2 Syntactic Level

The stability of the referential function causes the grammatical attributes of the whole word to be basically consistent with the components. 97.47% of N+N antonymous compounds is nominal.

The binomial semantic features of N+N compounds make the reference function of the whole word change from individual quantity to collective quantity, which require syntactic components with the same meaning to coexist with it in syntactical level, but the degree of this co-occurrence is different. For example:

(1)

    a. 前后两次模拟考试中，他都是第一名。
       [*Qianhou liang ci moni kaoshi zhong, ta dou shi di yi ming.*]
       He was the first in the two simulated exams.
  *b. 前后一次模拟考试中，他都是第一名。
       [*Qianhou yi ci moni kaoshi zhong, ta dou shi di yi ming.*]
       He was the first in the simulated exams.

The b sentence in the above example does not hold, because the syntactic component "一 [*yi*] (*one*)" does not match the binomial meaning of 前后 [*qianhou*] (*before and after*).

The word class category has prototypicality. Zhu pointed out the grammatical features of nouns: first, it can be modified by quantitative words, such as: 一支笔 [*yizhibi*] (*a pen*), 三本书 [*sanbenshu*] (*three books*), etc.; the second is not modified by adverbs, *很勇气 [*henyongqi*] (*very courage*) [3]. Prototype members or typical members have the above characteristics.

N+N antonymous compounds cannot be modified by adverbs, embodying the nominal side. The reference of a typical noun is independent and can act as a subject or an object independently in a sentence. However, antonymous compounds must be attached to other nouns to realize their reference function.

On the whole, the semantic integration degree of N+N antonymous compounds is very low. Their structure is analytical and the function is relatively stable. Compared with V+V and A+A compounds, the lexicalization degree of N+N compounds is not high, which means that the components independence of N+N antonymous compounds is stronger.

4   Function Migration of V+V Antonymous Compounds

V+V antonymous compounds present a functional migration as a whole, which is expressed in two aspects: syntactic function and utterance function.

4.1   Syntactic Function

V+V antonymous compounds composed of verbal morphemes are affected by the components, and their lexical semantics retain certain action meaning. After compounding, 73.91% of the compounds have no change in their grammatical properties and are verbs. The prototype syntactic function of verbs is to be predicate in the sentence, and the frequency of V+V compounds as predicates is different from typical verbs.

Through statistical data, in addition to the individual compounds such as 唱和 [changhe] (sing and echo), 臧否 [zangpi] (reward and punishment), 隐现 [yinxian] (hid and appear), etc., among the entire V+V compounds, 81.92% of the compounds have very low ability as predicate or predicate centers, below 30%, which downgrade to atypical syntactic functions, 42.05% of compounds can't be used as predicate.

Hopper and Thompson conducted a survey on the relationship between the grammatical features of nouns and verbs, semantics, and functional factors [4]. The typological evidence they quoted shows that the verb is more semantically visible in the actual action, and the more the utterance function is used to report the events that actually occur on the action participants in a particular scene, the more action it has. That is, there are more verb characteristics in form, such as changes in tense, aspect and modality; on the contrary, there are fewer formal characteristics of prototype verb.

When V+V antonymous compounds are used as predicates, the action is weakened, and the actual action that is visible is not reported. Instead, it tends to express a habitual, continuous, and repeated behavioral state. Therefore, V+V compounds that are not encoded as prototype verbs in language form have fewer formal characteristics of the prototype verbs.

V+V antonymous compounds generally have fewer objects, which are mainly restricted by the binomial semantic features.

(2)

    a. 当新的供应量上市时，产品必定不落/涨价。

       [*Dang xin de gongyingliang shangshi shi, chanpin biding bu luo/zhang jia.*]
       When the new supply goes on the market, the price of product must not fall/increse.

  *b. 当新的供应量上市时，产品必定涨落价。

       [*Dang xin de gongyingliang shangshi shi, chanpin biding bu zhangluo jia.*]
       When the new supply goes on the market, the price of product must not increase and fall.

Habitual, continuous, and repeated behavioral state of V+V compounds in semantic expression gives them a certain descriptive meaning, which do not have a negative form and cannot be modified by 不 [*bu*]/没(有) [*mei(you)*] (*not*): *不/没(有)沉浮 [*bu/mei(you)chenfu*] (*not up and down*).

The weak action semantic feature of V+V compounds is not co-occurring with the tense and aspect markers (着 [*zhe*], 了 [*le*], 过 [*guo*]) in the syntactic level: *往返 [*wangfan*] (*back and forth*) 着 [*zhe*]/了 [*le*]/过 [*guo*].

V+V antonymous compounds cannot be followed by directional verbs as complements, which is also affected by the semantic features of the whole word: *涨落上来 [*zhangluo shanglai*] (*increase and fall up*).

The loss of the typical syntactic function of the predicate means that the non-predicate position is obtained as a typical syntactic function, that is, V+V antonymous compounds have a syntactic migration in function. Moreover, when 94.57% of compounds acting as subjects or objects is not free, but conditional. The imperfection of semantics causes the implementation of their referential function to be attached to a nominal subject, that is, they usually appear in the structure of "NP + (的) [*de*] (*auxiliary word*) + V1V2". In this structure, the mobility of V1V2 is weakened or disappeared, and it gradually migrates to the noun category. But the functional migration does not lead to a cross-class or part-of-speech transition.

4.2   Utterance Function

The migration of V+V antonymous compounds syntactical function causes their utterance function to change. The utterance function of verbs is statement. In the process of compounding and using, the utterance function of components transforms from statement into reference at the lexical level, such as 开关 [*kaiguan*] (*switch*), 支出 [*zhichu*] (*expenditure*), which are nouns. But most V+V compounds have not completed this vocabulary transformation. 75% of compounds representing the concept of materialization occupy the syntactic position of subject and object, expressing reference.

In summary, the internal function of V+V antonymous compound is in a continuum, which we can express as follows:

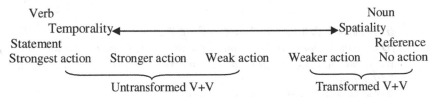

From this we can see that the independence of verb components is not very strong.

5 The Overall Transformation of A+A Antonymous Compounds

A+A antonymous compounds composed of adjectives, such as 大小 [daxiao] (size), 贵贱 [guijian] (nobleness), etc., can be transformed from attribute meaning to upper attribute category. There is a vertical relationship between the components and the compounds, which are in different semantic categories. Thus, it can be said that the semantics of the A+A compounds has changed as a whole.

The utterance function of A+A components is modification, but the utterance function of the whole word is reference. The transformation of utterance function has been completed at the lexicon level, which belongs to the lexical transformation [5].

The main syntactic function of adjectives in Chinese is to act as a predicate or an attribute, and can be modified by degree adverbs. When acted as an attribute, the opposite meanings of A1 and A2 show the asymmetry and the markedness both on the lexical level and the syntactic level, such as 大方 [dafang] (generous), *小方 [xiaofang]; 有多A1 [you duo A1] (how A1), *有多A2 [you duo A2]. A+A compounds can't be modified by degree adverbs. The lexical status of components is equal. The asymmetry and the markedness disappear. A+A compounds can act as the head, attribute, object in modifier-head construction to generalize and metonymy the whole property category, but can't take the tense, aspect marker and directional complements. For example: 大/小了 [da/xiao le] (getting larger/smaller), *大小了 [daxiao le]; 多起来 [duo qi lai] (getting more)/少下去 [shao xia qu] (getting less), *多少起来/下去 [duoshao qi lai/xia qu] [6].

Compared with the adjective components, the A+A compounds have taken place the fundamental change both at semantic and syntactic level. The details are as follows (Table 2).

Table 2. Comparison between A+A compounds and components [6].

	Syntactic structure	Syntactic position	Utterance		Overlapping	Negative	Markedness
			Modification	Reference			
Component	A+NP S+adv+A	Attributive, predicate	+	−	+	不[bu]+A	+
Compound	NP+(的) +A1A2 VP+A1A2	Subjective, objective	−	+	−	− 不[bu]+VP +A₁A₂	−

A+A antonymous compounds show the tendency of the whole transformation, and 90.16% of the compounds are transformed from the adjectives of components into the nouns with the function of the referential attribute. However, this referential attribute is different from the reference of a typical noun to a person or a specific thing. The semantics is not self-sufficient. At the syntactic level, there are the following performances:

A+A antonymous compounds cannot be used as an attributive alone. In modern Chinese, 68% of nouns can be used as attributives [5]. Therefore, acting as an attributive is also a syntactic function of nouns. But A+A compounds' semantics is not self-sufficient, which can only be attached to the subject entity or action behavior.

A+A antonymous compounds are not free to act as a subject or an object. They cannot implement the function of reference independently, and are mainly used in the structure of NP + (的) [de] (auxiliary word) + A1A2, for example:

(3)

    a. 图片的优劣，是此类图书成败的关键因素之一。

       [*Tupian de youlie, shi ci lei tushu chengbai de guanjian yinsu zhi yi.*]

       The pros and cons of the picture are one of the key factors for the success of such books.

  *b. 优劣，是此类图书成败的关键因素之一。

       [*Youlie, shi ci lei tushu chengbai de guanjian yinsu zhi yi.*]

       The pros and cons are one of the key factors for the success of such books.

Most A+A antonymous compounds cannot be directly modified by the quantity structure, such as *三种高低 [*san zhong gaodi*], *一个大小 [*yi ge daxiao*].

Based on the above analysis, A+A antonymous compounds are transformed into nouns as a whole, but they do not have the prototype or typical characteristics of nouns. The independence of components is the weakest.

6 Conclusion

Based on the analysis and comparison of the three types of antonymous compounds, it is not difficult to see that the independence degree of noun monosyllabic words, verbal monosyllabic words and adjective monosyllabic words as the components is different. Noun monosyllabic words are the strongest, followed by verbal monosyllabic words, and adjective monosyllabic words are the weakest. We can express the strength of independence as follows:

Noun monosyllabic words > Verbal monosyllabic words > Adjective monosyllabic words.

References

1. Haiman, J.: Iconic and economic motivation. Language **59**(4), 781–819 (1983)
2. Mithun, M.: The evolution of noun incorporation. Language **60**(4), 847–894 (1984)
3. Zhu, D.X.: Lectures on Grammar. The Commercial Press, Beijing (1982)
4. Hopper, P., Thompson, S.A.: The Iconic Basis of the Categories Noun and Verb. John Benjamins, Amsterdam (1985)
5. Guo, R.: Research on the part of speech in modern Chinese, pp. 84–101. The Commercial Press, Beijing (2002)
6. Zhang, J., Xiao, S.: The characteristics and the whole conversion analysis of quantitative antonymous compounds. In: Lu, Q., Gao, H. (eds.) CLSW 2015. LNCS, vol. 9332, pp. 380–391. Springer, Cham (2015). https://doi.org/10.1007/978-3-319-27194-1_38

Applications of Natural Language Processing

Microblog Sentiment Classification Method Based on Dual Attention Mechanism and Bidirectional LSTM

Wenjie Wei, Yangsen Zhang[(⊠)], Ruixue Duan, and Wen Zhang

Institute of Intelligent Information Processing, Beijing Information Science
and Technology University, Beijing 100192, China
wei_wjl228@163.com, zhangyangsen@163.com,
duanruixue@bistu.edu.cn, 1379853502@qq.com

Abstract. In the information age, the network technology continues to develop. As an emerging social media, Sina Weibo has a huge user base. Every day, hundreds of millions of users express their opinions on hot events, or share the joys and worries in life on the Weibo platform. Therefore, the analysis of the user's emotion has broad application prospects, which could also be used in the fields of public opinion monitoring, opinion guidance, and advertisement placement. This paper proposes a microblog sentiment classification method based on dual attention mechanism and bidirectional LSTM. Firstly, the bidirectional LSTM model is used to semantically encode the microblog text, then the self-attention and sentiment word attention are introduced into the bidirectional LSTM model. Finally, the Softmax classifier is used to classify the sentiment of microblog. In order to verify the validity of the model, several groups of comparative experiments are carried out, which use NLPCC2013 and NLPCC2014 evaluation task datasets as experimental data sets. The results show that the proposed microblog sentiment classification model based on dual attention mechanism and bidirectional LSTM is superior.

Keywords: Microblog · Sentiment classification · Deep learning · Attention mechanism · Long-short term memory

1 Introduction

According to the results of the China Internet Development Statistics Report 2018, the number of Chinese Internet users has reached 772 million, the total number of websites has reached 5.33 million, and the total number of web pages has exceeded 260.4 billion. With the continuous development and popularization of Internet technology, social media has become an indispensable part of daily life. The emergence of social media has made information flow and sharing more convenient. Among them, Sina Weibo, as an emerging information dissemination platform and social network platform, stands out among many new media in China because of its unique real-time interaction, free and open features. In recent years, China's social reform and social transformation have continued to deepen, social contradictions have gradually intensified, people's desire to express personal opinions and attitudes through the internet

© Springer Nature Switzerland AG 2020
J.-F. Hong et al. (Eds.): CLSW 2019, LNAI 11831, pp. 309–320, 2020.
https://doi.org/10.1007/978-3-030-38189-9_33

has gone from strength to strength, meanwhile, hundreds of millions of users have expressed their opinions and attitudes on social hot events on Weibo, some of which are positive, while others are pessimistic and negative. If pessimistic negative attitudes dominate the mainstream then form a bad public opinion situation, it is very likely to affect social security and stability. Sentiment analysis of microblog published by hundreds of millions of users to judge the user's sentiment tendency towards social hot events, which will be helpful for government agencies to predict the direction of the development of the situation, and take relevant measures in advance to guide positive public opinion to prevent the impact of social security and stability. Therefore, the research on the sentiment of microblog texts is of great significance. At present, a large number of researchers have conducted relevant research and achieved certain results.

2 Related Work

At present, the research work on Chinese microblog sentiment analysis is still in its infancy, that many research methods refer to the existing foreign research methods. However, when the foreign methods are directly applied to the Chinese language field, the effect is not very ideal because of the great differences between Chinese and English in grammar rules, language habits etc. After a lot of literature research, there are three kinds of sentiment analysis methods: (1) sentiment analysis method based on sentiment dictionary and rules; (2) sentiment analysis method based on machine learning; (3) sentiment analysis method based on deep learning.

2.1 Sentiment Analysis Method Based on Sentiment Dictionary and Rules

The construction of the sentiment dictionary brings great convenience to the sentiment analysis work. By comparing the words with the words in the sentiment dictionary, the polarity of the words can be directly determined, then the sentiment polarity of the microblog is determined. Based on HowNet, Zhu [1] proposed two methods for lexical semantic tendency calculation: a method based on semantic similarity the other one based on semantic correlation field, besides, proved through experiments that the effect of the method in Chinese common words preferably, the accuracy rate after word frequency weighting can reach more than 80%. Hou [2] proposed a method of semantic dictionary and rules to construct a phrase sentiment dictionary. The dictionary contains more phrases with sentiment polarity. The dictionary is used to classify the sentiment polarity of the microblog text, which has achieved better results. Wu [3] used the sentiment dictionary to select sentiment features. In order to reduce the influence of microblog corpus imbalance on feature selection, feature item frequency factors were used to reduce the dimension of feature items, which improved the accuracy of sentiment classification. Hatzivassiloglou [4] found that there is a specific semantic relationship between the sentiment texts before and after the related words. For example, the general sentiment polarity is the same before and after "and", while the general sentiment polarity is opposite before and after "but", but the limitation of this research is that it can only identify and analyze adjectives. Song [5] research the sentiment polarity in hot events based on the polarity dictionary. The above sentiment analysis method based on sentiment dictionary and rules depend too much on sentiment dictionary, so the result of sentiment analysis depends on the quality of the sentiment dictionary.

2.2 Sentiment Analysis Method Based on Machine Learning

At present, most of the research text sentiment classification adopts the method based on machine learning, using large-scale microblog corpus for model training, using the trained model for microblog sentiment classification. He [6] constructed a vector space model after calculating the weight of text features based on TF-IDF, which the dimension of feature vectors is reduced by using the method of information increment and document frequency, at the same time sentiment classification is carried out by using various methods. Finally, it is found that the classification effect based on machine learning is the best. Pang [7] conducted an in-depth study on the sentiment analysis of microblog. He proposed an unsupervised sentiment classification method based on sentiment knowledge, which used sentiment word and expression pictures as learning knowledge, and used machine learning method to construct microblog sentiment classifiers. The experimental results show that the performance of the classifiers is very good. Zhang [8] first constructed a sentiment knowledge base suitable for Chinese using the expression images and sentiment words in microblog, then used the information entropy to optimize the corpus. Finally, the microblog sentiment was classified by constructing Bayesian classifiers, which obtained a very good classification effect. Liu [9] used three machine learning algorithms, three feature selection algorithms and three feature item weight algorithms to conduct sentiment classification research on movie reviews. The results show that the three machine learning algorithms have their own advantages, and the performance of sentiment classification depends on the style of comment. Li [10] proposed a sentiment analysis method which combines SVM with CRF. It is proved by experiments that the performance of SVM model is better than that of CRF model when selected to use part-of-speech (POS), sentiment word and negative words as features. When sentiment word, negative word, degree adverbs and special symbols are selected as features, the performance of CRF model is better than that of SVM model.

Through a large number of literature investigations, the effect of sentiment analysis based on machine learning algorithm is better than traditional sentiment analysis method, but machine learning algorithm requires researchers to have some experience in text feature and classifier selection. This is also the focus and difficulty of the machine learning method.

2.3 Sentiment Analysis Method Based on Deep Learning

Since Hinton [11] proposed a deep learning method in 2006, both academia and industry have done a lot of research on it. At present, deep learning has made remarkable achievements in the fields of speech, image, online advertising, and natural language processing. Deep learning has also been widely used in the field of sentiment analysis. Yang [12] improved the Kim model based on convolutional neural network theory and verified the superior performance of convolutional neural networks for sentiment classification of twitter information. Socher [13] proposed a method based on tensor recurrent neural network, which greatly simplifies the scale of the model by introducing the concept of tensor to reduce the model parameters. Sun [14] proposed a multi-tag sentiment classification method for microblog. First, a large number of microblog data is used to train word vectors, using the CNN model to carry out supervised multi-sentiment classification learning. The learned CNN model is used to combine the word vectors in microblog sentences into sentence vectors. Finally, these sentence vectors are used as feature training multi-label classifiers to complete the multi-label sentiment classification of microblog. Wei Meng [15] designed a sentiment analysis method that uses CRT mechanism to fuse CNN and LSTM. Experiments based on three public data sets prove that the mixed model has higher accuracy in sentiment analysis in specific fields.

On the whole, sentiment analysis is a very practical research task, its research trend is mainly manifested in the continuous refinement of sentiment analysis objects, the deepening of sentiment analysis levels, and the enrichment of sentiment analysis content. The research technology adopted has gradually changed from traditional sentiment dictionary and machine learning model to various deep learning models.

3 Construction of Microblog Sentiment Classification Model

3.1 LSTM Model

Although the Recurrent Neural Network (RNN) solves the problem of dependency before and after the sequence, when the sequence is long, the influence of the distant state on the current state will be negligible, which will lead to the poor ability of RNN to learn the historical states. The emergence of LSTM is to solve the long-term dependence of RNN. The Long Short-Term Memory can effectively capture the context information of the sentences. The output h_t of the hidden node of the LSTM is as shown in the formulas (1)–(6):

$$i_t = \sigma(W_i \cdot [h_{t-1}, x_t] + b_i). \tag{1}$$

$$f_t = \sigma(W_f \cdot [h_{t-1}, x_t] + b_f). \tag{2}$$

$$f_t = \sigma(W_f \cdot [h_{t-1}, x_t] + b_f). \tag{3}$$

$$f_t = \sigma(W_f \cdot [h_{t-1}, x_t] + b_f). \tag{4}$$

$$f_t = \sigma(W_f \cdot [h_{t-1}, x_t] + b_f). \tag{5}$$

$$f_t = \sigma(W_f \cdot [h_{t-1}, x_t] + b_f). \tag{6}$$

Where σ denotes the sigmoid function, $W_i, W_f, W_o, W_c, b_i, b_f, b_o, b_c$ are parameters to be learned in the LSTM, h_{t-1} is the output of the hidden node at time $t-1$, C_{t-1} is the output of the memory unit at time $t-1$.

Since the unidirectional LSTM can only capture the above information at the current moment but does not consider the following information, this paper uses bidirectional LSTM (BiLSTM) to obtain the context information of the microblog text, that is, using forward and reverse LSTM to extract the semantic information of the sentence at the same time, finally splicing their output as the output of BiLSTM, and then pooling it. The context information that can represent microblog text is obtained, that is, semantic information representation vector.

3.2   A Method of Obtaining Self-attention Vector

Although LSTM is widely used in the field of text classification, its encoder structure encodes the input sequence as a fixed-length internal representation, which severely restricts the length of the input sequence and leads to poor classification performance of the long text sequence. Introducing the attention mechanism into the LSTM model can solve the above problems well. Considering the time complexity of each layer of deep learning and the long distance dependence, this paper uses the self-attention mechanism which proposed by the Google Machine Translation Team in June 2017 [16], which has the following advantages: (1) Ability to learn the internal connections of the sequence; (2) Reduce the computational complexity of each layer; (3) Effectively reduce the minimum number of units for parallel computing, thereby reducing training costs. The acquisition process of the self-attention vector v_s is shown in Fig. 1:

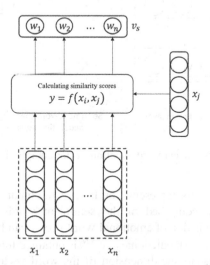

Fig. 1. Self-attention vector calculation diagram.

In Fig. 1, $x_1, x_2, \cdots, x_i, x_j, \cdots, x_n$ all represent input data, which is the word vector corresponding to each word in the microblog text, y represents the corresponding query. The similarity is calculated by Additive attention calculation method, $w_1, w_2, \cdots, w_n$ as the corresponding weight, and finally get the self-attention vector v_s.

3.3 A Method of Obtaining Sentiment Word Attention Vector

If we only use the self-attention of microblog text to strengthen the semantic representation of the text, there are still some shortcomings. In order to highlight the role of the sentiment words in the microblog text, according to the Emotional Ontology Library compiled by Dalian University of Technology [17] and HowNet, we constructed and built a sentiment word base. Because each sentiment word in microblog text only has the function of sentiment strengthening and weakening, if the sentiment word of the same polarity appears at the same time, the sentiment will be strengthened, and the sentiment of the opposite polar sentiment word will weaken or even reverse at the same time, because there is no semantic dependence between the sentiment words, so this paper uses the full connection network to carry on the nonlinear transformation to the sentiment word set. Finally, the sentiment attention vector v of the microblog is obtained, and the synthesis process is shown in Fig. 2:

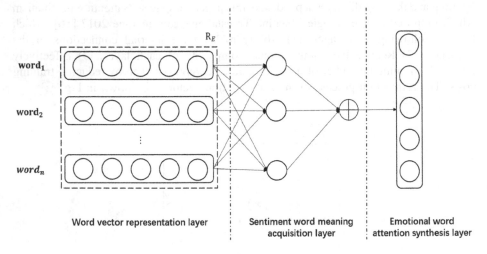

Fig. 2. Sentiment word attention vector composition graph.

In Fig. 2, the word vector representation layer is the input of the model, which is a word vector matrix R_E composed of the sentiment words contained in a single microblog text. The acquisition of emotional word vector can be regarded as a process of looking up a dictionary. The dictionary $R^{d \times N}$ is obtained through large-scale corpus training. Where d represents the dimension of the word vector and N represents the number of words in the dictionary. For a text sequence $T = \{w_1, w_2, \cdots, w_n\}$, the word vector of the sentiment word is spliced together, after that the word vector representation of the sentiment word set can be obtained. The splicing method is as shown in formula (7):

$$R_E = v_1 \oplus v_2 \oplus \cdots \oplus v_n. \tag{7}$$

Where $v_i \in R^{d \times N}$ indicates that w corresponds to an element in the dictionary, and $\oplus$ represents a row vector splicing operation. The dimension of R_T is the number of emotional words in the text of microblog.

3.4 Microblog Sentiment Classification Model Based on Dual Attention Mechanism and Bidirectional LSTM (Dual-Attention-BiLSTM)

The Dual-Attention-BiLSTM sentiment classification model is composed of bidirectional Long Short-Term Memory after adding self-attention and sentiment word attention. The first layer of the framework is a word vector representation layer, which changes the data after the word segmentation into a word vector matrix R_T, the column of the matrix is the dimension of the word vector, and the row of the matrix is the length of the sequence. The second layer of the framework is a bidirectional LSTM layer, which is used to encode the semantic features of microblog text. The word vector dimension of the layer is 200 dimensions, that the size of the hidden layer is 128. The third layer of the framework is the semantic composition layer. The self-attention vector v_s and the sentiment word attention vector v_e obtained in the two Sects. 3.2 and 3.3 are applied to the output of the bidirectional LSTM, and the result is input to the fourth layer of sentiment calculation layer, which is a softmax classifier, then the classification results are obtained by the classifier. The frame diagram is shown in Fig. 3:

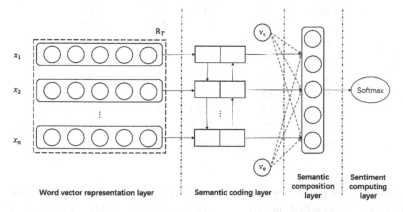

Fig. 3. Microblog sentiment classification framework based on dual attention and bidirectional LSTM.

4   Experiment and Analysis

4.1   Selection of Data Sets

The experimental data of this paper adopts the public data set NLPCC2013 and NLPCC2014 of Chinese microblog sentiment evaluation task. The microblog text is from Sina Weibo platform. Each data set is divided into two parts: training set and test set. The text is labeled with a primary sentiment tag and multiple secondary sentiment tags. The sentiment label is divided into eight categories, using the classification criteria of the literature [17]. Specifically, the eight emotional tags are: happiness, like, anger, sadness, fear, disgust, surprise, none. In view of the multiple sentences below each Weibo data, each sentence may have different sentiment tags, only the same microblog sentence as the main sentiment labels of microblog are extracted as effective data. Because the task of this experiment is used to classify the positive and negative polarities, we regard happiness and like as positive sentiment microblog, anger, sadness, fear, disgust and surprise as negative sentiment microblog, and delete the microblog data with the main sentiment tag as none. Through the parsing of XML, 14085 positive data and 15316 negative data are obtained, so that the positive and negative data are well balanced. The number of microblog text with each category as the main sentiment label is shown in Table 1:

Table 1. Number of microblog texts in NLPCC2013 and NLPCC2014 data sets

Sentiment	NLPCC 2013		NLPCC 2014	
	Training set	Test set	Training set	Test set
Happiness	370	1116	1459	441
Like	595	1558	2204	1042
Anger	235	436	669	128
Sadness	385	744	1173	189
Fear	49	102	148	46
Disgust	425	935	1392	389
Surprise	113	236	362	162
None	1828	4873	6591	3603
Total	4000	10000	13998	6000

Microblog data processing is mainly the filtering of abnormal data to achieve data formatting and standardization. According to the linguistic and structural characteristics of microblog data, we use the following three methods to preprocess: (1) The repeated letters, repeated symbols and repeated Chinese characters in the evaluation data are eliminated; (2) Delete non-text information from the evaluation data, such as URL links, emoticons, special characters; (3) Construct a user vocabulary and stop words list for microblog, then the evaluation data are segmented and removed based on jieba word segmentation system. The processed data is shown in Fig. 4:

```
0  图书馆 书架 间 逡巡 一本 熟悉 书 回忆起 初中 高中 六年 里 中午 做 图书馆 管理员 时 一本 看过 书 时间
0  只许州官放火 不许百姓点灯 中石化 中 石油 一再 涨价 联合利华 提 涨价 遭 万 罚款 康师傅 遭受 警告
1  圆点 长款 套头 针织衫 可爱 波点 超级 爱 感觉 迷人 长款 随身 感觉 超级 百搭 感觉
1  休闲 舒适 不失 可爱 俏皮 一身 很赞
0  累死 陪 逛街 买 件 薄 风衣 打底 吊带衫 床上 四件套 想 买 东西 买 一件 小衫 一件 打底 衫
1  明白 包容 岁月 痕迹 性感 皱纹 里 我俩 泪水 喜欢 这句 贯中 写诗 才华
1  西苑 医院 一种 神奇 面膜 早上 点去 排队 做 一周 痘子 米 块钱
1  找到 帮 排队 老公 做 美女 小护士 花心
0  爱情 走 尽头 绝情
0  金梦 想起 腰椎 结核 卧床不起 前男友 精神 浮躁 异 常人 他会 打电话 骂 骂 停机 帮 缴费 发泄 接片
0  讨厌 那种 拍 自拍 试妆 照说 渣 要死 艾特 一堆 还轮 那张 照片 版聊 死命 反驳 说 不萌 暗喜 妹子
```

Fig. 4. Pre-processed evaluation data.

4.2 Experimental Parameter Setting

The selection of hyper parameters is very important for the performance of neural networks. Therefore, the Hyperopt library is used for distributed parameter adjustment to obtain the optimal parameter set of the model. The model parameter adjustment list is shown in Table 2:

Table 2. Model parameter adjustment list

Parameter	Ranges	Value
Optimizer	Adadelta, RMSprop	Adadelta
Lstm_units	100, 150, 200, 250, 300	200
Batch_size	8, 16, 32, 64, 128	64
Dropout	[0.05, 0.95]	0.2
Learn_rate	[0.01, 0.1]	0.05

4.3 Experimental Contrast Model

MCNN Model (multichannel convolutional neural network, MCNN) [18]: The model uses multi-channel convolution neural network to learn the word vectors of microblog text, so as to construct the semantic representation of the text and realize the sentiment classification of the text, which is one of the early sentiment model using the deep learning model.

EMCNN Model (emotion-semantic-enhanced multichannel convolutional network, EMCNN) [19]: The EMCNN model constructs a semantic feature representation matrix for common microblog emoticons. Through matrix product operation, the semantic feature representation matrix of emoticons is used to semantically enhance the microblog text. Multi-channel convolutional neural network is used for feature learning to realize sentiment classification.

ESM Model [20]: The ESM model realizes the mapping of words to sentiment space by calculating the word vector of words in microblog text and the cosine function value of microblog emoji word vector, and uses SVM model to classify sentiment.

BiLSTM Model: The model does not introduce attention mechanism, directly uses bidirectional LSTM model to model microblog text, and carries on sentiment analysis, which is the benchmark model of this paper.

BiLSTM+self-attention Model: The model introduces the self-attention mechanism on the basis of the bidirectional LSTM model, and then carries on the emotion analysis.

BiLSTM+sentiment-attention Model: The model introduces the sentiment-attention mechanism on the basis of the bidirectional LSTM model, then carries on the emotion analysis.

4.4   Experimental Results and Analysis

In this paper, the experiment of classification tasks of emotion positive and negative polarity is designed by using the method of cross verification, and the accuracy is used as the evaluation index. The specific experimental results are shown in Fig. 5. From the graph, we can find that the accuracy of BiLSTM+self-attention and BiLSTM+sentiment-attention model with single attention is higher than that of BiLSTM, MCNN, EMCNN, ESM model, and the accuracy of BiLSTM+sentiment-attention model is higher than that of BiLSTM+self-attention model, and the accuracy of EMCNN model is also higher than that of BiLSTM, MCNN, ESM model, which fully shows that sentiment words make a great contribution to the task of sentiment positive and negative polarity classification. The accuracy of the Dual-Attention-BiLSTM model with self-attention and sentiment-word attention is higher than that of all other models, which shows that it is feasible to improve the performance of microblog sentiment classification by adding dual attention.

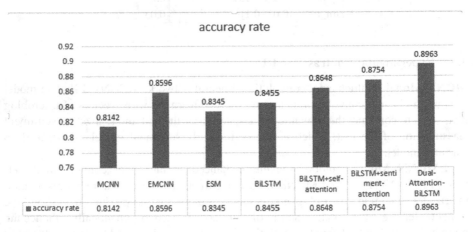

Fig. 5.   Contrast experimental results.

5   Conclusion

In this paper, NLPCC2013 and NLPCC2014 are used as experimental data for microblog sentiment classification task, and a microblog sentiment classification model based on dual attention mechanism and bidirectional LSTM is designed. The model first uses the bidirectional LSTM model to code the microblog data, then adds self-attention and sentiment word attention to the output of bidirectional LSTM, then carries on the sentiment classification through Softmax classifier, finally setting up several groups of contrast experiments. The experimental results show that the model proposed in this paper has excellent performance.

In the next work, we will focus on the following aspects: (1) Due to the limitation of data, the sentiment words collected in the sentiment lexicon of this paper are not comprehensive enough, and the sentiment symbols are not considered. Therefore, the next step is to further improve the sentiment lexicon and add the sentiment symbols to the sentiment lexicon; (2) To study whether other attention will improve the performance of the model, try to add other text elements as new attention, and carry out related experiments; (3) To replace the data set to test the universality of the model proposed in this paper.

Acknowledgements. This work was supported by grants from National Nature Science Foundation of China (No. 61772081).

References

1. Zhu, Y.L., Min, J., Zhou, Y.Q., et al.: Semantic orientation computing based on HowNet. J. Chin. Inf. Process. **20**(1), 16–22 (2006). https://doi.org/10.3969/j.issn.1003-0077.2006.01. 003. (in Chinese)
2. Hou, M., Teng, Y.L., Li, X.Y., et al.: Study on the linguistic features of the topic-oriented microblog and the strategies for its sentiment analysis. J. Appl. Linguist. **86**(2), 135–143 (2013). https://doi.org/10.16499/j.cnki.1003-5397.2013.02.019. (in Chinese)
3. Wu, Q.L., Wang, Y.: Research on the emotional feature selection method in the Chinese microblog. J. Inner Mongolia Normal Univ. **45**(1), 84–88 (2016). https://doi.org/10.3969/j. issn.1001-8735.2016.01.020. (in Chinese)
4. Hatzivassiloglou, V., Mckeown, K.R.: Predicting the semantic orientation of adjectives. In: Proceedings of the ACL, pp. 174–181 (1997). https://doi.org/10.3115/979617.979640
5. Song, S.Y., Li, Q.D., Lu, D.Y.: Hot event sentiment analysis method in micro-blogging. J. Comput. Sci. **39**(Z6), 226–228 (2012). (in Chinese)
6. He, Y., Deng, W.R., Zhang, D.: Study on sentiments recognition and classification of Chinese micro-blog. J. Intell. **2**, 136–139 (2014). https://doi.org/10.3969/j.issn.1002-1965. 2014.02.026. (in Chinese)
7. Pang, L., Li, S.S., Zhou, G.D.: Sentiment classification method of Chinese micro-blog based on emotional knowledge. J. Comput. Eng. **38**(13), 156–168,162 (2012). https://doi.org/10. 3969/j.issn.1000-3428.2012.13.046. (in Chinese)
8. Zhang, S., Yu, L.B., Hu, C.J.: Sentiment analysis of Chinese micro-blogs based on emoticons and emotional words. J. Comput. Sci. **39**(Z11), 146–148 (2012). https://doi.org/ 10.3969/j.issn.1002-137x.2012.z3.041. (in Chinese)
9. Liu, Z.M., Liu, L.: Empirical study of sentiment classification for Chinese microblog based on machine learning. J. Comput. Eng. Appl. **48**(1), 1–4 (2012). https://doi.org/10.3778/j. issn.1002-8331.2012.01.001. (in Chinese)
10. Li, T.T., Ji, D.H.: Sentiment analysis of micro-blog based on SVM and CRF using various combinations of features. J. Appl. Res. Comput. **32**(4), 978–981 (2015). https://doi.org/10. 3969/j.issn.1001-3695.2015.04.004. (in Chinese)

11. Hinton, G.E., Salakhutdinov, P.R.: Reducing the dimensionality of data with neural networks. Science **313**(5786), 504–507 (2006)
12. Yang, X., Macdonald, C., Ounis, I.: Using word embeddings in Twitter election classification. Inf. Retrieval J. **21**(2–3), 183–207 (2018). https://doi.org/10.1007/s10791-017-9319-5
13. Socher, R., Perelygin, A., Wu, J., et al.: Recursive deep models for semantic compositionality over a sentiment treebank. In: Proceedings of the 2013 Conference on Empirical Methods in Natural Language Processing, pp. 1631–1642 (2013)
14. Sun, S.T., He, Y.X.: Multi-label emotion classification for microblog based or CNN feature space. J. Adv. Eng. Sci. **49**(3), 162–169 (2017). https://doi.org/10.15961/j.jsuese.201600780 . (in Chinese)
15. He, Y.X., Sun, S.T., Niu, F.F., et al.: A deep learning model enhanced with emotion semantics for microblog sentiment analysis. Chin. J. Comput. **40**(4), 773–790 (2017). https://doi.org/10.11897/sp.j.1016.2017.00773. (in Chinese)
16. Meng, W., Wei, Y.Q., Liu, W.F.: Target-specific sentiment analysis based on CRT mechanism hybrid neural network. J. Appl. Res. Comput. (2019). https://doi.org/10.19734/j.issn.1001-3695.2018.08.0538. (in Chinese)
17. Vaswani, A., Shazeer, N., Parmar, N.: Attention is all you need. In: Advances in Neural Information Processing Systems, pp. 5998–6008 (2017)
18. Xu, L.H., Li, H.F., Pan, Y., et al.: Constructing the affective lexicon ontology. J. China Soc. Sci. Tech. Inf. **27**(2), 180–185 (2008). https://doi.org/10.3969/j.issn.1000-0135.2008.02.004. (in Chinese)
19. Wawre, S.V., Deshmukh, S.N.: Sentiment classification using machine learning techniques. Int. J. Sci. Res. **5**(4), 819–821 (2016)
20. Jiang, F., Liu, Y.Q., Luan, H.B., et al.: Microblog sentiment analysis with emoticon space model. J. Comput. Sci. Technol. **30**(5), 1120–1129 (2015). https://doi.org/10.1007/s11390-015-1587-1

High Order N-gram Model Construction and Application Based on Natural Annotation

Qibo Wang[1], Gaoqi Rao[2(✉)], and Endong Xun[1(✉)]

[1] College of Information Science, Beijing Language and Culture University, Beijing, China
{wangqibo,xunendong}@blcu.edu.cn
[2] Research Institute of International Chinese Language Education, Beijing Language and Culture University, Beijing, China
raogaoqi@blcu.edu.cn

Abstract. The language model based on the n-gram grammar plays an important role in NLP tasks. In this paper, language models based on language boundary are proposed to conquer the challenge of the very big language data: intra-sentence boundary model and inter-sentence boundary model. We developed a training tool on the Hadoop platform based on MapReduce programming, and conducted the prefix tree to compress and store the model. We implemented our model in identifying the boundary in the syntactic parsing, achieving a good result.

Keywords: Boundary language model · N-gram · Prefix tree · Boundary recognition

1 Introduction

N-gram model is designed to conquer the difficulties in computing long strings, depending on Markov assumption. Along with the increasing of the computing complexity, higher-order n-gram model was required. Also inspired by Li [1] and Rao [2], we used the boundary information to construct a high-order n-gram model based on massive scale of Chinese corpus. Our work focuses on: (1) Boundary N-gram Model. We conducted a new kind of language model based on the boundary information in natural annotations. (2) Optimizing loading speed and memory utilization by designing shared prefix of 2 Chinese characters, which is based on the structure of prefix tree. (3) Finally, the experiment on binary syntactical tree generation was conducted, and has achieved good performance in precision.

This paper is organized as: related works are described in Sect. 2. We describe the construction of boundary language model in Sect. 3. The compressing and loading method is described in Sect. 4. The experiment on binary syntactical tree generation lays in Sect. 5. Section 6 is the conclusion.

© Springer Nature Switzerland AG 2020
J.-F. Hong et al. (Eds.): CLSW 2019, LNAI 11831, pp. 321–328, 2020.
https://doi.org/10.1007/978-3-030-38189-9_34

2 Related Work

2.1 Language Modeling

N-gram modeling has widely adopted in various tasks in NLP and is updated in many aspects. Rosenfeld [3], Huang [4] and Ney [5] proposed skipping n-gram model, focusing on computing the property of a word by the words with the same space, instead of the neighbor ones. Brown [6] proposed the class-based model. It was then improved by Goodman [7]. This model was actually a cluster of n-gram models classified by word features like POS or other semantic features. Kuhn [8–10] proposed cache-based model, focusing on dynamic self-adaption in cache and memory.

The training and loading of a language model is the basic step of utilization of language model. SRILM [11] is a widely-used tool for training language model. But it cannot be adapted to distributed computing. Federico [12] and MSRLM [13] applied divide-and-conquer strategy to reduce memory usage, and conducted IRSTLM. Zhang [14] combined various smoothing algorithms to optimize complexity of the language modeling, used in machine translation. Otherwise, many scholars proposed distributed and parallel computing in language modeling. Zhang [15] applied Client-Server model in LM training. Dean [16] and Yu [17] combined Good-Turning algorithm and Fool Back-off to build the training tool based on MapReduce.

2.2 Boundary Computing

The research in boundary information mainly focuses on phrase recognition, which includes base NP recognition and PP recognition. Zhou [18] used the length distribution to compute the property of phrase boundary, in order to predict base NPs. Zhao [19] combined the boundary co-occurrence with traditional phrase recognition methods to mine base NPs. Li [20] and Ma [21] selected boundary information as a feature of HMM model to optimize the effect in NP extraction.

In the research on chunks, Liu [22] defined a computational Chinese chunk system, which can be automatically optimized by error driving methods. Huang [23] proposed a hybrid model on SVM and error driving method in chunk recognition. Li [24] used Stacking algorithm building a hierarchical framework to contain various context features including boundary information. Liu [25] viewed the chunk recognition as classification problem, combining HMM and a rule system.

3 The Construction of Boundary Language Model

The natural boundaries, especially punctuation marks among words, phrases and sentences can be used to train a boundary language model, because these punctuation marks can offer the signal of splitting strings. The intra-sentence boundary language model includes two statistical models focusing on the left and right sides of punctuation marks respectively. These models only calculate the probability of n-gram strings with punctuation marks at the beginning or the end, which can avoid the problem of size explosion of the model. Therefore, the order (value n) of the intra-sentence language model can be larger, by occupying the same space. More background information can also be used in model training, leading to better robustness of the model.

The corpus used in this paper was collected from the Internet, including 2T data of news, blog, literary novel, weibo and other domains. After format processing, it achieved sentence-level non-repetition, word segmented and part-of-speech annotated. The amount of data reached 280 billion words. We used CDH5 to implement Hadoop platform on five servers. All the contents of the file were treated as one long string to get the n-gram count. We did not count all the links between the beginning and the end of the content, but achieved the same effect by setting temporary variables.

3.1   Language Model of Boundary Information

Punctuation is one of the most important natural annotation information in text. Because punctuation marks have the function of segmenting each language unit of a sentence, we used punctuation marks to construct the boundary language model, and applied the boundary language model to predict the punctuation position in a sentence, so as to achieve the purpose of recognizing the boundary of each unit in a sentence.

3.2   Boundary Modeling in Sentence

The intra-sentence language model is an n-gram model of the string between two punctuation substitutes. Due to the limited length of sentences, the scale of the model can be significantly reduced. In the process of implementation, we constructed the right and left intra-sentence boundary language models according to the punctuation marks at the beginning or end of the n-gram string. The models can be viewed as a simple class-based language model. In our work, the comma, period, exclamation mark, semicolon and other symbols mentioned above were divided into SEPARATOR-like symbols, shorted for SEP. Other symbols were set as CIRCLE symbols, shorted for CIR. Arabic numbers were set as NUMBER class, referred as NUM.

In the process of calculation, the probability value was calculated by maximum likelihood estimation (formulas 3-1, 3-2). The backoff value was calculated as formulas 3-3 and 3-4. The denominator is the frequency of the n-grams with the same suffix and SEP in the end. The numerator is the frequency of the n-grams selves.

$$\text{Prob(left)} = \frac{c(w_1 w_2 \ldots w_{n-1} \text{SEP})}{\sum_i^j c(w_1 w_2 \ldots w_{n-1} w_i)} \tag{3 - 1}$$

$$\text{Prob(right)} = \frac{c(\text{SEP} w_2 \ldots w_n)}{\sum_i^j c(w_i w_2 \ldots w_n)} \tag{3 - 2}$$

$$\text{Back(left)} = \frac{c(w_1 w_2 \ldots w_{n-1} \text{SEP})}{\sum_i^j c(w_i w_2 \ldots w_{n-1} \text{SEP})} \tag{3 - 3}$$

$$\text{Back(right)} = \frac{c(SEPw_2 \ldots w_n)}{\sum_i^j c(SEPw_2 \ldots w_{n-1})} \qquad (3-4)$$

3.3 Word Boundary Language Modeling in Short Sentence

Most of the n-gram models are based on words. However, the accuracy of Chinese word segmentation varies greatly among domains. Instead of words, we used n-gram based on characters, which could solve the problem caused by word segmentation errors.

Boundary language model is mainly used to assist short sequence segmentation. When the length of the sequence is 3 or 4, the segmentation result can be viewed as the result of word segmentation. The training of the above model is based on characters. When the sequence to be segmented is three characters long, the result may be 1 + 2 or 2 + 1 characters. When the sequence to be segmented is four characters, the result may be 1 + 3 or 3 + 1 or 2 + 2 characters. Their frequency was taken as the numerator. The denominator was the statistics of 4 or 5 characters containing the sequences.

4 Compress and Loading of the Boundary Models

The fastest way to use a language model is to load the model into memory. However, it will take up too much time and memory. Therefore, it's necessary to optimize the data structure for the compression storage and speeding up the loading of the models.

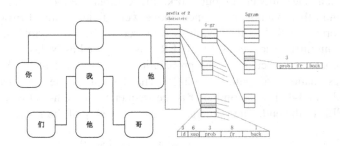

Fig. 1. Prefix tree and prefix-like tree (right)

In this paper, we designed a prefix tree-like data structure to compress, store and invoke the text-formatted language model. As Fig. 1 shows, the memory optimization strategy in the compression is as follows: The content of the shared prefix string array is composed of two characters, each of which takes up 2 bytes of storage space. By encoding and compressing characters, two characters occupy three bytes, which further saves the storage space of the shared prefix string array. In addition, the probability and backoff data are discretized to reduce the spatial complexity of the algorithm. In this way, the model trained on 280 billion words can be compressed on a server with 16G memory. The sizes change as shown in the following table (Table 1):

Table 1. Compression of the language models

Models	Before compression	After compression	Data size	Model size	Compress rate
Left intra-sentence	111.8G	106.8M	17G	17.1G	15.2%
Right intra-sentence	117.9G	106.8M	17.6G	17.7G	15.02%
Inter sentence	281.8G	106.8M	42.1G	42.2G	14.98%
Word boundary	28.8G	106.8M	4.3G	4.4G	15.3%

5   Experiments

We used the above language models to predict punctuation symbols in sentences, so as to achieve the goal of boundary recognition, and then generate a syntactic binary tree based on the results of boundary recognition.

Suppose s denotes a sentence to be recognized, BRS denotes a set of boundary recognition results, and BR denotes a result in the set. The process of finding the best boundary BR recognition result is to find the most probabilistic boundary recognition result from the set BRS which generates the boundary recognition result from the sentence. Namely:

$$BR = \max \arg_{BR \in BRS} P(BR, s) = \min \arg_{BR \in BRS} PPL(BR|S) \qquad (5-1)$$

In Formula 5-1 $P(BR, s)$ denotes the probability of the boundary recognition results of sentence s. PPL denotes the perplexity.

$$p(BR, s) = \begin{cases} p_1 * p_2 + p_3 & \text{if} \quad len(s) > 4 \\ p_1 * p_2 + p_3 + p_4 & \text{if } 2 < len(s) \leq 4 \end{cases} \qquad (5-2)$$

In 5-2 $p_1 = P_{left}(HZ_1 \ldots HZ_{root})$ and $p_2 = P_{right}(HZ_{root} \ldots HZ_{len(s)})$ denote the probabilities of the context calculated by the left and right models. $p_3 = P_{ngram}(s)$ denotes the probabilities of the current sentence in the inter-sentence model. p_4 denotes the probabilities of short sentence (phrase) boundary model.

$$p_1 = \sum_1^n \left(prob\left(HZ_1^i\right) * back\left(HZ_1^i\right) \right) \qquad (5-3)$$

$$p_2 = \prod_n^1 \left(prob\left(HZ_n^i\right) * back\left(HZ_n^i\right) \right) \qquad (5-4)$$

$$P_3(s) = \prod_{i=1}^m p(w_i | w_{i-n+1}^{i-1}) \qquad (5-5)$$

$$p(w_i | w_{i-n+1}^{i-1}) = \begin{cases} p(w_i | w_{i-n+1}^{i-1}) \text{if } w_{i-n+1}^{i-1} \text{ in A} \\ p(w_i | w_{i-n+1}^{i-1}) + back\left(w_{i-1} | w_{i-n+1}^{i-1}\right) \text{ if B} \end{cases} \qquad (5-6)$$

In 5-3 HZ_1^i denotes the string of first i characters, n denotes the offset of the current boundary in the sentence. In formula 5-4 HZ_n^i denotes the string of the first i characters. l denotes the length of the current sentence. $P_3(s)$ denotes the probability of the segmentation result in inter-sentence model. In 5-5 and 5-6 w_{i-n+1}^{i-1} denotes the string of $i - n + 1$ character to the $i - 1$ character. A denotes the string w_{i-n+1}^{i-1}, which exists in the model, and B denotes w_{i-n+1}^{i-1}, which does not exists.

In this experiment, one-best method was used to find the root node of binary tree. The specific method was as follows. First, ▲ was inserted after each Chinese character. The content on the left side of ▲ represented the left subtree of the segmentation result at this time, and the content on the right represented the right subtree of the segmentation result. We applied the boundary language model to calculate the perplexity, and found out the tree of minimum perplexity. Then the sentences were segmented at the ▲ to get the left and right subtrees. Then the subtrees were recursively segmented by the above steps, until final results were obtained.

Table 2. Precision of the chunk recognition

Data Set	Precision
msr_test	91.33%
pku_test	92.33%

The data set provided by Peking University and Microsoft in Bakeoff 2005 was used as test data to test the results of boundary recognition. The experimental results are shown in the Table 2. The experimental results show that the boundary language model based on large-scale corpus has good effect in the application of boundary recognition, and it can generate accurate binary parsing tree.

6　Conclusion

In this paper, we proposed the intra-sentence boundary language model and the word boundary language model. Aiming at the characteristics of boundary recognition task, a set of training tools for boundary language model was implemented with MapReduce. A char-based 10-gram boundary language model was trained on 560G corpus on Hadoop. By using two characters as shared prefix string and prefix tree, we designed a novel data structure to reduce the memory occupancy of the model loading and improved the speed significantly. Finally, we used the boundary information obtained from the boundary language model to generate binary tree in parsing. The experimental results show that our models has high accuracy in boundary recognition.

Acknowledgements. This paper is supported by Research Project of National Language Committee (YBI135-90), MOE Key Research Center Project (16JJD740004) and Beijing Language and Culture University Research Project (19YJ130005).

References

1. Li, Z., Sun, M.: Punctuation as implicit annotations for Chinese word segmentation. Comput. Linguist. **35**(4), 505–512 (2009)
2. Rao, G., et al.: Natural annotation research in large-scale corpora with a focus on Chinese word segmentation. Acta Sci. Nat. Univ. Pekin. **49**(1), 140–146 (2013)
3. Rosenfeld, R., Carbonell, J., Rudnicky, A., et al.: Adaptive statistical language modeling: a maximum entropy approach. A maximum entropy approach (1994)
4. Huang, X., Alleva, F., Hon, H.W., et al.: The SPHINX-II speech recognition system: an overview. Comput. Speech Lang. **7**(2), 137–148 (1992)
5. Ney, H., Essen, U., Kneser, R.: On structuring probabilistic dependences in stochastic language modelling. Comput. Speech Lang. **8**(1), 1–38 (1994)
6. Brown, P.F., Desouza, P.V., Mercer, R.L., et al.: Class-based n-gram models of natural language. Comput. Linguist. **18**(4), 467–479 (1992)
7. Goodman, J.T.: A bit of progress in language modeling. Comput. Speech Lang. **15**(4), 403–434 (2001)
8. Kuhn, R.: Speech recognition and the frequency of recently used words: a modified Markov model for natural language. In: Proceedings of ACL, pp. 348–350 (1988)
9. Kuhn, R., De Mori, R.: A cache-based natural language model for speech recognition. IEEE Trans. Pattern Anal. Mach. Intell. **14**(6), 219–228 (1992)
10. Kuhn, R., Mori, R.D.: Correction to: a cache-based natural language model for speech reproduction (1992)
11. Stolcke, A.: SRILM-an extensible language modeling toolkit. In: Interspeech, pp. 17–43 (2002)
12. Federico, M., Cettolo, M.: Efficient handling of n-gram language models for statistical machine translation. In: Proceedings of the 2nd WSMT, pp. 88–95. ACL (2007)
13. Nguyen, P., Gao, J., Mahajan, M.: MSRLM: a scalable language modeling toolkit. Microsoft Research MSR-TR-2007-144 (2007)
14. Zhang, R.: Research on Large Model and Its Application in Machine Translation, Ph.D thesis of Xiamen University (2009)
15. Zhang, Y., Hildebrand, A.S., Vogel, S.: Distributed language modeling for N-best list re-ranking. In: EMNLP, pp. 216–223 (2007)
16. Dean, J., Ghemawat, S.: MapReduce: simplified data processing on large clusters. Commun. ACM **51**(1), 107–113 (2008)
17. Yu, X.: Estimating language models using Hadoop and HBase. Ph.D thesis of University of Edinburgh (2008)
18. Zhou, Q., Sun, M., Huang, C.: Automatic identification of Chinese maximal noun phrases. J. Softw. **11**(2), 195–201 (2000)
19. Zhao, J., Huang, C.: Chinese basic noun phrase recognition model based on conversion. J. Chin. Inf. Process. **13**(2), 1–7 (1999)
20. Li, H., Yang, F., Zhu, J.: Transductive HMM based text chunking. Comput. Sci. **31**(2), 152–154 (2004)
21. Ma, Y., Liu, Y.: Base noun phrase identification based on HMM and candidates sorting by weighted templates. In: Proceedings of CCL (2005)
22. Liu, F., Zhao, T., Yu, H.: Statistics based Chinese chunk Parsin. J. Chin. Inf. Process. **14**(6), 28–32 (2000)
23. Huang, D., Wang, Y.: Chunk parsing based on SVM and error-driven learning methods. J. Chin. Inf. Process. **20**(6), 17–24 (2006)

24. Li, Y., Zhu, J., Yao, T.: Combined multiple classifiers based on a stacking algorithm and their application to Chinese text Chinese text chunking. J. Comput. Res. Dev. **42**(5), 844–848 (2005)
25. Liu, S., Li, Y., Zhang, L.: Chinese text chunking using co-training method. J. Chin. Inf. Process. **19**(3), 73–79 (2005)

A Printed Chinese Character Recognition Method Based on Area Brightness Feature

Yonghong Ke[(✉)]

Research Center for Folklore, Classics and Chinese Characters,
Beijing Normal University, Beijing, China
kyh@bnu.edu.cn

Abstract. This paper proposes a method for printed Chinese character recognition based on the area brightness feature, which is simple and has a low computational cost. It can achieve over 93% accuracy in recognizing printed Chinese characters equal to or greater than 10.5 pt which can meet the needs of certain situations (such as screen capture). The disadvantage of this method is its poor anti-distortion handling ability, and the recognition accuracy of smaller images still needs to be improved.

Keywords: Chinese character recognition · Printed Chinese character recognition · Area brightness feature

1 Introduction

1.1 History of Chinese Character Recognition

Chinese character recognition is an important branch of pattern recognition [1–3]. Major areas of Chinese character recognition can be divided into two categories: handwritten Chinese character recognition (HCCR) and printed Chinese character recognition (PCCR). This paper focuses on the latter. PCCR usually includes three main steps: image preprocessing, feature extraction and recognition. Image preprocessing generally includes image denoising, skew correction, image-text separation, text segmentation, binarization and so on. The purpose of image preprocessing is to improve image quality and prepare for feature extraction. The result of preprocessing directly affects the accuracy of feature extraction. Feature extraction is designed to extract useful data or information (such as strokes, structural points, projection features, contour features, histogram features, gray features, pixel brightness, etc.) from an image and obtain the representation or description of the text image, such as values, vectors, and symbols. The third step is to use the extracted eigenvalues to recognize the input text images and output candidate Chinese characters. In the three steps mentioned above, the first step of image preprocessing is the most foundation. This paper mainly focuses on the second step feature extraction and the third step Chinese character recognition. In the past few decades, much effort has gone into PCCR, which is mainly based on statistical and structural pattern recognition methods, such as those based on the complexity features [4], grid features [5], the indivisible wavelet and Zernike moments features [6], and the radical analysis network (RAN) features [7]. The accuracy of some methods is over 95%. All these explorations and research have provided us with valuable references and inspiration.

© Springer Nature Switzerland AG 2020
J.-F. Hong et al. (Eds.): CLSW 2019, LNAI 11831, pp. 329–336, 2020.
https://doi.org/10.1007/978-3-030-38189-9_35

1.2   Problems Faced With

As printed Chinese characters are distinguished from other printed characters use across the world, Chinese character recognition has been considered an extremely difficult issue. First, Chinese is used in daily communications by over one quarter of world's population, and Chinese characters are a super large set. China's national standards GB2312-80, GBK and GB18030-2000 contain 6763, 20902 and 27533 Chinese characters. The 'General Standardized Chinese Character List' [8] issued by China's State Council in June 2013 contains 8105 characters. Second, printed Chinese characters have many font styles, and the corresponding image features of different fonts vary greatly (Fig. 1).

永　永　永　永　永　永　永

Fig. 1. 永 in Seven Font Styles (Left to right: songti, fangsong, heiti, yahei, kaiti, lishu and youyuan)

Third, there are many similar characters, such as 已--己--巳, 盲--肓, 人--入 and 为 --办, etc. Some Chinese characters have complex structures and numerous strokes, for example, 麓, 霭, 鏖, 瀚, 瀣 and 瀛 in the second level of the 'General Standardized Chinese Character List'. In the case of small image size or low image quality, it is difficult to extract features of these Chinese characters. These factors indicate that PCCR has many difficulties to overcome.

1.3   Proposed Solutions

Area brightness was defined as the ratio of white pixels in an area to the total pixels in a binarized Chinese character image and which was used for training and recognition. This paper mainly includes two parts: 1. to divide the Chinese character image into certain areas and verify the accuracy of each area's brightness as a feature in various fonts and sizes; and 2. to verify the recognition accuracy of multiple feature combinations by a single classification method.

2   Area Brightness Feature of Chinese Characters

To maximize the extraction of useful features and minimize the computational complexity, the left-most, top-most, right-most and bottom-most tangent rectangular boxes were selected for each input Chinese character image, and each tangent rectangular box was divided into ten areas, which are the whole, middle part, upper half, lower half, left half, right half, upper left corner, upper right corner, lower left corner and lower right corner. Taking 阿 as an example, the feature areas and corresponding names are as follows (Fig. 2):

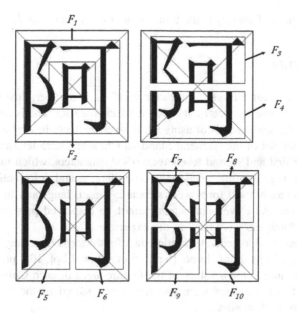

Fig. 2. Feature areas and corresponding names

For brevity, this paper numbered the areas 1–10, and the feature set is:

$$F = \{F_1, F_2, F_3, F_4, F_5, F_6, F_7, F_8, F_9, F_{10}\}$$

The X-Y coordinate system is established on the plane where the Chinese characters are located. If x and y are the abscissa and ordinate coordinates of the pixel points, then:

$$t(x, y) = \begin{cases} 0 & white\ pixel \\ 1 & black\ pixel \end{cases}, \quad 0 < x < i,\ 0 < y < j \tag{1}$$

where i is the width of the area and j is the height of the area. If T is the set of black and white values in an area, then:

$$T = \{t(x, y)\} \tag{2}$$

and the area eigenvalue of η is

$$\eta = \frac{\eta_1}{\eta_2} \tag{3}$$

η_1 is the '1' value in T and η_2 is the total number of elements in T.

3 Experiments

The 'General Standard Chinese Character List' is a number, level and structure specification of Chinese characters. It contains 8105 characters and is divided into three levels. To meet the actual needs of using Chinese characters, the first and second level Chinese Character Set of the 'General Standard Chinese Character List' were used in experiment. The first and second levels total 6500 characters, which mainly meets the needs of publishing, printing, dictionary compilation, and information processing. Seven most commonly used fonts were selected: songti, fangsong, heiti, kaiti, yahei, lishu and youyuan. As eleven characters cannot be displayed properly in the above fonts, so 6489 characters were used in the experiment.

Font files were used to generate training data for thirteen commonly used font sizes: 42 pt, 36 pt, 26 pt, 24 pt, 22 pt, 18 pt, 16 pt, 15 pt, 14 pt, 12 pt, 10.5 pt, 9 pt and 7.5 pt. According to formula 1, the η values of the feature areas of each Chinese character in seven fonts and thirteen sizes were calculated and stored in the template libraries corresponding to the font size.

According to formula 3, the η value of 6489 Chinese characters in seven font styles and thirteen font sizes were calculated as eigenvalues and stored them in files as template libraries.

3.1 Recognition Accuracy of a Single Area Brightness Feature

To verify the recognition accuracy of a single area brightness feature, the eigenvalues of Chinese characters to be recognized were calculated, and then compared the eigenvalues with those in the template libraries to calculate the recognition accuracy of each area.

The concrete calculating methods were as follows:

$$\Delta\eta = \left(\eta'_n - \eta_n\right)^2, \; 0 < n \leq N \tag{4}$$

$\Delta\eta$ is the difference between the eigenvalue η'_n of area n and the corresponding eigenvalue η_n of a candidate character in the template libraries. Then, $\Delta\eta$ were sorted and the Chinese characters corresponding to the feature values with the smallest difference in the template libraries were outputted. If the output Chinese character is the same as the Chinese character to be recognized, the recognition will be considered successful; if it is different, the recognition will be considered failed. If there are two or more minimum values of $\Delta\eta$, recognition will be also considered failed. Taking the songti font as an example, P_i is the accuracy of feature area i, and the recognition accuracy of each feature area is shown in Table 1.

Table 1. The accuracy of each feature area in songti (%)

Size	P_1	P_2	P_3	P_4	P_5	P_6	P_7	P_8	P_9	P_{10}
7.5 pt	1.83	0.69	0.89	1.14	0.96	1.28	0.35	0.52	0.51	0.62
9 pt	2.76	1.34	1.34	1.54	1.39	1.68	0.54	0.65	0.54	0.69
10.5 pt	3.44	1.09	1.60	2.08	1.88	2.20	0.60	0.91	0.82	1.00
12 pt	4.89	1.97	2.48	2.76	2.39	3.11	1.05	1.29	1.03	1.19
14 pt	7.03	1.71	3.21	3.68	3.50	4.22	1.51	1.82	1.53	1.65
15 pt	7.03	2.80	3.61	5.09	3.99	4.62	1.45	2.07	2.02	2.17
16 pt	7.60	3.45	4.59	4.45	4.67	4.50	1.96	2.10	1.97	1.82
18 pt	12.53	4.78	6.32	7.87	7.44	7.58	2.94	3.16	3.34	3.34
22 pt	18.65	7.43	11.02	10.86	10.68	11.16	4.61	5.10	5.21	4.65
24 pt	18.23	6.41	11.77	11.85	11.22	11.99	4.99	5.75	5.66	5.09
26 pt	22.38	9.40	13.58	14.95	14.67	14.53	5.86	6.86	7.41	6.36
36 pt	41.87	15.29	26.24	30.24	28.40	29.79	12.48	14.55	15.52	14.15
42 pt	53.38	23.30	35.12	40.58	37.99	37.22	16.40	18.72	22.02	19.08

We can see from the table above: (1) the recognition accuracy of each feature area increases with an increase in font size. Taking feature 1 as an example, the recognition accuracy of feature 1 in 7.5 pt font is 1.83%, in 42 pt is 53.38%. This shows that the higher the image quality is, the more accurately the features can be determined. Figure 3 shows the recognition accuracy of feature 1, feature 3 and feature 7 in different font sizes.

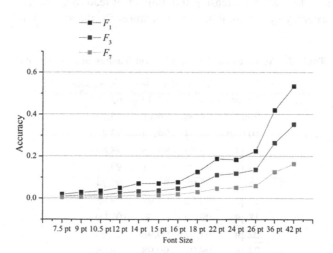

Fig. 3. The accuracy of F1, F3, F7 in different font sizes

(2) The average recognition accuracy of a single feature in the songti font from high to low is as follows: F_1, F_4, F_6, F_5, F_3, F_2, F_9, F_8, F_{10} and F_7. If classified according to the size of the feature area, the features can be divided into three groups: F_1 (100% of the whole character recognition area) is group one, F_3, F_4, F_5 and F_6 (approximately 50% of the whole character recognition area) are group two, and F_2, F_7, F_8, F_9 and F_{10} (approximately 25% of the whole character recognition area) are group three. The experimental data show that the recognition accuracy of group one is higher than group two, and group two is higher than that of group three. In different fonts, the recognition accuracy ranking of each feature within a group will be different, but the accuracy ranking between the groups is the same. (3) For common font sizes, the recognition accuracy of a single feature is insufficient to meet the needs of application.

3.2   Recognition Accuracy of Brightness Feature Combination

The accuracy of a single area brightness feature is not sufficiently high. Therefore, features were gradually increased in order to improve the recognition accuracy. When n features are used simultaneously, the formula for calculating $\Delta\eta$ of a candidate Chinese character in the template libraries and Chinese characters to be recognized is as follows:

$$\Delta\eta = \sum_{1}^{n} \left(\eta'_n - \eta_n\right)^2 \tag{5}$$

When the number of features reaches seven, the added improvement of recognition accuracy is very low when increasing the number of feature areas. Table 2 shows the recognition accuracy of the combination of features for the songti font:

Table 2. Accuracy of the combination features in songti font (%)

Size	P_{123456}	$P_{1234569}$	$P_{12345678910}$
7.5 pt	49.78	62.09	62.09
9 pt	74.06	86.11	86.13
10.5 pt	86.24	94.24	94.24
12 pt	96.59	98.83	98.83
14 pt	98.49	99.71	99.71
15 pt	98.98	99.74	99.74
16 pt	99.58	99.98	99.98
18 pt	99.89	100.00	100.00
22 pt	99.97	99.98	99.98
24 pt	99.98	99.98	99.98
26 pt	99.98	100.00	100.00
36 pt	100.00	100.00	100.00
42 pt	100.00	100.00	100.00

Most of the unrecognized Chinese characters are caused by the consistency of brightness eigenvalues calculated by different Chinese characters. For example, 旁 and 攻 brightness eigenvalues in songti and 7.5 pt are shown in Table 3.

Table 3. 旁 and 攻 Brightness Eigenvalues in songti and 7.5 pt

Characters	Brightness Eigenvalues				
	η_1	η_2	η_3	η_4	η_5
旁	0.7818	0.7143	0.7800	0.7833	0.7955
攻	0.7818	0.7143	0.7800	0.7833	0.7955
	η_6	η_7	η_8	η_9	η_{10}
旁	0.7727	0.8000	0.7667	0.7917	0.7778
攻	0.7727	0.8000	0.7667	0.7917	0.7778

When the minimum value of ΔT is greater than one, a new feature area can be randomly generated, calculate the brightness eigenvalues of the new one, and carry out a secondary recognition.

The following table shows the recognition accuracy of seven common fonts using all ten feature areas (Table 4).

Table 4. Recognition accuracy of seven fonts using all ten feature areas (%)

$P_{12345678910}$	7.5 pt	9 pt	10.5 pt	12 pt	14 pt
lishu	54.54	85.93	93.10	97.57	99.31
songti	62.09	86.13	94.24	98.83	99.71
fangsong	69.18	82.23	92.08	97.95	99.28
kaiti	72.92	87.27	96.87	99.06	99.77
heiti	80.58	94.88	96.16	99.51	99.78
youyuan	86.61	92.83	97.57	99.29	99.88
yahei	89.86	95.92	98.47	99.80	99.95

4   Conclusions and Future Work

This paper proposes a method for PCCR based on the area brightness feature, which is simple and has a low computational cost. It can achieve more than 93% accuracy in recognizing printed Chinese characters equal to or greater than 10.5 pt which can meet the needs of printed Chinese character recognition in certain situations (such as screen capture). The disadvantage of this method is its poor anti-distortion handling ability, and the recognition accuracy of smaller images still needs to be improved. Specifically, it can be improved in the following aspects. First, this paper only discussed a single feature type. To improve recognition accuracy, other features can be added for practical application, and a hybrid classifier can be used. Second, classify Chinese characters according to their complexity [4] to improve the retrieval speed.

Acknowledgments. This research is sponsored by the Major Project of National Social Science Foundation of China (15ZDB096) and Major Projects of the Key Research Base of Humanities and Social Sciences of the Ministry of Education (19JJD740003).

References

1. Stallings, W.: Approaches to Chinese character recognition. Pattern Recogn. **8**(2), 87–98 (2004)
2. Mori, S., Yamamoto, K., Yasuda, M.: Research on machine recognition of handprinted characters. IEEE Trans. Pattern Anal. Mach. Intell. **6**(4), 386–405 (1984)
3. Liu, C.-L., Jaeger, S., Nakagawa, M.: Online recognition of Chinese characters: the state-of-the-art. IEEE Trans. Pattern Anal. Mach. Intell. **26**(2), 198–213 (2004)
4. Liu, J.: Study and realization on printed Chinese character recognition system. Dalian University of Technology, Dalian (2011)
5. An, R., Zhang, S., Chen, H.: Method for removing burr to optical character recognition. Comput. Technol. Dev. **17**(9), 136–138 (2007)
6. Liu, B., Xiao, H.: Printed Chinese character recognition based on non-separable wavelet transform and Zernike moments. Comput. Appl. Softw. **35**(4), 229–236 (2018)
7. Zhang, J., Zhu, Y., Du, J., Dai, L.: Radical Analysis Network for Zero-shot Learning in Printed Chinese Character Recognition. arXiv:1711.01889 (2018)
8. State Council of the PRC: General Standard Chinese Character List. http://www.gov.cn/zwgk/2013-08/19/content_2469793.htm

"Love Is as Complex as Math": Metaphor Generation System for Social Chatbot

Danning Zheng[1(✉)], Ruihua Song[2], Tianran Hu[1,2], Hao Fu[1,2], and Jin Zhou[1,2]

[1] Computer Science Department, Columbia University, New York, USA
dzheng2@u.rochester.edu
[2] Microsoft, Beijing, China
song.ruihua@microsoft.com

Abstract. As the wide adoption of intelligent chatbot in human daily life, user demands for such systems evolve from basic task-solving conversations to more casual and friend-like communication. To meet the user needs and build emotional bond with users, it is essential for social chatbots to incorporate more human-like and advanced linguistic features. In this paper, the usage of a commonly used rhetorical device – metaphor – is investigated for social chatbot. Our work first designs a metaphor generation framework, which generates topic-aware and novel figurative sentences. Human annotators validate the novelty and properness of the generated metaphors. More importantly, we evaluate the effects of employing metaphors in human-chatbot conversations. Experiments indicate that our system effectively arouses user interests in communicating with our chatbot, resulting in significantly longer human-chatbot conversations.

Keywords: Metaphor generation · Social chatbot · User experience

1 Introduction

In recent years, we have witnessed the emergence of a new type of automatic conversational system – social chatbots, such as Microsoft XiaoIce[1]. Different from traditional task-oriented bots [1], social chatbots are designed to "communicate" and build "emotional bonds" with users [2]. To enrich the expressions of social chatbots, a natural approach is to introduce more human-like and advanced linguistic features. Previous studies [3, 4] suggest that figurative language such as metaphors and sarcasm are key to interesting and engaging conversations. Therefore, in this work, a framework is proposed to automatically generate novel metaphors and further apply such metaphors to enrich the interaction between a chatbot and users. As shown in Fig. 1, instead of returning relevant answers to a user's query, our method extends topics such as life and love by metaphors. Moreover, users might be impressed by how a chatbot can think, since our method can explain why the target and the source are similar.

---

[1] https://en.wikipedia.org/wiki/Xiaoice

© Springer Nature Switzerland AG 2020
J.-F. Hong et al. (Eds.): CLSW 2019, LNAI 11831, pp. 337–347, 2020.
https://doi.org/10.1007/978-3-030-38189-9_36

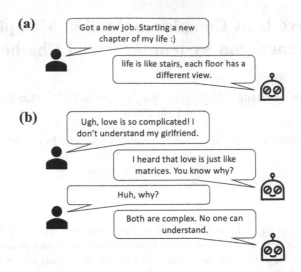

Fig. 1. Two examples of our social chatbot using metaphors in conversation with users. (a) demonstrates a one-round conversation in which the chatbot directly says the whole metaphor sentence. (b) demonstrates a two-round conversation in which the chatbot first says a novel comparison to interact with user, followed by the explanation in the second round.

Our framework starts from a randomly selected target-source pair, such as "love" and "math". The system then quantitatively finds proper connections between source and target. For example, "(being) complex" is considered as a feature shared by both "love" and "math". Based on the target-source pair and the discovered connection, the framework generates metaphorical sentences (e.g. "Love is as complex as math"). We validated our system from two perspectives: (1) the quality of generated metaphors in terms of properness and creativity; and (2) how do users react to these metaphors in real human-computer conversations. Evaluation results show that our framework is capable of generating novel and proper metaphors, and the users are significantly more willing to respond in human-computer conversations when a chatbot uses metaphors.

The contributions of this study are threefold:

1. We propose an automatic metaphor generation system for social chatbots. To the best of our knowledge, this is the first work that considers generating metaphors for conversational systems.
2. We evaluate the quality of generated metaphors. Results show that the system is able to generate novel and proper metaphors.
3. We systematically evaluate the effect of using metaphors in human-computer conversations. The results reveal that metaphors make users more willing to respond.

2 Related Work

Metaphors (e.g., Love is like chocolates, sweet and bitter at the same time) are a figure of speech involving the comparison of one thing with another thing of a different kind and are used to make a description more emphatic or vivid [9].

Previous studies [3, 4] suggest that the use of figurative language such as metaphors and sarcasm are important for creating interesting and engaging conversations. Roberts et al. [5] report that among all major figurative language types, metaphors are most able to make conversations more interesting. Specifically, 71% of participants indicated that they use metaphor to add interest to conversations and 12% used metaphors to get attention from their conversational partner.

Early works [1, 6, 8] on human-computer conversation systems mainly focused on task-completion, such as customer service, making recommendations and answering questions. Researches on task-oriented systems are mainly focusing on addressing users' queries and generating informative answers. In recent years, more and more attention has been paid to non-task-oriented chatbots [2], which aim to hold casual and engaging conversations with users in open domains. A number of studies [10, 11] have been done to meet users' emotional needs and make conversation systems more engaging.

Despite the pervasiveness of figurative language in human conversations, little attention has been paid to integrating figurative language with chatbots. Inspired by Roberts et al.'s study [5], we propose a metaphor generation system that is capable of generating metaphors and enhancing users' engagement with chatbot systems.

3 Metaphor Generation System

3.1 Targets and Sources Selection

Cognitive linguistic studies show that target and source are usually of different types of concepts: target are usually from abstract domains, while sources from concrete domains [12]. In other words, by utilizing metaphors, people manage to explain and express less-understood and abstract concepts (i.e., targets) using well-understood and concrete concepts (i.e., sources) [13]. Therefore, to select suitable targets and sources, we applied two different approaches in our system.

To select targets, we first collected 122 poetic themes and extended the candidate set by adding the closest five concepts of each poetic theme[2] [14, 15]. To ensure that the concepts are actually being used in human-computer conversations, we further filtered out the concepts that are rarely used (frequency lower than 0.001%) and obtained 96 concepts. These concepts spanned many diverse topics, such as romance (e.g., "love", "heart"), history (e.g., "war", "peace"), and nature (e.g., "earth", "spring"). Table 1 shows the top ten most popular concepts.

---

[2] The closest concepts are extracted from a pre-trained word embedding space learned from millions of social media (Weibo) posts.

To select sources, we first extracted the top 10,000 frequently used nouns from our chatbot conversation logs. We then learned the concreteness scores for these words from a concreteness database introduced by Brysbaert et al. [16]. We took the most concrete 3,000 nouns as source candidates for our system. Table 1 also shows the top ten most concrete concepts and their scores.

Table 1. Top 10 most frequent abstract concepts and concrete concepts in our chatbot conversation log. Conc. R. is the abbreviation for concreteness rating.

Target	Freq.	Source	Freq.	Conc. R.
Parting	0.68%	Food	0.92%	4.80
Love	0.38%	Signal	0.28%	3.86
Heart	0.21%	Game	0.27%	4.50
World	0.20%	Father	0.22%	4.52
Mother	0.16%	Robot	0.21%	4.65
Beauty	0.12%	Wife	0.20%	4.13
Man	0.11%	Picture	0.17%	4.52
Dream	0.10%	Brother	0.16%	4.43
Life	0.10%	Photos	0.16%	4.93
Happiness	0.09%	Phone	0.15%	4.86

3.2 Discovering Connections Between Targets and Sources

Besides a target and a source, a metaphor also requires a connection between these two concepts. In our framework, we quantitatively discovered words linking targets and sources semantically, and refer to these words as *connecting words*.

We first located targets and sources in a word embedding space. Since the distance in the space represents the semantic similarities of words [17], we can quantify how good a word is in terms of connecting target and source from two perspectives: (1) connectivity: the semantic distances from a connecting word to target and source should be smaller than the semantic distance between target and source; and (2) balance: a connecting word should maintain a balanced distance to target and source, thus drawing the target and source together. These two aspects can be clearly visualized in Fig. 2. Combining these two aspects, we designed a *connecting score*. Formally, given a target T and a source V, the connecting score of a word X for T and V is defined as:

$$\text{connecting}(X|T, V) = \text{dist}(T, X) + \text{dist}(V, X) + \log(|\text{dist}(T, X) - \text{dist}(V, X)| + \beta)$$

$$(1)$$

The lower the connecting score, the better a word could link the two concepts.

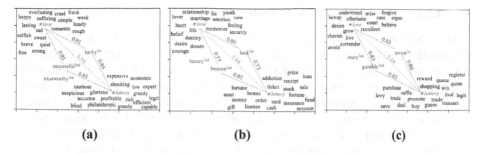

Fig. 2. An illustration of the connecting words (in blue) for target *love* and source *lottery* (in red) by different part of speech (POS) tags. Plots (a), (b), and (c) respectively show adjectives, verbs, and nouns in the underlying word vector space. Numbers on the dotted lines represent the semantic distance (defined as $1 - cosine(.,.)$) between a pair of words. (Color figure online)

3.3   Identifying Similarities by Different POS

As connecting words should convey enough information, we consider content words (i.e., adjectives, verbs, and nouns) as candidates. Connecting words semantically links a target and a source, but connecting words of different part of speech (POS) link the two concepts in different ways. We summarize the most representative case in each category: (1) The connecting word is an adjective and it is a common attribute or property shared by the target and the source. For example, "complex" is a proper connecting word for target "love" and source "math". (2) The connecting word is a verb and it can modify both the target and the source. For example, "scream" is a proper connecting word for target "soul" and source "football fans". (3) The connecting word is a noun and its relationship with the target is the same as its relationship with the source. For example, "gamble" is a proper connecting word for target "love" and source "lottery".

In the next three sections, we report the methods for generating metaphors from connecting words of different POS.

3.4   Generating Metaphors with ADJ Connecting Words

Ortony et al. [18] argue that metaphors project high-salience properties of a descriptive source term (the source) onto a target term (the target) for which those properties are not already salient. In other words, a proper connecting word for a target-source pair can be (1) used to describe the target, and (2) a salient attribute of the source.

To validate condition 1, we send two queries *adjective T* (e.g. "sweet love") and *T is adjective* (e.g. "love is sweet") to a web search engine and recorded the total number of returned web pages. Similarly, to validate condition 2, we queried *as adjective as (a| an) V* (e.g. "as sweet as apples"), and recorded the number of returned web pages. We considered an adjective as proper if both numbers are larger than certain thresholds. To generate complete metaphor sentences, we then manually constructed a few templates: *T is adjective, just like V.*, *T is as adjective as (a|an) V.*, and *T is like (a|an) adjective V.* For example, *Time is sweet, just like a tangerine.*

3.5   Generating Metaphors with Verb Connecting Words

A key observation is that verb and noun *connecting words* tend to exhibit diverse relationships with targets and sources. Therefore, we do not identify and handle all possible relationships, but rather handle the most representative case: subject-verb associations. Subject-verb is the most fundamental sentence structure. It is also an effective feature in metaphor detection [19] and metaphor generation [7]. Thus, to validate whether a verb exhibits the same relation with a target and source pair, we verified if target-verb and source-verb each demonstrate subject-verb relations.

Starting from a verb connecting word and a pair of target T and source V, we sent two queries *verb + T* and *verb + V* to a search engine and retrieved the top 10,000 web snippets for each query. We analyzed the syntactic dependency structure of each sentence and filtered out those sentences in which the target-verb or source-verb relation was not subject-verb. We ranked all sentences of targets by their semantic distance to the source word, in which the distance was calculated as the average distance of every word (excluding stopwords) in the sentence to the source word. We used the sentence with smallest distance as the explanation and generated *T is like V, [explanation].* metaphors. For example, *Soul is like a football fan, silently screaming.*

3.6   Generating Metaphors with Noun Connecting Words

Similar to the verb case, noun connecting words exhibit diverse relations with targets and sources. Therefore, we only handled the most representative relation: subject-predicate-object patterns. The idea is that if there exists a certain predicate such that both target-predicate-noun and source-predicate-noun frequently occur as subject-predicate-object in a large text corpus, then we know *predicate + noun* is a phrase that can modify both target and source.

We followed the same procedure to collect sentences for targets and sources from the search engine. We identified subject-predicate-object structure in each sentence via dependency parsing. We then followed the same approach to generate *T is like V, [explanation].* metaphors. For example, *Life is like a stairway, has a direction, not confused.*

4   Metaphor Generation System Evaluation

4.1   Connecting Words Evaluation

From all 96×3000 pairs of target and source, we randomly sampled 500 pairs and used Eq. (1) to retrieve the top 5 adjective connecting words, verb connecting words, and noun connecting words, respectively. Each of the 500×15 samples were annotated on a 3-point scale: 0 (not proper), 1 (proper), and 2 (very proper). Each sample was labeled by 3 human judges, and its average was used as the final rating.

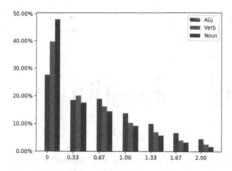

Fig. 3. The properness score distribution of connecting words by different POS categories. X-axis shows the properness scores, ranging from 0 (not proper) to 2.0 (very proper). Y-axis shows the percentage.

Figure 3 shows the properness score distribution of connecting words by different POS categories. If we consider a connecting word with score ≥ 1 as proper, then overall 26.2% of the connecting words are proper. An important observation is that adjective connecting words achieve higher scores than verb and noun connecting words. This result aligns with our previous analysis that verb and noun connecting words exhibit more diverse relations with targets and sources.

4.2   Generated Metaphors Evaluation

Metaphor generation was evaluated with 1965 proper (target, source, connecting word) triplets. Each generated sentence was evaluated by three human annotators from the following perspectives: (1) smoothness: if a sentence is clear and grammatically correct; (2) properness: if the comparison and explanation are understandable and make sense; and (3) novelty: if the comparison or explanation is novel or surprising. Smoothness is a binary metric: 0/1 (smooth/not smooth); properness and novelty are annotated on a 3-point scale: 0 (not at all), 1 (moderately), and 2 (very strongly). The average of three annotators is used as the final score.

Figure 4 shows the score distribution of each metric by different categories of connecting words. A sentence with a smoothness score ≥ 0.7 is considered clear, and a metaphor with a properness score ≥ 1 is considered proper. From plot (a) we can see that 92.02% of metaphors generated with adjectives are smooth while only 71.43% and 76.6% of the metaphors generated with verbs and nouns, respectively, are smooth. This result is not surprising since the former approach applies a far more restrictive template. It is also worth noting in plot (b) that 20% more metaphors generated with verbs are inappropriate as compared to the metaphors generated with adjectives and nouns. This is probably because subject + verb is a relatively loose constraint. Finally, plot (c) reveals that the novelty distribution of adjective metaphors is evenly distributed between 1.00 and 2.00, while the novelty distribution of noun metaphors is continuously increasing from 1.00 to 2.00. This is because noun and verb sentences are longer and tend to provide richer explanations.

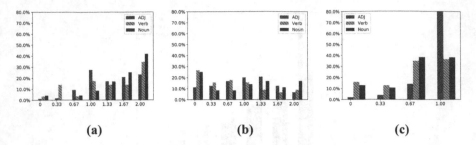

Fig. 4. Plots (a), (b), and (c) respectively show the smoothness score, properness score, and novelty score distribution of metaphor sentences by different categories of connecting words: adjective, verb, and noun.

5   Testing on Real Chatbot Users

To robustly evaluate the effect of our system, we integrated metaphors with an existing social chatbot and analyzed how different expressions affect real chatbot users' *follow-up* rate. *Follow-up* is defined as: the user responds to the chatbot and the user's response is related to the content generated by the chatbot. Therefore, if a user doesn't respond or a user's response is unrelated to the content generated by the chatbot, then it is not considered as a follow-up.

5.1   Experiment Design

Our data consists of 278 randomly sampled metaphor sentences with properness score ≥ 1. We implemented two different approaches to integrate our metaphor generation system into a chatbot. In the first approach, a chatbot directly says the complete metaphor sentence, e.g., "Heart is shining like a diamond". In the second approach, a chatbot first says a comparison, e.g. "I heard that heart is like a diamond. Do you know why?", and then follows with the explanation in the second round, e.g. "Because both are shining." These two approaches are demonstrated in Fig. 1(a) and (b), respectively. Both approaches are compared to the baseline where a chatbot simply says the literal sentence, e.g., "Heart is shining." Therefore, for each metaphor sentence, we generated three different types of expressions: one-round metaphor, two-round metaphor, and literal sentence (Table 2).

When integrating metaphors with social chatbot systems, we sought to make the integration context-aware and fit in the conversation flow. For example, if a user is talking about their boyfriend or girlfriend, a metaphor for love or marriage could be a good fit in the conversation. We used question-answer relevance, keyword matching, and topic similarity as input features and trained a classifier to predict whether a metaphor should be triggered.[3] When a metaphor was a good fit in the conversation flow, we randomly triggered one of the three expressions.

---

[3] Details of this classifier are beyond the scope of this paper and were omitted.

Table 2. Examples of generated metaphors in decreasing order of the smoothness, properness, and novelty scores. Targets (in red), sources (in orange), and connecting words (underlined) are highlighted.

Generated Metaphor	Smooth	Proper	Novel	POS
Time is sweet, just like a tangerine. 时光很甜，就像柑橘一样。	1.0	2.0	1.7	Adj.
Love is like salary, has a goal and is not blind. 爱情就像工资，都是有点目标的，不能盲目。	1.0	2.0	1.7	noun
Relationship is like a park, need to be operated and maintained. 感情就像园区，需要经营和维护。	1.0	2.0	1.0	verb
Loneliness is like an empty station. 孤独就像空无一人的车站。	1.0	1.3	0.7	Adj.
Soul is like a football fan, silently screaming. 灵魂就像球迷一样，在无声地呐喊。	0.7	1.0	1.0	verb
Marriage is like a guide, has its own set of rules. 婚姻就像指南，有自己的一套法则。	0.7	1.0	1.0	noun
Childhood is very cute, just like a dolphin. 童年很可爱，就像海豚一样。	0.7	0.7	1.0	Adj.
Work is as outstanding as ballet. 工作像芭蕾一样出色。	0.7	0.0	0.0	Adj.
Life is like a stairway, has a direction, not confused. 人生就像楼梯，有方向，再难不迷茫。	0.3	0.0	0.0	noun
Time is like sportswear, will not fade, memory will shine. 时光就像运动服，不会褪色，记忆也会发光。	0.0	0.0	0.0	verb

5.2   Evaluation Result

We tested our system on 924 users within a 3-week period. Users' follow-up rates are 22%, 27%, and 41% for literal sentences, one-round metaphors, and two-round metaphors, respectively. Overall, the results show that both metaphor expressions achieve more follow-ups than literal expressions. Importantly, we found that the follow-up for two-round metaphors was the highest among all three expressions, which indicates that users prefer such an interactive way with chatbots in human-computer conversations.

6   Discussion and Future Work

In this paper, we propose computational approaches to generate metaphors and report the effect of our system in the context of human-computer conversations. According to human evaluation results, our system is able to generate proper and novel metaphors. More importantly, integrating metaphors with an existing social chatbot increased users' follow-up rates.

There are many interesting and valuable directions for future work, including studying how the properness and novelty of metaphors affect users' experiences and engagement in human-computer interactions. In the meantime, it will also be important to study possible improvements to our proposed metaphor generation model, such as enhancing the percentage of proper metaphors.

References

1. López, G., Quesada, L., Guerrero, L.A.: Alexa vs. Siri vs. Cortana vs. Google assistant: a comparison of speech-based natural user interfaces. In: Nunes, I. (ed.) AHFE 2017. AISC, vol. 592, pp. 241–250. Springer, Cham (2018). https://doi.org/10.1007/978-3-319-60366-7_23
2. Shum, Y., He, D., Li, D.: From Eliza to XiaoIce: challenges and opportunities with social chatbots. Front. Inf. Technol. Electron. Eng. 19(1), 10–26 (2018)
3. Glucksberg, S.: Metaphors in conversation: how are they understood? Why are they used? Metaphor Symb. 4(3), 125–143 (1989)
4. Kaal, A.: Metaphor in Conversation, 1st edn. BOXPress, Conneautville (2012)
5. Roberts, M., Kreuz, J.: Why do people use figurative language? Psychol. Sci. 5(3), 159–163 (1994)
6. Xu, A., Liu, Z., Guo, Y., Sinha, V., Akkiraju, R.: A new chatbot for customer service on social media. In: Proceedings of the 2017 CHI Conference on Human Factors in Computing Systems, Denver, Colorado, USA, pp. 3506–3510. ACM (2017)
7. Harmon, S.: FIGURE8: a novel system for generating and evaluating figurative language. In: International Conference on Computational Creativity, Park City, Utah, USA, pp. 71–77. ACC (2015)
8. Li, J., Monroe, W., Ritter, A., Galley, M., Gao, J., Jurafsky, D.: Deep reinforcement learning for dialogue generation. In: Proceedings of the 2016 Conference on Empirical Methods in Natural Language Processing, Austin, Texas, USA, pp. 1192–1202. ACL (2016)

9. Richards, A.: The Philosophy of Rhetoric, 1st edn. Oxford University Press, New York (1965)
10. Hu, T., et al.: Touch your heart: a tone-aware chatbot for customer care on social media. In: Proceedings of the 2018 CHI Conference on Human Factors in Computing Systems, Montréal, Canada, pp. 415. ACM (2018)
11. Huber, B., McDuff, D., Brockett, C., Galley, M., Dolan, B.: Emotional dialogue generation using image-grounded language models. In: Proceedings of the 2018 CHI Conference on Human Factors in Computing Systems, Montréal, Canada, p. 277. ACM (2018)
12. Lakoff, G., Johnson, M.: Metaphors We Live By, 2nd edn. University of Chicago Press, Chicago (2008)
13. Gentner, D., Bowdle, B., Wolff, P., Boronat, C.: The Analogical Mind: Perspectives from Cognitive Science. MIT Press, Cambridge (2001)
14. Gagliano, A., Paul, E., Booten, K., Hearst, A.: Intersecting word vectors to take figurative language to new heights. In: Proceedings of the Fifth Workshop on Computational Linguistics for Literature, San Diego, California, USA, pp. 20–31. ACL (2016)
15. Gero, K., Chilton, L.: Challenges in finding metaphorical connections. In: Proceedings of the Workshop on Figurative Language Processing, New Orleans, Louisiana, USA, pp. 1–6. ACL (2018)
16. Brysbaert, M., Warriner, B., Kuperman, V.: Concreteness ratings for 40 thousand generally known English word lemmas. Behav. Res. Methods **46**(3), 904–911 (2014)
17. Mikolov, T., Yih, T., Zweig, G.: Linguistic regularities in continuous space word representations. In: Proceedings of the 2013 Conference of the North American Chapter of the Association for Computational Linguistics: Human Language Technologies, Atlanta, Georgia, USA, pp. 746–751. ACL (2013)
18. Ortony, A.: Metaphor and Thought, 2nd edn. Cambridge University Press, Cambridge (1979)
19. Shutova, E., Sun, L., Korhonen, A.: Metaphor identification using verb and noun clustering. In: Proceedings of the 23rd International Conference on Computational Linguistics, Beijing, China, pp. 1002–1010. ACL (2010)

Research on Extraction of Simple Modifier-Head Chunks Based on Corpus

Wang Chengwen, Zhang Zheng, Rao Gaoqi, Xun Endong[⊠],
and Miao Jingjing

Institute of Big Data and Language Education, Beijing Language
and Culture University, Beijing 100083, China
edxun@126.com

Abstract. The purpose of this study is to automatically extract a set of simple modifier-head chunks from a large-scale corpus. By analyzing the distribution of simple modifier-head chunks in usage, a set of formal rules of chunks extraction are formulated and a rule-based automatic extraction algorithm is designed. In the experiment of random sampling, the precision of extraction result with this method reaches 82.63%, which casts light on knowledge extraction based on large-scale corpus.

Keywords: Automatic extraction · Simple modifier-head chunks · Collocations · Corpus

1 Introduction

Based on a cognition from Miller that the experimental subjects had only a maximum memory of 7 ± 2 units in a short period of time [1], Becker proposed the concept of "chunk" for the first time in 1975 [2].

After the concept of "chunk" had been introduced to the field of Chinese linguistics, researchers did a lot of research on the identification and construction of Chinese chunks system [3–6].

In tasks of language information processing, there are many practices of automatic chunk extraction based on large-scale corpus [7, 8]. On the basis of high frequency Ngram strings, Jiang [9] uses statistical methods and linguistic rules to evaluate the internal tightness and external independence of candidate chunks, so as to obtain possible chunks and achieve ideal results.

The statistical method for the recognition of linguistic knowledge is based on the probability calculation of language facts, which obtains examples of multi-word co-occurrence rather than syntactically and semantically related chunks. Further development of natural language processing requires the guidance of deep linguistic knowledge, which must be given functional and semantic recognition in form. Thus this paper intends to explore a method of automatically extracting simple modifier-head chunks from large-scale corpus on a basis of language structure rules.

© Springer Nature Switzerland AG 2020
J.-F. Hong et al. (Eds.): CLSW 2019, LNAI 11831, pp. 348–358, 2020.
https://doi.org/10.1007/978-3-030-38189-9_37

2 Distribution and Composition of Simple Modifier-Head Chunk

2.1 Definition of Modifier-Head Chunk and Simple Modifier-Head Chunk

Whether units such as "白马(white + horse, a white horse)" and "大树(big + tree, a big tree)" are words or phrases does not affect the process of communication. In communication, many multi-word expressions are used as a whole, without their internal structure being analyzed or emphasized.

Based on this, this paper proposes a general concept of chunk. Any multi-word units of expression that tend to be used as a whole, when its internal constituents are paired, and the unit is used in a steady state of high frequency, it is considered to be a chunk.

We also define a simple modifier-head chunk as a modifier-head structure with the characteristics of binary collocation appearing in a steady state of high frequency without recursion in its syntactic structure. "Simple" here mainly means that the syntactic structure is not to recurse, and the most basic constituents of a simple chunk are words rather than phrasal components.

The examples from Group A in Table 1 are of the types of subjects in our current study. The examples in Group B are structurally modifier-head structures, whereas their internal structures are more complicated than those in Group A. Hence the modifier-head expressions in Group B are not covered within this paper. For each of them is a result of nesting two pairs of modifier-head structures. The purpose of this study is to extract collocations of simple modifier-head chunks, thus emphasizing the binary collocation of a simple modifier-head chunk becomes necessary. Given this, the research object is determined as the simple modifier-head chunks.

2.2 Distribution Analysis of Simple Modifier-Head Chunk Based on Pennsylvania Tree Bank

With the object of this study being the simple modifier-head chunks (binary modifier-head collocations), we need to first figure out the specific distribution and the internal structures of the simple modifier-head chunks in Chinese language facts. The simple modifier-head chunk is a kind of nominal phrase. We choose the Pennsylvania Tree Bank as the basis of statistics to figure out the internal part-of-speech sequences of NP-OBJ marked language units with the number of words being 2, in order to provide statistical support for the formulation of knowledge extraction rules for simple modifier-head chunks in the later stage. The specific statistical results are shown in the table below.

As shown in Table 2, in simple binary NP-OBJ marked language units, the frequency of occurrence with the part-of-speech sequence of "n + n" is the most in quantity with the frequency of occurrence being 2101, accounting for 34.7% of the total simple NP-OBJ marked language units. The "a + n" sequence ranks secondly with 999 occurrence, accounting for 16.5% of all. In the third place, "DEC" in Pennsylvania Tree Bank is a part-of-speech tag for the attributive marker "的", so the

Table 1. The compared examples

Group A	Group B
宏伟建筑, *magnificent + building, a magnificent building*	一件美丽的衣服, *one piece + beautiful + de + dress, a beautiful dress*
远大前程, *great + prospect, a great prospect*	美丽的鞋子和帽子, *beautiful + de + shoe + and + hat, beautiful shoes and hats*
漂亮[的]花朵, *beautiful + de + flower, a beautiful flower*	

third sequence mainly refers to the De-phrase in modern Chinese. There are also cases where quantifiers and pronouns are used as modifiers, such as those of "q + n", "r + n" and "r + v".

In order to confirm the internal structure of the NP-OBJ whose part-of-speech sequence is "n + n", 100 matching pairs were selected by random sampling survey to be manually labeled. A given language unit whose part-of-speech sequence is "n + n" will be marked as 1, and 0 if it is not. For instance, "科研单位(*research + unit, a scientific research institution*)" is marked as 1 and "爸爸妈妈(*father + mother, father and mother*)" is marked as 0. The following table lists some examples (Table 3).

According to the random sample survey, only 10 non-modifier-head structures appear in 100 binary "n + n" pairs. Such findings are much consistent with general linguistic knowledge. When two nominal constituents appear closely, without conjunctions in the middle, being a modifier-head structure is the main case.

2.3 Analysis of the Internal Composition of the Simple Modifier-Head Chunks

On the basis of the above statistics on the part-of-speech sequences of simple modifier-head chunks, some helpful information for the automatic extraction of binary modifier-head chunks knowledge could be summarized.

(1) Simple Binary Modifier-Head Chunks of "n + n(*noun + noun*)"

It is the most common case that noun constituents modify their following noun constituents to form modifier-head structures, such as "人民福祉(*people + well-being, people's well-being*)", "社会利益(*society + interest, social interests*)", etc.

(2) Simple Binary Modifier-Head Chunks of "a + n(*adjective + noun*)"

A preceding adjective can modify its following nominal component, implying its properties or states, such as "锦绣前程(*bright + prospect, bright prospect*)", "伟大前程(*great + prospect, a great prospect*)", etc.

(3) Simple Binary Modifier-Head Chunks of "b + n(*distinguishing words + noun*)"

Table 2. Statistics on sequences of part of speech in simple NP-OBJ units

Part of speech sequences of NP-OBJ with the number of words being 2	Frequencies of occurrence
n + n	2101
a + n	999
n + DEC	593
n + v	565
q + n	209
n + LC	152
r + n	50
r + v	47

Table 3. The annotated data (partial)

Language Examples with a Part-of-Speech Sequence of "n+n"	Modifier-Head/Non-modifier-head
司法单位, *judicial + unit, a judicial authority*	1
外交情势, *diplomacy+ situation, the diplomatic situation*	1
土地价格, *land + price, the land price*	1
省市, *province + city, provinces and cities*	0
飞机大炮, *plane + artillery, planes and artillery*	0

In combination with the research of previous scholars, it is found that modifier-head chunks forming from the combination of distinguishing words and nouns also make considerable proportions, such as "金链子(*gold + chain, a gold chain*)", "大型比赛 (*big + competition, a big competition*)", etc.

(4) Simple Binary Modifier-Head Chunks of "v + n(*verb + noun*)"

When a verb is used as an attributive, it is usually necessary in Chinese to add a functional marker "de(的)" to modify the head constituent. In this study, the binary collocation between the modifying component and the modified component is emphasized, so this kind of situation is also considered. Some verbs can also directly modify nouns. In these cases, the nominalization tendency of the verb components is relatively strong. For example: "爱怜的眼神(*affect + look, a look of affection*)", "合作情况(*collaborate + condition, the collaborating condition*)", etc.

(5) Simple Binary Modifier-Head Chunks of "q + n(*quantifier + noun*)"

In general, a numeral and a quantifier constitute a quantitative phrase to modify a nominal constituent. This study emphasizes the binary collocation characteristics of chunks so as to focus on "q + n" chunks.

(6) Simple Binary Modifier-Head Chunks of "r + n(*pronoun + noun*)"

Usually, the part-of-speech sequence is "r + 的 +n". And there are pronouns that directly modify nouns to form modifier-head structures.

(7) Simple Binary Modifier-Head Chunks of "s/t + n(*locative words/time words + noun*)"

It indeed exists in Chinese language facts that locative words or time words modify nouns to form simple modifier-head chunks.

(8) Simple Binary Modifier-Head Chunks of "n + v/a(*noun + verb/adjective*)"

The head constituents of this kind of chunks are mainly predicative constituents and the nominalization of the head constituents is relatively strong. For example, "语法研究(*grammar + research, grammar research*)", etc.

(9) Simple Binary Modifier-Head Chunks of "n/s/t + f(*noun/locative words/time words + nouns of locality*)"

Nouns, locative words or time words modify nouns of locality to form modifier-head chunks. For example, "道路两旁(*road + both sides, both sides of the road*)", etc.

(10) Simple Binary Modifier-Head Chunks of "n/s + s (*noun/locative words + locative words*)"

A noun or a locative word modifies another locative word to form a simple modifier-head chunk, indicating a location. For example, "活动现场(*activity + site, the event site*)".

(11) Simple Binary Modifier-Head Chunks of "n/t + t (*noun/time words + time words*)"

Nouns or time words modify other time words to form binary collocations, indicating some point or some period of time. For example, "昨天上午(*yesterday + morning, yesterday morning*)", etc.

3 Extractions of Simple Modifier-Head Chunks

3.1 Chunk-Extraction System

Based on the BCC system and serving the needs of knowledge extraction of chunk collocation [10], we developed a retrieval and extraction system with stronger retrieval capabilities.

The retrieval queries in BCC consist of basic queries and advanced queries. A basic query consists of strings, attribute symbols and wildcards. An advanced query adds conditional statements or output statements on the basis of a basic query. Statements are separated by ";" and written in "{ }" after the basic search query, in such a form as:

$$Query\{cond-1; cond-2; ...; cond-i; print(\$i)\}. \tag{1}$$

"Query" denotes the basic search query; the conditional statements in "{ }" restricts the content of a basic query; the output statement restricts the output content, and only one output statement can be used in an advanced search query. Restricted parts of a query will have to be separately enclosed in "()", and according to the order in which "()" appears, they can be obtained by using a "$" symbol followed by a serial number for conditional or output restriction. The component(s) in the first "()" is denoted by "$1" and are analogous in turn.

One example:

$$(a)的(n)\{len(\$1)=2; len(\$2)=2; print(\$1\ \$2)\}. \tag{2}$$

The "(a)的(n)" before "{len($1) = 2; len($2) = 2; print($1 $2)}" is the basic search query, which means that a string in the order of "a+的+n" (*adjective + de + noun*) is to be searched. "len($1) = 2" in the "{ }" means that the restricted adjective is disyllabic, while "len($2) = 2" means that the restricted noun is disyllabic, and "print ($1 $2)" means that the output will simply be "adjective + noun" without the "的(*de*)" in between.

3.2 Formulation of Extraction Rules

(1) Make Full Use of Natural Annotation Information

As a complete language unit, a simple modifier-head chunk is complete and self-sufficient both syntactically and semantically. Given this, natural annotation information is used in the process of extraction rules making.

The query of "W a 的 n W" means that punctuations occur before and after the language unit "a 的 n". This ensures the independency and self-sufficiency of the unit to a relatively considerable extend. It can greatly reduce results of pseudo-modifier-head collocations.

(2) Make Full Use of Prosodic Structure for Restriction

When it comes to the processing of written text, the pitch, intensity and length in prosody may have little to do with natural language text processing, while the number of syllables plays an important role. Therefore, we try to make full use of the syllable rules in collocations to formulate search queries for the extraction.

From the perspective of prosody, it is found that for binary "n + n" or "a + n", the occurrence of double syllables is the most frequent. For example, when initially conduct a search on BCC for strings of "monosyllabic noun + 的 + disyllabic noun", a large number of collocations will be found to be against language sense. And that is why conditional restricting expressions of BCC are needed to restrict syllable length of strings to be extracted. For example, the query of "的(n) (n) W{len($1) = 2;

len($2) = 2; print($1$2)}" indicates that two nouns appear immediately after the attributive form mark "的", and are located at the end of a punctuation sentence.

3) Make Use of Grammatical or Semantic Subcategory Information

In Chinese, the same part-of-speech combination sequence may correspond to different types of syntactic structures. This makes it difficult to extract ideal modifier-head chunks by relying solely on part-of-speech sequences. Therefore, it becomes necessary to introduce syntactic or semantic subcategory information to restrict the combination of part of speech sequences syntactically and semantically, so as to improve the correctness of rule matching.

For modifier-head chunks with a part-of-speech sequence of "n + v", the verb must be able to be directly modified by a noun to form modifier-head structures. Peking University's Modern Chinese Grammar Information Dictionary (hereinafter referred to as the Grammar Information Dictionary) gives a detailed description of the grammatical attributes of 11 notional words, 5 functional words, various set phrases, abbreviations, idioms and punctuation marks in modern Chinese. For example, for the verb "研究(to research)", the Grammar Information Dictionary defines an attribute named "前名(pre-noun)", under which the verb is marked with the value of "可(positive)". The "pre-noun" attribute under the verb means that the noun is able to appear immediately in front of some verbs to form a modifier-head structure. Therefore, in front of the verb "研究(research)", nouns can immediately appear to form a modifier-head collocation structure, such as "语法研究(grammar + research, grammar research)". In order to make full use of the grammatical subcategory information and the resources labeled with the grammatical subcategory information to extract knowledge, we have processed the Grammar Information Dictionary appropriately to construct a set of thesauri that satisfy the attributes of a certain grammatical subcategory, and named them in a form of "S_POS_Key_Value". Combining the conditional restriction statements mentioned above, we can effectively combine the grammatical subcategory vocabulary with the rules of chunk knowledge extraction, so as to improve the accuracy of chunk retrieval.

3.3   The Algorithm for Simple Modifier-Head Chunks Extraction

The extraction algorithm for this research can be introduced as follows:

Step 1: Queries Storage

Large-scale queries for extracting modifier-head chunks of different POS collocation sequences are put into a specified file named Extract_RuleSetto facilitate later processing. Some search rules for modifier-head chunk types are shown in Table 4.

Step 2: Extract in Batch with the Use of Web API Provided by BCC.

By Perl coding, we can read the retrieved form lines that have been stored in the file, and later use the cloud service to extract collocation knowledge from BCC.

Step 3: Sort and Accumulate the Result of Similar Modifier-Head Chunks Extraction

Extraction queries of each type will output the matching results to a file with the name form of "ChunckType_R1" after calling BCC. The result of each extracting instance of ChunckType_Rn is accumulated in frequency. A set of quaternion (W1,

Table 4. Partial modifier-head chunks searching rules

Types	Searching Rules
a n Chunks	W (a) (n){$1=[P_A_重叠_AA];$2=[S_N_名词]；print($1+$2)}
	(a) 的 (n){$1=[P_A_重叠_AABB];$2=[S_N_名词]}
	(a) 的 (n) W{$1=[P_A_重叠_ABB];$2=[S_N_名词]}
	(a) 的 (n){$1=[P_A_重叠_A 里 AB];$2=[S_N_名词]}
n n Chunks	(n) (n) W{len($1)=2;len($2)=2;print($1+$2)}
	W(n) 的 (n) {len($1)=2;len($2)=2}
	W(n) 的 (n) {len($1)>2;len($2)>2}
	的(n) (n) W{len($1)=2;len($2)=2;print($1+$2)}

W2, Sum_Frquency, ChunckType) will then be established, in which "W1" represents the attributive modifier, "W2" represents the head word, "Sum_Frquency" represents the frequency of this instance collocation appearing in corpus, and "ChunckType" represents the type of sequence of modifier-head chunk collocations. For example, there are 15 extraction rules for the "b + n" type of modifier-headchunks, which correspond to a set of {bn_R1, bn_R2, bn_R3… bn_R15} results. Files are extracted and the frequencies of their collocation instances are accumulated. The final set of quaternion (W1, W2, Sum_Frquency, bn) is thus obtained.

Step 4: Rank Instances W2 of the Same Modifier W1 from High to Low

In a certain ChunckType, the modifier W1 is used as a key, and its collocation instances W2 are ranked in the order of frequencies from high to low. Both W1 and W2 are output into chunk collocation files name-formed "ChunckType_W1".

Step 5: To Restrict the Threshold of the Collocation Frequency of Modifier-Head Chunks According to Zipf's Law

According to Zipf's Law, for all the modifier-head collocation chunks (as well as other language units in usage), only 20% of the high frequency parts are retained as the final result of the collocations. Hence the matching Wi instances under modifier W1 are sorted from high to low in order of their frequencies. And with a formula:

$$\text{Freq1} + \text{Freq2} + \ldots + \text{Freqi} = 80\% * \text{Total\_Count}. \tag{3}$$

The threshold is set to the Freqi spot, and the matching result is extracted if the frequency is higher than the threshold.

4   Chunk Extraction Assessment

4.1   Random Sampling Assessment

Considering the workload of evaluation, only the type of "adjective + noun" modifier-head chunk[1] is selected as the object of quality evaluation.

According to the frequencies of adjectives, 100 disyllabic adjectives were selected by stratified sampling to have their "adjective-noun" type of chunks manually verified. Two postgraduate students of linguistics make judgments on them. When the two judgments are consistently being "modifier-head", a collocation is considered as correct to be a modifier-head collocation, and otherwise, wrong.

The stratified sampling method is introduced as below:

Firstly, the frequencies of all disyllabic adjectives are counted in BCC corpus, and the frequency table of disyllabic adjectives is obtained by arranging the frequencies from high to low.

Secondly, on the basis of the frequency distribution, the frequency table is divided into five levels: levels of ten million, one million, one hundred thousand, ten thousand and ten thousand below. Proportions of words in each level are counted.

Finally, according to the proportion of each level, the words (100 * proportion) are randomly selected as part of the sample in this level. The sum of the samples at each level makes the final random detection samples.

The average correctness rate of the 100 modifier-head chunks with sample adjectives is 82.63%. Given the fact that no similar previous work has been found, the results of our present work can provide a benchmark for future work.

4.2   Analysis of Factors Affecting Experimental Results

Following are some factors affecting the experimental results.

(1)   The rigor of partial extraction rules is not district enough.

The validity and rigor among the 50 rules we made is not equal to one another. For locatively constrained queries such as "W a 的 n W", the collocation efficiency of extraction is higher. For rough extraction rules without locative constraints, the possibility of extracting pseudo-collocations will be faced. For example, for the part of speech sequence "a 的 n n" in a sentence, it is possible that the first noun component is actually the modifier of the second noun component, and the adjective is actually the modifier of the second noun component. For such contexts, a general extraction rule of "a 的 n" may extract collocations of pseudo-modifier-head chunks.

(2)   Errors in word segmentation and part of speech tagging in corpus.

The formulation of retrieval rules is mostly based on the restriction of position as well as part-of-speech sequences. Thus the errors of word segmentation and part-of-

---

[1] "Adjective + noun" modifier-head chunks in Chinese mainly consist of collocations of "disyllable adjective + [的] + disyllable noun", such as "宏伟的建筑"(*magnificent + de + architecture–a magnificent building*).

speech tagging in corpus will make a great impact on the results of knowledge extraction. Like the string of "突出重围(*to get through a tight encirclement*)是 (*is*)当务之急(*an urgent priority*)" could be segmented as "突出(*prominent*)/a 重围(*a tight encirclement*)/n 是(*is*)/v 当务之急(*an urgent priority*)/i", which can make a huge negative difference in the extraction results.

(3) The frequency calculation strategy needs to be improved.

Different rules in the modifier-head chunk extraction rule set (retrieval set) have different contributions to the modifier-head chunk extraction. This paper simply adds the frequency as the number of occurrences of a certain collocation. In the later judgment, collocations from different rules should be given different values of weight, and the normalized frequency should be evaluated.

5 Summary

We have, in this paper, summarized the internal types of simple modifier-head chunks from the perspective of linguistic ontology, then formulated formal rules for chunk knowledge retrieval, so as to realize automatic extraction of chunk knowledge.

For further work, besides the method that is currently used, we can also employ the method of mutual information measurement to determine the collocation of lexical chunks, and then choose the collocation of lexical chunks identified by the two methods as the final correct result.

Acknowledgement. This study was supported by National Natural Social Foundation of China (16AYY007), Beijing Advanced Innovation Center for Language Resources (TYR17001), Graduate Innovation Foundation in 2019 (19YCX117), the funds of Major Projects of Key Research Bases of Humanities and Social Sciences of the Ministry of Education (16JJD740004).

References

1. Miller, G.A.: The magical number seven, plus or minus two: some limits on our capacity for processing information. Psychol. Rev. **63**, 81–97 (1956)
2. Becker, J.: The phrasal lexicon. In: Shank, R., Nash-Webber, B.L. (eds.) Theoretical Issues in Natural Language Processing, pp. 60–63. Bolt Beranek & Newman, Cambridge (1975)
3. Zhou, J.: Reinforce the language chunk teaching to foster the intuition of Chinese. Jinan J. (Philos. Soc. Sci. Ed.) **1** (2007). (in Chinese)
4. Qian, X.: A preliminary study on Chinese chunk. J. Peking Univ. (Philos. Soc. Sci. Ed.) **5** (2008). (in Chinese)
5. Lu, B.: Defined, classify and teaching research of Chinese practical chunk. Guangzhou University (2012). (in Chinese)
6. Xue, X., Shi, C.: The nature of lexical chunks and the hierarchical relationship of the Chinese lexical chunk system. Contemp. Rhetor. **3** (2013). (in Chinese)
7. Zhan, H.: Methods and tools of retrieving lexical chunks from Corpora. Foreign Lang. Teach. (2011). (in Chinese)

8. Zhan, H.: Psychological reality of L2 phraseologisms: evidence from phoneme monitoring. Foreign Lang. Foreign Lang. Teach. (2012). (in Chinese)
9. Jiang, B.: Chinese multi-word chunks extraction for computer aided translation. Chin. J. Inf. Technol. **21**(1) (2007). (in Chinese)
10. Xun, E., Rao, G., Xiao, X., Zang, J.: The construction of the BCC corpus in the age of big data. Corpus Linguist. **3**(1), 93–109 (2016). (in Chinese)

Incorporating HowNet-Based Semantic Relatedness Into Chinese Word Sense Disambiguation

Qiaoli Zhou[1(✉)], Gu Yue[2], and Yuguang Meng[1]

[1] School of Computer, Shenyang Aerospace University, Shenyang, China
zhou_qiao_li@hotmail.com
[2] Foreign Language College, Beijing Wuzi University, Beijing, China
513672587@qq.com

Abstract. This paper presents a semi-supervised learning method that incorporates sense knowledge into a Chinese word sense disambiguation (WSD) model. This research also effectively exploits HowNet-based semantic relatedness in order to leverage system performance. The proposed method includes Sense Colony task for improving context expansion and semantic relatedness calculating for sense feature representation. To incorporate sense knowledge into WSD, this paper employs the Semantic relatedness in a semi-supervised label propagation classifier. This research demonstrates state-of-the-art results on word sense disambiguation tasks.

Keywords: Chinese word sense disambiguation · HowNet · Semi-supervised learning · Semantic relatedness

1 Introduction

Word Sense Disambiguation (WSD) has been a hard nut ever since the earliest days of computer-based treatment of language in the 1950s. WSD is the task to identify the intended sense of a word in a computational manner based on the context in which it appears [1].

Many algorithms devote to WSD by exploiting two powerful properties of human language: "one sense per collocation" and "one sense per discourse" [2,3]. In the "one sense per collocation", the adjacent words provide clues to the sense of the target word. "One sense per discourse" represents the sense that a target word is consistent with a given document. In the WSD research literature, currently, these two assumptions are widely accepted by natural language processing community. There are, however, several difficulties to WSD we need to face.

Firstly, one sense per collocation, the strong tendency for words which exhibits only one sense in a given collocation was observed and quantified in the paper [3]. Collocation refers to a group of practical words that habitually go together, whereas the sense of a word is figured out by accompanying words.

Supported by Humanities and Social Sciences of Ministry of Education Planning Fund (18YJA870020).

In this case, words are to be classified in terms of co-occurrence relation as well as sense. The co-occurrence relation means the constraints shown in the sense combination relation, which is called collocation constraint or selection constraint. When we make use of these assumptions, it is easy to identify the sense of common expressions or idioms containing a target word. For example, the word *place* means general location. But, the meaning of the idiom *take place* is quite different from the meaning of *take her place*. The idiom *take place* means that something occurs or happens at a particular time or place. Thus, an idiom is a group of words in a fixed order and has a particular meaning which is different from the meanings of the individual word regardless of the context of the word to be disambiguated. Although there are many researches aiming to solve WSD problem using the phrase in WordNet and idiom dictionary, when we take into consideration the overall occurrence in the target corpus, there still remain some cases where a dictionary may not cover some of the idioms that exist in the target corpus. Therefore, the effect of using collocations to resolve lexical ambiguities depends on the type of collocation. It is strongest for immediately adjacent collocations, and weakens with distance. It is much stronger for words in a predicate-argument relationship than for arbitrary associations at equivalent distance. It is very much stronger for collocations with content words than those with function words.

Secondly, one sense per discourse, the sense of a target word is highly consistent within any given document. The observation that words strongly tend to exhibit only one sense in a given discourse or document was stated and quantified in Gale, Church and Yarowsky [4]. For example:

A The *stock* would be redeemed in five years, subject to terms of the company's debt.
B Our soups are cooked with vegan *stock* and seasonal vegetables.
C In addition, they will receive *stock* in the reorganized company, which will be named Ranger Industries Inc.

Because the contexts between sentence A and sentence C are similar, we can identify that the *stock* in sentence 1 and in sentence C has the same meaning because they are both used in a "company" setting. Yet to date, the full power of this property has not been exploited for sense disambiguation. Many work derived from this assumption into statistical models based on local and topical features surrounding a target word to be disambiguated. However, even when we make use of these assumptions, it is difficult to identify the sense of common expressions or idioms containing the ambiguous term. For example:

(1) In an age when personal grievances is all the *rage*, this tale is uninstructive.
(2) In an age when *rage* powers personal grievances, this tale is uninstructive.

In these above two sentences, there are a word *rage* and an idiom *all the rage*, as the same time, the context of two sentences are almost the same. If we want to identify the meaning of *rage* in these two sentences using similarity context, the result of identification must have the same meaning. According to the above mentioned research, we find supervised systems for WSD often rely upon word collocations (i.e., sense-specific keywords) to provide clues on the most likely sense for a word within the given context. Collocation makes these features more obvious, so supervised learning techniques have generally been found to perform more accurately than knowledge-based methods in *one sense per collocation*. As for *One sense per discourse*, WSD, a knowledge-based method could obtain more accuracy than other methods. Similarity-based methods determine the sense of a polysemous word (a word with more than one possible meaning) by computing the relatedness between each of its possible senses and the terms in the surrounding context. The correct sense of the ambiguous term is then assumed to be that for which the relatedness is the greatest.

Finally, we find the supervised methods which use an annotated training corpus inducing the appropriate classification models in terms of *one sense per collocation*. We also find the relatedness-based method enables it to utilize a higher degree of semantic information, and is more consistent with the properties of *One sense per discourse*; that is, by considering the greater context in which the word appears. Because relatedness-based disambiguate all words in a text fragment simultaneously by exploiting semantic relatedness across word senses, it usually achieves higher performance than their supervised alternatives which usually do not considering the senses assigned to surrounding words. To overcome above problems and combine advantages, we propose a hybrid approach for WSD, which combines supervised and relatedness-based methods. This is achieved by combining supervised method and semantic relatedness measures. This approach integrates a diverse set of knowledge sources to disambiguate word sense, including POS, labeled training data, corpora of unlabeled data, salient neighboring words, and glosses of ambiguous words.

The rest of this paper is organized as follows. Section 2 presents a short review of some earlier works. Section 3 refer to HowNet we have used for our study and which calculates semantic relatedness. Section 4 explains the approach used in this paper and presents the results and corresponding explanations. A Discussion of the experimental results is given in Section 5. Finally, Sect. 6 concludes the proposed method.

2 Previous Work

Studies have shown that the word, the word n-gram and the traditional orthographic features, and the POS are the base for WSD, but are poor at representing semantic background. In order to incorporate semantic knowledge into supervised methods, Semi-Supervised Learning (SSL) techniques have been applied to WSD. SSL is a supervised approach that typically uses an unlabeled from knowledge bases and a small amount of labeled data to build a more accurate classification model than would be built using only labeled data. SSL has gained significant attention for two reasons. First, preparing a large amount of data for training requires a lot of time and effort. Second, since SSL exploits knowledge bases, the accuracy of classifiers is generally improved. There have been two different directions in SSL methods: (1) semi-supervised learning approaches, which randomly select a subset of a large unlabeled dataset and classify these samples using one (self-training) or two (co-training) classifiers, trained on a smaller set of labeled samples. After assigning labels to the new samples, these methods select the samples that were classified with a high confidence (according to a selection criterion) and add them to the set of labeled data, and (2) supervised model induction with unsupervised, possibly semantic knowledge, feature learning [13]. The approaches in the second research direction induce better feature representation by learning from knowledgebase.

The method that we use in this paper is a semi-supervised learning method which incorporates semantic knowledge from unlabeled datasets by using HowNet. To our knowledge, there currently exist three previous studies of WSD using HowNet. Yang et al. pioneered this work by using sememe co-occurrence information in sentences from a large corpus to achieve an accuracy of 71% [14]. Yang and Li collect sememe co-occurrence information from a large corpus, transferred the information to restricted rules for sense disambiguation [15]. Wong and Yang presented a maximum entropy method for the disambiguation of word senses as defined in HowNet [16]. Concepts meanings in HowNet are constructed by a closed set of sememes, the smallest meaning units, which can be treated as semantic tags. Wang used the sense tagged HowNet corpus with different approaches [17]. He applied a sense pruning method to reduce the complexity of word senses. The strategy of the current study reduces the complexity of sense tag by using the categorical attributes (first or the first two sememes) as semantic tags.

This study extends our previous work in the following ways. First, we propose a hybrid approach for WSD, which combines supervised methods and semantic relatedness. Second, we induce concept relevance field from HowNet obviateing that the context of the target word tends to be fairly short and does not equip enough vocabulary to make fine-grained distinctions. Third, we explore semantic relatedness, when more than two words are considered. This approach could reduce computational intensity. Finally, we take a small amount of labeled data to build supervised models to avoid being entrapped in the problem of creating annotated corpora. These changes lead to markeable improvement outperforming the official entries for SemEval-2007multilingual Chinese-English lexical sample task.

3   HowNet-Based Semantic Relatedness

3.1   Brief Introduction of HowNet

Why we use HowNet for Chinese semantic relatedness? HowNet is an on-line extra linguistic knowledge system for the computation of meaning in human language technology. HowNet unveils inter-concept relations and inter-attribute relations of the concepts as connoting in lexicons of the Chinese and their English equivalents. In contrast to WordNet [18], HowNet adopts a constructive approach of meaning representation [19]. Basic meaning units called sememe, which cannot be decomposed further, combine to construct concepts in HowNet.

HowNet is an extralinguistic knowledge system. Compared with some existing linguistic knowledge resources, such as Roget's thesaurus, WordNet, BarbelNet and EDR dictionaries, HowNet is unique in the following aspects:

(1) The definitions of concepts are based sememes and are described in a structured mark-up language which is easy to compute by a computer;
(2) It reveals not only the concept relations within the same part-of-speech (POS) categories, but also those cross-POS categories, especially the semantic-role relations between nouns and verbs;
(3) The representation is based on concepts denoted by words and expressions in both Chinese and English;
(4) It can be self-tested, self-evaluated by using HowNet-based tools, including Concept Relevance Calculator and Concept Similarity Measure.

The comparison of a HowNet entry record with the word definitions in LODCE and WordNet my give us a rough picture of the uniqueness of HowNet. Let's take "doctor" (only one sense) for example:

(1) LODCE: Doctor n 2 a person whose profession is to attend to sick people (or animals;) an animal doctor
(2) WordNet: (n) doctor#1 (doctor%1:18:00::), doc#1 (doc%1:18:00::), physician#1 (physician%1:18:00::), MD#2 (md%1:18:00::), Dr.#2 (dr.%1:18:00::), medico#2 (medico%1:18:00::) (a licensed medical practitioner) "I felt so bad I went to see my doctor"
3) HowNet:
NO.=162206 W_C=医生 G_C=N[yi1 sheng1]
E_C=
W_E=doctor
G_E=N
E_E= DEF={human—人:HostOf={Occupation—职位}, domain={medical—医}, {doctor—医 治:agent={~}}}

W_C, G_C and E_C would be entries for the words, the parts-of-speech and the examples respectively in Chinese, whereas W_E, G_E and E_E are the corresponding entries for English. The concept definition(DEF) in HowHet is not written in natural language but in a mark-up language, whose basic units are sememes such as "human—人", "Occupation—职位", "medical—医", "doctor—医治" and semantic roles and features such as "HostOf", "agent", "modifier" and "domain".

The definition of *doctor* can be literally paraphrased as follows: *A doctor is a human being, who has the attribute of occupation; he doctors (gives medical treatment to); he belongs to the domain of medicine.*

Obviously the definition in HowNet is formal or machine-tractable. We deliberately name HowNet a system because it is not merely a static database, but a dynamic and computer-tractable implement and because it does not only include a lexical database, but also includes a set of taxonomies and documents of axioms which are all capable to be manipulated and used for various kinds of the computation of meaning.

One of the initial objectives of HowNet is to offer special resources for word sense ambiguity [20]. HowNet is a knowledge base which was released on Internet. In HowNet, the concepts which were represented by Chinese or English words were described and the relations between concepts and the attributes of concepts were revealed. In this paper, we use Chinese knowledge base, which is an important part of HowNet, as the resource of our disambiguation. Please refer to Zhendong et al. and Gan et al. for details [20,21].

3.2 Sense Colony

Sense Colony is a group of words and expressions which are closely related in multiple sememe connections. HowNet is different from ordinary dictionaries and WordNet, and HowNet does not use natural language to define its words and expressions, but uses a formal language based on sememe and semantic roles. The context of the target word is extracted from a given sentence, which do not equip enough vocabulary to make fine grained distinctions. Therefore, we use Sense Colony to expand context. In this paper, we construct Sense Colony by semantic relatedness based on Concept Relevance Calculator and Examples in HowNet records.

Concept Relevance Calculator (CRC) is a device, which is capable to give a related concepts field of any senses of any words and expressions both in English and Chinese included in HowNet. CRC can be used to construct Sense Colony, or called *bag of the concepts*, by means of Concept Relevance Rules. To take *buy*(only for meaning as purchase) as a keyword, the Sense Colony will cover: buy → purchase location → buyer → buying manner → sell →select → pay → price → cheap/expensive → money → other concepts of commercial domain, it almost forms a script with an event as its centre. CRC is one kind of tools of meaning computation provided by HowNet [20]. We construct Sense Colony using the API which is provided by HowNet.

For example, *treatment* has fore senses {doctor—医治}, {handle—处理}, {treat—对待}, {fact—事情: CoEvent={treat—对待}} in HowNet, and each sense of Sense Colony, $F(sense)$, are given in Table 1. Sense Colony is a fully disambiguated list, because some ambiguous words are disambiguated during the construction of Sense Colony. Each sense of a plysemous word or expression in HowNet should normally be provided with as sufficient examples as possible for the purpose of disambiguation. The examples should be gestured by broad sense distinction for computer processing rather than for explanation for humans. Let us take one of sense of the word *take* as an example. One of sense is *conduct*, with its concept definition as {conduct—实施} in HowNet. Suppose we have an English text with the word *take* which is to be disambiguated: *Take vengeance on the Midianites for the Israelites*. Then we may find the exact match between the target phrase with *vengeance* and an instance in Record 018081. The words in examples were also added in Sense Colony to expand context.

Table 1. Sense Colony of each sense

sense	{doctor—医治}	{handle—处理}	{treat—对待}	{fact—事情: CoEvent={treat—对待}}
F(sense)	medical, condition, doctor, pharmacist, bloodletting, curative, inject, immunity, dose ···	conduct, solvable, finance, fiscal, deal, thorny, address, care···	reception, approach, handle, regard, treat, envisage ···	abstract, arise, begin, commence, initiate, open, rise, start, adopt, apply, assume, inauguration, apply, execute···

3.3 Semantic Relatedness Measures

We regard semantic relatedness as sense representation feature. So, this section introduces the semantic relatedness measure between a given sentence and word senses based on HowNet. In the following, let $W = \{w_0, w_1, w_2, \ldots, w_{n-1}\}$ be a set of n words given by a sentence of context of length n and w_0 be the target word to be disambiguated. Suppose word w_0, has m possible senses $def(w_0) = \{s_{01}, s_{02}, s_{03}, \ldots s_{0m}\}$. If we disambiguate w_0 by means of semantic relatedness, then the objective function is.

$$\max_{1 \leq j \leq m} Rele(W, s_{0j}). \tag{1}$$

Where $Rele(W, s_{0j})$ denotes the semantic relatedness between W and s_{0j}, s_{0j} denotes one of the senses of w_0. When we get Sense Colony of each sense s_{0j}, Eq. 1 is translated into Eq. 2.

$$\max_{1 \leq j \leq m} Rele(W, F(s_{0j})). \tag{2}$$

Where $F(s_{0j})$ denotes Sense Colony of jth sense of w_0. We get Sense Colony by using the method as mentioned in Sect. 3.2. It consists of K terms t_l and the sense s_{t_l} of t_l, $1 \leq l \leq k$. Hence, let $T = t_1, t_2, t_3, \ldots, t_k$ be a set of k terms given by Sense Colony, and $F(s_{0j})$ is described as follows:

$$F(s_{0j}) = \{(t_1, s_{t_1}), (t_2, s_{t_2}), (t_3, s_{t_3}) \ldots (t_k, s_{t_k})\}. \tag{3}$$

To define this semantic relatedness function, we utilize the weight of w_i.

$$Rele(W, F(s_{0j})) = \sum_{i=1}^{n} \frac{Wei(w_i)}{D(w_i, w_0)}. \tag{4}$$

Where, $w_i \in W$, $0 \leq i < n$, D stands for the length from w_i to w_0 in a given sentence and Wei is a numeric value that stands for the weight assigned to w_i. To define this weight, we utilize the below terms which will be used further in the paper:

$$t_l = w_i \bigcap T. \tag{5}$$

t_l, denotes a common word between w_i and T, and s_{t_l} denotes sense corresponding to t_l in $F(s_{0j})$. $Sem(s_{0j})$ denotes the jth sense of w_0, and it consists of a
group of $sememes(se_k)$, $1 \le k \le p$, and $sem(s_{0j})$ is described as follows:

$$sem(s_{0j}) = \{se_1, se_2, se_3 \& se_p\}. \tag{6}$$

$C(s_{0j}, s_{t_l})$ denotes the number of common sememes between $sem(s_{0j})$ and
$sem(s_{t_l})$ and is given as follows:

$$C(s_{0j}, s_{t_l}) = sem(s_{0j}) \bigcap sem(s_{t_l}). \tag{7}$$

$$Wei(w_i) = \begin{cases} 0, & (w_i \bigcap T = \phi) \\ \lambda, & \begin{array}{l}(w_i \bigcap T \ne \phi) \\ \&(B(s_{0j}) = 1)\end{array} \\ \lambda + (1 - \lambda) * \frac{C(s_{0j}, s_{t_l})}{B(s_{0j}) - 1}, & \begin{array}{l}(w_i \bigcap T \ne \phi) \\ \&(B(s_{0j}) > 1)\end{array} \end{cases} \tag{8}$$

In formula (8) $B(s_{0j})$ is the total number of se_k in $sem(s_{0j})$, and λ equal to 0.6.

4   Methods

The machine learning tool that the supervised system used for word sense disambiguation is Conditional Random Fields (CRFs). We extract types of features
and then use CRFs as the classifier. The features and templates implemented
in our system are explained below Sect. 4.2. After extracting these features, the
classifier (CRFs) is used to train a model for the same number sense of target
word. In the test phase, the model is used to classify test samples and to assign
a sense tag to each sample.

4.1   Preprocessing

In this paper, the training and testing corpus were provided by SemEval2007 task
5. First, the text data is cleansed by removing non informative characters and
eliminating the stop list. In the corpus, each sentence has one ambiguous term
for which there are multiple senses. Therefore, the proposed sense representation
feature method sees a sentence as its basic research units.

4.2   Feature and Template

The word representation feature is essential to classifier, but it is poor because it
only carries some morphological and shallow-syntax information of words. However, the sense representation features can be extracted by calculating semantic
relatedness and may be capable of introducing sense knowledge background to
the WSD model.

Table 2. Features and temples

Feature	Temple
t	t_4, t_3, t_2, t_1, t_0,t_1, t_2, t_3, t_4
SR	SR_1, SR_2, ...SR_{j-1}, SR_j
POS	p_0
t+POS	$t_0 p_0$
t+SR	$t_0 SR_1, t_0 SR_2, ..., t_0 SR_{j-1}, t_0 SR_j$
others	$w_0 p_0 SR_1 SR_2 ... SR_{j-1} SR_j$

Table 3. Tagging format of CRFs training set

Term	POS	SR_1	SR_2	SR_3	SR_4	Tag
information	NOUN	0	0	0	0	O
about	ADP	0	0	0	0	O
your	PRON	0	0	0	0	O
medical	ADJ	0	0	0	0	O
condition	NOUN	0	0	0	0	O
or	CONJ	0	0	0	0	O
your	PRON	0	0	0	0	O
treatment	NOUN	1.68	0	0	0	SR_1
,	W	0	0	0	0	O

In this paper, we take each semantic relatedness score as a semantic feature, and then an ML algorithm is employed to build a model for WSD. We applied features and templates in ML algorithm as Table 2.

In Table 2, $1 \leq j \leq m$, 't' represents term in a given sentence W, '$Rele$' represents semantic relatedness score, '$t + Rele$' represents the combination feature, 't_0' represents an ambiguous term, 't_{-1}' is the word preceding 't_0', 't_1' is the word following 't_0', 's_{0j}' represents one sense of t_0, $F(s_{0j})$ represents Sense Colony of s_{0j}.

4.3   Semi-supervised Learning

We apply the CRFs tools to build WSD model. In this paper, tagging format of CRFs training is shown in Table 3. In Table 4, $SR1$ denotes the score of $Rele(W, F(doctor|医治))$. $SR2$, $SR3$ and $SR4$ denote scores of other three $Rele$ respectively.

4.4   Experiments Results

We evaluated the system for WSD dataset provided by the SemEvel2007Task #5 (Multilingual Chinese-English Lexical Sample Task) organizers. The goal of this task is to create a framework for the evaluation of word sense disambiguation in Chinese-English machine translation systems. The dataset consists of 40 Chinese polysemous words: 20 nouns and 20 verbs, and each sense of one word will be provided at least 15 instances, in which around 2/3 is used as the training data and 1/3 test data. The "sense tags" for the ambiguous Chinese target words are given in the form of their English translations. The translator comes from the Chinese Semantic Dictionary (CSD) developed by the Institute of Compu-tational Linguistics, Peking University (ICL/PKU). The texts will be extracted from the corpus of People's Daily News, which have been word-segmented and POS-tagged. The semantically ambiguous target words will be manually sense tagged with their English equivalents. In Table 4, we refer to HowNet-based sense relatedness as 'HBSR'. According to this table, HBSR leads to improvements on this task. The improvements obtained through HBSR with CRF over the base-line are significant in SemEvel2007Task 5. We have included the scores obtained by the first and the second best participating systems in SemEvel2007Task #5 [22,23]. SRCB-WSD adopted a supervised learning approach with Maximum Entropy classifier. The features used were neighbouring words and their POS, as well as single words in the context, and other syntactic features based on shallow parsing. In addition, SRCB-WSD used word category information of a Chinese thesaurus as features for verb disambiguation. SRCB-WSD obtained precision of 71.66% in Micro-average, which is the best among all participated systems. Out best result is about 1.1% which is higher than SRCB-WSD. For comparison pur-poses, we included the result reported by the state-of-the-art knowledge-based systems in SemEvel2007Task 5, namely Semantic Primitive Embedding [24].

Table 4. Comparison of our system with other systems

System	Micro-average
CRF (baseline)	71.34
HBSR	42.35
CRF+BHSR	**72.73**
SRCB-WSD	71.66
I2R	71.23
Sematic Primitive Embedding	58.00

5   Discussion

Supervised word sense disambiguation systems usually treat words as discrete entities and consequently ignore the concept of relatedness between words. How-ever, by adding semantic relatedness, some of the samples that cannot be dis-criminated basing on the original features (surrounding words, long distance dependency) have more chances to be classified correctly. Moreover, semantic relatedness contains valuable linguistic information too. Hence, adding representations of semantic relatedness can provide valuable information to the classifier and the classifier can learn better discriminative criteria based on such information. Our approach to WSD does not rely on large labeled data sets. Instead, it leans on supervised models learned from small labelled data sets, on representations of semantic relatedness learned from structured semantic resources and on medium unlabeled corpora. This enables us to exploit how to integrate the semantic knowledge of word in the framework of WSD.

6   Conclusion

This paper proposes an approach to WSD using a semi-supervised learning method. The semantic relatedness is treated as the semantic feature in the model. This paper puts forward a method of improving context expansion by constructing Sense Colony of ambiguous terms presented in a given sentence. In order to incorporate sense knowledge into a WSD model, this paper proposes a semi-supervised learning method that exploits HowNet-based semantic relatedness effectively. The key feature of the method is using semantic representations from HowNet-based semantic relatedness. HowNet is sense inventory resource, in which structured information is provided about all senses (meanings) of each word. The information includes the relationships among the words (thesauri), words glosses and examples (dictionaries), or taxonomies, and a set of semantic relationships (ontologies). Meanwhile, corpora resource provides collocation occurrences of a target word that cannot be gained from sense inventory resources. Our system that solves WSD would enable corpora resource to take full advantage of their enlightened decision to incorporate sense knowledge into supervised learning method.

References

1. Navigli, R.: Word sense disambiguation: a survey. ACM Comput. Surv. **41**(2), 1–69 (2009)
2. Gale, W.A., Church, K.W., Yarowsky, D.: One sense per discourse. In: Proceedings of the Workshop on Speech and Natural Language. HLT 1991, Stroudsburg, PA, USA, pp. 233–237. Association for Computational Linguistics (1992)
3. Yarowsky, D.: One sense per collocation. In: Proceedings of the Workshop on Human Language Technology. HLT 1993, Stroudsburg, PA, USA, pp. 266–271. Association for Computational Linguistics (1993)
4. Gale, W., Church, K., Yarowsky, D.: One sense per discourse. In: Proceedings of the 4th DARPA Speech and Natural Language Workshop, pp. 233–237 (1992)
5. Miller, G.A.: WordNet: a lexical database for English. Commun. ACM **38**(11), 39–41 (1995)
6. Pal, A.R., Saha, D.: Word sense disambiguation: a survey. Int. J. Control. Theory Comput. Model. (IJCTCM) **5**(3), 1–16 (2015)

7. Parameswarappa, S., Narayana, V.N.: Kannada word sense disambiguation using decision list. Int. J. Emerg. Trends Technol. Comput. Sci. (IJETTCS) **2**(3), 272–278 (2013)
8. Singh, R.L., Ghosh, K., Nongmeikapam, K., Bandyopadhyay, S.: A decision tree based word sense disambiguation system in Manipuri Language. Adv. Comput. Int. J. (ACIJ) **5**(4), 17–22 (2014)
9. Le, C., Shimazu, A.: High WSD accuracy using Naive Bayesian classifier with rich features. PACLIC **18**, 105–114 (2004)
10. Aung, N.T.T., Soe, K.M., Thein, N.L.L.: A word sense disambiguation system using Naive Bayesian algorithm for Myanmar Language. Int. J. Sci. Eng. Res. **2**(9), 1–7 (2011)
11. Yuan, D., Doherty, R., Richardson, J., Evans, C., Altendorf, E.: Word sense disambiguation with Neural Language models. Eprint arXiv:1603.07012 (2016)
12. Navigli, R., Litkowski, K.C., Hargraves, O.: Semeval-2007 task 07: coarse-grained English all- words task. In: Proceedings of the 4th International Workshop on Semantic Evaluations, pp. 30–35. Association for Computational Linguistics (2007)
13. Taghipour, K., Ng, H.T.: Semi-supervised word sense disambiguation using word embeddings in general and specific domains. In: The 2015 Annual Conference of the North American Chapter of the ACL, pp. 314–323 (2015)
14. Yang, E., Zhang, G., Zhang, Y.: The research of word sense disambiguation method based on co-occurrence frequency of HowNet. In: Proceedings of the Second Chinese Language Processing Workshop, ACL 2000 Conference, pp. 60–65 (2000)
15. Yang, X., Li, T.: A study of semantic disambiguation based on HowNet. Int. J. Comput. Linguist. Chin. Lang. Process. **7**(1), 47–78 (2002)
16. Wong, P.W., Yang, Y.: A maximum entropy approach to HowNet-based Chinese word sense disambiguation. In: Proceedings of the ACL 2002 Workshop on Effective Tools and Methodologies for Teaching Natural Language Processing and Computational Linguistics (2002)
17. Wang, C.-Y.: Sense Pruning by HowNet - a knowledge-based word sense disambiguation, MPhil Thesis, Hong Kong University of Science and Technology (2002)
18. Miller, G.A., Beckwith, R., Fellbaum, C., Gross, D., Miller, K.J.: Int. J. Lexicogr. **3**(4), 235–244 (1990)
19. Miller, G.: Nouns in WordNet: a lexical inheritance system. Int. J. Lexicogr. **3**(4), 245–264 (1990)
20. Dong, Z., Dong, Q.: HowNet and the Computation of Meaning. World Scientific Publishing Co., Inc. (2006)
21. Gan, K.-W., Wong, P.-W.: Annotating information structures in Chinese text using HowNet. In: Proceedings of the 2nd Chinese Language Processing Workshop, Association for Computational Linguistics 2000 Conference, pp. 85–92(2000)
22. Xing, Y.: SRCB-WSD: supervised Chinese word sense disambiguation with key features. In: Proceedings of the 4th International Workshop on Semantic Evaluations (SemEval-2007), pp. 300–303 (2007)
23. Niu, Z.-Y., Ji, D.-H., Tan, C.-L.: Three systems for word sense discrimination, Chinese word sense disambiguation, and English word sense disambiguation. In: Proceedings of the 4th International Workshop on Semantic Evaluations (SemEval-2007), pp. 177–182 (2007)
24. Sun, M., Chen, X.: Embedding for words and word senses based on human annotated knowledge base: a case study on HowNet. J. Chin. Inf. Process. **30**(6), 1–5 (2016)

Protein/Gene Entity Recognition and Normalization with Domain Knowledge and Local Context

Weihong Yao[1], Xuefei Li[1], Zongze Li[2(✉)], Zhe Liu[1], and Shixian Ning[1]

[1] School of Computer Science and Technology,
Dalian University of Technology, Dalian, China
weihongy@dlut.edu.cn,
{lixuefei,njjnlz,ningshixian}@mail.dlut.edu.cn
[2] School of Electronic and Information Engineering,
Dalian University of Technology, Dalian, China
lizongze@mail.dlut.edu.cn

Abstract. Biomedical named entity recognition and normalization aim at recognizing biomedical entity mentions from text and mapping them to their unique database entity identifiers (IDs), which are the primary task of biomedical text mining. However, name variation and entity ambiguity problems make this task challenging. In this paper, we leverage domain knowledge by a novel knowledge feature representation method to recognize more entity variants, and model important local context through a dual attention mechanism and a gating mechanism to perform entity normalization. Experimental results on the BioCreative VI Bio-ID corpus show that our proposed system achieves the new state-of-the-art performance (0.844 $F1$-score for protein/gene entity recognition and 0.408 $F1$-score for normalization).

Keywords: Entity recognition · Entity normalization · Domain knowledge · Local context

1 Introduction

The unprecedented growth in biomedical literature necessitates perpetual reformations of automated text mining to uncover the knowledge contained in the vast biomedical text. The first step of biomedical text mining is biomedical named entity recognition and normalization, which facilitates downstream applications such as relation extraction [1] and knowledge base completion [2]. However, manually annotating all of them is time-consuming. New recognition methods and tools need to be developed to support more effective extraction of biomedical entities and their identifiers (IDs).

To address these needs, the Bio-ID track in BioCreative VI focuses on accurately recognizing entities and associating them with their corresponding database IDs [3]. Two subtasks are involved in this task: (1) biomedical named entity recognition (BioNER) and (2) normalization (BioNEN), also known as disambiguation.

J.-F. Hong et al. (Eds.): CLSW 2019, LNAI 11831, pp. 371–380, 2020.
https://doi.org/10.1007/978-3-030-38189-9_39

BioNER is usually considered as a sequence labeling task. Traditional machine learning (ML)-based methods [4, 5] use the BIO (Begin, Inside, Outside) labeling scheme to tag each word for entity recognition based on one-hot represented linguistic features. However, these methods require complicated feature engineering, which is labor intensive. In recent years, deep learning techniques have been proposed to learn low-dimensional feature representations of words for entity recognition without manual feature engineering [6–8]. Among them, bidirectional long short-term memory with conditional random field model (BLSTM-CRF) exhibits promising results. In addition, Devlin et al. [9] propose a new language model framework Bidirectional Encoder Representations from Transformers (BERT), which has achieved the highest performance in entity recognition tasks.

Moreover, large-scale knowledge bases (KBs), such as UniProt [10] and NCBI gene [11], usually contain rich domain knowledge, which is quite useful for BioNER task. Therefore, how to represent knowledge and introduce it to the recognition model deserves further exploration.

Compared with BioNER, BioNEN is a more challenging task, whose purpose is to normalize each recognized entity to its database ID. Prior work uses dictionary matching and develops a set of heuristic rules [5, 8] to resolve ambiguous mentions. These methods are simple and effective, but rely heavily on the integrity of the dictionary and the design of the rules. Recently, deep learning-based methods have achieved considerable success in entity normalization task [12, 13]. These methods use neural networks to learn context representations of entity mentions, and then calculate the similarity between the candidate IDs and the context representations to determine which candidate ID is correct.

This paper aims at protein/gene named entity recognition (PNER) and normalization (PNEN). We propose a pipeline identification system, which leverages entity knowledge from biomedical KBs by a BLSTM-CRF model for PNER and captures important local context by a dual attention-based Convolutional Neural Network model for PNEN. Specifically, UniProt and NCBI gene are used as a form of domain knowledge to generate n-gram boolean knowledge features to help BLSTM-CRF model to recall more protein/gene mentions. Then, we employ a dual attention mechanism and a gating mechanism to capture important local context for mention disambiguation.

The contributions of this paper can be summarized as follows: (1) knowledge features are effectively introduced to BLSTM-CRF model for improving PNER performance, and (2) important local context is explicitly captured by our disambiguation model for promoting PNEN performance.

2 Entity Recognition

As in previous work, we treat PNER as a sequence labeling problem whose goal is to assign a label to each token in a sentence. It can be divided into two steps: (1) feature extraction; (2) BLSTM-CRF model training.

2.1 Feature Extraction

Besides word and character features, we use GENIA Tagger tool [14] to extract linguistic features, such as part of speech (POS) and chunking features, to enrich the information of each token.

Furthermore, a large amount of common sense information is difficult to reflect in the sample data, resulting in the feature representation that the neural network learns lacking the general expression of the text. This paper extracts the language model features through the pre-trained language model BERT, and indirectly introduces common sense information into the entity recognition model.

Finally, UniProt and NCBI gene are used as a form of domain knowledge for knowledge features extraction. The knowledge features proposed for PNER are represented as n-gram boolean vectors, which are called KF. Specifically, text segments based on the context of each token x_i are constructed using the pre-defined feature templates which are listed in Table 1. For a text segment that appears in the feature templates, we can generate a binary value to indicate whether the text segment is an entity in the UniProt and NCBI gene databases or not. A 7-dimensional boolean vector containing the entity boundary information is generated as KF.

Table 1. KF feature templates for each token.

Type	Template
1-gram	x_i
2-gram	$x_{i-1}x_i$, x_ix_{i+1}
3-gram	$x_{i-2}x_{i-1}x_i$, $x_ix_{i+1}x_{i+2}$
4-gram	$x_{i-3}x_{i-2}x_{i-1}x_i$, $x_ix_{i+1}x_{i+2}x_{i+3}$

2.2 BLSTM-CRF Mode

The architecture of BLSTM-CRF mainly consists of three components: an embedding layer, a BLSTM layer and a CRF layer.

Embedding Layer: This layer is used to map features to distributed representations. For a sentence with n tokens, each token is represented as a d-dimensional pre-trained word embedding. Character-level feature learned by a character-level BLSTM model is also used to address the out-of-vocabulary (OOV) problem. The POS, chunking features are all embedded into randomly initialized vectors and language model features are obtained by pre-trained BERT model. Through the embedding layer, these obtained feature vectors are concatenated as an input sequence $I = \{I_1, I_2, \ldots, I_n\}$ to predict an output label sequence $y = \{y_1, y_2, \ldots, y_n\}$.

BLSTM Layer: This layer is beneficial to have access to both past (forwards) and future (backwards) context information. We concatenate a hidden state of forward LSTM $\vec{h}_t = LSTM(I_t, \vec{h}_{t-1})$ and that of backward LSTM $\overleftarrow{h}_t = LSTM(I_t, \overleftarrow{h}_{t+1})$ as the final output $h_t = [\vec{h}_t; \overleftarrow{h}_t]$ of BLSTM at the t-th time step.

Then the output $\boldsymbol{h} = \{h_1, h_2, \ldots, h_n\} \in \mathbb{R}^{d \times n}$ of BLSTM feed to a two-layer perceptron. We take a hyperbolic tangent function as the activation function. The transformation is as follows:

$$P = V(\tanh(\boldsymbol{W}\boldsymbol{h} + \boldsymbol{b})) \tag{1}$$

where $\boldsymbol{W} \in \mathbb{R}^{(d/2) \times d}$, $\boldsymbol{V} \in \mathbb{R}^{k \times (d/2)}$, $\boldsymbol{b} \in \mathbb{R}^{(d/2) \times n}$ are the trainable parameters, and k is the number of labels.

CRF Layer: To make dependence between output tags, a linear chain CRF is added on top of the BLSTM layer. The linear chain CRF is given by:

$$P(\boldsymbol{y}|\boldsymbol{I}) = \frac{1}{Z(\boldsymbol{I})} \exp\left(\sum_{i=1}^{n} \left(T_{y_{i-1}, y_i} + P_{i, y_i}\right)\right) \tag{2}$$

where P_{i, y_i} represents the score of the label y_i of the i-th token in the sequence, and T_{y_{i-1}, y_i} represents the transition probability from the label y_{i-1} to label y_i. $Z(\boldsymbol{I})$ is a normalization factor defined as follows:

$$Z(\boldsymbol{I}) = \sum_{\boldsymbol{y}} \exp\left(\sum_{i=1}^{n} \left(T_{y_{i-1}, y_i} + P_{i, y_i}\right)\right) \tag{3}$$

During the training period, our objective is to maximize the log-likelihood of the correct label path. During the prediction period, CRF uses Viterbi decoding to find the best path with the highest probability:

$$y^* = \underset{y' \in \boldsymbol{Y}}{\operatorname{argmax}} P(y'|\boldsymbol{I}) \tag{4}$$

where $\boldsymbol{Y}$ denotes the set of possible label sequences for the input sequence $\boldsymbol{I}$.

2.3 Post-processing

In order to correct some errors that may exist in the prediction results and further improve the recognition performance, we use the following three heuristic rules to post-process the prediction results:

(1) Tagging consistency: the same entity mentions appeared in the same document should be tagged with the same labels. To meet this requirement, we assume that if the same entity mentions appear more than twice in the same document and the length of the mention is not less than three, these mentions should be tagged with the same label as the first entity mention.
(2) Bracket balance: entity mentions with an odd number of brackets should be corrected until the brackets are balanced.
(3) Remove uninformative terms: some mentions known for being non-informative or unwanted terms should be removed.

3 Entity Normalization

In this section, we explain how to normalize the recognized entity mentions to their standard database IDs. PNEN consists of two steps: (1) candidate generation, (2) entity disambiguation.

3.1 Candidate Generation

For each recognized entity mention, we use the following two methods to generate the corresponding candidate IDs:

Dictionary Matching: This method uses the exact string matching to find candidate IDs from a name dictionary, which is built by mapping all annotated entities that occur in the training set to the list of IDs that entities have been linked to. If this method fails to return any results, we then turn to the second method to continue retrieving candidates.

API Lookup: This method uses the APIs provided by UniProt and NCBI gene databases to search for candidate IDs. For memory saving considerations, the top 5 results returned by API lookup are taken as the candidate IDs of the entity mention.

3.2 Entity Disambiguation

Due to the entity ambiguity, many mentions may be linked to multiple candidate IDs. To solve this problem, we design a dual attention-based Convolutional Neural Network model (Att-CNN) to eliminate the ambiguity of these entities. The architecture of Att-CNN consists of four components: an embedding layer, an attention layer, a gating layer and a softmax layer, as shown in Fig. 1.

Embedding Layer: In this layer, the left local context $\{x_{i-w}, \ldots, x_{i-2}, x_{i-1}\}$ and right local context $\{x_{i+1}, x_{i+2}, \ldots, x_{i+w}\}$ of the entity mention x_i are represented as the corresponding context word embeddings $\{e_{i-w}, \ldots, e_{i-2}, e_{i-1}\}$ and $\{e_{i+1}, e_{i+2}, \ldots, e_{i+w}\}$, where $e_i \in \mathbb{R}^d$ is a d-dimensional word embedding and w is the window size of context. Then we apply shared convolution kernels to perform convolution operations over the left and right context embedding matrixes to capture semantic features o^l and o^r respectively.

Attention Layer: In this layer, a dual attention mechanism takes the candidate ID as a control signal to capture important context clues. The following equations provide details of our attention layer

$$\alpha_t = \text{softmax}(f(o_t, e_{cand})) \tag{5}$$

$$o_{att} = \sum_t \alpha_t o_t \tag{6}$$

where the similarity score between the output of embedding layer o_t at the t-th time step and the candidate entity ID embedding e_{cand} is calculated by a function

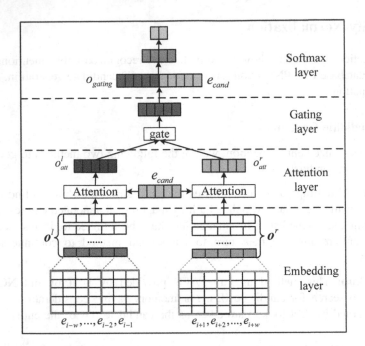

Fig. 1. Architecture of the Att-CNN model.

$f(o_t, e_{cand}) = W_f o_t + V_f e_{cand} + b_f$, where W_f, V_f and b_f are the parameters that need to be trained. Then a softmax function is used to normalize the similarity score to get the attention weight α_t. The context representation o_{att} is the weighted sum of the output vectors of each time steps of the semantic features.

Gating Layer: In this layer, a gating mechanism is employed to dynamically combine left context representation o^l_{att} and right context representation o^r_{att} from a dual attention mechanism. The gate is computed as follows:

$$g = \sigma\left(W_g o^l_{att} + V_g o^r_{att} + b_g\right) \tag{7}$$

where W_g, V_g, b_g are the model parameters that need to be trained, and σ denotes an element-wise sigmoid function.

The final context representation o_{gating} is computed using a gating mechanism,

$$o_{gating} = g \odot o^l_{att} + (1 - g) \odot o^r_{att} \tag{8}$$

where $\odot$ denotes element-wise product between two vectors.

Softmax Layer: In this layer, the output of gating mechanism o_{gating} and the candidate ID embedding e_{cand} are concentrated as the final feature representation, which is then fed to a 2-layer perceptron using softmax function. The cross-entropy loss function is used as the training objective. During the test, the candidate ID which gets the highest probability will be selected as the final result for the mention.

4 Experiments

4.1 Experimental Settings

Dataset: The Bio-ID corpus released by the BioCreative VI Track 1 [3] is used to evaluate the performance of our system. The training set consists of 570 documents with 58,853 annotations for gene/protein entities, and the test set consists of 196 documents with 16,293 annotations for gene/protein entities. We train our model using the training set, and evaluate it on the test set.

Network Details: In our experiments, words are initialized with 200-dimensional pre-trained word embeddings [15]. The vector dimensions for character, POS, chunking, and KF features are 50, 25, 10, and 15 respectively. The size of the local context window w is set to 10 for the disambiguation model. RMSProp [16] is used to optimize all parameters of the BLSTM-CRF model and the Att-CNN model. Mini-batch size is set to 32 for both models.

Evaluation Metrics: During the testing, we evaluate our system by official evaluation scripts. For PNER, the scoring tool calculates the precision (P), recall (R) and $F1$-score ($F1$) at two levels: strict span match and span overlap match. The strict span match requires that span in byte level must exactly match the golden annotation, including all entities with or without IDs. Span overlap match slightly relaxes this requirement, allowing the span to completely overlap or partly override the golden annotation. For PNEN, the levels of scoring include micro-average and macro-average, which calculate P, R and $F1$ at the sentence level and document level respectively.

4.2 PNER Results

We explore the effects of linguistic features, BERT and KF on the performance of our recognition model in this experiment. We design a baseline method that uses the concatenation of the word embedding and character embedding as input. Table 2 lists the results of different combinations of these features on the test set.

Table 2. Official PNER results on the test set.

System	Strict span match			Span overlap match		
	P	R	F1	P	R	F1
Baseline	0.749	0.816	0.781	0.800	0.871	0.834
+ linguistic	0.817	0.764	0.789	0.868	0.812	0.839
+ BERT	0.802	0.809	0.805	0.778	0.878	0.825
+ KF	0.753	0.831	0.790	0.803	0.886	0.842
+ linguistic + BERT + KF	0.822	0.827	**0.825**	0.841	0.846	**0.844**
Sheng et al. [8]	0.509	0.613	0.556	0.686	0.826	0.749
Kaewphan et al. [5]	0.729	0.739	0.734	0.825	0.836	0.831

From Table 2, we can see that linguistic features improve $F1$-score by 0.5% under the overlap criterion over the baseline. The main reason is that some entity boundary errors can be revised by the linguistic information. Knowledge feature KF bring 1.5% increase in recall under the overlap criterion over the baseline, which demonstrates that the entity name knowledge of protein/gene provided by UniProt and NCBI gene KBs are effective for PNER. When linguistic features, knowledge feature KF and Language model feature BERT are added, the best performance (an improvement of 1.0% in overlap $F1$-score) is achieved. This shows that these three features are complementary.

To further demonstrate the effectiveness of our PNER method, the results of the other two top-ranked systems developed by [5, 8] are also shown in the Table 2. Sheng et al. [8] use the concatenation of word embedding and character embedding as the input of BLSTM-CRF model, but without using other external knowledge or linguistic features. Kaewphan et al. [5] propose a CRF-based method for PNER, in which complex feature engineering is employed to improve the performance. As can be seen from the results, our recognition method (BLSTM-CRF model with all the proposed features) acquires a better $F1$-score than both of them.

4.3 PNEN Results

For the protein/gene entity normalization, we consider the following variant methods to verify the validity of our disambiguation model. (1) w/o attention: This variant removes the attention mechanism from our Att-CNN model, that is, directly using the CNN to obtain the left and right context representations. (2) w/o gating: This variant removes the gating mechanism from our Att-CNN model, and directly concatenate the left context representation, the right context representation and the candidate ID embedding for classification. (3) w/o attention and gating: This variant removes both attention mechanism and the gating mechanism from our Att-CNN model, that is, directly using the CNN to obtain the left and right context representations, then concatenates them with the candidate ID embedding for classification. Table 3 presents the official protein/gene normalization results on the test set.

Table 3. Biomedical Entity Normalization results on the test set.

System	Micro-averaged			Macro-averaged		
	P	R	F1	P	R	F1
Att-CNN	0.526	0.333	**0.408**	0.615	0.376	**0.353**
w/o attention	0.481	0.329	0.391	0.566	0.379	0.343
w/o gating	0.505	0.321	0.392	0.595	0.363	0.338
w/o attention and gating	0.483	0.314	0.381	0.585	0.355	0.337
Sheng et al. [8]	0.170	0.224	0.193	–	–	–
Kaewphane et al. [5]	0.472	0.343	0.397	–	–	–

For w/o attention, when we remove the attention mechanism, the $F1$ of PNEN drops 1.7% under the micro-averaged criterion. The plausible reason is that the proposed attention mechanism allows the system to focus on finding useful information in the context that is more relevant to the candidate ID, thereby improving the quality of the output context embeddings.

For w/o gating, when we remove the gating mechanism, the performance of PNEN also drops under the micro-averaged criterion. By introducing the gating mechanism to control the update and fusion of the left and right context representations, the disambiguation model could dynamically determine how much information can be delivered to the final context semantic representation to find the balance between the left and right representations.

For w/o attention and gating, when we remove both attention mechanism and the gating mechanism, the performance of PNEN further drops. This proves that the two mechanisms are effective for PNEN.

Finally, we compare our PNEN system with the same two top-ranked systems [5, 8] as well. Sheng et al. [8] compile a contextual dictionary based on the training set and then check if the entity mention is in this contextual dictionary. If in the dictionary, they normalize the mention to the known ID that shares the most contextual words with the sequence the entity belongs to. For cases without matched IDs in the compiled dictionary, they use the same API sources as ours, to search for candidate IDs and directly assign the first ID match to ambiguous mentions. Kaewphane et al. [5] apply exact string matching to retrieve candidate IDs of protein/gene mentions based on KBs. For the ambiguous mentions with multiple candidate IDs, some heuristic rules are developed for disambiguating them and uniquely assigning an ID. Since Sheng et al. [8] and Kaewphane et al. [5] use heuristic rules rather than disambiguation models to normalize candidate IDs, their approach achieves a relatively low precision and recall on the PNEN subtask. As can be seen from Table 3, our normalization approach outperforms the above related approaches and achieves a state-of-the-art result (0.408 micro-averaged $F1$-score). We attribute this to the validity of our disambiguation model, which accurately models the important context representation through the attention and gating mechanisms.

5   Conclusion

This paper develops a pipeline identification system for protein/gene entity recognition and normalization. On the one hand, we leverage domain knowledge from biomedical KBs by a BLSTM-CRF model for PNER. The n-gram boolean knowledge features are designed and explored, meanwhile, language model feature is also introduced to realize the feature transfer of common sense information. On the other hand, we propose a dual attention-based Convolutional Neural Network model for PNEN and investigate the effect of the dual attention mechanism and gating mechanism. Experimental results on the BioCreative VI Bio-ID corpus show that our system achieves the new state-of-the-art performance on both PNER and PNEN.

In the future work, we will explore the usage of joint learning of the PNER and PNEN to bridge the gap between them.

Acknowledgments. This work was supported by the grants of the Ministry of education of Humanities and Social Science project (No. 17YJA740076) and the National Natural Science Foundation of China (No. 61772109). Comments from the audience of CLSW2019 and the reviewers are also acknowledged.

References

1. Lin, Y., Liu, Z., Sun, M.: Neural relation extraction with multi-lingual attention. Proc. Assoc. Comput. Linguist. **1**, 34–43 (2017)
2. Rudolf, K., Ondrej, B., Jan, K.: Knowledge base completion: baselines strike back. In: Proceedings of the Association for Computational Linguistics, pp. 69–74 (2017)
3. Arighi, C., et al.: Bio-ID track overview. In: Proceedings of BioCreative Workshop, pp. 482–376 (2017)
4. Sheikhshab, G., Starks, E., Karsan, A., Sarkar, A., Birol, I.: Graph-based semi-supervised gene mention tagging. In: Proceedings of the 15th Workshop on Biomedical Natural Language Processing, pp. 27–35 (2016)
5. Kaewphan, S., Mehryary, F., Hakala, K., Salakoski, T., Ginter, F.: TurkuNLP entry for interactive Bio-ID assignment. In: Proceedings of the BioCreative VI Workshop, pp. 32–35 (2017)
6. Ma, X., Hovy, E.: End-to-end sequence labeling via bi-directional LSTM-CNNs-CRF. arXiv preprint arXiv:1603.01354 (2016)
7. Chiu, J., Nichols, E.: Named entity recognition with bidirectional LSTM-CNNs. Trans. Assoc. Comput. Linguist. **4**, 357–370 (2016)
8. Sheng, E., Miller, S., Ambite, J., Natarajan, P: A neural named entity recognition approach to biological entity identification. In: Proceedings of the BioCreative VI Workshop, pp. 24–27 (2017)
9. Devlin, J., Chang, M., Lee, K.: Bert: Pre-training of deep bidirectional transformers for language understanding. arXiv preprint arXiv:1810.04805 (2018)
10. Apweiler, R., et al.: UniProt: the universal protein knowledgebase. Nucl. Acids Res. **32** (suppl_1), D115–D119 (2004)
11. Edgar, R., Domrachev, M., Lash, A.: Gene expression omnibus: NCBI gene expression and hybridization array data repository. Nucl. Acids Res. **30**(1), 207–210 (2002)
12. Eshel, Y., Cohen, N., Radinsky, K., Markovitch, Y., Levy, O.: Named entity disambiguation for noisy text. arXiv preprint arXiv:1706.09147 (2017)
13. Ganea, O., Hofmann, T.: Deep joint entity disambiguation with local neural attention. arXiv preprint arXiv:1704.04920 (2017)
14. GENIA Tagger tool Homepage. https://omictools.com/genia-tagger-tool. Accessed 12 Aug 2019
15. Moen, S., Ananiadou, T.: Distributional semantics resources for biomedical text processing. In: Proceedings of the 5th International Symposium on Languages in Biology and Medicine, pp. 39–43 (2013)
16. Tieleman, T., Hinton, G.: Lecture 6.5-rmsprop: divide the gradient by a running average of its recent magnitude. COURSERA Neural Netw. Mach. Learn. **4**(2), 26–31 (2012)

Sentence-Level Readability Assessment for L2 Chinese Learning

Dawei Lu[1], Xinying Qiu[2(✉)], and Yi Cai[3]

[1] School of Liberal Arts, Renmin University of China, Beijing, China
[2] School of Information Science and Technology,
Guangdong University of Foreign Studies, Guangzhou, China
xy.qiu@foxmail.com
[3] School of Software Engineering, South China University of Technology,
Guangzhou, China

Abstract. Automatic assessment of sentence readability level can support educators in selecting sentence examples suitable for different learning levels to complement teaching materials. Although there exists extensive research on document-level and passage-level Chinese readability assessment, the sentence-level evaluation remains little explored. We bridge the gap by providing a research framework and a large corpus of nearly 40,000 sentences with ten-level readability annotation. We design experiments to analyze the influence of 88 linguistic features on sentence complexity and results suggest that the linguistic features can significantly improve the predictive performance with the highest of 70.78% distance-1 adjacent accuracy. Model comparison also confirms that our proposed set of features can reduce the bias in prediction without adding variances. We hope that our corpus, feature sets, and experimental validation can provide educators and linguists with more language resources, enlightenment, and automatic tools for future related research.

Keywords: Readability assessment · Feature engineering · Language resources

1 Introduction

Estimating the scale of sentence readability is often an important component in NLP related tasks including text simplification, text summarization, controlled language validation, and more populously L2 education research. L2 language instructors often need to complement teaching materials with additional sentence examples either synthetically designed or from authentic resources. These sentences, even if they may be tagged with grammatical cues, may not be readily indicated with an overall readability (i.e. complexity or understandability) level. Automatic assessment of sentence readability level can support educators in exploring, analyzing, and selecting proper sentences for exercise, for example, as in the Swedish L2 sentence selection module. The goal of our research is to automatically identify sentences suitable for foreign speakers of different learning levels, and to study features that influence readability predictive models.

© Springer Nature Switzerland AG 2020
J.-F. Hong et al. (Eds.): CLSW 2019, LNAI 11831, pp. 381–392, 2020.
https://doi.org/10.1007/978-3-030-38189-9_40

To build sentence readability predictive models, we first need to construct an annotated corpus of sentences. Currently, digitized Chinese L2 corpora are available mostly as text-book articles or sentences tagged with grammatical rules. Sentences from textbooks of different grades, however, may contain grammatical rules at different levels. We propose to annotate sentence readability levels by considering both the grammatical learning difficulty and the grade of source text-books, as illustrated with examples in Tables 1 and 2.

Table 1. Example one to illustrate challenges and methods for annotating the sentence readability level

Sentence example one. 我这次是专为了别他而来的。(I came here to say goodbye to him.)				
Source Textbook Level	Grammar Sets	Grammatical Cues	Grammar Difficulty	Sentence Readability
Level 5	复句 (Complex sentence)	为了+动词 （in order to + verb.)	3	4

Table 2. Example two to illustrate challenges and methods for annotating the sentence readability level

Sentence example two. 其实刚开始我来中国留学不是为了学经济，而是为了学中国功夫。 (In fact, at first, I came to China not to study economics, but to learn Chinese Kungfu.)				
Source Textbook Level	Grammar Sets	Grammatical Cues	Grammar Difficulty	Sentence Readability
Level 3	复句 (Complex sentence)	为了+动词 （in order to + verb.)	3	4
		不是+动词+而是+动词 (not to + verb ... but to + verb)	5	

In example one, the sentence contains a grammatical cue of level 3 difficulty, but it is extracted from a textbook of grade five. This is reasonable because of the complex semantic meaning of the words in the sentence. We assign a readability level 4 by average. In example two, the sentence contains two grammatical cues of level 3 and 5 respectively and we use level 5 as its grammatical difficulty. Since the source textbook of the sentence is of grade 3, however, we take an average score of 4 as its final sentence-level readability. Further details on corpus construction are provided in the Methodology section.

We experiment with a ten-level sentence readability categorization system based on the grammatical annotation of nearly 40,000 sentences and their source text books. We evaluate the influence of an enriched and comprehensive set of 88 features covering four linguistic categories on the performance of different machine learning models in predicting sentence readability. Our findings suggest that linguistic features can significantly increase most models' predictive performance and achieve the highest of 70.78% distance-1 adjacent accuracy for classification algorithms. By comparing performances of different classifiers, we further validate that our proposed set of features can reduce the bias in predicting readability levels without adding variances.

2   Related Research

Readability research has a 70-year history and many formulas and models have been applied in developing educational theories, supplementing instruction manuals, and publishing easy-to-read newspapers, reports and professional documents. Traditional readability assessment measures such as Flesch-Kincaid formula adopt simpler, easy-computed statistics from a combination of shallow and lexical features. In recent years, with the need to evaluate texts from the web and other media, methodology has advanced to adopting NLP and machine learning approaches as shown by Collins-Thompson and Callan [2] using language modeling for passage-level readability assessment.

Sentence-level readability needs to be addressed separately because of its shorter length and syntactic structure. For example, Woodsend and Lapata (2011) [3] incorporated two components from Flesch-Kincaid formula (i.e. words per sentence and syllabus per sentence) into objective functions of integer linear program to evaluate sentence readability for sentence simplification. The following table summarizes some representative work in sentence-level readability (Table 3).

Table 3. Summary of representative work on sentence-level readability

References	Target/Domain	Features/Heuristics	Models
Husak 2010 [4]	Sentence selection for English lexicography	Sentence length, common words, punctuation, etc.	Scores with manual weights
Segler 2007 [5]	German L2	Vocabulary, syntax	Logistic regression
Vajjala and Meurers 2014 [6]	English sentence simplification	Lexical, POS, syntactic, psycholinguistic	Binary classification
Pilan 2014 [7]	Swedish L2	Lexical, semantic, syntactic, shallow	Binary classification
Schumacher et al. 2016 [8]	English sentence pair-wise ranking	Lexical, syntactic	Logistic regression

The works in the above table differ in their targeted applied domain, the features or heuristic metrics used for characterizing sentences, and the models for predicting readability. With this background, we propose to build predictive models for Chinese L2 sentence-level readability assessment by evaluating the influence of a comprehensive set of linguistic features tailored for Chinese L2 education.

3 Methodology

In this study, we explore two questions, in addition to providing corpus and feature resources for sentence-level readability study. First, we compare and analyze performances of different machine learning algorithms for sentence-level readability prediction; second, we evaluate the influence of linguistic features on the performance of different models.

3.1 Corpus Preparation

Sentences and Grammatical Difficulty Level Annotation. We collected sentence data from "Corpus of Teaching Chinese as Second Language" website[1] and extracted 39,959 sentences annotated with 546 grammatical rules. Grammatical rules were categorized into four groups: complex sentence, fixed structure, preposition or prepositional phrases, and special construction. We aligned the 546 grammatical rules with grammatical learning objectives specified in the "International Curriculum for Chinese Language Education", a national standard for L2 Chinese proficiency test (i.e. HSK). According to the HSK Curriculum, we assigned one of the six HSK levels to each of the grammatical rules. Therefore, each of the sentence is annotated with one or more grammatical rules with each assigned with an HSK level.

Source Text-Books Grades. From the sentences data, we identified 73 text-books from which the sentences were extracted. We classified the text-books into five grades according to the semester that they are used. Specifically, text-books used in the first, second, third, and the fourth semester were assigned grade one to grade four respectively. Text-books used for students learning Chinese for more than two years were grade five. In deciding in which semester each text-book is used, we surveyed seven universities about their usage of the 73 text-books before we finalized the grade for each text book.[2]

Sentence Readability Level. In most cases, the sentence grammatical difficulty level is consistent with the text-book grade. There are exceptions, however, as illustrated in the two examples in Tables 1 and 2 in the Introduction section. We used the arithmetic mean of the sentence's grammatical level (on a scale of 1–6) and its source textbook

---

[1] http://www.aihanyu.org/.

[2] The seven universities participating in text-books survey are Renmin University of China, Beijing Language and Culture University, Sun Yat-sen University, Jinan University, South China Normal University, Huaqiao University, and Fujian Normal University.

grade (on a scale of 1–5) as the final sentence readability level. Thus, we have a ten-level readability scaling system ranging from 1.0 to 5.5. We performed manual check and statistical distribution analysis to validate that our scaling method would give an accurate score that matches human judgement. Data distribution is shown in Table 4.

Table 4. Corpus data statistics

Readability level	# of sentences	Percentage
1.0	503	1.26%
1.5	2571	6.43%
2.0	3271	8.19%
2.5	6426	16.08%
3.0	8532	21.36%
3.5	8247	20.64%
4.0	5355	13.4%
4.5	3141	7.86%
5.0	1351	3.38%
5.5	562	1.41%
Total	39959	100%

3.2 Feature Engineering

We provided a set of 88 linguistic features organized into four groups: shallow features, POS features, syntactic features, and discourse features summarized in Table 5. Please refer to the Appendix for complete description. The metrics include percentage and the number of features per sentence, and other statistics based on length, maximum, average, and total. For pre-processing, we used LTP[3] platform for word segmentation, POS tagging, and named entity recognition, and NiuParser[4] for syntactic parsing, grammatical labeling, and clause annotation.

The 88 features were designed by referring to previous research findings in readability study for different languages. We further adapted the feature set in particular for L2 Chinese learning as follows.

Shallow Features. We designed 29 shallow features for Chinese characters, words and sentences. The new features that have not been used widely are character statistics based on stroke-count, and the proportion of characters and words at different HSK (Chinese Proficiency Test) levels.

POS features are also labelled as "semantic" or "lexical" features in many readability studies. We focused on seven categories of tags: adjectives, nouns, verbs, adverbs, idioms, functional words, and content words.

---

[3] http://www.ltp-cloud.com/.

[4] http://www.niuparser.com/.

Syntactic Features. Parse tree features have been studied to show the grammatical complexity of text. We focus on three aspects: phrases, clauses, and sentence features. We implemented three new features based on punctuation clauses defined by Song [9], and two features based on dependency distance defined by Lin [10] and Liu [11] respectively.

Discourse Features. Inspired by previous work (Feng [12]) on evaluating text difficulty, we implemented two types of 24 discourse features: entity-density features, and cohesion features.

Table 5. Summary of sentence-level linguistic features

Feature category	Sub-category	Features
Shallow features	Character	Common characters, stroke-counts, characters by HSK levels
	Words	N-gram, words by HSK levels
	Sentence	Sentence length
POS features		Adjective, functional words, verbs, nouns, content words, idioms, adverbs
Syntactic features	Phrases	Noun phrases, verbal phrases, prepositional phrases
	Clauses	Punctuation-clause, dependency distance
	Sentences	Parse tree, dependency distance
Discourse features	Entity density	Entities, named entities
	Coherence	Conjunctions, pronouns

3.3 Models and Evaluation

To represent sentences, we use the following approaches:

- BOW: This is the default baseline representation where each sentence is a vector of terms in the sentence weighted with ltc variant of TF*IDF.
- BOW+88: We calculated the 88 feature scores for each sentence and appended the 88 scores to the BOW vector for each sentence.
- BOW+39: We performed logistic regression on the 88 features and identified 39 features that were coefficient with readability at 95% confidence interval. We appended the scores of the 39 features to the BOW vector for each sentence.
- BOW+25: We identified 25 features that were coefficient with readability at 99% confidence interval. We appended the scores of these features (also a subset of the 39 features) to the BOW vector for each sentence.

We used multi-class classifiers and regression to build predictive models for sentence-level readability. Specifically, we chose the following models:

- SVM. This is one of the default classifiers commonly used. It applies Kernel to project data into high-dimensional space to find a linear separation.
- Xgboost. eXtreme Gradient Boosting is an implementation of gradient boosted decision trees with ensemble methods. It is based on weak learners with the assumption that data is of high bias and low variance.
- Light GBM. Similar to Xgboost, LGBM is also a high-performance gradient boosting framework based on decision trees, but it splits the tree leaf wise and can reduce more loss very fast. It is also the most promising for data with lower noise but more complex.
- Extra-trees Classifier (ETC). Also known as extremely randomized tree, this algorithm is similar to random forest (RF), using bagging methods to aggregate fully grown decision trees. ETC only evaluates a random few break points. It is more suitable for noisy data by decreasing variance while increasing bias at the same time.
- Logistic regression (LR). We experimented with LR for fitting a linear model to explain the categorical output given the input.
- Support Vector Regression (SVR). Similar to SVM, support vector regression (SVR) maps data into high-dimensional space but fits a function with tolerable error and as flat as possible.

To evaluate multi-class classification, we used Accuracy and Distance-1 Adjacent Accuracy. Adjacent Accuracy is often used in computational linguistics when predicting a text to be within one level of the true level label is still considered accurate (Sung et al.) [13]. According to our data distribution shown in Table 4, the Majority Vote accuracy is 21.36%, and adjacent accuracy is 58.05%. With Uniform Random evaluation for ten-level classification, we have a baseline of 10% accuracy and 30% adjacent accuracy. We used R2 to evaluate Regression model. We performed stratified 10-fold cross-validation and presented average results and their paired t-test statistics analysis in the following section.

4   Results and Discussion

Our research purpose is two-fold, (1) to compare predictive models for sentence complexity, and (2) to evaluate the influence of linguistic features. We address these questions jointly with results in Table 5 and Fig. 1.

From the results in Table 5 and Fig. 1 of classifier models, sentence representation augmented with the linguistic features can significantly improve the BOW representation in four out of five classifiers except for the ETC classifiers. As we discussed in the previous section, the classifiers work differently from each other in prediction. The SVM and LR aim at finding a separable hyper plane or fitting a linear function to predict the output. The Xgboost and the LGBM reduce error mainly by reducing bias. The ETC model decreases variance while increasing bias. We see that adding linguistic features to data representation helps improve the boosting algorithms (Xgboost and LGBM) by achieving a lower model bias, but is not helpful for ETC which works best on BOW model by reducing data variance in itself. We may draw conclusions that augmenting sentence representation with the linguistic features are effective to improve readability prediction by reducing bias without increasing data variances.

We also experimented with fitting linguistic features into logistic regression model with LR and SVR, and present results in Table 6. We found that the features significantly correlated with readability at 95% and 99% confidence intervals perform similarly as the full set data in improving readability prediction. Although we are not able to fit a linear formula with high level of interpretation, and the R2 is rather small over all regression models, we see that adding linguistic features can significantly improve R2 for both logistic regression and SVR.

Table 6. Model performance for classifiers and regression

Models	BOW		BOW+88		BOW+39		BOW+25	
	Accu.	Adj. Accu.	Accu.	Adj. Accu.	Accu.	Adj. Accu.	Accu.	Adj. Accu.
SVM	_27.24%_	_63.86%_	**28.44%*****	**65.66%*****	**28.41%*****	**65%*****	**28.36%*****	**65.51%*****
Xgboost	_29.65%_	_67.95%_	**30.50%****	**70.78%*****	**30.35%***	**69.72%*****	**30.46%****	**69.65%*****
LGBM	_30.17%_	_68.57%_	**30.85%*****	**70.78%*****	**30.44%***	**69.99%****	30.55%	**70.67%****
ETC	**31.92%*****	**70.28%*****	28.76%	66.94%	30.03%	68.16%	30.53%	68.82%
LR	_30.03%_	_68.87%_	**30.87%****	**70.31%*****	30.38%	**69.83%*****	30.28%	**69.73%****
LR	_0.2893 (R2)_		**0.3324***** (R2)		**0.3256***** (R2)		**0.3222***** (R2)	
SVR	_0.4614 (R2)_		**0.477***** (R2)		**0.4726**** (R2)		**0.469*** (R2)	

Note: (1) Italic and underlined scores are baseline performance to be compared by models of the same row. (2) Bolded performances are significantly better than baselines. $* p < 0.1$; $** p < 0.05$; $*** p < 0.01$. (3) For ETC classifier, the BOW model performs significantly better than BOW model of SVM. (4) Logistic Regression (LR) is evaluated with accuracy, adjacent accuracy, and R2 scores.

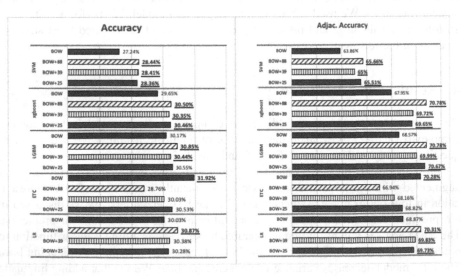

Fig. 1. Model performance based on accuracy and adjacent accuracy for five classifiers (note: significant performance are bolded and underlined.)

5 Conclusions

We present in this work an annotated corpus, and a comprehensive set of linguistic features to explore predictive models for Chinese L2 sentence-level readability assessment. Our findings suggest that linguistic features can significantly increase the performances of some state-of-the-art classifiers and regression algorithms by achieving the highest of 30.85% accuracy and 70.78% distance-1 adjacent accuracy. By analyzing the classifiers' performances, we further validate that our proposed set of features can reduce the bias in predicting complexity levels without adding variances. We hope that our corpus, feature sets, and experimental validation can provide educators and linguists with more language resources, enlightenment, and automatic tools for future related research.

Acknowledgements. This work was supported by National Social Science Fund (Grant No. 17BGL068). We thank undergraduate students Zhiwei Wu, Yuansheng Wang, Xu Zhang, Yuan Chen, Hanwu Chen, Licong Tan, and Hao Zhang for their helpful assistance and support.

Appendix

Linguistic Features

Category	Sub-category	Feature definition
Shallow features	*Character features*	1. Percentage of most-common characters per sentence
		2. Percentage of second-most-common characters per sentence
		3. Percentage of all common-characters per sentence
		4. Percentage of low-stroke-count characters per sentence
		5. Percentage of medium-stroke-count characters per sentence
		6. Percentage of high-stroke-count characters per sentence
		7. Average number of strokes per word per sentence
		8. Percentage of HSK1 to HSK3-characters per sentence
		9. Percentage of HSK4 to HSK5-characters per sentence
		10. Percentage of HSK6-characters per sentence
		11. Percentage of not-HSK-characters per sentence
	Word features	12. Average number of characters per word per sentence
		13. Average number of characters per unique word per sentence
		14. Number of two-character words per sentence
		15. Percentage of two-character words per sentence
		16. Number of three-character words per sentence
		17. Percentage of three-character words per sentence
		18. Number of four-character words per sentence
		19. Percentage of four-character words per sentence
		20. Number of five-up-character words per sentence

(continued)

(continued)

Category	Sub-category	Feature definition
		21. Percentage of five-up-character words per sentence
		22. Percentage of HSK1 to HSK3-words per sentence
		23. Percentage of HSK4 to HSK5-words per sentence
		24. Percentage of HSK6-words per sentence
		25. Percentage of Not-HSK-words per sentence
	Sentence features	26. Number of multi-character words per sentence
		27. Number of words per sentence
		28. Number of characters per sentence
		29. Number of characters (including punctuations, numerical, and symbols) per sentence
POS Features	*Adjectives*	30. Percentage of adjectives per sentence
		31. Percentage of unique adjectives per sentence
		32. Number of unique adjectives per sentence
		33. Number of adjectives per sentence
	Functional words	34. Percentage of functional words per sentence
		35. Percentage of unique functional words per sentence
		36. Number of unique functional words per sentence
		37. Number of functional words per sentence
	Verbs	38. Percentage of verbs per sentence
		39. Number of unique verbs per sentence
		40. Percentage of unique verbs per sentence
		41. Number of verbs per sentence
	Nouns	42. Percentage of nouns per sentence
		43. Number of unique nouns per sentence
		44. Percentage of unique nouns per sentence
		45. Number of nouns per sentence
		46. Percentage of All-Nouns per sentence
		47. Number of unique All-Nouns per sentence
		48. Percentage of unique All-Nouns per sentence
		49. Number of All-Nouns per sentence
	Content words	50. Percentage of content words per sentence
		51. Number of unique content words per sentence
		52. Percentage of unique content words per sentence
		53. Number of content words per sentence
	Idioms	54. Percentage of idioms per sentence
		55. Number of unique idioms per sentence
		56. Percentage of unique idioms per sentence
		57. Number of idioms per sentence
	Adverbs	58. Percentage of adverbs per sentence
		59. Percentage of unique adverbs per sentence

(continued)

(continued)

Category	Sub-category	Feature definition
		60. Number of unique adverbs per sentence
		61. Number of adverbs per sentence
Syntactic features	*Phrases*	62. Total number of noun phrases per sentence
		63. Total number of verbal phrases per sentence
		64. Total number of prepositional phrases per sentence
		65. Average length of noun phrases per sentence
		66. Average length of verbal phrases per sentence
		67. Average length of prepositional phrases per sentence
	Clauses	68. Number of punctuation-clauses per sentence
		69. Average dependency distance per sentence
		70. Maximum dependency distance per sentence
	Sentences	71. Height of parse tree per sentence
		72. Total number of dependency distances per sentence
		73. Average number of dependency distances per sentence
Discourse features	*Entity density*	74. Total number of entities per sentence
		75. Total number of unique entities per sentence
		76. Percentage of entities per sentence
		77. Percentage of unique entities per sentence
		78. Percentage of named entities per sentence
		79. Percentage of named entities against total number of entities per sentence
		80. Percentage of Not-NE nouns per sentence
		81. Number of Not-NE nouns per sentence
		82. Number of Not-Entity nouns per sentence
	Cohesion	83. Percentage of conjunctions per sentence
		84. Number of unique conjunctions per sentence
		85. Percentage of unique conjunctions per sentence
		86. Percentage of pronouns per sentence
		87. Number of unique pronouns per sentence
		88. Percentage of unique pronouns per sentence

References

1. Flesch, R.: A new readability yardstick. J. Appl. Psychol. **32**, 221–233 (1948)
2. Collins-Thompson, K., Callan, J.: A language-modelling approach to predicting reading difficulty. In: Proceedings NAACL-HLT, Boston, pp. 193–200 (2004)
3. Woodsend, Lapata: Learning to simplify sentences with quasi-synchronous grammar and integer programming. In: Proceedings of EMNLP 2011, pp. 409–420 (2011)
4. Husák, M.: Automatic retrieval of good dictionary examples. Bachelor Thesis, Brno (2010)

5. Segler, T.M.: Investigating the selection of example sentences for unknown target words in ICALL reading texts for L2 German. PhD Thesis. University of Edinburgh (2007)
6. Vajjala, Meurers: On improving the accuracy of readability classification using insights from second language acquisition. In: Proceedings of the ACL 2012 BEA 7th Workshop, pp. 163–173 (2012)
7. Pilán, et al.: Rule-based and machine learning approaches for second language sentence-level readability. In: Proceeding of the ACL 2014 BEA 9th Workshop, pp. 174–184 (2014)
8. Schumacher, E., et al.: Predicting the relative difficulty of single sentences with and without surrounding context. In: Proceedings of EMNLP 2016, pp. 1871–1881 (2016)
9. Song, R.: Stream model of generalized topic structure in Chinese text. Stud. Chin. Lang. **357** (6), 483–494 (2013). (in Chinese)
10. Lin, D.: On the structural complexity of natural language sentences. In: Proceedings of COLING 1996, pp. 729–733 (1996)
11. Liu, Haitao: Dependency distance as a metric of language comprehension difficulty. J. Cogn. Sci. **9**(2), 159–191 (2008)
12. Feng, L.: Automatic readability assessment. Ph.D. thesis, The City University of New York (2010)
13. Sung, Y., et al.: Leveling L2 texts through readability: combining multilevel linguistic features with the CEFR. Mod. Lang. J. **99**(2), 371–391 (2015)

Text Readability Assessment for Chinese Second Language Teaching

Shuqin Zhu[1,2(✉)], Jihua Song[1(✉)], Weiming Peng[1(✉)],
Dongdong Guo[1], and Gu Wu[1]

[1] College of Information Science and Technology, Beijing Normal University,
Beijing, China
sftzhushuqin@buu.edu.cn,
{songjh,pengweiming}@bnu.edu.cn,
dongdongguo@mail.bnu.edu.cn, 494689444@qq.com
[2] Teachers' College of Beijing Union University, Beijing, China

Abstract. This paper proposes a multi-type and multi-granularity text readability feature set for Chinese second language teaching, which takes into account the dynamic and static features of the texts and integrates three linguistic units: character, word and sentence. On this basis, this paper analyses and compares various text readability assessment methods, and discusses how to effectively use various features for text readability assessment.

Keywords: Text readability · Chinese second language · Classification algorithms

1 Introduction

Famous input theory of second language acquisition holds that effective input should be comprehensible, interesting and relevant, enough and not grammatically sequenced [1]. Comprehensibility means that the difficulty of second language input should be slightly higher than the current level of learners, which is a necessary condition for language acquisition. Only by providing the second language input the difficulty is slightly higher than learners' current level, can learners understand information according to the context and by means of certain non-linguistic ways, and will second language acquisition occur. Enough input means that learners must receive sufficient input training of language materials. Once the amount of reading materials is limited, the effect of language acquisition will be greatly reduced. These two characteristics put forward requirements for "input content" and "input quantity" of the second language, that is, the "input content" should be comprehensible and the "input quantity" should be sufficient.

Input theory of second language acquisition has also tremendously enlightened Chinese second language (CSL) teaching. It is necessary not only to have a comprehensive understanding of the level of CSL learners, but also to provide learners with more comprehensible input and carry out graded teaching. This requires an assessment of the text readability to match the difficulty of texts with the level of learners. Text readability is also called text complexity. Text readability assessment gives the difficulty level of the text or determines which level of readers the text is suitable for.

J.-F. Hong et al. (Eds.): CLSW 2019, LNAI 11831, pp. 393–405, 2020.
https://doi.org/10.1007/978-3-030-38189-9_41

This paper proposes a multi-type and multi-granularity text readability feature set for CSL teaching. On this basis, we analyze and compare various text readability assessment methods, and discuss how to effectively use various features for text readability assessment.

2   Related Work

The study of English readability has a relatively early start. There are many classical readability formulas, which generally calculate the weighted sum of some indicators. For example, The Flesch-Kincaid formula and the SMOG formula used the average number of words per sentence, the average number of syllables per word, and the average number of words with three or more syllables per sentence [2, 3]. In addition to these classical readability formulas, many researchers have introduced Machine Learning methods into readability assessment. For example, the Multi-class Logistic Regression method was used to measure the readability, which is an extension of traditional readability formulas [4]; the SVM method was used to assess the readability [5]; Decision Tree, Bagging Decision Tree, Linear Regression, SVM Regression and Gauss Process Regression were compared to various text readability features [6].

Text readability involves various features, including word features, sentence features, etc. In traditional readability formulas, word features such as word length and syllable number are often used to measure lexical difficulty. In [7], word features included Type-Token Ratio, Lexical Density, Lexical Word Variation, Verb Variation, Noun Variation, Adjective Variation, and Adverb Variation. In terms of sentence features, early studies mainly focused on the measurement of sentence length. In the current text readability assessment, attempts have been made to parse sentences to acquire grammatical structure information. For example, in [8], the grammatical features mainly include average grammar tree height, average clause number, average noun (verb, adjective) phrase number.

The readability features of Chinese texts are similar to those of other languages. Because of the particularity of Chinese, Chinese texts have their own characteristics. In [9], the factors affecting the complexity of Chinese texts were investigated through questionnaires for CSL learners. The results show that the Chinese character is one of the main factors affecting text complexity. Based on the Chinese Level Vocabulary and Chinese Character Level Syllabus (Revised Version, 2001), a text readability formula including the complexity of Chinese characters and words was established [10]. Among the readability formulas for intermediate European and American students, the function word was designated as one of the important indicators [11].

3   Text Readability Assessment

3.1   Multi-type and Multi-granularity Text Readability Features

In this paper, two kinds of features are extracted, one is domain-related dynamic features, and the other is static features determined only by the text itself. Dynamic features are extracted from texts based on the existing knowledge bases in the field of CSL teaching. When constructing these knowledge bases, not only the laws of second language acquisition, but also the development and change of language in current social life are taken into account. Static features, based on the inherent characteristics of language itself, don't change with the domain or application requirements, and they belong to the stability characteristics of texts. These two kinds of features are measured in multi-granularity at the level of character, word and sentence. When measuring features at the sentence level, sentences are divided to two types for processing: big sentences and small sentences. A big sentence refers to the sentence that is formally complete and separated with "。", "?" or "!". A small sentence refers to the internal segment of a big sentence that is relatively complete in structure and paused in tone. For example, in "国有国法，家有家规。" (Every country has its laws, and every family has its rules.), there is only one big sentence, but there are two small sentences. The reason for introducing small sentences is that understanding small sentences is the basis of understanding big sentences, and learners need to understand each small sentence and combine them to understand the whole sentence.

Dynamic Features

For the measurement of dynamic features, two major knowledge bases in the field of CSL teaching are introduced: the hierarchical Chinese character list of International Curriculum for Chinese Language Education (2014, compiled by Confucius Institute Headquarters/Hanban) and the hierarchical vocabulary of HSK Test Syllabus (2015, compiled by Confucius Institute Headquarters/Hanban). International Curriculum for Chinese Language Education combs and describes the curriculum objectives and contents of CSL teaching, aiming to provide reference frames and reference standards for Chinese teaching institutions and teachers in the formulation of teaching plans, assessment of learners' abilities and compilation of teaching materials. HSK Test Syllabus is a Chinese proficiency test syllabus formed by the experts in the fields of Chinese language teaching, linguistics, psychology and educational measurement, who fully investigated and understood the actual situation of overseas Chinese language teaching, in order to adapt to the new situation of the international Chinese promotion. Both of them are the official standards with a guiding status in the field of CSL teaching.

Dynamic features are shown in Table 1. A total of 22 indicators are extracted. According to the hierarchical Chinese character list, Chinese characters are classified into 1–6 levels. The proportions of 1–6 level Chinese characters and super-outline characters are counted. According to the hierarchical vocabulary, words are classified into 1–6 levels. The proportions of words at 1–6 levels and super-outline words are counted.

Table 1. Dynamic features

Unit	Dynamic features	Total
Character	Proportions of Chinese characters at each level (7) Average number of Chinese character level (1) Standard deviation of Chinese character level (1)	9
Word	Proportions of words at each level (7) Average number of word level (1) Standard deviation of word level (1)	9
Sentence	Average number of Chinese character levels in big sentences (1) Standard deviation of Chinese character levels in big sentences (1) Average number of word levels in big sentences (1) Standard deviation of word levels in big sentences (1)	4

Static Features

Static features of Chinese texts are classified from three perspectives: surface features, structural features and functional features, as shown in Fig. 1.

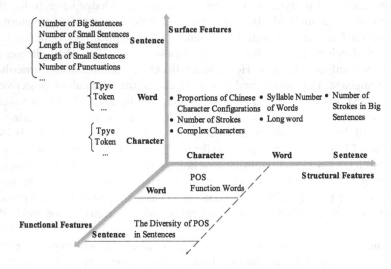

Fig. 1. Three-dimensional diagram of static features

Surface features refer to direct statistics of various language units, or simple transformations of the statistical results. A total of 20 indicators are extracted for surface features. Some of these indicators are basic indicators, and some are composite indicators. Composite indicators are calculated based on basic indicators, such as "Average Frequency of Chinese Characters in the Document = Number of

Characters/Number of Character Categories", which reflects the diversity of Chinese characters in the document.

In terms of structural features, there are great differences between Chinese and alphabetic languages. In writing form, Chinese characters are planar blocks, and their components are combined to form a certain configuration. Chinese characters with different structures and strokes have different effects on cognitive processing. This paper calculates the proportions of Chinese character configurations, the number of character strokes and the number of complex characters (Chinese characters with more than 14 strokes). There are 14 Chinese character configurations: left-right, up-down, left-middle-right, up-middle-down, encirclement, left-upper encirclement, left-lower encirclement, right-upper encirclement, lower-open-frame encirclement, right-open-frame encirclement, upper-open-frame encirclement, variant, overlap, and singleton. Chinese takes monosyllabic words as its core and disyllabic words as its basis. The syllable number of the word is one of the factors of word structure, which reflects the difficulty of the word to a certain extent. The more syllables of a word has, the more difficult it is for CSL learners to memorize and recognize. The structural features in this paper calculate the syllable numbers of words and the number of long words (words with more than 3 syllables).

Different linguistic units have different functions in the text. The word is the smallest meaningful unit in a sentence that can be used independently. Therefore, when measuring the functional features of texts, only word and sentence indicators are considered. Part of speech (POS) is an abstract category of words that has the same syntactic function and can appear in the same position, which will directly affect the complexity of sentences. For example, adjectives and adverbs generally act as modifiers. If there are too many modifiers, sentences may be more complex. The use of function words often involves the syntactic structure and affects the meaning of the whole sentence. The correct use of function words is also one of the difficulties for CSL learners. The functional features mainly calculate the proportions of POS, the number of function words and the diversity of POS in sentences. POS contains 14 categories, including nouns, time words, location words, numerals, quantifiers, pronouns, verbs, adjectives, adverbs, prepositions, conjunctions, auxiliary words, interjections and onomatopoeias. Function words include conjunctions, prepositions and auxiliary words.

This paper also considers the practicability when choosing various features. Most of the indicators are easy to obtain and can reflect the readability of the text. This paper does not use the current analysis tools which can't reach the actual application effect. For example, the parsing tool is not used to analyze sentence structure, which avoids secondary errors brought by the parsing tool and facilitates the later practical application. Tables 2, 3 and 4 below illustrate the detailed features.

Table 2. Surface features

Unit	Surface features	Total
Character	Number of characters (1) Number of character categories (1) Average frequency of Chinese characters in the document (1) The proportion of Chinese characters that appear only once in the document (1)	4
Word	Number of words (1) Number of word categories (1) Average frequency of words in the document (1) The proportion of words that appear only once in the document (1) Number of idioms (1)	5
Sentence	Average length of small sentences (by word) (1) Average length of big sentences (by word) (1) Standard deviation of the small sentence length (by word) (1) Standard deviation of the big sentence length (by word) (1) Average length of small sentences (by character) (1) Average length of big sentences (by character) (1) Standard deviation of the small sentence length (by character) (1) Standard deviation of the big sentence length (by character) (1) Number of small sentences (1) Number of big sentences (1) Number of punctuations (1)	11

Table 3. Structural features

Unit	Structural features	Total
Character	Proportions of Chinese character configurations (14) Average stroke of Chinese characters (1) Standard deviation of Chinese character strokes (1) Number of complex characters (1) The proportion of complex characters (1)	18
Word	Proportions of words with different syllables (10) Average syllable of words (1) Standard deviation of word syllables (1) Number of long words (1) The proportion of long words (1)	14
Sentence	Average number of strokes in big sentences (1) Standard deviation of strokes in big sentences (1)	2

Table 4. Functional features

Unit	Functional features	Total
Word	Proportions of POS (14) Number of function words (1) The proportion of function words (1)	16
Sentence	Average number of POS in big sentences (1) Standard deviation of POS in big sentences (1)	2

3.2 Assessment Methods

In this paper, text readability assessment is regarded as a classification problem. Based on the above text readability features, a variety of classification algorithms are used to assess text readability. The classification algorithms used include Logical Regression, K Nearest Neighbor, Naive Bayesian Classifier, CART Decision Tree and Support Vector Machine.

In addition, integrated algorithms including Bagging and Boosting are introduced to improve the accuracy of classification by aggregating the prediction results of multiple classifiers. Bagging constructs classifiers with random and replayed training data, and then combines them. Boosting is a method to improve the accuracy of weak classification algorithm. This method constructs a series of predictive functions and then combines them into a predictive function in a certain way. In this paper, the algorithms of Random Forest, Extreme Random Tree, AdaBoost and Gradient Boosting Machine are used to further improve the accuracy of the text readability assessment.

4 Experiment and Evaluation

4.1 Corpus

This paper uses eight sets of planning textbooks as experimental corpus, which consist of 36 books and 807 documents. These eight sets of textbooks cover a wide range of learners at different levels. The textbooks include not only elementary textbooks for beginners of Chinese, but also intermediate and advanced textbooks for higher-level learners.

At present, the HSK syllabus divides the learners' level into 6 levels, and the corresponding vocabulary & language point syllabus divides the difficulty into 6 levels. However, these experimental textbooks have different standards of difficulty classification. Some textbooks cover only some of the levels. Therefore, before the experiment of readability assessment, it is necessary to re-label and adjust the difficulty level of the experimental corpus, and divide the difficulty of the corpus texts from easy to difficult into 1 to 6 levels. Three front-line teachers in the field of CSL teaching were invited to independently annotate the 807 documents. For consistent annotations, their results are taken directly. In view of the inconsistent annotations, three annotators reached an agreement on the levels of the documents upon discussion. The annotation results are shown in Table 5.

Table 5. Annotation results of the experimental corpus

Number of books	Number of documents	Level 1	Level 2	Level 3	Level 4	Level 5	Level 6	Standard deviation
36	807	161	173	308	106	45	14	105.5

4.2 Experiments and Analysis

In the following experiments, for each classification algorithm, 20% of the above experimental corpus is randomly selected as test data in each experiment, and the rest as training data. A total of 100 experiments are carried out for each classification algorithm, and the results of 100 experiments are averaged for the final results. The experimental results are measured by three indicators.

- Accuracy: the proportion of documents predicted to be correct.
- Proximity Accuracy: the proportion of documents with prediction error less than 1 level.
- Pearson Correlation Coefficient: the index to measure the correlation between the predicted level and the actual level of the document.

Experiment 1 Comparison of Single Classification Algorithms
This experiment compares five classification algorithms: Logical Regression (LR), K Nearest Neighbor (KNN), Naive Bayesian Classifier (NB), CART Decision Tree (CART) and Support Vector Machine (SVM). The experimental results are shown in Table 6. It can be seen that the SVM algorithm is superior to other algorithms in each indicator, the LR algorithm takes second place, and the other three classification algorithms have similar results.

Table 6. Results of single classification algorithms

Classifier	Accuracy	Proximity accuracy	Pearson correlation coefficient
LR	0.58	0.95	0.78
KNN	0.57	0.95	0.77
NB	0.57	0.95	0.77
CART	0.57	0.95	0.77
SVM	0.63	0.96	0.78

In order to check the stability of the classifier performance, the accuracy of each classifier is displayed in the box-plot. It can be seen that the median of SVM is larger than that of other classifiers, and the performance of SVM is more stable than that of other classifiers, as shown in Fig. 2.

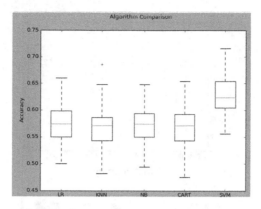

Fig. 2. The accuracy box-plot of single classification algorithms

Experiment 2 Comparison of Integrated Classification Algorithms

In order to further improve the classification effect, we use integrated algorithms including Bagging and Boosting to synthesize multiple weak classifiers to construct a strong classifier in this experiment. Four integrated algorithms are compared: AdaBoost (AB), Gradient Boosting Machine (GBM), Random Forest (RF) and Extreme Random Tree (ET). As shown in Table 7, the GBM is the best classification algorithm. The accuracy of GBM is 67%, the proximity accuracy is 97%, and the Pearson correlation coefficient is 84%. Compared with single classification algorithms, GBM shows better classification effect.

Table 7. Results of integrated classification algorithms

Classifier	Accuracy	Proximity accuracy	Pearson correlation coefficient
AB	0.54	0.94	0.78
GBM	0.67	0.97	0.84
RF	0.64	0.97	0.83
ET	0.64	0.97	0.83

From the box-plot (shown in Fig. 3), the median of the GBM classifier is larger than that of other classifiers, and the performance of the GBM classifier is more stable than that of other classifiers.

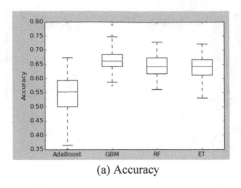

(a) Accuracy

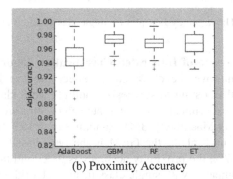

(b) Proximity Accuracy

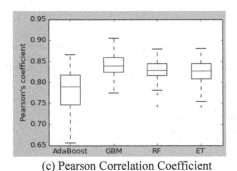

(c) Pearson Correlation Coefficient

Fig. 3. The box-plots of integrated classification algorithms

Experiment 3 Experiment after Equalizing Data

In the experimental corpus, the number of documents in level 5 and 6 is small, while the number of documents in level 3 is large. In order to solve the problem of data imbalance, documents of level 5 and level 6 are removed, leaving only the documents of level 1 to 4. Besides, 160 documents are randomly selected from all of level 3 documents. In the end, the average number of samples at each level is 150, and the standard deviation of samples is reduced from 105.5 to 29.9, as shown in Table 8.

Table 8. The experimental corpus after equalizing data

Level 1	Level 2	Level 3	Level 4	Average	Standard deviation
161	173	160	106	150	29.9

Based on the equalized data, the GBM classifier in experiment 2 is adopted in this experiment. The experimental results are shown in Table 9. The Classifier divides documents into four levels with an accuracy of 68%, which is 1% higher than that before the data is equalized. Therefore, the features designed for Chinese readability assessment in our work have been proved effective.

Table 9. Results of GBM after equalizing data

Classifier	Accuracy	Proximity accuracy	Pearson correlation coefficient
GBM	0.68	0.98	0.83

Experiment 4 Comparisons of Various Features

This experiment mainly compares the effects of different types of features on classification results, and adopts the GBM classifier which has the best effect in experiment 2. The experimental results are shown in Table 10. If only one kind of feature is used, the surface feature has the highest accuracy, which is consistent with previous studies. The accuracy of using the dynamic feature alone is relatively low, which is different from the expected result. The reason may be that the hierarchical vocabulary of HSK Test Syllabus contains only 5000 words. The 5000 words can only cover part of the vocabularies in the corpus, and the vocabularies outside the syllabus can only be regarded as super-syllabus words, which greatly affects the readability assessment.

Table 10. Results of different types of features

Features	Accuracy	Proximity accuracy	Pearson correlation coefficient
The surface feature	0.65	0.96	0.81
The dynamic feature	0.55	0.94	0.75
The functional feature	0.62	0.95	0.79
The structural feature	0.55	0.93	0.75
The surface feature + the dynamic feature + the functional feature	0.66	0.97	0.84
The surface feature + the dynamic feature + the structural feature	0.66	0.97	0.84
The surface feature + the functional feature + the structural feature	0.66	0.96	0.83
The dynamic feature + the functional feature + the structural feature	0.63	0.97	0.82
All	0.67	0.97	0.84

The accuracies of the functional feature and the structural feature reach 62% and 55% respectively, indicating that these features are still effective. However, the best classification result can be obtained only when the four types of features are combined. When all the features are adopted, the accuracy of 67% and the proximity accuracy of 97% are achieved.

5 Conclusion

This paper proposes multi-type and multi-granularity text readability features which combine dynamic and static features for CSL teaching. Dynamic features mainly reflect the domain particularity of texts; and static features reflect the stability characteristics of texts. Static features can be divided into surface features, structural features and functional features. Both dynamic and static features are measured in multi-granularity at the level of character, word and sentence. Four groups of experiments show that SVM is the best when a single classifier is used; the GBM classifier is the most effective in integrated classification algorithms; the classification effect is better when data is equalized; and the classification result based on all the features is better than those based on a single group or some of the features.

The next steps are to further analyze the current text readability features, analyze the importance of each feature, and try to filter the features. In addition, some new text readability features will be explored, such as semantic features. Based on these features, we will further improve the text readability evaluation model.

Acknowledgements. This study was supported by National Natural Science Foundation of China (61877004) and Natural Science Foundation for the Higher Education Institutions of Anhui Province of China (KJ2019A0592).

References

1. Krashen, S.: Second Language Acquisition and Second Language Learning. Prentice-Hall International, Upper Saddle River (1988)
2. Kincaid, J.P., Fishburne, J.R.P., Rogers, R.L., Chissom, B.S.: Derivation of new readability formulas for navy enlisted personnel. Adult Basic Educ. 1(1), 8–75 (1975)
3. Mclaughlin, G.H.: Smog grading-a new readability formula. J. Reading 12(8), 639–646 (1969)
4. Heilman, M., Collinsthompson, K., Eskenazi, M.: An analysis of statistical models and features for reading difficulty prediction. In: Proceedings of the Third ACL Workshop on Innovative Use of NLP for Building Educational Applications, Columbus, pp. 71–79. Association for Computational Linguistics (2008)
5. Schwarm, S.E., Ostendorf, M.: Reading level assessment using support vector machines and statistical language models. In: Proceedings of the 43rd Annual Meeting of the ACL, Ann Arbor, pp. 523–530. Association for Computational Linguistics (2005)
6. Kate, R.J., et al.: Learning to predict readability using diverse linguistic features. In: Proceedings of the 23rd International Conference on Computational Linguistics, Beijing, pp. 546–554. Association for Computational Linguistics (2010)

7. Vajjala, S., Meurers, D.: On improving the accuracy of readability classification using insights from second language acquisition. In: NAACL HLT 2012 Proceedings of the Seventh Workshop on Building Educational Applications Using NLP, Montréal, pp. 163–173. Association for Computational Linguistics (2012)
8. Feng, L., Jansche, M., Huenerfauth, M., Elhadad, N.: A comparison of features for automatic readability assessment. In: COLING 2010 Proceedings of the 23rd International Conference on Computational Linguistics, Beijing, pp. 276–284. Association for Computational Linguistics (2010)
9. Wanghao, G.: Research on the Readability Formula of Chinese Texts as a Foreign Language. Shanghai Jiao Tong University, Shanghai (2010)
10. Wanghao, G., Feihong, S.: The influencing factors of Chinese text complexity and the determining of its weight by subjective observation. Mod. Chin. (Lang. Res. Ed.) **481**(11), 104–106 (2012)
11. Hong, Z., Yong, Z.: Research on the Chinese text readability formula of intermediate European and American students. Chin. Teach. World **23**(2), 263–276 (2014)

Statistical Analysis and Automatic Recognition of Grammatical Errors in Teaching Chinese as a Second Language

Yingjie Han, Mengjie Zhong, Lijuan Zhou, and Hongying Zan[(✉)]

School of Information Engineering, Zhengzhou University, Zhengzhou 450001,
Henan, China
{ieyjhan,ieljzhou,iehyzan}@zzu.edu.cn,
1837361628@qq.com

Abstract. Foreigners make various grammatical errors when learning Chinese due to the negative transfer of their mother tongue, learning strategies, etc. At present, the research on grammatical errors mainly focuses on a certain word or a certain kind of errors, resulting in a lack of comprehensive understanding. In this paper, a statistical analysis on large-scale data sets of grammatical errors made by second language learners is conducted, including words with grammatical errors and their quantities. The statistical analysis gives people a more comprehensive understanding of grammatical errors and have certain guiding significance for teaching Chinese as a second language (TCSL). Because of the large proportion of grammatical errors of "的[de](of)", the usages of "的[de] (of)" are integrated into automatic recognition of Chinese grammatical errors. Experimental results show that the performance is overall improved.

Keywords: Automatic recognition · Function words · Grammatical errors · TCSL · Usage

1 Introduction

With the improvement of China's international prestige and overall national strength, there are an increasing number of foreigners who are learning Chinese. Chinese is a very complex language, lacking changes in morphological and singular/plural. Second language learners tend to make various grammatical errors in learning Chinese.

Chinese grammatical errors diagnosis (CGED) is one of the important tasks in natural language processing. Research on automatic recognition of English grammatical errors began earlier, and scholars have proposed automatic recognition methods, including rule-based methods [1, 2], statistics-based methods [3–5], and neural network-based methods [6, 7]. Compared with automatic recognition of English grammatical errors, research on CGED started late. Since 2014, the CGED shared tasks [8] organized by natural language processing techniques for educational applications (NLPTEA) have greatly boosted the research on automatic recognition of Chinese grammatical errors. Researchers have proposed a variety of methods for diagnosing Chinese grammatical errors, such as rule-based methods [9, 10], statistics-based

© Springer Nature Switzerland AG 2020
J.-F. Hong et al. (Eds.): CLSW 2019, LNAI 11831, pp. 406–414, 2020.
https://doi.org/10.1007/978-3-030-38189-9_42

methods [11, 12], and deep learning-based methods [13, 14]. However, they don't work well in diagnosing Chinese grammatical errors, their types and positions.

A statistical analysis is conducted on Lang-8 data set and HSK data set, which contain grammatical errors made by second language learners. The statistics is conducted on content words and function words with grammatical errors. According to the statistical results, the function word "的[de](of)" is chosen and its usage is integrated as a feature into our automatic recognition model, bidirectional long short-term memory (BiLSTM)-conditional random field (CRF). The experimental results show that false positive rate (FPR), accuracy and F1 value are improved.

2 Statistical Analysis

2.1 The Statistical Analysis of Chinese Grammatical Errors in Lang-8 Data Set

Lang-8 [15] is a website that allows people to learn multiple languages for free. The lang-8 dataset we studied is provided by the NLPCC2018 shared task [16], which contains 717,241 records. Each record contains a sentence with grammatical errors with several candidate corrections. Samples are shown in Table 1.

Table 1. Sample of Lang-8 data set

No.	Original sentence	Candidate correction
(1)	在垃圾站我碰见的邻居的大妈。	在垃圾站，我碰见了邻居家的大妈。 在垃圾站前，我碰见了邻居家的大妈。 在垃圾站我碰见了邻居大妈。
(2)	其实今天我去大阪看演唱会。	其实今天我要去大阪看演唱会。 其实今天我去了大阪看演唱会。
(3)	她一定就很高兴了！	她一定会很高兴的！
(4)	在三月我可能去东京。	三月，我可能去东京。
(5)	我饿了不得了。	我饿得不得了。
(6)	他从来不会抱怨什么。	
(7)	然后我们一起观看了期待亿久的舞龙舞狮表演。	今天中国是猴年，今天是大年初一。

The statistical analysis is performed with the following steps.

Step 1: The data set is preprocessed by deleting the sentences without grammatical errors (sentence (6) as shown in Table 1) and those sentence pairs of which the originals and the candidates don't match (sentence (7) as shown in Table 1). In addition, only the sentences with the least modification are retained to avoid counting grammatical errors repeatedly.

Step 2: The originals and the candidate corrections are compared to identify words with grammatical errors and their frequencies are counted via a self-designed program.

The total number of Chinese grammatical errors in Lang-8 data set is 1,333,489, among which the number of Chinese grammatical errors caused by improper use of function words is 428,602, accounting for 32.14%, and the number of Chinese grammatical errors caused by improper use of content words is 904,887, accounting for 67.86%. Although the number of content words with grammatical errors is much larger than that of function words, the number of content words is a lot larger than that of function words. Therefore, the grammatical errors caused by the improper use of function words make up a high proportion despite its small quantity. Moreover, Lv pointed out that the function of a content word is limited to itself, whereas the function of a function word is beyond itself [17].

Table 2 shows the top ten words with grammatical errors in adverb, preposition, conjunction, and auxiliary words. We do not list localizer and modal particle due to limited space of this paper. As can be seen, the number of grammatical errors caused by improper use of auxiliary word "的[de](of)" is 94,189, accounting for 7.1% of the total.

Table 2. The top ten function words with grammatical errors

Adverb		Preposition		Conjunction		Auxiliary word	
Word	Number	Word	Number	Word	Number	Word	Number
也[ye](also)	11,046	在[zai](at)	24,138	和[he](and)	10,168	的[de](of)	94,189
就[jiu](as far as)	7,669	对[dui](to)	5,863	所以[suoyi](so)	5,043	了[le]( )	43,231
不[bu](not)	7,154	把[ba]( )	4,137	因为[yinwei](because)	4,274	得[de]( )	5,421
都[dou](all)	7,150	从[cong](from)	4,106	可以[keyi]( )	4,164	着[zhe]( )	3,993
很[hen](too)	6,207	给[gei](to)	3,991	但是[bu](but)	3,726	过[guo]( )	3,248
还[hai](also)	3,710	跟[gen](with)	3,551	而[er](where)	2,815	地[de]( )	3,096
又[you](again)	3,448	被[bei](by)	3,174	可是[keshi](however)	2,626	的话[dehua]( )	2,163
就是[jiushi](exactly)	2,661	用[yong]( )	3,110	而且[erqie](but also)	2,385	等[deng](etc)	964
太[tai](extremely)	2,113	为了[weile](for)	2,172	然后[ranhou](then)	2,176	来说[laishuo]( )	670
再[zai](again)	2,103	为[wei](by)	2,122	还是[haishi](nevertheless)	2,087	之[zhi]( )	395

2.2 The Statistical Analysis of Chinese Grammatical Errors in HSK Data Set

The HSK dynamic composition corpus is a corpus for non-native speakers of Chinese to participate in HSK [18]. The HSK data set we studied is provided by CGED shared tasks [19–22], annotating the types and positions of Chinese grammatical errors. It contains 22,658 records. The statistical analysis is performed with the following steps.

Step 1: The data set is preprocessed by deleting the sentences without grammatical errors and the sentences with incorrect tags.
Step 2: The number of words with grammatical errors is counted and their types are counted via a self-designed program.

There are totally 54,352 Chinese grammatical errors, among which the number of function words with grammatical errors is 20,312, accounting for 37.4%, and the number of content words with grammatical errors reach 34,040 with the proportion of

62.6%. Table 3 shows the top ten function words with grammatical errors. Obviously, the grammatical errors caused by improper use of "的[*de*](*of*)" is 5,139, which account for 9.5% of the total.

Table 3. The top ten function words with grammatical errors

Adverb		Preposition		Conjunction		Auxiliary word	
Word	Number	Word	Number	Word	Number	Word	Number
也[*ye*](*also*)	588	在[*zai*](*at*)	1,008	而[*er*](*where*)	331	的[*de*](*of*)	5,139
就[*jiu*](*as far as*)	523	对[*dui*](*to*)	861	和[*he*](*and*)	242	了[*le*]( )	1,844
都[*dou*](*all*)	514	给[*gei*](to)	233	可以[*keyi*]( )	170	得[*de*]( )	251
不[*bu*](*not*)	236	从[*cong*](*from*)	219	还是 [*haishi*](*nevertheless*)	128	地[*de*]( )	182
就是 [*jiushi*](*exactly*)	144	跟[*gen*](*with*)	143	不是[*bushi*]( )	90	着[*zhe*]( )	156
很[*hen*](*too*)	130	为[*wei*](*by*)	136	因为[*yinei*](*because*)	74	过[*guo*]( )	156
还[*hai*](*also*)	128	用[*yong*]( )	115	只有[*zhiyou*](*only*)	51	来说 [*laishuo*]( )	118
更[*geng*](*even*)	87	以[*yi*]( )	113	或[*huo*](*or*)	49	之[*zhi*]( )	74
才[*cai*](*only*)	80	为了[*weile*](*for*)	111	如果[*ruguo*](*if*)	49	的话[*dehua*]( )	73
又[*you*](*again*)	79	把[*ba*]( )	96	并[*bing*](*and*)	43	等[*deng*](*etc*)	44

It can be seen from the statistical results of the two data sets that the proportion of different categories of grammatical errors caused by improper use of function words is similar despite different sources of the data sets, so is the distribution of high-frequency function words. Furthermore, the number of grammatical errors of "的[*de*](*of*)" is the largest in the above two data sets. Therefore, we incorporate the usage of "的[*de*](*of*)" in automatic recognition of Chinese grammatical errors to improve the performance of automatic recognition of grammatical errors.

3   Automatic Grammatical Errors Recognition Incorporated with the Usage of "的[*de*](*of*)"

3.1   The Basic Model

The BiLSTM-CRF [23] model is used for Chinese grammatical errors recognition. First, the sentences with grammatical errors are converted into vector sequences. Then, the vector sequences are encoded by BiLSTM. Finally, the encoded sequences are input into CRF and the corresponding tag sequences are obtained.

3.2   Integration of the Usage of "的[*de*](*of*)"

The usage of "的[*de*](*of*)" is described in Chinese function word usage knowledge base (CFKB) [24]. There are 39 usages of "的[*de*](*of*)" in CFKB, which are known as fine-grained usages. In the literature [25], there are 8 merged usages of "的[*de*](*of*)", which are known as coarse-grained usages. The automatic annotation tool [26] is applied to annotate the usage of "的[*de*](*of*)" in the data sets.

The features of character, word, part-of-speech (POS), and usage are added to BiLSTM-CRF model. For "的[de](of)", its usage ID is embedded as a feature, as is shown in Fig. 1. For the rest, "<b>" is used as its usage to maintain the consistency of the dimension of the vectors, as shown in Fig. 2.

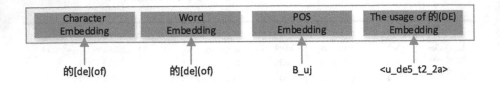

Fig. 1. The feature of "的[de](of)"

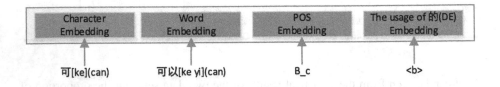

Fig. 2. The feature of other words

3.3 The Performance Metrics

Evaluating metrics of CGED include false positive rate (FPR), accuracy, precision, recall and F1 Value, which are shown in formulas (1) to (5), where *TN* (True Negative) is referred to as the number of sentences without grammatical errors and the system predicts that there are no errors in the sentences, *FN* (False Negative) is referred to as the number of sentences with grammatical errors while the system predicts that there are no errors in the sentences, *TP* (True Positive) is referred to as the number of sentences with grammatical errors and the errors are correctly recognized, and *FP* (False Positive) is referred to as the number of sentences without grammatical errors while the system predicts that there are errors in the sentences.

$$\text{False Positive Rate} = \text{FP} / (\text{FP} + \text{TN}) \tag{1}$$

$$\text{Accuracy} = (\text{TP} + \text{TN}) / (\text{TP} + \text{FP} + \text{TN} + \text{FN}) \tag{2}$$

$$\text{Precision} = \text{TP} / (\text{TP} + \text{FP}) \tag{3}$$

$$\text{Recall} = \text{TP} / (\text{TP} + \text{FN}) \tag{4}$$

$$\text{F1} = 2 \times \text{Precision} \times \text{Recall} / (\text{Precision} + \text{Recall}) \tag{5}$$

3.4 Experiment Setting and Results

The data set of CGED2016 is chosen as the training set, and the test set of CGED2018 is chosen as the test set, called data set 1. At the same time, the sentences with grammatical errors caused by improper use of "的[*de*](*of*)" are extracted from data set 1 and constitute a new data set for comparison experiments, called data set 2.

There are three experimental methods, namely *Baseline*, *Baseline+FG* and *Baseline +CG*. The *Baseline* method integrates the feature of word and POS into BiLSTM-CRF. The *Baseline+FG* method incorporates the fine-grained usages of "的[*de*](*of*)" into the *Baseline* method. The *Baseline+CG* method incorporates the coarse-grained usages of "的[*de*](*of*)" into the *Baseline* method. There are three sub-tasks for CGED share task, including: (1) detection-level-based sub-task for checking whether a sentence contains grammatical errors, (2) identification-level-based sub-task for identifying the type of the grammatical errors, and (3) position-level-based sub-task for determining the position of the grammatical errors.

The experimental results are shown in Table 4. It is apparent that *Baseline+FG* and *Baseline+CG* outperform *Baseline* in detection-level and identification-level sub-tasks. Therefore, the method that incorporates the usage of "的[*de*](*of*)" further improves the accuracy, precision and F1 value, and reduces FPR. Moreover, *Baseline+CG* achieves better performance than *Baseline+FG* because *Baseline+CG* uses fewer types of usage, which makes more training samples available for the corresponding usage and is conducive to training and testing. In contrast, *Baseline+FG* uses more usages, resulting in too few or no samples for a certain usage in the training set. Meanwhile, the result on data set 2 is better than that on data set 1, because the training data set 2 contains only the sentences with grammatical errors caused by improper use of "的[*de*](*of*)". However, *Baseline+FG* and *Baseline+CG* do not work well in position-level. This may be due to the fact that the usage of "的[*de*](*of*)" does not provide location information in determining the location of grammatical errors.

Table 4. The experimental results

Data set		Data set 1			Data set 2		
Method		Baseline	Baseline+FG	Baseline+CG	Baseline	Baseline+FG	Baseline+CG
FPR		0.6645	**0.6058**	0.6735	0.6899	**0.6786**	0.7232
Detection-level	Accuracy	0.6007	**0.6202**	0.61	0.6256	0.6263	**0.6289**
	Precision	0.6071	**0.6256**	0.6108	0.6361	**0.6381**	0.633
	Recall	0.81	0.7984	**0.8337**	0.8472	0.8404	**0.8763**
	F1	0.694	0.7015	**0.705**	0.7267	0.7254	**0.7351**
Identification-level	Accuracy	0.4189	0.4341	**0.4451**	0.4282	**0.4332**	0.403
	Precision	0.3846	0.3903	**0.406**	0.4028	**0.4072**	0.376
	Recall	0.493	0.4673	**0.4993**	**0.5397**	0.5305	0.5302
	F1	0.4321	0.4253	**0.4478**	**0.4613**	0.4608	0.44
Position-level	Accuracy	0.1581	0.1682	**0.1716**	**0.1379**	0.1337	0.1275
	Precision	**0.0928**	0.0852	0.0857	**0.088**	0.0808	0.0829
	Recall	**0.1082**	0.0912	0.0986	**0.1102**	0.0999	0.1074
	F1	**0.0999**	0.0881	0.0917	**0.0978**	0.0894	0.0936

There are two causes for the slight improvement in the performance of CGED incorporated the usage of "的[de](of)". One cause is that errors caused by word segmentation and POS will propagate to the recognition, and the other one is that the accuracy of automatic recognition of the usage of "的[de](of)" is less than 30% [27], which affects the feature extraction, model training and recognition results.

4 Conclusion and Future Work

We conduct statistical analysis of the words with grammatical errors and their quantity for two large-scale data sets. The statistical results give people a more comprehensive understanding of grammatical errors made by second language learners. It can also guide users to teach Chinese as a second language. At the same time, the statistical results show that the number of grammatical errors caused by improper use of "的[de] (of)" is the main cause for the errors. The usages of "的[de](of)" is beneficial for improving the overall performance of grammatical errors recognition. Therefore, we incorporate the usage of "的[de](of)" into the automatic recognition and conduct experiments. The experimental results show that the performance is improved, up to 3%.

The future work includes improving accuracy of automatic annotation and studying the merging strategy of "的[de](of)" to improve the performance. In terms of improving the performance of the position-level of CGED, boundary identification results of phrases involving "的[de](of)" [28] will be used.

Acknowledgments. This research is supported by the National Social Science Fund of China (No. 18ZDA315), the Key Scientific Research Program of Higher Education of Henan (No. 20A520038), the Science and Technology Project of Science and Technology Department of Henan Province (No. 192102210260), the China National Social Science Foundation Mega-Project (17ZDA318), the Henan Provincial Project of Soft Science (No. 182400410454), and the Henan Project of Philosophy and Social Science (No. 19BYY016).

References

1. Heidorn, G.E., Jensen, K., Miller, L.A., et al.: The EPISTLE text-critiquing system. IBM Syst. J. **21**(3), 305–326 (1982)
2. Bustamante, F.R., León, F.S.: GramCheck: a grammar and style checker. In: 16th Conference on Computational Linguistics, pp. 175–181. ACL, Stroudsburg (1996)
3. Han, N.R., Chodorow, M., Leacock, C.: Detecting errors in English article usage by non-native speakers. Nat. Lang. Eng. **12**(2), 115–129 (2006)
4. Rozovskaya, A., Roth, D.: Algorithm selection and model adaptation for ESL correction tasks. In: 49th Annual Meeting of the Association for Computational Linguistics, pp. 924–933. ACL, Stroudsburg (2011)
5. Lafferty, J., McCallum, A., Pereira, F.: Conditional random fields: probabilistic models for segmenting and labeling sequence data. In: 18th International Conference on Machine Learning 2001, pp. 282–289. ACM, New York (2001)

6. Sun, C., Jin, X., Lin, L., Zhao, Y., Wang, X.: Convolutional neural networks for correcting English article errors. In: Li, J., Ji, H., Zhao, D., Feng, Y. (eds.) NLPCC 2015. LNCS (LNAI), vol. 9362, pp. 102–110. Springer, Cham (2015). https://doi.org/10.1007/978-3-319-25207-0_9

7. Yuan, Z., Briscoe, T.: Grammatical error correction using neural machine translation. In: 2016 Conference of the North American Chapter of the Association for Computational Linguistics: Human Language Technologies, pp. 380–386. ACL, Stroudsburg (2016)

8. Yu, L.C., Lee, L.H., Chang, L.P.: Overview of grammatical error diagnosis for learning Chinese as foreign language. In: 1st Workshop on Natural Language Processing Techniques for Educational Applications, pp. 42–47. Asia-Pacific Society for Computers in Education, Japan (2014)

9. Han, Y., Lin, A., Wu, Y., Zan, H.: Usage-based automatic recognition of grammar errors of conjunctions in teaching Chinese as a second language. In: Liu, P., Su, Q. (eds.) CLSW 2013. LNCS (LNAI), vol. 8229, pp. 519–528. Springer, Heidelberg (2013). https://doi.org/10.1007/978-3-642-45185-0_54

10. He, J.: Automatic identification and application research of conjunction errors for HSK library. Master thesis, Zhengzhou University, Zhengzhou (2014). (in Chinese)

11. Yeh, J.F., Yeh, C.K., Yu, K.H., et al.: Condition random fields-based grammatical error detection for Chinese as second language. In: 2nd Workshop on Natural Language Processing Techniques for Educational Applications, pp. 105–110. ACL & AFNLP (2015)

12. Zhang, Y.S.: Statistical Language Modeling and Chinese Text Automatic Proofreading Technology. Science Press, Beijing (2017). (in Chinese)

13. Zheng, B., Che, W.X., Guo, J., et al.: Chinese grammatical error diagnosis with long short-term memory networks. In: 3rd Workshop on Natural Language Processing Techniques for Educational Applications, pp. 49–56. COLING 2016 Organizing Committee (2016)

14. Li, C.L., Qi, J.: Chinese grammatical error diagnosis based on policy gradient LSTM model. In: 5th Workshop on Natural Language Processing Techniques for Educational Applications, pp. 77–82. ACL, Stroudsburg (2018)

15. Lang-8 homepage. http://www.lang-8.com/. Accessed 10 Apr 2019

16. Zhao, Y., Jiang, N., Sun, W., Wan, X.: Overview of the NLPCC 2018 shared task: grammatical error correction. In: Zhang, M., Ng, V., Zhao, D., Li, S., Zan, H. (eds.) NLPCC 2018. LNCS (LNAI), vol. 11109, pp. 439–445. Springer, Cham (2018). https://doi.org/10.1007/978-3-319-99501-4_41

17. Lv, S.X., Zhu, D.X.: A Talk on Grammatical Rhetoric. Liaoning Education Press, Shenyang (2002). (in Chinese)

18. HSK dynamic composition corpus. http://hsk.blcu.edu.cn. Accessed 10 Apr 2019

19. Lee, L.H., Yu, L.C., Chang, L.P., et al.: Overview of the NLP-TEA 2015 shared task for Chinese grammatical error diagnosis. In: 2nd Workshop on Natural Language Processing Techniques for Educational Applications, pp. 1–6. ACL & AFNLP (2015)

20. Lee, L.H., Rao, G.Q., Yu, L.C., et al.: Overview of NLP-TEA 2016 shared task for Chinese grammatical error diagnosis. In: 3rd Workshop on Natural Language Processing Techniques for Educational Applications, pp. 40–48. COLING 2016 Organizing Committee (2016)

21. Rao, G.Q., Zhang, B.L., Xun, E.D., et al.: IJCNLP-2017 task 1: Chinese grammatical error diagnosis. In: 8th International Joint Conference of Nature Language Processing, pp. 1–8. Asian Federation of Natural Language Processing (2017)

22. Rao, G.Q., Gong, Q., Zhang, B.L., et al.: Overview of the NLP-TEA 2018 shared task for Chinese grammatical error diagnosis. In: 5th Workshop on Natural Language Processing Techniques for Educational Applications, pp. 42–51. ACL, Stroudsburg (2018)

23. Liu, Y.J., Zan, H.Y., Zhong M.J., et al.: Detecting simultaneously Chinese grammar errors based on a BiLSTM-CRF model. In: 5th Workshop on Natural Language Processing Techniques for Educational Applications, pp. 188–193. ACL, Stroudsburg (2018)

24. Zhang, K.L., Zan, H.Y., Chai, Y.M., et al.: A survey on the Chinese function word usage knowledge base. J. Chin. Inf. Process. **29**(3), 1–8 (2015). (in Chinese)

25. Liu, Q.H., Zhang, K.L., Xu, H.F., et al.: Research on automatic recognition of auxiliary "DE". Acta Scientiarum Naturalium Universitatis Pekinensis **54**(3), 11–19 (2018). (in Chinese)

26. Yuan, Y.C., Zan, H.Y., Zhang, K.L., et al.: The automatic annotation algorithm design and system implementation rule-base function word usage. In: 11th Chinese Lexical Semantics Workshop, pp. 163–169. Soochow University Press, Suzhou (2010). (in Chinese)

27. Han, Y.J., Zan, H.Y., Zhang, K.L., et al.: Automatic annotation of auxiliary words usage in rule-based Chinese Language. J. Comput. Appl. **31**(12), 3271–3274 (2011). (in Chinese)

28. Feng, X.B.: Studies on function words phrase boundary identification and its application in syntactic parsing. Master thesis, Zhengzhou University, Zhengzhou (2016). (in Chinese)

Tibetan Case Grammar Error Correction Method Based on Neural Networks

Cizhen Jiacuo[1,2], Secha Jia[1,2], Sangjie Duanzhu[1,2],
and Cairang Jia[1,2(✉)]

[1] Qinghai Normal University, Xining, Qinghai, China
czjcaiyaogun@hotmail.com, sechajia@126.com,
sangjeedondrub@live.com, zwxxzx@163.com
[2] Tibetan Information Processing and Machine Translation Key Laboratory
of Qinghai Province, Xining, Qinghai, China

Abstract. Grammar Error Correction (GEC) is an important researching subject among Nature Language Processing tasks. In this work, aiming at tackling with genitive and ergative grammatical errors in Tibetan formal text, we collect 1793563 consecutive sentence pairs as training set and 5000 sentence pairs with the same distribution as well as 1159 sentence pairs in different distributions as testing sets. In our approach, we firstly preprocess Tibetan text data with compositional rules and then build a neural network architecture which is a combination of BERT and Bi-LSTM, to estimate the probability of given token being genitive or ergative. In experiments, 98.38% and 86.16% in terms of accuracy are observed respectively in testing the proposed model on two different testing sets.

Keywords: Tibetan · Grammar error · Neural networks

1 Introduction

Spelling and grammar errors presented in text has a huge impact on reading comprehension for human. Similarly, Natural Language Processing (NLP) systems might also suffer from spelling error and grammatical misuse issues in text data, by affecting the performance some fundamental tasks as well as many downstream tasks, including tokenization, semantic tagging and parsing, etc. Tibetan grammar correction is a key technique to detect the existence of grammar errors in a given Tibetan text and also a crucial component in many Tibetan NLP applications and domains, such as corpus construction, text auto-correction, speech recognition and OCR applications.

In this work, our main focus lies in auto-correction technique to deal with genitive and ergative grammar errors in Tibetan text. Virtual words and verbs are at the core of Tibetan language itself, which reveals a grammatical system with dominant structural features and characteristics. In particular, the auxiliary words (རྣམ་དབྱེ་ཚིག་ཕྲད) among virtual words, has four functionalities in Tibetan sentence [1], including (1) being one of the main components to construct a sentence, (2) keeping semantic information from change, (3) providing the ability to enhancing rhetoric style and (4) sub-dividing fine sentence structures. Both genitive and ergative belong to the stipulations, in which the

J.-F. Hong et al. (Eds.): CLSW 2019, LNAI 11831, pp. 415–422, 2020.
https://doi.org/10.1007/978-3-030-38189-9_43

genitive is the sixth case among eight in Tibetan grammatical theory, and its manifestation has five variants, namely, "ཀྱི་ཀྱི་གི་ཡི་ཡི". Its matching rules depend on the post
addition of the previous word and it is mainly used to express restriction or relationship
between people or other entities. Being the third case according to Tibetan grammar
theory, ergative case is built upon the genitive by appending addiction ས to the latter
which has five variants as well, namely, "ཀྱིས་ཀྱིས་གིས་ཡིས་ཡིས", its matching rules also depend
on the post addition of the previous word, which is a grammar structure to signify
agent-recipient relationship of verbs [2]. Matching rules for both genitive and ergative
cases are as shown in Table 1.

Table 1. Matching rules for genitive and ergative cases

Post additions and Grammatical variants	Matching rules			
Post addition	ན་མ་ར་ལ	ད་བ་ས	ག་ང	འ
Genitive variants	ཀྱི	ཀྱི	གི	ཡི ཡི
Ergative variants	ཀྱིས	ཀྱིས	གིས	ཡིས ཡིས

As we can learn from Table 1, there are 10 assistive words to tackle with both
genitive and ergative cases in Tibetan text, even though they function differently (as
shown in Table 2), the matching rules for these two cases are exactly identical.

Table 2. Example for differences in grammatical variants

Tibetan sentences	English translation
བཀྲ་ཤིས་ཀྱི་ཨ་ཞང་གིས་རླངས་འཁོར་སྐྱོར།	*Tashi's uncle drives the car.*
བཀྲ་ཤིས་ཀྱིས་ཨ་ཞང་གི་རླངས་འཁོར་སྐྱོར།	*Tashi drives his uncle's car.*

Like many other alphabetic languages, in Tibetan, a sentence is formed by words
and words itself are combinations of syllables, and then syllables can be further divided
into basic Tibetan characters. Due to this hierarchical building structure, proper
grammar error correction also requires the capability to tackle with all syllable-level,
word-level and semantic-level nuances. In this work, we implement a hybrid model
architecture to pre-train a large language model based on BERT and fine-tuning this
model in downstream domain specific task, which is modeled by a Bi-LSTM network
to detect and correct grammatical errors in forms of genitive and ergative misuses.
Experiments indicate our model can outperform two carefully designed baseline
models, showing great potential in Tibetan grammar error correction applications and
research.

Table 3. Grammar error examples in Tibetan

Error types	Errors	Corrections
Genitive errors	དགེ་རྒན་གི་དཔེ་ཆ།	དགེ་རྒན་གྱི་དཔེ་ཆ། (*Teacher's book*)
Ergative errors	སློབ་གྲོགས་གིས་ལས་བྱ་བྱས།	སློབ་གྲོགས་ཀྱིས་ལས་བྱ་བྱས། (*The classmate have done homework.*)
Mutual errors	ཁྱེད་ཀྱི་དོ་སྣང་བྱ་རོགས།	ཁྱེད་ཀྱིས་དོ་སྣང་བྱ་རོགས། (*Please pay attention.*)
Absence errors	ཕ་མ་རེ་བ་ལྟར་བསྒྲུབ།	ཕ་མ་འི་རེ་བ་ལྟར་བསྒྲུབ། (*Do as parents wished*)
Multi-cases errors	ཁྱེད་ཀྱིས་ནམ་འགྲོའམ།	ཁྱེད་ནམ་འགྲོའམ། (*When will you leave?*)

2   Related Works

Due to the world-wide trend of learning main-stream languages such as Chinese or English as a second language, researching works in spell check for these languages has drawn a lot of attention. Many applications were built to help non-native second language learner. Early spell checking researches can be categorized into dictionary-based, rule-based and statistical approaches [3–5] to detect and correct specific types of errors such as the absence of words, wrong words or word order errors. In recent years, neural networks based approach has been proven to be much more effective in many NLP tasks in term of performance as well as implementation than traditional approaches. In spell checking scenario, the most commonly used network architectures includes Long Short-term Memory Networks (LSTM) [6], Convolutional Neural Networks (CNN) or stacked combinations of these two [7, 8]. The availability of public dataset for languages such as Chinese and English has eased the researching work for these languages. The most commonly known datasets include NUCLE, CLEC and CoNLL (2013, 2014) for English, CLS (Chinese as Second Language) and NLPCC2018-GEC for Chinese. There is a recent tendency that the research not only focuses on the text error in local context but also the fluency of more than two consecutive sentences, which often turn to be a difficulty even for native speakers [9].

Research on Tibetan related spell check started relatively late, and currently the research in the field mainly focuses on word-level methods, which can be categorized into two types, namely, dictionary matching and rule-based approaches [5]. It is reported that the accuracy of detecting ill orthography reach 99.99% [5]. This level of performance is attributed to the fact that the orthographic words are limited in number and Tibetan language has very strictly designed orthographic rules. The studies in terms of word-level and semantic level remain to be explored in Tibetan NLP researching community. In this work we presented a unified hybrid method which is a combination of rule and neural network approaches, to address detection and auto-correction of errors in form of misuses in the genitive and ergative cases in Tibetan text.

3 Related Works

3.1 Methods

The main intention in this work is that we design a model to predict proper grammatical case between genitive and ergative options with correct Tibetan binding rules, giving contextual information of a Tibetan sentence. The idea itself is compatible with original modeling scheme of BERT [10] (as shown in Fig. 1). As reported in many NLP benchmarks, BERT is empowered with excellent capacity to capture global contextual information. In the meanwhile, RNN variants such as LSTM, especially Bi-LSTM model, due to their recursive nature, is inherently capable of sequence modeling. Therefore, choosing pre-trained BERT for representation learning and Bi-LSTM for further capturing sequential structure is an effective composition of our hybrid model (as shown in Fig. 2).

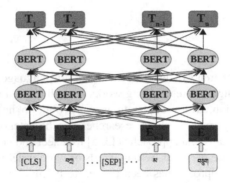

Fig. 1. Model architecture with BERT

Unlike traditional language modeling techniques, BERT proposed a new multi-task training scheme, in which rather than next token estimation, it predicts randomly masked tokens and next consecutive sentence. Transformer [11] lies at the core of BERT, and provides BERT with powerful representation learning capacity.

In our work, considering the close relation between ergative case and verbs in Tibetan, an input tag marked as [CLS] is added for denoting given sentence is presented with verbs to ease introducing post-verb context into grammar error detection and auto-correction objectives. By following BERT's original implementation, another tag marked as [SEP] is also appended to denote separation of two consecutive sentences. As shown in Figs. 1 and 2, hidden states produced by Transformer in BERT is directly fed into Bi-LSTM networks. Due to the fact that BERT is pre-trained, the general training process only needs to optimize parameters in Bi-LSTM networks, which makes our model much smaller and easier to converge. An example of sequence representation as shown in Table 3, in which input Tibetan sentence is "བཀྲ་ཤིས་ཀྱིས་ཨ་ཞང་གི་རླངས་འཁོར་བཏང་སྟེ། ཨ་ཞང་ལ་ཤེས་སུ་མ་བཅུག" (*Tashi drove his uncle's car without letting him know.*). In Fig. 3, *Input* denotes input sequence, *Token Embeddings* denotes corresponding representation for input sequence, *Segment Embeddings* denotes

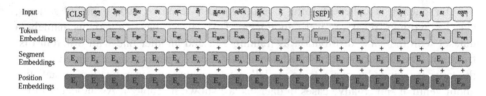

Fig. 2. An example of learning input representation

representation for consecutive sentence, and *Positional Embeddings* denotes positional representation for tokens in a sequence. The final input to the network is simply the concatenation of these aforementioned hidden state representations. Due to the fact that we are not well equipped with abundance of training data, in actual implementation, we manage to limit the model's parameter space by decreasing Transformer layer number to 2, setting model dimension to 256, position number of positional embedding to 50, multi-head number in multi-head attention layer to 4, initial learning rate to 0.2, vocabulary size to 6803 (Tibetan syllables as tokens), and Bi-LSTM layer number to 2.

3.2 Datasets

To the best of our knowledge, there is no availability of public dataset for Tibetan grammar error correction task. In our work, we use a simple yet feasible technique to collect positive and negative samples only for genitive and ergative grammatical structures. Sentence pairs presented with genitive and ergative cases are extracted from large-scaled Tibetan text corpus to serve as positive data samples and then randomly select a certain portion of sentence pairs, subsequently replace the grammatical structure with some noisy tokens or [MASK] tag to serve as negative samples. Due to grammar error correction itself requires a large amount of contextual dependencies, all sentence pairs are all consecutive sentences. What needs to be noted is that we refer consecutive sentence pairs as two sentences separated by Tibetan punctuation "ༀ" (*wylie: shad*) and these sentence pairs are by all mean not necessarily two separate sentences in a linguistic point of view. To ensure the effectiveness of sequence modeling, we delete all sentences less than 10 tokens and constrain sequence length by add "ༀ" (*shad*) after special virtual characters such as "དེ་ནི་" which usually indicate a semantic transition in Tibetan.

Table 4. Data distribution

Datasets	Training Data	Test Data1	Test Data2
Sentence	1793563	5000	1159
Genitive	1685819(93.99%)	4682(93.64%)	1059(91.37%)
Ergative	662371(36.93%)	1801(36.02%)	775(66.87%)
Genitive & Ergative	571313(31.85%)	1537(30.74%)	676(58.33%)

Our dataset is collected from a multi-domain resource, which includes news, classical writings, poetry, proverb, motto, etc. Test datasets has two variants, namely, a dataset with the same distribution as training set, and another one with different distribution. The former one is intended to evaluate model's performance on inference and the latter is intended to evaluate on model's performance on generalization. A statistics about genitive and ergative cases in datasets are shown in Table 4. As indicated in the table that there is very high frequency of occurrence for genitive grammatical structure in Tibetan text, however, co-occurrence of both genitive and ergative is relatively low which only reaches 31.85% totally.

3.3   Model Training

The training process of BERT + Bi-LSTM model follows common practice in recent large language model pre-training research [12] and contains two steps. The First one is pre-training BERT model separately on mono text data, aiming at empowering model with a good representation of language itself, which avoids to model sequence from scratch. The second one is fine-tuning the pre-training model on subsequent task-specific Bi-LSTM network. The first can be divided into 5 sub-steps, which includes:

1. Randomly sample 15% non-genitive and non- ergative syllables in training set
2. Randomly replace 80% syllables in the sampled subset with [MASK] tag
3. Randomly replace 10% syllables in the sampled subset with another syllable
4. Leave left 10% syllables as it is
5. If a sentence is presented with the ergative case, then there must be verbs in the given sentence. Therefore, sentences contain ergative case are denoted as 1 and others are denoted as 0

The step (5) corresponds to [CLS] tag in the model design, to help learn whether a sentence is presented with transitive verb which indicates there usually an ergative structure in sentence, and provides the model with valuable information to improve performance on grammar error detection and correction.

In the meanwhile, the second steps contains 3 sub-procedures, which includes:

1. Select all genitive and ergative cases in training set
2. For 30% of selected data, exchange genitive and ergative cases with each other to serve as negative samples
3. For 70% of selected data, no further processes are required, due to they serve as positive samples
4. In experiments, we design strong baselines by implementing two extra models, namely, sole Bi-LSTM and BERT + FNN (Feed Forward Network). To ensure comparability between models in term of performance, the data itself, preprocessing and training procedures are all identical

In experiments, we design strong baselines by implementing two extra models, namely, sole Bi-LSTM and BERT + FNN (Feed Forward Network). To ensure comparability between models in term of performance, the data itself, preprocessing and training procedures are all identical.

Results on our model and baseline models are as shown in Table 5. In the table, $BERT_{nopre}$ and $BERT_{pre}$ denote pre-trained and non-pertained BERT language model. Positive tests indicate all genitive and ergative cases are detected and corrected (no grammatical error at all). Negative tests indicate genitive or ergative misuse (there are grammatical errors). Both tests indicate a combination of two aforementioned test criteria. All test accuracy scores are evaluated 10 times via cross-validation.

Table 5. Model performance comparisons

Models	Test Data1			Test Data2		
	Pos	Neg	Both	Pos	Neg	Both
BiLSTM	97.99	97.50	97.93	85.37	84.36	85.18
$BERT_{nopre}$ + FNN	95.99	93.56	95.64	86.08	82.00	85.49
$BERT_{pre}$ + FNN	96.67	95.15	96.46	86.03	83.29	85.56
$BERT_{nopre}$ + BiLSTM	97.65	96.84	97.53	85.73	84.12	85.40
$BERT_{pre}$ + BiLSTM	**98.45**	**98.13**	**98.38**	**86.23**	**85.53**	**86.16**

In experiments, $BERT_{pre}$ + Bi-LSTM model outforms our baselines in all test criteria, indicating the effectiveness of our presented model. As we dive deeper into the failure instance of $BERT_{pre}$ + Bi-LSTM on test data 1, we notice that the proposed model find it difficult to deal with detection of abbreviated syllables. For example, it is able to detect sentence "དབེ་ཀློག་ལ་མོས་པ་ཡོད་" (Interested in reading) as incorrect sentence "དབེ་ཀློག་ལ་མོ་འིས་པ་ཡོད་" which makes auto-correction much difficult. As we may also learn from Table 5, the $BERT_{pre}$ performs better than $BERT_{nopre}$ indicating integration of language model pre-training techniques help model generalize better. Furthermore, even Transformer can learn sequence representation very effectively, RNN variants such Bi-LSTM can be beneficial to further improve model's capacity on capturing local positional information.

4 Conclusions

In this work, we firstly analyze extensively five dominant error types in Tibetan genitive and ergative grammatical cases. Subsequently, we collect a total number of 1793563 sentences which contain those aforementioned grammatical structures in Tibetan text corpus to serve as training set. In the meanwhile, two test sets are also collected, which includes 5000 sentences with the same distribution as training set and 1159 sentences with different distributions, respectively. Finally, we design a hybrid architecture which combines Tibetan genitive rules and BERT + Bi-LSTM, to detect and auto-correct Tibetan text. Experiment on test set with same distribution indicates that our model can tackle in-case error, and case interchange misuse. Furthermore, the experiment on test set with different distribution shows that the proposed model can generalize better than baseline models. Even though our model can process sequence to sequence task with fixed length, e.g. in-case errors, and interchange misuse with promising performance, unsatisfactory results are observed on cases in dynamic length situation, e.g. multiple case error and case absence error. The possible reason for this issue is that the training dataset itself is based on certain kind of grammatical structure. In the future, we are planning to further explore a wider range of collection of grammatical structures to deepen the research on Tibetan grammatical error correction.

Acknowledgement.   National Key R&D Program of China (2017YFB1402200), The National Natural Science Foundation of China (61063033, 61662061), The National Social Science Fund of China (14BYY132).

References

1. Ji, T.: Tibetan Syntactic Research. China Tibetology Press, Beijing (2013). (in Tibetan)
2. Gesang, J., Gesang, Y.: Practical Tibetan Grammar Tutorial. Sichuan Nationalities Press, Chengdu (2008). (in Chinese)
3. Zhu, J., Li, T., Liu, S.: The algorithm of spelling check base on TSRM. J. Chin. Inf. Process. **28**(3), 92–98 (2014). (in Chinese)
4. Cai, Z., Sun, M., Cairang, Z.: Vector based spelling check for Tibetan characters. J. Chin. Inf. Proess. **32**(9), 47–55 (2018). (in Chinese)
5. Zhu, J., Li, T., Liu, S.: An approach for Tibetan text automatic proofreading and its system design. Acta Scientiarum Naturalium Universitatis Pekinensis **50**(1), 142–148 (2014). (in Chinese)
6. Hochreiter, S., Schmidhuber, J.: Long short-term memory. Neural Comput. **9**, 1735–1780 (1997)
7. Luo, W., Luo, Z., Gong, X.: Study of techniques of automatic proofreading for Chinese texts. J. Comput. Res. Dev. **41**(4), 244–249 (2004). (in Chinese)
8. Zhang, Y., Yu, S.: Summary of text automatic proofreading technology. Appl. Res. Comput. **23**(6), 8–12 (2006). (in Chinese)
9. Chollampatt, S., Ng, H.T.: Connecting the dots: towards human-level grammatical error correction. In: Proceedings of the 12th Workshop on Innovative Use of NLP for Building Educational Applications. Association for Computational Linguistics, Copenhagen (2017)
10. Fu, K., Huang, J., Duan, Y.: Youdao's winning solution to the NLPCC-2018 task 2 challenge: a neural machine translation approach to Chinese grammatical error correction. In: Zhang, M., Ng, V., Zhao, D., Li, S., Zan, H. (eds.) NLPCC 2018. LNCS (LNAI), vol. 11108, pp. 341–350. Springer, Cham (2018). https://doi.org/10.1007/978-3-319-99495-6_29
11. Vaswani, A., et al.: Attention is all you need. Computation and Language (cs.CL); Machine Learning (cs.LG). arXiv:1706.03762 (2017)
12. Devlin, J., Chang, M., Lee, K., Toutanova, K.: BERT: pre-training of deep bidirectional transformers for language understanding. arXiv preprint arXiv:1810.04805 (2018)

Chinese Text Error Correction Suggestion Generation Based on SoundShape Code

Hanru Wang, Yangsen Zhang[✉], Lipeng Yang, and Congcong Wang

Institute of Intelligent Information Processing, Beijing Information Science
and Technology University, Beijing 100192, China
hanru-wang@foxmail.com, zhangyangsen@163.com

Abstract. Text error correction is an essential part of text proofreading. This paper presents a method for generating text error correction suggestion based on SoundShape Code. By converting the target words into SoundShape Code and using an improved editing distance algorithm to make an ambiguous match with the words in the vocabulary, a set of candidate words whose similarity exceeds a certain threshold are obtained. Based on the contextual relevance model, each words in the candidate words set is scored, and then reasonable error correction suggestions are given according the score. In this paper, four types of errors are marked: substitution error in words with two-character, missing error in words with more than three-character, inserting error in words with more than three-character, substitution error in words with three-character. In total, 617 errors are tested and analyzed. Experiments show that the similarity calculation based on SoundShape Code can provide reasonable error correction suggestions.

Keywords: Text error correction · SoundShape Code · Improved editing distance · Context relevance model

1 Introduction

With the rapid development of Internet technology and the increasing popularity of electronic publications, the field of text proofreading faces great challenges. To cope with the huge amount of documents in the Internet era, it is very important to adopt automatic text proofreading method. There are two main reasons: first, the traditional manual proofreading method is slow and inefficient; second, it's too difficult to find the error in the text because of authors' insufficient vocabulary or carelessness. Text error correction is a very important research in the field of Chinese automatic text proofreading. Reasonable error correction suggestions not only verify the correctness of the proofreading results, but also help users understand the causes of errors and solutions.

In Chinese, words are composed of one or more Chinese characters. Chinese text is composed of many words. The analysis of Chinese text is related to the words itself, words order and words collocation of Chinese text. There are two reasons for text errors: (1) the errors of words, the text errors caused by missing Chinese characters, inserting Chinese characters and wrong Chinese characters; (2) the semantic errors, the text errors caused by the inappropriate position of words in the text. The algorithm proposed in this paper mainly aims at correcting words errors and gives reasonable error correction suggestions by combining SoundShape Code with improved editing distance.

© Springer Nature Switzerland AG 2020
J.-F. Hong et al. (Eds.): CLSW 2019, LNAI 11831, pp. 423–432, 2020.
https://doi.org/10.1007/978-3-030-38189-9_44

The similarity of Chinese characters consists of sound similarity and shape similarity. Sound similarity compares the Pinyin differences of Chinese characters, so it is also called "Pinyin similarity". Sound similarity is used to solve the problem of pronunciation confusion caused by homonyms or dialects: (1) two distinct Chinese characters in Chinese text may also have the same pronunciation; (2) Dialects lead to the Chinese pronunciation varies with the change of provinces. In some areas which have strong accents cannot recognize the differences between some pinyin. Considering the above two reasons, Sound similarity is important for the calculation of Chinese character similarity. Shape similarity compares structural differences of Chinese Characters. Considering Chinese characters are pictographic characters, Chinese characters with similar structures are likely to express similar meanings, so Shape similarity is reasonable for the calculation of Chinese character similarity. In practice, it is unrealistic to separate the similarity between pronunciation and shape of Chinese characters. Therefore, this paper uses SoundShape Code [1] (SSC) to solve the similarity problem in Chinese.

2 Previous Research

At present, there are a few theories on the automatic error correction of Chinese text. The most common method is to construct an error correction knowledge base and correct the text by traversing the error correction knowledge base. For the research of the error correction method, the existing documents are the construction method of the error correction knowledge base and the sorting method of the error correction suggestion.

Zhang [2, 3] proposed the construction method of error correction knowledge base and the automatic error correction algorithm based on the error correction knowledge base for the first time. According to the characteristics of pronunciation and shape of Chinese characters, this paper established many candidate word sets, such as homophonic, similar pronunciation, similar shape and so forth. Combined with context information, this paper provides error correction suggestions for different words error types in the text. Liu [4] used sound and shape to establish confusing sets of error-prone words; then he used context features to sort candidate words in the error-correcting knowledge base, and gave suggestions for revising errors in the text according to ranking. For the construction of candidate word set, Shi [5] collected a large number of confusing words manually; He regarded Chinese characters as nodes, and the confusion relation of Chinese characters as the connection between nodes. Then he constructed confusion graph and designed an extended algorithm to expand the confusion set. Shi [6] constructed a homophone dictionary based on Chinese dictionary and Chinese phonetic alphabet, and used the homophone dictionary as a candidate word set. For the ranking method of error correction proposals, in addition to using the context of the text, Cai [7] constructed the candidate similar word set, and used the concept of membership function in ambiguous mathematics to calculate the ambiguous importance of the context information of words, and selected the candidate words with the highest value as the error correction proposals.

3   Error Correction Algorithm Based on Tone Shape Code

Text error correction suggestion proposed in this paper uses the SoundShape Code to calculate the similarity of Chinese characters, comprehensively considering the similarities and differences between Chinese characters in sound and shape, and gives reasonable and effective error correction suggestions according to contextual relevance model.

3.1   Error Correcting Thought

Confused dictionary include words that people tend to make mistakes in their lives. When providing error correction suggestions, we first look for error-prone words in confused dictionary, which greatly improve the efficiency of error correction. However, the error correction method based on confused dictionary is not comprehensive: compared with right words, the number of wrong words is quite high. Therefore, the algorithm proposed in this paper combine SoundShape Code with confused dictionary. The idea of error correction in this paper includes the following steps:

(1) Segmenting the text and extracting continuous single word in the text;
(2) Positioning error location preliminarily;
(3) Using the improved editing distance to find the list of candidate words in the vocabulary.
(4) Computing similarity using SoundShape Code;
(5) Ranking candidate words according to similarity between candidate words and error words and coherence between candidate words and context;
(6) Generating Error Correction Suggestions based on Sorting Results.

3.2   Error Location

The algorithm flow for obtaining a list of suspected errors from the text is shown in Algorithm 1:

Algorithm 1: suspected error location algorithm
Input: Text to be proofread
Output: List of suspected errors
Process:
  Step1: Get the position of all the continuous single Chinese character in the result of text segmentation, and add them to the error list, and process the continuous single Chinese character in the error list one by one.
  Step2: If a single Chinese character is not a words consisting of one Chinese character, leave this error in the error list.
  Step3: If the part of speech (pos) of single Chinese character is Chinese function words or person's name, remove this error from the error list.
  Step4: If a continuous Chinese character is a stop word, remove this error from the error list.
  Step5: If the Chinese character and the subsequent Chinese character can form words in the wrong word list, leave this error in the error list.
  Step6: If the Chinese character with the character before or after it satisfies the continuation relationship, remove this error from the error list.
  Step7: If there is a continuous Chinese character in the error list that has not been processed, continue to step 2 to step 6; otherwise, turn to step 8.
  Step8: The algorithm is over.

3.3 Improved Editing Distance Algorithm

Editing distance is proposed by Levenshtein [8]. It is an algorithm to measure the similarity between source and target strings. It refers to the minimum editing cost required to convert a source string into a target string. Editing operations in Chinese strings include inserting one character, missing one character and modifying one character. In this paper, an improved editing distance algorithm is used to find all candidate words in the vocabulary whose similarity with the words to be corrected is greater than a certain threshold.

Assuming that there are string $S = s_1 s_2 \ldots s_m$ and $T = t_1 t_2 \cdots t_n$, the matching matrix D between S and T is established, where S is the first column of the matrix and T is the first row of the matrix. Based on the dynamic programming method, the minimum number of operations is calculated as follows.

$$D(i,j) = \begin{cases} 0, & i = 0 \\ 0, & j = 0 \\ Min \begin{pmatrix} D(i-1,j)+1 \\ D(i,j-1)+1 \\ Di-1,j-1+M(i,j) \end{pmatrix}, & otherwise \end{cases} \tag{1}$$

$$M(i,j) = \begin{cases} 1, & Replace\ s_i\ with\ t_j \\ 0, & if\ s_i = t_j \end{cases} \tag{2}$$

$D(i,j)$ is the minimum number of times the string S is converted to T requiring insertion, deletion, and modification operations. $M(i,j)$ is used to measure whether s_i and t_j are the same character.

After the minimum number of operations is obtained, the path of matrix D is traced to find the operation of each step from string S to T. The similarity between the two strings is calculated as follows:

$$Sim(S,T) = 1 - \frac{1}{n}\sum_{i=1}^{n} P \tag{3}$$

Where n is the number of paths from string S to T, and P is the operations performed by each path, including inserting a character, deleting a character, or modifying a character. When inserting or deleting a character, the value of P is 0. When modifying a character, the similarity between Chinese characters x and y is calculated, $P = sim(x,y)$.

3.4 Similarity Algorithm of Chinese Characters Based on SoundShape Code

Coding Method. The SoundShape Code converts a Chinese character into a ten-digit character sequence, and calculates the similarity of Chinese characters through the character sequence. To a certain extent, this coding method retains the pronunciation and shape characteristics of Chinese characters. The encoding method of SoundShape Code is shown in Fig. 1.

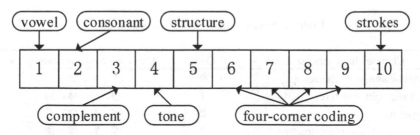

Fig. 1. SoundShape Code

The SoundShape Code consists of two parts: the first part is the sound code, which mainly includes vowel, consonant, complement code and tone; the second part is the shape code, including Chinese character structure, Chinese character four-corner coding and the number of Chinese character strokes. Each part is briefly introduced below.

The first bit is the vowel. Through simple substitution rules, the 24-bit vowel of Chinese characters is mapped to characters, some of which are replaced by the same characters. The matching table is shown in Table 1.

Table 1. Vowel matching table

A	1	O	2	E	3	I	4
U	5	V	6	ai	7	Ei	7
Ui	8	Ao	9	ou	A	Iu	B
Ie	C	Ve	D	er	E	an	F
En	G	In	H	un	1	ven	J
Ang	F	Eng	G	ing	H	ong	K

The second bit is the consonant. Similarly, the 23-bit consonant is converted into characters by substitution rules, and the consonant matching table is shown in Table 2.

Table 2. Consonant matching table

B	1	P	2	m	3	F	4
D	5	T	6	n	7	L	7
G	8	K	9	h	A	j	B
Q	C	X	D	zh	E	ch	F
Sh	G	R	H	z	E	c	F
Sh	G	Y	I	w	J		

The third bit is the complement, which is usually used when there is a consonant between the initial and the vowel. The replacement rule is the same as the vowel bit.

Table 3. Structure table of Chinese characters

Chinese characters structure	Code	Chinese character examples
Single-component character	0	上、人、土、一
Left and right structure	1	挣、伟、休、姐
Upper and lower structure	2	志、苗、字、胃
Left, middle and right structure	3	湖、脚、溅、谢
Upper, middle and lower structure	4	奚、禀、亵、髻
Upper right surrounding structure	5	句、可、司、式
Upper left surrounding structure	6	庙、病、房、尼
Lower left surrounding structure	7	建、连、毯、尴
Upper three surrounding structure	8	同、问、闹、周
Lower three surrounding structure	9	击、凶、函、画
Upper left three surrounding structure	A	区、巨、匝、匣
Full surround structure	B	囚、团、因、囹
Embedded structure	C	坐、爽、夹、噩
"品" structure	D	品、晶、森
"田" structure	E	燚、田

The fourth bit is tone, using 1, 2, 3, 4 to represent four tones in Chinese characters.

The fifth bit is the structure. There arc many structural types of Chinese characters. Each structure is represented by a code, and each Chinese character is classified into only one structure. The structure of Chinese characters is shown in Table 3.

The sixth to ninth bits are the four-corner coding of Chinese characters. It encodes the upper left corner, the upper right corner, the lower left corner and the lower right corner of the Chinese character according to the four-corner coding formula.

The tenth bit is the number strokes of Chinese characters: 1–9 means the number of strokes of Chinese characters is from 1 to 9; A–Z means the number of strokes of Chinese characters is from 10 to 35, and all Chinese characters with more than 35 strokes are represented by Z.

The Similarity Calculation of Chinese Characters. In fact, the similarity of Chinese characters is transformed into the similarity between SoundShape Code, and the SoundShape Code of each Chinese character is $(c_1c_2c_3c_4c_5c_6c_7c_8c_9c_{10})$. $c_1c_2c_3c_4$ is sound code, and $c_6c_7c_8c_9c_{10}$ is shape code. The similarity calculation of a Chinese character is divided into sound code similarity *VSim* and shape code similarity *XSim*. The similarity calculation formula between Chinese character x and Chinese character y is demonstrated as follows.

$$sim(x, y) = \alpha_1 \times VSim + \alpha_2 \times XSim \tag{5}$$

Among them, α_1 and α_2 represent the weights of *VSim* and *XSim* respectively, which can be adjusted according to practical application [9]. In this paper, 0.7 and 0.3 are selected. If text error correction is applied to documents identified by OCR technology, the weight of shape similarity should be greater. If it is applied to documents input by Pinyin Input Method, the weight of sound similarity should be greater.

The sound similarity is the mathematical expectation of the two Chinese characters' SoundShape Code after bitwise "and", and the calculation formula is demonstrated as follows.

$$VSim = \frac{1}{4\left(\sum_{i=0}^{4}(x_i \& y_i)\right)} \tag{6}$$

Shape code similarity consists of three parts: structure bit, four-corner coding and stroke similarity. The calculation formula is as follows.

$$XSim = \beta_1(x_4 \& y_4) + \beta_2\left(1/4 \sum_{i=5}^{9}(x_i \& y_i)\right) + \beta_3\left(1 - \frac{|x_{10} - y_{10}|}{max(x_{10}, y_{10})}\right) \tag{7}$$

Among them, $\beta_1, \beta_2, \beta_3$ represent the weight of structure bit, four-corner coding and stroke respectively. In this paper, $\beta_1 = 0.3$, $\beta_2 = 0.4$, $\beta_3 = 0.3$.

3.5 Generate Error Correction Suggestions

For the list of candidate words provided above, this paper uses contextual relevance model to rank the candidate words. The model examines the relevance between candidate words and context to determine whether the error correction proposals are correct and reasonable.

In context T_k with sliding window k, the relationship between word w_i and other words in context is called its "contextual relevancy degree". The calculation formula is demonstrated as follows:

$$Re(w_i, T_k) = P(w_i | w_{i-k} \cdots w_{i-2} w_{i-1} w_{i+1} w_{i+2} \cdots w_{i+k}) \tag{8}$$

$$Re(w_i, T_k) = \sum_{j=i-k}^{i+k} \frac{d_{|i-j|}}{\sum_{m=-k}^{k} d_m} \times P(w_i | w_j) \tag{9}$$

Based on the contextual relevance model and the similarity between the candidate words and the word to be corrected, the candidate words are scored and ranked according to the score. The weighting function is shown as follows:

$$Weight(w_i) := \eta_1 Re(w_i, T_2) + \eta_2 sim(S, T) \tag{10}$$

4 Experimental Results and Analysis

4.1 Experimental Data

The data set in this paper is based on the Internet web page documents captured by Sogou Laboratory in 2012, and 400 of them are randomly selected for manual marking. The error types of marks are shown in Table 4.

Table 4. Error marks

Error type	Substitution error in words with two-character	Missing error in words with more than three-character	Inserting error in words with more than three-character	Substitution error in words with three-character	Total
Error marks number	116	121	113	267	617

Table 5. Comparison of calculation results between editing distance and improved editing distance similarity algorithm

(Wrong words, right words)	Editing distance	Improved editing distance
(吉详, 吉祥)	0.5	0.976
(瘾匿, 隐匿)	0.5	0.896
(和睦睦, 和和睦睦)	0.75	0.75
(一生一世案, 一生一世)	0.8	0.8
(良秀不齐, 良莠不齐)	0.75	0.808
(珊珊来迟, 姗姗来迟)	0.4	0.772
(出类拔粹, 出类拔萃)	0.75	0.94
(独劈奚迳, 独辟蹊径)	0.25	0.685

4.2 Experimental Results and Analysis

Taking eight groups of words pairs as the experimental object, the similarity calculation results between the editing distance and the improved editing distance are compared, and the experimental results are shown in Table 5.

Table 6. Candidate word sets for error-correcting words

Wrong words	(Candidate words, similarity)	Right words
擦绕	(缭绕, 0.97) (裛绕, 0.81)	缭绕
躁热	(燥热, 0.96) (招惹, 0.76)	燥热
爆乱	(暴乱, 0.89) (饱暖, 0.78)	暴乱
陷井	(陷阱, 0.91) (仙境, 0.81) (先进, 0.73)	陷阱
爆发户	(暴发户, 0.92) (爆发力, 0.72) (爆发星, 0.7) (爆发音, 0.7)	暴发户
爱屋及鸟	(爱屋及乌, 0.89)	爱屋及乌
功亏箦	(功亏一篑, 0.750)	功亏一篑
开诚不步公	(开诚布公, 0.759)	开诚布公

For the data labeled in Table 4, an improved editing distance algorithm is used to match similar words from the correct lexicon. In this paper, the threshold of similarity is set to 0.68. In Table 6, eight groups of results are selected to present.

The accuracy, recall and F values of the labeled data in Table 5 are shown in Table 7.

Table 7. Error types and experimental results

Error types	Accuracy	Recall	F values
Substitution error in words with two-character	94.5%	81.5%	87.5%
Missing error in words with more than three-character	91.2%	78.4%	84.3%
Inserting error in words with more than three-character	90.6%	89.%	89.8%
Substitution error in words with three-character	95.3%	83.5%	89.0%

Experimental analysis: from Tables 5, 6 and 7, it can be seen that the improved editing distance algorithm based on SoundShape Code can effectively query words with similar sound or shape in the lexicon, thus giving reasonable error correction suggestions.

5   Conclusion

In this paper, the similarity of Chinese characters is calculated by SoundShape Code, which consists of sound similarity and shape similarity, and the words to be corrected are ambiguous matched with the correct vocabulary by using an improved editing distance algorithm, which greatly improves the accuracy of error correction suggestions.

Acknowledgements. This work was supported by grants from National Nature Science Foundation of China (No. 61772081).

References

1. Chen, M., Du, Q.Z., Shao, Y.B., et al.: Chinese characters similarity comparison algorithm based on phonetic code and shape code. Inf. Technol. **11**, 73–75 (2018)
2. Zhang, Y.S.: The structuring method of correcting knowledge sets and the producing algorithm of correcting suggestion in the Chinese text proofreading system. J. Chin. Inf. Process. **15**(5), 33–39 (2001)
3. Zhang, Y.S., Cao, Y.D., Xu, B.: Correcting candidate suggestion algorithm and its realization based on statistics. Comput. Eng. **30**(11), 106–109 (2004)
4. Liu, L.L., Cao, C.G.: Research on automatic correction of Chinese true word errors based on combination of local context features. Comput. Sci. **43**(12), 30–35 (2016)
5. Shi, H.L., Liu, L.L., Wang, S., et al.: Research on method of constructing Chinese character confusion set. Comput. Sci. **41**(8), 229–232 (2014)
6. Shi, M.: Chinese text automatic proofreading system. Jiangsu University of Science and Technology (2015)
7. Cai, D.F., Bai, Y., Yu, S., et al.: A context based word similarity computing method. J. Chin. Inf. Process. **24**(3), 24–28 (2010)
8. Levenshtein, V.I.: Binary code capable of correcting deletions, insertions and reversals. Dokl. Akad. Nauk SSSR **163**(4), 708–710 (1966)
9. Liu, L.L., Cao, C.G.: Research on automatic proofreading method for Chinese non-multiple word errors. Comput. Sci. **43**(10), 200–205 (2016)

Extracting Hierarchical Relations Between the Back-of-the-Book Index Terms

Ning Li[✉], Meng Tian, and Shuqi Lv

Beijing Information Science and Technology University, Beijing 100101, China
ningli.ok@163.com

Abstract. Aiming at solving the problem that the single level back-of-the-book index system is not enough to fully explore the semantics relations between the index terms, a method to extract the hierarchical relations between the index terms based on combination of lexical-syntactic analysis and text structure features is proposed in this paper. It first organizes index terms according to the text structure features, and constructs the indexed term pairs with hierarchical relations step by step. Then based on word vectors, the semantic similarity of paired index terms is calculated to eliminate the misidentified pairs. Finally, the index term pairs with hierarchical relations are optimized in the direct graph to remove redundant and conflict relations, and the hierarchical index system is built at last. Compared with the other results, our method improves precision rate and F value by 11.44% and 5.65% respectively.

Keywords: Hierarchical relation extraction for index terms · Bob index system · Hierarchical relations of index terms

1 Introduction

Books remain the main resource of knowledge dissemination. As an important part of books, back-of-the-book index system (bob index system) is essential in acquiring knowledge quickly and precisely. Index terms are knowledge units of books, and there are rich semantic relations between them. The hierarchical bob index system plays the role of local knowledge ontology of the book. It can not only help to reveal the semantic relations between index terms, help readers to find the internal relationship between the words in the content, but also increase the ability to retrieve and navigate the content [1]. In the future, it is expected that these knowledge ontologies can be interlinked to form a large knowledge base.

The key to establish the bob hierarchical indexing system is the extraction of hierarchical relations of the index terms. However there is little research on this work so far. Conceptual classification researches the hierarchical relations between concepts, which is close to this topic, and can be used as a reference. At present, relation extraction methods used in conceptual classification are mainly divided into: dictionary-based method [2, 3], statistical based method [4, 5], lexical-syntactic pattern based method [6, 7]. The accuracy of dictionary-based methods is higher, but the coverage is lower because of the limited size of dictionaries. Statistical-based methods use machine learning algorithms such as clustering, classification and association rules

© Springer Nature Switzerland AG 2020
J.-F. Hong et al. (Eds.): CLSW 2019, LNAI 11831, pp. 433–443, 2020.
https://doi.org/10.1007/978-3-030-38189-9_45

to extract relations. These methods can process large-scale data with high coverage, but their accuracy is not as good as those based on dictionaries and lexical-syntactic patterns. The lexical-syntactic pattern based method is the typical method to extract conceptual hierarchical relations, which has a high accuracy. However, it has some limitations when is applied directly in extraction of hierarchical relations of bob index terms. For example, in literature [8], the method combining lexical analysis with dependency parsing is proposed to extract conceptual hierarchical relations in power field, and good results have been achieved. However, this method only extracts the conceptual hierarchical relations according to the existing "is-a" pattern, and cannot extract the conceptual hierarchical relations other than this pattern. Whereas hierarchical bob index system is designed to better show the semantic relations between index terms with more patterns. As the hierarchical structure of a book is composed of titles at all levels, chapters and subchapters are organized around the same topic from different aspects, they embodies the internal logic of books [9] and is more conducive to the identification of hierarchical relations of index terms in terms of structure and semantics. Therefore, in addition to extracting the hierarchical relations of index terms by combining lexical-syntactic analysis, we use book's text structure features to extract the hierarchical relations of index terms. Moreover, the candidate term pairs are verified to eliminate those with errors so as to achieve good results.

2   Extracting Hierarchical Relations of Index Terms

Books contain a huge amount of information. The average number of index words accounts for about 0.42% of the total number of books only [10]. Effective integration of relevant index terms and full mining of semantic association between the terms will enhance the navigation and retrieval ability of the bob index system. The index terms are extracted based on the method proposed in literature [11], we then recognize the hierarchical relations in the extracted terms. First, we use the method based on text structure features, as well as the method combining lexical-syntactic analysis to identify the candidate term pairs with hierarchical relations. As these candidates may have errors, we further filter the pairs by calculating the semantic similarity of them, and eliminate the wrong ones.

2.1   Extracting Hierarchical Relations of Index Terms Based on the Text Structure

The process of extracting hierarchical relations of index terms includes: extracting the candidate index terms with hierarchical relations and dividing them into term sets and then identifying hierarchical relations between the index terms.

The extracted index terms are organized according to the text structure of the book. An index term set consists of index terms from the same chapter/subchapter. The term sets as a whole imply the hierarchical relation between the index terms. Figure 1 shows the hierarchical structure of index term sets located within different headings and text parts.

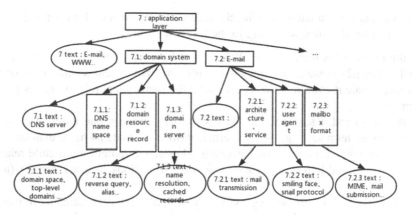

Fig. 1. The hierarchical structure of index term sets

In the hierarchical structure of index term sets, each non-root node is a sub-topic of the upper node. In order to identify the hierarchical relation between index terms, it is necessary to associate each non-root node with its parent node and ancestor nodes.

The associated nodes are chained together, namely the chain of index term sets. They are constructed by taking non-root nodes as the starting point and traversing the tree, see Fig. 2. The specific steps are as follows:

1. Starting from the non-root node, traverse upwards layer by layer until it reaches the root node, record the traversal path, and flips the traversal path to get the chain of index term sets.
2. Jump to the next non-root node that is not accessed, repeat 1. until all non-root nodes are accessed.
3. Collecting all the chains and remove duplications.

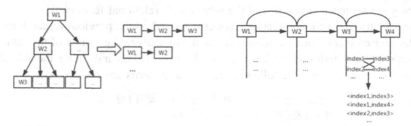

(a) Construct the chains of index term sets.   (b) Identify relations between index terms.

Fig. 2. The process to decompose the hierarchical index term set.

When chains of index term sets become available, pairs of index terms can be constructed as following: (1) each node in the chain pairs with its adjacent sub-node starting from the head node; (2) each term in the node pairs with every term in the paired node. As shown in Fig. 2(b), three term pairs, <W1, W2>, <W2, W3>,

<W3, W4>, are constructed from the chains, Finally the relation between the terms in the pair can be identified according to the following rules:

1. If the upper index term set, e.g., W3 = {term1} and the lower index term set, e.g., W4 = {term2} contain one index term only, the hierarchical relation is recognized as single parent-child relation. It constitute the hierarchical relation set as {(term1, term2)}.
2. If the upper index term set, e.g., W3 = {term1} contains one index term and the lower index term set, e.g., W4 = {term2, term3,...} contains more than one index terms, the hierarchical relation is recognized as single- parent-multi-child relations. It is decomposed into several single parent-child relations, {(term1, term2), (term1, term3), (...)}.
3. If the upper index term set, e.g., W3 = {term1, term2,...} contains more than one index terms, the lower index term set, e.g., W4 = {term3} contains one index term, the hierarchical relation is identified as a multi-parent-single-child relation. It is decomposed into several single parent-child relations, e.g. {(term1, term3), (term2, term3), (...)}.
4. If the upper index term set, e.g., W3 = {term1, term2,...} contains more than one index terms, the lower index term set W4 = {term3, term4,...} also contains more than one index terms, the hierarchical relation is identified as multi-parent-multi-children relations. It is decomposed into multi-parent-single-child relations, and then further decomposed into multiple single parent-child relations, e.g., {(term1, term3), (term1, term4), (term2, term3), (term2, term4)...}.

2.2 Extracting Hierarchical Relations Between Index Terms Combining Lexical-Syntactic Analysis

Syntax in books is more formal than in free texts on the Internet. It should be easier to recognize the index terms with hierarchical relations in the sentences. Therefore, we can quickly identify the hierarchical relations between the terms in the sentences, by analyzing the sentence structure and the words with relational features.

First, we try to obtain the relational patterns. Based on the previous research results in pattern-based hierarchical relation extraction, we analyze the syntax of the sentence where hierarchical index terms exist, as well as the words indicating the relational features. For this purpose, the following relational patterns are constructed:

```
Pattern1：<?C1><包括|包含|囊括|涵盖|分为|主要有|分成><?C2>
          {<或|及|以及|和|与|、><?C3>}*<等>
Pattern2：<?C1><由|是由><?C2>{<或|及|以及|和|与|、><?C3>}*<组成|构成>
Pattern3：<?C1><的><?C2>{<或|及|以及|和|与><?C3>}*<是|为|称为><?C3>
```

The sentences where the index terms are located are called supporting sentences, which are divided into simple sentences and complex sentences. The identified supporting sentences are segmented by the words with relational features, namely relational feature words. Hypernyms are located in the left text of relational feature words and recorded as S1. Hyponyms are located in the right text of relational feature words and recorded as S2. Corpus analysis tells that sentences with commas in S1 usually contain multiple subject-predicate structures or multiple hypernyms. In this case, it is rather difficult to obtain the correct hierarchical relation of the index terms. That is why we regard the latter as a complex sentence.

Second, the hierarchical relation of index terms is identified. For simple sentences, the defined patterns are directly matched with supporting sentences to identify hierarchical relations. For example, the sentence "resource sharing mainly includes hardware sharing, software sharing and data sharing." The results obtained by matching the relational pattern are as follows: {resource haring} <?C1> mainly includes {hardware sharing} <?C2>, {software sharing} <?C3> and {data sharing} <?C4>. Three hierarchical relations are identified directly: Hierarchy (hardware sharing, resource sharing), Hierarchy (software sharing, resource sharing) and Hierarchy (data sharing, resource sharing).

For complex sentences, when there are modifiers or multiple subject-predicate structures, literature [7] uses dependency parsing to get the subordinate concept of "is a" relation. By analyzing complex sentences, it is found that hypernyms are usually the subject components of relational feature words. We apply the Language Technology Platform (LTP) developed by Harbin Institute of Technology to analyze complex sentences using dependency parsing and obtain the subject components by relational feature words, so as to obtain the correct hypernyms. For example:

`Coaxial cable is a widely used transmission medium, which has inner conductor and insulation layer.`

According to the pattern matching method, there are two indexing items "coaxial cable" and "transmission medium" before the relational feature word "has", thus the two hypernyms can be extracted, then the whole sentence is parsed with dependency grammar to determine the correct hierarchical relation. The parsing result is shown in Fig. 3.

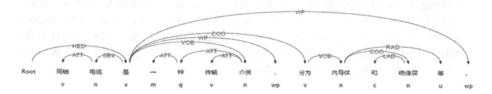

Fig. 3. Example of dependency parsing

From the result of dependency parsing, it is found that although LTP separates the index terms "coaxial cable" and "transmission medium" in the process of word segmentation, it can correctly find the predicate relation of "cable" and "is" (SBV), and the juxtaposition relation indicated by "has", so "cable" is also the subject of "has". Thus, it can be judged that the word "cable" is the hypernyms. However, "coaxial cable" should be the real hypernyms of "inner conductor" and "insulator". In order to solve the problem of index item separation, the method of partial matching is applied. By combining with the words appearing before and after the index item and comparing the result with terms in the knowledge dictionary like Wikipedia, more accurate index terms can be obtained. For the above example, we can get the superordinate index item "coaxial cable".

2.3   Validating Hierarchical Relations Between Index Terms

The foregoing extraction method uses hierarchical relational features, sentence structure features and text structure features to extract hierarchical relations, taking into account both global and local features. Although hierarchical relations extracted are rather complete, some wrong hierarchical relations may be extracted too. Literature [11] points out that if the concepts C1 and C2 have hyponymy, C1 and C2 should have similar semantics. Therefore, we further verify the hierarchical relation of index terms using semantic features, and eliminate the pairs with wrong hierarchical relations whose semantic similarity looks too low. The main steps to verify the hierarchical relations of index terms are as follows:

1. Preprocessing. The index terms are added to the user dictionary where the Chinese Wikipedia and the book text are used as corpus. Next, the corpus is segmented and the stop words are removed.
2. Getting word vectors. The Skip-gram model in Word2vec [13] is used to train the pre-processed corpus to obtain the word vectors of all the index terms. Skip-gram model mainly involves two parameters in the training process, namely word vector dimension and window size. After training, we get the optimized parameters, vector dimension is 300, and window size is 5.
3. Calculating Semantic similarity. After the word vectors of index terms are obtained, the cosine semantic similarity of each pair is calculated by the formula:

$$CosDis(w_1, w_2) = \frac{w_1 \cdot w_2}{\|w_1\| \times \|w_2\|} = \frac{\sum_{i=1}^{n} w_{1i} \times w_{2i}}{\sqrt{\sum_{i=1}^{n} (w_{1i})^2} \times \sqrt{\sum_{i=1}^{n} (w_{2i})^2}} \qquad (1)$$

Formula (1) shows the cosine semantic similarity of word vectors W_1 and W_2, where n is the dimension of word vectors. The threshold is set according to the experimental results. The index term pairs whose semantic similarity exceeds the threshold are considered with correct hierarchical relations and are added in to the final set, otherwise, they are discarded.

3   Construction of Bob Index System

Index terms that are validated by semantic similarity are organized in directed graph, which is called index term hierarchical graph. Since the bob index system needs to show the multi-level semantic relations of index terms, we chooses to create a three-level index. Therefore, we need to get all paths with length of 3 in the subsequent processing of the hierarchical graph. The index term hierarchical graph can be expressed as $G(V, E)$, where V is the node, representing the index terms; E is the directed edge, representing the hierarchical relations between the index terms. We use adjacency matrix to represent the relations between index terms, and use weighted voting to score the strength of hierarchical relations of index terms. Assuming that there are N indexing terms, then we will have an $N \times N$ adjacent matrix $A^{N \times N}$, where $a_{i,j}$ represents the strength score of the hierarchical relation between index terms numbered i and j.Score(i,j) is determined by weighted voting. The calculation formula is as follows:

$$\text{Score}(i,j) = \alpha Score_c(i,j) + \beta Score_p(i,j) \tag{2}$$

In the formula, the values of α and β are determined by experiment, and are set 0.6 and 0.4 respectively. If the hierarchical relation pairs of index terms i and j are extracted based on the text structure, the value of $Score_p(i,j)$ is 1, otherwise is 0. If the hierarchical relation pairs of index terms i and j are extracted by combining lexical-syntactic analysis, the value of $Score_p(i,j)$ is 1, otherwise is 0. The optimization method of weighted voting is shown in Fig. 4. In Fig. 4(a), "local area network" and "Ethernet" are hyponymy of each other, resulting in conflicts. Let's get all the 3-length paths passing through the nodes of "local area network" and "Ethernet". The length of the path "network-local area network-Ethernet" (the sum of weights is 1.6) is longer than the length of the path "network-Ethernet-local area network" (the sum of weights is 1.4), thus the directed edge "network-Ethernet-local area network" is removed. In Fig. 4(b), there are two paths from node "IEEE" to its superordinate index term "network", i.e., the path "network-local area network-IEEE" and path "network-IEEE". To remove the redundancy, the weights of the two paths are calculated and the directed edge with less weight sum is removed, i.e., the path "network-IEEE".

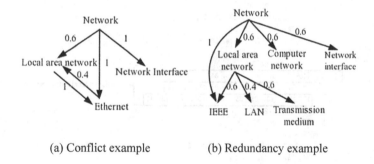

(a) Conflict example　　　　(b) Redundancy example

Fig. 4. Optimization method of weighted voting for hierarchical relation strength

The specific optimization steps are as follows:

1. Obtain the weight of each index term pair with hierarchical relations.
2. Get all paths between any two nodes i and j in the graph. The set of paths is denoted as $C(i, j)$. Calculate the length of each path separately using:

$$w_{sum}(p) = \sum_{e \in C(i,j)} w(e) \tag{3}$$

3. Retain the path with the biggest sum of weights between nodes *i* and *j* and remove the remaining paths.

When the index term hierarchical graph is available, the bob hierarchical index system can be constructed accordingly. The major work includes: term sorting, hierarchical construction and annotating the source of index. The specific steps are as follows:

1. Sort index terms according to Pinyin or strokes
2. Traverse the index terms. Assuming that the initial index term is W_1, all paths with the starting point of W_1 and the length of 3 are searched. The second node of the path acts as the second-level label corresponding to the index term W_1, and the third node of the path acts as the third-level label corresponding to the index term W_1. They are placed below the index word W_1 accordingly.
3. Extract the page number of each index item by Microsoft Word SDK, and mark the position information for all index terms.

We establish the index system for the textbook *University Computer Foundation* as an example. Partial results are shown in Fig. 5.

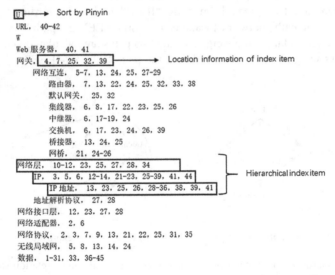

Fig. 5. Part of index system for the textbook *University Computer Foundation*

4   Experiment and Analysis

In this research, 10 Chinese textbook in computers and the offline package of Chinese version of Wikipedia as used as the experimental data. The index terms are extracted by the method of literature [11]. After manual review, the index terms are added into the set for subsequent extraction of hierarchical relations. The word segmentation tool used is ICTCLAS Chinese word segmentation system. According to general principles of index compilation listed in GB/T 22466, the index system we created is compared with the manually constructed system, and we have the following results.

4.1 Experiment on Index Term Hierarchy Relation Extraction

In order to illustrate the influence of hierarchical relation verification on the results, the method without verification are compared with the method with verification. Meanwhile, in order to verify the effectiveness of our method, the method proposed in literature [14] is adopted for comparison. Precision rate, recall rate and F value are used to measure the extraction effect. The experimental results of each group are shown in Table 1.

Table 1. Statistics of results from index term hierarchical relation extraction

	Precision rate	Recall rate	F value
Comparison with result of literature [14]	71.13%	77.27%	74.07%
Lexical-syntactic analysis plus text structure features	68.29%	87.16%	76.58%
Lexical-syntactic analysis plus text structure features plus hierarchical relation verification (our method)	82.57%	77.06%	79.72%

From the experiment results, we can find that the combination of lexical-syntactic analysis and textual structure features achieves a higher recall rate, because the combination of the two methods covers the index terms with hierarchical relations more comprehensively; however, the precision rate is lower, because there are some wrong hierarchical relation pairs in the extraction results. After hierarchical relation validation, the precision rate is improved by 14.28%, which indicates that hierarchical relation validation eliminates most of the wrong index item pairs. Compared with literature [15], precision rate of our method is improved by 11.44% and F value by 5.65%. This is due to the fact that more hierarchical pairs of index terms are identified based on text structure features. The experiment results show that the proposed method is effective in extracting hierarchical relations of index terms.

4.2 Experiment on Construction of Hierarchical Index System

Based on the index term pairs obtained above, the index term hierarchical graph is used to organize the index terms. It is found that the best result can be obtained when α is 0.6 and β is 0.4. Comparing with the hierarchical graph of index terms constructed manually, our method achieves 79.81% precision rate and 87.32% recall rate in the construction of three-level index, which indicates that our method is useful to construct the bob index system.

5 Summary

This paper proposes a method to extract hierarchical relations of index terms in the bob index system. This method organizes index terms according to the text structure, and constructs the indexed term pairs with hierarchical relations step by step. Based on word vectors, the semantic similarity of paired index terms is calculated to eliminate the misidentified pairs. The index term pairs with hierarchical relations are optimized in the direct graph to remove redundant and conflict relations, and finally, the bob hierarchical index system is established. Experiments show that our method can extract hierarchical relations of index terms efficiently, and improves the precision rate and F value significantly compared with the existing method. However, there are still some limitations in this research. In the future, more knowledge dictionaries, especially domain knowledge dictionaries, will be used to further improve the accuracy of domain-related hierarchical relation extraction. In the calculation of semantic similarity of index terms, more dimensions will be introduced to further improve the result of hierarchical relation validation of index terms and increase the recall rate.

Acknowledgements.   This paper was supported by the National Natural Science Foundation of China - The Intelligent Analysis and Optimization Method for Re-flowable Documents (61672105) and the National Key R&D Program of China (2018YFB1004100).

References

1. Guo, L.F., Wen, G.Q.: Comparative research of index software between English and Chinese. Library **4**, 47–48 (2010)
2. Suchanek, F.M., Kasneci, G., Weikum, G.: YAGO: a large ontology from Wikipedia and WordNet. In: Web Semantics Science Services & Agents on the World Wide Web, vol. 6(3), pp. 203–217 (2008)
3. Rebele, T., Suchanek, F., Hoffart, J., Biega, J., Kuzey, E., Weikum, G.: YAGO: a multilingual knowledge base from Wikipedia, Wordnet, and Geonames. In: Groth, P., et al. (eds.) ISWC 2016. LNCS, vol. 9982, pp. 177–185. Springer, Cham (2016). https://doi.org/10.1007/978-3-319-46547-0_19
4. Tian, F., Ren, F.: Hyponymy acquisition from Chinese text by SVM. In: International Conference on Natural Language Processing & Knowledge Engineering, Dalian, pp. 1–6. IEEE (2009)
5. Wang, S., Liang, C., Wu, Z., et al.: Concept hierarchy extraction from textbooks. In: ACM Symposium on Document Engineering, pp. 147–156. ACM (2015)
6. Sang, E.T.K., Hofmann, K., de Rijke, M.: Extraction of hypernymy information from text*. In: van den Bosch, A., Bouma, G. (eds.) Interactive Multi-modal Question-Answering. NLP, pp. 223–245. Springer, Heidelberg (2011). https://doi.org/10.1007/978-3-642-17525-1_10
7. Tang, Q., Lv, X.Q., Li, Z.: Research on domain ontology concept hyponymy relation extraction. Microelectron. Comput. **31**(6), 68–71 (2014)
8. Ruan, D.R., He, X.Y., Li, D.Y.: Modeling and extracting hyponymy relationships on Chinese electric power field content. In: 8th International Conference on Modelling, Identification and Control (ICMIC), Algiers, pp. 439–443. IEEE (2016)
9. Jing, C., Bo, X., et al.: A research on internal hierarchical topic organization model of the book based on hLDA. Libr. Inf. Serv. **60**(18), 140–148 (2016)

10. Wu, Z.H., Li, Z.H., Mitra, P., et al.: Can back-of-the-book indexes be automatically created? In: CIKM 2013 Proceedings of the 22nd ACM International Conference on Information & Knowledge Management, San Francisco, pp. 1745–1750. ACM (2013)
11. Tian, M., Li, N., et al.: Extraction of index terms for Chinese books. Comput. Eng. Des. **40**(1), 261–267 (2019)
12. Liu, L., Cao, C.G.: Hyponymy relation verification method based on hybrid features. Comput. Eng. **34**(14), 12–13 (2008)
13. Mikolov, T., Chen, K., Corrado, G., et al.: Efficient estimation of word representations in vector space. arXiv:1301.3781v3, pp. 1–12 (2013)
14. Lv, S.Q.: Research on the Method of Automatically Generating Back-of-the-Book Index. Beijing Information Science & Technology University, Beijing (2017)

A Situation Evaluation System
for Specific Events in Social Media

Bojia Li[1], Ruoyu Chen[1,2(✉)], and Yangsen Zhang[1,2]

[1] Institute of Intelligent Information Processing,
Beijing Information Science and Technology University, Beijing, China
berger.bojia@gmail.com, zhangyangsen@163.com
[2] Beijing Key Laboratory of Internet Culture and Digital Dissemination Research,
Beijing Information Science and Technology University, Beijing, China
ruoyu-chen@foxmail.com

Abstract. With the widespread use of social media, social networks have become an important information carrier and platform for users to explore the world. Social networks not only reflect the hot events in society but also influence the trends and evaluation of events through user network behaviors. In this paper, a situation evaluation system is established for social event trends. We use web crawlers to collect multi-source data for a series of events of interest to form a basic knowledge base, and based on this, we extract statistical data and language features. Then, we use the Analytic Hierarchy Process (AHP) algorithm to calculate the weight of important features that can capture the development of the event situation and establish a hot social event situation evaluation system. Finally, we apply this system and stream computing technology to achieve situational awareness of events in real-time.

Keywords: Situation evaluation · Social media · Social events

1 Introduction

With the wide use of social media, social networks have become an important topic. As early as the 1990s, Schuler pointed out in his article [1] that software acted as the focus and intermediary of social relations in all types of computing applications. Dryer [2] considered social computing as the interplay among people's social behaviors and computing technologies. It is precise because it can reflect social relations and social concerns to a certain extent that social computing is concerned by more and more scholars. Today, online social media is so widely used and has become an important channel and carrier for social relations and information dissemination in human society. According to Twitter's third quarter financial report of 2018, the number of its monthly average active users was 326 million, of which the number of monthly average active users in the United States was 67 million, accounting for about 20% of US's total population [3]. China's largest social media platform Sina Weibo's third-quarter

© Springer Nature Switzerland AG 2020
J.-F. Hong et al. (Eds.): CLSW 2019, LNAI 11831, pp. 444–455, 2020.
https://doi.org/10.1007/978-3-030-38189-9_46

earning report of 2018 showed that its monthly active users reached 446 million, accounting for about 32% of China's total population [4]. Early researches focused on the relationship between event situations in online social media and event situations in the real world, e.g. Becker et al. [5] used a classifier to identify event situations in the real world on Twitter.

In recent years, technologies of situation evaluation have been studied extensively. Researchers from various angles and professional fields have devised various situation evaluation systems. From the point of view of network infrastructure security and network traffic security, key research works are as follows: Snort [6] is a lightweight, cross-platform network intrusion detection tool. With the data collected by the rule-based traffic collection engine of Snort, Roesch realized the evaluation of the security situation based on the order of magnitude of the traffic and the importance of the attacked target. Porras [7] constructed an alert aggregation strategy for heterogeneous information security devices based on topology analysis, alert priority analysis and general attributes, which could automatically fuse the alarm information that poses the greatest security threat to the system. Hariri et al. also designed a framework for analyzing network vulnerabilities based on proxy facts, quantifying the level of risk at the time of a network attack through network performance [8]. In particular, network security evaluation, which depends on network nodes, is added to the monitoring, analysis, and quantification of failures and network attacks.

Among all the security incidents, information content security incident refers to those security incidents that use the network to publish and disseminate content that endangers national security, social stability, and the public interest [9]. Nowadays, researches in the field of information content security involve the following aspects: the research on network information penetration technology which involves information confidentiality and information isolation between networks [10], the research on Access Control Strategies to suppress the spread of harmful information [11], the analysis of emotional tendentiousness, public opinion evolution volatility analysis and related technical means of public opinion mining [12] and research of security risk detection through speech recognition and topic extraction [13] and so on.

The following are representative works in the research of multi-level and multi-angle situation measurement, and in the research of quantitative methods in the weight assignment problem. Among them, Chen [14] proposed an evaluation method based on massive alarm information and network performance indicators, following a bottom-up and global-to-local manner. With corresponding importance factors, different influencing factors are weighted to calculate the threat indicator of the service, host or network system as a whole. Liu [15] obtained the fuzzy maximum membership vector through the weight of each indicator to the previous indicator in the situation system and combined with the multi-level comprehensive evaluation method, so as to quantitatively analyze the network security situation. Chen et al. constructed a network device security situation analysis system [16] based on Analytic Hierarchy Process (AHP) by using a simple network management protocol, and in the final quantification, they used the central limit theorem to determine the confidence, so as to carry out quantitative monitoring. The Delphi method was utilized by Xu et al. to analyze various influencing factors of network security [17]. After establishing the evaluation system, they used an artificial neural network to model and analyze the safety evaluation.

Based on the above research results, we find that the existing research on situation evaluation has mainly gone through the research on network transmission and network structure security threats, and then move towards the research process on the security of transmission information content. Through the analysis of these studies, it is not difficult to find that in these studies, there are more studies on network security events using vulnerabilities and system risks, but relatively few on the situation evaluation of security events at the information content level. According to the characteristics of data in network social media, this paper constructs the situation evaluation system of specific security events and studies the situation evaluation model to realize the event situation evaluation.

2   Microblog Data Collection and Preprocessing

2.1   Microblog Data Collection

Traditionally, there are generally two methods to obtain Weibo data, including Weibo data acquisition method based on Sina Weibo API and Weibo data acquisition method based on a breadth-first crawling strategy. The first method is to use the API accessible on Sina Weibo's open platform. This paper constructs a real-time data acquisition system without login requirement. The basic flow of data acquisition for the system is shown in Fig. 1. The process is as follows:

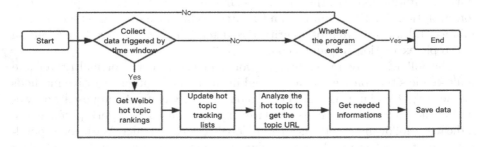

Fig. 1. Weibo data collection process

(1) At the beginning of the program, to determine whether the time window trigger is in effect. The system is updated synchronously with the hot topics of Weibo, and the time window is set at the hour level. If the time window trigger is already in effect, move on to step (2), otherwise, wait for it to take effect;

Table 1. Weibo data structure

Fields name	Example	Fields name	Example
id	4293247910014519	event	公安消防部队转为行政编制
content	#公安消防部队转为行政编制# 公安消防部队转为行政编制公安消防部队移交应急管理部交接仪式正式举行，53年的消防现役成为历史再见了，中国最后的消防战士你好，崭新的中国消防员	like count	15
comments count	8	reposts count	6
create time	1539075602	user id	1578569790
user name	安徽中公教育	user gender	f
device	360安全浏览器	fans count	26256
follower count	913	keywords	部队 中国 再见 公安 历史
topics	公安消防部队转为行政编制	topics count	1
at count	0	URL count	0

(2) Get the Weibo hot topic list, update the Weibo hot topic tracking list, get the Weibo content under the hot topic;

(3) Extract the topic keywords corresponding to the Weibo content and update the topic tracking list of the information publishing platform;

(4) Store the acquired data in the basic resource database;

(5) Exit the program manually.

2.2 Microblog Data Preprocessing

In view of the data obtained in the network social media, this paper has carried on the following basic preprocessing work:

(1) Collation of release time

In order to facilitate the use of the subsequent time information, the information of the release month, the day of the week and the release time is extracted.

(2) Traditional Chinese Characters to Simplified Chinese Characters

In order to facilitate the extraction of keywords and the calculation of text similarity in the subsequent research.

(3) Topic acquisition and quantitative statistics

According to the characteristics of the topic in Weibo, (that is, the topic is in two symbols "#"), the system uses "regular expression" to obtain and count the topic name and the number of topics corresponding to Weibo text.

(4) Remind users of quantity statistics and replacements

According to the characteristics of Weibo usage, when a Weibo user is mentioned in Weibo text, a special tag "@" is used to identify the user. The system uses "regular expression" to obtain and count the number of users who are reminded of the text of Weibo, and replaces these users with special symbol "<USER>".

(5) Hyperlink quantity statistics and replacement

Using regular expression and according to the characteristics of Weibo, the number of hyperlinks in Weibo text is counted and replaced with special symbol "<URL>".

About microblogs that are too short to filter and filtering advertising promotion types. Because the crawler system crawls from the hot Weibo under the hot topic, the Weibo content obtained is the original Weibo with long text content and non-promotional nature. Therefore, in the preprocessing of this paper, it does not involve filtering too short Weibo and promotional advertisements of Weibo. The obtained microblog raw data structure is shown in Table 1.

Table 2. Statistical function rule

Statistics	Calculation rules ($rule_i$)	Description
C_{wb}	$count(w_{id})$	Count the number of Weibo ID
C_{users}	$dedup(w_{userid})$	Count the number of independent users
C_{att}	$sum(w_{attitudes})$	Sum of Weibo attitudes
C_{com}	$sum(w_{comments})$	Sum of Weibo comments
C_{rep}	$sum(w_{reposts})$	Sum of Weibo reposts
C_{avgrep}	$avg(w_{reposts})$	Average reposts count of microblog text
C_{fans}	$sum(w_{fans})$	Sum of Weibo fans
$C_{avgfans}$	$avg(len(w_{content}))$	Average fans count of microblog text
C_{fol}	$sum(w_{followers})$	Sum of Weibo followers
L_{cont}	$avg(len(w_{content}))$	Average length of microblog text
C_{pos}	$count(filter(w_{sent})\{sent = positive\})$	Get the number of positive emotions in Weibo
P_{pos}	C_{pos}/C_{wb}	Proportion of positive Weibo
C_{neg}	$count(filter(w_{sent})\{sent = negative\})$	Get the number of negative emotions in Weibo
P_{neg}	C_{neg}/C_{wb}	Proportion of negative Weibo
C_{neu}	$count(filter(w_{sent})\{sent = neutral\})$	Get the number of neutral emotions in Weibo
P_{neu}	C_{neu}/C_{wb}	Proportion of neutral Weibo

3   Selection of Situation Evaluation Indicators

When selecting indicators that describe the situation of events, it includes the number of microblogs, the number of comments, the number of forwarded microblogs, the number of positive and negative sentimental tendencies, and so on. The basic functions used in this paper is defined as follows:

Deduplication function ($dedup(\cdot)$): Calculate the number of independent non-repeating elements.

Summation function ($sum(\cdot)$): Calculate the sum of all elements.

Counting function ($count(\cdot)$): Calculate the number of elements.

Filter function ($filter(\cdot)$): Filter functions that satisfy the condition.

Length function ($len(\cdot)$): Learn the length of each element.

Average function ($avg(\cdot)$): Filter functions that satisfy the condition.

The combined application of these functions together constitutes a conditional function.

Conditional function ($rule_i$): A function of the statistical calculation rule, which contains the constraints and calculation rules. Common rules are shown in the Table 2.

4 Construction of Situation Evaluation System

In order to evaluate the situation of social security events on the network social media more comprehensively and objectively, this paper constructs a multi-level situation evaluation system. The system is composed of several second-level indicators and three first-level indicators and uses network attention, emotional orientation, network diffusion to analyze and evaluate the current situation of the event. At the same time, in order to better explain the development and changes of the event, the system traces the situation of the event on the timeline and makes a better prediction of the situation developed on the basis of the situation prediction model. The definitions of timeline and time slice are described below

(Time Line): It refers to the time period $[t_1, t_n + \tau]$ covered by social media data related to a particular social security event. The time line can be divided into multiple time slices according to the granularity of time τ.

(Time Slice): It refers to a specific social security event in the timeline. According to the time granularity τ, the timeline is divided into a sequence of time points $[t_1, \ldots, t_n]$. Every two adjacent time points in the composition of the time segment as a time slice, represented as $T = [t_i, t_{i+1}]$, and $t_{i+1} = t_i + \tau$.

In order to reflect the development and change of situation indicator more accurately, in the following section, we set the time granularity τ to 1 h, that is to say, the granularity of situation indicator analysis of specific social security events is 1 h.

4.1 Distribution of Index Weight in Social Security Event Situation Evaluation System

By referring to the relevant research on event situation evaluation at home and abroad, combined with the social media data available in the existing system, and referring to the characteristics of definition changes in the process of event development found in the course of the study, this paper puts forward a situation rating system of social security events in online social media. The system is shown in Table 3.

Based on the two-level indicator composed of basic statistics, the situation evaluation system proposed in this paper constructs the degree of attention by the network, so as to measure the attention of a certain social security event in the range of netizens on the network. Emotional orientation is used to indicate the emotional tendency of network public opinion to a certain social security event; Network diffusion is used to measure the spread of a social security event

Table 3. Event situation evaluation system

Event situation	Netizens' attention	$C_{wb}, C_{users}, C_{com}, C_{att}, L_{cont}, C_{fol}$
	Sentiment orientation	$C_{pos}, P_{pos}, C_{neu}, P_{neu}, C_{neg}, P_{neg}$
	Diffusion degree	$C_{fans}, C_{avgfans}, C_{rep}, C_{avgrep}$

in the network social media. Through this trinity situation evaluation system, the evaluation and analysis of the situation of specific social security events are realized.

4.2 Judging Matrix for Constructing Primary Indicators Based on Secondary Indicators

After constructing the situation evaluation system of social security events, the relatively complex problem of situation evaluation has been decomposed into multiple sub-problems composed of multiple indicators. In this paper, the Analytic hierarchy process (AHP) is used to solve this problem. AHP is proposed by operational research scientist Saaty [18], which is a decision-making method for quantitative analysis of qualitative problems.

In order to calculate the degree of network attention, emotional tendency, and network diffusion, after obtaining the situation evaluation system constructed above, it is necessary to give the numerical judgment of relative importance through the pairwise comparison between the two-level indicators. In order to construct the matrix of importance judgment, a six-tuple reflecting the network attention indicator, a six-tuple reflecting the emotional tendency indicator, and a quaternion reflecting the network diffusion indicator are defined. As shown in Eqs. 1, 2 and 3:

$$E_n = (C_{wb}, C_{users}, C_{com}, C_{att}, L_{cont}, C_{fol}) \tag{1}$$

$$E_s = (P_{neg}, P_{pos}, P_{neu}, C_{neg}, C_{pos}, C_{neu}) \tag{2}$$

$$E_d = (C_{fans}, C_{rep}, C_{avgfans}, C_{avgrep}) \tag{3}$$

The relative importance judgment matrix of network attention indicator elements, the relative importance judgment matrix of emotional tendency indicator elements, and the relative importance judgment matrix of network diffusion indicator elements, as shown in Eq. 4:

$$A_n = \begin{bmatrix} 1 & 3 & 4 & 5 & 8 & 9 \\ 1/3 & 1 & 3 & 4 & 7 & 8 \\ 1/4 & 1/3 & 1 & 2 & 6 & 7 \\ 1/5 & 1/4 & 1/2 & 1 & 5 & 6 \\ 1/8 & 1/7 & 1/6 & 1/5 & 1 & 1 \\ 1/9 & 1/8 & 1/7 & 1/6 & 1 & 1 \end{bmatrix} \quad A_s = \begin{bmatrix} 1 & 2 & 4 & 6 & 7 & 9 \\ 1/2 & 1 & 3 & 5 & 6 & 8 \\ 1/4 & 1/3 & 1 & 3 & 4 & 6 \\ 1/6 & 1/5 & 1/3 & 1 & 2 & 4 \\ 1/7 & 1/6 & 1/4 & 1/2 & 1 & 3 \\ 1/9 & 1/8 & 1/6 & 1/4 & 1/3 & 1 \end{bmatrix} \quad A_d = \begin{bmatrix} 1 & 3 & 5 & 7 \\ 1/3 & 1 & 3 & 5 \\ 1/5 & 1/3 & 1 & 3 \\ 1/7 & 1/5 & 1/3 & 1 \end{bmatrix}$$

$$\tag{4}$$

All eigenvalues were obtained from the judgment matrix of the above Eq. 4, and the eigenvectors of all corresponding eigenvalues were obtained. At the same time, the eigenvectors were formed into a column vector matrix V_n, V_s, V_d, and the corresponding eigenvectors were obtained according to the maximum eigenvalues, and then normalized. And, the weight w_n (Eq. 5), w_s (Eq. 6), w_d (Eq. 7) of each secondary indicator was obtained.

$$w_n = (0.4364, 0.2617, 0.1434, 0.1005, 0.0303, 0.0276) \tag{5}$$
$$w_s = (0.4152, 0.2888, 0.1472, 0.0725, 0.0495, 0.0268) \tag{6}$$
$$w_d = (0.5650, 0.2622, 0.1175, 0.0553) \tag{7}$$

Table 4. Consistency Ratio table

CR_n	CR_s	CR_d
0.062894	0.044065	0.043327

4.3 Consistency Test of Judgment Matrix

The importance judgment matrix is a completely consistent pairwise comparison matrix in the best case, but it is difficult to satisfy the strict condition of pairwise consistency when constructing the matrix, so it is necessary to ensure its consistency when constructing the judgment matrix. The Consistency Ratio (CR) is solved by calculating Consistency index (CI), combined with Random index (RI). The degree of inconsistency is calculated as shown in Eq. 8: λ_{max} is the maximum eigenvalue of the judgment matrix and n is the order of the judgment matrix.

$$CI = \frac{\lambda_{max} - n}{n - 1} \tag{8}$$

The calculation of the CR is shown in Eq. 9, which indicates that the judgment matrix is acceptable when it is less than 0.1.

$$CR = \frac{CI}{RI} \tag{9}$$

The consistency of the above three importance judgment matrices is checked, and the results are shown in Table 4. The calculated results meet the acceptance requirements of the judgment matrix.

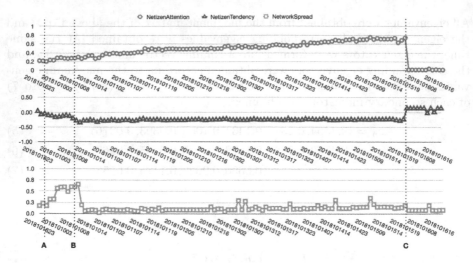

Fig. 2. Specific event situation

5   Event Situation Evaluation

The three first-level indicators were calculated according to the weights corresponding to the secondary indicators. For the event x of interest, there was a sequence of statistics $C_{rule_i x}^{T_k}$ on a time interval of interest $T = (T_1, T_2, ..., T_n)$. The value of the primary indicator was calculated from the secondary indicator weight determined above. And these statistics were normalized to get normalized statistics $N_{rule_i x}^{T_k}$ and a sequence of normalized statistics over time. According to Eq. 12, the weight w_* of the corresponding secondary indicator and the normalized statistical matrix were multiplied to obtain the primary indicator A_*. The obtained primary indicators were visualized and the relationship between the objective conditions of the events and the primary indicators was analyzed.

$$C_{rule_i x}^T = (C_{rule_i x}^{T_1}, ..., C_{rule_i x}^{T_n}) \tag{10}$$

$$N_{rule_i x}^T = (N_{rule_i x}^{T_1}, ..., N_{rule_i x}^{T_n}) \tag{11}$$

$$A_* = w_* \times N_{rule_i x}^T \tag{12}$$

The first-level indicator network attention degree and the network diffusion degree value range obtained are between $[0, 1]$. In order to better describe the sentiment orientation index, this article will take the weight negative when it comes to the statistics of negative emotions and the sentiment orientation index thus obtained is $[-1, 1]$.

6   Experimental Results

This experiment selects Weibo from 2018/10/09 21:00 to 2018/10/16 23:00, about 7 days, pays attention to the event of "anchorman insulting the National Anthem" and selects 48,700 related microblogs.

In the process of the development of the event, the three first-level indicators of the situation rating system have undergone more obvious changes. Among them, the three first-level indicators are network attention, emotional tendentiousness, network diffusion. There are three obvious indicator changes in Fig. 2. For ease of explanation, it is defined as the Time point A, Time point B, and Time point C, event timelines are divided into four intervals by these three points. Event occurs to time point.

(1) At the beginning of the incident, it had just spread in online social media, the relevant information was less, the degree of network attention was relatively low, the degree of emotional tendency was slightly negative, and the degree of network diffusion was not enough.

(2) Time point A: Discussed the incident from a number of topics, and the cause of the incident became clear.

(3) In the process of continuous fermentation of events from Time point A to Time point B: With the disclosure of event information, the degree of network attention continued to increase, the negative direction of emotional tendency was obviously intensified, and the degree of network diffusion was increased. As a result, this topic had gradually formed a heated discussion on the online social platform.

(4) Time point B: The parties to the incident publicly apologized. The facts of the event had been basically determined, the degree of network attention continued to rise, the degree of emotional tendency once again negative decline, the degree of network diffusion continued to maintain a high. The situation of the incident had not eased because of the apology of the parties, on the contrary, it was more intense in the reflection of some indicators.

(5) Time point B to Time point C: The network attention of the event continued to rise over a period of about five days, the emotional tendency remained in a negative low position, and the network diffusion decreased to a lower position. Although the incident was still under discussion on social media, as the spread decreased, the incident began to subside.

(6) Time point C: The police published the decision on punishment and imposed criminal detention on the person concerned. At the end of the development of the incident, netizens began to lose interest in the incident, and the degree of network attention began to decline rapidly. The emotional orientation of the public had also changed from the negative emotional attitude towards the behavior of the parties to the positive emotion recognized by the police. The network diffusion of the event dropped again.

(7) Time point C to the end of the incident: The incident had little online attention and diffusion, emotional attitude remained positive, the incident finally subsided.

Through the analysis of the case, it is obvious that the objective development process of the event is basically consistent with the evaluation of the situation evaluation system proposed in this paper, which also shows the effectiveness of the situation evaluation system.

7   Conclusion

This paper establishes a situation evaluation system for specific events in social media. Through the establishment of real-time crawler system, web crawlers collect social media data, and use statistical and neural network methods to extract corresponding features. Based on these features, AHP is used to construct a following event situation evaluation system. The results obtained in the situation evaluation system are combined with objective facts to illustrate the situation of the event. In future research, attention will be paid to how to assess the way to mitigate the situation and provide solutions to government agencies and people at the time.

Acknowledgements. This study was jointly supported by National Natural Science Foundation of China (No. 61772081), Scientific Research Project of Beijing Educational Committee (No. KM201711232014). We thank anonymous reviewers for their detailed and constructive comments.

References

1. Schuler, D.: Social computing. Commun. ACM **37**(1), 28–29 (1994)
2. Dryer, D.C., Eisbach, C., Ark, W.S.: At what cost pervasive? A social computing view of mobile computing systems. IBM Syst. J. **38**(4), 652–676 (1999)
3. WorldBank: World bank (2018). https://data.worldbank.org.cn/country/US
4. Communique, S.: Statistical communique of the People's Republic of China on 2017 national economic and social development (2018). http://www.gov.cn/xinwen/2018-02/28/content_5269506.htm
5. Becker, H., Naaman, M., Gravano, L.: Beyond trending topics: real-world event identification on Twitter. In: ICWSM (2011)
6. Roesch, M.: Snort - lightweight intrusion detection for networks. In: Proceedings of the 13th USENIX Conference on System Administration, pp. 229–238 (1999)
7. Porras, P.A., Fong, M.W., Valdes, A.: A mission-impact-based approach to INFOSEC alarm correlation. In: Wespi, A., Vigna, G., Deri, L. (eds.) RAID 2002. LNCS, vol. 2516, pp. 95–114. Springer, Heidelberg (2002). https://doi.org/10.1007/3-540-36084-0_6
8. Qu, G., Raghavendra, C.S., Dharmagadda, T., Hariri, S., Ramkishore, M.: Impact analysis of faults and attacks in large-scale networks. IEEE Secur. Priv. **1**(05), 49–54 (2003). https://doi.org/10.1109/MSECP.2003.1236235
9. General Administration of Quality Supervision, I., Quarantine of the People's Republic of China, S.A.o.t.P.R.o.C.: Information security technology_guidelines for the category and classification of information security incidents (2017)
10. Xun-xun, C., Bin-xing, F., Ming-zeng, H., Lei, L.: A new field in security of internet information and content-network information penetration detection technology. J. China Inst. Commun. **25**(7), 185–191 (2004)
11. Bin-xing, F., Yun-chuan, G., Yuan, Z.: Information content security on the internet: the control model and its evaluation. Sci. China Ser. F Inf. Sci. **53**(1), 30–49 (2010). https://doi.org/10.1007/s11432-010-0014-z
12. Yuan, W.: Research on mining of internet public opinion based on semantic and statistic analysis. Ph.D. thesis, Wuhan University of Technology (2012)

13. Barroso, N., de Ipiña, K.L., Ezeiza, A., Hernández, C.: An ontology-driven semantic speech recognition system for security tasks. In: 2011 Carnahan Conference on Security Technology, pp. 1–6 (2011). https://doi.org/10.1109/CCST.2011.6095948

14. Xiu-zhen, C., Qing-hua, Z., Xiao-hong, G., Chen-guang, L.: Quantitative hierarchical threat evaluation model for network security. J. Softw. **17**(4), 885–897 (2006)

15. Lei, L., Hui-qiang, W., Ying, L.: Evaluation method of service-level network security situation based on fuzzy analytic hierarchy process. J. Comput. Appl. **29**(9), 2327–2331, 2335 (2009)

16. Hong-xing, C.: Research and simulation of network security evaluation based on AHP weights calculation. Comput. Simul. **30**(8), 266–269 (2013)

17. Fu-yong, X., Jian, S., Jian-ying, L.: Study on comprehensive assessment method for network security based on Delphi and ANN. Microcomput. Dev. **15**(10), 11–13, 15 (2005)

18. Saaty, T.L.: Decision making with the analytic hierarchy process. Int. J. Serv. Sci. **1**(1), 83 (2008)

Automatic Recognition of Chinese Separable Words Based on CRFs

Ning Dong and Weiming Peng[✉]

College of Information Science and Technology, Beijing Normal University,
Beijing, China
dongning93@163.com, pengweiming@bnu.edu.cn

Abstract. Currently, most of the automatic recognition tasks of separable words adopt a rule-based method, which relies on automatic word segmentation results and lexical patterns generated from common inserted constituents. However, they suffer from incorrect word segmentation results and inaccurate and limited rules. Moreover, they ignore the rich information contained in the context. To address these issues, this paper proposes a CRFs-based method which employs nine features, such as character, POS tag, punctuation, word boundary, keyword and POS sequential rule. Experimental results on real-world datasets show that our approach can make full use of rich information and achieve significant improvements on recognition efficiency compared to all the baselines.

Keywords: Separable words · Separated form · Conditional Random Field · Sequence labeling · Automatic identification

1 Introduction

In Chinese, if two morphemes are weakly connected and can be separated by a limited number of other constituents without causing change of meaning, the words they form are called Chinese separable words (Lihe words). The task of separable words automatic recognition is that computers can automatically extract separated cases from given contexts[1]. More formally, given string "*A*B*"[2], computers have the ability to decide whether characters A and B can be combined together to serve as a proper usage of separable word "AB", or, in other cases, A and B are two different words[3] with two different meanings. For example, string "*当*面*" in (1) shows a positive case in which morpheme "当" and morpheme "面" are combined to express the inalienable meaning of word "当面 (in the presence of)", while (2) demonstrates a negative one where "当" functions as a pronoun, meaning "when" in English and "面" is a morpheme of noun "面纱", meaning "veil".

---

[1] As combined cases can be directly recognized by word segmentation.

[2] * refers to characters other than separated morphemes.

[3] More precisely, in some cases, A and B are joined with other morphemes to form different words, as shown in (2). No matter A/B is a word or XA/BX is a word, it has a meaning distinct from that of separable word "AB".

© Springer Nature Switzerland AG 2020
J.-F. Hong et al. (Eds.): CLSW 2019, LNAI 11831, pp. 456–465, 2020.
https://doi.org/10.1007/978-3-030-38189-9_47

(1) 他当着你的面对我动手动脚,[4] *(He molests me in the presence of you,)*
(2) 可当她揭开面纱的一刹那, *(But when she lifted the veil,)*

The recognition task has significant practical values in natural language processing. The downstream applications are as diverse as word segmentation, POS task, tagged corpus construction, machine translation and syntactic parser. In recent years, researchers have shown an increasing interest. As much work has focused on rule-based methods that explicitly attempt to generate typical lexical patterns appearing from those inserted constituents between morpheme A and morpheme B. For the reason that A and B are certain characters, a natural way to represent those patterns is provided by POS tag sequences. For example, given the rule ① "A + r + B", which means "left morpheme A + pronoun + right morpheme B", the string "帮/v 他/r 忙/n *(do him a favor)*" will be recognized as a positive case of the separable word "帮忙 (to help)". Although this rule-based approach has its advantages, it also has limitations.

First, the efficiency directly suffers from word segmentation errors. For separable word "AB", A and B could be word-forming morphemes, which could cause combinational ambiguity to Chinese automatic word segmentation tools. Approaches like [1–4] set a filter to restrict the segmentation result of A/B to only one character. Segmentation results like "XA" and "BX" would be left out. However, cases like (3), where segmentation errors constantly occur, would be missed, leading to reduced performance.

(3) 并使/c 您/r 记起/v 您/r 当着/p 他/r 的/uj 面向/v 我/r 施加/v 的/uj 侮辱/v,/x[5] *(It reminds you that you insulted me in his presence,)*

Second, the matching rules, no matter generated by linguists or computers, are finite and limited compared to the abundant real-world usage. Under the effort of matching typical grammar patterns of typical separable words, existing rule sets are mostly optimized in a common sense. However, a relatively tricky example (4) shows some negative cases can also satisfy the rule ①, leading to reduced accuracy. Besides, language in use is intricate and subtle. It is difficult to generate clear rules from separated forms with long and complex inserted constituents [1]. Furthermore, such approaches only extract information within A and B, which will lose a large amount of rich information contained in those neglected context, thus having failed to address distinctions among similar cases like (5) and (6). POS tag sequences of "A*B" are similar in form but different in meaning.

(4) 但/c 见/v 此人/r 面/n 如/v 美玉/nr,/*(It only can be seen that this person has a face as beautiful as jade,)*
(5) 一年/m 只见/v 几次/m 面/n,/x *(Only meet several times a year,)*

---

[4] All examples are collected from [11]. Positive separated forms are underlined, whereas negative ones are not.

[5] (3) is given by the following Chinese automatic word segmentation tools: jieba (https://pypi.org/project/jieba/), Thulac (THU Lexical Analyzer for Chinese, http://thulac.thunlp.org/), CUCBst (中国传媒大学文本切分标注系统). All three tools made wrong segmentations on "面向/v (to face)", whereas the correct one should be "面/n (presence) 向/p (to)". The former is a verb. The latter are two words: a noun and a preposition.

(6) 只见/v 几个/m 面/n 带/v 病容/n 的/uj 船客/n,/x *(There are several guests on board who look sick to their faces,)*

To address these issues, this research models the automatic recognition task as a sequence labeling task based on Conditional Random Fields (CRFs). Here we reconsider the word boundary results given by automatic segmentation tools. After that, to utilize all information of inserted constituents such as lexical patterns, grammar rules, keywords and POS tag sequences. And then, within the window of CRFs, we extract useful context features such as punctuations, keywords, POS features and collocations. Finally, we aggregate all those features to obtain the CRFs feature template. The ability to leverage a large number of input features of CRFs enables our approach to improve robustness as well as to maintain efficiency.

The experimental results on datasets collected from large-scale corpus including BCC and CCRMA[6] show that our approach achieves significant and consistent improvements compared to baselines by rule-based methods, reaching higher precision and recall rate. In particular, we show that our model can achieve comparable performance on those long and complex separated forms as well as on those separable words with higher segmentation ambiguity.

2 Related Work

The idea of automatic recognition of separable words was first articulated by [5] and popularized later by [4], which is to extract typical grammar patterns shared among massive separated forms collected from large-scale corpus, either by linguists or computers. An application of this idea is carried out within the automatic identification system [1], which relies on the words segmentation results and the typical structure types of extended components. However, the system only selects the higher discrete frequency of separable words and neglects those extended constituents longer than three words. It loses a large amount of positive cases and suffers from segmentation errors. More systems under similar ground can be found in [6, 7]. Hence, an advanced system attempts to improve the system by sorting rules into four sets: the characteristic set, the part-of-speech set, the reduplication set and the multiple extended constituents set [3]. By matching rules in order, it successfully improves the recognition efficiency, especially the recall rate. However, the system still suffers from word boundary issue. For example, in string "*碍*事儿* (be in the way; hinder)", due to the "儿 (pro-suffix of noun)" in the right boundary word "事儿 (things)", this positive case would be neglected. Besides, the matching window is restricted to up to 5 characters, still quite limited compared to the real-world usage of 7 or more characters. Different from these methods, our model employs a CRFs classifier to obtain boundary and context information, which can make full use of rich information of longer inputs.

Our work is also inspired by the idea of sequence labeling tasks in NLP. [8] uses a CRFs model to identify prepositional phrases and uses left and right word boundaries, e.g. "除......外 (except for)", as training features. [9] regards keywords and class

---

[6] Please refer to Sect. 4.1 for details.

sequential rules as features in CRF classifier to mine subjects and objects in comparative relations. According to the fact that separated forms share common explicit grammar features with prepositional phrases and comparative relations, our model also employs word boundary, keywords and class sequential rules as CRFs features.

3 Methodology

CRFs is a statistical method for sequence tagging and segmentation proposed by [10].

Define G = (V, E) as a non-directional graph (Y = {Y_v |v ∈ V}). That is, each node in V corresponds to the ingredient Y_v of mark sequence expressed by a random variable. If every random variable Y_v for G complies with condition independence attribute (Markov property), then CRF (X, Y) is a non-directional graph model taking observation sequence X as the condition:

$$P(Y_v | X, Y_w, w \neq v) = p(Y_v | X, Y_w, w \sim v) \tag{1}$$

When modeling for sequence labeling tasks, we assume that the input sequence is X = $X_1 X_2 \cdots X_n$, the output sequence is Y = $Y_{1\ 2} \cdots Y_n$, X and Y can be seen as random variables. Then we can simply get a first-chain CRFs as shown in Fig. 1:

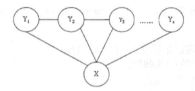

Fig. 1. The structure sketch of the first-chain CRFs mark sequence

4 Experiment

4.1 Datasets

To our best knowledge, most of existing lexical analyzing tools and tagged corpus have not paid much attention to separable words. So we create two datasets to meet our training purpose.

1. BaseSep. A dataset collected from BCC corpus [11]. It contains 8,314 positive cases and a total of 14,664 cases of 13 separable words. BaseSep adopts a word list according to [12]. Then sentences containing the form as "*A*B*" are retrieved from search engine of BCC corpus.
2. CCRMA. A dataset retrieved from Corpus of Language and Character Resources Research Center, Beijing Normal University [13]. It contains 5,777 positive cases and a total of 16,450 sentences of 786 separable words. CCRMA adopts word list from [14] and sentences involving "*A*B*" are collected by a tree-structure-parser [15] (Table 1).

Table 1. The distribution of experimental datasets

Datasets	Training set (BaseSep*80%)	Closed test set (BaseSep*20%)	Open test set (CCRMA*100%)
Amount of sentences	11731	2933	16450

Then experimental datasets are mainly processed by the following three steps:

1. Manual annotation. After preliminary collected from search engine, each sentence is annotated by two human linguists to decide if it is a proper separated form of the target separable word. Using the eligibility criteria of [16], the mark "1" is given at the end of the case if it is positive, "0" otherwise.
2. Sentence segmentation. Sentence segmentation is performed every time there appear punctuations, including but not limited to comma, period, question mark, colon, exclamation mark, ellipsis mark.
3. BIEO marks. Finally, the marks of "1/0" are annotated by computers to each character with BIEO marks[7]. Because the final result is highly sensitive to word segmentation results near A and B, BIEO marks are employed to character-level instead of word-level. Examples of annotations are shown in Table 2.

Table 2. Examples of preprocessing of datasets

Raw state	Step1	Final state
1754:,高雅大方,简直是梦寐以求的情人,虽然只<U>见过几面</U>,但是我对她的印象非常深刻。 [a]	1	虽/O 然/O 只/O 见/B 过/I 几/I 面/E,/O[b]
1645:宛如一盆冷水,当头泼下任风萍目光一转,<U>见到他面</U>上的神态,目中暗露喜色……[c]	0	见/O 到/O他/O面/O上/O 的/O 神/O 态,/O[d]

[a]1754:, Elegant and generous, she is a dream lover. Though we have only met for several times, I'm very impressed with her.
[b]Though we have only met for several times,
[c]It's like a basin of cold water pouring right over her head. Fengping Ren turned her eyes and saw the look on his face. Her eyes glowed with joy.
[d]Saw the look on his face.

4.2 Features

After several attempts with different feature templates, along with integration of previous works, nine features are formally selected, including character, part of speech, punctuation, word segmentation boundary, keyword and POS sequential rule. The meanings of features are as follows (Table 3):

---

[7] BIEO marks: "B" represents the beginning or the left boundary of the separable word; "I" represents the inserted constituents in the middle; "E" represents the end or the right boundary of the separable word; "O" represents the outside or non-separable constituents.

Table 3. Features employed in the CRFs experiment

Wk, Pk, Punck, CutAk, CutBk, Keyk, Rulek, PAk, PBk
Wk·Wk + 1, Wk·WK + 1·Wk + 2, Pk·Pk + 1, Pk·CutAk, Pk·CutBk, PAk·CutAk, PBk·CutBk, Keyk·Rulek, Wk·Rulek·Cutk

① Wk: the current character; ② Pk,: the current POS tag; ③ PAk: the POS tag of left boundary (LB in short) of the target separable word (verb, locative preposition, noun as top three); ④ PBk: the POS tag of right boundary (RB in short) of target separable word (noun, quantifier, verbal morpheme as top three); ⑤ Punck: the punctuation following RB; ⑥ CutAk: LB. There are three types of LB indicating positive cases. The first is that the LB is composed by only one character, as the LB "洗 (to wash)" in the case (8) "洗/v 个/q 澡/ng 再/d 回家/n 怎么样/r ?/x *(How about taking a shower before going back home?)*". The second is that LB takes the form of "AX", where X could be a complement to A, e.g., the "洗完 (have washed)" in the case (9) "洗完/v 澡/n,/x *(After taking the shower,)*". The third type is segmentation as "XA" where X may be the adverbial modifier of A, such as case (10) "没洗/v 上/f 澡/n,/x (have not taken a shower,)". ⑦ CutBk. Similarly, CutBk indicates three types of RB: "B", "BX" and "XB". The "XB" is more likely to indicate positive cases as case (11) "下/f 着/uz 小雨/n,/x *(It is raining a little bit,)*", while "BX" tends to be negative as case (12) "见/v 到/u 我/r 面纱/n 下/f ......*(to see under my veil...)*"; ⑧ keywords: keywords = {了,过,着,个,什,么,的,上,完,好,起,成,得,一,大,高,闷,透,尽,碎,足}   (all Chinese adverbs or function words); ⑨ Rulek: POS sequential rules.

4.3 Baselines

In the annotation result of CRFs model, only if every mark of a positive case is correct, the result is right. The negative cases do not account for accuracy.

Currently there is no state-of-the-art classifier published on this issue. To measure the results of our approach, we apply several rule-based methods of earlier researches [1–4] on our datasets to serve as baselines. We also compare our final model to its initial state, where only two features, characters and POS tags, are employed.

Table 4. Experiment results compared to baselines

Dataset	Baselines				Our model		
	[4]	[2]	[1]	[3]	Our initial model (BaseSep)	Closed Test (BaseSep)	Open Test (CCRMA)
Accuracy rate	0.987	0.828	0.954	0.930	0.976	**0.984**	**0.958**
Recall rate	0.423	0.117	0.321	0.912	0.972	**0.989**	0.889

4.4   Overall Results

Table 4 shows that our approach significantly outperforms all the baselines, reaching higher accuracy and recall rate. Particularly, the recall rate is dramatically improved compared to baselines. The closed test dataset BaseSep achieves ideal efficiency both in accuracy and recall rate. However in the open test dataset CCRMA, the recall rate has relatively worse performance.

4.5   Experimental Results Analysis

1. Our approach shows a better recognition result on those words with higher risk of segmentation errors. To examine the impact of combinational ambiguity near morphemes A and B, we measure the ambiguity extent using the following formula:

$$\text{Ambiguity extent} = \frac{\text{the amount of words whose LB and RB are composed by single character}}{\text{the total amount of target separable words}}$$

$$(2)$$

As can be seen in Fig. 2, the accuracy of baselines decreases with the increase of the ambiguity extent, whereas our model keeps a consistent better performance. More specifically, CRFs classifier made correct judgment over all four segmentation errors in closed test dataset BaseSep. For instance, in case (13) "您/zg 何妨/r 想法子/l 见/v 上/f 一/m 面试/v 上一/b 试/vn,/x *(Why don't you manage a meeting and have a try,)*", the RB "面" ought to be combined with LB "见" to deliver an inàlienable meaning of "见面 (to meet)". The Chinese automatic word segmentation tool made a mistake by forming "BX" as a different word "面试 (to interview)". Challenged by such errors, the rule-based methods would certainly fail as we discussed earlier, while our model still recognized the separated forms in spite of errors. It demonstrates that our CRFs classifier can alleviate the influence of wrong word segmentations to some extent and by doing so, improve the robustness of the model.

2. Our CRFs model has a relatively better performance on long and complex separated forms as illustrated in Fig. 3. What stands out in the polygonal chart is the recognition ability over separated forms longer than 5 characters, which has not been discussed in earlier researches.

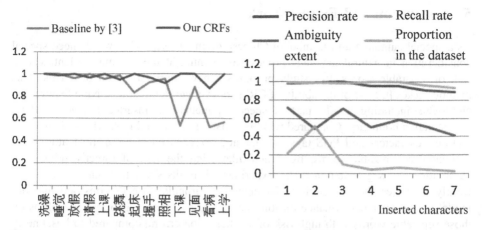

Fig. 2. Comparison of accuracy rates of our model and baselines[8]

Fig. 3. Recognition performance on different length of separated forms

4.6 Error Analysis

In the investigation there are mainly two sources of errors. The first is the ambiguity arose by polysemy of separated forms. Take the word "看病" for example, the typical meaning of "*看*病*" is the doctors to deliver medical treatment or the patient to see doctors [17], as shown in positive case (14). But there is also another meaning of "*看*病*", in which "看" means "to observe and judge", as shown in negative case (15). Moreover, (14) and (15) have similar contexts and both fit the grammar rule of "A+(r) +的+B". Our model annotated both of them as positive. The responses relating to polysemy are subjective and are therefore susceptible to recall bias.

(14) 是哪个大夫给她看的病? *(From which doctor did she get medical treatment?)*
(15) 杨杏园道: "我看她的病, 这时候好得多, 暂时不搬到医院, *(Qingyuan Yang says: "It seems to me that she has restored much to good health. No need to move to the hospital currently.")*

Another major source of error is that our approach relaxes the restriction of the segmentation results of LB and RB. As a result, some verb-object phrases explicitly raise the uncertainty and diminish the distinctions with separated forms. For instance, case (16) is a verb-object phrase to most of human linguists, whereas our CRFs classifier regarded it as a separated form.

(16) 紧张的高考结束了,我们班开了一个告别晚会。 *(After the stressful college entrance examination, our class held a farewell party.)*

---

[8] The meanings of words are as follows: 洗澡 to take a shower; 睡觉 to sleep; 放假 to have a vacation; 请假 to ask for leave; 上课 to go to class; 跳舞 to dance; 起床 to get up; 握手 to shake hands; 照相 to take a picture; 下课 to get out of class; 见面 to meet; 看病 to see a doctor; 上学 to go to school.

5   Conclusion and Future Work

As a special language unit in modern Chinese grammar, separable words need special consideration in natural language processing. Aiming at the automatic identification task of separable words, this study proposes a new approach based on CRFs. This method makes full use of the ability of CRFs to leverage rich features. Nine features are employed, including words, part of speech, punctuation, inserted keywords, POS, segmentation boundary. Compared with the existing rule-based method using only two features, characters and POS tags, our method converts more linguistic features into computer-recognizable forms, by which alleviating the risk of directly relying on automatic word segmentation tools. Experimental results show that our model significantly improves the recognition efficiency. In particular, we demonstrate that our model has a better performance on long and complex separated forms, as well as on those separable words with high risk of segmentation errors, compared to baselines.

Two datasets are created in this survey. But due to the limitation of authors' ability, only a small scope of separable words is studied. The small size of the dataset severely reduces the recall rate in open test dataset CCRMA. In the future, we will explore this issue in following directions:

1. A further data collection. We will explore more types of separable words as well as separated forms. The inversion form "*B*A*" is worthy of discussion.
2. Further research in natural language processing perspective. With combined forms and separated forms, separable words may reveal some correspondences between the Chinese morphology and the syntax. The automatic recognition task of separable words may shed light on not only the task of Chinese automatic word segmentation, but also on downstream tasks like shallow parsing and verb-object collocation.

Acknowledgments. This research is supported by: National Natural Science Foundation of China (61877004); Natural Science Foundation for the Higher Education Institutions of Anhui Province of China (No: KJ2019A0592).

References

1. Bo, L.: Research on automatic recognition of separable words based on corpus. Master thesis, Hebei University, Baoding (2015)
2. Haibo, R., Gang, W.: The analysis of modern Chinese separated forms based on corpus. Lang. Sci. **06**, 75–87 (2005)
3. Jiaojiao, Z., Endong X.: Automatic recognition of separable words based on BCC. J. Chin. Inf. Sci. **31**(01), 75–83+93 (2017)
4. Weihua, Z.: Information processing oriented researches on the semantic collocation between the verbs and objects in modern Chinese language. Master thesis, Central China Normal University, Beijing (2007)
5. Aiping, F.: Word recognition in source language analysis of Chinese-English machine translation. J. Chin. Inf. Process. **05**, 7–13 (1999)

6. ChunXia, W.: Researches of separable words based on corpus. Master thesis, Beijing University of Language and Culture, Beijing (2001)
7. Haifeng, W.: The study on separable words' separated form function of modern Chinese. Master thesis, Beijing Language and Culture University, Beijing (2008)
8. Silei, H.: Automatic identification of Chinese prepositional phrase based on CRF. Master thesis, Dalian University of Technology, Dalian (2008)
9. Gaohui, H., Tianfang, Y., Quansheng, L.: Mining Chinese comparative sentences and relations based on CRF algorithm. Appl. Res. Comput. **27**(06), 2061–2064 (2010)
10. Lafferty, J., Mccallum, A., Pereira, F., et al.: Conditional random fields: probabilistic models for segmenting and labeling sequence data. In: Proceedings of the 18th International Conference on Machine Learning 2001, pp. 282–289. Morgan Kaufmann Publishers Inc., San Francisco (2001)
11. BCC. http://bcc.blcu.edu.cn/. Accessed 21 May 2019
12. Office of Examination Center of National Chinese Proficiency Examination Committee: Modern Chinese Dictionary, 5th edn. Economic Science Press, Beijing (2001)
13. CCRMA. http://www.jubenwei.com/. Accessed 21 May 2019
14. Institute of Linguistics: CASS: Modern Chinese Dictionary, 6th edn. The Commercial Press, Beijing (2012)
15. Song, T., Peng, W., Song, J., Guo, D., He, J.: The construction of sentence-based diagrammatic treebank. Chinese Lexical Semantics. LNCS (LNAI), vol. 10085, pp. 306–314. Springer, Cham (2016). https://doi.org/10.1007/978-3-319-49508-8_29
16. Daoqin, W., Zhongchu, L.: On grammatical features and distinctive principles of split words. Soc. Sci. J. Xiangtan Polytechnic Univ. **03**(03), 47–50 (2001)
17. Qinghui, Y.: Dictionary of the Usage of Separable Words, 1st edn. Beijing Normal University Press, Beijing (1995)

Tibetan Sentence Similarity Evaluation Based on Vectorized Representation Techniques

Zhou Maoxian[1,2,3], Cizhen Jiacuo[1,2,3], and Cai Rangjia[1,2,3(✉)]

[1] Qinghai Normal University, Xining, China
zoe988223@sina.com, 543819011@qq.com, zwxxzx@163.com
[2] Tibetan Information Processing and Machine Translation Key Laboratory
of Qinghai Province, Xining, China
[3] Key Laboratory of Tibetan Information Processing, Ministry of Education,
Xining, China

Abstract. Sentence similarity evaluation is an essential subject among the researching fields of Natural Language Processing (NLP), however, Tibetan related research on this subject is fairly inactive and has rarely drawn attention. In this work, we proposed an approach by leveraging vectorized representation techniques to tackle this problem by implementing two methods, namely, Euclidean distance evaluation and Jaccard similarity evaluation. Experiments indicated the performance of presented methods is satisfactory.

Keywords: Similarity evaluation · Tibetan sentences · Word embedding

1 Introduction

Sentence similarity is a relatively complex task. The meaning of similarity varies in different professional fields. Researchers approach of this task with different models or metrics to evaluate text similarity in different domains and applications. Similarity evaluation between sentences plays an important role in the field of Natural Language Processing [1, 2], and it has applied to example-based machine translation, computer-assisted translation, corpus-based language teaching, automatic summarization, question answering system, language comparison, information retrieval, information filtering, etc. The evaluation of sentences similarity is one of the key technologies [3]. Back in the 1960s, researches on text similarity were carried out abroad, and various models and algorithms were proposed. In 1969, the Vector Space Model (VSM) was proposed by Gerard Salton and McGill. The basic idea of this work is aiming at compress the document to a vector representation of the feature items' weight. Although domestic research on text similarity is later than foreign countries, many of the results have been applied to practice.

There are very few related studies on Tibetan text similarity evaluation in the field of Tibetan information processing. The study of Tibetan sentence similarity algorithm proposed in the literature [3], which using hashed words inverted index and quickly selecting the set of candidate sentences from the corpus based on rough selection of sentence length similarity algorithm, and then using the multi-strategy selection algorithm based on the similarity of the word form and continuous word sequence.

J.-F. Hong et al. (Eds.): CLSW 2019, LNAI 11831, pp. 466–473, 2020.
https://doi.org/10.1007/978-3-030-38189-9_48

They selected 166 Tibetan sentences from the Tibetan textbook. It is divided into 8 categories, each of which has 19 to 21 sentences as a standard set. Each sentence has 17 to 21 similar sentences in the standard set, and then the similarity is calculated for the eight types of sentences. The average accuracy is 83.6%. This method can effectively measure the similarity between two Tibetan sentences. In 2016, the literature [4] proposed a similarity evaluation method for Tibetan segmentation fusion. This method treats each paragraph in the text as a short text in units of paragraphs, by calculating the similarity between every two short texts. Then the similarity values between long text and long text are obtained, so that the similarity value of two Tibetan texts is obtained, and the accuracy rate is 76%.

In this work we introduce vectorized representation techniques to Tibetan sentence similarity evaluation task by designing and implementing two evaluation metrics, including Euclidean distance evaluation and Jaccard similarity evaluation. A satisfactory result was obtained through experimental comparisons.

2 Word Embedding and Sentence Similarity Evaluation

2.1 Word Embedding and Language Model

The establishment and training of language models in the field of Natural Language Processing are very important components, such as the common N-Gram model and the deep learning model that have been widely discussed and explored recently [5]. In machine learning, text needs to be vectorized to ease the processing before feeding into machine learning pipelines. The word embedding in the deep learning model is a way to represent the semantic information of the words by converting discrete symbolic form to a continuous dense vectorized representation. A simple word embedding approach is one-hot representation which takes the size of the vocabulary as the length of the word embedding. There are only two components in embedding vector, namely 1 and 0, where 1 indicates the index of corresponding word in the vocabulary. For example, the word embedding of word a is [0, 0, 1, 0, 0, ...], then the 1 in vector indicates that the word a is the third word in the vocabulary. The representation of such word embedding is easily plagued by dimensional disasters and cannot express the similarity between words. In 1986, Hinton proposed the use of the words distributed representation to represent semantic information, this method is also known as word representation or word embedding [6]. Word2Vec[1] is an effective tool provided by Google to ease training and inference word embedding, which does not use the discrete One-hot representation, but distributed representation to represent the word semantic information. Through training, each word in its given language is mapped into a dense vector of fixed length, such as ཤིང་སྡོང་ [0.414029 0.130032 0.191205 0.163355 0.078749......]. The Word2vec tool mainly includes the Continuous Bag of Word (CBOW) model and the Skip-gram model. Both models are proposed by Mikolov [7]. The model architecture is as shown in Fig. 1. The former is modeled based on continuous vocabulary co-occurrence. And the latter is modeled by skip certain characters, which takes longer time to train than the former.

---

[1] https://code.google.com/archive/p/word2vec.

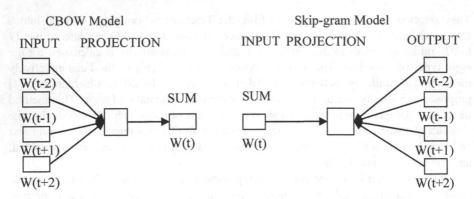

Fig. 1. Schematic diagram of CBOW model and Skip-gram model.

2.2 Sentence Similarity Evaluation

From the perspective analysis of sentences, the evaluation of sentence similarity is basically divided into two categories: the evaluation based on the similarity of surface information and the evaluation based on syntax and semantic similarity. The similarity evaluation based on surficial information of the sentence, such as word frequency, Part-of-Speech, word order, word length, etc., to consider a sentence as a linear sequence of words, without analysis on grammatical structure of the statement. This method cannot manage to capture the similarity of the whole sentence structure when evaluating similarity between sentences. The syntax and semantic similarity based evaluation carry out in-depth syntactic and semantic analysis of the sentence, compare the relationship between the two sentences from the deep structure, and perform similarity evaluation on this basis.

3 Tibetan Sentence Similarity Evaluation

3.1 Tibetan Sentence Representation

Since there is no effective tool to learning vectorized representation for Tibetan sentence vector. In this work we uses the word embedding to indirectly obtain the sentence vector. Therefore, the sentence similarity evaluation method based on the surface information is adopted, which is, the word similarity in the sentence is equivalent to the sentence similarity. On the basis of the training results of Tibetan word embedding, the Tibetan sentence vector is obtained by adding and averaging the word embedding. For example, a Tibetan sentence, after the word segmentation, four words are denoted as A, B, C, D, assuming that the training word embedding is a 3-dimensional vector, and the word embedding of the four words A, B, C and Dare [1, 2, 3], [2, 3, 4], [1, 1, 2], [2, 2, 3] then the vector of the sentence is [(1 + 2 + 1 + 2)/4, (2 + 3 + 1 + 2)/4, (3 + 4 + 2 + 3)/4]. Experiments show that this method can better preserve sentence information.

3.2   Euclidean Distance Evaluation

The Euclidean metric is a distance definition that refers to the true distance between two points in multi-dimensional space or the natural length of the vector (i.e. the distance from the point to the origin). In terms of its meaning, the smaller the Euclidean distance indicating the greater the similarity between the two Tibetan sentences, and vice versa.

The n-dimensional Euclidean space is a set of points, and each of its points X or vector x can be expressed as $(x_{[1]}, x_{[2]}, \cdots, x_{[n]})$, where $x_{[i]}$ $(i = 1, 2, \ldots, n)$ is a real number, called the ith coordinate of X.

The distance ρ between two points $A = (a_{[1]}, a_{[2]}, \ldots, a_{[n]})$ and $B = (b_{[1]}, b_{[2]}, \ldots, b_{[n]})$ (A, B) is given by the following formula (1):

$$\rho(A, B) = \sqrt{\left(\sum (a_{[i]} - b_{[i]})^2\right)}, (i = 1, 2, \ldots, n) \ . \tag{1}$$

The two Tibetan sentences A and B have respective vector representations, and the similarity can be judged by calculating the Euclidean distance between the two sentence vectors. Shorter the distance, the greater the similarity; the longer the distance, the less the similarity.

3.3   Jaccard Similarity Evaluation

For sets A and B, the Jaccard similarity is given by formula (2):

$$Jaccard(A, B) = |A \text{ intersect } B| \, / \, |A \text{ union } B| \tag{2}$$

The similarity value takes a value between [0, 1], and when A = B, the similarity is 1. The Jaccard similarity can be converted into a Jaccard distance, as shown in formula (3):

$$Jaccard\,distance(A, B) = 1 - Jaccard(A, B) \tag{3}$$

The disadvantage of Jaccard similarity is that the value of the element can only be 0 or 1, and no more information can be utilized. Since each component of the previously obtained sentence vector is a real number, the generalized Jaccard similarity definition is used here so that the value of the element is an arbitrary real number. This similarity is also called the Tanimoto coefficient and is expressed by EJ. The evaluation method is given by formula (4):

$$EJ(A,B) = (A * B) / \left(||A||^2 + ||B||^2 - A * B \right) \tag{4}$$

A and B are respectively represented as two vectors. Each element in the set is represented as a dimension of the vector. In each dimension, the value range is [0, 1], and A * B represents the vector product, $||A||^2$ represents the modulus of the vector, i.e. $||A||^2 = \operatorname{sqrt}(a_{12} + a_{22} + a_{32} + \ldots)$.

Different from the similarity evaluation method in 3.2, the result calculated here is similarity rather than distance, so when the similarity is obtained for two Tibetan sentences, the larger the EJ (A, B) value, the higher the similarity; The smaller the EJ (A, B) value, the lower the similarity.

4 Experiment

4.1 Word Embedding

The corpus used in the experiments is a Tibetan corpus of 500 MB in size, including news, textbooks, government documents, legal texts, and spoken language, with a total number of 2.84 million sentence as a word embedding training set. Use the word embedding training tool Word2vec to get the Tibetan word embedding. In order to compare the effects of different word embedding models on the similarity evaluation, in this work we use the CBOW model and the Skip-gram model to train two sets of Tibetan word embedding files. The window size is 5 and the dimension is 200. The training obtained a Tibetan word embedding file with the 98122 words, as shown in Table 1:

Table 1. Partial Tibetan word embedding generated by training.

No.	Tibetan Word	word embedding
98115	ཞེན་མེན་	[-0.064197,0.078042,...]
98116	ཐུངས་གཟུགས་	[0.193874,0.136975...]
98117	ཤེས་དྲུང་	[-0.124637,-0.052450,...]
98118	གན་གྱུའ་	[0.187092,0.226460,...]

Table 2. Tibetan vocabulary similarity test results. Table shows the lexical similarity test results of Word2vec training using the skip-gram model:

Enter word or sentence (EXIT to break): སྐྱག་ཡུང་	
Word: སྐྱག་ཡུང་ Position in vocabulary: 4401	
Word	**Cosine distance**
བརྟན་སྐྱིས་	0.693358
རྟེས་འབོར་	0.679643
ཚོས་ལེད་	0.671659
མཐེན་ཚས་	0.669562
སྐྱགས་འརྫེག་	0.655946

4.2 Tibetan Sentence Similarity Evaluation

4.2.1 Dataset

9,700 Tibetan sentences were randomly selected from the Tibetan corpus, 100 of which were selected and matched with three synonymous or near-semantic sentences for each sentence. We generated 100 groups of 400 sentences corpora, as shown in Table 3. The first sentence in each group is selected as the base sentence, and the other three sentences are used as similar sentences of the first sentence for testing. The remaining 9600 Tibetan sentences constitute a noise set, which is mixed with 400 synonymous sentences to form a test corpus with a size of 10,000 sentences. According to the word embedding of two different models trained before, the sum is obtained by adding and averaging the two sets of sentence vectors for this 10,000 sentence.

Table 3. A group of thesaurus.

NO.	sentence
1	ཁྱེད་ཀྱི་སྐྱག་ཏུ་ང་ལ་ཉིན་ཞིག
2	ཁྱེད་ཀྱི་སྐྱོ་ཁག་ང་ལ་ཉིན་ཞིག
3	ཁྱེད་ཀྱི་སྐྱོ་སོ་ང་ལ་ཉིན་ཞིག
4	ཁྱེད་ཀྱི་དུས་རྐང་ང་ལ་ཉིན་ ཞིག

4.2.2 Experimental Metrics

For the Tibetan corpus, the corresponding test criteria is given. For each of the Tibetan sentences, there are three similar sentences, so the following test methods are used for testing:

Calculate the percentage of the number of similar sentences in the four sentences with the highest similarity to each of the benchmark sentence, i.e., formula (5) to calculate the accuracy:

$$\text{Accuracy} = \sum \text{Number of Hits per Group}/300 \times 100\%. \tag{5}$$

4.3 Similarity Evaluation

The experiment uses two similarity evaluation methods, each of which uses the Tibetan word embedding trained by the two models, that is, four sets of experiments are performed.

The experimental results of one benchmark sentence are as follows:

Table 4. The similarity evaluation based on the word embedding and Euclidean distance (skip-gram model).

No.	Benchmark sentence: ཁྱེད་ཀྱི་སྐྱག་ཏུ་ང་ལ་ཉིན་ཞིག	Euclidean distance
1	ཁྱེད་ཀྱི་སྐྱག་ཏུ་ང་ལ་ཉིན་ཞིག	2.25E-07
2	ཁྱེད་ཀྱི་སྐྱོ་ཁག་ང་ལ་ཉིན་ཞིག	0.0290994
79	ང་ལ་དུ་བ་འཇེན་སྐྱད་ཉིན་ཞིག	0.116304
4	ཁྱེད་ཀྱི་དུས་རྐང་ང་ལ་ཉིན་ ཞིག	0.122168

The reference sentence number in Table 4 is 1, and the similar sentence number is 2, 3, 4. Based on the similarity between the word embedding (skip-gram model) and the Euclidean distance, the hit rate is 2/3 in the four sentences with the closest distance, that is, the highest similarity.

Table 5. Jaccard similarity evaluation based on word embedding (skip-gram model).

No.	Benchmark sentence: ཆོད་ཀྱི་སྐྱག་ཊྱ་ང་ལ་ཅིན་ཞིག	Jaccard similarity
1	ཆོད་ཀྱི་སྐྱག་ཊྱ་ང་ལ་ཅིན་ཞིག	1.36E+00
2	ཆོད་ཀྱི་རྟོར་ཁྲག་ང་ལ་ཅིན་ཞིག	0.842694
77	ང་ལ་ཁྱག་ས་ཞིག་སྐྱེར་རོགས།	0.841035
4	ཆོད་ཀྱི་ཏུབ་ཚོད་ང་ལ་ཅིན་ ཞིག	0.785952

The reference sentence number in Table 5 is 1, and the similar sentence number is 2, 3, 4. The hit rate of the four sentences with the highest similarity calculated based on the Jaccard similarity of the word embedding (skip-gram model) is 2/3.

The accuracy of the final 100 benchmark sentences using the two similarity evaluation methods is shown in Table 6. The experimental results show that the accuracy of the similarity evaluation method based on the word embedding (skip-gram model) and Jaccard similarity is reached 85.6%, which is more accurate than the other three groups of experiments, and can obtain better results in Tibetan sentences within 15 words. The more the number of words, the higher the accuracy of sentence similarity evaluation.

Table 6. Comparison of similarity accuracy evaluation s by different methods.

word embedding Model	Evaluation Method Evaluation of Euclidean distance	Jaccard similarity
Skip-gram Model	83.3%	85.6%
CBOW Model	72.6%	60.3%

5 Conclusion

The similarity evaluation of sentences is a promising basic research work in the field of natural language processing. With the development of natural language processing at home and abroad, compared with the continuous evolution of foreign language and Chinese sentence similarity evaluation methods, research on Tibetan sentences similarity evaluation methods is rare. This paper analyzes the concept of word embedding and sentence similarity evaluation and common evaluation methods. The Word2vec tool is used to train two Tibetan sentence vectors based on CBOW model and skip-gram model respectively. On this basis, the vector of Tibetan sentences is obtained. Two methods for calculating the similarity of Tibetan sentences based on word embedding are designed (the similarity evaluation method based on word embedding and Euclidean distance and the Jaccard similarity evaluation method based on word embedding). The implementation steps and related evaluation formulas are described. Finally, four sets of comparative experiments are carried out using different word embedding models and two sentence similarity evaluation methods. The accuracy of each set of experiments is obtained, and the accuracy of the evaluation method based on word embedding (skip-gram model) and Jaccard similarity reaches 85.6%, which shows that the method adopted in this paper is feasible.

Based on this work, we will further try to use different similarity evaluation methods to test texts in different genres, and we will add more semantic and grammatical information to the similarity evaluation method of sentence vectors to further improve the accuracy.

Acknowledgments. This work was supported by National Natural Science Foundation of China (grant numbers: 61063033, 61662061), the National Key Research and Development Program of China and Qinghai Normal University Scientific Research Fund Project.

References

1. Li, B., Liu, T., Qin, B., Li, S.: Chinese sentence similarity calculation based on semantic dependence. Comput. Appl. Res. **20**(12), 15–17 (2003). (in Chinese)
2. Zhou, W.: Research on the calculation method of Chinese sentence similarity and its application. Master thesis, Henan University, (2005). (in Chinese)
3. Anjian, C.: Research on the algorithm of Tibetan sentence similarity. Chin. J. Inf. **25**(4), 2–6 (2011). (in Chinese)
4. Yan, M.: Research on the calculation method of Tibetan text similarity based on segmentation fusion. Master thesis, Northwest University for Nationalities (2016). (in Chinese)
5. Li, F., Hou, J., Zeng, R., Ling, C.: Research on multi-character sentence similarity calculation method of fusion word vector. Comput. Sci. Explor. **11**(4), 608–618 (2017). (in Chinese)
6. Hinton, G.E.: Learning distributed representations of concepts. In: Proceedings of the Eighth Annual Conference of the Cognitive Science Society, vol. 1, p. 12 (1986)
7. Mikolov, T., Chen, K., Corrado, G., Dean, J.: Efficient estimation of word representations in vector space. arXiv preprint arXiv:1301.3781 (2013)

An Easier and Efficient Framework to Annotate Semantic Roles: Evidence from the Chinese AMR Corpus

Li Song[1,2], Yuan Wen[1], Sijia Ge[1], Bin Li[1(✉)], and Weiguang Qu[3]

[1] School of Chinese Language and Literature, Nanjing Normal University,
Nanjing 210097, China
songli1105@sina.com, libin.njnu@gmail.com
[2] Department of Chinese Language and Literature, Tsinghua University,
Beijing 100084, China
[3] School of Computer Science and Technology, Nanjing Normal University,
Nanjing 210023, China

Abstract. Semantic role labeling (SRL) is a fundamental task in Chinese language processing, but there are three major problems about the construction of SRL corpora. First, disagreements occurred in previous studies over the definition and number of semantic roles. Second, it is hard for static predicate frames to cover dynamic predicate usages. Third, it is unable to annotate the dropped semantic roles. Abstract Meaning Representation (AMR) is a new method which provides a better solution to the above problems. The researchers use 5,000 sentences in the Chinese AMR corpus to make a comparison between AMR and other SRL resources. Data analysis shows that within the framework of AMR, it is easier to annotate semantic roles based on simplified distinction between core and non-core roles. In addition, 1,045 tokens of dropped roles are annotated under this new framework. This study indicates that AMR offers a better solution for Chinese SRL and sentence meaning processing.

Keywords: Abstract Meaning Representation · Predicate framework · Semantic role · Language knowledgebase

1 Introduction

In semantic representation, semantic relations between predicates and their semantic roles form the backbone of the sentence structure. Constructing a predicate frame lexicon, combining with labeling semantic roles of predicates in corpus, has become a research paradigm. Thus, building the predicate frames which contain such information becomes an important issue in linguistics and Natural Language Processing. Many semantic role labeling (SRL) systems and SRL resources in different languages already exist, but there are several problems in these SRL corpora.

First, the definition and number of the semantic role labels of predicates are still under discussion. The specific semantic roles in different SRL resources range from very general role labels to labels that are meaningful to a specific situation to predicate-specific labels in terms of levels of abstraction [1], which leads to quite different numbers of role labels in

© Springer Nature Switzerland AG 2020
J.-F. Hong et al. (Eds.): CLSW 2019, LNAI 11831, pp. 474–485, 2020.
https://doi.org/10.1007/978-3-030-38189-9_49

different SRL resources. VerbNet [2], Sinica Treebank [3] as well as NetBank [4] all define several predicate-general role labels such as *agent*, *theme* and *instrument* to represent semantic relations. The difference is that the former does not distinguish core and non-core roles, while the latter two resources make the distinction. FrameNet [5] and its continuers define semantic roles on a per-frame basis, so they avoid determining how many semantic roles are needed for a language, leading to a large quantity of role labels. PropBank [6] and its continuers such as Chinese PropBank (CPB) [7] deal with core and non-core roles in different specialized ways. They define 13 non-core role labels which are consistent across predicates, while 5 core role labels in a predicate-specific manner (Arg0–Arg4). Each sense of each verb has a specific set of core roles.

Second, it is hard for static predicate frames to cover dynamic predicate usages. Predicate frames which do not distinguish core and non-core roles are difficult to indicate whether a semantic role is necessary for the predicate. Additionally, resources that define core roles in a predicate-general manner just as non-core roles neither could solve the collision between core and non-core roles nor could they represent multi-functional semantic roles. (See Sect. 3 for details.)

Third, being limited to the annotating mechanism, most SRL systems are unable to annotate the dropped semantic roles of predicates. For example, it is hard for most SRL systems to correctly represent the meaning of the nominal phrase *the injured* whose central word is dropped and *one of which...* which drops the noun appeared in the preceding clause.

Table 1. Main SRL resources in English and Chinese

Resource[a]	Language	Definition of Role Labels	Number of Role Lables
VerbNet	English	predicate-general	36 (VerbNet 3.3)
Sinica Treebank	Traditional Chinese	predicate-general (core & non-core)	60 (12 core + 43, 5 for nouns)
NetBank	Simplified Chinese	predicate-general (core & non-core)	22 (10 + 12) (latest release)
FrameNet	English	frame-specific (core & non-core)	N/A (1,224 frames)
CFN	Chinese	frame-specific (core & non-core)	N/A (323 frames)
PropBank	English	predicate-specific (core), predicate-general (non-core)	18 (5 core + 13)
CPB	Chinese	predicate-specific (core), predicate-general (non-core)	18 (5 core + 13)
AMR	English	predicate-specific (core), predicate-general (non-core)	45 (5 core + 40)
CAMR	Chinese	predicate-specific (core), predicate-general (non-core)	49 (5 core + 44)

[a]VerbNet: https://verbs.colorado.edu/verbnet/.
Sinica Treebank: http://turing.iis.sinica.edu.tw/treesearch/.
NetBank: http://ccl.pku.edu.cn/973_sem_spec/pkunetbank/pku_netbank.htm.
FrameNet: https://framenet.icsi.berkeley.edu/fndrupal/.
CFN: http://sccfn.sxu.edu.cn/portal-zh/home.aspx.
PropBank: http://propbank.github.io/.
CPB: http://www.cs.brandeis.edu/~clp/ctb/cpb/.
AMR: https://amr.isi.edu/index.html.
CAMR: http://www.cs.brandeis.edu/~clp/camr/camr.html.

Abstract Meaning Representation (AMR), a new method to represent meaning of sentences [8], defines semantic roles in a way different from other SRL systems. Similar to PropBank, AMR defines core roles in a predicate-specific way while non-core roles are defined in a predicate-general way. The core role labels are the same as PropBank, and AMR adopts the predicate frame lexicon extracted from PropBank. However, the number of non-core role labels is up to 40. Besides, AMR allows annotators to add back dropped semantic roles in the sentences. Through the dynamic mechanisms, AMR can provide a better solution to the above problems. The AMR Sembank[1] has become an important semantic resource.

Referring to the guidelines of AMR and taking linguistic characteristics of the Chinese language into account, [9] has developed annotation specifications for Chinese AMR (CAMR), using the same 5 core role labels and 44 non-core role labels (the extra four are added for the needs of Chinese annotation) as AMR. The predicate frame lexicon [10] of CAMR is extracted from the corpus of CPB. In addition, [11] designs a framework for aligning the concepts and relations to word tokens in a sentence for CAMR, which is helpful for annotating dropped semantic roles. Since AMR can better address the above three problems, the researchers try to discuss whether CAMR can provide better solutions to these problems in Chinese.

Table 1 summarizes the definition and number of role labels of the main SRL resources in English and Chinese.

In the rest of this paper, Sect. 2 introduces the core and non-core role labels of CAMR and its corpus, Sect. 3 discusses the rationality of the core role labels of CAMR, Sect. 4 discusses the rationality of the non-core role labels of CAMR, Sect. 5 discusses the benefits of AMR permitting adding back dropped roles, and Sect. 6 shows the conclusions and future work.

2   Chinese AMR

2.1   Core and Non-core Roles of Chinese AMR

Following the annotation scheme of OntoNotes [12] adopted by AMR, CAMR uses predicate senses and core argument frames in CPB, and annotates semantic relations with core and non-core semantic relation labels. Core semantic relations refer to the inevitable semantic relations in the predicate-specific event framework of the predicates. For core arguments in CPB, they have 3 main attributes [10]: (1) obligate, meaning of a predicate will be incomplete if it lacks a core argument; (2) different, the core argument frames of predicates differ from one another, so each sense of each predicate has a specific set of roles; (3) exclusive, multiple core arguments do not serve as the same semantic role. Table 2 shows the 5 core semantic relations adopted from CPB.

---

[1] https://catalog.ldc.upenn.edu/LDC2017T10.

Table 2. Core semantic relations in CAMR

arg0	external argument (Proto-Agent)
arg1	internal argument (Proto-Patient)
arg2	indirect object/beneficiary/instrument/attribute/end state
arg3	start point/beneficiary/instrument/attribute
arg4	end point

Non-core semantic relations refer to the semantic relations which are predicate-general outside the core semantic relations. AMR defines 40 predicate-general non-core semantic relations so that they are fine-grained, and CAMR adds four new non-core semantic relations taking the characteristics of Chinese into account. In order to be compatible with AMR, CAMR still uses English words to represent labels of non-core relations. Table 3 shows non-core relations in CAMR.

Table 3. Non-core semantic relations in CAMR

accompanier	direction	mod	quant
*aspect	domain	mode	range
beneficiary	duration	name	source
cause	example	ord	subevent
compared-to	extent	part-of	subset
consist-of	frequency	path	superset
condition	instrument	*perspective	*tense
cost	li	polarity	time
*cunit	location	polite	topic
degree	manner	poss	unit
destination	medium	purpose	value

* are the added relations in CAMR

Since core semantic roles are defined with respect to an individual verb sense, AMR and CAMR need support from predicate frame lexicons. The frame lexicon of CAMR is extracted from the CPB corpus, consisting of 26,650 senses of 24,510 predicates.

2.2 The Chinese AMR Corpus

Take the following example to show how CAMR represents the meaning of a sentence.

受伤[1]	的[2]	想[3]	回家[4]	。
injure	DE	want	go back home	
The injured (person) wants to go back home				

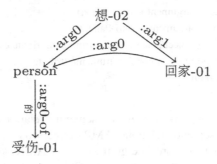

In this sentence, 想 (want) is the root, and *02* means the second sense of 想, which has other senses such as *think* and *miss* in the predicate frame lexicon. 受伤的 (the injured) drops the agent of the predicate 受伤 (injure), so CAMR adds a concept *person* for it and uses an inverse relation *:arg0-of* to represent the relation between *person* and 受伤.

According to the CAMR annotation specifications developed by [9], the researchers extracted and annotated 5,088 Chinese sentences from Penn Chinese TreeBank (CTB) 8.0[2]. The inter-agreement smatch score [13] of 500 randomly selected sentences between the 2 annotators is 0.83. The annotated sentences cover a wide range of fields and rich topics. Most sentences are long and complicated, containing rich semantic information. Before annotating, the researchers deleted invalid sentences which are ill-formed or contain messy codes, and then carried on automatic word segmentation and artificial proofreading. The final corpus consists of 5,000 Chinese sentences[3]. Table 4 shows the basic data of these sentences. Compared with the Chinese version of *the Little Prince* AMR corpus [14], whose average sentence length is 12.90 words and average number of concepts is 9.48, sentences in this corpus are longer and more complicated.

Table 4. Basic data of the CAMR corpus

Sentences	5,000	Characters (AVG)	34.34
Characters	171,703	Words (AVG)	22.46
Words	112,348	Concepts (AVG)	18.36
Concepts	91,808	Added Concepts[a] (AVG)	3.02

[a]There are three main kinds of added concepts in CAMR: (1) added semantic roles, (2) types of named entities which are used to identify the names of an entity, like *country* for *China*, (3) discourse relations such as *condition*, *temporal*.

---

[2] http://amr.isi.edu/download.html.

[3] The current CAMR corpus contains 10,149 sentences, and has been published at https://catalog.ldc.upenn.edu/LDC2019T07.

3 Core Roles in Chinese AMR Cover Dynamic Problems

The definition of core arguments in PropBank has been controversial in linguistics field. Some scholars consider it too broad and not conducive to classification of semantic roles [4], thus the predicate frame of AMR failed to be generally approved by linguistics field. Therefore, the researchers try to explore whether the predicate framework AMR adopts can represent core semantic roles of predicates more reasonably.

Before the exploration, the researchers consider that there are two inescapable problems in predicate frameworks whose core role labels are consistent across predicates: (1) the core and non-core roles may conflict when annotating concepts of location, cause, instrument and so forth. For example, a concept of location is indispensable to the meaning of *appear*. (2) It is difficult to properly annotate the multi-functional roles. For example, a concept of agent or cause can both serve as the subject of *change*.

These problems are common in Chinese and they can be solved by the predicate framework of CAMR, whose predicate-specific frame lexicon is extracted from the CPB corpus. The lexicon contains 26,650 senses of 24,510 Chinese predicates. CPB is a corpus adding information of semantic roles of predicates to CTB [15], a syntactically annotated Chinese corpus that is word-segmented, POS-tagged and syntactically bracketed with phrase structures [7]. Here the researchers elaborate how CAMR resolves the collision between core and non-core roles and annotates multi-functional roles based on data analysis of the predicate frame lexicon and CAMR corpus.

3.1 Resolve the Collision Between Core and Non-core Roles

Each sense of each predicate in the predicate framework of CAMR has a specific set of roles. If a concept is essential for the meaning of the predicate, it serves as a core role of the predicate, even if it represents the location or cause of the predicate, which is a kind of collision between core and non-core roles. If inessential, the concept serves as a non-core role of the predicate. For example, the concept of location is indispensable in the meaning of 遍布-01 (the first sense of 遍布, be spread throughout somewhere), so it is a core role of 遍布-01:

> 遍布-01 (be spread throughout somewhere)
> arg0: theme
> arg1: **location**

It is even possible that 4 of the 5 core roles of a predicate are in conflict with its non-core roles, such as 引进-01 (introduce something from one place to another):

> 引进-01 (introduce something from one place to another)
> arg0: agent / **cause**
> arg1: entity imported
> arg2: **location** arg1 is imported from
> arg3: predicate, **purpose**
> arg4: **destination**

Data shows that 2,453 (9.20%) senses of predicates in the CPB lexicon have collision between their core and non-core roles. Among them, 5.99% have more than one non-core-entering-core roles. Moreover, by analyzing all the descriptions of core roles in the lexicon, the researchers find there are 24 kinds of non-core roles which might conflict with the core roles. It means that over half of the categories of non-core roles are able to enter the core argument frame of predicates. Table 5 shows the top 5 non-core-entering-core roles in order of occurrences in the lexicon.

Table 5. Top 5 non-core-entering-core Roles

Roles	Freq
cause	1,454
location	934
destination	140
time	134
source	124

It can be seen that cause is used most frequently, which shows that concepts which represent the reason of a predicate are very easy to enter the predicate's core argument frame. Location and time take 2nd and 4th place. The destination and source are often used to represent initial and end point of location or time.

3.2   Representation of Multi-functional Roles

Although the number of core arguments of a predicate is limited to 5, CPB does not limit the types of concepts that can act as core roles of predicates. As long as a component is indispensable to the meaning of a predicate, it can act as a core role of the predicate no matter what semantic relation it has with the predicate. Taking 她缓解了我的痛苦 (she relieved my pain) for example, 她 (she) can serve as the agent as well as the cause of the predicate 缓解-01 (relieve), so the concept which represents agent and cause both can serve as the arg0 of 缓解-01 in CPB.

> 缓解-01 (relieve)
> arg0: **cause, agent**
> arg1: theme

As the descriptions of core roles in CPB lexicon are mostly in natural language, the researchers cannot exactly count how many predicate frames contain multi-functional core roles. However, data shows there have been 1,146 (4.30%) senses whose arg0 can be acted by both concepts of agent and cause, which means it is common for Chinese predicates having multi-functional roles, and the core argument framework of CPB lexicon can better represent this type of roles. That is to say, the CAMR's definition of core roles is reasonable for semantic representation.

4   Discrimination of Non-core Roles of Chinese AMR

In spite of the fact that AMR and CAMR have the same core role labels as PropBank and CPB, there is a great difference between them in terms of the quantity of non-core role labels. CAMR has 44 non-core role labels (Table 3), much more diversified than the 13 non-core role labels in CPB (Table 6). The researchers calculate the using frequency of each non-core role label in the corpora of CPB and CAMR (top 20 for lack of space), which are showed in Tables 6 and 7. Their mean deviations are 7,271.53 and 440.08 respectively, which means the degree of difference in using frequencies of non-core role labels is much higher in CPB corpus than in CAMR corpus.

Table 6. Freq of non-core role labels in CPB

Labels	Freq	%
ADV (adverbial)	38,262	46.63
TMP (temporal)	16,876	20.57
DIS (discourse maker)	10,270	12.52
LOC (locative)	7,104	8.66
MNR (manner)	3,793	4.62
PRP (purpose/reason)	2,344	2.86
DIR (direction)	874	1.07
CND (condition)	864	1.05
TPC (topic)	605	0.74
EXT (extent)	521	0.63
BNF (beneficiary)	470	0.57
FRQ (frequency)	49	0.06
DGR (degree)	21	0.03

Table 7. Top 20 Freq of non-core labels in CAMR

Label	Freq	%	Label	Freq	%
beneficiary	2,804	19.21	location	335	2.30
mod	2,098	14.38	domain	154	1.06
polarity	1,615	11.07	duration	146	1.00
*aspect	1,432	9.81	instrument	103	0.71
manner	1,164	7.98	frequency	99	0.68
mode	1,097	7.52	compared-to	86	0.59
time	1,045	7.16	condition	81	0.56
degree	1,012	6.93	*tense	76	0.52
cause	366	2.51	range	73	0.50
purpose	362	2.48	*perspective	57	0.39

5 AMR's Solution to Dropped Roles

Compared with other methods of meaning representation such as Dependency Graph, a big advantage of AMR is allowing annotators to re-analyze and add back dropped concepts of the sentences in order to represent the meaning of sentences more completely. For example, the nominal phrase *the injured* drops the agent of *injure*, so AMR adds back a virtual node *person*, just as the example shows in Sect. 2.2. Another example, *one of which...* drops the noun which appeared in the preceding clause, so AMR adds back a *thing* for it. Dropping semantic roles is common in Chinese. Data shows that CAMR annotates 1,045 tokens of dropped roles for the 5,000 sentences, which cannot be annotated in other SRL resources. For example, CAMR adds a virtual node *person* for 受伤的 (the injured). Among these added roles, 619 (59.23%) have core semantic relation with the predicates.

5.1 Adding Back Core Roles for Predicates

Core roles of predicates are very significant for the semantic analysis of a sentence. The researchers try to explore whether AMR's permission of adding back roles can help to annotate core roles of predicates more completely by comparing the 2 corpora of CAMR and CPB.

According to the difference value of the number of core roles annotated in the corpus and the number of core roles in the predicate framework lexicon, the annotation extent of core roles for each sense of each predicate can be classified into 3 categories: core roles annotated completely (difference = 0), core roles annotated incompletely (difference < 0), lexicon lacks core arguments (difference > 0)[4].

The researchers extract predicate frames from the 2 corpora[5] and calculate the difference per sense. Table 8 shows the distribution of quantity of senses in diverse difference value between the quantity of core roles in the 2 corpora and the lexicon of CPB.

It can be seen that the proportion of senses in CAMR corpus whose core roles are annotated completely or inadequate in the lexicon is higher than that in CPB corpus. In addition, the percentage of senses whose core roles are annotated incompletely is almost lower than the CPB corpus. It means that CAMR can annotate core roles of predicates more completely. The main reason is that AMR allows annotators to re-analyze and add back dropped concepts, so that CAMR is not restricted by the words of sentences, but to represent the core roles as completely as possible.

The proportion of senses in Table 8 shows that there are also many predicates whose core roles are annotated incompletely. The researchers consider it is mainly because AMR is a method to represent meaning of sentences, not the whole text, so that much information between sentences is missed. In the future, the researchers will attempt to extend AMR to discourses.

---

[4] The difference < 0 also contains the case of core roles being dropped, and the difference > 0 also contains the case of core roles having not being annotated. But these cases are negligible because they are few in number.

[5] Predicates without core roles in CAMR corpus are hard to be separated from other words, so the researchers ignore them currently.

Table 8. Difference value between the quantity of core roles in the two corpora and the lexicon of CPB

Corpus	Difference value (corpus minus lexicon)	−4	−3	−2	−1	0	1	2	Total
CAMR	Tokens of senses	23	272	1,527	6,862	**11,037**	**99**	**3**	19,823
	% of senses	0.12%	1.37%	7.70%	34.62%	**55.68%**	**0.50%**	**0.02%**	100%
CPB	Tokens of senses	344	1,260	10,060	36,539	**52,735**	**383**	**5**	101,326
	% of senses	0.34%	1.24%	9.93%	36.06%	**52.04%**	**0.38%**	**0.00%**	100%

* If the predicate in CPB has semantic relations with multiple roles, it just counts as one tokens of sense.

5.2 Adding Back Dropped Roles of 3 Types of Special Structures in Chinese

There are quite a few nominal structures dropping core roles of the predicates in Chinese. The researchers choose 3 types of special structures in Chinese: DE structure, SUO structure and SUO...DE structure, to analyze. The function of adding concepts of AMR can represent their meanings correctly. For example, a DE structure 受伤的 (the injured) drops the agent of the predicate 受伤 (injure), so CAMR adds a virtual node *person* and uses a triple *person: arg0-of* 受伤 to represent the relation between the added role and the predicate. Taking 我说的 ((what) I said) for another instance, it drops the theme of 说 (say), so CAMR adds a *thing* for it. The 所 and the predicate in a SUO structure form a nominal structure. As 我说的, 所说 ((what) is said) drops the theme of 说 and CAMR adds back a *thing*. A SUO...DE structure is a combination of a SUO structure and a DE structure. 所共有的 ((thing) shared by some people) drops the relative of 共有 (share), so CAMR also adds back a *thing*.

There are 309 tokens of DE structures, 9 SUO structures and 7 SUO...DE structures in the 5,000 CAMR sentences. Though not numerous, they are important and not negligible in Chinese. The researchers also statistically get the types and proportions of the dropped roles of each kind of the special structures, but lack of space forbids further treatment of the topic in this paper. From the data analysis, it can be seen that AMR's permission of re-analyzing and adding roles are helpful for annotating the meanings of predicates more completely.

The CAMR's function of adding dropped roles also benefits from the framework designed by [11] that can align the AMR concepts and relations to word tokens in a sentence. It uses the index of a word token as the ID of its aligned concept or relation in the AMR representation. When adding a role that is dropped, the added role will be assigned an ID which is greater than the length of the sentence, so that it is impossible to confuse the added roles with the words in the given sentence.

6 Conclusion and Future Work

In this paper, based on data analysis of the 5,000 sentences from Chinese AMR corpus, the researchers find that the AMR's definition of core roles can solve the collision between core and non-core roles and represent the multi-functional roles more concisely. Furthermore, the 44 non-core role labels from CAMR have a satisfactory discrimination to non-core semantic roles. In addition, benefited from the permission of re-analyzing and adding concepts, AMR can solve the problem of dropping semantic roles in the sentences, which is especially helpful for annotating special structures in Chinese such as DE structures. Therefore, as a method of representing meaning of sentences, AMR has unique advantages in semantic role labeling and it is suitable for representing meanings of Chinese sentences, so a larger AMR corpus should be built to serve the Chinese semantic processing.

A high-quality predicate framework lexicon is significant for ensuring the quality of the annotation. However, since it is directly extracted from the CPB corpus, the lexicon used by CAMR at present is still encountered with some problems, such as lack of senses and arguments. In the future, the researchers will try to re-build a high-quality lexicon for CAMR, and release more data for NLP applications and linguistics studies.

Acknowledgements. We thank the reviewers. This work is partially supported by project 18BYY127 under the National Social Science Foundation of China, project 61472191 under the National Science Foundation of China, and Project Funded by the Priority Academic Program Development of Jiangsu Higher Education Institutions.

References

1. Xue, N.: A Chinese semantic lexicon of senses and roles. Lang. Resour. Eval. **40**(3–4), 395–403 (2006)
2. Kipper, K., Dang, H., Palmer, M.: Class-based construction of a verb lexicon. In: Proceedings of the 17th National Conference on Artificial Intelligence and 12th Conference on Innovative Applications of Artificial Intelligence, pp. 691–696 (2000)
3. Chen, K., Luo, C., Chang, M., et al.: Sinica Treebank. In: Abeillé, A. (ed.) Treebanks. Text, Speech and Language Technology, vol. 20, pp. 231–248. Springer, Dordrecht (2003). https://doi.org/10.1007/978-94-010-0201-1_13
4. Yuan, Y.: The fineness hierarchy of semantic roles and its application in NLP. J. Chin. Inf. Process. **21**(4), 10–20 (2007)
5. Baker, C.F., Fillmore, C.J., Lowe, J.B.: The Berkeley framenet project. In: Proceedings of the 36th Annual Meeting of the Association for Computational Linguistics and 17th International Conference on Computational Linguistics, pp. 86–90 (1998)
6. Palmer, M., Gildea, D., Kingsbury, P.: The proposition bank: an annotated corpus of semantic roles. Comput. Linguist. **31**(1), 71–106 (2005)
7. Xue, N., Palmer, M.: Adding semantic roles to the Chinese Treebank. Nat. Lang. Eng. **15**(1), 143–172 (2009)
8. Banarescu, L., Bonial, C., Cai, S., et al.: Abstract meaning representation for sembanking. In: Proceedings of the 7th Linguistic Annotation Workshop and Interoperability with Discourse, pp. 178–186 (2013)

9. Li, B., Wen, Y., Bu, L., et al.: Annotating the Little Prince with Chinese AMRs. In: Proceedings of the 10th Linguistic Annotation Workshop (2016)
10. Bai, X., Xue, N.: Generalizing the semantic roles in the Chinese proposition bank. Lang. Resour. Eval. **50**(3), 643–666 (2016)
11. Li, B., Wen, Y., Song, L., et al.: Construction of Chinese abstract meaning representation corpus with concept-to-word alignment. J. Chin. Inf. Process. **31**(6), 93–102 (2017)
12. Weischedel, R., Hovy, E., Marcus, M., et al.: OntoNotes: a large training corpus for enhanced processing. In: Handbook of Natural Language Processing and Machine Translation (2011)
13. Cai, S., Knight, K.: Smatch: an evaluation metric for semantic feature structures. In: Proceedings of the 51st Annual Meeting of the Association for Computational Linguistics, pp. 748–752 (2012)
14. Li, B., Wen, Y., Bu, L., et al.: A comparative analysis of the AMR graphs from english and Chinese corpus of the Little Prince. J. Chin. Inf. Process. **31**(1), 50–57 (2017)
15. Xue, N., Xia, F., Chiou, F.D., et al.: The Penn Chinese Treebank: phrase structure annotation of a large corpus. Nat. Lang. Eng. **11**(2), 207–238 (2005)

Linguistic Knowledge Based on Attention Neural Network for Targeted Sentiment Classification

Chengyu Du and Pengyuan Liu[✉]

Beijing Language and Culture University, Beijing, China
Du_chengyu@163.com, liupengyuan@pku.edu.cn

Abstract. Deep learning approaches for targeted sentiment classification do not fully exploit linguistic knowledge. In this paper, we propose a Linguistic Knowledge based on Attention Neural Network (LKAN) to employ linguistic knowledge (e.g. sentiment lexicon, negation words, intensity words) to benefit targeted sentiment classification. Firstly, we extract linguistic knowledge words (e.g. sentiment lexicon, negation words, intensity words) in sentences by HowNet vocabulary. Then, we design an attention mechanism which drives the model to concentrate on such words and get a weighted combination of word embeddings as the final representation for the sentences. We evaluate our proposed approach on SemEval 2014 Task 4, whose performance as shown reaches the most advanced level.

Keywords: Linguistic knowledge · Targeted sentiment classification · Attention

1 Introduction

Targeted sentiment classification is a fine sentiment classification task. It aims to identify the sentiment polarity for a particular target. For example, the sentence *"Great food but the service was dreadful!"* would be assigned with *positive* polarity for target *"food"* while *negative* polarity for target *"service"*. Early studies for targeted sentiment classification focused on designing a set of features to train a classifier [1–3]. However, such kind of feature engineering work often relied on human ingenuity, which was a time-consuming process and lacked generalization.

Recently, many neural network based models have been proposed, such as Recursive NN [4], Recursive NTN [5] and Tree-LSTM [6]. In addition, there were more and more works tending to focus on fusing the information of the target and the contexts [7–12].

Although above deep neural network models have achieved great success on targeted sentiment classification, there are few studies using linguistic knowledge such as sentiment words, negation words and intensity words to benefit targeted sentiment classification. Sentiment lexicon offers the prior polarity of a word which can be useful in determining the sentiment polarity of longer texts such as phrases and sentences. Negation words are typical sentiment shifters [1], which constantly change the polarity

© Springer Nature Switzerland AG 2020
J.-F. Hong et al. (Eds.): CLSW 2019, LNAI 11831, pp. 486–495, 2020.
https://doi.org/10.1007/978-3-030-38189-9_50

of sentiment expression. Intensity words change the valence degree of the modified text, which is important for fine sentiment classification. The above three types of resources can contribute to the task, for example, *"the sangria was pretty tasty and good on a hot muggy day"* and *"sashimi was not fresh"*, if we give more attention to sentiment words *"tasty"*, *"good"* and *"fresh"*, intensity words *"pretty"* and negation words *"not"* in the sentences, the performance of sentiment classification will be improved.

In this paper, we propose a model employing linguistic knowledge (e.g. sentiment lexicon, negation words, intensity words) to benefit targeted sentiment classification. Firstly, We extract linguistic knowledge (e.g. sentiment lexicon, negation words, intensity words) in sentences by HowNet[1] vocabulary. Then, an attention mechanism is used to capture the most indicative words in the context using linguistic knowledge and generate the weighted sum embeddings as final sentence representation.

Our contribution is 3-fold:

(1) we are the first to address targeted sentiment classification with linguistic knowledge (e.g. sentiment lexicon, negation words, intensity words).
(2) We propose a Linguistic Knowledge based on Attention Neural Network (LKAN).
(3) the performance of our approach is on a par with the state-of-the-art.

The rest of the paper is organized as follows: In Sect. 2, we present some related works and in Sect. 3, the model is introduced. Experiments are presented in Sect. 4, and discussion follows in Sect. 5. Finally, we draw a conclusion in Sect. 6.

2   Related Work

There are many methods focusing on the task of targeted sentiment classification. Traditional approaches include rule based methods [13] and statistic based methods [14, 15]. These methods focused on converting a set of classification clues into feature vectors [1–3], which need either laborious feature engineering work or massive extra-linguistic resources.

In recent year, there are many neural network based models for targeted sentiment analysis, such as Recursive NN [4], Recursive NTN [5] and Tree-LSTM [6]. Many researchers focused on fusing the information of the targets and contextual. For example, Tang et al. developed two target-dependent LSTM to model the left and right contexts with targets [7]. Wang et al. proposed an attention based LSTM to explore the relation of targets and sentiment polarities [8]. Tang et al. designed a deep memory network to integrate the targets information [9]. Chen et al. also proposed a deep memory network to integrate the targets information, but the results of multiple attentions were non-linearly combined with a recurrent neural network [16]. Ma et al. proposed an interactive attention network which interactively learned attentions in the contexts and targets [10]. Zheng et al. introduced a rotatory attention mechanism to

---

[1] http://www.keenage.com.

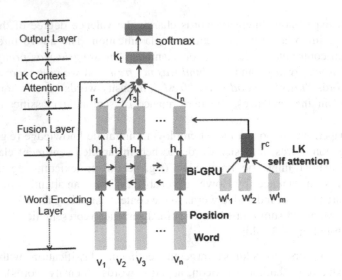

Fig. 1. The architecture of the proposed model.

achieve the representations of the targets, the left context and the right context [17]. Huang et al. introduced an attention-over-attention network modeling the targets and sentences in a joint way [18]. Li et al. proposed a hierarchical attention based position-aware network [11]. Wang et al. used both word-level and clause-level attentions to incorporate the knowledge of word-level and clause-level text information [12].

Although above deep neural network models have achieved great success on targeted sentiment classification, there are few works using linguistic knowledge to solve sentiment classification task. Qian et al. proposed a simple model which can address the sentiment shifting effect of sentiment, negation and intensity words [19]. Kumar et al. built a two-layered attention network taking advantage of the external knowledge to improve the sentiment prediction [20]. Lei et al. tried to integrate some linguistic knowledge for sentence-level sentiment classification [21]. These researches are all sentence-level sentiment classification. To the best of our knowledge, we are the first to address targeted sentiment classification with linguistic knowledge (e.g. sentiment lexicon, negation words, intensity words).

3   Model

The architecture of our model is shown in Fig. 1, which consists of four parts: Word Encoding Layer, LK Self Attention Layer, LK Context Attention, and Output Layer. We suppose that a sentence consists of N words $W = \{w_1, w_2, \ldots, w_n\}$, a target has M words $w^t = \{w_1^t, w_2^t \ldots w_m^t\}$ and the sentence has j knowledge words

$w^k = \{w_1^k, w_1^k \ldots w_j^k\}$, which is extracted from HowNet vocabulary. w^t and w^k is a subsequence of W. The goal of our model is to predict the sentiment polarity of the sentence over the target.

3.1 Word Encoding Layer

Let $W_w \in \mathbb{R}^{dw \times Vw}$ be a word embedding lookup table generated by an unsupervised method such as GloVe [22] or CBOW [23], where d_w is the dimension of the word embeddings and v_w is the size of word vocabulary. Inspired by Li et al. [11], we encode the position information into the representation of each word. Similar as the word embedding layer, the position embedding layer is a $Wp \in \mathbb{R}^{dp \times Vp}$, where d_p is the dimension of the position embeddings and v_p is the number of possible relevant positions between each word and the target. The position embedding lookup table is initialized randomly and tune in the training phase.

The vector representation $v_i \in \mathbb{R}^{dw+dp}$ of word w_i consists of its word embedding and position embedding, which is calculated as $v_i = w_w \cdot w_i \oplus w_p \cdot w_i$. Then, to learn a more abstract representation of the sentence, we apply a Bi-GRU [24] which can efficiently make use of past features (via forward states) and future features (via backward states) for a specific time frame. We obtain the representation for a given word w_i as follows:

$$hi = \overrightarrow{hi} \oplus \overleftarrow{hi}, h_i \in R^{2d_h} \tag{1}$$

d^h is the dimension of hidden states.

3.2 Linguistic Knowledge Based on Attention Neural Network

In this part, a clear-cut mechanism is illustrated which aims to fuse the information of linguistic knowledge. In detail, A LK self-attention is used to capture the most important clues in the linguistic knowledge words. Then, we fuse the weighted linguistic knowledge representation and the contextual words by a LK context attention.

LK Self-attention: There are numbers of sentences containing at least two linguistic knowledge words, so we also conduct a self-attention mechanism to generate the representation. The score function and the normalized importance weights are computed as follows:

$$fk\left(w_i^k\right) = \tanh\left(wk \cdot w_i^k\right), \quad w_k \in \mathbb{R}^{dw} \tag{2}$$

$$\alpha_i^k = \frac{\exp\left(fk\left(w_i^k\right)\right)}{\sum_{i=1}^{j} \exp(fk(w_i^k))} \tag{3}$$

Then, a weighted combination of word embeddings is considered as the representation for the linguistic knowledge.

$$r^c = \sum_{i=1}^{j} \alpha_i^k \cdot w_i^k \tag{4}$$

Information Fusion: Inspired by Li et al. [11], after achieving the linguistic knowledge representation, we then further make use of the achieved representation to construct the knowledge-specific representation of each word in the sentence by the following equation:

$$ri = wf \cdot [hi, r^c] \tag{5}$$

where $w_f \in \mathbb{R}^{(2dh+dw) \times dw}$ is a weight matrix. r^i denotes the linguistic knowledge-specific representation of the i-th word in the sentence.

LK Context Attention: The knowledge-specific representation of each word is used to learn attentions and further generate the sentence representation. A weighted hidden state is computed as the following equations:

$$fs([ri, hi]) = \tanh(w_s \cdot [ri, hi]), w_s \in R^{2dh+dw} \tag{6}$$

$$\alpha_i^s = \frac{\exp(f_s([ri, hi]))}{\sum_{i=1}^{n} \exp(f_s([ri, hi]))} \tag{7}$$

$$kt = \sum_{i=1}^{n} \alpha_i^s \cdot hi \tag{8}$$

which is considered as the final representation of the current sentence.

3.3   Output and Model Training

After we get the final representation k_t of the current sentence, we feed it into a softmax layer to predict the target sentiment. Then the model is trained by minimizing the cross entropy:

$$l = -\sum_{(x,y) \in D} \sum_{c \in C} y^c \log f^c(x; \theta) \tag{9}$$

where C is the sentiment category set, D is the number of training samples, $y \in \mathbb{R}^{|C|}$ is a one-hot vector where the element for the true sentiment is 1, $f^c(x; \theta)$ is the predicted sentiment distribution of the model. We also adopt dropout and early stopping to ease overfitting.

4 Experiments

4.1 Experiment Settings

We evaluate our proposed approach for targeted sentiment analysis on SemEval 2014 Task 4 [25], as shown in Table 1. The SemEval 2014 dataset contains reviews of restaurant and laptop domains, which are widely used in previous works. The evaluation metric is classification accuracy. We use 300-dimension word vectors pre-trained by GloVe [22] (whose vocabulary size is 1.9 M) for our experiments, as previous works did [9, 16, 17]. All out-of-vocabulary words are initialized as zero vectors, and all biases are set to zero. The dimensions of hidden states and fused embeddings are set to 300. The dimension of position embeddings is set to 50. In model training, we set the learning rate to 0.001, the batch size to 64, dropout rate to 0.5 and the model parameters are optimized via RMSprop proposed by Tieleman and Hinton [26].

4.2 Compared Methods

In this section, we give some state-of-the-art methods for comparison in order to evaluate the performance of our proposed approach.

Majority assigned the sentiment polarity with most frequent occurrences in the training set to each sample in test set.

Bi-LSTM and Bi-GRU adopted a Bi-LSTM and a Bi-GRU network to model the sentence and used the hidden state of the final word for prediction respectively.

TD-LSTM adopted two LSTMs to model the left context with target and the right context with target respectively (Tang et al. 2016a); It took the hidden states of LSTM at last time-step as the representation of sentence.

Mem Net (Tang et al. 2016b) applied attention multiple times on the word embedding, and the output of last attention was fed to softmax for prediction.

IAN (Ma et al. 2017) interactively learned attentions in the contexts and targets, and generated the representations for targets and contexts separately.

RAM (Chen et al. 2017) was a multi-layer architecture where each layer consists of attention-based aggregation of word features and a GRU cell to learn the sentence representation.

LCR-Rot (Zheng et al. 2018) employed three Bi-LSTMs to model the left context, the target and right context. Then they proposed a rotatory attention mechanism which modeled the relation between target and left/right contexts.

AOA-LSTM (Huang et al. 2018) introduced an attention-over-attention (AOA) based network to model targets and sentences in a joint way and explicitly captured the interaction between targets and context sentences.

HAPN (Li et al. 2018) proposed a hierarchical attention based position-aware network. They made position embedding as a part of inputs and proposed a hierarchical attention based mechanism to fuse the information of targets and the contextual words.

Table 1. Comparison with baselines on SemEval 2014 dataset.

Dataset	Restaurant (%)	Laptop (%)
Majority	65.00	53.45
Bi-LSTM	78.57	70.53
Bi-GRU	80.27	73.35
TD-LSTM	75.63*	68.13*
Mem Net	79.98	70.33
IAN	78.60	72.10
RAM	80.23	74.49
LCR-Rot	81.34	75.24
AOA-LSTM	81.20	74.50
HAPN	82.23	77.27
LKAN-SE	**82.69**	76.02
LKAN-IN	81.61	77.12
LKAN-NE	81.96	76.33
LKAN-SE+IN+NE	80.71	77.00

4.3   Experiment Result

Table 1 shows the performance comparison of our method with the state-of-the-art methods on the same test dataset. From the table, we can see that RNN based models are better than the Majority method and Bi-GRU achieves 80.27% and 73.35% accuracies which are 1.7% and 2.82% higher than those of Bi-LSTM on the Restaurant and Laptop dataset respectively. This result indicates that Bi-GRU is more suitable to this task than Bi-LSTM. In addition, the accuracy achieved by TD-LSTM is 2.94% and 2.4% lower than those by Bi-LSTM on the two datasets respectively. This results show that introducing target clues only by splitting the sentence according to the position of target is inadequate. Moreover, AOA-LSTM, RAM, IAN, LCR-Rot and HAPN, all perform better than TD-LSTM, which confirm the helpfulness of using attention mechanism to model target information in targeted sentiment classification.

In the Table 1, we can see that the performance of the model has been greatly improved after adding the linguistic knowledge information compared with Bi-GRU. Moreover, compared with the other state-of-the-art methods, the performance of our model is on a par with the state-of-the-art and LKAN-SE achieves the best performance on restaurant dataset, which highlights the importance of employing linguistic knowledge. In addition, we find that the performance of using different linguistic knowledge is different and the more linguistic knowledge may cause the worse performance. This depends on what the model learns before fusing linguistic knowledge information, that is to say, the addition of linguistic knowledge may become noise, if the model has learned this information.

EX1: I was pleasantly surprised at the taste

EX2: they really provide a relaxing, laid-back atmosphere.

EX3: the teas are great and all the sweets are homemade.

Fig. 2. Attention visualizations of some sentences.

5 Discussion

In order to get a better understanding of our Linguistic Knowledge based on Attention Neural Network and validate that this model is able to select informative words corresponding to a specific target in a sentence, we visualize **LK context attention layer** according to the obtained attention weight in Eq. (6).

Figure 2 shows the attention visualizations for some sentences as cases study. In the Figure, the color depth indicates the importance degree of attention weight for a specific target, the darker the more important. For example, in EX1, "*I was pleasantly surprised at the taste*" and the sentiment polarity of the target "*task*" is positive. We apply our approach to model the sentence and the target. Then, we can see that the model selects the "*pleasantly*", "*surprised*" as important information. This shows that our approach can effectively find out the informative words corresponding to a specific target in a sentence and make accurate predictions.

6 Conclusion

In this paper, we propose a novel Linguistic Knowledge based on Attention Neural Network (LKAN) to enhance the performance of targeted sentiment analysis. The model mainly aims to integrate linguistic knowledge into the deep neural network and get a weighted combination of word embeddings as the final representation to predict the target polarity. The experimental result shows that the performance of our approach is on a par with the state-of-the-art. In the future, we will explore different ways of merging linguistic knowledge to get a better integration into neural networks.

Acknowledgments. This work is supported by Beijing Natural Science Foundation (Project No. 4192057).

References

1. Kiritchenko, S., Zhu, X., Cherry, C., et al.: NRC-Canada-2014: detecting aspects and sentiment in customer reviews. In: Proceedings of SemEval, pp. 437–442 (2014)
2. Wagner, J., Arora, P., Cortes, S., et al.: DCU: aspect-based polarity classification for SemEval task 4. In: Proceedings of SemEval, pp. 223–229 (2016)
3. Vo, D.-T., Zhang, Y.: Target-dependent twitter sentiment classification with rich automatic features. In: Proceedings of IJCAI, pp. 1347–1353 (2015)

4. Socher, R., Pennington, J., Huang, E.H., et al.: Semi-supervised recursive autoencoders for predicting sentiment distributions. In: Proceedings of the 2011 Conference on Empirical Methods in Natural Language Processing, pp. 151–161 (2011)

5. Socher, R., Perelygin, A., Wu, J., et al.: Recursive deep models for semantic compositionality over a sentiment treebank. In: Proceedings of the 2013 Conference on Empirical Methods in Natural Language Processing (2013)

6. Tai, K.S., Socher, R., et al.: Improved semantic representations from tree-structured long short-term memory networks. In: Proceedings of ACL-2015, pp. 1556–1566 (2015)

7. Tang, D., Qin, B., et al.: Effective LSTMs for target-dependent sentiment classification. In: Proceedings of COLING-2016, pp. 3298–3307 (2016)

8. Wang, Y., Huang, M., et al.: Attention-based LSTM for aspect-level sentiment classification. In: Proceedings of EMNLP-2016, pp. 606–615 (2016)

9. Tang, D., Qin, B., et al.: Aspect level sentiment classification with deep memory network. In: Proceedings of EMNLP-2016, pp. 214–224 (2016)

10. Ma, D., Li, S., Zhang, X., et al.: Interactive attention networks for aspect-level sentiment classification. In: Proceedings of IJCAI-2017, pp. 4068–4074 (2017)

11. Li, L., Liu, Y., Zhou, A.Q.: Hierarchical attention based position-aware network for aspect-level sentiment analysis. In: Proceedings of the 22nd Conference on Computational Natural Language Learning, pp. 181–189 (2018)

12. Wang, J., Li, J., Li, S., et al.: Aspect sentiment classification with both word-level and clause-level attention networks. In: IJCAI, pp. 4439–4445 (2018)

13. Ding, X., Liu, B., Yu, P.S.: A holistic lexicon-based approach to opinion mining. In: Proceedings of the 2008 International Conference on Web Search and Data Mining, pp. 231–240 (2008)

14. Jiang, L., Yu, M., Zhou, M., et al.: Target-dependent Twitter sentiment classification. In: Proceedings of the 49th Annual Meeting of the Association for Computational Linguistics: Human Language Technologies-Volume 1, pp. 151–160 (2011)

15. Zhao, W.X., Jiang, J., Yan, H., et al.: Jointly modeling aspects and opinions with a MaxEnt-LDA hybrid. In: Proceedings of the 2010 Conference on Empirical Methods in Natural Language Processing, EMNLP 2010, pp. 56–65 (2010)

16. Chen, P., Sun, Z., Bing, L., et al.: Recurrent attention network on memory for aspect sentiment analysis. In: Proceedings of the 2017 Conference on Empirical Methods in Natural Language Processing, pp. 452–461 (2017)

17. Zheng, S., Xia, R.: Left-center-right separated neural network for aspect-based sentiment analysis with rotatory attention. arxiv preprint arXiv:1802.00892 (2018)

18. Huang, B., Ou, Y., Carley, K.M.: Aspect level sentiment classification with attention-over-attention neural networks. arxiv preprint arXiv:1804.06536 (2018)

19. Qian, Q., Huang, M., Lei, J., et al.: Linguistically regularized LSTMs for sentiment classification. arXiv preprint arXiv:1611.03949 (2016)

20. Kumar, A., Kawahara, D., Kurohashi, S.: Knowledge-enriched two-layered attention network for sentiment analysis. arXiv preprint arXiv:1805.07819 (2018)

21. Lei, Z., Yang, Y., Yang, M., et al.: A multi-sentiment-resource enhanced attention network for sentiment classification. arXiv preprint arXiv:1807.04990 (2018)

22. Pennington, J., Socher, R., Manning, C.: Glove: global vectors for word representation. In: Proceedings of the 2014 Conference on Empirical Methods in Natural Language Processing (EMNLP), pp. 1532–1543 (2014)

23. Mikolov, T., Sutskever, I., Chen, K., Corrado, G.S., Dean, J.: Distributed representations of words and phrases and their compositionality. In: Advances in Neural Information Processing Systems, pp. 3111–3119 (2013)

24. Chung, J., Gulcehre, C., Cho, K.H., Bengio, Y.: Empirical evaluation of gated recurrent neural networks on sequence modeling. arXiv preprint arXiv:1412.3555 (2014)
25. Wagner, J., Arora, P., Cortes, S., et al.: DCU: aspect-based polarity classification for SemEval task 4. In: Proceedings of SemEval, pp. 223–229 (2014)
26. Tieleman, T., Hinton, G.: Lecture 6.5-RMSprop. In: COURSERA: Neural Networks for Machine Learning (2012)

A Method of Automatic Memorabilia Generation Based on News Reports

Sun Rui[✉], Zhang Hongyi, Zhang Benkang, Zhao Hanyan, and Tang Renbei

School of Computer Science, Leshan Normal University, Leshan, China
ruisun@whu.edu.cn, 767699685@qq.com,
1305594195@qq.com, 824257826@qq.com, 912217728@qq.com

Abstract. This paper proposes a method of automatic memorabilia generation based on news reports, aiming to generate the memorabilia in a certain time period for specific enterprises or departments via machine learning technologies. Firstly, the nonparametric clustering algorithm DBSCAN is used to cluster news reports based on text similarity. Then, we propose a salience ranking model to calculate the salience score of each cluster from different aspects, such as news coverage, report forwarding and source website importance etc. Finally, time normalization and description generation are performed on the TOP-K clusters so as to generate the final memorabilia. Several experiments are carried out based on news reports crawled from the related website. Experimental results show that the proposed method can effectively discover important events from the corpus. This paper explores memorabilia generation and provides a baseline system for this task.

Keywords: Memorabilia generation · Clustering · DBSCAN

1 Introduction

As an important archive, memorabilia record important activities or events by different parties, governments, enterprises or other social organizations. Generally speaking, it chronologically outlines important events by gathering, arranging, and revising the relevant information in business development or activities [1–3]. On one hand, memorabilia provide a special search way for archives of institutions. Salient events at certain time point or in a specific time period can be obtained quickly and easily; on the other hand, memorabilia provide some clues and evidence for related historical reviews and experiences, which is beneficial to promote the culture development of institutions.

At present, the way of manual gathering and arranging is mostly adopted in memorabilia generation. Most related studies focus on principles and methods in manual compilation [1, 2], or the computer-assisted database construction of memorabilia [4]. There are two obvious shortcomings in this way. It achieves high accuracy but is inefficient. At the same time, the granularity and the time period of major events are more difficult to master. With the development of computer technology, machine learning is capable of liberating human beings from tedious labor. It is significant to

J.-F. Hong et al. (Eds.): CLSW 2019, LNAI 11831, pp. 496–506, 2020.
https://doi.org/10.1007/978-3-030-38189-9_51

automatically discover and generate memorabilia. To our knowledge, there is no open research on automatic generation of memorabilia.

Inspired by the efficiency of machine learning, this paper presents an automatic memorabilia generation method based on news reports. We aim at introducing machine learning technology into memorabilia generation and discovering T scientific problems in this task. As shown in Fig. 1, our framework is divided into two stages, namely, salient events ranking and memorabilia generation. In the first stage, the news reports are preprocessed by word segmentation and word frequency statistics, and then a nonparametric clustering algorithm DBSCAN [5] is used to cluster the news reports automatically. To capture salient events, a ranking model is proposed to evaluate the importance of news cluster. In the second stage, the time standardization and the description generation of each salient cluster are carried out respectively. At last, descriptions of salient events are sorted by timestamp to generate the final memorabilia.

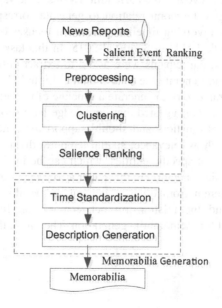

Fig. 1. Our framework for automatic memorabilia generation.

2 Related Work

Generally speaking, the memorabilia is a list consisting of some important events in certain time period, when each event contains two parts: a time and a description. These events are arranged in chronological order. The importance of events and the accurate timeline should be paid more attention to the memorabilia. The description of salient events should be prominent and brief [1, 2]. Traditional methods exploit manual collection or computer-aided recording of relevant events for memorabilia generation [6, 7], most of which propose the card arrangement method to collect and filter salient events, and use computer-aided timeline formulation.

In recent years, some work, such as timeline summarization [8–10] and timeline generation [11–14], have gotten more and more attention. Timeline summarization is based on news reports or microblogs, and uses the temporal features to identify the representative sentences. These sentences are used to construct the summary which can reflect the event or topic evolution. Based on the dependency relationship between sentences at different time, Yan et al. [8] suggested to model the summary relevance on time dimension to generate the timeline of news events. Ng et al. [9] adopted supervised learning to improve the quality of multi-document summarization based on different characteristics of news events in time dimension. Tran [10] exploited various relations between news reports, such as topic relevance, temporal relation and causal relation, to mine the evolutionary process of events. Above work focuses on organizing major news reports on specific topics. However, memorabilia generation does not care about topic.

In timeline generation, event is the basic unit. Events in news reports or microblogs are organized according to temporal relation to generate corresponding timelines. In recent years, most work involving timeline generation focuses on the English corpus. A typical task is the Task 4 [11] in SemEval-2015. In this task, given a set of documents and target entities, participants were asked to extract entity-related events and assign the timestamp, and generate the final timeline. Li and Cardie [12] proposed an unsupervised learning framework to generate a timeline of important events for entities in Twitter. Althoff et al. [13] suggested that knowledge base should provide the entity relationship and event information, and should support the timeline generation from multiple perspectives such as relevance, spatial-temporal diversity and content diversity. Above works focus on specific entities, which can be institutions or individuals. Similarly, the memorabilia generation in this paper also focuses on specific institutions. While our work is different from timeline generation, we do not take into account entity-related events and the relationship between events. In other words, event extraction is not focused for memorabilia generation. Therefore, this task is broader and more coarse-grained.

3   Method

Given a set of news reports for specific institution, we divide them into different clusters according to the similarity of reports using a nonparametric clustering algorithm. Then, the salience evaluation model is designed to capture important news clusters; the time standardization and description generation are conducted on these important clusters. The final memorabilia is generated by sequentially arranging the salient description chronologically.

3.1   News Reports Clustering

Usually, salient events are reported by different news media. Similar news reports are usually relevant to the same event. Naturally, these reports should be in the same cluster. The information of news reports cluster can be used to measure the event salience. Therefore, how to cluster news reports effectively is crucial to selecting salient events. In this paper, we adopt the Vector Space Model (VSM) to represent each news report and perform the clustering based on this representation.

Preprocessing of News Reports. First, the caption and content of each report in dataset are segmented by word segmentation tool. The dictionary is constructed after eliminating the stop words. Then, we calculate the TFIDF for each word in the dictionary, and the top 50 words with high TFIDF values are called feature words. Next, the VSM is adopted to represent each report $doc_i = (tfidf_{i,1}, \ldots, tfidf_{i,50})$, which $tfidf_{i,1}, \ldots, tfidf_{i,50}$ denotes the TFIDF value of the corresponding feature words in news report doc_i.

Nonparametric Clustering using DBSCAN. This paper suggests to exploit the nonparametric clustering method DBSCAN to cluster the dataset D and acquire the news clusters C. DBSCAN is a density-based clustering algorithm [5, 15], which discovers the clusters with different shapes based on the principle of density connectivity. It is capable of partitioning reports in the region with high density into clusters, and is able to detect clusters with arbitrary shape in noisy data space. Furthermore, previous studies have demonstrated that DBSCAN is efficient.

3.2 Salience Ranking of News Clusters

Intuitively, results of news clustering can be used to guide the memorabilia generation. Salient events in memorabilia can be obtained by evaluating the importance of news clusters. We propose a model of ranking salience for each news cluster. This model measures the salience of news cluster from three aspects: the coverage of news reports, the transmission quantity of reports and the importance of report source.

Coverage of News Reports. Generally speaking, the more news reports in the cluster is, the more salient the event is. We count up the number of news reports $Count(C_i)$ in the cluster C_i. The coverage of news reports is computed as follows:

$$RptScore(C_i) = Count(C_i)/|D| \tag{1}$$

where $|D|$ denotes the total number of data-centralized news reports.

Reports Forwarding. Usually, the salient event will be transmitted rapidly. In the same cluster, one event may be reported more times. Here, we focus on the transmission quantity, namely the number of websites reporting the same event. Obviously, the more salient event will be forwarded with high probability. We count up the number of report forwarding $FwdCount(C_i)$ in each cluster C_i. The score of report forwarding is calculated as follows:

$$FwdScore(C_i) = FwdCount(C_i) \Big/ \sum_j FwdCount(C_j) \tag{2}$$

Importance of Source Website. If an event is reported by the important website, to some extent, it is more salient. Firstly, we calculate the importance of each website $site_i$ in the news report dataset $|D|$, and it is calculated as follows:

$$siteScore(site_i) = -log(\frac{|\{doc : doc \in site_i\}|}{|D|}) \tag{3}$$

where $\{doc : doc \in site_i\}$ denotes the number of news reported by website $site_i$. Then, we add together the importance of all the source website in the cluster C_i and normalize the importance. It is computed as follows:

$$SrcScore(C_i) = \sum_{site_k \in C_i} siteScore(site_k) \bigg/ \sum_{site_j \in D} siteScore(site_j) \tag{4}$$

Salience Ranking of News Clusters. As above mentioned, three aspects are considered to measure the salience of news clusters C_i. Our objective function consists of above three score function, and aims at capturing important events. The overall score of a cluster is computed by:

$$Sal(C_i) = \alpha RptScore(C_i) + \beta FwdScore(C_i) + \gamma SrcScore(C_i) \tag{5}$$

where the parameters α, β, γ are used to scale the range of coverage of news reports, reporting forwarding and importance of source website ($0 \le \alpha, \beta, \gamma \le 1$), respectively. In this paper, their values are determined by grid search algorithm (The step size is set as 0.1). The top k clusters can be regarded as salient events (k is set to 100 in out experiment).

3.3 Memorabilia Generation

After capturing the salient event, we need to describe these events and arrange them in chronological order to develop the final memorabilia. In this paper, we design some simple but effective approaches on time standardization and description generation for each salient cluster.

Time Standardization of Salient Clusters. Each salient event in the final memorabilia comes from important news clusters, but the reporting time of each news report in the same cluster may be various. Therefore, we perform time standardization for each cluster. To be brief, the reporting time of each news is regarded as the occurrence time of the event reported in the news. If there is no reporting time, we extract the time which occurs firstly in the news reports as the event occurrence time. Intuitively, the time when the news was reported earliest can be regarded as the time when the event occurred. So, for each cluster, the minimum timestamp is taken as the standard time of the salient event in this work.

Description Generation of Salient Clusters. Every salient event should be summarized with simplicity and informative text. Some abstractive summarization methods are beneficial for description generation. Here, we design a simply method to acquire the description. We suggest that the center of the cluster is representative. For each cluster, the caption of the center news report is directly taken as the description of the event. So, the center report in each cluster should be identified firstly. Details are presented as follows:

Step1: For each salient cluster, we firstly compute the average of each feature of all reports. The center feature vector is composed of these average dimension feature.
Step2: Then, we calculate the cosine similarity between the center feature vector and each news report in cluster.
Step3: Finally, the caption of the news report with highest cosine similarity is directly taken as the representative description of this cluster.

The Final Memorabilia Generation. After time standardization and description generation, we acquire the standard time and representative description for each salient cluster. All the salient events are ordered by time so as to generate the final memorabilia.

4   Experiment

4.1   Settings

Dataset. Our experiments are carried out on the self-constructed dataset. Taking a college as an example, we crawl the news reports related to the college in the past three years from several websites, such as the official website of the university (UOW), the local news website (LNW), the official website of local municipal bureau of education (LMBE), the Window of Chinese Colleges and Universities (WCCU), the Chinese education news website (CENW) and the official website of Ministry of Education of China (CMEOW). After data cleaning, more than 2300 related news reports are retained. There are 500 clusters in the dataset after manual filtering. In our experiments, we acquire the top 100 salient clusters to generate the final memorabilia.

Baselines. Clustering algorithm is crucial to memorabilia generation in this paper. Given a set of news reports, the number of salient clusters is unknown. So the non-parametric algorithm is more suitable for aggregating the news which has reported the same event. Here, we select KMeans algorithm and APCluster algorithm as the baseline. The former is a parametric clustering algorithm, while the latter is non-parametric.

Evaluation Metric. Entropy and Purity are used for automatic evaluation of different clustering algorithms in this work. The better the clustering algorithm is, the higher the purity is, while the lower the entropy is. So far, there is no standard automatic evaluation method for memorabilia generation. Therefore, we perform human evaluation to measure the performance of memorabilia generated by different clustering algorithms. Five participants (all of them have 3 years of college experience) are asked to rate the generated memorabilia. Each event in the memorabilia is given a subjective score from 1 to 5, with 1 being the worst and 5 being the best. The average is taken as the final score.

4.2   Results

Table 1 shows results of different clustering algorithms, and Table 2 shows results of manual evaluation on the final memorabilia. At last, the three algorithms, KMeans, APCluster, and DBSCAN have generated 500, 475 and 493 clusters respectively. As shown in Tables 1 and 2, experimental results demonstrate that DBSCAN method outperforms KMeans and APCluster methods. Obviously, the performance of clustering algorithm directly affects the quality of final memorabilia.

Table 1. Performance comparison on different clustering algorithms for automatic evaluation.

Clustering method	Entropy	Purity
KMeans	0.1005	42.85%
APCluster	0.1129	35.57%
DBSCAN	0.0868	44.19%

Table 2. Results from the manual evaluation.

Clustering method	Score
KMeans	4.45
APCluster	4.11
DBSCAN	4.52

Compared with the performance of different clustering algorithms, APCluster has the worst performance and DBSCAN achieves the best performance. By analyzing clustering results, we find that some clusters generated by APCluster only contain a single sample. The noise data impairs the performance of APCluster method. KMeans outperforms APCluster in clustering performance. It also introduces some noise data while clustering samples correctly. As shown in Table 1, both the highest scores of purity and entropy are achieved by DBSCAN which clusters samples based on density and recognizes noise more effectively.

Observing the implementation details of three algorithms, KMeans is based on the geometric distance between samples, and APCluster is based on the graph distance. Both of them focus on the distance between samples, so they can generate better clusters based on spherical distribution. Different from them, DBSCAN is based on data density, does not follow the assumption of spherical distribution. When clustering, each sample is not always allocated to the corresponding cluster. Therefore, the algorithm can reduce the effect of noise and discover the clusters with arbitrary shape, so as to promote the clustering performance.

In addition, KMeans is fast and flexible. It achieves slightly lower performance than DBSCAN in our experiments. Unfortunately, the number of clusters needs to be specified in KMeans algorithm. In our experimental settings, it is difficult to determine the number of clusters when the dataset is not annotated. Obviously, the nonparametric algorithm is more suitable for the clustering in the memorabilia generation.

Table 3. Comparison of the clusters generated by different methods.

Method	Example (News Headlines)	Website
Reference	某网球代表队获中国大学生网球联赛总决赛资格 (The tennis team won the qualification of the finals of the Chinese University Tennis League)	CENW
	某网球队进军中国大学生网球联赛总决赛 (The tennis team entered the finals of the Chinese University Tennis League)	LMBE
	某网球队进军中国大学生网球联赛总决赛 (The tennis team entered the finals of the Chinese University Tennis League)	WCCU
	我校网球代表队获中国大学生网球联赛总决赛资格 (Our tennis team won the qualification of the finals of the Chinese University Tennis League)	UOW
APCluster	某网球代表队获中国大学生网球联赛总决赛资格 (The tennis team won the qualification of the finals of the Chinese University Tennis League)	CENW
	某网球队进军中国大学生网球联赛总决赛 (The tennis team entered the finals of the Chinese University Tennis League)	LMBE
	某网球队进军中国大学生网球联赛总决赛 (The tennis team entered the finals of the Chinese University Tennis League)	WCCU
KMeans	某教工在省高校网球比赛中获佳绩 (The faculty got good results in tennis competitions in provincial colleges and Universities) (×)	UOW
	某网球代表队获中国大学生网球联赛总决赛资格 (The tennis team won the qualification of the finals of the Chinese University Tennis League)	CENW
	某网球队进军中国大学生网球联赛总决赛 (The tennis team entered the finals of the Chinese University Tennis League)	LMBE
	某网球队进军中国大学生网球联赛总决赛 (The tennis team entered the finals of the Chinese University Tennis League)	WCCU
	我校网球代表队获中国大学生网球联赛总决赛资格 (Our tennis team won the qualification of the finals of the Chinese University Tennis League)	UOW
	某乒乓球队在"文理杯"2015年省高校大学生乒乓球比赛中获得优异成绩 (The table tennis team achieved excellent results in the "Wenli Cup" table tennis competition for college students in 2015) (×)	UOW
DBSCAN	某网球代表队获中国大学生网球联赛总决赛资格 (The tennis team won the qualification of the finals of the Chinese University Tennis League)	CENW
	某网球队进军中国大学生网球联赛总决赛 (The tennis team entered the finals of the Chinese University Tennis League)	LMBE
	某网球队进军中国大学生网球联赛总决赛 (The tennis team entered the finals of the Chinese University Tennis League)	WCCU
	我校网球代表队获中国大学生网球联赛总决赛资格 (Our tennis team won the qualification of the finals of the Chinese University Tennis League)	UOW

As shown in Table 2, the overall tendency of human evaluation on the final memorabilia is similar to the automatic evaluation of clustering. The memorabilia generation based on DBSCAN algorithms achieves the best results. Generally, our salient event ranking module alleviates the negative impact of clustering performance to some extent. In specific, the long tail clusters are discarded after ranking. However, the clustering performance affects the choice of salient cluster and its center report. An intuitive explanation is that the higher the purity of salient cluster is, the more robustly the memorabilia can be generated. The noise sample will inevitably affect the final performance of the system.

4.3   Example Outputs and Discussion

Table 3 presents an example of different clustering methods (The mark * denotes that the sample is wrongly clustered). This example is consistent with the experimental results. Reports that ought to be belonged to one cluster, are partitioned into different clusters by APCluster algorithm. Therefore, some samples are lost in the cluster. As shown in Table 3, for KMeans algorithm, the noise samples that should be in other cluster are introduced to the cluster. Relatively speaking, DBSCAN algorithms is more effective.

The experimental results demonstrate that our method is capable of generating memorabilia effectively. In our method, the time period of memorabilia is determined by the news reports in dataset, and the number of salient events in memorabilia is adjustable by the threshold of salient news clusters. So, our method is simple but flexible. Some issues in this work are shown as follows:

(1) Our work is a pioneering exploration on the memorabilia generation. Because of lack of news reports relevant to the college, experimental samples are insufficient so as to affect the performance of clustering algorithm to some extent.

(2) In this work, we simply choose the caption of center report as the description of salient event, which has some defects. Abstractive summarization methods are more helpful in promoting the performance, and our method can be regarded as a baseline.

(3) Human rating is used for the evaluation of memorabilia generation in this paper. Rouge or BLEU may be more optimal, which has been widely used in summarization tasks. In addition, external evaluation method is also a direction to be explored.

5   Conclusion and Future Work

In this paper, we propose an automatic memorabilia generation method based on network news reports. Firstly, news reports are clustered using a nonparametric algorithm. Then, the salience score of each cluster is calculate from three aspects, such as coverage, report forwarding and importance of source website. Next, the time standardization and description generation are performed on salient news clusters. Finally, the salient event descriptions are ordered by the time to generate the memorabilia.

Experimental results on dataset crawled from the related website show the proposed method is capable of discovering important events.

This work studies memorabilia generation. The proposed method can be regarded as a baseline for this task. In the stage of clustering, we compare different clustering algorithms based on a simple text representation VSM. In the stage of description generation, the caption of center report in the cluster is directly used for event description.

So, more research is needed on the generation of memorabilia from many perspectives, such as the measurement of event importance and the generation of event description. For future work, we plan to explore two directions. Firstly, we plan to introduce the semantic structure and distributed representation of documents to improve the performance of clustering. In addition, we plan to investigate natural language generation technology to generate more meaningful descriptions for each salient event, such as abstractive summarization methods.

Acknowledgments. We thank all reviewers for their detailed comments. This work is supported by the National Natural Science Foundation of China (Grant No. 61373056), the Key Projects of Sichuan Provincial Education Department of China (Grant No. 18ZA0239), and the Key Projects of Science and Technology Bureau in Leshan, China (Grant No. 18JZD116, 18JZD117), the Scientific Research Projects of Leshan Normal University (Grant No. XJR17001, ZZ201822), and the National College Students Innovation and Entrepreneurship Training Program of China (Grant No. 201710649005).

References

1. Liu, Y.: Discussion on the establishment of historiography museums in higher learning institution. J. Changchun Univ. Sci. Technol. (Soc. Sci.) **18**(4), 139–140 (2005)
2. Yang, C.: Archives work of colleges and universities on constructing harmonious campus culture. J. Quanzhou Norm. Univ. (Nat. Sci.) **26**(4), 130–132 (2008)
3. Fan, H.: Study on the writing of the events. J. Guangdong Ind. Tech. Coll. **9**(1), 68–69 (2009)
4. Xing, Q., Zhong, Y., Li, R., et al.: The assessment of culture diffusion impact intensity based on memorabilia temporal and spatial database: the case study of Taijiquan Culture in Yongnian County. J. Hebei Norm. Univ. (Nat. Sci. Ed.) **36**(3) (2013)
5. Ester, M., Kriegel, H.-P., Xu, X.: A density-based algorithm for discovering clusters in large spatial databases with noise. In: Proceedings of the 2nd International Conference on Knowledge Discovery and Data Mining (KDD 1996), pp. 226–231. AAAI Press, Portland (1996)
6. Ma, Y.: University's major events compilation: a perspective of university history research. J. Arch. Shanxi **1**, 70–72 (2013)
7. Zhao, R.: Research on biography generation based on events of character roles. Dalian University of Technology (2015)
8. Yan, R., Kong, L., Huang, C., et al.: Timeline generation through evolutionary transtemporal summarization. In: Proceedings of the 2011 Conference on Empirical Methods in Natural Language Processing (EMNLP 2011), pp. 433–443. ACL, EdinBurgh, UK (2011)

9. Ng, J.-P., Chen, Y., Kan, M.-Y., et al.: Exploiting timelines to enhance multi-document summarization. In: Proceedings of the 52th Annual Meeting of the Association for Computational Linguistics (ACL2014), pp. 923–933. ACL, Baltimore, Maryland, USA (2014)

10. Tran, G.B.: Structure summarization for news events. In: Proceedings of the 22th International conference on World Wide Web (WWW 2013), pp. 343–348. ACM, Republic and Canton of Geneva, Switzerland (2013)

11. Minard, A.-L., Speranza, M., Agirre, E., et al.: SemEval-2015 task 4: timeline: cross-document event ordering. In: Proceedings of the 9th International Workshop on Semantic Evaluation (SemEval 2015), pp. 778–786. ACL, Denver, Colorado, USA (2015)

12. Li, J., Cardie, C.: Timeline generation: tracking individuals on Twitter. In: Proceedings of the 23rd International Conference on World Wide Web (WWW 2014), pp. 643–652. ACM, Seoul, Korea (2014)

13. Althoff, T., Dong, X.L., Murphy, K., et al.: TimeMachine: timeline generation for knowledge-base entities. In: Proceedings of the 21th ACM SIGKDD International Conference on Knowledge Discovery and Data Mining (KDD2015). ACM, Sydney, Australia (2015)

14. Wei, X., Bin, Z., Genlin, J.: Microblog timeline summarization algorithm based on sliding window. J. Data Acquis. Process. 32(3), 523–532 (2017)

15. Feng, S., Xiao, W.: An improved DBSCAN clustering algorithm. J. China Univ. Min. Technol. 37(1), 105–110 (2008)

Lexical Resources

A Case Study of Schema-Based Categorized Definition Modes in Chinese Dictionaries

Hongyan Zhang[1], Lin Wang[2(✉)], and Wuying Liu[3]

[1] Center for Linguistics and Applied Linguistics, Guangdong University of Foreign Studies, Guangzhou 510420, Guangdong, China
20192520014@gdufs.edu.cn
[2] Xianda College of Economics and Humanities, Shanghai International Studies University, Shanghai 200083, China
lwang@xdsisu.edu.cn
[3] Laboratory of Language Engineering and Computing, Guangdong University of Foreign Studies, Guangzhou 510420, Guangdong, China
wyliu@gdufs.edu.cn

Abstract. Traditional category theories including classical category and prototype category haven't provided enough theoretical and practical support to category-based definition in dictionaries. Based on the schema category theory, this paper attempts to demonstrate how the schema-based categorized definition models work in Chinese Dictionaries.

Keywords: Category-based definition · Schema category · Chinese dictionaries

1 Introduction

As the main way of knowing and interpreting the world, categorization is a mental process of classifying things or objects, and the product of it is called cognitive categories [1]. According to the economy principle of language, words of the same category had better be defined in the same definition modes in dictionaries in order to improve users' learning efficiency.

Linguistic theories have long been studied to help compilers optimize their way of defining words in dictionaries, among which category theories have provided theoretical foundation for the category-based definition. However, these theories have not aroused compilers' enough attention so that they fail to organize the categorized definition systematically for words of the same category, thus causing confusion to dictionary users. Therefore, based on the schema category theory, this paper attempts to propose uniform definition modes for example words of the same category and provide some implications for future lexicographical work.

© Springer Nature Switzerland AG 2020
J.-F. Hong et al. (Eds.): CLSW 2019, LNAI 11831, pp. 509–516, 2020.
https://doi.org/10.1007/978-3-030-38189-9_52

2 Relevant Studies on Schema Category

The prototype theory concludes that there are two types of prototypes: prototype examples and prototype features. But the prototype theory itself has limitations on semantics interpretation because instances are categorized based on the resemblance with prototype examples or prototype features, causing instances to lose their own features. In fact, there exists a more abstract and general concept that can represent the relations of the prototype and instances - schema.

2.1 The Development of Schema

Schema is first proposed by German psychologist Kant. Piaget treats schema as a basic concept and thinks it is both the category of knowledge and the process of acquiring that knowledge [2]. Barlett defines schema as "an active organization of past reactions or experiences" [3]. We can find that schemas provide us frameworks to organize the information about the world, events, people and actions.

Later, schema theory becomes one of the most important parts in cognitive linguistics. Lakoff and Johnson regard a schema as an abstract construct through the interaction between people' experience and the world and a unit of knowledge in humans' mind [4, 5]. Langacker maintains that categorization comes from the prototype as the prototype mirrors the basic level of human beings' cognition so it is regarded as the reference or criterion to interpret other instances [6–8]. Langacker makes a clear distinction between the two terms of "prototype" and "schema": "A prototype is a typical instance of a category, and other elements are assimilated to the category on the basis of their perceived resemblance to the prototype; there are degrees of membership based on degrees of similarity. A schema, by contrast, is an abstract characterization that is fully compatible with all the members of the category it defines...; it is an integrated structure that embodies the commonality of its members..." [6]. Wang Yin displays the relations among the schema, prototype and instances [9]:

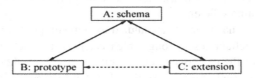

Fig. 1. Relations among the schema, prototype and extension

From Fig. 1, we can see that a schema is the commonality that emerges from distinct structures when one abstracts away from their points of difference by portraying them with lesser precision and specificity [8]. The solid arrow indicates that a schema can directly denote prototype and extension or instances. The dashed arrow between the prototype and extension means that instances are influenced not only by the schema but also by the prototype, but indirectly.

2.2 Studies on Schema-Based Categorized Definition

It is rare to see the schema category theory applied into the category-based definition. Yao Ximing and Zhang Yuening reckon that dictionary definitions are the reconstruction of schema in human mind, therefore the structures and defining words will largely influence the understanding of the definitions [10]. Zhang Yihua explains very clearly the development of schema category theory, makes clear the necessity and feasibility and proposes some definition modes by giving some examples in English and Chinese dictionaries [11, 12]. Liu Wei and Feng Tingting demonstrate the philosophical basis for the existence of the identity of modern Chinese morphemes on the basis of the schema category theory and discusses the means of reflecting the characteristic of the schema and instances in dictionary definition and example installation [13]. From the relevant studies mentioned above, we can find that the category-based definition from the perspective of schema category theory is far from enough and much has to be done to improve the quality of dictionary definition.

3 Construction of Schematic Network and Cognitive Domain

Since a schema is the commonality abstracted from the prototype and instances, and the instances are assimilated to the category on the basis of the resemblance to the prototype, similarities imply that there exist discrepancies. So how should we interpret the distinct instances different from the prototype? Here comes to the question of the cognitive domain, because all concepts are closely related to the cognitive domain to represent meanings.

The cognitive domain includes the basic domain and abstract domain [9]. A basic domain, including time, space and color, derives from humans' basic experience and can not be further simplified. An abstract domain refers to any nonbasic domain, i.e. any concept or concept complex that functions as a domain for the definition of a higher-order concept [9] and is of encyclopedia. That is to say, once characterized relative to a basic domain, a concept creates the potential for an array of higher-order concepts and thus functions as their domain; these latter concepts in return provide the domain for the emergence of still further concepts [9]. Therefore, we can think of the basic domain as the schematic features and the abstract domain as distinct features of instances in a schema category. The following figure can clearly demonstrate how the cognitive domain works in the schema of "egg" (Fig. 2):

In terms of defining words of the same category, we, except for indicating their schematic features, need to make certain abstract domains salient in the definitions of different words.

Now let's take the Chinese word "鸟" (bird) as an example. When learning the word "鸟" in Chinese culture, we may regard "麻雀" (sparrow) as the prototype of the bird category and abstract the schematic features as such: having feathers; having wings; being able to fly. When we see birds like "燕子, 乌鸦, 喜鹊" (swallow, raven, magpie), we will speak out without any hesitation that they all belong to birds. Based on the basic cognition, we form the basic cognitive schema of "鸟" (BIRD1), which we call them "鸣禽" (singing bird); When we see birds like "鹦鹉, 啄木鸟" (parrot,

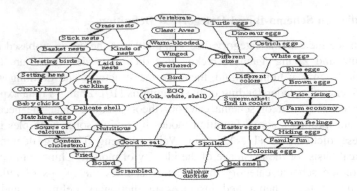

Fig. 2. Schema of "Egg" (From http://etec.ctlt.ubc.ca/510wiki/Schema_Theory.)

woodpecker), we can recognize that they also belong to birds because they are similar to common ones though they are a little bit different such as "good at climbing trees, having short but strong legs". Thus, we can form a new schema for these kinds of birds (BIRD2), and we can call them "攀禽" (climbing bird); When we see birds like "鹰, 雕" (hawk, vulture), we can form a new schema for such birds as we call "猛禽" (BIRD3); Likewise, birds having features like "having long necks, bills and legs, living in reedy areas, swamps or shadow waters, feeding on fish" are called "涉禽" (wading bird) and they form the a new schema (BIRD4), such birds include "鹤, 鹭, 鹳" (crane, heron, heron); Birds like "鸭, 天鹅, 海鸥" (duck, swan, seagull) are "游禽" (swimming bird), which have webbed feet, thick and dense feathers and fluff and good at swimming". We can form the other new schema (BIRD5); Birds like "鸡, 孔雀, 鸵鸟" (chicken, peacock, ostrich) which are "陆禽" (terrestrial birds), are usually with big size, short legs and good at running. Thus we form a new schema (BIRD6). Finally, we can form a common schema for "鸟" (bird): 有羽毛 (having feathers), 有翅膀 (having wings), 多会飞 (most can fly), 能生蛋 (laying eggs). What needs to be mentioned is, for each subcategory, it also has its own schematic features more or less different from the others. For example, "鹤" can include "丹顶鹤" (red-crowned crane), "灰鹤" (grey crane), "白鹤" (white crane), "黑颈鹤" (black-necked crane); "鹰" includes "赤腹鹰" (Chinese goshawk), "苍鹰" (goshawk), "雀鹰" (sparrow hawk), and they all have distinct features different from schematic features of "鹤" and "鹰". Thus, based on Fig. 1 and taking "鹰" as an example, we can develop such a schema category network of "鸟" as is presented below (Fig. 3):

4   Problems of Definitions in Current Chinese Dictionaries

Although the importance of category-based definition is obvious, dictionary compilers still fail to comply to it as words of the same category are organized alphabetically and defined by different people, thus causing definition inconsistent and chaotic. Now, let's take a look at how the members of "hawk" and "crane" are defined in the *Contemporary Chinese Dictionary (the Seventh Edition)* [14]. In order to show the problems of the definitions of "鹰" category, definitions of the "鹤" category are also provided. But

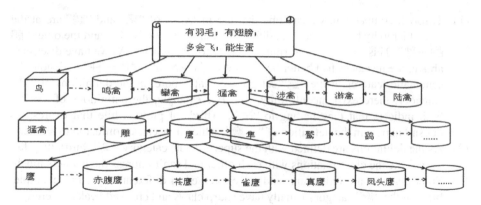

Fig. 3. Part of the schema category network of Chinese "鸟"

due to the limitation of space, the problems of the "鹤" category in the dictionary will not be discussed here.

Definitions of "鹰" category:

a. 鹰: 鸟, 上嘴呈钩状, 颈短, 脚部有长毛, 足趾有长而锐利的爪。是猛禽, 捕食小兽及其他
   鸟类。种类很多, 如苍鹰, 雀鹰, 老鹰等。
b. 苍鹰: 鸟, 身体暗褐色, 上嘴弯曲, 爪尖锐, 视力强, 性凶猛。捕食鼠, 小鸟, 野兔等。
c. 雀鹰: 鸟, 体形小, 羽毛灰褐色, 腹部白色, 有赤褐色横斑, 脚黄色。雌的比雄的稍大。是
   猛禽, 捕食小鸟。通称鹞子。

Definitions of "鹤" category:

a. 鹤: 头小颈长, 嘴长而直, 细长, 后趾小, 高于前二趾, 羽毛白色或灰色, 群居或双栖, 常
   在河边或沼泽地带捕食鱼和昆虫。种类很多, 常见的有丹顶鹤, 白鹤, 灰鹤等。
b. 丹顶鹤: 鹤的一种, 羽毛白色, 翅膀大, 末端黑色, 覆在短层上面, 能高飞, 头顶朱红色,
   颈和腿很长, 常涉水吃鱼, 虾等。叫的声音高而响亮。也叫仙鹤。
c. 灰鹤: 鹤的一种, 身体羽毛灰色, 颈下黑色, 头顶有红色斑点, 脚黑色。生活在芦苇丛中或
   河岸等地, 吃植物的浆果和昆虫等。
d. 白鹤: 鹤的一种, 羽毛白色, 只有翅膀上一部分是黑色的。尾短, 嘴和腿都很长。叫的声音
   很响亮。常生活在水边, 捕食鱼, 虾等。

From the definitions presented above, we can find there exist the following questions:

(1) Through comparison, we can find that the instances of "鹰" and "鹤" are at the same hierarchy but defined with different categories, one is "鸟" and the other "鹤的一种". This causes inconformity of the whole dictionary. As we have discussed above, the instances had better be defined in the way of "鹰的一种" to denote the schematic features of the "鹰" category. What's more, in the definitions of "鹰" category, there exist inconformity, too. For example, the definitions of "雀鹰" both indicate that they belong to "猛禽" (birds of prey), while that expression doesn't appear in the definition of "苍鹰", which may cause confusion to learners.

(2) As the definitions are defined with the superordinate category "鸟" (bird), they fail to indicate the close relations with each other and can't impress the learners much.

(3) Some features are redundant, and are expressed in different ways. For example, birds of the "鹰" category usually have sharp claws and crooked beaks. Therefore, expressions like "上嘴弯曲, 爪尖锐", "雌的比雄的稍大" don't need to appear in the definition of "苍鹰". Expressions like "也叫, 通称" are actually the same and don't need to appear in different ways.

(4) Some distinct features that are important in differentiating the different instances don't show up in the definitions. For example, in the domain of "size", the definition of "中小型猛禽" should appear in the definition of "苍鹰" to differentiate the others; "苍鹰" are usually trained to play its role in humans' activities, and this feature should exist in the definition.

(5) Some features are against facts and don't conform to humans' cognitive experience. For example, "老鹰" is usually a general expression of a big family, so it is improper to place it together with "苍鹰", "雀鹰" in the definition of "鹰".

(6) As a Chinese Dictionary for Chinese language learners, it should be to some extent encyclopedic. But only few instances like "苍鹰" and "雀鹰" are included, rare species in China like "赤腹鹰", "凤头鹰" are seen nowhere.

5   Construction of Schema-Based Categorized Definition Modes

According to the analysis above, we can find that the schema-based categorized definitions can really facilitate systematic and conform definition and help improve learners' efficiency. Therefore, in order to construct the schema-based categorized definition mode for words of "鹰" category, what we need to do first is abstract the schematic features of "鹰". As for the definitions of specific instances in this category, their distinct features will be presented according to their respective salient domains. Figure 4 can illustrate very clearly the relations of the schematic and distinct features between "鹰" and its instances "苍鹰", "雀鹰" and "凤头鹰":

Thus, the definitions of the four words can be organized in this way:

a. 鹰: 猛禽的一种。嘴强大呈勾状, 翼大善飞, 脚强而有力, 趾有锐利勾爪, 肉食性动物。种
类很多, 如苍鹰, 雀鹰, 凤头鹰等。

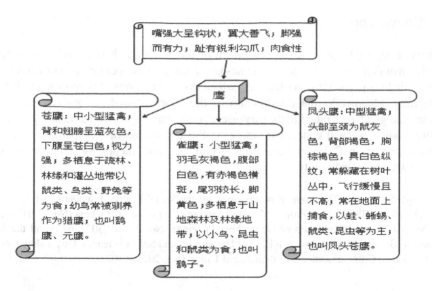

Fig. 4. Relations between schematic features of "鹰" and distinct features of its instances

b. 苍鹰: 鹰的一种, 属中小型猛禽。背和翅膀呈蓝灰色, 下腹呈苍白色。多栖息于疏林, 林缘和灌丛地带。视力强, 以鼠类, 鸟类, 野兔等为食。幼鸟常被驯养作为猎鹰。也叫鹞鹰, 元鹰。

c. 雀鹰: 鹰的一种, 属小型猛禽。羽毛灰褐色, 腹部白色, 有赤褐色横斑, 尾羽较长, 脚黄色。多栖息于山地森林及林缘地带; 以小鸟, 昆虫和鼠类为食; 也叫鹞子。

d. 凤头鹰: 鹰的一种, 属中型猛禽。头部至颈为鼠灰色, 背部褐色, 胸棕褐色, 具白色纵纹。常躲藏在树叶丛中, 飞行缓慢且不高。常在地面上捕食, 以蛙, 蜥蜴, 鼠类, 昆虫等为主。也叫凤头苍鹰。

From the definitions shown above, we can see that the definitions of the "鹰" category are systematic and uniform. The schematic features of the "鹰" category can be shown obviously. And for each instance, features different from the schematic features are salient in the definitions.

However, there are still questions we need to pay attention to when defining words of the same category based on the schema-based categorized definition modes. For instance, We, as lexicographers, will have to possess an encyclopedic knowledge of the defined words in order to differentiate words of the same category, which has something to do with the cognitive domains mentioned above. The language resources may come directly from lexicographers' mind or be taken from books, websites and so on when defining the categorization and extracting the features for each concept. Besides, the tentative definition models illustrated above are simply made for the *Contemporary Chinese Dictionary*, catering to the Chinese language learners. Nowadays, dictionaries are mainly user-based, which means different types of dictionaries should be defined in different ways to suit the levels of dictionary users. Unfortunately, Chinese dictionaries designed for different purposes are not as finely distinguished as the English dictionaries and we still have a long way to go.

6 Conclusion

Schema category theory is the development of the prototype theory and it can include all the members of categories. Schema-based categorized definition can reveal the schematic features and distinct features, which provide theoretical support to dictionary compiling. Meanwhile, the construction of the schema-based categorized definition modes can greatly facilitate dictionary users' understanding of word meanings. In the future studies, we will concentrate more on the application of schema-based categorized definition in Chinese dictionaries, trying to make dictionary definitions more systematic and user-friendly.

Acknowledgements. The research is supported by the Social Science Foundation of Shanghai (No. 2019BYY028), the Key Project of State Language Commission of China (No. ZDI135-26), the Natural Science Foundation of Guangdong Province (No. 2018A030313672) and the Key Project of Guangzhou Key Research Base of Humanities and Social Sciences: Guangzhou Center for Innovative Communication in International Cities (No. 2017-IC-02).

References

1. Ungerer, F., Schmid, H.J.: An Introduction to Cognitive Linguistics. Foreign Language Teaching and Research Press, Beijing (2001)
2. Piaget, J.: The Child's Conception of the World. Routledge and Kegan Paul, London (1928)
3. Bartlett, F.C.: Remembering: A Study in Experimental and Social Psychology. Cambridge University Press, Cambridge (1932)
4. Lakoff, G.: Women, Fire and Dangerous Things. What Categories Reveal about the Mind. The University of Chicago Press, Chicago (1987)
5. Johnson, M.: The Body in the Mind. The Bodily Basis of Meaning, Imagination, and Reason. The University of Chicago Press, Chicago (1987)
6. Langacker, R.W.: Foundations of Cognitive Grammar: Theoretical Prerequisites, vol. 1. Stanford University Press, Stanford (1987)
7. Langacker, R.W.: Concept, Image and Symbol: Cognitive Basis of Grammar. Mouton de Gruyter, New York (1990)
8. Langacker, R.W.: Grammar and Conceptualization (Cognitive Linguistics Research). Mouton de Gruyter, New York (2000)
9. Wang, Y.: The three category theories: classical, prototype, schema—the contributions of CL to post-modern philosophy. Foreign Stud. 1(1), 20–26 (2013)
10. Yao, X., Zhang, Y.: On schematic structure of definitions in english dictionary. J. Hefei Univ. Technol. (Soc. Sci.) 21(5), 129–132 (2007)
11. Zhang, Y.: Defining the Commonality and difference based on the relation between category schemata and instantiations: categorized lexical definitions for the english learner's dictionary. Foreign Lang. Teach. Res. 2, 240–253 (2017)
12. Zhang, Y.: Lexical relatedness and modes of definition based on the prototype and schema category theories: a case of categorization-based definition. Lexicogr. Stud. 5, 1–12 (2017)
13. Wei, L., Tingting, F.: A study on the identity of modern Chinese morphemes based on the schema category theory. Lexicogr. Stud. 5, 13–20 (2017)
14. Institute of Linguistics, CASS: Contemporary Chinese Dictionary, 7th edn. The Commercial Press, Beijing (2016)

The Construction and Analysis of Annotated Imagery Corpus of *Three Hundred Tang Poems*

Xingyue Hao[1], Sijia Ge[1(✉)], Yang Zhang[1], Yuling Dai[1], Peiyi Yan[1], and Bin Li[1,2]

[1] School of Chinese Language and Literature, Nanjing Normal University, Nanjing, China
sijiage007@gmail.com
[2] Institute for Quantitative Social Science, Harvard University, Cambridge, USA

Abstract. Imagery is one of the core elements in understanding and appreciating ancient poetry. Lack of imagery data leads to subjective researches in traditional imagery theory. Some quantitative studies are recently proposed but such studies are in lack of annotated corpora. This paper reports the construction of a richly annotated imagery corpus compiled from *Three Hundred Tang Poems*, a classic poetry anthology. The analysis of 4,496 imageries is made, showing that the use of imagery is a long tail distribution, which conforms to Zipf's law, and that poets prefer to use natural imageries with metaphorical meanings. The use of imagery reflects a poet's writing style to some extent, but it cannot be the golden standard for evaluating the quality of poetry.

Keywords: Imagery · *Three Hundred Tang Poems* · Poem corpus · Quantitative analysis

1 Introduction

As an essential element in ancient poetry, imagery has multi-layered metaphorical meanings, which is of great significance for understanding and appreciating ancient poetry. Traditional studies usually use case analysis and deductive reasoning methods, so conclusions are vulnerable to subjective experience of researchers. Besides, an overall and macroscopic analysis of imagery and their meanings is in shortage, for previous studies mainly concern individual cases.

With the improvement of digital humanity research methods, corpus technology and quantitative methods are involved in literature research. However, only a few corpora of ancient poems contain word segmentation and part-of-speech information at present, and the statistics gleaned from them unveil shallow ideas. Imagery is less considered, since these studies focus more on lexical items [1, 2]. Moreover, imagery units and lexical units are still blurred, leaving data and conclusions less convincing. Therefore, it is very important to construct a richly annotated imagery corpus.

Taking *Three Hundred Tang Poems*, a classic poetry anthology, as the study object, this paper constructs an annotated imagery corpus, which fully considers forms and meanings of imageries and covers their complete semantic information. On this basis,

J.-F. Hong et al. (Eds.): CLSW 2019, LNAI 11831, pp. 517–524, 2020.
https://doi.org/10.1007/978-3-030-38189-9_53

this paper conducts a quantitative research, investigates the distribution of imageries, and explores relations between poets and the use of imageries.

2 Related Work

2.1 Imagery Theory Research

Without any uniform and recognized standard, it is very difficult to define and categorize imagery although it is widely used. Ai thinks that imagery is a kind of concrete feeling [3]. Min and Yuan believe that imagery mainly contains physical objects and subjective emotions [4, 5]. Chen proposes that it is a poetic image with words as its surface [6]. Though slightly different, all these statements can be summarized into a formula given below:

$$Imagery = object + emotion \tag{1}$$

Imageries can be classified into dynamic and static imageries from the perspective of linguistics [6]. The former usually express narrative scenes, mainly composed of nouns and verbs, such as "雲破" (break through clouds) and "月來" (the moon appears), while the latter category is mostly descriptive, composed of nominal words or phrases. Static imageries can be further divided into two categories: specific and general imageries, depending on whether it refers to a particular object. For example, "黃河" (the Yellow River) and "岳陽樓" (Yueyang Tower) are specific imageries, while "河" (river) and "樓" (tower) are general ones. Static imageries can also be divided into simple and compound static imageries according to the number of lexical items. Compound static imageries have more than two words and are mainly in two forms. One is adjective + noun, like "白雲" (white cloud), and the other is noun + noun, like "燕草" (grassland in Yan).

2.2 Resources of Annotated Imagery Corpora

Constructing annotated imagery corpus is the basis for quantitative research. At present, resources of imagery corpora are relatively scarce, let alone those semantically annotated. Some representative annotated imagery corpora are shown in Table 1.

Table 1. Researches on semantically annotated imagery corpora.

Scholar	Target	Annotation objects	Scheme
Huang	Ontology	Plants, animals and artifacts in *Three Hundred Tang Poems*	SUMO/WordNet [7, 8]
Bi	Imagery similarity	Homesick and farewell imageries in *Three Hundred Tang Poems*	*Tongyici Cilin* [9]
Luo	Literature	Poems in Tang and Song Dynasty	Self-built system [10]
Lee	Lexical analysis	*Complete Tang poems*	Wang Li's scheme [11]

The corpora in Table 1 cover all imageries with different granularities and categories. But there are still some problems to be solved. Firstly, the definition of "imagery" in various studies lacks a normative explanation, so lexical items are often wrongly regarded as imageries. Secondly, most studies focus on literal meanings of imagery yet ignore metaphorical meanings. Luo attaches great importance to metaphorical meanings of imagery, but she equates lexical units with imagery units, and the semantic category scheme of imagery is not clearly explained [10]. Thirdly, the number of annotated imageries is quite small and they only annotate certain semantic information. A greater and more balanced corpus is needed.

3 The Construction of Annotated Imagery Corpus Based on *Three Hundred Tang Poems*

3.1 Data Source

Tang Dynasty is the golden age in the development of Chinese poetry. *Three Hundred Tang Poems*, a representative anthology of Tang Poems, contains 320 poems by 79 poets, all of which are written in traditional Chinese characters. This paper chooses it as the raw material and divides it into 1,597 sentences according to punctuations.

3.2 Annotation Scheme

(1) **Annotation Target.** It is difficult to define imageries due to many different opinions, while static imageries can be defined and annotated more easily because they are mainly nominal words. So this paper regards them as main study objects. Dynamic imageries are more complex with verbs involved, so most of them are ignored. As to a few dynamic imageries with specific metaphorical meanings, such as "斷腸" (anguish), "搗砧" (miss husband far away), they are also annotated in this paper because they are well-recognized.
Therefore, imageries annotated in this paper are nouns, noun phrases, predicate-object phrases and subject-predicate phrases that can arouse readers' association and senses in poetry.

(2) **Word Segmentation and Parts of Speech Annotation.** Segmenting imagery and annotating parts of speech are helpful to internal composition analysis at the lexical level, but there are few relevant studies. This paper refers to the "Pre-Qin Chinese word segmentation and part-of-speech annotation stipulations" formulated by Nanjing Normal University [12] and merges some noun and verb labels according to actual annotation. The whole process is carried out by machine combined with manual proofreading.

(3) **Scheme of Semantic Categories.** HowNet version 2007 is chosen as the semantic category scheme in this paper. It is a computer-oriented formal and computable semantic knowledge base, which has great application value [13]. It depicts meanings as "concepts", which are denoted by sememes. In other words, each concept can be decomposed into several sememes. As a basic and minimal

semantic unit in HowNet, each sememe represents a unique and specific meaning without ambiguity, functioning as a semantic feature in semantics [14].

In HowNet, 1,503 sememes are divided into seven categories, such as entity, event, attribute and attribute value, each having a complete and hierarchical sememe structure. The upper sememe indicates general characteristics, while the lower shows individual features. Each concept in HowNet is represented as "DEF = main feature + the secondary feature + other attributes or relations", in which the main feature is its literal meaning, that is, the most basic meaning, also known as "basic sememe". For example, "芳草" (flower and grass) in HowNet is defined as:

$$DEF = \{FlowerGrass|花草: modifier = \{fragrant|香\}\} \tag{2}$$

The basic meaning of "花草" is *flower and grass*. Emphasizing its odor, "modifier" indicates that the feature of "fragrance" modifies flower and grass. The main object of this paper is to annotate static imageries, so sememe label {FlowerGrass|花草} is annotated to "花草".

(4) **Semantic Annotation of Imageries.** In most studies, only literal meanings of imageries are considered, while internal semantic compositions, especially meanings of modifiers and subordinates, are neglected. However, imageries have multi-layered meanings. An imagery may have literal and metaphorical meanings, and one meaning can be expressed by different imageries. They show complex many-to-many relations. Therefore, we need to annotate literal meanings, modifiers' meanings and metaphorical meanings.

Literal meanings are unique and basic. We use the first sememe in the definition of a concept as its literal meaning. For example, "风" (wind) is defined in HowNet as {wind|风} {gas: modifier = {StateGas|气态}, {WeatherBad|坏天: experiencer = {~}}. So, {wind|风} is treated as its semantic category and basic meaning.

Modifiers refer to words that modify literal meanings, mainly being adjectives and nouns. "春風" (spring wind) is a signal of spring, symbolizing hope, while "秋風" (autumn wind) implies depression. Semantic differences can be reflected by modifiers. Meanings will be further clarified with subordinate modifiers.

Metaphorical meanings mean specific roles of imageries, which can be grouped into three categories. Firstly, emotion or action. Some imageries express feelings or refer to certain actions, which can be annotated directly. For example, the metaphorical meaning of "愁雲" (depressing cloud) is to create a melancholy atmosphere and express sad feelings. Secondly, metonymy. In specific contexts, some can imply certain objects that are different from their literal meanings, such as "觴" (goblet) referring to wine. This category is distinct by adding "!". The third category is nickname. For example, "諸葛亮" (Zhu Geliang, a famous ancient statesman) is often called "孔明" (Kong Ming), so these two proper names are annotated as one entity, which is conducive to constructing dictionaries of proper names in poetry.

3.3 Annotation Sample

Taking Li Bai's *Spring Thoughts* as an example, the imagery to be annotated is "春風" (spring wind) in sentence "春風不相識" (spring wind is a stranger). The annotation is as follows (Table 2):

Table 2. A sample to annotate meanings of "春风".

Sentence	春/t 風/n 不/d 相/d 識/v, /w
The first imagery	春風
Literal meaning	風/{wind|风};
Modifiers' meaning	春/{time|时间}; {spring|春};
Metaphorical meaning	怀春/{expect|期望}; {love|爱恋};

3.4 Quality Control

This paper mainly avoids the inconsistency of annotation and data sparsity caused by subjectivity from two aspects: annotation stipulations and error correction mechanism, since subjective factors can easily influence our understanding to imageries.

(1) **Annotation stipulation.** This paper designs a detailed instruction, stipulates common annotation problems one by one and provides abundant annotation samples for reference.
(2) **Error correction mechanism.** This paper requires interpretations from annotators when annotating imageries for later error correction. Besides, after multiple manual checks, annotated information is cross-checked.

4 Quantitative Analysis of Annotated Imagery Corpus

4.1 Basic Statistics of Imagery

We annotate 4,496 imageries in 1,597 sentences from 320 tang poems, including 2,857 imagery types. Each sentence contains an average of 2.82 imageries. Such distribution density shows that imageries are used frequently, which further confirms the necessity of annotation and analysis efforts of the current study.

Table 3. The top 10 high-frequency imageries.

Rank	1	2	3	4	5	6	7	8	9	10
Imageries	月	夜	風	山	花	淚	酒	水	雨	天
Frequency	51	39	37	33	30	27	25	25	22	22

All imageries in Table 3 are simple imageries, which are monosyllabic nouns in terms of language form. Most of them are general imageries, referring to a kind of objects, such as "山" (mountain) and "花" (flower). This shows simple imageries that refer to general objects weigh more compared to specific and compound ones, which is consistent with previous statistical results [15, 16].

Basic sememes in HowNet determine core meanings of concepts. Therefore, we can get semantic categories of imageries' literal meanings by their annotated sememes. 130 basic sememes are obtained from 4,496 imageries. The top 10 and their imageries are shown in Table 4.

Table 4. The top 10 frequently occurred semantic categories.

Basic sememe (semantic category)	Frequency	Ratio	Imagery sample
{part|部件}	445	10.12%	淚
{human|人}	397	9.02%	客
{place|地方}	232	5.27%	長安
{tool|用具}	230	5.23%	鐘
{time|时间}	194	4.41%	夜
{waters|水域}	171	3.89%	江
{山}	159	3.61%	山
{tree|树}	130	2.96%	樹
{FlowerGrass|花草}	117	2.66%	花
{bird|禽}	112	2.55%	鳥

In Table 4, sememes that express time and space appear frequently, because these two elements are necessary for cognitive scenes. Once they are defined in poems, the atmosphere and emotions are created. For example, "夜" (night) tends to be homesick and lonely, while "晨曉" (dawn) tends to be warm and glorious. "春天" (spring) is joyful and pleasant, yet "秋" (autumn) is lonely and desolate.

Landscapes, flowers and plants are also frequently used, reflecting poets' preference for natural imageries. A poet often chooses what he sees around him to create imageries. It is interesting to note that although they are both natural imageries, inanimate natural landscapes such as mountains and rivers are used more frequently than plants and animals. This is related to poets' cognitive characteristics: animals and other creatures are cognitive foci, while mountains and rivers are backgrounds.

In summary, imagery shows a long tail distribution as to its use and its semantic category, consistent with Zipf's law [17].

4.2   Poets and Their Typical Imageries

This paper proposes to use different imageries to distinguish poet's styles, that is, to find out typical imageries of different poets, which are calculated by TF-IDF. TF-IDF is a common method in document feature representation. It can quantitatively represent the ability of a word to distinguish current document from other documents. In TF-IDF,

TF denotes Term Frequency, while IDF denotes Inverse Document Frequency. The formulas are:

$$TF = \frac{frequencies\ of\ one\ item\ in\ current\ document}{total\ frequencies\ of\ items\ in\ current\ document} \tag{3}$$

$$IDF = \log \frac{number\ of\ total\ documents}{number\ of\ documents\ that\ own\ this\ item} \tag{4}$$

The TF-IDF value is the product of the two. In this paper, each imagery type is regarded as an item, and all works of each poet a document. Table 5 is the result arranged in descending order.

Table 5. Six famous poets and their typical imageries.

Poets	Imageries with high TF-IDF
Du Fu (杜甫)	將軍 風塵 三峽 先帝 王孫 長安 秋水 行人 骨肉 丞相
Li Bai (李白)	青天 碧山 秋風 蜀道 春風 長安 秋月 長風 天姥 幽徑
Bai Juyi (白居易)	君王 琵琶 春風 仙樂 絲竹 都門 魂魄 秋月 雲鬢 顏色
Li Shangyin (李商隱)	東風 巴山 新知 聖相 蓬山 垂楊 夜雨 世界 丹墀 二喬
Wang Wei (王維)	古木 寒山 桃花 清川 漁舟 窮巷 野老 青溪 深林 白雲
Han Yu (韓愈)	清風 鬼物 丘軻 丹柱 九疑 二雅 佛寺 佛畫 使家 侯王將相

Limited to the size of corpus, extracted imageries cannot fully reflect poet's artistic style. But they do reflect characteristics of each poet to a certain extent according to common sense of literature. For example, Du Fu's typical imageries are related to social reality, such as "王孫" (noble's offspring), "行 人" (pedestrian) and "骨肉" (bones and flesh), embodying his realistic feelings. As to Li Bai, he loves travelling and is more romantic, his typical imageries being full of nature scenes; Bai Juyi and Li Shangyin's imageries tend to be gorgeous and create extremely luxurious atmosphere; while those of Wang Wei are simple and elegant, conforming to his style that "integrating poems with paintings". And Han Yu pursues more peculiar and strange imageries, such as "鬼物" (ghost) and "丹柱" (scarlet pillar).

5   Conclusion and Future Work

This paper constructs an annotated imagery corpus based on *Three Hundred Tang Poems*. Quantitative analyses of the corpus show that the use of imagery is a long tail distribution, conforming to Zipf's law [17], and general imageries are used frequently, such as "月" (the moon) and "夜" (night). The distribution of imagery's semantic category levels out, and natural imageries that easily stimulate symbolic meanings are favored. The use of imagery reflects a poet's writing style to some extent, but it cannot be the golden standard for evaluating the quality of poetry.

In the future, we will make a better annotation scheme, and consider static and dynamic imageries separately. Besides, we will further expand the scale of annotation, extending it to diachronic poetry texts and making a diachronic analysis of imageries. Theoretically, more detailed statistics will be collected to dig out individual's characteristics and diachronic semantic features of imageries.

Acknowledgments. We are grateful for the comments of reviewers. This work is a staged achievement of the projects supported by National Language Committee Project of China (YB135-61), National Language Committee Project of China (WT135-24), and Higher Institutions' Excellent Innovative Team for Philosophy and Social Sciences Project of Jiangsu (2017STD006).

References

1. Jiang, S.Y.: "The Moon" in Li Bai and Du Fu's poems–how computers are used in classical poetry appreciation. Lan. Pl. **52**(11), 4–9 (2008). (in Chinese)
2. Song, W.: "The Wind" and "The moon" in Qi Liang's poems–a case study for imagery statistics in poems by a refined corpus. J. Chin. Lan. C. NNU. **16**(1), 173–178 (2014). (in Chinese)
3. Ai, Q.: The formal problem of poetry–against the formalist tendency of poetry. Litr. **5**(3), 22–37 (1954). (in Chinese)
4. Min, Z.: Chinese classical imagery theory. Litr. Art. Sd. **4**(3), 54–62 (1983). (in Chinese)
5. Yuan, X.Y.: The imagery of Chinese classical poetry. Litr. Hrt. **29**(4), 9–15 (1983). (in Chinese)
6. Chen, Z.E.: The Imagery of Poetry. China Social Sciences Press, Qinhuangdao (1990)
7. Huang, C.R.: Sinica BOW: integrating bilingual WordNet and SUMO ontology. In: Proceedings of International Conference on Natural Language Processing and Knowledge Engineering, pp. 825–826. IEEE, Beijing (2003)
8. Huang, C.R., Lo, F., Chang, R.Y. et al.: Reconstructing the ontology of the Tang dynasty: a pilot study of the Shakespearean-garden approach. In: Workshop of the Language Resource Conference (LREC), Lisbon (2004)
9. Bi, X.: Research on Imagery Retrieval Based on Tang Poetry Corpus. Dalian University of Technology, Dalian (2006). (in Chinese)
10. Luo, F.Z.: Research on semantic concept classification system based on the construction of Chinese Poetry language characteristics. J. Lib. Inf. S. **9**(4), 63–86 (2011). (in Chinese)
11. Lee, J., Kong, Y.H.: A dependency Treebank of classical Chinese poems. In: Proceedings of NAACL-HLT, pp. 191–199, Montreal (2012)
12. Chen, X.H.: Pre-Qin Literature Information Processing. World Book Publishing Company, Beijing (2013)
13. Dong, Z.D., Dong, Q., Hao, L.L.: The theoretical discovery of HowNet. J. Chin. Inf. Pro. **21**(4), 3–9 (2007). (in Chinese)
14. Li, B.: Acquisition and analysis of cognitive attributes based on the internet. Appl. Lng. **56**(3), 134–143 (2012). (in Chinese)
15. Zhou, F.X.: The specificity of natural imagery in Tang Poetry–discussing with Watson et al. J. Theor. Sd. Litr. Art. **3**(4), 42–48 (1983). (in Chinese)
16. Watson, B.: Chinese Lyricism, Shih Poetry from the Second to the Twelfth Century. Columbia University Press, New York (1971)
17. Zipf, G.K.: Human behavior and the principle of least effort. Addison-Wesley Press, Boston (1949)

Building Semantic Dependency Knowledge Graph Based on HowNet

Siqi Zhu, Yi Li, Yanqiu Shao$^{(\boxtimes)}$, and Lihui Wang

School of Information Science, Beijing Language and Culture University,
Beijing, China
123115@stu.blcu.edu.cn, blcu2014liyi@gmail.com,
yqshao163@163.com, timywang@163.com

Abstract. This paper introduces a method of constructing a semantic dependency knowledge graph (SDKG) by using the rich semantic knowledge in HowNet. The establishment of SDKG depends on correspondence between the lexical dependency labels in semantic dependence bank of BLCU-HIT and the event roles in HowNet. For words with few event roles or those which are not included in the knowledge graph, sememes are recommended to them based on SPWE and SPASE algorithms to extend the SDKG. The paper demonstrates that the experiments achieve an accuracy of 86% when the sememe recommendation is conducted. Considering the establishment of the dependency relationship, a correspondence table in this paper including 87 pieces of data of event role labels mapping to dependency labels is designed. The constructed SDKG has nearly 500000 nodes that contains rich dependency information, which can be used to assist the analysis of the Semantic Dependency Parser. Besides, the results of Semantic Dependency Analysis can be drawn on to supplement the SDKG.

Keywords: HowNet · Dependency knowledge graph · Relation correspondence table · Sememe recommendation

1 Introduction

The Semantic Dependency Analysis refers to breaking the shackles of the surface syntactic structure in the sentence structure, analyzing the semantic relationship between words, and reaching the essence of deep semantic expression [1]. The existing Semantic Dependency Analysis relies heavily on the manual semantic relationship annotation which is time-consuming and labor-intensive. Meanwhile, semantic dependency labelled by different people are prone to be inconsistent. Labeling errors will result in accumulative errors during the process of Semantic Dependency Analysis, thus causing the performance of Semantic Dependency Analyzer to be degraded. To improve the performance of the Semantic Dependency Analyzer, better analysis algorithms or additional data from the external semantic knowledge base can be used. It is necessary to construct a SDKG and realize the pre-labeling of sentence components by means of the semantic information it contains, which can reduce the workload of

© Springer Nature Switzerland AG 2020
J.-F. Hong et al. (Eds.): CLSW 2019, LNAI 11831, pp. 525–534, 2020.
https://doi.org/10.1007/978-3-030-38189-9_54

manual annotation, avoid possible errors that may occur, and improve the accuracy of automatic labeling.

At present, there are two main approaches to build a knowledge base. One is to pre-define the ontology and data schema and then add the entity to the knowledge base (top-down approach), such as Freebase. The other is to extract entity from open link data. Higher entities are added to the knowledge base to build ontology (bottom-up approach), such as Google's Knowledge Vault [2, 3]. Nowadays, existing ontology libraries (such as HowNet and WordNet) are utilized as an ontology to build most of the large-scale knowledge bases, and then valid information from a large amount of data are extracted to build these knowledge base.

Building a SDKG requires rich semantic information. It is very difficult and time-consuming to conduct without annotated corpus. HowNet [4] is a large-scale language knowledge base constructed by Mr. Dong Zhendong and Mr. Dong Qiang. HowNet contains rich semantic information, which makes it the first choice to build a SDKG. In recent years, with the emergence of technologies such as deep learning, how to integrate the semantic information of HowNet with existing technologies has gradually become a research hotspot. Most of the existing research have considered part of information of HowNet such as items or sememes. Despite breakthroughs that have been made, most research ignored the event role information. Using the rich semantic information in HowNet to build a SDKG can save a lot of manpower and material resources spent in the preliminary work, so that the construction can be carried out quickly. The SDKG completely inherits the features of HowNet and describes each concept in detail.

The structured data of HowNet are utilized to construct the SDKG, and sememes are recommended to the words with fewer sememes in the knowledge graph. Besides, the SDKG restores the dependency labels between sememes. Comparing with HowNet, it contains extra dependency information. Therefore, downstream applied research can get sufficient knowledge from the knowledge graph.

2   Related Work

HowNet uses sememes, the basic indivisible semantic unit to express the meaning of words. Meanwhile, the event role and semantic information are marked between the sememes, so that the meaning of words can be accurately described. In the structure of HowNet, a word consists of one or more meaning items, and meaning items are defined by sememes and event roles. At present, HowNet contains more than 800 sememes, 212541 meaning items, and 118347 words [4].

Relies on the Chinese semantic dependence graph theory, the semantic dependence bank (SDB) of BLCU-HIT includes 61 coarse-grained dependency labels, 170 fine-grained dependency labels, and 45000 annotated sentences. The SDB of BLCU-HIT has basically covered the semantic dependency phenomena in modern Chinese.

The SDKG constructed in our research relies on the semantic dependence bank of BLCU-HIT. The semantic dependency information between the words in the Semantic Dependence Analysis and different meanings of the same word are stored in SDKG. The SDKG contains 88 relationships (87 semantic dependency labels and 1 lexical-sememe relationship) and stores nearly 500000 nodes. In the process of building this SDKG, the graph database Neo4j is utilized to store the data of SDKG (vocabulary and semantic dependency labels). Neo4j [5] is a high-performance graph database that stores structured data in graphs. Because of its convenience, high performance and other advantages, it is increasingly popular among developers. Neo4j stores knowledge through nodes and relationships, supports the customization of node attributes. Furthermore, Cypher is utilized by Neo4j to query the database, making data retrieval convenient.

3   Construction of SDKG Based on HowNet

In HowNet, each concept is described by using Knowledge System Description Language (KDML). The illustration of the structure of HowNet is as follows (the information shown is from the 2008 version of HowNet's data file) [6].

As shown in Fig. 1, "NO" represents the identifier number of the word; "W_C" is the Chinese name of the word; "G_C" defines the part of speech of the word; "E_C" gives a specific example of the word (The example of "便桥 *bian qiao* temporary bridge" is not given); "W_E" gives an English explanation of the word; "G_E" indicates the part of speech in the English interpretation; "E_E" is a vacant part; "DEF" is a detailed description of the word by a combination of event roles and sememes.

```
NO.=008153
W_C=便桥
G_C=N [bian4 qiao2]
E_C=
W_E=makeshift bridge
G_E=N
E_E=
DEF={bridge|桥梁:modifier={temporary|临时}}

NO.=008154
W_C=便桥
G_C=N [bian4 qiao2]
E_C=
W_E=temporary bridge
G_E=N
E_E=
DEF={route|道路:modifier={temporary|临时},
    {cross|越过:LocationThru={waters|水域},location={~}}}
```

Fig. 1. The description of the word "便桥 (*bian qiao* temporary bridge)" in HowNet data file.

Figure 1 shows the description of words in HowNet. Different meanings of the same word are described by different data items.

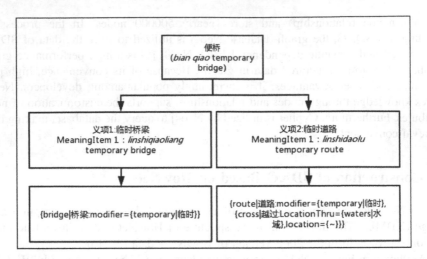

Fig. 2. The meanings of the word "便桥 (*bian qiao* temporary bridge)".

As shown in Fig. 2, the term "便桥 (*bian qiao* temporary bridge)" has two meaning items. The meaning item 1 is a temporary bridge and the meaning item 2 is a temporary road. The specific description of meaning 1 consists of two semantics and an event role: the basic sememe is "bridge|桥梁 (*qiao liang* bridge)", and the event role is "modifier", which is the further explanation of the basic sememe. Moreover, the "modifier" links the other sememe "temporary|临时 (*lin shi* temporary)", adding additional information, namely, the meaning of the bridge, so the word can be interpreted as a temporary bridge. The more complicated description of meaning 2 is composed of five sememes and two attributes. The basic sememe is "route|道路 (*dao lu* route)", and the inner two attributes are separated by commas. The two attributes further explain the basic sememe. The first attribute consists of the internal sememe "temporary|临时 (*lin shi* temporary)" and the event character "modifier" that follows. The second attribute is composed of sememes "cross|越过 (*yue guo* cross)", "waters|水域 (*shui yu* waters)", "~"and the two event characters "LocationThru", "location", which constitute the location description. And sememe "~" describes the relationship between the "cross|越过 (*yue guo* cross)" and "waters|水域 (*shui yu* waters)". The meaning of item 2 is described as a temporary road through a waters.

The ontology construction of the SDKG relies on HowNet. By comparing the results of the "DEF" and semantic dependency analysis of the concept in the HowNet, we find that if there is an event role between sememes in HowNet, there is a great similarity between the event roles of the two semantics and the semantic relationships between the semantic units in the semantic dependency analysis. In response to this phenomenon, we measure the fine-grained labels of 113 features of event roles in the event role and pragmatic feature table of the HowNet and the semantic dependency labels in semantic dependence bank of BLCU-HIT. The result has shown that 87 event roles have corresponding relationships with dependency labels. Therefore, this paper designs a correspondence table to assign dependency labels to the existence of the dependency relationship, and realize the transformation from concept interpretation to dependency.

In order to build a SDKG, the structure of the SDKG is first designed. The structure of the constructed knowledge graph is shown in Fig. 3. Considering the structure of HowNet, this paper retains the Chinese words, part of speech, the sememes of the words and the event relationship information in each concept interpretation, and use the data to construct the knowledge graph.

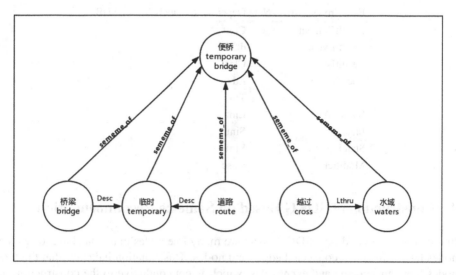

Fig. 3. The structure visualization of SDKG built in this paper.

In this experiment, the data of HowNet is split according to the above structure, and the information of each part is obtained. The ontology is reconstructed in the database based on the following steps:

(1) This paper uses the Neo4j database as the carrier of the SDKG, and build the SDKG in the graph database. In order to meet the needs of Chinese semantic dependence analysis, our knowledge graph uses the "W_C", "G_C" and "DEF" parts of the word information description in the HowNet data and eliminates the serial number information and English information.

(2) Since the meaning of the words in "DEF" is also presented in the form of Chinese words, when the knowledge graph is constructed, this paper defines the word and the original sememe of the word as the node, and part of speech of the word contained in the "G_C" part is assigned to each node as an attribute.

(3) For the definition of the relationship between the words and the sememes, the "sememe_of" relationship is defined in the knowledge graph, since there is an explanatory relationship between the original and the sememes. In order to distinguish between the main sememe and other sememes, this paper adds a "main sememe" property to the main sememe node. In the Neo4j database, variety of relationships among nodes can be defined, and that is convenient to expand the relationships between words and sememes.

(4) As for the event role, this paper manually converts the event role into a dependency labels, since the event roles and the dependency labels have correspondences. Table 1 is the correspondence table. Therefore, a knowledge graph with semantic dependency information is constructed.

Table 1. Partial correspondence between event roles and dependency labels

Event roles in HowNet	Dependency labels in our SDB
ResultContent	Cont
ResultEvent	dCont
ResultIsa	Clas
LocationThru	Lthu
TimeIni	Tini
SourceWhole	Lini
StateIni	Sini
StateFni	Sfni
Modifier	Desc

4 Construction of SDKG Based on Sememe Recommendation

In the process of building a SDKG, there are many free nodes in the database, of which nodes have only a few connected edges and nodes. This situation indicates that the free nodes have no sememe and event roles, which is not conducive to the construction of SDKG. To solve this problem, this paper utilizes the open source code developed by the natural language processing laboratory of Tsinghua University [7, 8] to recommend sememes for our SDKG. The sememes recommendation algorithms of Tsinghua University [9] adopts two main methods, both of which are based on the assumption that similar words have the similar sememes. For example, the word "orange" rather than "blacksmith" will be recommended for the word "apple", because apples and oranges are both fruits. The algorithm descriptions are as follows:

(1) **Sememe Prediction with Word Embeddings (SPWE).** When a new word is given, this method finds the most relevant words in HowNet according to the word vector of the word, and recommends sememes of these related words to the new word.

(2) **The Sememe Prediction with Aggregated Sememe Embeddings (SPASE).** This method constructs a "word-sememe" matrix by using the original definitions of existing words, and establishes a semantic vector that matched the word vector by decomposing the matrix. When a new word is given, the word vector obtained by the new word in the large-scale text data is obtained, and the dot product of the word and the original word vector is used to directly measure the correlation between the word and sememes. Therefore, the most relevant sememes can be recommended for new words based on above steps.

The experimental results show that SPWE can learn the complex structure of sememes according to related words, and SPASE can provide the potential relationship between words and sememes. These two methods are complementary and can improve the prediction performance. Therefore, the method used in this paper is the SPWE +SPASE hybrid model. Because the recommendation of dependency is in complexity, these free nodes are not supplemented by dependency labels.

5   Experimental Results and Analysis

Through the above process, this paper establishes a SDKG with 88 relationships (87 semantic dependency relationships and 1 lexical-sememe relationship) and nearly 500000 nodes. Among them, 431375 nodes are obtained by transferring the event role of HowNet to the dependency labels of semantic dependence bank of BLCU-HIT, and 51624 nodes are generated by sememe recommendation.

Since there are few studies related to the metrics of the semantic dependency knowledge graph, this paper considers the method of random sampling, in measuring the accuracy of the constructed SDKG information. Part of the samples are randomly selected as test samples, and the accuracy is determined by comparing the results of the dependency analysis with the dependency labels in the knowledge graph.

Since the number of each dependency labels is different in the knowledge graph, this paper randomly selects about 10% of the samples in the semantic dependency library for each label, and the total number of selected samples is 49368, which ensures the correctness of the sampling result. By observing the accuracy of various relationships and conducting a comparative study, this paper finds that the accuracy of the one-to-one correspondence can reach 99%, which is significantly higher than the one-to-many accuracy of the correspondence. In the one-to-many relationship correspondence, the accuracy of only 55% can be achieved. For a one-to-many relationship correspondence, a more in-depth research is needed to build a SDKG with a high accuracy.

In response to the above situation, the results are further analyzed as follows:

(1)   When examining the knowledge graph, it is found that the event roles in the knowledge graph and the dependency labels in the semantic dependency analysis are not only one-to-one, but have three correspondence relations, namely, one-to-one, one-to-many, and many-to-one. For the correspondence between one-to-one and many-to-one, the knowledge graph constructed by the corresponding relationship has better accuracy. As shown in Fig. 4, for each component in the sentence, "临时 (lin shi temporary)", "道路 (dao lu route)", "越过 (yue guo cross)", and "水域 (shui yu waters)" are the sememes of "便桥 (bian qiao temporary bridge)". The dependency tags of "越过 (yue guo cross)" and "水域 (shui yu waters)", "临时 (lin shi temporary)", and "道路 (dao lu route)" are consistent with the corresponding targets of "LocationThru" and "modifier" in the table, both of which are "Lthu" and "Desc". Accuracy can be as high as 99%. However, for a one-to-many correspondence relationship, a high accuracy is not achieved in the performance of such relationships. In response to this problem, this paper constructs a more accurate correspondence table based on the fine-grained semantic dependency labels in our SDB. The correspondence relationship is shown in Table 2. After that, this paper utilizes the results of the semantic dependency analyzer to modify the corresponding dependency labels in the knowledge graph.

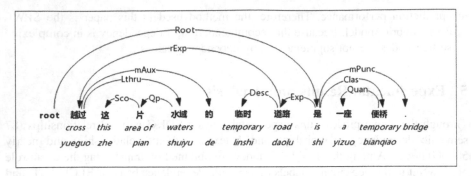

Fig. 4. Example of Semantic Dependency Analysis.

Table 2. Partially modified event role relationship and semantic dependency label correspondence table

Event roles in HowNet	Dependency labels in our SDB
Agent	Agt
	Exp
	Tool
Patient	Pat
Possession	Belg
	Pat
PatientPart	Pat
PatientProduct	Prod
Modifier	Desc

(2) During the Semantic Dependence Analysis, the part of speech significantly determines the dependency relationship. Therefore, when the knowledge graph is constructed, the part of speech of the word with dependency relations in the knowledge graph should be considered. After research, the result shows that for the core of the outermost sememe of each word, part of speech should be consistent. Because of the existence of event role relationship, a large part of sememes is the description of the main sememe. Therefore, part of speech of the word is directly used as the attribute of the node for the main sememe, which improves the accuracy of the corresponding.

(3) Since the relationship between the recommended words that are joined to the knowledge graph is not defined, the quality of the entire knowledge graph is lowered and the storage and retrieval are facing pressures. In the future, we will recommend the relationship, supplement the relationship between the nodes by using the related techniques of knowledge graph, and adopt the Semantic Dependency Analyzer to perform data cleaning and error correction.

In this paper, 470000 300-dimensional word vectors trained on 10 Gigabyte corpus of Wikipedia and People's Daily newspaper in the past 30 years are used in sememe recommendation. On the SPWE+SPASE model, the recommended sememes achieved mean average precision of 86%. This shows that word embedding vectors trained on large-scale corpus have better quality and the comprehensive use of matrix decomposition and collaborative filtering can effectively carry out the recommendation of sememe. Moreover, using a higher quality word vector can make the accuracy of the sememe recommendation higher.

Through the above steps, the sememes of free nodes have been completed. The specific experimental results are shown in Table 3:

Table 3. The experimental results of the sememe recommendation algorithms (* is the experimental result of the original paper)

Sememe recommendation algorithms	Mean average precision
SPWE	0.676
SPASE	0.506
SPWE+SPAWE*	0.713
SPWE+SPAWE (ours)	0.863

6   Conclusion and Future Work

This paper examines the way to use a large amount of semantic information in HowNet. It uses the event roles information between the sememes to establish a SDKG. The event role information between sememes is transformed to semantical dependency information, which ensures the dependency information is contained.

In the future, we will continue to expand the knowledge graph, carry out a further study on correspondence table, and further modify the one-to-many relationships in the corresponding table, free nodes and edges to modify the entire knowledge graph. Not only can capacity be expanded, the overall quality can also be enhanced to meet the needs of higher-level research.

Acknowledgments. This research project is supported by the National Natural Science Foundation of China (61872402), the Humanities and Social Science Project of the Ministry of Education (17YJAZH068), Science Foundation of Beijing Language and Culture University (supported by "the Fundamental Research Funds for the Central Universities") (18ZDJ03), the Fundamental Research Funds for the Central Universities and the Research Funds of Beijing Language and Culture University (19YCX122). In this way, we are sincerely grateful for the foundations.

References

1. Shao, Y., Qiu, L., Liang, C., Mao, N.: Construction and analysis technology of Chinese semantic dependency TreeBank. In: Sun, M., Chen, Q. (eds.) The 11th Chinese National Conference on Computational Linguistics, Advances of Computational Linguistics in China, pp. 236–241. Tsinghua University Press, Luoyang (2011). (in Chinese)
2. Liu, Q., Li, Y., Duan, H., Liu, Y., Qin, Z.: Knowledge graph construction techniques. J. Comput. Res. Dev. **53**(3), 582–600 (2016). (in Chinese)
3. Xu, Z., Sheng, Y., He, L., Wang, Y.: Review on knowledge graph techniques. J. Univ. Electron. Sci. Technol. China **45**(04), 589–606 (2016). (in Chinese)
4. HowNet. http://www.keenage.com. Accessed 14 May 2019
5. Neo4j. https://neo4j.com. Accessed 14 May 2019
6. Zhu, J., Yang, Y., Xu, B., Li, J.: Semantic representation learning based on HowNet. J. Chin. Inf. Process. **33**(3), 33–41 (2019). (in Chinese)
7. Sun, M., Chen, X.: Embedding for words and word senses based on human annotated knowledge base: a case study on HowNet. J. Chin. Inf. Process. **30**(6), 1–6 (2016)
8. Niu, Y., Xie, R., Liu, Z., Sun, M.: Improved word representation learning with sememes. In: Proceedings of the 55th Annual Meeting of the Association for Computational Linguistics, vol. 1 (Long Papers), pp. 2049–2058. Association for Computational Linguistics, Vancouver (2017)
9. Xie, R., Yuan, X., Liu, Z., Sun, M.: Lexical sememe prediction via word embeddings and matrix factorization. In: IJCAI 2017 Proceedings of the 26th International Joint Conference on Artificial Intelligence, pp. 4200–4206. AAAI Press, Melbourne (2017)

Study on the Order of Vocabulary Output of International Students

Xi Wang[2] and Zhimin Wang[1,2(✉)]

[1] Research Institute of International Chinese Language Education,
Beijing, China
wangzm000@qq.com
[2] Faculty of International Education of Chinese Language,
Beijing Language and Culture University, Beijing, China
wanx08biu@sina.com

Abstract. In the process of teaching Chinese as a foreign language, vocabulary output is an important standard to measure effects of learners' language acquisition. This paper collects 100 questionnaires from junior and senior international students, which require respondents to list the 300 most important daily words they think should be mastered when learning Chinese. Through the induction and statistical analysis of the questionnaire results, it is found that the vocabulary output of international students follows scene clues, word category clues and part of speech clues. At the same time, the vocabulary output of international students also has certain rules and characteristics. The output vocabulary has imageability and two-syllable words are dominant. Vocabulary output has a preliminary sense of morpheme, which embodies scene concept. In addition, it also has gender difference, and follows the acquisition order.

Keywords: Basic vocabulary · Vocabulary output order · Psychological wordlist · Teaching Chinese as a foreign language

1 Introduction

The vocabulary output of language learners plays a critical role in measuring effects of learners' language acquisition.

Previous studies have reported the effect of vocabulary output on language acquisition. Surveys such as that conducted by Bao [1] showed that the output task independently affected the vocabulary acquisition of English language learners. He conducted an experiment to test the output task, mainly by sentence formation and gap fill exercises. Huang [2] investigated the effect of output task types on attention and incidental acquisition of second language vocabulary in relevant input. Through different output tasks, such as writing, reconstruction and retelling, it was noted that all output tasks could effectively improve learners' attention to target words and promote their receptive vocabulary acquisition. Meanwhile, time had a moderating effect on task effectiveness. Kuai [3] through a word table experiment found that cognitive style and gender can influence memory. Zhang [4] proposed that the establishment of adult second language mental lexicon was often dependent on the mental lexicon of the mother tongue. If more second language words can be learned, it is possible to establish

J.-F. Hong et al. (Eds.): CLSW 2019, LNAI 11831, pp. 535–545, 2020.
https://doi.org/10.1007/978-3-030-38189-9_55

a separate second language mental lexicon. Wang [5] pointed out that morpheme teaching was the basis of vocabulary teaching in Chinese as a foreign language. Some monophonic morphemes with strong word-formation ability should be taught first, and morpheme teaching should be emphasized when discriminating and analysing words with the same pronunciations and synonyms.

Although some previous studies have focused on English vocabulary output, less research has been conducted into Chinese vocabulary output. These studies provided inspiration for this research, but how can second language vocabulary be obtained? What are the characteristics of psychological vocabulary output? Is there a certain rule? These are questions that need to be explored in depth.

The overall structure of the study takes the form of five sections, including this introductory section. The second section begins by designing and giving out questionnaires, and aims at obtaining respondents' psychological vocabulary. The third section is concerned with the inspection of foreign students' vocabulary output clue by statistics of the survey results. The fourth section presents the findings of the research, focusing on the six key aspects about analysis of vocabulary output characteristics of international students. Finally, we come to the conclusion and outline the further research.

2   A Survey of Basic Vocabulary

The respondents of this questionnaire are international undergraduates studying in China at Beijing Language and Culture University. We collect what these students considered to be the most important words in their daily life and studying, and then conduct statistical analysis using the collected results. By observing the order and frequency of words in the vocabulary output of international students, we can consider their psychological wordlist and therefore provide some references for teaching Chinese as a foreign language. The exact wording of the questionnaire is as follows:

> Dear classmates,
> With your experience in learning Chinese, what do you think are the 300 most commonly used words for international students? These words may relate to learning, survival, daily communication, daily activities and so on. Please fill in your ideas on the form.
> Thank you for your cooperation.

Different from the traditional questionnaire, this questionnaire is open-ended and subjective, what's more, it does not include any leading questions, which aims to encourage respondents to write the first words that come to their minds without being influenced by a choice of options. In this way, we are able to understand the cognitive and language level of international students to the maximum extent and thereby ascertain their real language needs.

The questionnaire is conducted in class and lasts about 60 min. The respondents are mainly undergraduates in the third and fourth grades, a total of 100 people. Instead of freshmen and sophomores, juniors and seniors are chosen, for that they have mastered a wide range of vocabulary therefore they have a greater capacity to express themselves and their responses to the questionnaire would not be limited by their level of study.

At the same time, we also collect data on the gender of respondents. Among the 100 respondents, males account for 22% of the total number and females 78%, so females are the main subjects of this study. Most of the students come from countries of which languages have some relations to Chinese, such as Japan, South Korea and Singapore. Other students come from countries of which languages have no connection to Chinese characters, such as the United States, Norway, Russia, Malaysia and Turkey.

The duration of foreign students learning Chinese is also an important index that affects vocabulary output. Therefore, the duration of foreign students learning Chinese is also investigated. From the survey results, we know that the students who have been learning Chinese for 2 to 3.5 years (including 3.5 years) account for 53% of the total number, which is the largest. Next are those who have been learning Chinese for 3.5 to 5 years, accounting for 31%. The rest makes up 16% of the total, the shortest time of learning Chinese is only one year. These students do not have any experience of learning Chinese prior to coming to China. The longest time of learning Chinese is 11 years. We speculate that this student starts to learn Chinese in his own country before his undergraduate study. Most students learn Chinese for the same amount of time as their grade level.

3   The Inspection of International Students' Vocabulary Output Clue

In the process of observing and analysing multiple questionnaires, we find that the vocabulary output of many overseas students follows a certain order. We take the vocabulary output sequence as the research focus and mark multiple questionnaires to explore the rules as much as possible and understand their learning situation and psychological state. We discover that a common problem is that there are fewer mistakes in the words exported first and more mistakes in the words later. This phenomenon is universal and can be found in students with lower or higher Chinese proficiency. In order to avoid the influence of students' vocabulary output errors and the negative emotions caused by too long thinking time, we only select the first 100 words in each questionnaire. Unlike words listed later in the questionnaire, the first 100 words are generally written correctly and are real words in Chinese.

Before exploding the vocabulary output clues of overseas students, it is necessary to define precisely what we mean by the term *vocabulary output clue*. We first establish our own annotation system, and divide a set of our own word categories, then annotate and classify the first 100 words in 100 questionnaires.

In each sample, where there are 3 or more words to appear consecutively and these words all have certain common features, for example, they all belong to the same word category, can appear in the same scene, or contain the same morpheme, we present them as one continuous paragraph, represented by dashes between words in a row.

For example, if an output order such as *Waiter – chicken – vegetarian – bill* appears in a questionnaire, we believe that these words belong to the *Dining* scene, and this output sequence is a clue about *Dining*. It is worth noting that the words must be continuous if they are to form a sequence to represent a clue. For instance, if in the 100 words the first and second words are *Father* and *Mother* respectively, but *Brother* is the tenth word, these three words cannot be classified into the same category of *Family, kinship*, nor can they be regarded as a clue.

There are many clues in the vocabulary output of each international student. In the 100 questionnaires, we make statistics according to the frequency of each clue, and learn that the clues can be roughly classified. Meanwhile, for each questionnaire, we calculate the length of each clue (the number of words), including the maximum length and the average length (we stipulate that the minimum of forming a sequence is 3, so the minimum length is 3 by default, which is not included in the table). The classified statistical results are as follows.

3.1 International Students' Vocabulary Output Clue Inspection

We observe that some words of continuous output can be summarised into a scene, so the scene clues are classified as follows (Table 1):

Table 1. Vocabulary output situation table following scene clues

Scene	Output sequence example	Frequency of occurrence in 100 questionnaires	Proportion	Maximum length	Average length
School, class, study	Teacher – student – classroom – lesson – textbook – study Text – examination – pronunciation	88	88%	12	4.45
Traffic, travel	Directions – traffic lights – maps – police Change – direction – return – drive – one way	48	48%	10	4.50
Daily communication	Hello – thank you – goodbye Good morning – good night – hello	33	33%	5	3.52
Shopping	Suitable – small size – large size – color – shopping mall Refund – supermarket – sell – buy	25	25%	6	3.72
Dining	Order – check – dumplings – chicken Waiter – spoon – chopsticks – fork	24	24%	10	4.54
Go to a doctor	Hospital – doctor – sick – cold Fever – cold – injections – medicine – conditioning – rest	11	11%	6	4.10
Concerning foreign affairs or foreign nationals	International students – foreigners – embassies Passport – wallet – embassy – China – visa	8	8%	5	4.25

As can be learned from the table above:

(1) Output sequences which appear in the *School, class, study* scene are of the highest frequency and the maximum length of these output sequences, which is greater than output sequences relating to other scenarios. It can be deduced that campus is still the main living environment of students, and they depend strongly on the school scene. Of course, it must nevertheless be recognised the fact that this questionnaire was completed in the classroom may have influenced this result.

(2) The words most employed by international students are included in the scenes of interpersonal communication, learning, travel, dining, medical treatment, etc. Although words in each category have their own characteristics, they still have strong regularity. Using this data, we can create a list of basic vocabulary to use when teaching Chinese as a foreign language.

(3) 300 words can comprehensively inspect the Chinese learning and language ability of overseas students. In a certain scene, such as ordering food, seeing a doctor, shopping or transportation, we need to observe what students can do and whether they have the ability to engage in normal communication, and the sequence of words output can reflect these contents to some extent. The more comprehensive the output is, the more detailed and specific the words extracted from this scene are. So we can further conclude that the more language international students master, the more they can complete the communicative tasks in this scene.

3.2   Word Category Clues

According to observation and statistics, it is demonstrated that some words with continuous output can be grouped into one category, and they all have the same superposition word.

Table 2. Vocabulary output table following word category clues

Word category	Output sequence example	Frequency of occurrence in 100 questionnaires	Proportion	Maximum length	Average length
Family, kinship	Father – mother – parents Brother – elder sister – younger sister – grandfather	76	76%	7	5.59
Personal feelings	Happy – sad – delightful Sad – comfortable – happy	36	36%	3	3.00
Dress up	Perm – dye – manicure Leather shoes – hoodies – skirts	7	7%	10	5.71
Colors	Red – blue – black – orange – white Brown – red – purple – green – gray	5	5%	6	4.20

The data in Table 2 reveals that the frequency of words such as *Family, kinship* is high, which indicates that people attach great importance to the social relation, and rely on intimate relationships and kinship. The frequency of *Mother* is the highest among all words relating to relatives, and the order is higher in this sequence, which indicates that *Mother* has the highest status in people's minds. The length of the appellation of relatives is mostly even. We can observe that international students often use the method of matching words in the output of vocabulary, which is also related to the characteristics of the pair of appellation words.

Kinship terms actually represent a kind of social relations. In the questionnaire, we discover many words that represent people in social relations, some of which can form a sequence, such as *Teacher – classmate – student* mentioned above, and some of which are not in a sequence and scattered in the vocabulary output list of international students, such as *Roommate, Friend, Lover* and so on. The study reveals that the social relation is also an important part of their lives when studying abroad.

Table 3. Vocabulary output table following part of speech clues

Part of speech	Output sequence example	Frequency of occurrence in 100 questionnaires	Proportion	Maximum length	Average length
Temporal noun	Yesterday – today – tomorrow – the year after Morning – afternoon – evening	78	78%	14	4.65
Pronouns	I – you – he – they This – that – there	75	75%	6	3.20
Orientation	East – west – north – south Up – down – left – right – in front – beside – in	43	43%	8	5.16
Contains the same morpheme	Expression (表情) – performance (表演) – praise (表扬) TV (电视) – movie (电影) – telephone (电话)	24	24%	6	4.00
Location noun	Dining hall – dormitory – restaurant Bus stop – subway station – mosque	23	23%	5	3.83
Positive adjectives	Pretty – lovely – nice Pretty – smart – cute – brave	21	21%	5	3.19
Numerals	One – two – three – four – five Ten – a hundred – a thousand – ten thousand	21	21%	20	8.62

3.3   Part of Speech Clues

This category is classified from the perspective of linguistic ontology (for the purpose of statistics, the clue *Containing the same morpheme* is also included in this category, since both are related to linguistic ontology). We observe that some output words follow part of speech clues.

From Table 3, we can draw the following conclusions:

(1) In all questionnaires, there are more content words than function words. This is in line with the characteristics of student learning and language teaching. Many of these content words are nouns, especially concrete nouns, most likely because the meaning is simple, practical and easy to learn. There is a correlation between Chinese proficiency and the output frequency of abstract nouns: the more advanced a respondent's Chinese is, the more likely they output abstract nouns in their questionnaires. We speculate that as the complexity of the syntactic structure of learning increases, the amount of function words learned and used will also increase, thereby causing the proportion of function words in the output vocabulary to increase.

(2) During the process of analysing the statistics, we also observe that the vocabulary output of international students is related to their culture. Of the 100 questionnaires collected, 17 are filled out by students from Malaysia, a country of which state religion is Islam. Among the 17 questionnaires, 11 of them mention words such as *Halal* or *Mosque*, indicating that the vocabulary output of overseas students is closely related to the culture and religion of their home country.

3.4   Investigation on the Distribution Quantity of Lexical Output Clues

The above tables present the clues of foreign students' vocabulary output. Each international student exports their words selectively. Therefore, for each questionnaire, there must be a variety of output clues. We collect data on 100 questionnaires, and sort the number of clues output from each questionnaire. The number of copies of the clues and the proportion of 100 questionnaires are as follows.

Table 4. Statistical table of clues output from each questionnaire

Number of clues for each questionnaire output	Frequency of occurrence in 100 questionnaires	Proportion
3	6	6%
4	17	17%
5	20	20%
6	13	13%
7	19	19%
8	4	4%
9	6	6%
10	5	5%
11	7	7%
12	3	3%

Table 4 shows us that the number of clues in each questionnaire is at least 3 and at most 12. Questionnaires with 5 clues are the most frequent, accounting for 20%, and questionnaires with 7 clues are the second most frequent, accounting for 19%. Based on this, we calculate that the average number of clues contained in each questionnaire is 6.46. It can be inferred that international students will use 6 or 7 complete clues when filling in the questionnaire for vocabulary output.

4 Analysis of Vocabulary Output Characteristics of International Students

Through the statistical description of the questionnaire above, the clues of the vocabulary output of international students are explored and summarised. We discover that there are some rules and characteristics in the vocabulary output of international students. The specific contents are as follows.

(1) The output words have imageability

Imageability refers to the ease and speed with which a word can evoke a mental image. Generally speaking, the imageability of a word refers to whether its image (including vision and sound) is easy to appear in people's mind when they see a word. In Chinese vocabulary, concrete nouns and proper nouns have strong imageability. Consequently, the frequent occurrence of concrete nouns in the questionnaire is related to the specific images that international students see when filling in the questionnaire, such as *Desk*, *Chair*, *Student*, *Teacher*, *Key*, *Computer* and so on. Similar examples include *Textbook*, *Glasses*, *Bicycle*, *Library*.

(2) Disyllabic words are dominant in output words

By analyzing the questionnaire results, we learn that the majority of output words are disyllabic, while monosyllabic words and words with three syllables or more account for a small proportion, which is because of the syllabic characteristics of Chinese. In modern Chinese, monophonic morphemes are abundant and disyllabic words are dominant. In the process of teaching Chinese as a foreign language, the teaching of vocabulary is inevitably dominated by two-syllable words.

(3) Preliminary morpheme awareness of vocabulary output

In the cognition and learning process of foreign students, it is still worth discussing whether they have a clear understanding of morphemes, whether the boundaries between morphemes and words are clear and to what extent, based on the existing data and corpus. According to the results of this study, the boundary between morpheme and word is not clear. Apart from most two-syllable words and a small number of three-syllable words and above, the remaining monosyllabic words actually include some monosyllabic morphemes, such as *Shi (师)*, *Yue (阅)*, *An (安)*, *Jian (荐)* and so on.

Students who have been learning Chinese for a long time and are proficient in the language, when output vocabulary, arrange words according to whether words contain the same morpheme. After they have reached the output level, they will choose new morphemes to combine and output. It can be assumed that foreign students with a higher learning level have preliminary morpheme awareness. In their cognition, they can extract morphemes and combine them with other morphemes as a block. However, in the Chinese vocabulary system, compound words are in the majority, and the actual meaning of words composed of the same morpheme can vary either greatly or only slightly.

For example, in *Rockery (假山)* – *wig (假发)*, although those two words both contain the morpheme *False (假)*, the meaning of the morpheme *Mountain (山)* and *Hair (发)* is not related to each other, so they are not in the same category. But the morpheme meaning of *False* is the same, meaning *Artificial*, not natural. In the output sequence of *Quintessence (国粹)* – *territory (国土)*, both contain the morpheme *Guo (国)*, which means Country, but there is not a strong link between the meaning of the two words. *Quintessence* means *Essence of Chinese culture and art*, while *Territory* means *Land of a country*. By putting these words together, we can see that international students are aware that these words have the same morpheme meaning even though the actual meanings of the two words are very different.

(4) Vocabulary output reflects the concept of the scene

For example, in the output sequence of *Fever* – *cold* – *injection* – *medicine* – *conditioning* – *rest*, it reveals that international students output words according to the scene of *Go to a doctor*. Similarly, *Order* – *WeChat* – *Alipay* – *KFC* – *chicken chops* – *ribs* is based on output relating to the restaurant dining scene and the output sequence of *Suitable* – *small size* – *large size* – *color* – *good-looking* – *shopping mall is* related to the scene of shopping. Thus, it can be learned that some international students with higher learning level can choose vocabulary according to different scenarios and group them together in a similar way.

(5) There are gender differences in vocabulary output

The gender difference of international students is also an important feature that affects their output of words. Different genders result in different concerns and different interests. In questionnaires filled out by female respondents, there are many common words related to diet, clothing, beauty and emotion, such as *Delicious, Bread, Skirt, Manicure, Love* and *Crush*, etc., while words included in male respondents' output shows high frequency of sports words, such as *Tennis* and *Basketball*. In terms of how different genders handle vocabulary relating to family members, female respondents tend to list more family members, including *Grandfather, Father, Mother, Grandmother, Brother, Sister*, etc., and these words appear more in the beginning of their questionnaires. Male respondents, on the other hand, output fewer words relating to family members, with priority given to *Mother* and *Father* above other relatives.

This suggests that the female participants in this study pay more attention to and rely more on family members and familial relationships.

Another difference between the responses of male and female participants is that the male respondents generally give priority to rational thinking and pay little attention to the expression of personal emotions. Unlike the female respondents, they are not particularly sensitive to psychological activities and emotions. Females, on the contrary, whose output words are related to rich and delicate personal feelings, pay attention to the expression of emotions. The gender difference is clearly reflected in the questionnaire. According to the statistics, 61.5% of females output *Personal emotion* words, such as *Happy*, *Sad*, and *Disappointed*, while only 27.2% of males output *Personal emotion* words.

(6) Vocabulary output reflects the order of acquisition

In the process of learning, foreign students tend to use some words or phrases more frequently than others. According to the law of human memory, the contents mastered first will be first entered into long-term memory. The words which are most deeply embedded in learners' long-term memories are often among the first learners that are able to extract. The first part of respondents' vocabulary output is actually the part with the highest level of mastery. In the teaching material of teaching Chinese as a foreign language, the order of knowledge points is related to communication needs. In the process of communication, communicative terms and high frequency vocabulary relating to everyday life will appear first. Therefore, the words involved in these contents will also appear first in some textbooks, which will be acquired by students first and output by students first. For example, it is reported that the output frequency of *Hello*, *Thank you*, *Goodbye* and *Excuse me* (phrases are misrepresented as words, here similar chunks are counted entirely as words) is relatively high because they are the most commonly used communicative terms, which are the first to be mastered when learning and the first to be entered into long-term memory.

5 Conclusion

We give out questionnaires to 100 international students to investigate their Chinese vocabulary output. By statistically summarising and analysing their vocabulary output, we find that the vocabulary output of international students follows certain clues, such as scene clues, word category clues, and part-of-speech clues. At the same time, the vocabulary output of international students also reflects certain rules and characteristics. The output words have imageability and two-syllable words are dominant. Vocabulary output has a preliminary sense of morpheme, which embodies scene concept. In addition, vocabulary output also has gender difference, and follows the acquisition order. Our research hopes to provide some reference for the integration of Chinese teaching in foreign languages, the compilation of textbooks and the basic vocabulary of Chinese learners.

Acknowledgments. The work was supported by Major Program of National Social Science Foundation of China (18ZDA295); Funding Project of Education Ministry for Development of Liberal Arts and Social Sciences (16YJA740036); Top-ranking Discipline Team Support Program of Beijing Language and Culture University (JC201902); the Fundamental Research Funds for the Central Universities (18YBT03); BLCU Supported Project for Young Researchers Program (supported by the Fundamental Research Funds for the Central Universities) (19YCX047).

References

1. Bao, G., Li, J.-Y.: Effects of output task and glossing on EFL learners' vocabulary acquisition. J. PLA Univ. Foreign Lang. **40**(06), 1–10 (2017). (in Chinese)
2. Huang, Y., Hu, X., Wang, H.-X.: The effect of task type on output-triggered incidental L2 lexical acquisition. Mod. Foreign Lang. **40**(05), 642–653 (2017). (in Chinese)
3. Kuai, Y.-J.: Verbal-imagery cognitive style to false memory in the DRM procedure influence. Master thesis, Yanbian University, Yanbian (2011). (in Chinese)
4. Zhang, X.-S.: Research on vocabulary teaching in the classroom in TCFL from a perspective of cognitive psychology. Master thesis, Central China Normal University, Wuhan (2012). (in Chinese)
5. Wang, Z.-Y., Qing, X.-H.: Morpheme teaching, the basis of vocabulary teaching in TCFL. J. Yunnan Norm. Univ. (Teach. Stud. Chin. Foreign Lang. Ed.), (05), 39–42 (2004). (in Chinese)

Construction of the Contemporary Chinese Common Verbs' Semantic Framework Dictionary

Tongfeng Guan[1], Kunli Zhang[1(✉)], Xuemin Duan[1], Hongying Zan[1], and Zhifang Sui[2]

[1] School of Information Engineering, Zhengzhou University, Zhengzhou, China
guantf@gs.zzu.edu.cn, {ieklzhang, iehyzan}@zzu.edu.cn,
xueminduan@stu.zzu.edu.cn
[2] Institute of Computational Linguistics, Peking University, Beijing, China
szf@pku.edu.cn

Abstract. Semantic lexicon and semantic framework are the primary support of natural language processing tasks such as information extraction, sentiment analysis, and machine translation. Therefore, it is essential to construct the contemporary Chinese common verbs' semantic framework dictionary that covers rich semantic knowledge. Based on an analysis of current research results, this paper defines the lexical framework of common Chinese verbs. According to the predicate thematic roles, the semantic framework is divided into the basic semantic framework and extended semantic framework. Frameworks which are automatically extracted, taking semantics as the processing unit, and summarized based on large-scale lexical and thematic roles labeling corpus. The complete and simplified versions of the verb framework is constructed with the help of manual proofreading. The final verb framework contains a detailed description and corresponding example sentences of 2,782 common verbs with 4,516 meanings.

Keywords: Contemporary Chinese common verbs · Semantic framework dictionary · Automatic extraction · Basic semantic framework · Extended semantic framework

1 Introduction

Natural Language Processing (NLP) is the technology used to aid computers to understand natural language statements and texts, and extract essential information for analysis, retrieval, reading, question and answer, machine translation and text generation. To achieve this goal, scholars have conducted in-depth research in the fields such as part-of-speech tagging, syntactic analysis, and semantic dependency parsing. Significant advances have also been made in related research areas.

There have been many lexical-semantic knowledge bases, such as WordNet [1], FrameNet [2], MindNet [3]. Representative achievements in Chinese language include the Chinese Semantics Dictionary (CSD) [4], the Chinese Concept Dictionary (CCD) [5], the Chinese Function words Knowledge Base (CFKB) [6, 7] and the

© Springer Nature Switzerland AG 2020
J.-F. Hong et al. (Eds.): CLSW 2019, LNAI 11831, pp. 546–553, 2020.
https://doi.org/10.1007/978-3-030-38189-9_56

Chinese Large-Scale Knowledge Base (CLSKB) [8]. In Chinese FrameNet (CFN) [9] and Mandarin VerbNet (MV) [10], verbs are taken as the primary research object. CFN is a FrameNet-style Chinese framework semantic frame net, which translates or creates a framework suitable for Chinese semantic content, and defines framework-framework relationships. However, it defines a small number of verb frames, only 309 frames. MV is the result of linguistic semantics research using a framework-based constructional approach. It has completed research on major verb categories such as 'caused-motion', 'cognition', 'communication', 'emotion', 'motion', 'perception' and 'social-interaction'. It proves that framework-based Chinese verb and verb class framework analysis methods have proper motivation and effect in language. However, due to the artificial construction method, the classification is limited. Currently, the construction of a semantic knowledge base mostly adopts the method of artificial construction and fails to put the semantic of words in a particular combination framework for observation. The static aggregation classification method is often used in the construction. But few attribute descriptions are added, so the construction scale and update speed cannot meet the needs of current NLP development.

The lexical semantic frame of verbs is a semantic relationship framework system composed of verbs and their thematic roles in the dictionary. Verbs exist in every complete sentence, expressing the movement, development, change, existence, or demise of a person or thing. Their thematic roles are divided into core roles and peripheral roles. The core roles, including an agent of the action, a patient of the action, and both. The peripheral roles that complement the sentence context and add a sense of picture to the sentence include instrument, material, manner, reason, purpose, a point of departure, a point of arrival, start and end time, etc.

This paper is supported by the National Key Basic Research and Development Program 973 (2014CB340504), in the context of the Broad-Spectrum semantic Dictionary of Contemporary Chinese (BCSD). It has integrated and absorbed current research results, and it defines the verb semantic framework dictionary structure. The paper separates the semantic framework into a Basic Semantic Framework (BSF) and an Extended Semantic Framework (ESF). With the semantics as the processing unit, the BSF and the ESF are automatically extracted and summarized from the large-scale lexical and thematic roles labeling corpus. Supplemented by manual proofreading, a lexical semantic framework dictionary has been constructed. This paper simplifies the framework when studying the differences in the verb implementation framework. We have constructed a detailed description of the semantic framework containing a total of 2,782 commonly used verbs (4,516 meanings) with corresponding example sentences.

2   Verbs Semantic Framework

Predicate verbs play a cohesive role in sentences. The construction of sentences must surround the core of verbs. Therefore, verbs are considered the soul of sentences. On the other hand, verbs have constraints on the semantic classes of their arguments, so constructing the BSF of verbs is conducive to grasping the human cognitive model of linguistic knowledge. In the construction of the lexical semantic dictionary, current research results are fully absorbed and borrowed. On this basis, the semantic attribute description framework of the existing word classes is expanded and improved.

The construction of the verb semantic framework refers to the principle of the collection of words and the semantic of the fifth edition of the Modern Chinese Dictionary (XH5). Meanwhile, fusion and inheritance of the modern Grammar Knowledge Base (GKB) [11], The Modern Chinese Semantic Dictionary and the word class system, semantic classification system and related semantic attribute description information of the large-scale lexical semantic database are involved to reduce the time and manpower required for the construction process, and to ensure the high quality and high credibility of the dictionary. The lexical semantic dictionary selects 2782 of the intersection of the fifth edition of Modern Chinese Dictionary and the Modern Chinese Grammar Information Dictionary, including 1865 single senses and 917 polysemous words, according to Modern Chinese predicate Semantic Role Labeling Corpus Specification (the 973 specification) in the 'predicate thematic roles hierarchy classification system' as shown in Fig. 1. It is a system of semantic relationship framework constructed by combining the verbs with their theoretic roles.

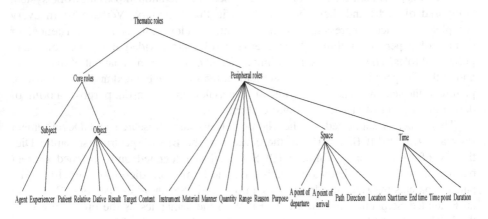

Fig. 1. The hierarchical classification system of predicate thematic roles

According to the different relationships between the roles of the verbs, roles can be divided into two types: core roles and peripheral roles. Core roles represent an essential component closely related to the verb. Without this part, the verb cannot be completely described. The core roles include agent, experiencer, patient, relative, dative, result, target, content. Peripheral roles represent a component that is more distant from the verb. If this part is missing, the verb of the core roles can still describe the event, but the content is not comprehensive. Adding the peripheral components makes the narrative vivid and stereoscopic. Peripheral roles consist of two subordinate themes of space and time and sixteen three-level theoretic roles.

2.1   Basic Framework of the Dictionary

The basic framework of the verb library includes three main parts: verb entity, semantic framework, and thematic roles. The verb entity includes verb term and phonetic, definition, and example of the word. The semantic framework contains the BSF and ESF. Moreover, the thematic roles consist of 8 core roles and 18 peripheral roles (a total of 24 three-level thematic roles and 2 secondary thematic roles).

2.2   Framework Description

The verb entity part is extracted directly from the fifth edition of the Modern Chinese Dictionary. Semantic framework refers to the semantic relationship framework composed of verbs and their thematic roles, including BSF and ESF. Verb semantic framework dictionary needs to match the corresponding example sentences with the verbs, and then automatically extracts the BSF and the ESF from the example sentences. Therefore, the dictionary needs to sort the verb necessary information, the semantic framework, and the thematic roles. Some verbs may be empty due to insufficient corpus size. The BSF only involves the semantic relationship framework that contains core roles, and the ESF involves the peripheral roles.

3   Semantic Framework Extraction

Both the BSF and the ESF have various implementation frameworks. Therefore, it is necessary to select a typical structural framework as a representative. Verb semantic framework dictionary should extract a typical structural framework.

The building of the verb semantic framework dictionary and semantic roles labeling corpus is an iterative process. Firstly, develop a verb framework and semantic roles labeling specifications for characteristics. Secondly, the automatic framework extraction method is determined. Then, the annotations are discussed weekly. As problems arise, adjust the labeling specification, and guide the labeling of the next corpus to sort out the new verb framework.

The semantic framework is extracted from Peking University Chinese Tree Library [12] and People's Daily's corpus. Among 55,764 sentences in the Chinese tree library of Peking University, 10,634 sentences were extracted according to rules of Chinese grammar. The semantic roles were marked in the corpus (referred to as 10,000 sentences). From the corpus of People's Daily on January 2000 and January 1998, we have extracted two batches of corpus (referred to as People's Daily and new, total 42,000 sentences) that have been marked with the target verb meaning and thematic roles.

In the process of processing the semantic framework, four main steps are included: data preprocessing, BSF processing, ESF processing, and simplification, as shown in Fig. 2. Monosyllabic words can be treated as polysemous words with only one meaning. In the process, when there are multiple example sentences, the default example sentences are selected according to the priority order of '10,000 sentences -> People's Daily -> new'.

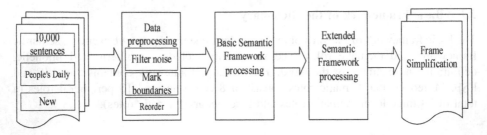

Fig. 2. Frame processing

3.1 Data Preprocessing

The main task of the data preprocessing is deleted useless example sentences, mark boundaries, reorder, and filter noise. Reordering is to align the same implementation framework. Noise is filtered according to the rationality of the framework before calculating the typical framework. In the 973 specifications, the correspondence between the subject and the object is constrained, as shown in Table 1. After processing, monosyllabic words only retain the words with more than three example sentences, and polysemous words delete the meanings without the example sentences.

Table 1. Subject and object correspondence relation

Object	Subject	Relation
Relative	Experiencer	Can't correspond to agent
Patient	Agent	Can't correspond to experiencer
Dative	Agent	Usually does not correspond to the experiencer
Result	Agent | experiencer	Can correspond to agent and experiencer
Content	Agent | experiencer	Can correspond to agent and experiencer
Target	Agent | experiencer	Can correspond to agent and experiencer

3.2 Basic Semantic Framework Processing

The position of the thematic roles in the actual statement is flexible. A semantic term of a verb may have multiple semantic structures that differ in the order in which the roles are located. The verb library regards it as the semantic framework of the implementation. The semantic framework abstracted from the implemented semantic framework is considered as a typical semantic framework. The implemented semantic framework and typical semantic framework are general and typical of the semantic framework. The typical semantic frameworks are arranged in the logical order in which the thematic roles appear.

The implemented framework can be regarded as a group member, and the typical framework is the squad leader selected from the group members. With the typical framework, the meaning item can be easily recognized. The processing of the BSF is divided into three steps. The first step is to select all the monosyllabic words to identify the typical framework. The second step is to replace the typical structure that is automatically identified in the first step by using a typical structure that has been manually selected. And the third step determines the example sentences of the corresponding implementation framework and the merged semantic roles for the results of the first step and the second step.

3.3   Extended Semantic Framework Processing

The ESF contains 18 peripheral roles, so there are multiple implemented semantic frameworks, from which the typical semantic framework needs to be extracted. The typical ESF is based on the typical BSF but contains peripheral roles. Also, some adjustments are made according to the actual situation.

Begin the judgment of whether the typical BSF is a subset of some implementation frameworks. If it is a subset, the next step is to consider whether there are multiple implemented semantic frameworks. If yes, merge these frameworks, and if not, choose the only framework. Intersections also need to consider whether the core roles are the same when merging. If they are different, consider choosing the one with the longest implementation structure and the largest number of examples. Eventually, all verbs can reach the end state.

3.4   Frame Simplification

After completing the automatic extraction of the BSF and the ESF, to compare the differences in the implementation framework of the verbs in different sentences, and explore the internal rules of the semantic framework, we need to simplify the results. The implementation framework is merged into a single cell, separated by '|', leaving only typical example sentences. The semantic roles are deduplicated and merged, and finally, one verb retains only two rows of records.

4   Results and Analysis

Statistical analysis of the results of a typical framework for the extraction of monosyllabic and polysemous words, as shown in Table 2. The total number of meanings in the table is the number of all the meanings contained in each type of verb. Since each polysemous word contains multiple meanings, 917 polysemous words have a total of 2651 meanings.

Among these meanings, 785 show semantic frameworks and extended semantic frameworks, 672 show basic semantic frameworks only, 208 show extended semantic frameworks only, and 986 meanings do not show any framework. Extracting parts from 986 meanings for analysis leads to the identification of two reasons. The first is that some verbs appear in single words, such as '上[shàng] (to board; to attend; to go to)', '下[xià] (to fall; to go down)', '交[jiāo] (exchange, communicate; to turn over; to make)', '令[lìng] (to order; to cause)', etc. Some of the meanings of these verbs are not used often enough, so it does not appear in the corpus, there are also verbs containing these words, such as '上台[shàngtái] (to go on stage (in the theater); to rise to power (in politics))',

'上学[shàngxué] (to go to school; to attend school)', '下船[xiàchuán] (disembark; boarding)', '下岗[xiàgǎng] (to lay off; to come off sentry duty)', '交换[jiāohuàn] (to exchange; to switch)', '交替[jiāotì] (to replace; alternately; in turn)', etc., respectively covering the meaning of '上[shàng] (to board; to attend; to go to)', '下[xià] (to fall; to go down)', and '交[jiāo] (exchange, communicate; to turn over; to make)'. When the corpus is segmented, it is taken as a whole. The second reason is that the corpus used is mainly news. Some words do not appear in the news of that period, such as '上升[shàngshēng] (to rise; to go up)', '临门[línmén] (facing the goalmouth (soccer); to be at the door)', '修养[xiūyǎng] (accomplishment; training)', etc. Subsequent work should consider more types of corpora, thus increasing the number of examples.

Table 2. Statistical semantic framework extraction results

Category	Num	Filter noise	Total of meanings	Only BSF	Only ESF	Both	Neither
Polysemous	917	–	2651	672	208	785	986
Monosyllabic	1865	190 (The number of examples is less than 3)	1865	236	37	1592	0
Total	2782	190	4516	908	245	2377	986

5 Conclusion

This paper defines a verb semantic framework dictionary, divides the framework into BSF and ESF, and automatically extracts and summarizes the BSF and ESF based on the large scale labeled lexical and thematic corpus. We have constructed a dictionary of contemporary Chinese common verbs semantic framework containing 917 polysemous words and 1865 monosyllabic words, which provide support for natural language processing applications such as machine translation, sentiment analysis, and information extraction. By analyzing the dictionary, we find that some meanings of polysemous words have neither the BSF nor the ESF. The main reason is that the source of the corpus is news. For the next step, coverage of other corpora with different data sources should be considered to expand the diversity of example sentences, extract a more representative typical framework from the diversified implementation frameworks to improve the quality of the verb semantic framework dictionary. Moreover, the scale of the verb semantic framework should be expanded to include more verb entries.

Acknowledgments. We thank the anonymous reviewers for their constructive comments, and gratefully acknowledge the support of the National Key Basic Research and Development Program under Grant No. 2014CB340504; The National Social Science Fund of China under Grant No. 18ZDA315; the Key Scientific Research Program of Higher Education of Henan under Grant No. 20A520038; the science and technology project of Science and Technology Department of Henan Province under Grant No. 192102210260; and the international cooperation project of Science and Technology Department of Henan Province under Grant No. 172102410065.

References

1. Fellbaum, C., Miller, G.: WordNet: An Electronic Lexical Database. MIT Press, Cambridge (1998)
2. Bake, C.-F., Fillmore, C.-J., Lowe, J.-B.: The Berkeley FrameNet project. In: Proceedings of COLING 1998, pp. 86–90 (1998)
3. Richardson, S.-D., Dolan, W.-B., Vanderwende, L.: MindNet: acquiring and structuring semantic information from text. In: ACL, pp. 1098–1102 (1998)
4. Hui, W., Zhan, W.-D., Yu, S.-W.: The specification of the semantic knowledge-base of contemporary Chinese. J. Chin. Lang. Comput. 13(2), 159–176 (2003). (in Chinese)
5. Liu, Y., Yu, S.-W., Yu, J.-S.: A study on the construction of CCD. J. Chin. Comput. Syst. 26 (8), 1411–1415 (2005). (in Chinese)
6. Zan, H.-Y., Zhang, K.-L., Zhu, X.-F., et al.: Research on the Chinese function word usage knowledge base. Int. J. Asian Lang. Process. 21(4), 185–198 (2011). (in Chinese)
7. Zhang, K.-L., Zan, H.-Y., Chai, Y.-M., et al.: Survey of the Chinese function word usage knowledge base. J. Chin. Inf. Process. 29(3), 1–8 (2015). (in Chinese)
8. Shi, J.-M., Zan, H.-Y., Han, Y.-J.: Specification of the large-scale Chinese lexical semantic knowledge base building. J. Shanxi Univ. Nat. Sci. Ed. 38(4), 581–587 (2015). (in Chinese)
9. You, L.-P., Liu, K.-Y.: Building Chinese FrameNet database. In: Proceedings of 2005 IEEE NLP-KE, pp. 301–306 (2005)
10. Liu, M.-C., Chang, J.-C.: Semantic annotation for Mandarin verbal Lexicon: a frame-based constructional approach. In: Proceedings of the 2016 International Conference on Asian Language Processing, pp. 30–36 (2016)
11. Yu, S.-W., Zhu, X.-F., Wang, H., et al.: The Grammatical Knowledge-Base of Contemporary Chinese—A Complete Specification. Tsinghua University Press, Beijing (2003). (in Chinese)
12. Zhan, W.D.: Encyclopedia of Chinese Language and Linguistics, vol. 3, pp. 332–336. Brill Publishing House, Leiden (2016)

Knowledge Graph Representation
of Syntactic and Semantic Information

Danhui Yan[1](✉), Yude Bi[2](✉), and Xian Huang[1]

[1] Luoyang Campus, Information Engineering University, Zhengzhou, China
`diem1987@163.com`, `a77huang852yi@126.com`
[2] Fudan University, Shanghai, China
`biyude@fudan.edu.cn`

Abstract. Representation of linguistic knowledge is one of the keys to helping machines understand natural languages. This paper follows the idea from linguistic data to linguistic knowledge and to knowledge representation. At the syntactic level, the syntactic structure and its variants in the corpus are summarized, and the syntactic functions undertaken by the arguments are analyzed. At the semantic level, the semantic roles and semantic types of arguments are analyzed. The purpose is to reveal the interaction between syntax and semantics. Finally, this paper explores a fusion representation method of linguistic data and linguistic knowledge, and carries out a case study.

Keywords: Syntactic structure · Semantic structure · Knowledge representation

1 Introduction

The study of the representation of linguistic knowledge is of great value to the study of natural language understanding. Linguistic knowledge is hidden in linguistic data. From the perspective of Rationalism, linguistic knowledge is summarized from the linguistic data, and the acquisition of linguistic knowledge relies on the study of language structures by linguists.

The process above is a process of forming linguistic knowledge from linguistic data. From the perspective of computational linguistics, to let a machine have certain "language ability", the first thing to do is to generalize the linguistic knowledge hidden in the language data. However, this is only the first step. Machines have to "learn" this knowledge, which requires knowledge representation. This highlights the importance of the representation of syntactic and semantic knowledge, and makes the representation the core content of knowledge base construction.

Starting from knowledge representation, this paper proposes a new representation architecture, which is a knowledge graph that fuses syntactic and semantic information, and discusses its architecture and construction methods. Finally, an exemplary research is carried out.

© Springer Nature Switzerland AG 2020
J.-F. Hong et al. (Eds.): CLSW 2019, LNAI 11831, pp. 554–562, 2020.
https://doi.org/10.1007/978-3-030-38189-9_57

2 Knowledge Representation

2.1 Knowledge Representation

Knowledge is the regular understanding and structured information of things that humans have summed up. The expressibility of knowledge is beyond doubt. From a broader perspective, language, literature, and performing arts are all subcategories of knowledge representation. In the light of computational linguistics, what knowledge representation essentially solves is the question of what form of knowledge is stored in the machine. This is an indispensable key component of a language information processing system.

Knowledge representation methods based on symbolic logic can be divided into statement representation and process sequence representation. The statement representation describes various concepts related to things and the interrelationships between things, mainly through predicate logic, semantic network [1], framework and object-oriented representation. It has the characteristics of being intuitive, readable, modular, and easy to modify and add new knowledge. The process sequence representation mainly emphasizes the objective law between things and the method of solving the problem. It is highly efficient, but it is not easy to modify or add new knowledge.

Semantic network representation can not only express the classification knowledge related to things, but also intuitively represent the various semantic relationships between concepts. It is in line with human cognitive habits and thinking processes. It can be easily used for machine translation [2], QA systems [3] and natural language understanding [4].

In view of these characteristics and the characteristics of linguistic knowledge, this paper selects knowledge graph belonging to the semantic network category as the representation method, however, the knowledge graph needs to be improved, which will be discussed below.

2.2 From Semantic Network to Knowledge Graph

Semantic Network is a psychological model proposed by J. R. Quillian in 1968. It is a kind of directed graph composed of directed lines between nodes.

Structurally, a semantic network consists of some basic semantic units, namely the triples represented by directed graphs as shown below:

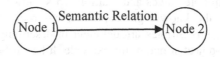

Fig. 1. The basic unit of a semantic network

Nodes represent entities or abstract concepts. The direction of the line represents the primary-secondary relationship between the nodes; the annotation of the line is the semantic relationship between the nodes. It can also be clearly seen from Fig. 1 that the semantic network and the predicate logic are equivalent in representation. A basic unit is equivalent to a binary predicate. A basic unit (node 1, line, node 2) can be directly rewritten as the predicate P (individual 1, individual 2) in the predicate logic. When multiple basic units are related by semantic relations, a semantic network is formed.

Semantic network is abstract when expressing events. It tends to abstract things into upper concepts and then generalize them. At the same time, a semantic network can explicitly reflect the attributes of entities (concepts) and the semantic connections between entities (concepts). It is a structured and intuitive representation of knowledge. It conforms to human habits and thus makes it easier to convert natural language into a semantic network. This is also the main reason why we use it as a means of knowledge representation.

At present, semantic network has revived in the form of knowledge graph, but there are also differences between the two. The focus of the knowledge graph is on the acquisition of triples (entities, relationships, entities) and the attribute-value pairs. Knowledge graphs constructed by companies in the industry often are composed of hundreds of millions of entities (nodes). However, they lack the abstractness of the semantic network, while the latter is not as good as the knowledge graph in terms of the number of entities and attributes.

We believe that linguistic knowledge is extracted from linguistic data. At the same time, linguistic knowledge should also return to the linguistic data and the two form an organic whole. A semantic network is suitable for expressing linguistic knowledge; and a knowledge graph is suitable for expressing linguistic data.

2.3   Knowledge Graph

Knowledge graph is a type of knowledge base proposed by Google in 2012. In general, a knowledge graph organizes knowledge primarily in the form of binary relationships. Entities, concepts, and their relationships in the text are described by binary relationships. Each entity has several attribute values, and the relationship is generally a predicate in the text, thus forming a triple as (entity, relationship, entity). Triples, as basic organizational units, are connected with each other to form a network.

Current studies related to knowledge graph mainly focus on entity relation extraction [5–8], entity alignment [9–12], relation classification [13–16], entity linking [17–20], knowledge graph embedding [21–23], etc. Studies similar to what this paper covers have not been found.

At present, t more and more entities are included in the knowledge graph, along with more and more attribute values and relation types. However, we believe that the current architecture of the knowledge graph still has room for improvement. The reasons are as follows: First, triples are deficient in describing some entity relationships. Take the double-object structure for example. It has to be split into two triples, which increases complexity artificially. Secondly, although a knowledge graph includes the design of the ontological layer in its structure, it is not the focus of the construction. Entities have attribute values, and there are relations between entities. This structure is only a shallow representation. Natural languages have a variety of expressions for the same relation between entities, but the syntactic and semantic structures behind them are stable and limited in number.

We mentioned above that linguistic knowledge should be integrated with linguistic data. Therefore, from the perspective of knowledge representation, this paper proposes a knowledge graph structure that integrates syntactic and semantic information, which can further enrich the semantic information of a knowledge graph.

3   Knowledge Graph Representation of Syntactic and Semantic Information

The knowledge graph representation of syntactic and semantic information can be regarded as a knowledge graph that combines syntactic and semantic information. It is language independent and is rich in semantic information.

3.1   Structure and Construction Method

In the structure of the knowledge graph proposed in this paper, N-tuples are used to describe entities and the relationships between entities. Starting from the perspective of syntactic and semantic research, the syntactic-semantic structure is taken as the semantic layer of the knowledge graph.

To obtain this semantic layer, categorization of entities and their relationships (verbs) is key issue to solve. Therefore, it is necessary to study categorization of the entities and relationships in depth, and thus form a syntactic-semantic structure based on the investigation of the corpus.

The structure of the knowledge graph that incorporates syntactic-semantic information is mainly divided into two parts:

(1)   Syntactic-Semantic Structure: it is mainly obtained through the analysis of the verb. The representation method is language independent. In Table 1, the Vietnamese verb gửi is taken as an example, and the cases where the arguments are omitted are not listed:

Table 1.   Syntactic-semantic information of verb gửi

Verb	Syntactic structure	Semantic types of arguments		
		Giver_NP1	Receiver_NP2	Things Given_NP3
gửi	NP1+V+cho +NP2+NP3	[+Human]/ [+ORG]	[+Human]/ [+ORG]	[+DoCon]
	NP1+V+NP3 +cho+NP2	[+Human]/ [+ORG]/ [+DoCon]	[+Human]/ [+ORG]/ [+LOC]	[+Human]/ [+Weapon]/ [+DoCon]
	NP3+mà+NP1 +V+cho+NP2	[+Human]/ [+ORG]	[+Human]/ [+ORG]	[+DoCon]
	NP3+được+V +cho+NP2	—	[+Human]/ [+ORG]	[+Weapon]/ [+DoCon]

(2) Entities, Attributes, Relations: this part is similar to the current knowledge graph. On the basis of syntactic-semantic structure, entities, the semantic types of the entities, attributes and relations are filled in for each syntactic-semantic structure. Semantic types, entities and attributes related to NP1 in NP1+V+NP3+cho+NP2 are shown in the following table:

After each syntactic-semantic structure is filled, the syntactic structures and the semantic types are connected with each other to form a graph. The syntactic-semantic layer constitutes the deep structure of the knowledge graph. With the deep structure, syntactic-semantic information of the natural language is integrated into the knowledge graph, forming a knowledge graph that integrates both syntactic and s information.

3.2   Data Storage and Representation

Data storage in knowledge graph can be divided into table structure storage and graph structure storage. Table structure is stored in a relational database, such as Microsoft SQL Server, Oracle, PostgreSQL, MySQL, etc.; and graph structure is stored in graph databases.

When the amount of data increases, under the table structure, whether it is a single table or multiple tables, the overhead of querying, adding, deleting, modifying of the data is large. In addition, there is a need to meet the requirements of relational databases for data types, integrity constraints and so on. This paper chooses the graph database storage method.

Common graph databases are Neo4j, OrientDB, InfoGrid, etc. In this paper we use the Neo4j database. In the case study we use the Neo4j database to illustrate the storage and presentation of graphs.

4   Case Study

4.1   Storage and Representation of a Single Structure

In Sect. 3, the knowledge graph representation of syntactic-semantic information is discussed. Firstly, linguistic knowledge (syntactic and semantic knowledge) is summarized from the linguistic data, and then the Neo4j database is used to fuse data and knowledge in the form of knowledge graph. In this section, the "NP1+V+NP3+cho +NP2" structure is taken as an example to illustrate the knowledge graph representation.

Syntactic structure and semantic types of the arguments listed in Table 2 can be visualized in the following figure:

First, Syntactic structure and semantic types of the arguments are shown in Fig. 2. In Fig. 2, a node represents a syntactic unit or a semantic category. The directed lines between nodes represent the relationship between the two. The label "giver_S" indicates that the semantic role of the node is the giver and the syntactic function is the subject; and the label "SemC" represents the semantic classes, such as an organization, people, place names, weapons, etc.

Table 2. Entity and attribute filling of NP1

Syntax	Semantic types	Entities	Attributes
NP1	ORG	Ma-lai-xi-a	atr 1-val 1; atr 2-val 2; ......
		tập đoàn Commercial Corp.	atr 1-val 1; atr 2-val 2; ......
		...	atr 1-val 1; atr 2-val 2; ......
	Human	Andi Arsana	atr 1-val 1; atr 2-val 2; ......
		Ôn Gia Bảo	atr 1-val 1; atr 2-val 2; ......
		...	atr 1-val 1; atr 2-val 2; ......

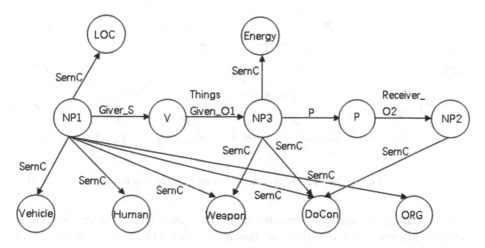

Fig. 2. Visualization of NP1+V+NP3+cho+NP2

It can be seen from Fig. 2 that the syntactic structure is NP1+V+NP3+cho+NP2; and arguments in NP1 position can be one of the following six semantic classes: [+Human], [+ORG], [+DoCon], [+LOC], [+Weapon], and [+Vehicle].

Second, entities and <attribute-value> pairs are added to the corresponding nodes. To distinguish different syntactic structures, sequence numbers are added to the arguments in each syntactic structure when actually storing the information. Thus, S1_NP1 refers to the first argument of the first syntactic structure. In addition, the start position BoS and the end position EoS of the syntax structure are added. An "Instance" tag represents an instance of a node. For example, under the "person" node, there are two instances of "Ôn Gia Bảo" and "Andi Arsana". The attributes of the entities are not displayed. In the visualized interface, when a node is selected, the attributes of the node can be displayed.

4.2   Storage and Representation of Multiple Structures

In Sect. 4.1 we describe the storage and representation of a single syntactic structure. Other syntactic structures are stored and represented in the same way.

The following figure shows the storage and visualization of the two syntactic structures NP1+V+NP2+v+NP3 and NP2+mà+NP1+V+v+NP3 of the verb gọi (Fig. 3):

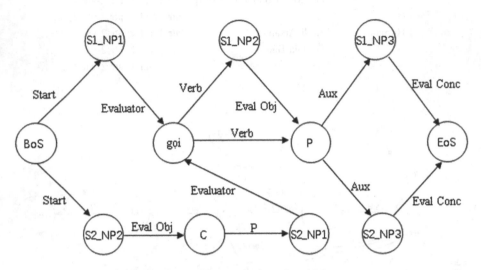

Fig. 3. Storage and representation of multiple structures

In the storage of multiple syntactic structures, syntactic units other than verbs, conjunctions, auxiliary verbs, BoS and EoS are marked with sequence numbers. For example, a node starting with "S1_" belongs to the first syntactic structure.

The following conventions are agreed on the transfer process between nodes:

Convention 1: The transition between nodes starts with BoS and ends with EoS;
Convention 2: The transition between nodes can only be made through semantic role labels or word class labels;
Convention 3: The first node with sequence number arrives from the start point BoS, and the sequence numbers of its subsequent nodes should be consistent with it;
Convention 4: Under the premise of not violating the above conventions, if the current node does not have a sequence number, it can reach any node it points to;
Convention 5: If other nodes can be transferred before the terminator EoS, the transfer operation is performed, and then go to the end EoS.

When multiple verbs and multiple structures are stored, a more complex network will be formed. Multiple structures of one verb share verbs, conjunctions, and auxiliary verbs nodes. According to the transfer conventions, different syntactic structures of verbs can be obtained. In addition, the syntactic-semantic information of the arguments in each syntactic structure can also be clearly displayed. For each semantic class, the number of instances is unlimited.

Through above construction methods, the goal of storing and representing syntactic semantic information is achieved. That is to say, the syntactic structure, the syntactic function of the argument, the semantic role of the argument, the semantic class of the argument, and the instances are organically combined to realize the integrated representation of the linguistic knowledge and linguistic data.

5 Conclusion and Prospects

This paper starts from linguistic knowledge representation and proposes a knowledge graph architecture that incorporates syntax and semantics. The deep syntactic-semantic structure describing entities and their relations is integrated into the knowledge graph. Thus, the semantic representation of the existing knowledge graph is strengthened and improved. At the same time, a general method of constructing the knowledge graph is given, that is, the categorization of the relations (verbs) is first carried out, next the syntactic-semantic structure possessed by each category is analyzed and summarized based on the linguistic data, and then the entities and attribute values are populated into each structure, eventually forming a knowledge graph that fuses syntax and semantics.

This paper shows a top-down construction method. Of course, it can also be constructed from entities, relations and attributes, which is a bottom-up construction method.

Transferring linguistic knowledge into knowledge representation is another important step. Linguistic knowledge is the outline of linguistic data. The representation of linguistic knowledge cannot be limited to the representation of knowledge itself, but should be integrated with linguistic data. This paper explores the fusion representation of linguistic data and linguistic knowledge and has reference value.

The deep structure of syntactic-semantic knowledge can be integrated into the existing knowledge graph. It can provide strong knowledge support for natural language processing applications such as natural language understanding, machine translation and information retrieval. In the subsequent research, we will analyze the corpus and try to form a knowledge graph of a relative large scale and explore the application of it in natural language processing tasks.

References

1. Sowa, J.F.: Principles of semantic networks: exploration in the representation of knowledge. Frame Probl. Artif. Intell. 135–157 (1991)
2. Simmons, R.F.: Technologies for machine translation. Future Gener. Comput. Syst. 83–94 (1986)
3. Simmons, R.F.: Natural language question answering systems. Commun. ACM **13**, 15–30 (1970)
4. Yu, Y.H., Simmons, R.F.: Truly parallel understanding of text. In: National Conference on 5th Artificial Intelligence, 29 July 29 – 3 August 1990, Boston, Massachusetts, USA, pp. 996–1001 (1990)

5. Quirk, C., Poon, H., Quirk, C., et al.: Distant supervision for relation extraction beyond the sentence boundary. In: Conference of the European Chapter of the Association for Computational Linguistics, vol. 1, pp. 1171–1182 (2017)
6. Ji, G., Liu, K., He, S., Zhao, J.: Distant supervision for relation extraction with sentence-level attention and entity descriptions. In: AAAI, pp. 3060–3066 (2017)
7. Miwa, M., Sasaki, Y.: Modeling joint entity and relation extraction with table representation. In: Conference on Empirical Methods in natural Language Processing, pp. 944–948 (2014)
8. Miwa, M., Bansal, M.: End-to-end relation extraction using LSTMs on sequences and tree structures. In: Annual Meeting of the Association for Computational Linguistics, pp. 1105–1116 (2016)
9. Zhuang, Y., Li, G., Zhong, Z., et al.: Hike: a hybrid human-machine method for entity alignment in large-scale knowledge bases. In: ACM on Conference on Information and Knowledge Management, pp. 1917–1926. ACM (2017)
10. Duan, S., Fokoue, A., Hassanzadeh, O., Kementsietsidis, A., Srinivas, K., Ward, M.J.: Instance-based matching of large ontologies using locality-sensitive hashing. In: Cudré-Mauroux, P., et al. (eds.) ISWC 2012. LNCS, vol. 7649, pp. 49–64. Springer, Heidelberg (2012). https://doi.org/10.1007/978-3-642-35176-1_4
11. Li, J., Wang, Z., Zhang, X., et al.: Large Scale instance matching via multiple indexes and candidate selection. Knowl.-Based Syst. **50**, 112–120 (2013)
12. Shao, C., Hu, L.M., Li, J.Z., et al.: RiMOM-IM: a novel iterative framework for instance matching. J. Comput. Sci. Technol. **31**, 185–197 (2016)
13. Xu, Y., Mou, L., Li, G., Chen, Y., Peng, H., Jin, Z.: Classifying relations via long short term memory networks along shortest dependency paths. In: EMNLP, pp. 1785–1794 (2015)
14. Yang, Y., Tong, Y., Ma, S., Deng, Z.: A position encoding convolutional neural network based on dependency tree for relation classification. In: EMNLP, pp. 65–74 (2016)
15. Hashimoto, K., Stenetorp, P., Miwa, M., et al.: Task-oriented learning of word embeddings for semantic relation classification. Comput. Sci. 268–278 (2015)
16. Lu, W., Dan, R.: Joint mention extraction and classification with mention hypergraphs. In: Conference on Empirical Methods in Natural Language Processing, pp. 857–867 (2015)
17. Blanco, R., Ottaviano, G., Meij, E.: Fast and space-efficient entity linking for queries. In: ACM International Conference on Web Search and Data Mining, pp. 179–188 (2015)
18. Zhang, W., Sim, Y.C., Su, J., et al.: Entity linking with effective acronym expansion, instance selection and topic modeling. In: International Joint Conference on Artificial Intelligence, pp. 1909–1914 (2011)
19. Shen, W., Wang, J., Luo, P., et al.: Linking named entities in tweets with knowledge base via user interest modeling. In: ACM SIGKDD International Conference on Knowledge Discovery and Data Mining, pp. 68–76. ACM (2013)
20. Huang, H, Heck, L, Ji, H.: Leveraging Deep Neural Networks and Knowledge Graphs for Entity Disambiguation. Computer Science. 1275–1284(2015)
21. Lin, Y., Liu, Z., Sun, M.: KR-EAR: knowledge representation learning with entities, attributes and relations. In: IJCAI, pp. 2866–2872 (2016)
22. Feng, J., Huang, M., Yang, Y., Zhu, X.: GAKE: graph aware knowledge embedding. In: Proceedings of the 26th International Conference on Computational Linguistics, pp. 641–651 (2016)
23. Ji, G., He, S., Xu, L., Liu, K., Zhao, J.: Knowledge graph embedding via dynamic mapping matrix. In: Proceedings of ACL, pp. 687–696 (2015)

Annotation Scheme and Specification for Named Entities and Relations on Chinese Medical Knowledge Graph

Donghui Yue[1,3], Kunli Zhang[1,3(✉)], Lei Zhuang[1,3], Xu Zhao[1,3],
Odmaa Byambasuren[2,3], and Hongying Zan[1,3]

[1] School of Information Engineering, Zhengzhou University, Zhengzhou, China
iedhyue@gs.zzu.edu.cn,
{ieklzhang, ielzhuang, iehyzan}@zzu.edu.cn,
zhaox917@163.com
[2] Key Laboratory of Computational Linguistics, Ministry of Education,
Peking University, Beijing, China
odmaa_b@pku.edu.cn
[3] Peng Cheng Laboratory, Shenzhen, China

Abstract. The medical knowledge graph describes medical entities and relations in a structured form, which is one of the most important representations for integrating massive medical resources. It is widely used in intelligent question-answering, clinical decision support, and other medical services. The key to building a high-quality medical knowledge graph is the standardization of named entities and relations. However, the research in annotation and specification of named entities and relations is limited. Based on the current research on the medical annotated corpus, this paper establishes an annotation scheme and specification for named entities and relations centered on diseases under the guidance of physicians. The specification contains 11 medical concepts and 12 medical relations. Medical concepts include the diagnosis, treatment, and prognosis of diseases. Medical relations focus on relation types between diseases and medical concepts. In accordance with the specification, a new Chinese medical annotated corpus of high consistency is constructed.

Keywords: Multi-source medical resources · Named entity · Entity relation · Annotation scheme · Annotation specification

1 Introduction

With the development of medical information, medical institutions have accumulated massive medical resources. Medical resources include electronic medical records (EMRs), medical papers, medical Wikipedia, medical textbooks, etc., which record diagnosis, treatment and other descriptions of diseases and cover lots of information closely related to the patient's health. How to effectively extract medical knowledge from medical resources is one of the most important tasks in information extraction. The Knowledge Graph [1] does well in organizing and managing medical data. Besides, it is the basis of intelligent semantic search. At present, how to construct a large-scale and accurate medical knowledge graph is a hot topic at home and abroad.

© Springer Nature Switzerland AG 2020
J.-F. Hong et al. (Eds.): CLSW 2019, LNAI 11831, pp. 563–574, 2020.
https://doi.org/10.1007/978-3-030-38189-9_58

Considering the privacy of doctors and patients, we select publicly available multi-source medical data as the corpus of Chinese medical knowledge graph. There are medical thesauruses, online medical resources, and national norms. The medical thesauruses include MeSH (Medical Subject Heading) [2], ICD-10 (The International Classification of diseases, 10th Revision), ICD-9-CM [3], and ATC (Anatomical Therapeutic Chemical). The online medical resources mainly refer to BMJ Best Practice, Medical Wikipedia and DXY Drugs Information. According to BMJ Best Practice[1], disease-related information is obtained. Medical Wikipedia is based on the classification of diseases and includes the definition, clinical features, diagnosis and treatment of the disease. DXY Drugs Information is the largest professional website of the pharmaceutical and biological industry, which contains ingredients, indications and adverse reactions of Chinese and western medicines. Clinical pathway is a set of standardized treatment patterns and procedures for diseases, formulated by the Ministry of Health of the People's Republic of China [4].

This paper focuses on the formulation of the annotation scheme and specification for named entities and relations on Chinese medical knowledge graph. The construction of Chinese medical corpus is briefly introduced, and it will be discussed separately. The rest of the paper is organized as follows. The related work is reviewed in Sect. 2. Section 3 presents the medical knowledge graph description scheme. Section 4 elaborates on the specification of medical named entities and relations. In Sect. 5, the Chinese medical knowledge graph corpus is constructed. Finally, the paper draws the conclusion in Sect. 6.

2 Related Work

The general knowledge graph has achieved its recent success, such as Google's Knowledge Graph, Baidu Bosom, and NELL [5]. In the medical domain, famous ontology libraries are SNOMED, CT [6] and IBM Watson Health. In 2010, the i2b2/VA Workshop on Natural Language Processing Challenges for Clinical Records presents three tasks: concepts extraction, assertions classification, and relations classification [7], and provides a total of 394 training reports, and 477 test reports. The entity types include medical problems, treatments, and tests. Relation classification aims to classify relations of pairs of given reference standard concepts from a sentence. The corpus annotation work is done by medical experts and evaluated by F1 measure. Although the annotation scheme is reasonable, it refers to diseases and symptoms collectively as a medical problem without distinction and ignoring the writing operation of EMRs. Besides, Mizuki et al. [8] organize the NTCIR-12 MedNLPDoc task which attempts to extract important information from medical reports written in Japanese. This task provides 50 medical reports. The annotation scheme contains patients' personal information and treatment. Because 50 documents are fabricated by doctors and only the complaints and diagnoses are marked, it is difficult to obtain more information on the real EMRs.

---

[1] https://bestpractice.bmj.com/.

Compared with English and Japanese medical data, the construct of Chinese medical corpus starts late. Lei et al. [9] construct EMRs dataset which contains 400 admission notes and 400 discharge summaries from Peking Union Medical College Hospital. For each note, they annotate four types of entity-clinical problems, procedures, laboratory tests, and medications. The annotation guidelines are developed by two doctors, similar to the 2010 i2b2 NLP challenge. Wang et al. [10] select 115 operation notes from Zhong Shan Hospital. Two doctors are recruited to annotate these 12 data elements manually and 961 entities are annotated. Yang et al. [11] propose an annotation scheme for named entities and relations on Chinese EMRs, involving five types of entity-disease, disease type, symptom, test, treatment, and six types of relation. The corpus comprises 992 medical text documents. Under the guidance of two doctors, the annotation process lasts for one and a half years. Although the annotation scheme is complete and large-scale, the type of medical entities less and it is difficult to cover key information in complex medical data. Xia et al. [12] describe three corpora that they created for clinical NLP projects and discuss various challenges they have encountered. Their studies show that it is necessary for medical experts and NLP researchers to design detail guidelines and discussion. Medical experts and NLP researchers are important for the success of annotation.

Throughout the annotation scheme for medical named entities and relations, the disadvantage is: (1) the type of text marked is relatively simple, which is mainly EMRs; (2) the classification scheme of named entities is not complete, which mainly includes disease, test, treatment and symptom; (3) relations description scheme is fewer. To overcome the limitations of the medical annotation scheme, this paper regards the medical knowledge graph covering treatment, diagnosis, and prognosis of the disease as the starting point. Under the guidance of physicians, a complete and detailed annotation specification for named entities and relations is formulated according to multi-source medical data, and an annotated corpus is constructed which is suitable for Chinese medical knowledge graph.

3   Medical Knowledge Graph Annotation Scheme

According to the analysis of multi-source medical data, this paper summarizes key information about diseases. It includes: (1) the basic knowledge of diseases: including the epidemiology, pathogenesis and cases. (2) the diagnosis process of diseases: detecting diseases according to some tests and diagnosing diseases referring to diagnostic criteria; (3) the treatment process of diseases: after the diagnosis of diseases, it is necessary to determine the treatment steps and implement treatment according to the patient's health, and guide patients on how to prevent diseases; (4) the follow up of diseases: after treatment of diseases, patients need to be followed up regularly to see the prognosis of diseases. The description framework of diseases is illustrated in Fig. 1. From this process, the disease mainly involves four types of information: basic knowledge, diagnosis, treatment, and follow-up, divided as follows.

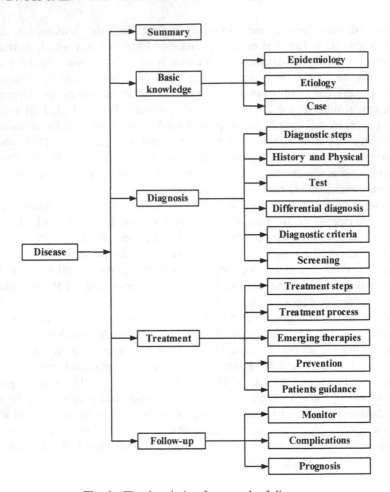

Fig. 1. The description framework of diseases

(1) The basic knowledge is subdivided into 3 types of entity: disease, anatomy (related parts of the disease) and epidemiology.
(2) Diagnosis is divided into 3 types of entity: symptom, test, and sociology.
(3) Treatment is split into 3 types of entity: drug, operation and other treatments (other than drug and operation).
(4) Follow-up is divided into 2 types of entity: prognosis and other (closely related to diseases).

Chinese medical relation extraction focuses on assigning relation types between diseases and other entities. Table 1 shows the relations pre-defined for Chinese medical knowledge graph. For example, there are 7 relations between two diseases. Distinguishing two diseases are called differential diagnosis, and one disease combined with another disease or multiple diseases is called the complication. Complications (drugs) mean one or more diseases are complicated by the usage of drugs.

Table 1. Medical entity types and relations between each other

Entity 1	Relation	Entity 2	Relation number
Disease	metastasis/invasion/pathogenic site	Anatomy	3
Disease	clinical symptoms/post-treatment symptoms/ .../clinical signs	Symptom	4
Disease	auxiliary examination/imaging/.../endoscopy	Test	6
Disease	differential diagnosis/complications/ .../pathological type	Disease	7
Disease	radiation therapy/adjuvant therapy/chemo	Other treatments	3
Disease	operative treatment	Operation	1
Disease	medication	Drug	1
Disease	onset age/onset sex/.../incidence rate	Epidemiology	6
Disease	prognosis situation/prognostic survival time/.../ prognosis survival rate	Prognosis	5
Disease	etiology/risk assessment factors/.../high-risk factors	Sociology	5
Disease	consulting departments/period/.../prevent	Other	5

4 Medical Named Entities and Relations Annotation Specification

This section describes the definition of named entities and relations, and the annotation specification. The specification refers to the entities and relations defined by the 2010 i2b2 NLP challenge.

4.1 Named Entity

The medical knowledge graph named entity annotation specification involves 11 types of entity. This work adopts 4 medical thesauruses to identify the boundary of entities, namely MeSH, ICD-10, ICD-9-CM and ATC. In this specification, other treatments can treat diseases or relieve symptoms, apart from operation and drug. The prognosis is the specific consequences of diseases, and the appearance or disappearance of other abnormalities, such as symptoms, signs, and complications. The other reflects important information that is closely related to diseases and is a part of clinical data. The remaining 8 common medical entities are described as follows. In order to distinguish entities with other words, correctly marked entities are noted with underlines.

4.1.1 Disease

The disease is the diagnosis made by the doctor, or an abnormal life activity caused by the self-regulation disorder. The MeSH and the ICD-10 define the boundary of diseases. What's more, diseases can be identified by Baidu Wikipedia, medical Wikipedia and the like. Annotated examples are shown in sentences (1) and (2).

(1) 患者有<u>高血压</u>和<u>糖尿病</u>的治疗历史。(The patient had a history of treatment for <u>hypertension</u> and <u>diabetes</u>).

(2) 本病被视为<u>先兆子痫</u>的一种严重形式，所以诱发因素仍是推测的。(The disease is considered as a serious form of <u>pre-eclampsia</u>, so the predisposing factors are speculated).

4.1.2 Anatomy

The anatomy is a functional structural unit consisting of a variety of tissues. The MeSH can define the boundary of anatomy. The MeSH includes body parts, sensory organs, tissues, cells, etc. Examples are shown in sentences (3) and (4).

(3) <u>肺癌</u>经常蔓延到<u>纵膈淋巴</u>。(Lung cancer often spreads to the mediastinal <u>lymph</u>).

(4) 侵袭到<u>胸膜</u>或<u>胸壁</u>导致胸痛。(Invading into the <u>pleura</u> or <u>chest wall</u> causes chest pain).

4.1.3 Symptom

A symptom is a departure from normal function or feeling which is apparent to a patient, reflecting the presence of an unusual state, or a disease. A symptom can be objective or subjective. A sign is a clue to a disease elicited by examiners or doctors. The subjective symptom is described by patients. Symptoms refer to as Diagnostics and Chinese symptom base. Chinese symptom base comes from OpenKG.CN[2] which is provided by East China University of Science and Technology. Examples are shown in sentences (5) and (6).

(5) 患者自述最近<u>食欲不振</u>、<u>嗜睡</u>、<u>怕冷</u>。(The patient reports <u>inappetence</u>, <u>drowsiness</u>, and <u>coldness</u>).

(6) 体格检查可触及<u>胆囊肿大</u>。(<u>Swollen gallbladder</u> can be found in physical examination).

4.1.4 Test

A test is a medical procedure which is performed to detect, diagnose, or monitor diseases and disease processes, or to determine a course of treatment. The MeSH and the medical insurance diagnosis and treatment catalog are used to recognize the boundary of test. The test can find more information about diseases and symptoms. However, the disease is not treated and the symptoms are not relieved either through tests. Examples are shown in sentences (7) and (8).

(7) <u>肝功能检查</u>可见碱性磷酸酶和胆红素水平略升高。(<u>Liver function</u> tests showed a slight increase in alkaline phosphatase and bilirubin levels).

(8) <u>血肌酐</u>为确定GFR是否异常而进行的筛查试验。(<u>Serum creatinine</u> is a screening test to determine if GFR is abnormal).

---

[2] http://www.openkg.cn/dataset/symptom-in-chinese/.

4.1.5 Operation

An operation uses instrumental techniques on a patient to treat a pathological condition such as a disease or injury. The MeSH and the ICD-9-CM can identify the boundary of operation. Sentences (9) and (10) are examples.

(9) 少数患者选择肝脏移植。(A small number of patients choose liver transplantation).

(10) 对于孕龄30至32周未临产者,可以考虑剖宫产。(Pregnant women with the gestational age of 30 to 32 weeks can consider cesarean).

4.1.6 Drug

Drugs are chemicals that prevent, diagnose, and treat diseases. The boundary of drugs is recognized by the ATC, the MeSH and DXY Drugs information. The examples are shown in sentences (11) and (12).

(11) 非甾体类抗炎药可以有效的减少疼痛。(NSAIDs can reduce pain effectively).

(12) 通常拉贝洛尔和硝苯地平可以作为备选药物。(Normodyne and nifedipine can be used as alternative drugs).

4.1.7 Epidemiology

Epidemiology is the study that analysis of the distribution and determinants of health and disease conditions in defined populations, thereby preventing the occurrence of diseases. In this paper, epidemiology focuses on vulnerable groups of the disease and its subsequent development. Examples are shown in sentences (13) and (14).

(13) 肺癌发生率在黑人和非西班牙裔白人中最高。(The incidence of lung cancer is the highest among Negro and non-Hispanic whites).

(14) 发病率为每10万人166例。(The incidence rate is 166 cases per 100,000 people).

4.1.8 Sociology

Sociology is the actual situation of a social phenomenon or the causes of related phenomena. In this paper, it focuses on the factors that drive the disease. Sentences (24) and (25) are examples.

(15) 垂体卒中是现存腺瘤突发梗死或出血引起。(Pituitary apoplexy is caused by sudden infarction or bleeding of an adenoma).

(16) 他既往健康, 无眼病史。(He is healthy and has no history of eye disease).

4.2 Relations

Relations are annotated based on entities. This paper focuses on relation types between diseases and other entities. There are 12 relations and 46 sub-relations, as shown in Table 1. These relations are not only in a sentence but also across sentences. The attribute description of relations and 8 relations are briefly introduced.

4.2.1 Attribute Description of Relations

When triples are not sufficient for relations, this paper adopts quads and regards attribute as the fourth element. The triple is represented as (Entity1, Entity2, Relation). The quad is defined as (Entity1, Entity 2, Relation, Attribute). The attribute is the interpretation and assertion of entities. The value is as follows.

(1) Number, indicating a partial relationship between diseases and other entities. For example, 乳头状癌是最常见的类型, 占了甲状腺癌的80%。 (*Papillary carcinoma is the most common type, accounting for 80% of thyroid cancer*). (thyroid cancer, papillary carcinoma, pathological type, 80%), "80%" is a property description.

(2) Neg, expressing no relationship between diseases and other entities. For example, 急性阑尾炎通常无发热现象。 (*Acute appendicitis usually has no fever*). Expressed as: (acute appendicitis, fever, clinical signs, no), "no" is an attributed description.

(3) String, representing a constraint on the occurrence of relations. For example, 病毒性脑膜炎在美国的年发病率为11/10万。 (*The annual incidence of viral meningitis is 11/10 million in the United States*). Expressed as (viral meningitis, 11/100,000, incidence, the United States), "the United States" for attribute description, indicating the incidence of viral meningitis in a certain place, the incidence varies from place to place.

4.2.2 Synonym

The synonym refers to a group of words with the same meaning, such as abbreviations, English names, proverbs and so on. In this paper, the synonym involves 11 entities. For example, 胸部X线检查 (CXR)。 (*Chest X-ray (CXR)*). Represented as: (chest X-ray, CXR, synonym).

4.2.3 Disease and Anatomy

The relationship between disease and anatomy includes metastasis, invasion, and pathogenic site. Two sub-relations are introduced.

(1) Metastasis is the spread of a pathogen from an initial or primary site to a different or secondary site in the host. Then, the new pathological site is the metastasis. For example, 甲状腺成为肺癌的远处转移部位。 (*The thyroid becomes a distant metastatic site for lung cancer*). Described as: (lung cancer, thyroid, metastasis).

(2) An invasion is the direct extension and penetration of cancer cells into neighboring tissues. For example, 胸膜是肺癌常见的侵犯部位。 (*The pleura is a common site of invasion of lung cancer*). Denoted as: (lung cancer, pleura, invasion).

4.2.4 Disease and Symptom

The relation between disease and symptom involves clinical symptoms, symptoms of an invasion of neighboring tissues, post-treatment symptoms and clinical signs. Two sub-relations are described.

(1) Symptoms of an invasion of neighboring tissues are caused by the metastasis or an invasion of other parts of the disease. For example, 睾丸癌患者,当腰大肌或神

经根侵犯时会出现腰背部疼痛。 (*For the patient with testicular cancer, muscle pain occurs when the psoas muscle or nerve roots invade*). Regarded as: (testicular cancer, muscle pain, symptoms of an invasion of neighboring tissues, psoas muscle or nerve root invasion).

(2) Post-treatment symptoms are caused by the treatment of diseases. For example, 慢性憩室炎患者在结直肠吻合术后第二天, 出现腹胀症状。 (*The patient with chronic diverticulitis presented with distension on the second day after colorectal anastomosis*). Denoted as: (chronic diverticulitis, distension, post-treatment symptoms, colorectal anastomosis).

4.2.5   Disease and Test

The relation between disease and test contains imaging, endoscopy, histology examination, screening, laboratory examination, and auxiliary examination. Two sub-relations are presented.

(1) Histological examination observes the pathological changes of gross specimens, and then removes specific size lesions for further examination with a microscope. For instance, 喉癌诊断中, 若肿物不可触及, 则可行B超引导下穿刺活检。 (*In the diagnosis of laryngeal cancer, if the tumor is inaccessible, it is feasible to perform a biopsy guided by B-scan ultrasonography*). Expressed as: (laryngeal cancer, biopsy, histological examination).

(2) The auxiliary examination is a physical examination by medical equipment in addition to imaging, endoscopy, histological, screening and laboratory examination. For example, 多导睡眠图结果有助于鉴别失眠症和 PLMD。 (*Polysomnography is beneficial for identifying insomnia and PLMD*). Defined as: (insomnia, polysomnography, auxiliary examination), (PLMD, polysomnography, auxiliary examination).

4.2.6   Disease and Disease

The relationship between disease and disease includes 7 sub-relations, namely differential diagnosis, complications, complications (postoperative), complications (drug), correlation (transformation), correlation (cause) and pathological type. Four sub-relations are described.

(1) Complications (postoperative) refer to other diseases caused by surgical procedures during treatment. For example, 甲状腺癌患者在全甲状腺切除术后可能出现一些并发症,其中喉返神经损伤或甲状腺功能减退的风险为 2%。 (*The patient with thyroid cancer may have some complications after thyroidectomy, and the risk of recurrent laryngeal nerve injury or hypothyroidism was 2%*). Represented as: (thyroid cancer, recurrent laryngeal nerve injury, complications (postoperative), thyroidectomy), (thyroid cancer, hypothyroidism, complications (postoperative), thyroidectomy).

(2) Complications (drug) are diseases that may be caused by the side effects of drugs. For instance, 确诊为未足月胎膜早破后, 不推荐使用阿莫西林, 因其可能会增加新生儿坏死性小肠结肠炎的风险。 (*Amoxicillin is not recommended for premature rupture of fetal membranes as it may increase the risk of NEC*).

Denoted as: (premature rupture of fetal membranes, NEC, complications (drug), amoxicillin).

(3) Correlation (transformation) is the correlation, transformation, and secondary among diseases. For example, 在 15% 的病例中,慢性淋巴细胞白血病可突然转化成高级别非霍奇金淋巴瘤。(*In 15% of cases, chronic lymphoblastic leukemia can be suddenly converted into high-grade non-Hodgkin lymphoma*). Expressed as: (chronic lymphocytic leukemia, high-grade non-Hodgkin's lymphoma, correlation (transformation)).

(4) Correlation (cause) means one disease leads to another. In the annotation, etiological diseases are in front and the caused diseases are in the back. For example, 皮肤破裂导致白喉感染。(*Skin rupture leads to diphtheria infection*). Described as: (skin rupture, diphtheria infection, correlation (cause)).

4.2.7 Disease and Operation

The relationship between disease and operation is operative treatment. For instance, 一些患者仍会发生进行性肾损伤, 最终需要肾脏替代治疗(透析,移植)。(*Some patients have progressive kidney damage and require renal replacement therapy (dialysis, transplantation)*). Represented as: (progressive kidney injury, renal replacement therapy (dialysis, transplantation), operative treatment).

4.2.8 Disease and Drug

The relation between disease and drug is the medication. For example, 对于复杂性膀胱炎患者, 一线治疗是连续 7–14 天的抗生素治疗。(*For patients with complicated cystitis, the first-line treatment is continuous 7–14 days of antibiotics*). Expressed as: (complex cystitis, antibiotics, medication).

4.2.9 Disease and Epidemiology

The relation between the disease and epidemiology includes 6 sub-relations, which are onset age, onset sex, incidence rate, mortality rate, multiple groups, and multiple regions. Two sub-relations are described.

(1) Onset age refers to the age in which the disease tends to occur. The values are numeric or strings.

(2) Multiple groups refer to groups with a high incidence of disease. The keywords are multiple, influential, prevalent, common, high frequency and concentrated.

4.2.10 Disease and Sociology

The relationship between the disease and sociology consists of 5 sub-relations, including etiology, risk assessment factors, high-risk factors, medical history, and genetic factors. Three sub-relations are discussed.

(1) Risk assessment factors are the evaluation of factors that may lead to disease. The keywords include inducement, exposure risks, risk sources and risk factors. For example, 导致危重患者出现胆道疾病的因素为胆囊运动障碍。(*Gallbladder dyskinesia causes biliary tract in critically ill patients*). Expressed as: (biliary tract, gallbladder dyskinesia, risk assessment factors).

(2) High-risk factors increase the probability of disease or death and have a certain causality with the occurrence of disease. When eliminating this factor, the probability of diseases will be reduced. For example, 血管损伤, 尤其是在曾有低血压发作的胆囊炎患者中，被认为是危险因素。 *(Vascular injury, especially in patients with cholecystitis who have had a low blood pressure episode, is considered as a risk factor).* Represented as: (cholecystitis, vascular injury, high-risk factors).

(3) Genetic factors are the effects of genes, specific genes and chromosomal aberrations on diseases. For instance, 麻风病可能的遗传因素为PACRG基因。 *(The possible genetic factor for leprosy is the PACRG gene).* Described as: (leprosy, PACRG gene, genetic factors).

5 Construction of Medical Knowledge Graph Corpus

The corpus of medical knowledge graph mainly consists of 3 parts, among which the medical Wikipedia covers about 23.74 million words. DXY Drugs Information includes about 31.49 million words. Clinical pathway contains about 60,000 words. An annotation tool is developed for medical texts. Under the guidance of Chinese medical annotation scheme and specification for named entities and relations, this paper adopts a multi-round iteration for tagging. Each disease is annotated by two people independently. The annotation results are evaluated by F1 score. This work discusses some problems existing in the annotation process and modifies the specification. Based on the new specification, the next round of annotation will be started.

At present, 106 diseases are annotated, and named entities and relations achieve 0.873 and 0.829 in F1 score, respectively. The overall process of annotation lasted 4 months and a total of 22 people participated. We annotated 3029,448 words and obtained 23,475 entities and 32,530 relations. The visual display is an important part of Chinese medical knowledge graph. For now, CMeKG 2.0 (Chinese Medical Knowledge Graph) has been released for trial. (See online supplements at http://zstp.pcl.ac.cn:8002/).

6 Conclusion

According to the characteristics of multi-source medical data and existing researches, this paper establishes Chinese medical annotated scheme and specification for named entities and relations under the guidance of professionals. Compared with the existing English and Chinese medical annotated scheme, this annotation scheme focuses on 11 entities, 12 relations and 46 sub-relations. The scheme is more detailed and builds a solid foundation for the construction of Chinese medical knowledge graph. This specification has a certain universality. According to the features of other departments (e.g., obstetrics and pediatrics), the annotated scheme and specification are able to be more suitable.

Acknowledgments. This research is supported by the National Social Science Fund of China (No. 18ZDA315), the Key Scientific Research Program of Higher Education of Henan (No. 20A520038), the science and technology project of Science and Technology Department of Henan Province (No. 192102210260), and the international cooperation project of Science and Technology Department of Henan Province (No. 172102410065).

References

1. Chen, Y., Liu, Z.: The rise of mapping knowledge domain. Stud. Sci. Sci. **23**, 149–154 (2005). [In Chinese]
2. Martín, R.W.: The use of medical terminology. A thesaurus. Medical subject headings. Aten Primaria **23**, 548–552 (1999)
3. Quan, H., Sundararajan, V., Halfon, P., et al.: Coding algorithms for defining comorbidities in ICD-9-CM and ICD-10 administrative data. Med. Care **43**, 1130–1139 (2005)
4. Li, M.: Basic concept and application of clinical pathway. Chin. J. Nurs. **45**, 59–61 (2010). [In Chinese]
5. Carlson, A., Betteridge, J., Kisiel, B., et al.: Toward an architecture for never, ending language learning. In: Twenty, fourth Aaai Conference on Artificial Intelligence (2010)
6. Stearns, M.Q., Price, C., Spackman, K.A., et al.: SNOMED clinical terms: overview of the development process and project status. In: Proceedings of AMIA Annual Symposium, vol. 8, p. 662 (2001)
7. Uzuner, O., South, B.R., Shen, S., et al.: 2010 i2b2/VA challenge on concepts, assertions, and relations in clinical text. J. Am. Med. Inform. Assoc. **18**, 552–556 (2011)
8. Mizuki, M., Yoshinobu, K., Tomoko, O., Mai, M., Aramaki, E.: Overview of the NTCIR-10 MedNLP task. In: Proceedings of the NTCIR-10 (2013)
9. Lei, J., Tang, B., Lu, X., et al.: Research and applications: a comprehensive study of named entity recognition in Chinese clinical text. J. Am. Med. Inform. Assoc. Jamia **21**, 808 (2014)
10. Wang, H., Zhang, W., Zeng, Q., et al.: Extracting important information from Chinese operation notes with natural language processing methods. J. Biomed. Inform. **48**, 130–136 (2014)
11. Yang, J.F., Guan, Y., He, B., et al.: Corpus construction for named entities and entity relations on Chinese electronic medical records. J. Softw. **27**, 2725–2746 (2016). [In Chinese]
12. Xia, F., Yetisgen Yildiz, M.: Clinical corpus annotation: challenges and strategies. In: Proceedings of the 3rd Workshop on Building and Evaluating Resources for Biomedical Text Mining of the International Conference on Language Resources and Evaluation (LREC), pp. 32–39 (2012)

Directionality and Momentum of Water in Weather: A Morphosemantic Study of Conceptualisation Based on Hantology

Sicong Dong[1]([⊠]) [iD], Yike Yang[2] [iD], Chu-Ren Huang[2,3] [iD],
and He Ren[4] [iD]

[1] The Hong Kong Polytechnic University - Peking University Research Centre
on Chinese Linguistics, Beijing, China
szechungtung@gmail.com
[2] Department of Chinese and Bilingual Studies,
The Hong Kong Polytechnic University, Hong Kong, China
yi-ke.yang@connect.polyu.hk, churen.huang@polyu.edu.hk
[3] Center for Chinese Linguistics, Peking University, Beijing, China
[4] Department of Chinese Language and Literature,
Peking University, Beijing, China
renhe1214@126.com

Abstract. We present in this paper a study of the conceptualisation of meteorological events involving water in Chinese based on Hantology, a SUMO-based ontology of Chinese orthography. Our comprehensive investigation of the morphosemantic behaviours of these weather words in both Mandarin and Sinitic languages reveals that they are predicted by the directionality and momentum of their formation and movement. We studied events involving water in both liquid and solid forms: such as rain, snow, hail, fog, dew and frost. They share the radical 雨, which can be linked to two SUMO nodes according to Hantology. This ontological bifurcation can be shown to bring about not only the diversity of direction expressions referring to these words for water, but also the differences of semantic features and PoS between them in Archaic Chinese. Moreover, the momentum of different water forms is proposed to be the physical basis for the differences of PoS, semantic features and node linking.

Keywords: Weather and language · Directionality · PoS · Mandarin Chinese · Sinitic languages · Archaic Chinese

1 Introduction

1.1 Scientific Felicity of Weather Verbs

Weather verbs are often verbs of movement or appearance but used in highly conventionalised and typologically different ways [1, 2, etc.]. Hence whether these weather

This work was partially funded by the Hong Kong Polytechnic University (Project 4-ZZHK) and the Hong Kong Polytechnic University - Peking University Research Centre on Chinese Linguistics.

© Springer Nature Switzerland AG 2020
J.-F. Hong et al. (Eds.): CLSW 2019, LNAI 11831, pp. 575–584, 2020.
https://doi.org/10.1007/978-3-030-38189-9_59

verbs embody or metaphorically represent the actual weather events are highly debated. For instance, a common expression in weather forecast to describe fogs is 普降大雾 *pǔ jiàng dà wù* 'widely-fall-big-fog' in Chinese, which uses the verb 降 *jiàng* 'to fall' to report that certain areas are covered in dense fogs. Interestingly, however, fog does not fall. According to Ahrens [3], fog is usually formed by cooling or by evaporation and mixing of moist air with relatively dry air. Different types of fog (e.g., radiation fog, advection fog or upslope fog) either do not have a vertical movement, or move upwards, but never go downwards. Hence it seems that this particular use of the weather verb is scientifically infelicitous and does not directly represent the actual movement of fog.

In fact, 降 *jiàng* 'to fall' can also be used with other weather events such as 降露 *jiànglù* 'to dew' and 降霜 *jiàngshuāng* 'to frost'. As stated by Ahrens [3], dew is formed when air cools to the dew point due to contact with cold surfaces, and frost is produced on the ground and other surfaces by direct change of water vapour into ice via cooling. Therefore, there is no downward movement during the formation of fog, dew and frost, which is also in accordance with our life experience. Unlike rain or snow, fog, dew and frost do not fall down from the sky. Thus, it seems that the use of the weather verb 降 *jiàng* 'to fall' is highly conventionalised and neither embodied nor scientifically felicitous. Then a question arises: why does the Chinese language use verbs with downward movement meaning to represent such water-related weather events without downward movements? Interestingly, besides Mandarin Chinese, many Sinitic languages[1] also exhibit such morphosemantic convention. One thing various Sinitic languages have in common is that they all use Chinese characters. It is worth mentioning that the characters for the aforementioned weather phenomena, together with 雨 *yǔ* 'rain', 雪 *xuě* 'snow' and 雹 *báo* 'hail', share the same radical 雨. Rain, snow and hail are also forms of water, and belong to precipitation that does fall under gravity which can co-occur with 降 *jiàng* 'to fall', too. So, a further question we want to address is: is the fact of radical sharing just a coincidence, or a key for us to explain such idiosyncratic usage?

1.2 Ontology and Hantology

Borrowed from philosophy, ontology refers to the 'explicit specification of a conceptualisation' [4]. Ontology generally concerns two questions [5]: (1) What are the basic concepts for knowledge representation? and (2) How are the concepts organised?

Recent studies show that the Chinese orthography has conventionalised a system of semantic relations of basic concepts [6, 7] and such ontology has its psychological reality among Chinese people [8]. Following this line of research, the current study investigates the directions in the expressions of different weather phenomena in various Sinitic languages based on Hantology.

---

[1] Sinitic languages are traditionally referred to as Chinese dialects. Both terms are used interchangeably in our study.

2   Method

Corpora and dictionaries were consulted to obtain comprehensive information of how Chinese people describe the formation and/or appearance of water (both in liquid and solid forms) in weather events, all with the radical 雨 in Chinese characters. After analysing the results, we provided an account from Hantology for the distribution of directional meanings. Further supporting evidence from Archaic Chinese and physics was also discussed.

The BCC Corpus [9] and the CCL Corpus [10] were consulted for the usage of weather expressions in Mandarin Chinese. As for the former, the balanced and diachronic corpora were used. We have searched combinations of verbs with directional meanings and weather phenomena 雾 wù 'fog', 露 lù 'dew' and 霜 shuāng 'frost' respectively in these corpora, and calculated the frequencies after manual check. The verbs include 降 jiàng 'to fall', 下 xià 'to fall', 起 qǐ 'to rise' and 上 shàng 'to rise'. In total, there were 12 possible forms of compounds ('降/下/起/上 + 雾/露/霜').

Besides Mandarin Chinese, related usage in other Sinitic languages was also investigated. We collected the expressions of eight weather phenomena from the dictionaries of Li [11], Xu and Miyata [12], Tao [13] and Zhang and Mo [14], which are usually represented by characters 雨 yǔ 'rain', 雪 xuě 'snow', 雹 báo 'hail', 雾 wù 'fog', 露 lù 'dew', 霜 shuāng 'frost', 雷 léi 'thunder' and 电 (traditional form: 電) diàn 'lightning', all containing the radical 雨. Compared to Mandarin, five more phenomena were added for Sinitic languages, since we planned to conduct a comparative study with the diverse expressions in these languages. Verbs that co-occur with weather nouns to indicate the occurrence of such phenomena were examined and grouped into four major types according to directional meanings: upward, downward, both, and no obvious vertical direction.

In our study, the verbs involving upward movement are 起 qǐ 'to rise' and 上 shàng 'to rise'. The ones referring to downward movement are 降 jiàng 'to fall', 落 luò 'to fall', 下 xià 'to fall' and 矺 zhé 'to press down'. It needs to be mentioned that most of the verbs are polysemes. Those implicating the downward direction clearly indicate a downward movement towards the ground, while the others, i.e., 起 qǐ and 上 shàng, may have different interpretations about which sense they are used in. For instance, 起 qǐ means to rise, to start or to occur, etc.; as a result, 起雾 qǐwù could be interpreted as fog rising, fog appearing, or both. Similarly, 上 shàng can mean to rise or to add, so 上雾 shàngwù may also indicate a non-directional meaning as fog being added.

3   Results

Distribution of the 12 forms ('降/下/起/上 + 雾/露/霜') in the BCC and CCL corpora (accessed 8 and 13 Dec 2018, respectively) is shown in Table 1.

Table 1. Corpus search results

	BCC balanced	BCC diachronic	CCL
降雾 *jiàngwù*	11	2	8
下雾 *xiàwù*	174	18	10
起雾 *qǐwù*	352	32	22
上雾 *shàngwù*	5	0	0
降露 *jiànglù*	7	0	2
下露 *xiàlù*	7	1	4
起露 *qǐlù*	1	0	0
上露 *shànglù*	0	1	0
降霜 *jiàngshuāng*	66	66	13
下霜 *xiàshuāng*	119	87	26
起霜 *qǐshuāng*	14	1	0
上霜 *shàngshuāng*	16	1	2

Among the fog-related compounds, the upward direction constitutes 65.9%, 61.5% and 55%, respectively in each corpus, which is larger than the downward direction, but there is no disparity. However, downward meanings are much more frequently expressed when describing the occurrence of dew and frost. The number of dew-related compounds is quite small, compared to the other two phenomena. It may be due to the fact that the visibility of dew is lower than fog and frost. Also, dew is not as relevant to human beings as fog and frost, because fog and frost could have an impact on transportation, agriculture and people's daily life.

Related expressions in 229 dialects/subdialects have been investigated. As previously mentioned, verbs were classified based on directional meanings and the percentage under each category was calculated. Since our focus is the directions the expressions denote, we did not consider the order of the verbs and the weather nouns. Besides, not all the eight weather phenomena in every dialect are included in the dictionaries, so instead of frequencies, percentages by directions concerning each phenomenon are provided in Table 2.

Table 2. Distribution of directions in Sinitic languages (%)

	Down	Up	Both	None
Snow	100	0	0	0
Rain	98.3	0	0	1.7
Hail	97.0	0	0	3.0
Fog	51.7	24.7	13.5	10.1
Dew	50.0	15.2	4.5	30.3
Frost	45.5	0	2.0	52.5
Thunder	0	6.8	0	93.2
Lightning	0	0	0	100

The eight phenomena exhibit three directional tendencies. Rain, snow and hail, which belong to precipitation, are highly inclined to be expressed with downward meanings. Thunder and lightning also show a clear tendency that they are seldom used with directional meanings. Fog, dew and frost, however, show more complicated patterns. First, downward meanings have greater proportions than upward ones. Secondly, both directions co-exist in some languages, which cannot be found in expressions of other phenomena. Thirdly, frost-related compounds or phrases cannot be solely expressed as moving upwards, and have more non-directional meanings than directional ones.

Please note that the two verbs we counted as with upward meanings may have ambiguous interpretations, as has been mentioned. Based on the results, we might state that not only can fog, dew and frost 'fall' in Mandarin and other Sinitic languages, they even 'fall' more than they 'rise'. This is in line with Dong's [15] brief observation on 雾 wù 'fog'.

4   Discussion

4.1   One Radical, Two Ontological Concepts

Hantology is the integration of both ontological and linguistic information [6]. In particular, morphosyntactic cues, such as word and compound formation generalisations, are preserved when mapped to formal ontological representations. As such, Hantology is a linguistic ontology which integrates shared world views of the speakers of Chinese. Thus the conceptual representation of the radical 雨 in Hantology reflects Chinese people's perception of the water-related weather phenomena and how it influences the morphosemantic behaviours of the different characters sharing this radical.

In Hantology, the radical 雨 represents both the ontology node of 'water' and that of 'weather process' in SUMO [16]. We propose that characters 雾 wù 'fog', 露 lù 'dew' and 霜 shuāng 'frost' are linked to the 'water' node, 雷 léi 'thunder' and 电/電 diàn 'lightning' are linked to the 'weather process' node, and 雨 yǔ 'rain', 雪 xuě 'snow' and 雹 báo 'hail' are linked to both. The reason why fog, dew and frost can share the same verbs with rain, snow and hail is that they share the same ontology node in the conceptual structure, which reflects that people would regard the former ones as kinds of precipitation. Also, the fact that fog, dew and frost appear near ground and rarely involve any obvious vertical direction change has facilitated such analogy, because their ground-level positions match the 'fallen' status and they do not show typical symptoms of 'rising'. Our claim can be bolstered by the following arguments.

First, similar 'atypical' taxonomical classifications are well attested in Chinese radical systems [17]. For example, 鱿 yóu 'squid', 鲸 jīng 'whale', 鳄 è 'crocodile', 鲍 bào 'abalone', 鲵 ní 'giant salamander', and 鳖 biē 'softshell turtle' are not fishes, yet all the characters have the component of 鱼 'fish', indicating that they are deemed to belong to the conceptual class of fishes. That is, these characters share most functional characteristics with the class of fishes.

Secondly, some Sinitic languages use both types of verbs with opposite directions to describe the occurrence of fog, dew and frost. For example, fog can both 'fall' and 'rise' in 13.5% of the languages. This means that speakers of those languages must have independent inference for each directional expression, otherwise they would conflict with each other. Fog in some cases does involve upward movement, as noted earlier, which may contribute to the usage of 起 qǐ 'to rise' and 上 shàng 'to rise'. On the other hand, the reason for using verbs with downward meanings is most likely that those phenomena are deemed as precipitation, since they do not actually move downwards.

Thirdly, the distributions of 降 jiàng 'to fall' and 下 xià 'to fall' in the CCL Corpus demonstrate differences in terms of text style. As illustrated in Table 3, 降 jiàng tends to be used in formal texts, while 下 xià in informal ones. The '降/下' contrast is a method to make a distinct style difference when they take rain, snow and hail as objects. When such method is applied to fog, dew and frost, we can infer that they should also be regarded as precipitation. Another piece of evidence is that, all the six cases of 降雾 jiàngwù are from newspapers after 2003, while cases of 下雾 xiàwù have a much wider time range, suggesting that 降雾 jiàngwù may be an emerging copy of 降雨 jiàngyǔ, 降雪 jiàngxuě, etc. As for the cause for the only exception 降霜 jiàng-shuāng, its reverse form 霜降 probably plays a role. 霜降 is one of the traditional solar terms (节气), and the common use of which in daily life is likely to have rendered 降霜 more informal.

Table 3. Distributions of 降 jiàng and 下 xià (CCL)

	降雾	下雾	降霜	下霜	降露	下露
Formal	6	1	6	7	2	0
Informal	2	9	7	19	0	4

4.2   Evidence from Archaic Chinese

Further evidence can be obtained from Archaic Chinese with the Hantology link to the radical 雨. The morphosemantic and grammatical differences of the characters sharing the radical in fact can be predicted according to the differences in the conceptualisation of perception of 'water' vs. 'weather process'.

First, the expressions of directional meanings concerning those weather phenomena have the same distribution in three different groups in both Archaic and Modern Chinese. According to Ren [18], 雨 yǔ 'rain', 雪 xuě 'snow' and 雹 báo 'hail' can function as weather verbs in Archaic Chinese, denoting to rain, to snow and to hail, respectively. They hence express downward movement themselves. The second group of weather nouns, namely 露 lù 'dew' and 霜 shuāng 'frost', need to combine with verbs with downward meanings, such as 降 jiàng and 陨 yǔn in (1) and (2) below, to indicate their occurrence.[2] The last group, 雷 léi 'thunder' and 电/電 diàn 'lightning', are not found to convey any directional meanings.

---

[2] No expression describing the occurrence of fog was found in the ancient documents we investigated.

(1) 凉风至, 白露降, 寒蝉鸣
    liáng__fēng__zhì__bái__lù__jiàng__hán__chán__míng
    cool__wind__arrive__white__dew__fall__cold__cicada__chirp
    Cool wind arrives. White dew appears. Winter cicadas chirp. (Yueling, in *Liji*)
(2) 驷见而陨霜
    sì__xiàn__ér__yǔn__shuāng
    Star-Si__appear__then__fall__frost
    There will be frost when Star Si appears. (Zhouyuzhong, in *Guoyu*)

Secondly, of the eight weather phenomena, the three groups can be attributed to their denotational differences. A weather phenomenon can be viewed as consisting of weather event and weather product. We can view the ontological class of 'water' as specifically the weather product of such phenomena. As analysed previously, in Hantology, 雾 *wù* 'fog', 露 *lù* 'dew' and 霜 *shuāng* 'frost' are linked to 'water' and refer to weather products exclusively. 雷 *léi* 'thunder' and 电/電 *diàn* 'lightning' refer to 'weather process' as events. And lastly, 雨 *yǔ* 'rain', 雪 *xuě* 'snow' and 雹 *báo* 'hail' refer to both. Based on ancient scholars' annotations and semantic facets of the words in actual use, Ren [18] argues that 雾 *wù*, 霜 *shuāng* and 露 *lù* are [+material], and have almost no verbal usage; 电/電 *diàn* and 雷 *léi* are [+process], and can be used as nouns and verbs; 雨 *yǔ*, 雪 *xuě* and 雹 *báo* are [+material, +process], and can also function as nouns and verbs. It shows that the ontology nodes and semantic features have one-to-one correspondence on this issue and can thus have mutual corroboration. Note that although the thunder group seems to be exceptional to have both N/V categories, their nominal usages are different from the rain group. In Mandarin Chinese, the rain group has deep unaccusative behaviours in appearing as subjects of intransitive constructions. However, the thunder group cannot typically appear in subject position. This indicates the strong referentiality and 'nouniness' of the rain group; while it is likely that the nominal usage of the thunder group is in fact nominalisation of the weather process. This difference can also be underlined by the wider range of modifiers the rain group can take, vs. the restricted group of manner-related modifiers allowable for the thunder group.

We can make one important observation about these three groups of linguistic representation with regard to Eriksen et al.'s [1] typology of weather and language. They propose that weather events are encoded in three types of linguistic representations: Predicate type, Argument type, and Argument-Predicate type. They also observe that precipitation events tend to be encoded by only one type in any language. We found that the three groups of weather words with radical 雨 *yǔ* attest the first two types of representation. However, true precipitation events are represented by either Predicate or Argument types, contrary to their prediction. This should be accounted for with closer examination of types of weather events in the future. In addition, recent study [19] shows that fog, dew and frost expressions exhibit intriguing typological behaviours across languages in terms of encoding type and directionality, thus also deserve further investigation.

According to Huang's [5, 20] research, a concept which can be defined independent of time is endurant, and a concept which must be defined dependent of time is perdurant. Furthermore, Huang claims that [+N] feature stands for endurant properties, and

[+V] feature represents perdurant properties. Since a process is dependent of time, while material is not, we can now connect the endurant/perdurant dichotomy with semantic features proposed by Ren [18], together with the ontology node linking and the directions of those weather phenomena, as illustrated in Table 4.

Table 4. Connection of related aspects concerning weather words in Archaic Chinese

	雾露霜	雨雪雹	雷电
Direction	Down	Down	None
Node	Water	Both	Weather process
PoS	N.	N./V.	N./V.
Feature	+Material	Both	+Process
Temporal dichotomy	Endurant	Both	Perdurant

4.3   A Momentum Account

We can know from the mechanisms of fog, dew and frost, as introduced before, that they involve a formation process, and some types of fog even involve movement. Then why can't they be linked to 'weather process' node, or be used as verbs, or have [+process] feature in Archaic Chinese? The answer may lie in physics.

In physics, momentum is a quantity expressing the motion of a body, which equals to the product of its mass and velocity, and is defined as below [21].

$$p = mv$$

We propose that the momentum of the weather phenomena is positively correlated with its activity. Large momentum brings about high activity, and high activity leads to more 'verby' properties. Based on the research of Houghton [22], Hughes and Brimblecombe [23], Matzner [24], Libbrecht [25] and Ahrens [3], fog droplets are around 0.01–0.02 mm in diameter; dew droplets at initial stage are about 0.035 mm in diameter, and their mean diameter at sunrise is about 0.2 mm; rain drops are equal to, or greater than 0.5 mm in diameter; most of the snow crystals are much larger than rain drops, e.g., the famous multi-branched stellar dendrite crystal could be 3 mm from tip to tip; hailstones are greater than 5 mm in diameter. We have not found data on frost size, but since it comprises ice crystals as snow and is usually in clusters, there are reasonable grounds to presume that frost crystals are not smaller than rain drops. Thus, the mass of the six phenomena can be roughly divided into two groups, fog and dew being 'smaller', and frost, rain, snow and hail being 'bigger'. Now we count in velocity. The velocities of fog, dew and frost are very small, even close to zero; therefore, their momentum is much smaller than the falling rain, snow and hail. It can thus be seen that the phenomena with smaller momentum have lower activity, lacking the capacity for being verbs in Archaic Chinese. In addition, momentum is also linguistically represented in other respects.

We just mentioned that frost has big mass but small velocity, which makes its momentum smaller than rain, snow and hail. On the other hand, this also makes its momentum greater than fog and dew. Such physical difference is reflected in two aspects. First, as shown in Table 2, for expressions describing frost, there is no Sinitic language with the 'upward' direction only, while there are 24.7% for fog and 15.2% for dew. In this regard, frost resembles rain, snow and hail. Second, 霜 *shuāng* can compound with verbs with high activity in 55.6% of the Sinitic languages, e.g., 打 *dǎ* 'to hit'/扯 *chě* 'to pull' /拉 *lā* 'to pull'. The percentage is 31.8% for 露 *lù* and 5.6% for 雾 *wù*. The results are in direct proportion to their momentum. In addition, more evidence can be obtained from our data of Sinitic languages that verbs with high activity also tend to take other high momentum weather phenomena as objects. For example, languages may use 打雨 *dǎyǔ* hit-rain to denote raining a heavy rain (e.g. Xingyi Chinese), use 打风 *dǎfēng* hit-wind to denote blowing strong wind (e.g. Guangzhou Cantonese), or use 打闪 *dǎshǎn* hit-lightning to mean lightning flashing (e.g. Harbin Chinese). Please see Dong [19] for more discussions on the relation between momentum and activity.

5   Conclusions

In this paper, we examined the correlation between eight weather phenomena and their linguistic representations in Mandarin and other Sinitic languages, aiming to account for two observations: that they share the radical 雨 in Chinese orthography, and that the Chinese language somehow contradicts science and allows fog, dew and frost to 'fall'. We argue that weather events consist of weather process and weather object; hence the radical 雨 is shown by Hantology to represent two corresponding ontology nodes. Such conceptual structure gives rise to the distribution of directional meanings, and also the differences in semantic features and PoS in Chinese. Their momentum is argued to be the physical basis for the differences in PoS, semantic features and node linking.

References

1. Eriksen, P., Kittilä, S., Kolehmainen, L.: Weather and language. Lang. Linguist. Compass **6**(6), 383–402 (2012). https://doi.org/10.1002/lnc3.341
2. Van Hoey, T.: Does the thunder roll? Mandarin Chinese meteorological expressions and their iconicity. Cogn. Semant. **4**(2), 230–259 (2018). https://doi.org/10.1163/23526416-00402003
3. Ahrens, D.: Essentials of Meteorology: An Invitation to the Atmosphere, 6th edn. Brooks/Cole, Cengage Learning, Belmont (2012)
4. Gruber, T.: Toward principles for the design of ontologies used for knowledge sharing. Int. J. Hum Comput Stud. **43**(5–6), 907–928 (1995). https://doi.org/10.1006/ijhc.1995.1081
5. Huang, C.-R.: Notes on Chinese grammar and ontology: the endurant/perdurant dichotomy and Mandarin D-M compounds. Lingua Sinica **1**(1), 1 (2015). https://doi.org/10.1186/s40655-015-0004-6
6. Chou, Y.-M., Huang, C.-R.: Hantology: conceptual system discovery based on orthographic convention. In: Huang, C.-R., et al. (eds.) Ontology and the Lexicon: A Natural Language Processing Perspective, pp. 122–143. Cambridge University Press, New York (2010). https://doi.org/10.1017/cbo9780511676536.009

7. Huang, C.-R., Yang, Y.-J., Chen, S.-Y.: Radicals as ontologies: concept derivation and knowledge representation of four-hoofed mammals as semantic symbols. In: Cao, G., et al. (eds.) Breaking Down the Barriers: Interdisciplinary Studies in Chinese Linguistics and Beyond, pp. 1117–1133. Institute of Linguistics, Academia Sinica, Taipei (2013)
8. Yang, Y., Huang, C.-R., Dong, S., Chen, S.: Semantic transparency of radicals in Chinese characters: an ontological perspective. In: Proceedings of the 32nd Pacific Asia Conference on Language, Information and Computation, pp. 788–797 (2018)
9. Xun, E., Rao, G., Xiao X., Zang, J.: The construction of the BCC corpus in the age of big data (in Chinese). Corpus Linguist. **3**(1), 93–109 (2016). http://bcc.blcu.edu.cn/
10. Zhan, W., Guo, R., Chen, Y.: The CCL corpus of Chinese texts: 700 million Chinese characters, the 11th century B.C. - present (2003). http://ccl.pku.edu.cn:8080/ccl_corpus
11. Li, R. (ed.): Great Dictionary of Modern Chinese Dialects (42 Volumes) (in Chinese). Jiangsu Education Publishing House, Nanjing (1993–2003)
12. Xu, B., Miyata, I. (eds.): A Comprehensive Dictionary of Chinese Dialects. Zhonghua Book Company, Beijing (1999). (in Chinese)
13. Tao, G. (ed.): Dictionary of Nantong Dialect. Jiangsu People's Publishing Ltd, Nanjing (2007). (in Chinese)
14. Zhang, W., Mo, C.: Dictionary of Lanzhou Dialect. China Social Sciences Press, Beijing (2009). (in Chinese)
15. Dong, S.: Directions of fog (in Chinese). Yuyan Wenzi Zhoubao, 2, 1 August 2018. https://doi.org/10.13140/rg.2.2.15796.81289
16. Niles, I., Pease, A.: Mapping WordNet to the SUMO ontology. In: Proceedings of the IEEE International Knowledge Engineering Conference, pp. 23–26 (2003)
17. Huang, C.-R., Hong, J.-F., Chen, S.-Y., Chou, Y.-M.: Exploring event structures in Hanzi radicals: an ontology-based approach. Contemp. Linguist. **15**(3), 294–311 (2013). (in Chinese)
18. Ren, H.: Noun-verb conversion in old Chinese: from the perspective of lexical semantic analysis (in Chinese). Doctoral dissertation, Peking University (2018). https://doi.org/10.13140/rg.2.2.16642.22724
19. Dong, S.: A study on meteorological words in Chinese (in Chinese). Postdoctoral report, Peking University (2019). https://doi.org/10.13140/rg.2.2.22648.29445
20. Huang, C.-R.: Endurant vs perdurant: ontological motivation for language variations. In: Park, J.C., Chung, J.-W. (eds.) Proceedings of the 30th Pacific Asia Conference on Language, Information and Computation, pp. 15–25. Hankookmunhwasa, Seoul (2016)
21. Walker, J., Halliday, D., Resnick, R.: Fundamentals of Physics, 10th edn. Wiley, Hoboken (2014)
22. Houghton, H.: The size and size distribution of fog particles. Physics **2**(6), 467–475 (1932). https://doi.org/10.1063/1.1745072
23. Hughes, R., Brimblecombe, P.: Dew and guttation: formation and environmental significance. Agric. For. Meteorol. **67**(3–4), 173–190 (1994). https://doi.org/10.1016/0168-1923(94)90002-7
24. Matzner, R. (ed.): Dictionary of Geophysics, Astrophysics, and Astronomy. CRC Press, Boca Raton (2001)
25. Libbrecht, K.: The physics of snow crystals. Rep. Prog. Phys. **68**(4), 855–895 (2005). https://doi.org/10.1088/0034-4885/68/4/r03

Construction of Adverbial-Verb Collocation Database Based on Large-Scale Corpus

Dan Xing, Endong Xun[(⊠)], Chengwen Wang, Gaoqi Rao,
and Luyao Ma

School of Information Science, BeiJing Language and Culture Univercity,
Xueyuan Road 15, Beijing 100083, China
xingdan1@126.com, edxun@126.com

Abstract. This paper constructs a high-quality adverbial-verb collocation database based on a large-scale corpus. First, we established a knowledge system of adverbial-verb collocations based on previous studies and linguistic rules. Then, we designed and implemented a knowledge acquisition model of adverbial-verb collocation based on a large-scale corpus. Finally, we evaluated and analyzed the extracted results. The main purposes are to obtain high-quality adverbial-verb collocations by formal means and to provide data support for natural language processing and theoretical and applied linguistic research.

Keywords: Large-scale corpus · Knowledge extraction · Adverbial-verb collocation

1 Introduction

Because of the lack of morphology in Chinese, phrases are combinations of multiple words with complex internal hierarchical relations, which makes it difficult to use word information effectively. However, collocation is a combination of two words, which lies between words and phrases, and builds a bridge between words and phrases. According to the syntactic structure between the components of collocation, collocation can also be divided into types such as subject-predicate collocation, verb-object collocation, modifier-head collocation, adverbial-verb collocation and verb-complement collocation [1]. In the chain of language composition, word collocation is one of the important links [2], especially in dependency grammar analysis, since collocation describes the combination of words and does not involve phrase structure, it has become the most basic knowledge source of many automatic dependency grammar analysis systems [3]. The construction of the collocation base can improve the ability of the computer to process language, and also provide collocation cases for language ontology, teaching and application research. In this paper, we divide the resource construction of the adverbial-verb database into three parts: First, we construct a knowledge system of collocation based on previous studies; Second, we design and implement knowledge acquisition from a large-scale corpus; Finally, we analyzed the extracted collocation knowledge statistically.

© Springer Nature Switzerland AG 2020
J.-F. Hong et al. (Eds.): CLSW 2019, LNAI 11831, pp. 585–595, 2020.
https://doi.org/10.1007/978-3-030-38189-9_60

2 Related Work

The word collocation was first introduced by Firth, who emphasized the concept of word pairing and pointed out that vocabulary in natural language is not used alone or in isolation, but in combination with other words [4]. Firth did not give a clear definition, however, Halliday defined collocation in the lexical framework [5]. Based on this definition, many scholars have inherited and developed the definition of collocation [6–9]. Benson et al. defined collocation as "an arbitrary and recurrent word combination", and classified collocations into a lexical group and a grammatical group [9]. Xu et al. gave a more specific definition: "a collocation is a recurrent and conventional expression containing two or more contents that hold syntactic and/or semantic relations [10]". This definition emphasizes the syntactic and semantic relations between words but excludes the function words [11]. The earliest study of Collocation in China mainly focuses on whether the essence of collocation is syntactic or semantic [12–16]. Besides, there are several practical achievements, such as Common Words Collocation, Concise Chinese Collocation Dictionary and Modern Chinese Substantive Collocation Dictionary [17–19]. However, researches at this stage have no consensus because of the lack of quantitative analysis [20].

The earliest collocation extraction was Choueka et al., who used the corpus of about 11 million words in the New York Times Weekly to extract word collocation by calculating the co-occurrence frequency of repeated adjacent word strings [21]. Church et al. used a news corpus of about 44 million words to calculate the mutual information between the words to get the ideal word collocation [22]. Smadja used the Xtract system to extract collocations from a stock market news corpus of about 10 million words and the results showed that the accuracy of collocation extraction can reach 80% [23]. The earliest research on collocation knowledge extraction in China was Sun et al., who referred to relevant foreign knowledge extraction and made a quantitative analysis of collocation in Chinese, introducing the definition and nature of collocation [24]. After that, Sun summarized 14 simple rules by using the method of statistics and rules to judge under what circumstances the "V + N" sequence in Chinese labeled corpus is a legitimate collocation combination [25]. However, these rules do not use any dictionary information; Qu et al. proposed a framework-based automatic word collocation extraction method [26]; Huang et al. extended Sketch Engine to Chinese and found POS-based rules were efficient in extracting grammatical information [27]. Sun et al. used the Maximum Entropy Markov Model to conduct basic block recognition experiments on CPTB and MSRA of 470,000 words scale [28]; Zhou developed a basic automatic analyzer for Chinese real text based on rule-based method [29]; According to Lin, rules-based and statistics-based methods have their strengths and weaknesses and can complement each other, but both rely on reliable linguistic knowledge. Previous studies on collocation knowledge extraction based on statistics, linguistic rules and the fusion of various statistical methods are usually based on a single lexical association method, which is based on the co-occurrence number of candidate collocations and constituent words in the corpus, often omitting the high-quality collocation with some co-occurrences, and the timeliness of data information leads to the extraction of "outdated" collocation combinations [3]. This paper is based on BCC corpus which is relatively new, and the extracted data has relative timeliness compared with previous studies. Meanwhile, this paper carries out the knowledge extraction project of adverbial-verb collocation based on the features of part of speech, word length, pause, rhythm and language rules, which is innovative in methods and provides a large-scale real data for further research based on statistics and machine learning and the examples of collocation for language ontology, teaching and applied research.

3 Construction of the Knowledge System of Adverbial-Verb Collocations

There have been some previous studies on adverbial-verb structures. For example, Zhu classified adverbials into two categories: adverbial modifier and adjectival modifier. Adverbial modifier includes adverbs that are transformed by the verb and nouns with adverbial suffixes 的 (de, auxiliary); adjectival adverbials include state adjectives, some compound words with state adjective suffixes can be used as adverbials. Zhu also thinks some substantives have the nature of the predicate so they are also modified by adverbials, such as numerals, quantifiers [30]. Xing thinks it usually defines adverbial as the modifier of verbs and adjectives. The adverbial verb can also be classified by noun or noun phrase. From the semantic point of view to classify the adverbial, it mainly includes the state adverbial, potential adverbial, degree adverbial, negative adverbial, condition adverbial, and object adverbial [31]. This paper establishes an adverbial-verb collocation system based on previous studies. The extracted adverbial collocations only refer to the simple non-recursive situations in which modifiers modify predicate headwords, such as 紧紧地抓住 (jin3jin3 de zhua1zhu4, hold tightly), 不断地提高 (bu2duan4 de ti2gao1, improve constantly), 很喜欢 (hen3xi3huan1, like very much) and so on. Because of the particularity of the data extraction by computer, we do not consider the complex adverbial-verb structure for the time being, and the nouns which modify the predicate components such as 群众的支持 (qun2zhong4 de zhi1-chi2, mass support) and modifier which modifies the nominals, such as 很淑女 (hen3shu1nv3, very lady) and 才周二 (cai2zhou1er4, only Tuesday).

The adverbial-verb collocations in this paper are mainly centered on the predicative headword. It can be divided into three types in the form: adverbial + verb, adverbial + adjective, adverbial + predicate pronoun. The adverbial-verb structure of complex adverbials and complex headwords is excluded for the time being. As shown in the Table 1 below:

Table 1. Classification table of adverbial-verb collocations

Adverbial collocation classification	Adverbial classification
Adverbial + Verb	Adverbial modifiers
	Adjective modifiers
	Verb modifiers
	Nominal modifiers
	Adverbial modifiers of predicative pronouns
	Quantitative phrase modifiers
	Modifiers of morpheme phrases
	Adverbial modifiers of preposition structures
Adverbial + Adjective	Adverbial modifier of adverbs
	Adverbial modifiers of adjectives
	Adverbial modifiers of predicative pronouns
	Quantitative phrase modifiers
Adverbial + Predicative Pronoun	Adverbial modifier

4   Acquisition of Adverbial-Verb Collocations

4.1   The Composition of Retrieval Formula

The acquisition of adverbial-verb collocation is based on the corpus of Beijing Language and Culture University, which contains 15 billion words [32]. To extract collocation from the large corpus, this paper used a series of retrieval methods. We added conditional or output statements based on basic retrieval. Statements are separated by ";", such as Query {cond1; cond2;...; condi; print ($i)}. "Query" denotes the basic search formula; the conditional statement in "{ }" restricts the query content; the output statement restricts the output content, and only one output statement can be used in a high-level search formula. The limited part of the search formula needs to be enclosed in "()", and according to the order in which "()" appears, it can be obtained by using "$" and serial number. The first "()" is represented by "$1", for example, (a) (n) {len ($1) > 1; $2 = [S_N_名词]; len ($2) > 1} denotes the definite collocation that the attributive adjectives precede the noun. Nouns and adjectives can be acquired by "$1" and "$2" respectively in the order in which "()" appears. The qualifications in "{ }" indicate that the length of adjectives and nouns is greater than 1 and that nouns are limited to the list of "S_N_名词" (*ming2ci2, the noun list*), with default output of "$1" and "$2". The accuracy of extraction results can be improved by restricting part of speech, length, the word list, and pause. Table 2 shows the partial extraction results based on large data.

4.2   The Construction of Adverbial-Verb Retrieval Formula

We constructed the retrieval formula of the collocation by four means. First, we make it through the existence of 地 (*de, auxiliary*), 着 (*zhe, auxiliary*) and other forms of markers; Second, we establish the word collocation table and the exclusive table; Third, we use rhythm structure for length restriction; Fourth, through pause, punctuation W restrictions and other forms of means to construct retrieval.

Table 2. The partial extraction results based on large data

Retrieval constitution	Results	Frequency
(d) 地 (v) {len ($1) = 2; len ($2) = 1}	默默地看 (*mo4mo4 de kan4, look silently*)	132
	大声地说 (*da4sheng1 de shuo1, speak loudly*)	111
	远远地看 (*yuan3yuan3 de kan4, look distantly*)	99
	慢慢地走 (*man4man4 de zou3, walk slowly*)	86
(a) 的 (n) {len ($1) > 1; $2 = [S_N_名词]; len ($2) > 1}	重要的作用 (*zhong4yao4 de zuo4yong4, important role*)	3952
	重要的意义 (*zhong4yao4 de yi4yi4, important significance*)	3101
	好玩的活动 (*hao3wan2 de huo2dong4, fun activities*)	2347
	最好的朋友 (*zui4hao3 de peng2you3, best friend*)	1337
	美好的回忆 (*mei3hao3 de hui2yi4, good memories*)	1094

A case study of adverbial-verb collocation retrieval in the form of verbs modified by the adverbial modifier: First, according to the existence and absence of "地" (*de, auxiliary*), the retrieval forms of adverbial modifiers modifying verbs are divided into two categories. For example, the retrieval formula *(d)地(v){print($1 $2)}* refers to querying the corpus of adverbs followed by "地" (*de, auxiliary*) modifying verbs from large data and outputting the adverbs and verbs; The retrieval formula *(d) (v) {print ($1 $2)}* refers to querying the collocation of adverbs followed by verbs from large data and outputting adverbs and verbs. Second, we construct the word collocation table or the exclusive table to construct retrieval formulas to make the retrieval results more accurate and comprehensive. For example, the retrieval formula *(~) 地(v){$1=[S_D_重叠_DD2];print($1 地 $2)}*. Table *S_D_重叠_DD2* is compiled and supplemented according to the Peking University Modern Chinese Grammar Information Dictionary, which indicates the double-syllable adverb overlap such as 常常 (*chang2chang2, frequently*), 渐渐 (*jian4jian4, gradually*) and so on [33].

Then, we construct the retrieval formula by restricting the length of the prosodic structure. Lv mentioned that the three-syllable structures are divided into two categories except for a few cases: adverbial-verb structure and verb-object structure [34]. For adverbial-verb combinations, 2 + 1 is much more than 1 + 2 in three-syllable combinations [35]. Wang believed that adverbial-verb combinations, 2 + 1 is the majority, such as "慢慢走" (*man4man4zou3, walk slowly*) and so on. The length of the phonetic affects the combination of words. According to this rule, a series of retrieval formulas are summarized, such as *(d) (v) {len ($1) = 2; len ($2) = 1; print ($1 $2)}* indicates the collocation of two syllables adverbs and one-syllable verbs.

Table 3. The partial extraction and evaluation results based on large data

Modifier	Head	The retrieval of adverb-modified collocations	Evaluation of results
Adverbial modifiers of adverbs	Verb	(d)地(v){len($1) > 1; print($1 地 $2)}	5
		(~ ~)地v{$1 = [DD_ABAB]}	4
		(~)地(v){$1 = [S_D_重叠_DD2]; print ($1 地 $2)}	4
		(~)地(v){$1 = [S_D_重叠_AABB3]}	4
		(~)(v){$1 = [S_D_情态词_情]; print($1 $2)}	4
		(~)(v){$1 = [S_D_重叠_AABB3]; print ($1 $2)}	4
		(~)(v){$1 = [S_D_重叠_DD2]; len ($2) = 2; $print($1 $2)}	4
		(~)(v){$1 = [S_D_重叠_DD2]; print($1 $2)}	4
		(~)(v){$1 = [S_N_兼类_d]; print($1 $2)}	3
		(d)(v){len($1) = 2; len($2) = 2; $1! = $2; print($1 $2)}	4
		(d)(v)W{len($1) = 3; $2! = [是 有]; print ($1 $2)}	4
		(d)(v){len($1) = 3; $2! = [是 有]; print ($1 $2)}	4
		(d)(p)W{$2 = [P_p兼v]; print($1 $2)}	5
		(d)(v){len($1) = 2; len($2) = 1; print($1 $2)}	4

Finally, we construct the retrieval formula by a pause. For example, the retrieval formula *(d) (v) W {len ($1) = 3; $2!= [是 有]; print ($1 $2)}* refers to adverbs with a length of 3, modifying the verbs except "是" (*shi4, be*) and "有" (*you3, have*), and the punctuation should be followed by verbs closely, such as "不由得哀号" (*bu4you2 de ai1hao2, couldn't help wailing*), "好容易安置" (*hao3rong2yi4an1zhi4, It's hard to settle down*), "背地里安抚" (*bei4di4li3an1fu3, appease in the back*).

4.3 Evaluation of Adverbial-Verb Retrieval Formula

As shown in Table 3, the number refers to the accuracy rate, for example, 5 refers to 90%, 4 refers to over 70%, 3 refers to over 50%, 2 refers to 30%, and 1 refers to 10%. The results of this evaluation are combined forces of machine and human. Machine verification mainly excludes low-frequency invalid retrieval formulas. Human verification mainly includes two aspects: one is to observe the retrieval results, delete and improve the retrieval formulas through formal markers, such as the length of the word, and the other is to screen the retrieval results manually.

In constructing retrieval formula and extracting adverbial-verb collocations, we find that linguistic rules summarized by linguists can help us get higher quality adverbial-verb collocations. Take the retrieval of adverbial-head collocations "v + v" as an example, Sun thinks there is a great difference between the direct modification of verbs and the post-modification of verbs with auxiliary words, and the direct modification is limited [36]. Verbs are often modified by adverbials with auxiliary words such as "地" (*de, auxiliary*) and "着" (*zhe, auxiliary*), which highlight the modifiability of adverbials. According to this rule, we construct the retrieval formula (v) {$1 = [P_feixinliv]; print ($1 $2)}, which improves the accuracy by 40%. Shen believes that verb modified as adverbials is very obvious, monosyllabic verbs as adverbials should be marked with "着" (*zhe, auxiliary*), disyllabic verbs and overlapping verbs should be marked with "着" (*zhe, auxiliary*) or "地" (*de, auxiliary*) [37]; Zhang believes that the number of verbs modified as adverbials is much more than the number of verbs modified adverbials directly [38]. Therefore, according to the rules summarized by linguists, we classify and refine the retrievable formulas, and classified the retrieval formulas of the verb modified by verb into three parts: V着V, V地V, and V-V. As shown in Table 4 below:

Table 4. The retrieval formulas of Adverbial-verb collocation and their evaluation

The retrieval formulas of verb modified verb	Evaluation	Notes
(~)地(v){$1=[S_V_单作状语_地]}	5	The table *S_V_单作状语_地* is summarized according to the grammar information dictionary [34].
(v)地(v){$1=[P_vdv_v1];$1!=$2;print($1 $2)}	4	The table *P_vdv_v1* is summarized according to Sun and Zhang [37][39].
(v)地(v){len($1)=2;len($2)=1}	4	
(v)地(v){len($1)=2;len($2)=2}	4	
(v)地 v{len($1)=2;print($1 地 v)}	4	
(~)(v){$1=[S_V_单作状语_可];print($1 $2)}	3	The table *S_V_单作状语_可* is summarized according to the dictionary of grammatical information [34].
(~)(v){$1=[S_V_单作状语_可];$2!=[P_notv1v2x];$1!=$2;print($1 $2)}	4	The table *P_notv1v2x* is summarized according to retrieval results from big data.
(~)(v){$1=[P_V_双音节_状];len($2)=2;$1!=$2;$2!=[P_notv2v2];print($1 $2)}	4	The table *P_V_双音节_状* is summarized according to the dictionary of grammatical information [34].
(~)(v){$1=[P_V_单音节_状];len($2)=1;$1!=$2;$2!=[P_notv1v1];print($1 $2)}	3	The table *P_notv2v2* and the table *P_notv1v1* are summarized according to Sun [37].
(~)(v){$1=[P_V_双音节_状];len($2)=1;$2!=[P_notv2v2];print($1 $2)}	4	

We can also use the retrieval results from large data to see whether the existing language rules are generally applicable. By observing the adverbial-verb retrievals of verbs modified by verbs, we can find that the retrievals restricted by part of speech or exclusive table and the retrievals with markers such as "地" (*de, auxiliary*) and "着" (*zhe, auxiliary*) are better than those without markers. In V-V retrievals, the retrieval effect of the query with first verb being two syllables is better than that with first verb being one syllable, which shows that there are a large number of double-syllable verbs used as adverbial modifiers to modify the verb head in V-V type and the regularity is obvious.

5 Statistical Analysis of Retrieval Results

Observing the statistical analysis Table 5 as followed, we find that the proportion of adverbial modifiers modifying verbs is the highest, regardless of the proportion of the number of headwords or the collocation items, the proportion of "adverbial modifiers + verb" collocations in the total number of collocation items is as high as 44%, followed by the prepositional structure as the adverbial modifier which reflects the grammatical function of adverbs as adverbials.

Table 5. The retrieval results of Adberbial-verb collocations

Classification of Adverbial collocations	Adverbial classification	Numbers of the headwords	The proportion of the headwords	Numbers of the items	The proportion of the item	Examples
Adverbial + Verb	Adverbial modifiers	18922	0.1561	3851121	0.444028	偷偷地哭 (*tou5tou5 de ku1, cry secretly*)
	Adjective modifiers	17032	0.1405	1213775	0.139946	早知道 (*zao3zhi1dao4, know early*)
	Verb modifiers	17982	0.1484	811103	0.093519	同情地看 (*tou2qing2 de kan4, look at it with sympathy*)
	Nominal modifiers	14195	0.1171	233582	0.026932	现场表演 (*xian4chang3biao3yan3, performing live*)
	Adverbial modifiers of predicative	15592	0.1287	140417	0.01619	怎样做 (*zen3yang4zuo4, how to do*)
	Quantifier modifiers	5320	0.0439	10648	0.001228	一天天地过 (*yi4tian1tian1 de guo4, go through life day by day*)
	Morpheme phrases modifiers	132	0.0011	132	0.000015	小口地吃 (*xiao3kou3 de chi1, eat with a small mouth*)
	Preposition structures modifiers	16465	0.1359	1385811	0.159782	把苍蝇赶跑 (*ba3cang1ying1gan3pao3, drive away the flies*)
Adverbial + Adjective Modifier	Adverbial modifier	4430	0.0366	823888	0.094993	尽情地放松 (*jin4qing2 de fang4song1, relax as much as you like*)
	Adjectives modifiers	4201	0.0347	165403	0.019071	惊人地相似 (*jing1ren2 de xiang1si4, similar astonishingly*)
	Predicative pronouns modifiers	4747	0.0392	26062	0.003005	那么美 (*na4me mei3, so beautiful*)
	Quantitative phrase modifiers	2155	0.0178	6440	0.000743	一抹红 (*yi4mo3hong2, hints of red*)
Adverbial + Predicative Pronoun	Adverbial Modifier	9	0.0001	4766	0.00055	究竟怎样 (*jiu1jing4zen3yang4, how exactly*)
Total number		121182	1	8673148	1	

6   Conclusion and Prospects

This paper extracted 259 5108 adverbial-verb collocations based on BCC corpus, which mainly improves the retrieval method by observing the retrieval results and combines machine verification and manual screening to acquire knowledge.

Based on the construction of a large-scale corpus, this paper mainly focuses on the knowledge system establishment, extraction, evaluation and data statistics of the adverbial-verb collocation knowledge system. We find that certain linguistic rules can guide knowledge acquisition, but the lack of linguistic formal markers, the ubiquitous grammatical and pragmatic ambiguities of words and sentences, and the habitual, dynamic and multi-domain characteristics of language increase difficulties in knowledge acquisition. The results of linguistic research, such as the part of speech, word length, and prosodic information, can help us extract simple adverbial collocations more accurately. Collocation resources can help us find new ideas, supplement or discover new rules. However, collocation knowledge acquisition based on a large-scale corpus also has drawbacks. First, some retrieved results might have errors, mostly in the low-frequency part, because of the imperfect classification of parts of speech, the part of speech tagging system and inaccurate segmentation. Second, the existing linguistic rules are not thoroughly studied, or the existing linguistic laws do not meet the needs of natural language processing. When retrieving large data, there are many collocation noises, which cannot guarantee the accuracy of acquired collocations. Finally, the lack of context, the limitations of collocation identification, overlap between different collocations could also influence the accuracy.

For the existing problems, the future work prospects are as follows: First, given the imperfect part of the part-of-speech tagging system, we are trying to screen and tag the verb lists of BCC corpus manually. Second, because of the incomplete laws of linguistic research, we can establish collocation tables or exclusion tables according to the results and then retrieve them again.

Acknowledgments. This work is supported by the Beijing Advanced Innovation Center for Language Resources (TYR17001), Graduate Innovation Fund (19YCX119) and National Natural Science Foundation of China (16AYY007).

References

1. Li, Y.D.: Modern Chinese Word Collocation. Commercial Press, Beijing (1998). (in Chinese)
2. Sun, H.L.: Inductive grammar rules from the annotated corpus: experimental analysis of the "V + N" sequence. Papers Collection of the Fourth National Joint Conference on Computational Linguistics. Tsinghua University Press, pp. 157–170 (1997). (in Chinese)
3. Lin, J.F.: Word collocation extraction and its application in information retrieval. Doctor, Harbin Industrial University (2010). (in Chinese)
4. Firth, J.R.: Modes of Meaning in Papers in Linguistics. Oxford University Press, Oxford (1957)
5. Halliday M.A.K.: Lexis as a linguistic level. In: Memory of JR Firth, pp. 148–162 (1966)
6. Sinclair, J.: Corpus, Concordance, Collocation. Oxford University Press, Oxford (1991)

7. Singleton, D.: Language and the Lexicon: An Introduction. Routledge, London (2016)
8. Kjellmer, G.: Some thoughts on collocational distinctiveness in recent developments in the use of computer corpora in English language research. Costerus. **45**, 163–171 (1984)
9. Ilson, M., Benson, E., Benson, R.: The BBI Combinatory Dictionary of English: A Guide to Word Combinations. John Benjamins, Amsterdam (1986)
10. Xu, R.F., Lu, Q., Wong, K.F., Li, W.J.: Building a Chinese collocation bank. Int. J. Comput. Process. Lang. **22**(01), 21–47 (2009). (in Chinese)
11. Hu, R.F., Chen, J.Y., Chen, K.H.: The construction of a chinese collocational knowledge resource and its application for second language acquisition. In: Proceedings of COLING 2016, the 26th International Conference on Computational Linguistics, Technical Papers (2016)
12. Xing, G.W.: Is collocation a grammatical problem? J. Anhui Norm. Univ. (Hum. Soc. Sci. Edn.) **4**, 77–84 (1978). (in Chinese)
13. Lin, X.G.: Research on Mr. Zhang Shoukang and the collocation of words. J. Capit. Norm. Univ. (Soc. Sci. Edn.) (1), 59–63 (1995). (in Chinese)
14. Lin, X.G.: The nature and research of word collocation. Chin. Lang. Learn. **1**, 7–13 (1990). (in Chinese)
15. Lin, X.G.: On semantic classification and word collocation. J. Renmin Univ. China **5**, 77–82 (1991). (in Chinese)
16. Zhu, Y.S.: The semantic basis of collocation and the practical significance of collocation research. Foreign Lang. 14–18 (1996). (in Chinese)
17. Zhang, S.K., Lin, X.G.: A Dictionary of Collocation of Common Words Used by Students. Hebei Children's Press (1989). (in Chinese)
18. Zhang, S.K., Lin, X.G.: Concise Chinese Collocation Dictionary. Fujian People's Press (1990). (in Chinese)
19. Zhang, S.K.: Modern Chinese vocabulary dictionary. Commercial Press, Beijing (1999)
20. Hu, Q.G., Gao, Q.Y.: Word collocation and teaching Chinese as a foreign language. Lang. Transl. (2017). (in Chinese)
21. Choueka, Y., et al.: Automatic retrieval of frequent idiomatic and collocational expressions in a large corpus. Literary Linguist. Comput. **4**, 34–38 (1983)
22. Church, K.W., Hanks, P.: Word association norms, mutual information, and lexicography. Comput. Linguist. **16**, 22–29 (1990)
23. Smadja, F.: Retrieving Collocations from Text: Xtract. Comput. Linguist. **19**(19), 143–177 (1993)
24. Sun, M.S., Huang, Ch.N., Fang, J.: A preliminary study on the quantitative analysis of Chinese. Chin. Lang. (1997). (in Chinese)
25. Sun, H.L.: Distribution characteristics of word collocation in text. In: Papers of the International Conference on Chinese Information Processing, pp. 230–236 (1998). (in Chinese)
26. Qu, W.G., et al.: Frame-based automatic word collocation extraction method. Comput. Eng. **30**(23), 22–24 (2004). (in Chinese)
27. Huang, Ch.R., et al.: Chinese sketch engine and the extraction of grammatical collocations. In: Proceedings of the Fourth SIGHAN Workshop on Chinese Language Processing, pp. 48–55 (2005). (in Chinese)
28. Sun, G.L., Huang, C.N.: Chinese chunking based on maximum entropy markov models. Comput. Linguist. Chin. Lang. Process. **11**(2), 115–136 (2006). (in Chinese)
29. Zhou, Q.: Automatic learning and extended evolution of chinese basic block rules. J. Tsinghua Univ. (Nat. Sci. Edn.) **48**(1), 88–91 (2008). (in Chinese)
30. Zhu, D.X.: Lectures on Grammar. Commercial Press, Beijing (2016). (in Chinese)

31. Xing, F.Y.: Three Hundred Questions on Chinese Grammar. Commercial Press, Beijing (2014). (in Chinese)
32. Xun, E.D., Rao, G.Q., Xiao, X.Y., Zang, J.J.: Development of BCC Corpus under the Background of Big Data. Corpus Linguistics, Beijing (2016). (in Chinese)
33. Yu, S.W., et al.: Detailed Explanation of Modern Chinese Grammar Information Dictionary. Tsinghua University Press, Beijing (1998). (in Chinese)
34. Lv, Sh.X.: A preliminary study on monosyllabic problems in modern Chinese. Chin. Lang. 10–22 (1963). (in Chinese)
35. Wang, L.J.: Rhythm form of Chinese. Beijing Language University Press (2015). (in Chinese)
36. Sun, D.J.: An investigation of verbs as adverbials in modern Chinese. Language Teaching and Research, 116–129 (1997) (in Chinese)
37. Shen, J.X.: Asymmetry and Tagging Theory, pp. 269–270. Jiangxi Education Press, NanChang (1999). (in Chinese)
38. Zhang, J.: The main forms and motivation analysis of modern Chinese verbs as adverbials. J. Central China Norm. Univ. (Hum. Soc. Sci. Edn.) 53(05), 100–108 (2014). (in Chinese)

On the Definition of Chinese Quadrasyllabic Idiomatic Expressions in Chinese-French Dictionaries: Problems and Corpus-Based Solution

Fang Huang[(⊠)]

Center for Lexicographical Studies, Guangdong University of Foreign Studies,
Baiyundadaobei 2, Guangzhou 510420, China
huangfang@gdufs.edu.cn

Abstract. To define Chinese quadrasyllabic idiomatic expressions in Chinese-French dictionaries is a demanding work for dictionary compilers. By comparing and analyzing the definition of quadrasyllabic idiomatic expressions in Chinese-French dictionaries, this study discussed 4 kinds of problems (wrong definition, inappropriate definition, omission of senses and absence of contextual information), and then gave some suggestions to improve the definition of these Chinese idioms by observing and analyzing their real use in large-scale corpus.

Keywords: Definition · Chinese quadrasyllabic idiomatic expressions · Chinese-French dictionaries · Problems · Corpus-based solution

1 Introduction

As the treasure of Chinese traditional culture, Chinese quadrasyllabic idiomatic expressions are considered to be significant expressions in dictionaries and "have gained increasing attention" [1] these years. It is also important to well describe in forms of definitions these idioms in bilingual dictionaries, because the accurate, appropriate and user-friendly definitions on such expressions help both Chinese and foreign language users know exactly what they mean and how to use them correctly.

But to define[1] these idioms in bilingual dictionaries is a demanding task for lexicographers due to various reasons, among which two are very critical.

Firstly, to fully define a quadrasyllabic idiomatic expressions is a complicated work. Most of these expressions have specific, conventionalized meanings, which are not predictable from the usual meanings of its constituents [1, 2]. They also have some

---

[1] The researcher of this paper uses the terms "define" and "definition" instead of "translate" and "translation", because for bilingual dictionaries, to well explain the meanings and the usages of an entry word, translational equivalents are not enough. Other methods of explanation used to define an entry word in monolingual dictionaries should also be used in the target language in bilingual dictionaries. The definition of Chinese quadrasyllabic idiomatic expressions refers to the explanation of all the possible meanings of each idioms as entry words.

© Springer Nature Switzerland AG 2020
J.-F. Hong et al. (Eds.): CLSW 2019, LNAI 11831, pp. 596–603, 2020.
https://doi.org/10.1007/978-3-030-38189-9_61

special features on grammatical and pragmatic rules. Therefore, the definition of the idioms concerns not only all the possible meanings, but also all the usages including the semantic, syntactic, stylistic rules and the context about the idioms. It is an impossible task to define all these meanings only by personal experience without resorting to others resources, such as corpus.

Secondly, to give definitions to an idiom in another language is not an easy job because compilers of bilingual dictionaries are not native speakers of both languages. The Chinese idioms could not be defined only by the intuition in their mother tongue. So great efforts should be put to render the "various aspects of meaning that an idiom or a fixed expression conveys into the target language" [3].

Related studies focused mainly on problems of the definition of Chinese idioms especially of quadrasyllabic idiomatic expressions in Chinese-Foreign dictionaries [4–7]. As Pinchuck [8] observed, most traditional bilingual dictionaries only offered a collection of equivalents chosen according to the arbitrary judgement of the compiler. Without careful or objective analysis, inappropriate or inaccurate equivalents were offered in dictionaries.

Based on their experience or intuition, researchers also proposed methods to improve the definition, like comparative studies between Chinese idioms and their counterparts in the target language, different translation strategies for different kinds of idioms [7, 9, 10]. While seldom are studies proposing to improve the definition of these idioms by analyzing their actual use in a large-scale corpus.

This research investigated Chinese quadrasyllabic idiomatic expressions in Chinese-French dictionaries, discussed problems on the definition of these idioms and made suggestions to improve their definition based on observations of the corpus data.

2 Methods

A comparative research was done between the definition of Chinese quadrasyllabic idiomatic expressions in two Chinese-French dictionaries and their actual use in large-scale corpus.

2.1 Research Object

The research object of this article was the definition of Chinese quadrasyllabic idiomatic expressions whose first character is 大 (Da).

2.2 Corpus and Dictionaries

Two large-scale corpus were chosen: Chinese web 2011 and French web 2012 from the corpus tool-Word Sketch Engine [11]. Two dictionaries were used in our investigation: Ricci Chinese-French dictionary (henceforth RCFD) published in 2014 by Commercial Press [12], and Chinese-French dictionary of idioms (henceforth CFDI) published in 2014 by Xiamen University Press [13].

3   Results

The results showed that some of the idioms presented in the two bilingual dictionaries under investigation were not well defined especially for their wrong, inappropriate, incomplete or low contextual information.

3.1   Basic Statistics

119 idioms were collected in total, among which 68 appeared in both dictionaries (see Table 1). To get a comparable data, the researcher of this paper only investigated the latter one.

Table 1. Idioms in CFDI and RCFD

Dictionary	CFDI	RCFD	Total	Idioms in both dictionaries
Number of idioms	98	89	119	68

3.2   Problems Found on the Definition of Idioms in Two Dictionaries

After an exhaustive and comparative research on the definition of the idioms in the two dictionaries and their real use in the corpus, the researcher of this paper found that 36 (the real number is 29 because some idioms had more than one problem, like 大摇大摆, 大腹便便, 大谬不然) out of 68 idioms (see Table 2) were not satisfactorily defined. The recurring problems were classified in the following four main types: wrong definition, inappropriate definition, omission of senses and absence of contextual information.

Table 2. Statistics on idioms with problems of definition

Problems	Total idioms	Percentage
Wrong definition	3	4.4%
Inappropriate definition	8	11.8%
Omission of senses	15	17.4%
Absence of contextual information	10	14.7%
Total	36	48.3%

3.2.1   Wrong Definition

4.4% (3 out of 68) of the idioms contained wrong information in their French definitions. For example, in CFDI, one of the equivalent expressions offered for 大摇大摆 "en avoir dans les mécaniques" was not found in French monolingual dictionaries. To be more confirmed, the researcher of this paper also tested this expression in French web 2012. 0 hit was found for this expression. That proved "en avoir dans les mécaniques" was a wrong definition and may not exist in French. This result was also confirmed by French native speakers.

Another example is one of the equivalents provided for 大彻大悟 by CFDI: Les écailles lui sont tombés des yeux. The French dictionary *Le Petit Robert* defined this French equivalent expression as "he suddenly recognizes his mistakes" [14]. But an analysis on the syntactic and semantic representation of 大彻大悟 in Chinese web 2011 showed that the meaning of the Chinese idiom was actually the following one: be totally awaken, be illuminated of something (especially one's value, meaning of life, methods, etc.) after others' instructions, or after one's own reflexions, or after prolonged trials and tribulations. Therefore, "Les écailles lui sont tombés des yeux" was a wrong definition for 大彻大悟. It would be strange to say: 他对自己的错大彻大悟".

3.2.2　Inappropriate Definition

11.8% of the idioms were not defined in an appropriate way especially in stylistic registers or in grammatical categories.

(1)　Mismatched Stylistic Register

Some equivalents did not have the same stylistic register as their counterpart idiom. Some idioms were frequently used, but their French equivalents in the two bilingual dictionaries were out of date, old or archaic. Some idioms were of formal or literary use, but their equivalents were used in an informal or vulgar situation. The mismatch of the registers between the Chinese idioms and their counterparts in French may lead to a misunderstanding and the wrong use of the Chinese idioms by dictionary users.

For example, in RCFD the idiom 大腹便便 had a French equivalent "entripaillé". But this word was not included in the two famous French dictionaries Le Petit Robert [14] and Le Petit Larousse [15]. In French Web 2012 the frequency of this word was quite low (only 13 hits, among which 3 were repetitive examples). The researcher of this paper made a detailed research for the examples, and found that the repetitive examples were fragments of Molière's works: L'impromptu de Versailles (in 17th century):

> …gros et gras comme quatre, un roi, morbleu ! qui soit entripaillé comme il faut, un roi d'une vaste circonférence, et qui puisse remplir un trône de la belle manière.
> from French Web 2012

A quite low frequency in French Web 2012 with examples from classical works could probably mean that "entripaillé" was out of date, old or archaic. But the Chinese idiom 大腹便便 had a very high frequency in Chinese Web 2011, which showed "entripaillé" was not an appropriate definition for 大腹便便 due to the mismatch of the time register.

Another example is 大谬不然. According to Chinese web 2011, when compared to its synonym 大错特错, 大谬不然 was less frequently used. 大错特错 had 1126 hits, while 大谬不然 had only 127 hits. What's more, 大谬不然 appeared mostly in a literary context, especially in Chinese classical context. But in the two dictionaries, among the equivalents for 大谬不然, three (see Table 3) (bourde, se mettre le doigt dans l'œil, se gourer jusqu'au trognon) were used in a very familiar and casual situation. The mismatch of the styles between the idiom and its equivalents could mislead the users to an inappropriate use of the idiom.

Table 3. Definition for 大谬不然

Dictionary	Definition
RCFD	erreur grossière; (fam.) bourde
CFDI	se mettre (se donner, se fourrer) le doigt dans l'œil (jusqu'au coude)/se gourer jusqu'au trognon Fam./se tromper complètement

(2) Mismatched Grammatical Category

4 out of 68 idioms were defined using equivalents with mismatched grammatical category. For example, in RDCF 大谬不然 "(sth or sb) be entirely wrong or mistaken" was translated as a noun: erreur grossière; bourde (big mistake) (see Table 3). While the distribution of this idiom in the Chinese web corpus showed it was used as a predicate. So the translation could be misleading because foreigners may wrongly use 大谬不然 like the following example: *这真是一个大谬不然.

3.2.3 Omission of Senses

17.4% (15 out of 68) of the idioms omitted one or more senses (see Table 2). For the idiom 大吃大喝, all the equivalents offered in CCFD and RCFD (like "faire bombance", "tuer le veau gras" etc.) concerned mainly one's behavior or habit, which meant "eat and drink a lot, even excessively; guzzle" (see Table 4).

Table 4. Definition for 大吃大喝

Dictionary	Definition
RCFD	Faire un repas plantureux; faire bombance; faire ripaille; manger et boire tout son content; manger et boire comme un glouton; ripailler à plaisir; tuer le veau gras; faire bombance; (fam.) la noce.
CFDI	manger et boire abondamment/manger et boire tout son content/ripailler/«Trente rudes gaillards s'escrimaient de la fourchette, buvaient comme des éponges» (J. Fréville)./Ce soir-là, ce commerçant de Floride décide de faire la bringue. (Le Point.fr, 30-11-2012)/Les dizaines d'invités mangent et boivent sans retenue. (Le Point.fr, 31-10-2013)/On ripaille, on trinque, on braille. (Le Point, 25-02-2014)

But the word sketch of 大吃大喝 in Chinese web 2011 (see Table 5) showed the idiom co-occured frequently with words like 公款 (public funds), 严禁 (to forbid), 刹住/制止/狠刹 (to stop), which brought a meaning of illegal behavior by using public funds to eat and drink. Some sentence examples which appeared frequently in the corpus are 严禁用公款大吃大喝 or 不准用公款大吃大喝. A detailed analysis on the concordances of the idiom revealed that 42% examples had the original meaning of guzzle, while 58% examples had the pejorative illegal meaning. Therefore, the latter sense was already very salient and distinctive in the corpus. Unfortunately, this meaning was not included in both Chinese-French dictionaries.

大腹便便 was the same case. Based on the analysis of the idiom in Chinese web corpus, it had also two meanings: (someone who is) fat because he/she has a big belly; women who is pregnant. But the two dictionaries only included the first sense.

Table 5. Salient sketch data for 大吃大喝 in Chinese web 2011

Grammar relation	Collocate	Freq	Score
Subject	647	41.08	
	公款	364	9
	餐	5	2.08
SentObject_of	644	40.89	
	刹住	7	6.88
	狠刹	3	6.25
	严禁	96	5.76
	挥霍	5	4.64
	制止	13	3.21

3.2.4 Context Missing

Context is very important for language learners to understand full usages of a word expression. As Baker and Kaplan [16] argues, in bilingual dictionaries, equivalence cannot be represented only by translation equivalents. "A better dictionary is one that provides information about context and usage". But in both dictionaries, 14.7% (10 out of 68) of the idioms were defined with low context.

For example, in both dictionaries, 大快人心 was explained as "People are all very satisfied and happy" (see Table 6). While Chinese Contemporary Standardized Dictionary (CCSD) defined 大快人心 as "People are all very satisfied and happy because bad people are punished for what they deserved" [17].

Table 6. Definition for 大快人心

Dictionary	Definition
RCFD	faire grand plaisir à tout le monde; égayer (réjouir) beaucoup les gens; contenter tout le monde; à la satisfaction générale; pour le plus grand bonheur de chacun
CFDI	à la grande satisfaction générale /égayer (réjouir) tout le monde/qui fait réjouir le public (toutes les parties concernées)

What's more, the data from Chinese web 2011 confirmed the definition of the idiom from the CCSD. In the corpus, 大快人心 always cooccurred with the following words which conveyed a strong negative stance: 被抓 (be arrested), 公安局 (police), 有罪 (be guilty), 罪犯 (criminals), 刑罚 (penalty), 惩罚 (punishment). The data from the corpus explained the reason why people were all very satisfied and happy was that bad man had been punished. While in the two bilingual dictionaries the context for the reason of the happy mood was missing.

4 Some Suggestions to Improve the Definition of an Idiom by Using Corpus

Based on previous analysis, some idioms presented in the two bilingual dictionaries under investigation were not properly defined because of the inaccurate definition, inappropriate equivalents, or decontextualized information offered to them. These definitions could not help second language learners use the idioms correctly and effectively in a special communication situation. While corpus, with its real, big and objective data, could help solve all these problems by presenting the real usage of the idioms through frequencies, word sketches and distributional information.

Lexicographers should base their analysis on corpus, know about the real use of an idiom in the Chinese corpus, its equivalent translations (idioms or expressions) in the French corpus and try to analyze problems of definition on idioms in bilingual dictionaries and suggest solutions based on corpus data.

To offer accurate, appropriate and highly contextual information for Chinese quadrasyllabic idiomatic expressions, the researcher of this paper gives the following suggestions for dictionary compilers.

Firstly, it is suggested that when compiling a bilingual dictionary on idioms, compilers analyze the real use of the idioms in a Chinese corpus, verify in a French corpus if their equivalents in the target language convey exactly the same meaning, have the same register, the same context or the same grammatical category as the idioms in the source language. If not, compilers could delete the incorrect and inappropriate information proved by corpus. For example, after a corpus analysis, compilers could delete the wrong equivalent "en avoir dans la mécanique" for the idiom 大摇大摆, delete the archaic equivalent "entripaillé" for the idiom 大腹便便, the vulgar and familiar equivalents "se mettre (se donner, se fourrer) le doigt dans l'œil (jusqu'au coude)", "se gourer jusqu'au trognon" for 大谬不然.

Secondly, compilers could give contextual information extracted from the corpus. They could add potential context (bad people are punished) for 大快人心 instead of only giving its literal meaning (people are all very happy).

Lastly, compilers could add the sense missing in the dictionary but well represented in the corpus. For example, a second meaning could be added for 大腹便便: (femme) enceinte/(woman) pregnant, and for 大吃大喝: <péj.> (les fonctionnaires) manger et boire beaucoup voire trop en profitant des fonds publiques/(public functionary) eat and drink extravagantly by using public funds.

5 Conclusion

A corpus-based investigation on Chinese quadrasyllabic idiomatic expressions showed that current Chinese-French dictionaries were not usage-based for their incomplete, insufficient or inappropriate definition.

Lexicographers, when making or revising a bilingual dictionary on Chinese quadrasyllabic idiomatic expressions, can analyze problems on their definition in dictionaries and provide correct, complete, appropriate and contextual definition by observing and extracting information from corpus data instead of doing it by intuition.

Funding Acknowledgments. This work was supported by Special Creative Project of Guangdong Universities (2017WTSCX026); project of Guangzhou social science (2019GZGJ65) and French subproject of CNCTST major project (ZD2019001).

References

1. Hsieh, S.-K., Chiang, C.-Y., Tseng, Y.-H., Wang, B.-Y., Chou, T.-L., Lee, C.-L.: Entrenchment and creativity in chinese quadrasyllabic idiomatic expressions. In: Wu, Y., Hong, J.F., Su, Q. (eds.) Chinese Lexical Semantics. LNCS (LNAI), vol. 10709, pp. 576–585. Springer, Cham (2018). https://doi.org/10.1007/978-3-319-73573-3_52
2. Zhang, D., Bai, J.: A cognitive semantic study of idioms from the book of songs. In: Lu, Q., Gao, H. (eds.) Chinese Lexical Semantics. LNCS (LNAI), vol. 9332, pp. 120–129. Springer, Cham (2015). https://doi.org/10.1007/978-3-319-27194-1_13
3. Baker, M.: In Other Words: A Coursebook on Translation, 2nd edn. Routledge, Abingdon & New York (2011)
4. Zeng, D.J.: Some problems in the translation of Chinese idioms in the Bilingual edition of modern Chinese dictionary (2002 supplement). J. Sichuan Foreign Lang. Inst. **2**, 116–120 (2005)
5. Gao, J.Z., Gao, J.: English translation of idiom allusions. Shanghai Transl. **1**, 29–32 (2010)
6. Mo, L.L., Ge, L.L.: Translation of Chinese idioms from the perspective of relevant translation theory. Hun. Soc. Sci. **2**, 189–191 (2012)
7. He, Y.B.: Analysis of the translation of chinese idioms from the perspective of cultural communication. Lang. Constr. **18**, 73–74 (2016)
8. Pinchuck, I.: Scientific and Technical Translation. Deutsch, London (1977)
9. Zhao, L., Wu, J.R.: On ways of translation of Chinese idioms into English based on the translations in dictionaries. Shanghai Transl. (S1), 53–54 (2005)
10. Hu, C.T., Sheng, P.L.: To optimize the translation of idioms: a review of the Chinese-English bilingual learners' dictionary. Dictionary Res. (03), 94–100 (2009)
11. Word sketch engine. https://www.sketchengine.eu/
12. Taipei Ricci Institute: Ricci Chinese-French Dictionary (RCFD). Commercial Press (2014)
13. Sun, Q.: Chinese-French Idioms Dictionary (CFDI). Xiamen University Press (2014)
14. Le Petit Robert Dictionary 2018. Robert Press (2017)
15. Le Petit Larousse Dictionary 2018. Larousse Press (2017)
16. Baker, M., Kaplan, R.: Translated! A new breed of bilingual dictionaries. Babel **40**(1), 1–11 (1994)
17. Li, X.J.: Chinese Contemporary Standardized Dictionary, 3rd edn. Foreign Language Teaching and Research Press, Beijing (2014)

Funding Acknowledgement. This work was supported by Special Creative Project of Guangdong Universities ([illegible]), project of Guangdong social science ([illegible]), and [illegible] project of GDUFS [illegible].

References

1. Hsieh, S.-K., Chiang, C.-Y., Tseng, Y.-H., Wang, B.-Y., Chou, T.-L., Lee, C.-L.: Constructionality and creativity in Chinese quadrasyllabic idiomatic expressions. In: Wu, Y., Hong, J.-F., Su, Q. (eds.) Chinese Lexical Semantics. LNCS (LNAI), vol. 10709, pp. 576–585. Springer, Cham (2018). https://doi.org/10.1007/978-3-319-73573-3_52
2. Zhang, D., Bai, L.: A cognitive semantic study of idioms from the book of songs. In: Lu, Q., Gao, H. (eds.) Chinese Lexical Semantics. CLSW 2015. LNCS (LNAI), vol. 9332, pp. 176–185. Springer, Heidelberg (2015). https://doi.org/10.1007/978-3-319-27194-1_18
3. Baker, M.: In Other Words: A Coursebook on Translation. Routledge, Abingdon, New York (2011)
4. Zhou, D.: Some problems in the translation of Chinese idioms in the bilingual edition of modern Chinese dictionary (2002 Supplement). J. Sichuan Foreign Lang. Inst. 2, [illegible]–120 (2005)
5. Gao, Z.: English translation of Chinese idioms. Shanghai J. Transl. 1, [illegible] (2010)
6. Mo, L.Y., Chen, L.: Translation of Chinese idioms from the perspective of relevant translation theory. [illegible] 2, [illegible] (2013)
7. Li, Y.J.: Analysis of the translation of Chinese idioms from the perspective of cultural communication. Lang. [illegible] 12, [illegible] (2016)
8. [illegible] Technique [illegible]. Dobson, London (1977)
9. Song, L., Wang, R.: On ways of translation of Chinese idioms into English based on the [illegible] relations [illegible]. Shanghai J. Transl. [illegible] (2005)
10. [illegible] Chinese [illegible] corpus [illegible] Dictionary [illegible] (2009)
11. [illegible]
12. [illegible] Chinese [illegible] Dictionary (CID). Commercial Press (20[illegible])
13. [illegible] Chinese-English Idioms Dictionary [illegible] University Press [illegible]
14. [illegible] Dictionary [illegible] Press ([illegible])
15. [illegible] Dictionary [illegible] Press (20[illegible])
16. [illegible] Kaplan, [illegible]: A new method of [illegible] dictionaries. [illegible] 40(1), 1–11 ([illegible])
17. Li, Y.L.: Chinese [illegible] Standard [illegible] Dictionary, 3rd edn. Foreign Language Teaching and Research Press, Beijing ([illegible])

Corpus Linguistics

A Metaphorical Analysis of Five Senses and Emotions in Mandarin Chinese

Jie Zhou[1], Qi Su[1(✉)], and Pengyuan Liu[2]

[1] School of Foreign Language, Peking University, Beijing, China
{jiezh, sukia}@pku.edu.cn
[2] School of Information Science, Beijing Language and Culture University, Beijing, China
liupengyuan@pku.edu.cn

Abstract. Emotions can be expressed by the five major external senses of human beings (i.e. vision, hearing, touch, smell and taste) via metaphors. Previous studies have mainly explored the relation between the five senses and emotions from the perspectives of physiology and cognition, and research on the five senses focuses on their semantic meanings. This paper attempts to investigate their relation based on corpus linguistics, centering on sensory verbs and emotional words. It is found that in Mandarin Chinese, five basic emotions (i.e., happiness, sadness, fear, anger, and surprise) can be expressed via olfactory, tactile, visual, and auditory modalities while among these five basic emotions, surprise cannot be expressed through taste.

Keywords: Sense · Corpus-based · Sensory modality · Mandarin

1 Introduction

Metaphor is a way of language rhetoric, cognition and thought. Many conceptual systems are construed through metaphors. Abstract, unfamiliar, complex and intangible concepts are usually cognized and experienced through specific, familiar and concrete concepts; thus, an object or experience can be known via other objects or experiences [1]. The five senses (i.e., vision, hearing, taste, smell and touch) are the most basic experiences of human beings. On this basis, human beings acquire and accumulate all other experiences including emotions. Physically, senses are dependent on sensory organs which are the direct channel connecting human bodies and the outside world. Therefore, sensory organs are of great significance in exploring emotions. The relation between senses, sensory organs and emotions can be partly manifested through metaphors. More specifically, emotion, an important and indispensable part of human life, though being abstract, can be expressed through specific, well-known sensory words. However, the five senses are in different levels [2], which reflects linguistically that words expressing the five senses are different and distinctive in terms of types and tokens. Consequently, there exist differences when using sensory words to express emotions. This paper attempts to explore the relationship between the five senses of the human body and emotions by studying the relationship between sensory verbs and emotional words from the perspective of corpus.

© Springer Nature Switzerland AG 2020
J.-F. Hong et al. (Eds.): CLSW 2019, LNAI 11831, pp. 607–617, 2020.
https://doi.org/10.1007/978-3-030-38189-9_62

2 Previous Research

Many studies concerning the five senses and emotions investigate the relationship between a certain sensory modality and emotions through physiological or psychological experiments from the perspective of physiology and cognition. When it comes to visual modality, by exploring the relationship between facial expressions and emotions [3], it has been found that negative emotions are more easily identified than positive and negative emotions [4] and that the recognition rate of happiness is the highest among different emotions [5]. With regard to auditory modality, the speaker's voice rhythm and semantic information can convey emotions [6], among which the most easily recognized emotions are sadness and fear, while the most difficult ones to identify are ridicule and disgust [7]. As for gustatory modality, [8] studies the emotional valence and arousal of various taste words by displaying different taste words (i.e., *acid*, *sweet*, *bitter*, *salty*) to participants. The results show that *sweet* has the highest emotional valence and the strongest arousal level while *bitter* has the lowest emotional valence and it is hard to evoke emotions. With reference to olfactory modality, smell can trigger positive emotions as well as negative emotions [9]. On the one hand, odor like the smell of sulfide is disgusting, and this particular physiological mechanism can help humans resist diseases [10]. On the other hand, fragrance such as pleasing perfume can evoke positive emotions [11]. Relating to tactile modality, [12] investigates what kinds of emotions can be transmitted through tactile modality (i.e. the touch of forearm in the experiment), indicating that certain emotions including anger, fear, disgust, love, gratitude and sympathy can be transmitted through touch, while some emotions cannot, and these emotions are happiness, surprise, embarrassment, jealousy and pride.

However, very little has been said on the relationship between sensory modalities and emotions from the perspective of linguistics, and the majority of them focus on the semantics of different sensory words in terms of metaphors. [13] calculates the distribution of adjectives whose entries are included in *the Modern Chinese Dictionary* (6th edition) describing the five sensory modalities in the Sinica corpus, summarizes the usage of linguistic synaesthesia and its mapping in different sensory domains, concludes the mapping model of modern Chinese linguistic synaesthesia and finally explains the model from the biological mechanism. [14] employs conceptual blending theory to study the linguistic synaesthesia in Tang poetry, and believes that linguistic synaesthesia in English and Chinese follow the same tendency from low-level sensory modalities to high-level ones. What's more, words related to temperature and taste express more emotions. Since the concept of "embodiment" has been introduced into cognitive linguistics [15], there has been an endless stream of researches on body parts [16], but still scarce research has focused on sensory modalities. Up to now, there is no scholar studying the relationship between the five senses and emotions from a corpus-based perspective. Therefore, this paper attempts to compare the distribution of different sensory verbs and different emotional words in sentences and obtain the directionality of modern Mandarin Chinese emotional metaphors via a corpus-based method.

3   Methods

This paper selects the multi-domain corpus of Beijing Language and Culture University Corpus Center (BLCU Corpus Center, BCC) [17]. First, we adopt the five basic emotions classified by [18], namely happiness, fear, anger, sadness, and surprise. Based on the emotional vocabulary of [19], we eliminate complex emotional words formed by a combination of basic emotions. Thus, only words expressing one basic emotion are kept (see Table 1).

Table 1. Mandarin words for five basic emotions

Type of emotions	Emotional words							
乐 *le4* "happiness"	畅快	放松	高兴	欢畅	欢快	欢乐	欢喜	欢欣	欢娱
	欢愉	欢悦	开心	康乐	亢奋	快活	快乐	快慰	如意
	舒畅	舒坦	舒心	爽心	顺心	松快	痛快	喜悦	闲适
	晓畅	欣慰	欣喜	兴奋	怡和	愉悦	振奋	自在	
惧 *ju4* "fear"	迟疑	发慌	害怕	害臊	害羞	惶恐	紧张	恐慌	恐惧
	苦恼	困惑	为难	畏惧	畏怯	心慌	心虚	羞怯	羞涩
怒 *nu4* "anger"	暴怒	悲愤	鄙薄	鄙视	鄙夷	冲动	敌视	烦乱	烦恼
	烦杂	烦躁	反感	忿恨	忿怒	愤恨	愤慨	愤懑	愤怒
	浮躁	激愤	蔑视	恼恨	腻烦	歧视	瞧不起	轻蔑	生气
	讨厌	痛恨	窝火	厌恶	厌烦	厌倦	怨恨	憎恶	自卑
哀 *ai1* "sadness"	哀戚	哀伤	哀痛	悲哀	悲怆	悲凉	悲凄	悲切	悲伤
	悲恸	悲痛	憋闷	憋气	怅惘	沉闷	沉痛	沉郁	乏味
	烦闷	感伤	灰心	沮丧	绝望	苦闷	丧气	伤感	伤心
	痛心	颓丧	颓唐	无聊	消沉	心酸	阴郁	忧郁	郁闷
	郁悒								
惊 *jing1* "surprise"	诧异	吃惊	愕然	骇怪	骇异	惊诧	惊愕	惊奇	惊讶
	奇怪	震惊							

Second, we extract sentences containing these emotional words in the corpus. After manual search, sentences containing both sensory verbs and emotional words and sentences containing both verbs meaning *feel* in English (i.e. 感到 *gan3-dao4*, 感觉 *gan3-jue2* or 觉得 *jue2-de2*) and emotional words are obtained, and the latter express emotions in a non-synaesthetic manner. The sensory verbs selected in this paper are Chinese characters frequently used for the five senses (see Table 2). Since 闻 *wen2* can express both senses of hearing and smell, we carry out corresponding semantic analysis and classification of sentences containing both 闻 *wen2* and emotional words.

Table 2. Mandarin verbs from five sensory modalities

Sensory domain	Sensory verbs
Vision	看 kan4 见 jian4 视 shi4 望 wang4
Hearning	听 ting1 闻 wen2
Touch	触 chu4 摸 mo1 碰 peng4
Smell	嗅 xiu4 闻 wen2
Taste	尝 chang2 吃 chi1 品 pin3

Here are some examples expressing emotions through sensory modalities.

(1) 不想别人看见自己的悲伤

Don't want others to see their sorrow.

(2) 你有没有听到你心底真正的愤怒

Have you heard the true anger in your heart?

(3) 吉中海再次敏锐地嗅到了小城中的恐惧

Jizhonghai once again keenly smelled the fear in the small town.

(4) 他终于尝到了快乐的滋味

He finally tasted the joy of happiness.

(5) 我立即触摸到他的轻松和亢奋

I immediately touched his ease and excitement.

4   Results and Discussions

4.1   Relation Between Language, Emotion and Senses

The distributions of every single type of emotion in all sentences containing emotional words are shown in Fig. 1(a). Among the five basic emotions, happiness has the largest percentage of 41%, followed by anger (19%), sadness (16%) and fear (15%), and at the bottom is surprise, which only takes up 9%. Previous research has discovered that children at the age of 2 are able to recognize happiness and sadness; after that, they are gradually capable of identifying anger, fear and surprise [20]. Moreover, surprise is the most difficult emotion to recognize, since apart from the understanding of happiness, the comprehension of expectation and belief is necessary for the recognition of surprise [21], which might account for the highest and the lowest proportions of happiness and surprise respectively.

The distributions of a specific sensory modality with emotional words in all sentences containing sensory verbs are shown in Fig. 1(b). Human beings can express various emotions through various sensory modalities, and the most frequently used is to express emotions in a non-synaesthestic way. In terms of five sensory modalities, the frequency order for them to express emotions is taste (27%), vision (21%), hearing (11%), smell (5%) and touch (3%). Therefore, the rank for Chinese sensory verbs from the perspective of emotion is: touch → smell → hearing → vision → taste, differing from the mapping model of [13], namely, (touch/taste) → (vision/hearing) → smell. The difference results from three factors. The first is that [13] concerns the mapping model of modern Chinese adjectives instead of sensory verbs. The second is that dictionary is chosen and that types are analyzed in [13] while our study concentrates on tokens in a certain corpus. Last but not least, we only focus on sensory verbs conveying emotions, so the metaphors of emotion may influence the directionality of Chinese sensory verbs.

(a)   (b)

Fig. 1. (a) The comparisons among five basic emotions in all sentences including emotional words. 41% means that sentences including emotional words for 乐 *le4* "happiness" account for 41% in sentences including emotional words for five basic emotions. (b) The comparisons among five senses and feeling (i.e. expressing emotions in a non-synaesthetic way). 11% means that the ratio of sentences with/without emotional words in all sentences including hearing verbs accounts for 11% of that of sentences with/without emotional words in all sentences including sensory verbs.

4.2 Relation Between Emotion and Senses

As shown in Fig. 2, happiness is most expressed through gustatory modality, and fear is most conveyed by tactile and visual modalities, anger by tactile and olfactory modalities, sadness by auditory and visual modalities, and surprise by olfactory modality. Happiness, fear, sadness, and anger can be expressed by five sensory modalities. Sentences expressing surprise via taste are not found in the corpus.

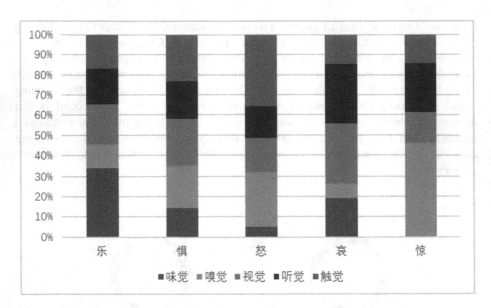

Fig. 2. The comparisons of five sensory modalities among five basic emotions

As shown in Fig. 3, psychological sense (i.e. feeling, or expressing emotions in a non-synaesthetic manner), smell, touch, vision and hearing can convey five basic emotions, while taste cannot show surprise. In terms of the biological mechanism, visual, olfactory, auditory, tactile, and gustatory modalities can trigger a variety of emotions, demonstrating the possibility of expressing various emotions from five sensory modalities. Among the five sensory modalities, the most expressed emotion through taste, vision, hearing and touch is happiness; the most expressed emotion through olfactory modality is surprise, followed by happiness; surprise cannot be expressed through gustatory modality and the least expressed emotion through taste is anger; the least expressed emotion through smell and touch is sadness; the least expressed emotion through vision and hearing is fear. Hence, from the perspective of similarity of expressing emotions, visual and auditory modalities are more closely related, and it is the same between tactile and olfactory modalities. This similarity is also closely related to the proportion of emotions expressed by different sensory modalities above. A more specific analysis will be carried out below from the relationship between each sensory modality and emotions.

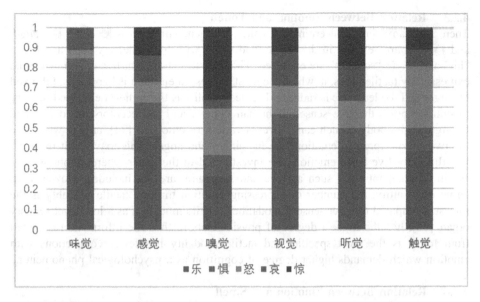

Fig. 3. The comparisons of five basic emotions among five sensory modalities

4.2.1 Relation Between Emotion and Vision

The sensory verbs that depict the action of seeing are mainly 看 kan4, 见 jian4, 视 shi4, and 望 wang4. The data show that five basic emotions can all be expressed by visually related verbs, in which the most expressed emotion is happiness and the least is fear. Among the five basic emotions, fear, anger, and sadness are negative, and surprise is not necessarily positive. In terms of proportion, visual expression of negative emotions exceeds 50%, reflecting the negative affect between vision and emotion [4], that is, in positive emotions and negative emotions, the recognition of negative emotions through vision is higher. Among the five emotions, the most expressed one is happiness, namely, pleasant emotion, which reflects the happy face recognition advantage [5]. In other words, humans have a higher accuracy rate of recognition and faster response for happy facial expressions than other emotional faces.

4.2.2 Relation Between Emotion and Hearing

Sensory verbs expressing the action of hearing are mainly 听 ting1 and 闻 wen2. Similar to the relation between vision and emotions, five emotions can all be expressed by aural verbs, in which the most expressed emotion is happiness, and the least is fear. [22] explores the relation between hearing and emotion in different languages and finds that the participants are able to recognize the five basic emotions expressed in English, German, Hindi and Arabic, indicating that in different languages studied, auditory modality can express the five emotions (i.e., happiness, fear, anger, sadness and surprise). Our study extends this conclusion from a corpus-based perspective to another language, namely Chinese, which once again verifies the argument that hearing can express a variety of emotions [23].

4.2.3 Relation Between Emotion and Touch

There are many verbs concerning the action of touch. Three verbs 触 *chu4*, 碰 *peng4*, and 摸 *mo1* are selected in the article, and verbs with high intensity such as 打 *da3* "hit" and 挤 *ji3* "squeeze" are excluded. The statistics show that five emotions can be expressed by tactile verbs, in which the most expressed emotion is happiness, followed by anger and the least one is sadness. Tactile modality is the earliest developed sensory modality among the five senses of human beings, and its receptors are distributed throughout the body. Touch can convey a variety of emotions [24], among which the top two most expressed emotions are happiness and anger. This may result from the fact that only five basic emotions are investigated in the paper where emotions concerning human intimacy such as love and sympathy are not included. However, in terms of quantity, the number of expressing emotion through tactile modality is the lowest. Compared to other sensory modalities, tactile modality, as a low-level sensory organ [25], has the highest degree of physicalization, thus the information obtained from touch is the most specific, and tactile modality has fewer connections with emotion which demands higher degree of cognition as a psychological phenomenon.

4.2.4 Relation Between Emotion and Smell

Sensory verbs that express the action of smelling are mainly 嗅 *xiu4* and 闻 *wen2*. The figures suggest that five emotions can be expressed by olfactory verbs, in which the most expressed emotion is surprise, and the least is sadness. In the aspect of expressing surprise, the difference between smell and taste is the biggest, because the most expressed emotion through olfactory modality is surprise which cannot be expressed through gustatory modality. This may be because at the time of discerning the smell, we perceive simultaneously the characteristics of this smell through olfactory modality, and thus the corresponding emotion is triggered [9]. In contrast, before distinguishing the taste, we can choose not to eat foods that are possibly very strange based on our experience, so as to avoid feeling surprised.

4.2.5 Relation Between Emotion and Taste

Sensory verbs that describe the action of tasting are mainly 尝 *chang2*, 吃 *chi1*, and 品 *pin3*. However, it is found that only four emotions including happiness, fear, anger and sadness can be expressed through taste, and the most expressed emotion is happiness. In comparison with other four sensory modalities, the majority of emotion expressed through taste is happiness. The vital function of taste is to help us choose food. Among different tastes, salty and sweet tastes increase appetite more easily than bitter and sour tastes. Therefore, salty and sweet tastes are more in line with our intentions, leading to their higher frequency. Sweetness is mainly related to positive emotions [26], with the highest emotional valence [8], and corresponds to happiness and surprise [27], which can explain why happiness is the most expressed emotion through olfactory modality.

No sentence expressing surprise through taste is found, which is not consistent with the above findings that sweetness can respond to surprise [27]. First of all, this may be related to the emotional words selected in this paper. It is common to use 吃惊 *chi1-jing1* in Chinese to express surprise. However, single character is not selected as emotional words in the paper; consequently, examples like 尝 *chang2*, 吃 *chi1*, or 品 *pin3* combined with 吃惊 *chi1-jing1* are not included. Secondly, since sweet tastes corresponding to both surprise and happiness, while the latter is the easiest to recognize [5] and has the highest frequency, sweetness may evoke the emotion of happiness more, leaving limited chances to express surprise through gustatory modality. Finally, due to the size of the corpus, no examples of using taste to express surprises have been found in this study.

5 Conclusion

Emotion is a vital and intriguing psychological phenomenon. It is a new endeavor to investigate emotions from the perspective of embodiment and cognition. As a connection between the body and the outside world, the five senses play a significant role in the perception of emotions, and the relation between emotions and five senses is very close. On the one hand, as a result of physiological mechanism, different sensory modalities can evoke different emotional responses from different external stimuli. On the other hand, according to the corpus-based study of sensory verbs and emotional words, we find that in Mandarin Chinese, emotional words can be collocated with various verbs that describe the five senses, demonstrating that senses are closely associated with emotions from the perspective of corpus.

Specifically, visual, tactile, olfactory, and auditory modalities can express five basic emotions, namely, happiness, fear, anger, sadness, and surprise; while taste can convey four emotions including happiness, fear, anger, and sadness, there is no example expressing surprise via taste. Except olfactory modality, the most expressive emotions of vision, touch, hearing, and taste is happiness. Although there are abundant commonalities in the relation between the five senses and emotions, each distinct modality is unique in expressing emotions. That is, the proportions of different emotions expressed by different sensory modalities are not the same. In addition to the commonalities and characteristics between sensory modalities and emotions, our research also indicates that in the connection between senses and emotions, visual and auditory modalities share more similarities, and that it is the same between olfactory and tactile modalities. There exist complicated relations among different sensory modalities since they are important tools for human's perception of the outside world. Also, sensory modalities are not isolated from emotions. Therefore, how emotions are expressed under the joint participation of different sensory modalities deserves a prominent place on the agenda for discussion, especially focusing on the relation between visual and auditory modalities as well as that between olfactory and tactile modalities.

There are still a few limitations in this study. The first is the choice of basic emotions and emotional words contained in each type of emotion. Although these emotional words certainly cannot cover all emotional words, most of common emotional words are included. Second, the choice and scale of the corpus. The corpus selected in this study is multi-disciplinary, and future research can be based on different genres and explore the impact of a particular genre on the research results. Finally, this study only focuses on Mandarin Chinese. Whether the results of this study are equally applicable to other languages or dialects requires other corpus-based studies in other languages and dialects.

Acknowledgments. We would like to thank The National Social Science Fund of China for the financial support on the project (No. 18YJA740030).

References

1. Lakoff, G., Johnson, M.: Metaphors We Live By. University of Chicago Press, Chicago and London (1980)
2. Ullmann, S.: Language and Style. Basil Blackwell, Oxford (1964)
3. Schirmer, A., Adolphs, R.: Emotion perception from face, voice, and touch: comparisons and convergence. Trends Cogn. Sci. **21**(3), 216–228 (2007)
4. Mogg, K., Bradley, B.P.: A cognitive-motivational analysis of anxiety. Behav. Res. Ther. **36** (9), 809–848 (1998)
5. Calvo, M.G., Nummenmaa, L.: Perceptual and affective mechanisms in facial expression recognition: an integrative review. Cogn. Emot. **30**(6), 1081–1106 (2016)
6. Dupuis, K., Pichora-Fuller, M.K.: Use of affective prosody by young and older adults. Psychol. Aging **25**(1), 16–29 (2010)
7. Wang, H., Li, A.: The construction of CASS-ESC (Putonghua emotion speech corpus) and its listening and recognizing experiment. In: Proceedings of the Sixth Modern Phonetics Workshop, pp. 127–132. Chinese Information Processing Society of China (2003). (in Chinese)
8. Wang, Q., Woods, A.T., Spence, C.: "What's your taste in music?" A comparison of the effectiveness of various soundscapes in evoking specific tastes. i-Perception **6**(6), 1–23 (2015)
9. Herz, R.S., Schankler, C., Beland, S.: Olfaction, emotion and associative learning: effects on motivated behavior. Motivat. Emot. **28**(4), 363–383 (2004)
10. Oaten, M., Stevenson, R.J., Case, T.I.: Disgust as a disease-avoidance mechanism. Psychol. Bull. **135**(2), 303 (2009)
11. Warrenburg, S.: Effects of fragrance on emotions: moods and physiology. Chem. Senses **30** (suppl_1), i248–i249 (2005)
12. Hertenstein, M.J., Keltner, D., App, B., Bulleit, B.A., Jaskolka, A.R.: Touch communicates distinct emotions. Emotion **6**(3), 528–533 (2006)
13. Zhao, Q., Huang, C.R.: Mapping models and underlying mechanisms of synaesthetic metaphors in Mandarin. Lang. Teach. Linguist. Stud. **1**, 44–55 (2018). (in Chinese)
14. Luo, Y., Zhang, H., Qin, X.: Conceptual integration analysis of synaesthetic metaphors: a case study on synaesthetic metaphors in three hundred poems of the Tang Dynasty. Foreign Lang. Learn. Theory Pract. **164**(4), 13–18 (2018). (in Chinese)
15. Lakoff, G.: Explaining embodied cognition results. Top. Cogn. Sci. **4**(4), 773–785 (2012)
16. Yu, N.: The body in anatomy: looking at "head" for the mind-body link in Chinese. In: Caballero, R., Díaz Vera, J. (eds.) Sensuous Cognition: Explorations into Human Sentience: Imagination, (E)motion and Perception, pp. 53–73. De Gruyter Mouton, Berlin (2013)
17. Xun, E., Rao, G., Xiao, X., Zang, J.: The construction of the BCC Corpus in the age of big data. Corpus Linguist. **1**, 93–109 (2016). (in Chinese)
18. Turner, J.H.: The evolution of emotions in humans: a Darwinian-Durkheimian analysis. J. Theory Soc. Behav. **26**(1), 1–33 (1996)
19. Chen, Y., Lee, S.Y., Huang, C.R.: A cognitive-based annotation system for emotion computing. In: Proceedings of the Third Linguistic Annotation Workshop, pp. 1–9. Association for Computational Linguistics (2009)

20. Flavell, J.H.: Theory of mind development: retrospect and prospect. Merrill-Palmer Q. **50**, 274–290 (2004)
21. Golan, O., Baron-Cohen, S., Hill, J.J., Golan, Y.: The "reading the mind in films" task: complex emotion recognition in adults with and without autism spectrum conditions. Soc. Neurosci. **1**, 111–123 (2006)
22. Pell, M.D., Paulmann, S., Dara, C., Alasseri, A., Kotz, S.A.: Factors in the recognition of vocally expressed emotions: a comparison of four languages. J. Phonet. **37**(4), 417–435 (2009)
23. Darwin, C.: The Descent of Man. John Murray, London (1871)
24. Thompson, E.H., Hampton, J.A.: The effect of relationship status on communicating emotions through touch. Cogn. Emot. **25**(2), 295–306 (2011)
25. Pham, T.T.H.: A study of lower senses sensation words between Chinese and Vietnames and its teaching. Ph. D. dissertation (unpublished). Wuhan University, Wuhan (2017). (in Chinese)
26. Wendin, K., Allesen-Holm, B.H., Bredie, W.L.: Do facial reactions add new dimensions to measuring sensory responses to basic tastes? Food Qual. Prefer. **22**(4), 346–354 (2011)
27. Robin, O., Rousmans, S., Dittmar, A., Vernet-Maury, E.: Gender influence on emotional responses to primary tastes. Physiol. Behav. **78**(3), 385–393 (2003)

Research on Chinese Animal Words Extraction Based on Children's Literature Corpus

Huizhou Zhao[1]([⊠]), Zhimin Wang[2], Shuning Wang[1], and Lifan Zhang[1]

[1] School of Information Science, Beijing Language and Culture University, Beijing, China
zhaohuizhou@blcu.edu.cn, wangshuning1997@yahoo.com, xkl8zlf@126.com
[2] Research Institute of International Chinese Language Education, Beijing Language and Culture University, Beijing, China
wangzm000@qq.com

Abstract. Categorized and graded vocabularies are an important aspect of children's graded reading. Taking animal words from the Thesaurus of Modern Chinese as the seed words, this paper studies a method of extracting animal words from the children's literature corpus and attempts to construct a word sequencing model. The method used is to match the results of automatic word segmentation with the seed words. There are 786 animal nouns extracted from the corpus, with an increasing rate of 39.36% compared to the 564 seed words, and there are 780 derivative animal words. The animal word sequencing model is based on word-work-popularity and word-writer-popularity, which resolves the problem of having an unbalanced number of characters and writer's works.

Keywords: Word sequencing · Animal words · Children's literature · Graded reading

1 Introduction

Vocabulary grading is important for readability measurements. Categorized and graded vocabularies are an important aspect of children's graded reading.

"Animal vocabulary" is closely related to children's literature. "In the process of children's growth, the natural attributes of animals give them an extremely important spiritual influence." [1] Therefore, among the word lists in children's reading, the animal word list is an important category.

The current animal word lists can be found in semantic classification dictionaries, such as the *Thesaurus of Modern Chinese* [2]. This dictionary provides a sound basis for the construction of a word meaning system [3], but the animal words in it cannot be directly used in assessing children's graded reading. The reasons for this are as follows: (1) There are more than a thousand words in the animal category that are not graded; (2) The words are widely collected from Chinese dictionaries, the large modern Chinese corpus, and the latest word lists. In the field of children's literature, some words are rare, such as "draught animal" (役畜), "Emu" (鸸鹋), and "shad" (鲥鱼); (3) The

© Springer Nature Switzerland AG 2020
J.-F. Hong et al. (Eds.): CLSW 2019, LNAI 11831, pp. 618–627, 2020.
https://doi.org/10.1007/978-3-030-38189-9_63

language of children's literature is childlike, having new forms for words that are different from the ones adults use, such as the use of reduplicated words [4] "Xiaoxiaoniu" (小小牛), "Maomaoxiong" (毛毛熊) etc., and thus word lists for children's reading should also contain childlike words that appear in high-frequency. In conclusion, based on the existing semantic classification dictionary, we can extract and sequence animal words from the children's literature corpus, and construct a word list for the animal category in children's graded reading.

Frequency and distribution rate are often used to extract common words from a corpus. *The Modern Chinese Frequency Dictionary* [5], published in 1986, for the first time calculated the usage of words based on their frequency and distribution. The first list contained 8,000 words which were edited according to their usage, [6] and mainly considered the "classification" and the number of "texts". Wang Zhimin posited that vocabulary should be acquired by combining frequency with time span, which considers the distribution of words in the time dimension [7]. Yu Shiwen advanced the concept of building a common vocabulary oriented towards information processing, which considers this common vocabulary as a function of the corpus and coverage coefficient. These methods provide a good reference for extracting general and basic words [8].

Categorized words are those in a specific semantic category. Compared to general and basic words, they have lower frequency and distribution in the large-scale corpus, and thus they are less likely to be listed in basic vocabulary. However, cognitive linguistics holds that vocabulary in each language is graded, and that there is a "basic level of vocabulary" in all categories of meanings [9]. Basic level words are more naturally and frequently used in daily life than words of other levels in the same category. In terms of word frequency, in the large-scale corpus, the word frequency of basic level words should be higher than that of other words in the same category [10]. Song [11] extracted basic level mammalian words from the 220 million words in the newspaper literature corpus for Chinese learning, and used the relative word frequency in the category of meaning to determine candidates for basic level words, in order to resolve the problem of having an unbalanced vocabulary between the categories of meaning.

When extracting animal words, there is no cross-semantic category imbalance, but there is an inherent imbalance in the children's literature corpus. First, children's literature covers a wide range of ages, and there are differences in the number of characters among the different works. Second, different writers have written a different number of works, and different writers have different preferences for animals in their works. This paper will study a method for extracting animal words from the unbalanced children's literature corpus.

2 A Survey of the Children's Literature Corpus

The children's literature corpus in this paper is selected from the non-translated Chinese children's literature corpus. In order to guarantee a 100% recall rate for sentences with animal words, we use the animal words in the *Thesaurus of Modern Chinese* [2] to match the non-segmented sentences, and selected 172 works with animal words. These works are as follows:

(1) *The 60-year collections of Chinese children's literature*, 46 works, short fairy tales by various writers, totaling 38,825 characters.
(2) *Selected Works of Wang Yimei's Fairy Tales*, 8 works, short fairy tales, totaling 8,241 characters.
(3) *The Scarecrow Man*, collection of Ye Shengtao's fairy tales, 32 works, written in the 1920s and 1930s, totaling 106,156 characters.
(4) Cao Wenxuan's works, 14 works, novels, totaling 795,415 characters.
(5) Shen Shixi's works, 35 works, animal novels, totaling 1,477,470 characters.
(6) Yang Hongying's *Diary Series of Laughing Cats*, 19 works, novels, totaling 700,617 characters.
(7) Zheng Yuanjie's works, *The Biography of Shuke and Beita* and *The Biography of Lu Xixi*, 2 works, totaling 620,404 characters.
(8) 16 other works, totaling 1,043,243 characters.

Zhu [1] considers animal literature as a distinct genre in children's literature, and that animal literature in a broad sense refers to all animal related literature, including: (1) fairy tales; (2) animal novels; (3) works about real animals in which human beings objectively observe as the protagonists. In our corpus, the first three items are works in the first category, the sixth and seventh items are works in the second category, and the fifth item includes works in the third category. There are more than 20 of each kind of work. The corpus has solid coverage of the animal literature genre. In addition, it also includes children's life stories and fantasy stories which have animal-related content. In short, the corpus has solid coverage in children's narrative literature.

These works are suitable for child readers, are recommended by primary schools or experts, or are listed among the best-selling books on the most popular book websites. It is a common classification method to classify children's books according to their corresponding grade level. Grades 1–2 are classified as the lower grades, Grades 3–4 as the middle grades, and Grades 5–6 as the higher grades. The number of works, writers, characters, and sentences with animal words are shown in Table 1.

Table 1. Suitable grades data table of Children's Literature Corpus.

	Number of works	Number of writers	Total characters	Number of sentences with animal words
Lower grades	89	20	235,882	2,007
Middle grades	26	7	1,600,137	17,843
Higher grades	57	10	2,954,352	34,711
Total	172	34*	4,790,371	54,561

*The different works of some writers belong to different grade levels. All the 172 works were written by 34 writers.

From Table 1, we can see that in the lower grades the numbers of works and writers are both more than half of the total number. There are 89 works, accounting for 51.7%

of the total works, and 20 writers, accounting for 58.8% of the total writers. However, due to the short lengths of the works, there are only 2,007 sentences with animal words, accounting for 3.68% of the total sentences with animal words. It can be predicted that the contribution of the frequency of animal words in these works is relatively small. However, the short-length works in the lower grades have a diverse range of writers, which should be considered in the design of the animal word sequencing model.

There are 26 works in the middle grades, accounting for 15.1% of the total number of works, and 7 writers, accounting for 20.6% of the total number of writers, with 17,843 sentences containing animal words, accounting for 32.7% of the total number of sentences with animal words. There are 19 works by Yang Hongying in the middle grades, which is more than half of the total. There are 57 works in the higher grades, accounting for 33.1% of the total number of works, and 10 writers, accounting for 29.4% of the total number of writers, with 34,711 sentences containing animal words, accounting for 63.6% of the total number of sentences with animal words. There are 35 works by Shen Shixi in the higher grades, which is more than half of the total. The diversity of writers in the middle and higher grades is not broad, so the problem of controlling the imbalance of writers' works should be considered when designing the animal word sequencing model.

Children's literary works can also be classified by the number of characters. The classification of the 172 works by grade level and character number is shown in Fig. 1.

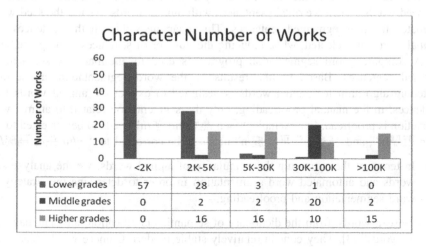

Fig. 1. Character number distribution of works in different grades.

From Fig. 1, it can be seen that all the works in the lower grades are short stories with less than 30,000 characters. The middle grades have mainly medium length works with 30,000–100,000 characters, while the long works are mainly in the higher grades. However, the works in the higher grades are divided into short stories, medium length works, and long stories. This is mainly because some of Shen Shixi's and Cao Wenxuan's works include several relatively independent works. Although the lengths of these works are shorter, considering their content, they still belong in the higher grades.

3 Animal Words Extraction

Animal words refer to words that record animals [12]. Zhou [13] divides animal words into two categories. The first category is used to name animals, such as lion, tiger, mouse, etc. The animal category words in the *Thesaurus of Modern Chinese* [2] belong to this category. The second category is based on animal name words, animal body parts, or the names of their products, and are called morpheme-derived animal words. These words can be used to refer to people, other animals, plants or fruits, utensils, constellation, disease, place or person name, human body parts, animal products or body organs and so on. For example, 蜂猴 ("bee monkeys", i.e. slow loris), or 斑马线 ("zebra line", i.e. pedestrian crosswalk).

The purpose of this paper is to construct a graded vocabulary for animal words in children's literature to support children's graded reading, in which the above two categories of animal words are extracted. When extracting animal words, the following principles are to be followed:

(1) The extraction of animal words should be comprehensive. The steps in automated computer processing should focus on the recall rate, and the steps in manually assisted processing should consider the accuracy, work objectives, and work efficiency.

We call the words in the animal category of the *Thesaurus of Modern Chinese* [2] the seed words. When we obtain sentences with animal words, we use the seed words to match the non-segmented sentences. This guarantees that all the sentences with animal words are selected, while reducing the number of sentences to be processed.

A computer word segmentation program is used to automatically segment the selected sentences. Based on the results of the word segmentation, an accurate matching algorithm for animal words is designed to extract the animal words from sentences. In the manual proofreading, only the segmentation related to animal word extraction is proofread. For example, "马/n 鹿野/nrj 驴/n" needs to be corrected to "马鹿/n 野驴/n", and "一/m 头马/n 鹿/n" needs to be corrected to "一/m 头/q 马鹿/n".

(2) In terms of the extracting the granularity of animal words, we the analyze seed words and automated word segmentation in order to determine the strategy for word segmentation and proofreading.

The seed words, from the dictionary of semantic classification, are "language and literacy words" [3]. They contain relatively stable modern Chinese words. They only include nouns used to name animals. The automated word segmentation computer program is based on newspaper corpus training. Over-segmentation, error segmentation, and inseparability segmentation can occur when segmenting the children's literature corpus. However, the automated word segmentation program, which is designed based on statistical methods, has a certain adaptability when segmenting new words, and the results of the segmentation are also reproducible. Considering that the vocabulary resources extracted in this work will support the automated assessment of graded texts for children, we thus respect the granularity of the results of the automated word segmentation as long as they are not contrary to the seed words.

For example, "小/a 公鸡/n" and "小/a 兔/n" are both results of the automated word segmentation. There are words in the seed vocabulary "公鸡" and "小兔". Therefore, "小/a 公鸡/n" is not corrected, and "小/a 兔/n" is corrected as "小兔/n".

The combination of seed words and automated word segmentation can match a large number of derivative animal words, such as "九牛二虎之力/n" ("the strength of nine bulls and two tigers/n", meaning "tremendous effort"), "虎毒不食子/nz" ("Even a vicious tiger will not eat its cubs/nz", meaning "No one is capable of hurting their own children"), "马戏团/n" ("circus/n"), "动物园/n" ("zoo/n"), "马倌/n" ("horseman/n"), "驯兽/v" ("taming animals/v").

Based on the above combination of seed words and automated word segmentation, we extracted 1,566 animal-related words from the children's literature corpus. We classify and annotate the extracted animal words by using two types of classification methods: **naming animal words** and **derivative animal words** [13].

564 animal words in the seed vocabulary belong to naming animal words. We consider that 56 words from the extracted results can be categorized as naming animal words. These words can be divided into three situations:

(1) Specific animal names: 鼹鼠 ("mole"), 鼬鼠 ("weasel") and 锦鲤 ("koi fish").

(2) General animal names, which belong to the automated word segmentation results, or isomorphic words that can be found in the seed vocabulary. For example, "珍稀动物, 野生动物, 灵长类动物, 两栖动物, 鱼类, 蛇类" ("rare animals, wildlife, primates, amphibians, fish, snakes") are all automated word segmentation results, and "鱼类, 蛇类" are isomorphic with the seed word "兽类".

(3) Animal-specific words belong to the automated word segmentation, and isomorphic words can be found in the seed vocabulary. For example, "羊群, 鱼群" are isomorphic with the seed words "蜂群, 马群"; "女猫" is isomorphic with the seed word "男猫"; "小鸟, 小猫, 小鱼, 小鹰, 小象, 小虾, 小天鹅, 小鼠, 小狮子, 小狮, 小山羊, 小鸟, 小蜜蜂, 小毛驴, 小猫" are isomorphic with the seed words "小鸡, 小猪"; "小狮子" is isomorphic with the seed word "小燕子"; "幼熊, 幼豹" are isomorphic with the seed words "幼虎, 幼鸟"; "野羊, 野牛, 野鸟, 野耗牛, 野骆驼, 野驴, 野鹿, 野狼" are isomorphic with the seed words "野鼠, 野马"; "雄鱼, 雄象, 雄鹿, 雄虎, 雄豹" are isomorphic with the seed word "雄狮".

Some animal words exist that cannot be classified by the above three situations, but they are used to name animals. We classify them as derivative animal words referring to other animals. There are 166 such terms and all of them are automated segmentation entries. For example, "白兔, 大熊, 大公鸡, 毛毛熊, 黑叶猴", among others.

In summary, we extracted **786 naming animal words** from the children's literature corpus, accounting for 50.2% of the total words; another **780** words belong to **derivative animal words** which are not used to name animals, accounting for 49.8% of the total words.

4 The Design of the Sequencing Model for Animal Words

After extracting the animal words, we need to sequence the 1,566 animal words extracted from the children's literature corpus. Frequency and distribution rate are two important parameters for word sequencing. For the animal words in this study, the frequency-related parameters includes the total word frequency, the word frequency in each work, and the total word frequency in all the works of each writer. The distribution-related parameters are the number of works containing a certain word, and the number of writers using a certain word.

In order to resolve the character-number imbalance in the corpus, we define *word-work-popularity*, which is shown in Formula (1) and represents the frequency parameters of the words in a work.

$$f_{i,j} = k \frac{FC_{i,j}}{FZ_j} \tag{1}$$

Where $f_{i,j}$ is the word-work-popularity of word i in work j, $FC_{i,j}$ is the frequency of word i in work j, FZ_j is the total number of characters in work j, and K is a constant whose function is to ensure that the range of the $f_{i,j}$ value is not too small. The *word-work-popularity* of a word can be understood as *the word frequency corresponding to the unit number of characters in a work*.

The total popularity of an animal word in the literature corpus is the sum of the word-work-popularity of each work in the corpus, as described in Formula (2).

$$\sum_j f_{i,j} = k \frac{FC_{i,j}}{FZ_j} \tag{2}$$

Table 2 shows the first 30 words and the popularity value of each word in descending order based on the word-work-popularity. In this experiment, the constant k = 10,000.

Table 2. The top 30 words (in descending order by the total popularity of words as per the word-work-popularity metric).

1–10	11–20	21–30
猫 2289.166483	动物 654.1816596	松鼠 459.0335541
狼 1904.984296	蜗牛 605.7345608	狐 447.9132898
狐狸 1773.467076	鱼 597.075746	豺 428.2569274
老鼠 1572.445574	大象 588.589844	大灰狼 407.5813084
狗 1084.048364	羊 544.8326726	小鸟 401.5427446
小熊 974.7443846	兔子 543.9215425	虎皮 396.1200276
熊 858.3838494	象 540.3039441	小猫 389.7929705
小猪 784.2135316	牛 496.4121095	马 371.943547
鸟 762.7401354	兔 479.2312622	小兔 350.2542461
老虎 662.9780725	青蛙 465.0850346	刺猬 330.3246476

In addition to the imbalance in the number of characters in the corpus, there is also an imbalance in the number of writers' works, that is, the number of works contributed by each writer to the collection of works is unbalanced. Formula (2) calculates the sum of the word-work-popularity of each work in the corpus, which causes the animals words in the works of high-volume writers to have an advantage in the ranking. In order to resolve this problem, when summing the popularity of a word, we only use one work from each writer to calculate the total word popularity. In this experiment, we choose the work with the greatest popularity value for a word. We call this word popularity the word-writer-popularity, that is, the word popularity is the greatest among the works of the same author. The word-writer-popularity metric is shown in Formula (3). Where, $f_{i,J}$ is the word-writer-popularity of word i in all works of author J, and Z_J is the collection of works of author J in the set of works.

$$f_{i,J} = \max_{j \in Z_J} \left(k \frac{FC_{i,j}}{FZ_j} \right) \tag{3}$$

The total popularity of an animal word based on word-writer-popularity is shown in Formula (4), which is the sum of the word-writer-popularity of each writer in the corpus.

$$\sum_J f_{i,J} = \max_{j \in Z_J} \left(k \frac{FC_{i,j}}{FZ_j} \right) \tag{4}$$

Table 3 shows the first 30 words and the popularity value of each word in descending order based on the word-writer-popularity. In this experiment, the constant k = 10,000.

Comparing Tables 2 and 3, it can be seen that the word-writer-popularity metric is improved. "豹" and "虎皮" are removed from the top 30 words. Compared to other animal words, "豹" is less basic, while "虎皮" belongs to derivative animal words and is not suitable for basic vocabulary.

The top 30 words listed in Table 3 all refer to naming animal words, and this word list has the following characteristics:

(1) It is highly consistent with previous research results. Song Fei [11] extracted basic mammal words based on the literary material in large-scale newspapers. That work extracted "象(大象), 马, 狗, 牛, 羊, 猪, 猫, 虎(老虎), 狼, 驴(驴子), 鼠(老鼠), 狮(狮子), 兔, 骆驼, 猴(猴子), 鹿, 熊, 熊猫(大熊猫), 豹(豹子), 狐(狐狸), 獾, 獭, 蝙蝠, 鲸(鲸鱼), 狸, 海豹, 袋鼠, 刺猬, 鼬, 貂" as the basic level mammal vocabulary. 21 of the words in Table 3 belong to Song Fei's basic mammalian words list. They are "象(大象), 狗, 牛, 羊, (小猪), 猫(小猫), (老虎), 狼(大灰狼), (老鼠), 兔(小兔, 白兔, 兔子), 熊(小熊, 大熊), (狐狸), 刺猬". Of the nine underlined words that do not belong to the basic mammalian words in Song Fei's list, only "松鼠" ("squirrel") and "斑马" ("zebra") are mammal words. The high correlation between our results and those of previous studies shows that the method of animal word extraction and the sequencing model in this paper are effective.

Table 3. The top 30 words (in descending order by the total popularity of words as per the word-writer-popularity metric).

1–10	11–20	21–30
狐狸 1496.183026	大象 506.5847906	小鸟 326.1634936
狼 1348.900531	鱼 494.4563183	斑马 304.180122
猫 1109.308394	蜗牛 490.618014	鹊 301.7933388
小熊 974.7443846	青蛙 446.3736771	小猫 287.2952268
小猪 775.45682	松鼠 408.6710506	白兔 279.6688984
熊 756.0862751	牛 408.5951738	刺猬 278.8668811
狗 719.2780114	兔 407.3089319	大熊 277.1532992
老鼠 638.0853704	兔子 400.7819874	小兔 276.0531936
鸟 590.1793644	羊 368.569084	大灰狼 268.1112944
老虎 566.7929598	动物 336.1857819	象 266.1123801

(2) There are prefixes such as "小" ("small"), "大" ("big") and "老" ("old") in many of the words. Regarding different words with the same central word, the ranking of words with "small" is higher than that of "big", which reflects the childlike nature of the words used in children's literature.

(3) "老鼠, 老虎, 大象" rank before "鼠, 虎, 象" respectively, which is consistent with the previous results for word frequency statistics based on the large-scale corpus [11]. However, the size of the corpus used in previous studies contains 220 million characters, while the children's literature corpus used in our work contains only 4.8 million characters, and there are still imbalance problems in the number of characters and the number of writers' works. These results again demonstrate that the word sequencing model in this paper is very effective in resolving these two types of imbalances.

5   Conclusion

This paper studies the extraction of animal words for children's graded reading. Taking animal words in the *Thesaurus of Modern Chinese* as the seed words, this paper studies the method and sequencing model for extracting animal words from the children's literature corpus. When extracting animal words, 786 animal nouns were extracted from the corpus by matching the seed words with the automated word segmentation, an increase of 39.36% from the 564 original seed words, and 780 derivative animal words. Designing an animal word sequencing model based on word-work-popularity and word-writer-popularity resolves the problem with imbalances in the number of characters and writer's works. The results of the sequencing experiments correspond to the previous studies based on the large-scale corpus, which verifies the validity of this method and sequencing model.

Acknowledgments. The research is supported by Science Foundation of Beijing Language and Culture University (supported by "the Fundamental Research Funds for the Central Universities") (19YJ040005); Major Program of National Social Science Foundation of China (18ZDA295); MOOC Project of Beijing Language and Culture University (FZ201911); Top-ranking Discipline Team Support Program of Beijing Language and Culture University (JC201902); Beijing College Student innovation and entrepreneurship training program (No: 18XKGJ05, 201910032045, 201910032046).

References

1. Zhu, Z.: An Introduction to Children's Literature, p. 161. Higher Education Press, Beijing (2009). (in Chinese)
2. Su, X.: Thesaurus of Modern Chinese. Commercial Press, Beijing (2013). (in Chinese)
3. Hong, G., Su, X.: A lexical classification system based on meaning: thesaurus of modern Chinese. Lexicogr. Stud. (1) (2015). (in Chinese)
4. Chen, X.: On the language of children's literature from the use of words. Sci. Educ. Collect. (Mid-Term J.) (11), 68–69 (2012). (in Chinese)
5. Institute of Language Teaching, Beijing Language and Culture University. Modern Chinese Frequency Dictionary. Beijing Language and Culture University Press (1986). (in Chinese)
6. Su, X.: Comparative study of GOTCFL and the glossaries in two textbooks. Appl. Linguist. (2) (2006). (in Chinese)
7. Wang, Z.: The statistical research on diachronic changes of the common wordlist for Chinese teaching. TCSOL Stud. (4) (2010). (in Chinese)
8. Yu, S., Zhu, X.: Quantitative lexicon study and knowledge base construction for commonly used words. J. Chin. Inf. Process. (03), 16–20 (2015). (in Chinese)
9. Wang, Y.: Cognitive Linguistics. Shanghai Foreign Language Education Press, Shanghai (2007). (in Chinese)
10. Song, F.: The method of extracting and classifying vocabulary in basic category in international teaching of Chinese language and its future application. Int. Dissem. Chin. Lang. (2) (2012). (in Chinese)
11. Song, F.: Research on large-scale corpus-based relative frequency location method for modern Chinese basic-level vocabulary. Appl. Linguist. (04), 77–84 (2014). (in Chinese)
12. Li, S., Ai, H.: The distribution of meaning-items about animal polysemous words in modern Chinese based on the typology. J. Ocean Univ. China (Soc. Sci. Ed.) (06), 99–104 (2017). (in Chinese)
13. Zhou, X.: A study of Chinese animal words. Jilin University (2012). (in Chinese)

The Concatenation of Body Part Words
and Emotions from the Perspective
of Chinese Radicals

Yue Pan[1], Pengyuan Liu[1($\boxtimes$)], and Qi Su[2]

[1] School of Information Science, Beijing Language and Culture University,
Beijing, China
viki.pan@qq.com, liupengyuan@pku.edu.cn
[2] School of Foreign Languages, Peking University, Beijing, China
sukia@pku.edu.cn

Abstract. Emotional stimuli can cause physical reactions in the body, and physiological responses further lead to emotional experiences. In the past, the study of emotional body response in linguistics mostly examined the emotions of the language structure of body parts, and it was mostly limited to the study of dictionary meanings, rarely conducting on the basis of corpus. This paper attempts to examine the concatenation of Chinese body part words and emotions in the microblog corpus from the perspective of Chinese radicals. The study found that each type of body radical can be used with any emotion, but the strength of the concatenation with emotion is not the same, such as "鼻(nose)" or "舌(tongue)" are most closely connected with the emotions of disgust; "舌 (tongue)" and "牙(齿, tooth)" can best express the feelings of disgust and surprise. This provides a new perspective for the study of body parts and emotions.

Keywords: Chinese radicals · Body part words · Emotional concatenation

1 Introduction

Emotions are not only the sensations in people's minds, but also inextricably linked with people's physical reactions. For example, when people are happy, they are more likely to stand upright, but when people feel sad or depressed, their bodies tend to drop [1]. Similarly, anger may manifest itself as blushing, which increases temperature and heart rate [2]. In this respect, the body is not only the place where we experience emotions, but also the medium to convey them.

Language is also a tool or medium for expressing emotions. In every language, there are lexical expressions describing human emotional response. These lexical expressions describing human emotions include not only the emotional words themselves, but also some words describing body parts to express emotions. For example, the English phrase "someone pulled a long face" means sad or unhappy mood [3]. In Chinese, it means anger or resentment with "咬牙切齿(gnashing teeth)".

The previous researches choose body part words as research objects from the perspective of cognitive linguistics, and perform manual analysis of the selected

© Springer Nature Switzerland AG 2020
J.-F. Hong et al. (Eds.): CLSW 2019, LNAI 11831, pp. 628–638, 2020.
https://doi.org/10.1007/978-3-030-38189-9_64

combinations through dictionary interpretation, to get the emotional types of language units where body part words belong [1, 4].

From the perspective of Chinese character radicals, this paper attempts to examine the association of body parts with emotions in corpus and the emotional tendencies they express, based on the emotional vocabulary. Chinese character component and radical play an important role in the meaning of Chinese characters. Studying the relationship between body parts and emotions from radicals provides a new perspective for previous studies based on dictionary interpretation. In addition to knowing whether radicals can be associated with various emotions, corpus-based research can also provide the "strength" radicals connected to each emotion, so as to supplement the lack of dictionary research methods.

2 Research Review

When people cognize the world, they mostly follow the principle of "nearly taken from the body, far from the objects (近取诸身, 远取诸物)", using physical experience to cognize the world. For example, humans first recognize their own organ "eyes" and then use it to refer to a similar part of the object: "eye of a needle", which in turn refers to the abstract thing: "eye of the soul" [5]. Therefore, many scholars [1, 6–8] study the metaphorical and metonymy cognitive mechanisms of body parts from the perspective of cognitive linguistics.

At the same time, Gibbs argues that "all aspects of the physical experience are shaped by cultural processes" [2]. Although human body shapes are similar, the link between physiological symptoms and internal states may change from culture to culture. This leads to some scholars conducting comparative studies on the vocabulary of body parts in different languages. For example, Los Angeles [8] compares the body parts of Britain and Japan. The results of the study show that the same emotion or feeling are associated with different parts of the body in English and Japanese. English uses "heart" to express "self". Japanese is expressed in "gut", such as "hara no uchi (inside of the belly)" and "hara o waru (crack the belly in half)", indicating that a person's true self (including feelings and emotions) are located in gut. Zhang and Cheng compare the "face" in Chinese and English [9, 10]. The former finds that the semantic derivation of English and Chinese meanings is similar in basic cognition, but it exists differences in the aspects of social culture and Noun-Verb transfer. The latter analyzes the metaphorical and metonymic words produced by "face" and "脸(face)" in English and Chinese, and describes the process of word meaning extension in metaphor and metonymy.

In addition, scholars study more specific body parts. The most research on "heart", such as Wang [11], Wang [12], Li [7] and so on. Among them, Li [7] takes the Chinese characters with radical "忄(heart)" appearing in Three Hundred Tang Poems as the research object, citing the ancient sentences to analyse the hidden emotional metaphor and metonymy of these Chinese characters from the perspective of cognitive linguistics. There are also cognitive studies on body part words as a whole. For example, Yu [4] divides body parts into 22 external body parts and 7 internal body parts or organs. At first, external body terms mostly use metaphors to express emotions, that is, describing emotions through physical events that can be observed from the outside. For example, “垂头丧气(be downcast)” and “翘首以盼(eagerly anticipated to)” are expressions of emotions such as discouragement and expectation through head movements. The internal body parts often use metaphors. The imagination of the body images is triggered by emotions, such as “肝肠寸断(Liver and intestines are cut into inches)” and “愁肠百结(the sad bowels were entangled)” which use the length of the intestines to describe the degree of sadness. Baş [1] uses 488 idioms containing body parts to systematically describe the habitual expression of body words in Turkish, and analyze them in the framework of metaphor and metonymy cognitive theory.

In addition to the analysis of the body part terminology, Xu [13] added the Chinese radicals to the expression model of the emotional vocabulary according to the emotional classification, and achieved good results, but this study did not specifically target the radicals containing the body parts.

3 Data Preparation

Because there are many body parts in Chinese characters, we focus on the analysis of head-related parts in this paper. Head is the main part of people’s emotional expression, including facial, eye, mouth and other important organs. In Chinese, there are corresponding character components and radicals. The main work of this paper is to select the radicals and Chinese characters related to the body parts of the head from the relevant dictionaries, classify and form the radical word list, count the frequency of the relevant radicals in the corpus containing the emotional words, and calculate the possibility of the concatenation of body parts and emotions. The specific process is as follows:

3.1 Data Collection

Radical Word List. First, look for the radicals related to the body parts from *Modern Chinese Dictionary* [14] and *Grand Chinese Dictionary* [15], and then climb the Chinese characters containing these radicals from the website of the guoxuedashi [16] to form the radical word list. There are 12 kinds of radicals including head body parts: 首(head), 面(face), 目(eye), 鼻(nose), 自(zi), 口(mouth), 舌(tongue), 牙(tooth), 齿(tooth), 耳(ears), 页(ye), 頁(ye) (Table 1).

Table 1. The radical word list (partial)

首(head): 首/馗/䭕/䭘/馛/䭙/䭓/䭚/䭜/馘	**面(face):** 面/靣/耐/靬/靭/靮/靰/靯/靶/靼;
目(eye): 目/盯/盲/看/眉/盼/省/相/眨/眄	**鼻(nose):** 鼻/鼿/齀/鼾/齁/齂/齃/齆/齇/齈
自(zi): 自/臬/臭/臮/臯/臰/臱/臲/臵/臱	**口(mouth):** 口/叼/叮/叫/叹/叶/召/叱/叽/叩
舌(tongue): 舌/舍/乱/舍/舐/括/舲/舑/舓/舔	**牙(tooth):** 牙/𤘅/犽/玡/跖/琅/呀/讶/邪/雅
齿(tooth): 齿/龀/龁/龂/龃/龅/龆/龄/龇/龈	**耳(ears):** 耳/耵/耶/耷/耸/耿/聃/聆/联
页(ye): 额/领/颉/项/顶/颅/顺/颓/颀/顷	**頁(ye):** 額/項/頡/頏/顏/顫/頭/頓/須/領

The above radicals are classified into 8 types according to their meanings: head (including 首, 页and 頁), face(面), eye(目), nose (鼻 and 自), mouth(口), tongue(舌), tooth (牙 and 齿), ears(耳). After that, the radical word list is used to match in the sentence containing the emotional words, and the obtained word table is filtered to form the final word list.

Emotion Word List and the Extraction of Emotional Sentence. The corpus is Sina Weibo [17] data from March 25 to April 13, 2013 for totally 20 days. First, we should filter out irrelevant ids, time, location, emoticons, and more. In addition, this paper adopts 6 classification emotion system [18], combining hownet and the affective lexicon ontology of Dalian University of Technology [19], and then counts the word frequency of "感到(feeling)" plus emotional words, taking the top 5 highest frequency of each class to form an emotion word list. The resulting emotion word list is as follows (Table 2):

Table 2. Emotion word list

喜(happiness): 快乐/高兴/欢乐/开心/愉快	哀(sadness): 难过/悲伤/伤心/悲痛/痛心
恶(disgust): 讨厌/仇恨/厌恶/痛恨/怨恨	怒(anger): 愤怒/生气/气愤/恼火/恼怒
惊(shock): 惊讶/震惊/大吃一惊/惊奇/难以置信	惧(fear): 害怕/恐惧/恐慌/畏惧/提心吊胆

On this basis, extracting the clauses of various emotional words, the emotions expressed by a sentence can be reflected by the emotional words appearing in the sentence. The number of various emotional sentences extracted in this experiment is: happiness (367,538 sentences), sadness (112,751 sentences), disgust(112,899 sentences), anger (56,701 sentences), shock (17,569 sentences), fear (84,408 sentences).

3.2 Statistical Result

For each type of emotional sentence, we use the radical word list above to match the word frequency of each body part word. The obtained result may have Chinese characters that are not related to the body part or whose meaning is blurred and are not related to the body part, such as "叶(口)", "醒, 靛(齿)", "朕(耳)", etc. We will manually screen these Chinese characters. The specific screening criteria are as follows:

i. a part of the body, such as "面, 口, 耳, 眼, 鼻, 额(face, mouth, ears, eyes, nose and forehead)"
ii. actions related to body parts: "舔, 吃, 看(licking, eating and watching)"
iii. ideological activities or mental states, feelings related to the body parts: "臭(smelly)".

Since some body radicals may include characters expressing other body parts, we will classify some typical body parts: add the "嗅(sniff)" to the radical "鼻(nose)", "听(listen)" to "耳(ear)", "脸/臉(face)" to radical "面(face)". In this way, we get the radical_emotion list. Table 3 is the word frequency of each radical in 6 kinds of emotional sentences and the top three highest frequency words in the radical-emotion word list:

Table 3. High-frequency words in radical_emotion word list and frequency of radicals in emotional sentences

Radical/Emotional tendency	喜 (happiness)	哀 (sadness)	恶 (disgust)	怒 (anger)	惊 (shock)	惧 (fear)	Characters of High frequency	Character frequency in total
首(页, 頁, head)	8435	3029	2349	1313	456	1826	首3720; 顾 2641; 领2135	17408
面(face)	10180	3155	4278	1792	720	3703	面18207; 脸 (臉)5566; 靥 50	23828
目(eye)	54067	18568	16824	9292	3301	13183	看55486; 睡 19038; 眼 12505	115235
鼻(自, nose)	1102	372	746	220	53	202	臭1391; 鼻 1199; 鼾61	2695
耳(ear)	14170	5756	3501	1747	519	2123	听17665; 聊 6751; 耳1718	27816
口(mouth)	64175	21034	21848	11016	2936	11272	吃24804; 哭 10387; 叫8251	132281
舌(tongue)	209	80	158	30	16	44	舌359; 舔162; 舐16	537
牙(齿, tooth)	1073	288	470	156	75	224	牙1878; 齿 343; 龈38	2286

Since the number of each emotional sentences and the frequency of each type of radical in use are different, we will normalize the results in order to compare the emotional tendency (Q) of each radical in various emotional sentences:

$$Q = \frac{b}{B * E} * c \tag{1}$$

Learning from mutual information formula, b refers to the frequency of characters that contain a certain body part in one kind of emotional sentences; B refers to the total word frequency of characters that contain such body part in all six types of emotional sentences; E refers to the total number of sentences that represent such emotions; c is the coefficient added to avoid the result being too small. This article takes 1000,000. We will draw the emotional connections of the various body radicals in Fig. 1:

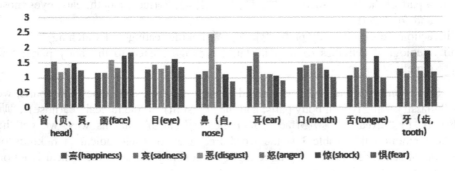

Fig. 1. The concatenation of the radicals and emotions of the head

4 Statistical Analysis

Figure 1 shows that each type of body part radicals can be associated with all emotions, but the emotional tendencies expressed by various body parts are different. The radical "鼻(nose)" and "舌(tongue)" are most closely linked to disgust. Radical "舌(tongue)" and "牙/齿(tooth)" are most able to express feelings of disgust and shock. Radical "页/頁(ye)" expresses sad and shocked emotions. "面(face)" department is more closely connected with the emotions of fear, surprise and disgust. Radical "目(eye)" expresses shock and sadness. "耳(ear)" expresses sadness and happiness. "口(mouth)" has no particularly prominent emotional expression.

4.1 Radical "首(Head, Including 首, 页 and 頁)"

"首" means "head", such as "昂首(head up)", but in modern Chinese, "首" is always quantifier, indicating the number of poems. Figure 1 tends to express sad emotions, because the "首" in the corpus is used as a quantifier and together with "sad songs", such as:

(1) 其实这是首悲伤的歌(In fact, this is a sad song)

This situation was not taken into account in this article. After retaining the "页/頁" section, the right side of Fig. 2 is the emotional tendency and the emotions of surprise and joy are most prominent.

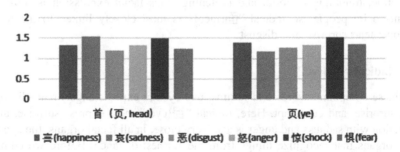

Fig. 2. Before and after deleting "首"

Yu [4] examined the compound words and idioms containing the terms of the Chinese head body parts and found that the head(头 or 首) can express sadness, joy, hope, and disgust, such as "垂头丧气(be downcast)" and "翘首以盼(eagerly antici-pated to)". Using the action to express emotions, in line with the cognitive mechanism of happy is up and unhappy/sad is down. In this article, the sad and shocked emotions are the closest to "首(head)". When "首" is removed, the emotions of shock and happiness are most prominent.

4.2   Radical "面(Face)"

The radical "面(face)" most expresses the feelings of fear, shock and disgust, but the connection with the other three emotions is also very close. This is because the highest frequency of characters in radical "面(face)" is "面" and "脸". The frequencies are 18207 and 5566. It accounts for 99.7% of the total character frequency of 23828, which is close to 100%.

(2) 看着那一张小脸上写满了恐惧(look at that little face full of fear)
(3) 他一脸惊讶的问我是怎么知道的(He looked surprised and asked me how I knew)
(4) 最讨厌这苦瓜脸了(I hate this sour face)
(5) 她还一脸高兴地冲过去(She rushed over with a happy look on her face)
(6) 满脸的伤心(A sad face)
(7) 吃醋生气那黑脸实在太棒(The black face with jealousy and anger is wonderful)

Yu [4] believes that the face is the most expressive body part, because it contains some physical features that express emotions such as eyebrows, eyes, mouth, etc., which can express happiness, sadness, anger, fear, disgust and surprise. The radical "面(face)" in this article best expresses fear, surprise and disgust, but the connection with the other three emotions is also very close, which may be related to the advantage of recognizing negative emotions [20]. When people express fear, surprise or disgusting emotions, they often cannot pass through only one facial organ. They may need a variety of facial expressions, such as fearful emotions: sluggish eyes, pupil dilation or contraction, trembling of tooth and twitching. This facial expression is most easily recognized by people, so radical "面(face)" is most closely linked to the negative emotions: fear, surprise and disgust.

4.3   Radical "目(Eye)"

In Chinese compound words and idioms, use eyes to express anger, fear, disappointment, surprise, and contempt. Here, radical "目(eye)" most express surprise, and the connection with sadness and anger is also very close. In all facial organs, the eye is one of the organs that recognize things from the furthest distance. Before we come into contact with something, or evaluate something, it is often observed through the eyes. For things that are unfamiliar or unconventional, of course, we could feel surprise.

(8) 惊奇地看着这世界(Look at the world in wonder)

In addition, the connection between radical "目" and sadness and anger is also very close: the highest frequency of "eyes" in the corpus is "看(look)", which can trigger various emotions:

(9)   每次看你哭得那么伤心妈妈好难受哦(Every time I see you cry so sad, i am so sad)
(10) 看着他那犟牛样很生气(I was very angry to see that he was so stubborn)
(11) 提心吊胆的看完了(Read it with fear)

This may be related to the negative tendency of vision, that is, in positive and negative emotions, the rate of visual recognition of negative emotions is higher [21].

4.4   Radical "鼻(自,Nose)"

In all emotional sentences, there are 2695 words belonging to radical "鼻(nose)", "臭 (smelly)" and "鼻(nose)" with the highest frequency is 1391 and 1199 respectively, which together account for the total frequency of the "鼻(nose)" 96%.

"鼻(nose)" refers to the disease associated with the nose. "臭(smelly)" itself is a disgusting smell. These are also related to negative emotions, causing others or themselves disgust, such as:

(12)   最讨厌那些有口臭的还在地铁里对着你狂说话的人(I hate people who are always talking to you with bad breath in the subway)
(13)   我讨厌给我用臭脸(I hate people giving me the dirty looks)
(14)   我最讨厌的就是鼻塞(What I hate most about colds is nasal congestion)
(15)   最讨厌春天鼻炎过敏(I most hate rhinitis allergy in spring)

Compared with the eyebrow, eye and mouth, the nose and ear are not prominent in expressing emotions. The compound words and idioms with radical "鼻(nose)" are mostly contemptuous and sad emotions, such as "嗤之以鼻(be contemptuous of)" and "令人酸鼻(cause sb.'s heart to ache)" [4]. This is different from the emotional results in this article. In daily life, when people smell a certain smell, they may produce positive or negative emotions. This physiological mechanism can help people stay away from these things and avoid themselves being hurt [22, 23]. Therefore, the "鼻(nose)" department is most closely linked to the disgusted emotions. Radical "耳(ear)"

4.5   Radical "耳(Ear)"

Wang and Li [24] pointed out that when people use their ears to listen to rhythms or information, the easiest emotions to be felt are sadness and fear, the most difficult is ridicule and disgust. The "ears" in idioms express more anxiety and unpleasant emotions [4]. In this article, radical "耳(ear)" tends to express sadness and happiness. The reason for the difference from the previous conclusions may be that in all emotional sentences, "听(listen)" and "聊(talk)" are the two words with the highest frequency. We may "听(listen)" to pleasing news, or sad news. Word "聊", in addition to simply chatting, often occurs in words "无聊(boring)", which may lead to sad mood.

(16)   刚听到一些很让人高兴的事情(I just heard something very happy)
(17)   这种无聊的日子好让人难过(Such a boring day makes me very sad)

4.6   Radical "口(Mouth)"

An idiom or compound that includes "口(mouth)" or "嘴(mouth)" can mean emotions such as surprise, praise, envy, contempt, and depression [4]. In this article, there is no particular type of connection between the "口(mouth)" and emotions.

(18) 值得高兴的事情是吃了两个好吃的包子(The happy thing was to eat two delicious buns)

(19) 伤心的晚上连饭都没吃(He was so sad that he did not eat dinner)

(20) 每次都好讨厌吃这个药(I hate taking this medicine every time)

(21) 让你生气吃醋的人(People who make you angry and jealous)

4.7 Radical "舌(Tongue)"

In all emotional sentences, only three words containing radial "舌(tongue)": "舌(tongue)", "舔(lick)" and "舐(lick)". The words it forms often express negative emotions, such as "毒舌(poison tongue)", "长舌妇(gossip woman)", "(油嘴)滑舌(glib-tongued)", "(嚼)舌根(gossip)", etc. E.g:

(22) 讨厌嚼舌跟的男生(I hate guys who gossips)

"舌tongue" also tends to express surprise with the action of the tongue:

(23) 惊讶得舌头都忘了收回去(I was so surprised that i forgot to withdraw my tongue)

In compound words and idioms, "舌(tongue)" can mean shock and fear, such as "咋舌(be left speechless or breathless)", but in daily life, people often use "咋舌" to express surprised emotions as in example (27). This may be related to people's habits.

4.8 Radical "牙(齿,Tooth)"

Figure 1 shows that radical "牙/齿(tooth)" tends to express surprise and disgust, such as:

(24) 讨厌一个人讨厌到咬牙地恨(I hate a person so much that i gnash my teeth with hatred)

(25) 今天惊讶的发现自己扎大牙了(Today, I was surprised to find that i got a big tooth)

Because the teeth are in the mouth and should not be observed easily, people are often surprised to find that the caries or teeth grow up. In daily life, people also express their disgust by the action of "咬牙(biting the teeth)" and "切齿(cutting the teeth)". In addition, "牙/齿"often refer to diseases associated with teeth, and these diseases may also cause people's disgust.

5   Conclusion

In summary, from the perspective of the radicals, the corpus is used to examine the concatenation of head parts words and emotions. The results are different from the previous researches. The former only examines compound words and idioms that contain body part words. This article uses radicals to include words other than the body part itself, such as the "看(look)" and "眶(eye socket)" of the radical "目(eye)", which is broader.

Based on the corpus, the results are more objective because of the research from the real data. In addition, the former mostly judges their emotional tendency and has more emotional types. This paper judges the connection with emotion through co-occurrence with emotional words. The types of emotions are divided into 6 categories, which are more systematic. Finally, we try to judge the emotional tendency by calculating the strength of the connection between each radical and emotion so as to obtain a more specific result.

Acknowledgments. This study was supported by the Ministry of education of Humanities and Social Science project (18YJA740030), and supported by Science Foundation of Beijing Language and Culture University (supported by "the Fundamental Research Funds for the Central Universities") (19YJ040003).

References

1. Baş, M.: Conceptualization of emotion through body part idioms in Turkish: a cognitive linguistic study. Master thesis, Hacettepe University, Ankara (2015)
2. Gibbs, R.W.: Embodiment and Cognitive Science. Cambridge University Press, New York (2005)
3. Yu, N.: What does our face mean to us? Pragmat. Cogn. **9**(1), 1–36 (2001)
4. Yu, N.: Body and emotion: body parts in Chinese expression of emotion. Pragmat. Cogn. **10** (1/2), 341–367 (2002)
5. Huang, B.: Research developments and problem analyses on study of body terms. J. Engl. Stud. (1), 7–12 (2012). (in Chinese)
6. Li, S.: Cognitive model and semantic analogy of human words. Sinogram Cult. (4), 8–12 (2004). (in Chinese)
7. Li, Q.: Emotional metaphor and metonymy in chinese characters with the radical of "heart"—a cognitive case study. Master thesis, Dalian University of Technology, Dalian (2006). (in Chinese)
8. Tsurumi, K.: Corpus-based analysis of body-part terms for emotions and feelings in English and Japanese. UCLA Electronic Theses and Dissertations, UCLA (2015)
9. Zhang, J.: Polysemous words: the English and Chinese "face". J. Foreign Lang. (4), 2003. (in Chinese)
10. Cheng, D.: Metaphor and metonymy of face——based on the Chinese - English corpus of face. J. East China Jiaotong Univ. **24**(3), 151–153 (2007)
11. Wang, Y.: A cognitive study on polysemy: a case study of HEART. Master thesis, Harbin Institute of Technology, Harbin (2006)
12. Wang, W.: The mental structures of spatial metaphorization of "Xin" (Heart) in Chinese. J. PLA Univ. Foreign Lang. **24**(1), 60–63 (2001)
13. Xu, L., Lin, H., Qi, R., Guan, J.: Sentiment lexicon embedding based on radical and phoneme. J. Chin. Inf. Process. **32**(06), 124–131 (2018)
14. Institute of Linguistics: Modern Chinese Dictionary. The Commercial Press, Beijing (2016)
15. Chinese Dictionary Editorial Committee: Grand Chinese Dictionary. Hubei Dictionary Publishing House, Wuhan (1987)
16. Guoxuedashi Homepage. http://guoxuedashi.com
17. Sina Weibo Homepage. http://weibo.com
18. Ekman, P.: Facial expression and emotion. Am. Psychol. **48**, 384–392 (1993)

19. Xu, L., Lin, H., Yu, P., Ren, H., Chen, J.: Constructing the affective lexicon ontology. J. China Soc. Sci. Tech. Inf. **27**(2), 180–185 (2008)
20. Fox, E., Lester, V., Russo, R., Bowles, R.J., Pichler, A., Dutton, K.: Facial expressions of emotion: are angry faces detected more efficiently? Cogn. Emot. **14**(1), 61–92 (2000)
21. Al-Mohizea, M.I.: The comprehension of body-part idioms by EFL learners: a cognitive linguistics-inspired approach. J. Cogn. Sci. **18**(2), 175–200 (2017)
22. Herz, R.S., Schankler, C., Beland, S.: Olfaction, emotion and associative learning: effects on motivated behavior. Motiv. Emot. **28**(4), 363–383 (2004)
23. Oaten, M., Stevenson, R.J., Case, T.I.: Disgust as a disease-avoidance mechanism. Psychol. Bull. **135**(2), 303–321 (2009)
24. Wang, H., Li, A.: Establishment of Putonghua emotional phonetic database and audio recognition Experiment. In: Papers of the Sixth National Academic Conference on Modern Phonetics, (Part I), pp. 119–124 (2003)

A Corpus-Based Study of Keywords in Legislative Chinese and General Chinese

Shan Wang[(✉)] and Jiuhan Yin

Department of Chinese Language and Literature, University of Macau,
Taipa, Macau
shanwang@um.edu.mo

Abstract. The study of keywords is a hot topic in corpus linguistics and plays an important role in investigating lexical features in specific contexts. For legislative Chinese, as a kind of Chinese for specific purposes, the comparative study of its keywords with those of general Chinese is of great significance for understanding its linguistic features. This study uses the legislative Chinese corpus and Chinese Web 2011 corpus mutually as the observation corpus and the reference corpus, and extracts 50 keywords with the highest keyness score respectively from each corpus. Through a comparative analysis of the semantic classification of these keywords, this study finds that legislative Chinese has the characteristics of focusing on political and economic meanings, showing strong professionalism, having more numerals and monosemous words. This study is of great significance for exploring the characteristics of legislative Chinese vocabulary and exploring the textual features of legislative Chinese.

Keywords: Keywords · Legislative Chinese · General Chinese · Corpora

1 Introduction

Keywords refer to words of unusual frequency in comparison with a reference corpus [1]. The study on keywords plays an important role in analyzing the characteristics of texts in corpus linguistics. Scott and Tribble [2] pointed out that keywords can avoid insignificant details and show the essential features of a text or a set of texts. Keywords are like a mirror that reflects the meaning of texts and a window to look into the culture [3]. Legal language is a language activity aimed at achieving the intent of legal acts [4]. Legislative Chinese, as a kind of legal Chinese, has words that are significant in the field of law, which has unique features that distinguish it from other types of Chinese.

This paper extracts the keywords from the legislative Chinese corpus [5, 6] and Chinese Web 2011 corpus[1] through using each other as an observation corpus and a reference corpus. By classifying the semantic categories of the keywords based on the semantic classification in *A Thesaurus of Modern Chinese* [7], this paper further summarizes the linguistic features of legislative Chinese. The research on keywords is of great significance for revealing the characteristics of legislative texts and can provide a reference for teaching legal Chinese vocabulary and compiling legal Chinese textbooks.

---

[1] https://www.sketchengine.eu/zhtenten-chinese-corpus/.

© Springer Nature Switzerland AG 2020
J.-F. Hong et al. (Eds.): CLSW 2019, LNAI 11831, pp. 639–653, 2020.
https://doi.org/10.1007/978-3-030-38189-9_65

2 Related Research

2.1 Legal Chinese Studies

The study of legal Chinese can be divided into research on legal Chinese language itself and legal Chinese for specific purposes. The former mainly includes research from the following aspects: (1) The classification of legal vocabulary. Pan [8] divided the vocabulary in laws into legal vocabulary and general vocabulary. Liu [9] pointed out that whether there is a legal meaning is an important basis for distinguishing legal words from general ones. Wang and Wang [10] further divided legal terms into specific lexical items and expressive function words. (2) The semantic evolution of legal vocabulary. Wu [11] proposed that the semantic evolution of legal vocabulary includes semantic expansion, semantic reduction, semantic transfer, metaphorical evolution and referential evolution. She pointed out that historical factors and foreign language influence are the main factors causing the semantic evolution. Liu [12] further pointed out that language factors, psychological factors, and the need for new concepts are important factors affecting the evolution of semantics. (3) The style of legal language, including standardization [13, 14], authority [15], conciseness [16], solemnity [17], and accuracy and ambiguity [18, 19]. (4) Language use of legal Chinese. Luo and Wang [20] investigated legal vocabulary by a self-constructed legislative Chinese corpus. They found the different usages between 但 dàn and 但是 dànshì, which can help to improve the standardization requirement of legislative language.

These studies focus on legal Chinese itself and there is a lack of comparative studies between legal Chinese and general Chinese.

The study of legal Chinese for specific purposes covers legal Chinese textbooks [21] and legal vocabulary in the vocabulary syllabus [22]. First, the study on textbooks compared two legal Chinese textbooks published so far. It pointed out that *The Specialized Chinese Course of Chinese Language* [23] is a transitional Chinese teaching material, whose main content is based on basic legal common sense. It paves the way for intermediate-level students to learn legal Chinese. The main content of *Legal Chinese-Commercial Chapter* [24] is the laws and regulations related to business activities, which are used by students with more than three years' experience of learning Chinese [21]. Second, the study of legal vocabulary in the vocabulary syllabus has a very limited scope [22], and thus cannot reflect the characteristics of legal Chinese.

2.2 Research on Keywords

The research on keywords focuses on various types of texts analysis, including English learners' corpus [25], literary works [26], news [27], speeches [28], political texts [29], and English tour guide commentaries [30]. These studies cover a variety of social and cultural phenomena, in which keywords play an important role in analyzing textual features.

In the research of language for specific purposes, there are many studies on keywords of English for specific purposes. Keyword research is regarded as the starting point of business English's stylistic research [31]. Scott and Tribble [2] analyzed the keywords in the corpus of business letters and found that vocabulary preference of people with different relationship distance is different. Liu [32] extracted the keywords from a business English corpus and found that there are more colloquial words and professional words. The semantic classification of keywords in his study indicated that words about quotations and job advertisements account for a large proportion. Kong [30] investigated keywords in English tour guides and suggested that the proper nouns of tourist attractions, historical and cultural nouns related to tourism, verbs expressing visits and appreciation, and positive adjectives are the lexical features of tour guides.

Research on keywords has great influences on textual feature analysis and the study on language for specific purposes. However, there is no comparative study of keywords between legislative Chinese and general Chinese at present.

3 Research Design

3.1 Corpus Selection

This paper selects the legislative Chinese corpus and Chinese Web 2011 corpus. The corpora's size is shown in Table 1. The legislative Chinese corpus is composed of 35 legal texts covering the seven legal categories of National People's Congress, including the constitution, administrative laws, criminal laws, civil and commercial laws, economic laws, social laws, and procedural laws. Chinese Web 2011 used as a general Chinese corpus in this paper. It contains 1.7 billion words, composed of texts from the Internet. It was annotated by the Stanford Log-linear Part-Of-Speech Tagger using

Table 1. The size of the corpora

Corpora	Word types	Tokens
The Legislative Chinese Corpus	6411	253287
Chinese Web 2011	9936706	1729867455

Chinese Penn TreeBank. It belongs to the TenTen corpus family, which is a set of the web corpora built using the same tagging methods and contains corpora in more than 30 languages.

3.2 Research Methods

The Chinese legislative corpus and Chinese Web 2011 are mutually used as the observation corpus and reference corpus. Sketch Engine [33] is then applied to extract the top 50 keywords from the two corpora respectively. A comparative study is conducted between the keywords of legislative Chinese and general Chinese.

For the extracted keywords, this paper classifies each word's semantic category based on *A Thesaurus of Modern Chinese*. This dictionary is arranged according to word meanings; the semantic classification of words in it is logical and hierarchical, which is of great significance to the study of Chinese vocabulary [34]. There are nine first-level semantic categories in the dictionary, which are Living Things, Concrete Objects, Abstract Things, Time and Space, Biological Activities, Social Activities, Movement and Change, Nature and State, and Auxiliary Words. In addition, this paper also analyzes the second-level semantic categories under Abstract Things and Social Activities. The following are the second-level categories under Abstract Things: Issues, Attributes, Consciousness, Society, Politics, Military, Economy, Science and Education, Culture, Sports and Health, and Quantity Units. The following are the second-level categories under Social Activities: Management, Trade, Production, Transportation, Culture and education, War, Justice, Faith, Social Intercourse, Help, and Strife. After the keywords in legislative Chinese and general Chinese are semantically classified, they are compared to analyze the lexical features.

4 Keywords' Statistics

4.1 Keywords in the Legislative Chinese Corpus

To extract the keywords of the legislative Chinese corpus, it is used as the observation corpus and Chinese Web 2011 as the reference corpus. The top 50 keywords are shown in Table 2. The keywords 三节 *sānjié*, 上一 *shàngyī* and 孳 *zī* is incompletely captured. By looking at the context, they are completed as 第三节 *(dì)sānjié* 'the third section', 上一(级)*shàngyī (jí)* 'upper (level)' and 孳(息) *zī (xī)* 'fructus'. In Table 2, *frequency* refers to the frequency of occurrences of the word in each corpus. *Frequency/million* is the frequency of occurrences of the word in each million words. The *keyness score* shows the saliency of the keywords. The higher the *keyness score*, the more important it is in the observation corpus compared with the reference corpus, and vice versa. The formula for calculating the *Keyness score* [35] is:

$$\frac{fpm_{focus} + n}{fpm_{ref} + n} \tag{1}$$

fpm_{focus} is the normalized (per million) frequency of the word in the focus/observation corpus, fpm_{ref} is the normalized (per million) frequency of the word in the reference corpus, and n is the simple Maths (smoothing) parameter (n = 1 is the default value).

There are a large number of professional nouns and terms in the keywords of legislative Chinese, such as 本法 *běnfǎ* 'this law', 有期 *yǒuqī* 'fixed term', 孳(息) *zī(xī)* 'fructus', and 标的 *biāodì* 'target'. In addition, there are a large number of verbs that express social activities, such as 拘役 *jūyì* 'criminal detention', 裁定 *cáidìng* 'rule', 候审 *hòushěn* 'awaiting trial', and 追索 *zhuīsuǒ* 'recourse'. These verbs are not commonly used in general Chinese. In addition to legal activities, the keywords of legislative Chinese also involves political and economic words, such as 罚金 *fájīn* 'fine' and 要约 *yàoyūe* 'offer', as well as numerals such as 第六十一 *dì liùshí yī* 'sixty-first' and 二十万 *èrshíwàn* 'two hundred thousand'. The 'sixty-first' refers to the

Table 2. Keywords in the legislative Chinese corpus

Number	Word	The Legislative Chinese corpus		Chinese Web 2011		Keyness score
		frequency	frequency/million	frequency	frequency/million	
1	不得 bùdé 'must not'	587	1921.4	2363	1.1	906.1
2	本法 běnfǎ 'this law'	955	3126	7938	3.8	655.8
3	拘役 jūyì 'criminal detention'	390	1276.6	2091	1	641.2
4	罚金 fájīn 'fine'	488	1597.3	6393	3	396.2
5	有期 yǒuqī 'fixed term'	843	2759.3	12917	6.1	387.1
6	各级 gèjí 'each levels'	117	383	21	0	380.2
7	徒刑 túxíng 'imprisonment'	959	3139	17172	8.2	343.1
8	未经 wèijīng 'have not yet'	115	376.4	245	0.1	338.1
9	承运人 chéngyùnrén 'carrier'	285	932.9	5906	2.8	245.5
10	及其 jíqí 'and (its/their/his/her)'	168	549.9	4667	2.2	171.3
11	债务人 zhàiwùrén 'obligor'	284	929.6	10622	5	154
12	不能 bùnéng 'cannot'	47	153.8	25	0	153
13	海损 hǎisǔn 'average; sea damage'	57	186.6	492	0.2	152.1
14	出租人 chūzūrén 'lessor'	92	301.1	2265	1.1	145.6
15	偿 cháng 'compensation'	82	268.4	1967	0.9	139.3
16	(第)三节 (dì)sānjié 'section three'	65	212.8	1212	0.6	135.7
17	尚未 shàngwèi 'not yet'	41	134.2	22	0	133.8
18	保管人 bǎoguǎnrén 'depositary'	55	180	753	0.4	133.4
19	承租人 chéngzūrén 'lessee'	138	451.7	5135	2.4	131.7
20	所得 suǒdé 'income'	301	985.2	13752	6.5	131
21	裁定 cáidìng 'rule; adjudicate'	323	1057.3	15294	7.3	128.1
22	无期 wúqī 'have no end in sight'	113	369.9	4112	2	125.6
23	第六十一 dì liùshí yī 'sixty-first'	58	189.8	1109	0.5	125
24	付款人 fùkuǎnrén 'drawee'	60	196.4	1221	0.6	125
25	要约 yàoyuē 'offer'	76	248.8	2200	1	122.2
26	五十万 wǔshí wàn 'five hundred thousand'	62	202.9	1546	0.7	117.6
27	上一(级)shàngyī (jí) 'upper (level)'	36	117.8	92	0	113.9
28	各种 gèzhǒng 'various'	34	111.3	0	0	112.3
29	债权人 zhàiquánrén 'creditor'	257	841.2	13862	6.6	111.1
30	重整 chóngzhěng 'reorganization'	102	333.9	4256	2	110.9
31	候审 hòushěn 'await trial'	39	127.7	353	0.2	110.2
32	汇票 huìpiào 'money Order'	152	497.5	7674	3.6	107.4
33	取保 qūbǎo 'bail out'	39	127.7	457	0.2	105.7
34	状 zhuàng 'shape'	49	160.4	1217	0.6	102.3
35	雇 gù 'hire'	36	117.8	360	0.2	101.5
36	依照 yīzhào 'according to'	820	2684.1	55378	26.3	98.4
37	孳(息) zī(xī) 'fructus'	29	94.9	4	0	95.7
38	清偿 qīngcháng 'to discharge'	128	419	7199	3.4	95.1
39	各方 gèfāng 'all parties'	30	98.2	128	0.1	93.5
40	不可抗力 bùkěkànglì 'force majeure'	30	98.2	147	0.1	92.7
41	标的 biāodì 'target'	261	854.3	17461	8.3	92.1
42	罪 zuì 'crime'	733	2399.3	53248	25.3	91.3
43	托运 tuōyùn 'consignment'	84	275	4288	2	90.9
44	赠与 zèngyǔ 'gift'	60	196.4	2611	1.2	88.1
45	各项 gèxiàng 'each and every item'	26	85.1	0	0	86.1
46	承兑 chéngduì 'acceptance'	70	229.1	3548	1.7	85.7
47	二十万 èrshíwàn 'two hundred thousand'	48	157.1	2037	1	80.4
48	背书 bèishū 'endorsement'	64	209.5	3531	1.7	78.7
49	农用地 nóngyòngdì 'agricultural land'	29	94.9	469	0.2	78.5
50	追索 zhuīsuǒ 'recourse'	48	157.1	2190	1	77.5

number of legal provisions, while 'two hundred thousand' refers to the total amount of fines stipulated in the law.

4.2 Keywords in Chinese Web 2011

In order to get the keywords of Chinese Web 2011, it is used as the observation corpus and the legislative Chinese corpus is used as the reference corpus. The top 50 keywords are shown in Table 3.

The keywords of general Chinese cover a lot of parts of speech, including content words and function words. Content words involve personal pronouns such as 我 *wǒ* 'I' and 她 *tā* 'she'; vocational nouns such as 教授 *jiàoshòu* 'professor', 老师 *lǎoshī* 'teacher', and 记者 *jìzhě* 'reporter'; daily life vocabulary such as 学院 *xuéyuàn* 'college' and 课程 *kèchéng* 'curriculum'; time words such as 现在 *xiànzài* 'now' and 今年 *jīnnián* 'this year'. Function words include auxiliary words, such as 着 *zhe* 'progressive aspect', 来 *lái* 'come; about'; adverb like 却 *què* 'but' and 很 *hěn* 'very'. The use of time words is also a major feature of general Chinese, which is different from legislative Chinese. Due to the accuracy of legislative Chinese, words such as 现在 *xiànzài* 'now' and 今年 *jīnnián* 'this year' rarely appear in it.

5 Semantic Analysis of Keywords

In order to examine the semantic preference, the keywords are further classified into different semantic categories according to *A Thesaurus of Modern Chinese*. By comparing the number of the keywords in each semantic category and the number of semantic categories each keyword belongs to (showing the polysemy and monosemy of the keywords), we further examine differences between legislative Chinese and general Chinese.

5.1 The Semantic Categories of Keywords

In the process of classifying the semantic categories of keywords according to *A Thesaurus of Modern Chinese*, we find that nearly half of the keywords in general Chinese appear in multiple semantic categories. For example, 看 *kàn* 'see' belongs to both Biological Activities and Social Activities in the first-level semantic category. Under Biological Activities, the second-level semantic category of 看 *kàn* 'see' can be Head Movements and Mental Activities. Under Social activities, the second-level semantic category of 看 *kàn* 'see' involves Culture and Education, Justice, Social Intercourse, and Help. In contrast, the keywords in legislative Chinese have only one semantic category. For example, the semantic category of 不得 *bùdé* 'must not' is Biological Activities (first level) and Psychological Activities (second level). The semantic category of 罪 *zuì* 'crime' is Abstract Things (first level) and Affairs (second level). It is obvious that keywords in legislative Chinese show a tendency of monosemy. The semantic category distribution of the keywords is shown in Table 4.

Table 3. Keywords in Chinese Web 2011

Number	word	Chinese Web 2011		Legislative Chinese corpus		Keyness score
		frequency	frequency/million	frequency	frequency/million	
1	我 wǒ 'I'	6688087	3174.7	0	0	3175.7
2	我们 wǒmen 'we'	4256560	2020.5	0	0	2021.5
3	大学 dàxué 'university'	2484425	1179.3	0	0	1180.3
4	你 nǐ 'you'	2303135	1093.3	0	0	1094.3
5	万 wàn 'ten thousand'	1700612	807.3	0	0	808.3
6	中心 zhōngxīn 'core'	1661154	788.5	0	0	789.5
7	学院 xuéyuàn 'college'	1565280	743	0	0	744
8	里 lǐ 'inside'	1373611	652	0	0	653
9	看 kàn 'see'	1279267	607.2	0	0	608.2
10	她 tā 'she'	1269915	602.8	0	0	603.8
11	教师 jiàoshī 'teacher'	1216334	577.4	0	0	578.4
12	一些 yīxiē 'some'	1194523	567	0	0	568
13	去 qù 'go'	1180940	560.6	0	0	561.6
14	那 nà 'that'	1163134	552.1	0	0	553.1
15	这样 zhèyàng 'such'	1111019	527.4	0	0	528.4
16	目前 mùqián 'at present'	1075119	510.3	0	0	511.3
17	什么 shénme 'what'	930158	441.5	0	0	442.5
18	教授 jiàoshòu 'professor'	898867	426.7	0	0	427.7
19	想 xiǎng 'think'	897345	426	0	0	427
20	现在 xiànzài 'now'	893990	424.4	0	0	425.4
21	美国 měiguó 'U.S.A'	866966	411.5	0	0	412.5
22	亿 yì 'a hundred million'	854330	405.5	0	0	406.5
23	大家 dàjiā 'everybody'	837319	397.5	0	0	398.5
24	老师 lǎoshī 'teacher'	830984	394.5	0	0	395.5
25	则 zé 'though'	814626	386.7	0	0	387.7
26	这 zhè 'this'	5966474	2832.2	2	6.5	375.4
27	却 què 'but; retreat'	783236	371.8	0	0	372.8
28	课程 kèchéng 'curriculum'	728737	345.9	0	0	346.9
29	所以 suǒyǐ 'therefore'	723666	343.5	0	0	344.5
30	今年 jīnnián 'this year'	693269	329.1	0	0	330.1
31	如何 rúhé 'how'	690065	327.6	0	0	328.6
32	很多 hěnduō 'quite a lot'	684245	324.8	0	0	325.8
33	应该 yīnggāi 'should'	679891	322.7	0	0	323.7
34	记者 jìzhě 'journalist'	666282	316.3	0	0	317.3
35	党员 dǎngyuán 'party member'	633913	300.9	0	0	301.9
36	优秀 yōuxiù 'excellent'	628682	298.4	0	0	299.4
37	而且 érqiě 'and'	625004	296.7	0	0	297.7
38	上海 shànghǎi 'Shanghai'	620851	294.7	0	0	295.7
39	深入 shēnrù 'thorough'	616703	292.7	0	0	293.7
40	着 zhe 'progressive aspect'	2622042	1244.6	1	3.3	291.5
41	学科 xuékē 'subject'	610553	289.8	0	0	290.8
42	孩子 háizi 'child'	610145	289.6	0	0	290.6
43	来 lái 'come; about'	2610872	1239.3	1	3.3	290.3
44	模式 móshì 'pattern'	606237	287.8	0	0	288.8
45	高校 gāoxiào 'colleges and universities'	590750	280.4	0	0	281.4
46	成功 chénggōng 'success'	586172	278.2	0	0	279.2
47	同学 tóngxué 'classmate'	583701	277.1	0	0	278.1
48	提升 tíshēng 'promote'	572288	271.7	0	0	272.7
49	很 hěn 'very'	2448235	1162.1	1	3.3	272.2
50	认识 rènshi 'know'	540601	256.6	0	0	257.6

Table 4. Semantic category distribution of keywords

Semantic distribution condition	Legislative Chinese		General Chinese	
	Quantity	Percent	Quantity	Percent
Words with more than one category	0	0%	18	36%
Words with one semantic category	50	100%	32	64%
Total	50	100%	50	100%

5.2   Contrast of the First-Level Semantic Categories of Keywords

Based on *A Thesaurus of Modern Chinese*, we manually classify the keywords into different semantic categories. For words that do not exist in the dictionary, we refer to similar words in the dictionary for manual classification.

Among the keywords in legislative Chinese, 14 words cannot be found in *A Thesaurus of Modern Chinese*, including three numerals: 第六十一 *dì liùshíyī* 'sixty-first', 二十万 *èrshí wàn* 'two hundred thousand', and 五十万 *wǔshí wàn* 'five hundred thousand'. By referring to 第一 *dìyī* 'the first' and 一 *yī* 'one' in the dictionary, all numerals are classified into Abstract Things - Quantity. The remaining 11 words are classified with reference to their synonyms: referring to the semantic category of 法律 *fǎlǜ* 'law' in the dictionary, 本法 *běnfǎ* 'this law' is classified into Abstract Things - Politics; referring to the semantic category of 无期 *wúqī* 'have no end in sight', 有期 *yǒuqī* 'fixed term' is classified into Nature and State - Nature; referring to the semantic category of 各方 *gèfāng* 'all parties', 各级 *gèjí* 'each level' and 各项 *gèxiàng* 'every item' are classified into Abstract Things - Attributes; referring to the semantic category of 承租人 *chéngzū rén* 'lessee', 承运人 *chéngyùn rén* 'carrier', 保管人 *bǎoguǎn rén* 'depositary', and 付款人 *fùkuǎn rén* 'drawee' are classified into Living Things - Person; referring to the semantic category of 章节 *zhāngjié* 'chapter', (第)三节 *(dì) sānjié* 'section three' is classified into Abstract Things - Science and Education; referring to the semantic category of 合同 *hétong* 'contract', 要约 *yàoyuē* 'offer' is classified into Abstract Things – Politics; referring to the semantic category of 级 *jí* 'level', 上一(级) *shàngyī(jí)* 'upper (level)' is classified into Abstract Things - Quantity; referring to the semantic category of 利息 *lìxi* 'interest', 孳(息) *zī(xī)* 'fructus' is classified into Abstract Things - Economy. Among the keywords in general Chinese, only one word cannot be found in *A Thesaurus of Modern Chinese*. With reference to the semantic category of 中国 *zhōngguó* 'China', 美国 *měiguó* 'U.S.A' is classified into the Abstract Thing - Politics. All the keywords are classified into different categories and the contrast results obtained are shown in Table 5.

Through comparing the quantitative differences, we find that the keywords' semantic categories in legislative Chinese are concentrated in the fields of Social Activities and Abstract Things, while those in general Chinese are concentrated in the fields of Living Things, Auxiliary Words, Abstract Things, and Biological Activities. The number of keywords in legislative Chinese and general Chinese differs most obviously in the three semantic categories of Social Activities, Abstract Things, and Living Things. To further analyze their differences, the second-level semantic categories of the keywords from these three first-level semantic categories are compared in the next section.

Table 5. Comparison of keywords' first-level semantic categories

First-level semantic category of the keywords	Number of keywords in legislative Chinese	Number of keywords in general Chinese	Quantity difference	Absolute value of quantity difference
Social Activities	10	1	9	9
Abstract Things	19	11	8	8
Living Things	7	11	−5	5
Time and Space	0	3	−3	3
Biological Activities	4	2	−2	2
Auxiliary Words	4	5	−1	1
Concrete Objects	2	1	1	1
Movement and Change	1	0	1	1
Nature and State	3	3	0	0
Abstract Things, Time and Space, Nature and State	0	1	−1	1
Abstract Things, Time and Space	0	1	−1	1
Biological Activities, Social Activities	0	1	−1	1
Living Things, Concrete Objects	0	1	−1	1
Nature and State, Concrete Objects	0	1	−1	1
Abstract Things, Biological Activities, Social Activities, Movement and Change	0	1	−1	1
Concrete Objects, Auxiliary Words	0	2	−2	2
Social Activities, Living Things	0	1	−1	1
Concrete Objects, Time and Space	0	1	−1	1
Social Activities, Auxiliary Words	0	1	−1	1
Biological Activities, Movement and Change, Auxiliary Words	0	1	−1	1
Abstract Things, Biological Activities	0	1	−1	1
Total	50	50	/	/

5.3   Contrast of the Second-Level Semantic Categories of Keywords

The three first-level semantic categories with the greatest difference in the number of keywords between legislative Chinese and general Chinese are Social Activities, Abstract Things, and Living Things. This section further compares their second-level semantic categories. In addition to keywords that only belong to one semantic category, there are keywords that belong to more than one semantic category and at least one of them is from the above three semantic categories. Such kind of keywords are also included, so the number of keywords in each category is 12, 15 and 5 respectively.

For the keywords whose first-level semantic category is Abstract Things, their second-level semantic categories are shown in Table 6. The words of legislative Chinese are more professional. Among the 18 keywords in Abstract things, nearly one-third of them are legal terms, such as 本法 *běnfǎ* 'this law', 徒刑 *túxíng* 'imprisonment', 要约 *yàoyuē* 'offer', 标的 *biāodì* 'Target', 罚金 *fájīn* 'fine', and 孳(息) *zī(xī)* 'fructus'. They belong to the semantic categories of Attributes, Politics, Affairs, and Economy. There are also numerals like 第六十一 *dì liùshí yī* 'sixty-first' and 二十万 *èrshíwàn* 'two hundred thousand' that express quantity. In contrast, the second-level semantic category Science and Education accounts for a large proportion of the keywords of Abstract Things.

All the keywords of Living Things in legislative Chinese and general Chinese are shown in Table 7. Their second-level semantic category all belongs to Human. However, the words representing "human" used in legislative Chinese are professional vocabulary with special legal definitions, which are general definitions of the bearers or implementers of a certain right or obligation in legal acts. For example, 承运人 *chéngyùnrén* 'carrier' means people who take responsibility for transportation and provide transportation services. 债务人 *zhàiwùrén* 'obligor' means people who are liable to repay creditors. In contrast, the general Chinese words for "human" are those used in daily life, such as personal pronouns 我 *wǒ* 'I' and 我们 *wǒmen* 'we', and professional nouns 教授 *jiàoshòu* 'professor' and 记者 *jìzhě* 'reporter'.

The second-level semantic categories of the keywords of Social Activities in legislative Chinese and general Chinese are shown in Table 8. It shows that the semantic categories of keywords in legislative Chinese are mostly Help, Justice and Management. The keywords in legislative Chinese reflect activities in different fields. For example, 拘役 *jūyì* 'criminal detention' reflects judicial activities, 承兑 *chéngduì* 'acceptance' reflects economic and trade activities, and 雇 *gù* 'hire' reflects management activities. In contrast, for the keywords of Social Activities in general Chinese, most of them belong to the second-level semantic category Culture and Education.

Table 6. Second-level semantic categories of the keywords that belong to Abstract Things

Legislative Chinese		General Chinese	
word	Second-level semantic category under Abstract Things	word	Second-level semantic category under Abstract Things
状 *zhuàng* 'shape'	Attributes	学院 *xuéyuàn* 'college'	Science and Education
各方 *gèfāng* 'all parties'	Attributes	课 程 *kèchéng* 'curriculum'	Science and Education
各级 *gèjí* 'each levels'	Attributes	学科 *xuékē* 'subject'	Science and Education
各项 *gèxiàng* 'each and every item'	Attributes	高校 *gāoxiào* 'colleges and universities'	Science and Education
不可抗力 *bùkěkànglì* 'force majeure'	Attributes	去 *qù* 'go'	Science and Education
本法 *běnfǎ* 'this law'	Politics	万 *wàn* 'ten thousand'	Quantity Units
徒 刑 *túxíng* 'imprisonment'	Politics	亿 *yì* 'a hundred million'	Quantity Units
(第)三节 *(dì)sānjié* 'section three'	Politics	上 海 *shànghǎi* 'Shanghai'	Politics
要约 *yàoyuē* 'offer'	Politics	美国 *měiguó* 'U.S.A'	Politics
第六十一 *dì liùshí yī* 'sixty-first'	Quantity Units	模式 *móshì* 'pattern'	Attributes
二十万 *èrshíwàn* 'two hundred thousand'	Quantity Units	大学 *dàxué* 'university'	Society
五十万 *wǔshí wàn* 'five hundred thousand'	Quantity Units	这样 *zhèyàng* 'such'	Affairs
上一(级) *shàngyī(jí)* 'upper (level)'	Quantity Units	成 功 *chénggōng* 'success'	Affairs
标的 *biāodì* 'target'	Affairs	中心 *zhōngxīn* 'core'	Attributes, Science And Education
罪 *zuì* 'crime'	Affairs	里 *lǐ* 'in'	Affairs, Quantity Units
所得 *suǒdé* 'income'	Affairs	-	-
罚金 *fájīn* 'fine'	Economy		
孳(息) *zī(xī)* 'fructus'	Economy		

Table 7. All the keywords of Living Things

Legislative Chinese	General Chinese
承运人 chéngyùn rén 'carrier'	我 wǒ 'I'
债务人 zhàiwù rén 'obligor'	我们 wǒmen 'We'
出租人 chūzūrén 'lessor'	你 nǐ 'you'
保管人 bǎoguǎn rén 'depositary'	她 tā 'she'
承租人 chéngzū rén 'lessee'	教师 jiàoshī 'teacher'
付款人 fùkuǎn rén 'drawee'	教授 jiàoshòu 'professor'
债权人 zhàiquán rén 'creditor'	大家 dà jiā 'everybody'
-	老师 lǎoshī 'teacher'
	记者 jìzhě 'reporter'
	党员 dǎngyuán 'party member'
	孩子 háizi 'child'
	同学 tóngxué 'classmate'

Table 8. Second-level semantic categories of the keywords in Social Activities

Legislative Chinese	Second-level semantic category of Social Activities	General Chinese	Second-level semantic category of Social Activities
偿 cháng 'compensation'	Help	提升 tíshēng 'promote'	Management
清偿 qīngcháng 'pay off'	Help	看 kàn 'see'	Social Intercourse, Justice, Culture and Education, Help
赠与 zèngyǔ 'gift'	Help	去 qù 'go'	Culture and Education
取保 qǔbǎo 'bail out'	Justice	教授 jiàoshòu 'professor'	Culture and Education
拘役 jūyì 'criminal detention'	Justice	却 què 'but; retreat'	War
裁定 cáidìng 'rule; adjudicate'	Management	-	-
雇 gù 'hire'	Management		
承兑 chéngduì 'acceptance'	Trade		
托运 tuōyùn 'consignment'	Transportation		
候审 hòushěn 'awaiting trial'	Social Intercourse		

6   Conclusion

This paper selects the legislative Chinese corpus and Chinese Web 2011 corpus, extracts the keywords from respective corpus by using each other as the observation corpus and the reference corpus, classifies and compares the semantic categories of the keywords, and explores the characteristics of legislative Chinese in comparison with general Chinese.

This study finds that the keywords of legislative Chinese present the following characteristics: first, the semantic categories that represent Abstract Things and Social Activities predominate in legislative Chinese. In Abstract Things, keywords in legislative Chinese account for a large proportion in terms of Attributes, Politics, and Quantity Units. It shows that the legislative Chinese is biased towards Politics and Economy in the semantic domain. Second, keywords in legislative Chinese show strong professionalism. In the categories of Abstract Things and Living Things, the keywords of technical terms in legislative Chinese are particularly prominent, especially the keywords in Living Things. Third, there are many keywords representing Quantity in legislative Chinese, such as 第X章 *dì X zhāng* 'Chapter X' and 第X节 *dì X jié* 'Section X' of the chapter classification, the legal article 第X条 *dì X tiáo* "Article X", and 二十万 *èrshíwàn* 'two hundred thousand' which represent the amount of fines. Fourth, the top 50 keywords in legislative Chinese all belong to one semantic category, while half of the top 50 keywords in general Chinese belong to two semantic categories. It reflects the accuracy requirement of legislative Chinese, which strives to be accurate in the use of words. In the future, we will carry out research on the collocation of keywords to further deepen the study of vocabulary comparison between legal Chinese and general Chinese.

Acknowledgement.   This research is supported by MYRG2019-00013-FAH, University of Macau.

References

1. Scott, M.: Analysis of keywords and key keywords. System, pp. 1–13 (1997)
2. Scott, M., Tribble, C.: Textual Patterns: Key Words and Corpus Analysis in Language Education. John Benjamins Publishing Company, Amsterdam (2006)
3. Wu, Z.: A corpus-based study of keywords—taking news reports on Christmas as an example (Jīyú yǔliàokù de zhǔtící yánjiū—yǐ guānyú shèngdànjié de xīnwén bàodào wéilì). Lang. Plan. **22**, 53–54 (2012)
4. Hu, F.: Some issues in the study of legal speech acts in China (Zhōngguó fǎlǜ yányǔ xíngwéi yánjiū de ruògān wèntí). Contemp. Rhetor. **4**, 1–7 (2006)
5. Luo, H., Wang, S.: The construction and application of the legal corpus of Mainland China (Zhōngguó dàlù fǎlǜ yǔliàokù jíqí yīngyòng yánjiū). In: The 18th Chinese Lexical Semantics Workshop (CLSW2017). Leshan Normal University, Leshan, Sichuan, China (2017)
6. Luo, H., Wang, S.: The construction and application of the legal corpus. Chinese Lexical Semantics. LNCS (LNAI), vol. 10709, pp. 448–467. Springer, Cham (2018). https://doi.org/10.1007/978-3-319-73573-3_41
7. Su, X.: A Thesaurus of Modern Chinese (Xiàndài hànyǔ fēnlèi cídiǎn). The Commercial Press, Beijing (2013)
8. Pan, Q.: Judgment of Chinese Forensic Language (Zhōngguó fǎlǜ yǔyán jiànhéng). Truth & Wisdom Press, Shanghai (2004)
9. Liu, H.: Forensic Linguistics (Fǎlǜ yǔyánxué). Peking University Press, Beijing (2007)
10. Wang, D., Wang, L.: The classification of legal terms in the context of the legislative standardization and the scientific pursuit of legal language (Lìfǎ guīfànhuà 、 kēxuéhuà shìjiǎo xià de fǎlǜ shùyǔ fēnlèi yánjiū). Appl. Linguist. **3**, 108–114 (2010)

11. Wu, Y.: The Semantic Changes of Chinese Legal Words (Zhōngguó fǎlǜ cíhuì de yǔyì yǎnjìn). Southwest University of Political Science & Law (2012)

12. Liu, Q.: On the causes of the semantic evolution of chinese legal vocabulary (Zhōngguó fǎlǜ cíhuì yǔyì yǎnbiàn de yòuyīn tàntǎo). Soc. Sci. Rev. **29**, 70–71 (2014)

13. Xu, R.: Standardization of legal language—a historical jurisprudence interpretation based on modern China context (Fǎyán fǎyǔ de guīfàn huà (xìng)—yīgè jīyú jìndài Zhōngguó yǔjìngde lìshǐ fǎxué jiěshì). Law Sci. **2010**, 30–44 (2010)

14. Zhu, T.: Standardization of legislative language in the compilation of civil code (Mínfǎdiǎn biānzuǎn zhōngde lìfǎ yǔyán guīfànhuà). China Leg. Sci. **2017**, 230–248 (2017)

15. Xu, M.: The challenges of scientific and technological development to the authority of legal language (Kējì fāzhǎn duì fǎlǜ yǔyán quánwēi xìng de tiǎozhàn). Lang. Educ. **6**, 36–39 (2018)

16. Lv, W., Yao, S.: Vocabulary regulation and the simplicity of legal language (Cíhuì guīzhì yǔ lìfǎ yǔyán de jiǎnmíng xìng). Appl. Linguist. **11**, 65–74 (2018)

17. Zheng, J.: A contrastive analysis of Chinese realizations of legal "solemnity": a study based on complex comparable corpora of legal Chinese varieties (Fǎlǜ hànyǔ zhuāngzhòng xìng yǔtǐ shíxiàn lùjìng duìbǐ fēnxī). J. Guangdong Univ. Foreign Stud. **28**, 78–86 (2017)

18. Zhang, N.: Analyzing the accuracy and ambiguity legal document (Qiǎnxī fǎlǜ wénshū yǔyán de jīngquè xìng yǔ móhū xìng). J. Wuhu Inst. Technol. **7**, 54–55 (2005)

19. Du, J.: The transition from uncertainty of legal language to certainty of the judicial result (Cóng fǎlǜ yǔyánde móhūxìng dào sīfǎ jiéguǒde quèdìngxìng). Mod. Foreign Lang. (Q.) **24**, 305–310 (2001)

20. Luo, H., Wang, S.: The Transitional Sentences in Legislative Language in the Cases of the dan and danshi sentences (Lìfǎ yǔyán zhōng de dànshū hé fēidànshū yánjiū). Contemporary Rhetoric (Dāngdài xiūcí xué) 6, 77–89 (2018). [Reproduced by China Social Science Excellence Réndà fùyìn bàokān zīliào, 2, 130–139, 2019]

21. Liu, Z.: Comparative Case of Legal Chinese Textbooks—Take Specialized Chinese Course of Chinese Law and Legal Chinese-Commercial Chapter as the Case. School of Chinese as a second language (Fǎlǜ hànyǔ jiàocái gè'àn bǐjiào—yǐ 'Zhōngguó fǎlǜ zhuānyè hànyǔ jiàochéng' hé 'fǎlǜ hànyǔ—shāngshì piān' wéilì). Beijing Postgraduate Forum on Teaching Chinese as a Foreign Language, vol. 10 (2013)

22. Yang, D.: Research on Legal Vocabulary in Teaching Chinese as a Foreign Language (Duìwài hànyǔ jiàoxué zhōngde fǎlǜ cíhuì yánjiū). Heilongjiang University (2012)

23. Wang, R.: Specialized Chinese Course of Chinese Law (Zhōngguó fǎlǜ zhuānyè hànyǔ jiàochéng). Peking University Press, Beijing (2007)

24. Zhang, T.: Legal Chinese—Commercial Chapter (Fǎlǜ hànyǔ—shāngshì piān). Peking University Press, Beijing (2008)

25. Li, W.: A CLEC-based analysis of key words and associates (Jīyú yīngyǔ xuéxízhě yǔliàokù de zhǔtící yánjiū). Mod. Foreign Lang. (Q.) **26**, 283–293 (2003)

26. Jiang, X.: A corpus-based representative analysis of language and keywords in Jane Eyre (Jīyú yǔliàokù de "Jiǎn ài" yǔyán tèdiǎn jí zhǔtící biǎozhēng fēnxī). J. Jiangsu Univ. Sci. Technol. (Soc. Sci. Ed.) **16**, 77–84 (2016)

27. Li, L., Bai, L.: A corpus-based analysis of keywords in VOA agriculture report (Jīyú VOA nóngyè xīnwén yīngyǔ yǔliàokù de zhǔtící fēnxī). J. Anhui Agri. Sci. **44**, 330–332 (2016)

28. Zhong, X., Feng, Z.: A corpus of Michelle Obama speeches-based analysis and research of key words (Jīyú yǔliàokù de Mǐxiēěr àobāmǎ yánjiǎng zhǔtící yánjiū). Sci. Educ. Artic. Collect. **7**, 125–127 (2013)

29. Di, Y., Yang, Z.: A corpus-based analysis of central key words in Chinese government work reports (Jīyú yǔliàokù de Zhōngguó zhèngfǔ gōngzuò bàogào héxīn zhǔtící yánjiū). Foreign Lang. Res. **6**, 69–72 (2010)

30. Kong, Y.: A corpus-based study on lexical analysis of English tour guide commentary. Overseas Engl. **23**, 246–248 (2016)
31. Hu, C.: Corpora and the Study of Business English Lexis (Yǔliàokù yǔ shāngwù yīngyǔ cíhuì yánjiū). J. Guangdong Univ. Foreign Stud. **22**, 55–58 (2011)
32. Liu, Y.: A corpus-based study of the keywords in BEC (Jīyú shāngwù yīngyǔ yǔliàokù de zhǔtící fēnxī). J. Hunan Inst. Eng. **24**, 48–51 (2014)
33. Kilgarriff, A., Rychly, P., Smrz, P., Tugwell, D.: The sketch engine. In: Williams, G., Vessier, S. (eds.) Proceedings of the 11th EURALEX International Congress, Lorient, France, pp. 105–116 (2004)
34. Wang, S.: The concept system construction: a case study on the "economy and trade" category (Cíhuì xìtǒng jiàngòu de xīn shíjiàn—yǐ "jīngjì màoyì" lèi wéilì). In: The 19th Annual Conference of the International Association of Chinese Linguistics (IACL-19). Nankai University, Tianjin (2011)
35. Kilgarriff, A., et al.: The sketch engine: ten years on. Lexicography **1**, 7–36 (2014)

The Semantic Prosody of *"Youyu"*: Evidence from Corpora

Zhong Wu[1(✉)] and Xi-Jun Lan[2]

[1] Jianghan University, Wuhan 430056, Hubei, China
zhongwu2000@163.com
[2] China University of Geosciences, Wuhan 430074, Hubei, China
120751192@qq.com

Abstract. Semantic prosody provides a new perspective to identify the affective meaning of a word. Based on two comparable corpora and an online parallel corpus, this paper explores the semantic prosody of a Chinese functional word *'youyu'* ('because'). The statistics in the Chinese corpus, TorCH2014, indicate that *"youyu"* has two colligations that exhibit obvious negative semantic prosodies. The evidence of its English equivalents in the English corpus and the parallel corpus also proves that the semantic prosody of *"youyu"* is negative. This study shows that the combination of comparable corpora and parallel corpus provides a powerful tool for language research.

Keywords: *Youyu* ('because') · Semantic prosody · Corpus

1 Introduction

Corpus research demonstrates that lexical collocation shows a certain semantic tendency: some words habitually attract a certain type of lexical items with the same semantic features. This kind of inclination is called semantic prosody as it deals with meaning and typically includes combinations of words in an utterance rather than being attached to one [1]. Semantic prosody is defined as "words occur in characteristic collocations, showing the associations and connotations they have and therefore the assumptions which they embody" and generally falls into three categories: positive, neutral and negative [2].

In modern Chinese, *"youyu"* (由于 'because') is a multi-class word. In the *Modern Chinese Dictionary* [3: 1646], *"youyu"* belongs to two word classes: ① Preposition, indicating reasons or causes; ② Conjunction, indicating reasons. Currently, few attempts have been made to explore the semantic prosodies of functional words like conjunctions or prepositions. However, some functional words also have certain semantic tendencies in language units. Zhang Bin believes that *"cong'er"* (从而 'thereby') is positive in meaning while *"zhishi"* (致使 'cause') is negative [4]. Qu Shaobin, by using questionnaire and examples, argues that *"youyu"* has a semantic preference for unpleasant, dissatisfying or negative events [5]. They are the same as the semantic prosody discussed in this research, only differing in terms.

© Springer Nature Switzerland AG 2020
J.-F. Hong et al. (Eds.): CLSW 2019, LNAI 11831, pp. 654–660, 2020.
https://doi.org/10.1007/978-3-030-38189-9_66

2 Research Design

The present study on semantic prosody relies on observation and quantitative analysis. This is quite different from traditional lexicology, which is mostly qualitative analysis. Therefore, the corpora, research methods and research questions are given as below:

2.1 Corpora

Two comparable corpora and one parallel corpus were adopted in this research. The comparable corpora were established by National Research Center for Foreign Language Education of Beijing Foreign Studies University. The sampling frame of Brown family corpora is closely followed by two comparable corpora: Texts Of Recent CHinese 2014 (ToRCH2014) and A Brown family corpus of written British English (CLOB), with the purpose of ensuring comparability with other corpora in the same family. Because they are similar in size and genre proportion, they are perceived as appropriate research objects in contrastive linguistics.

The parallel corpus is Collected Chinese Documents Aligned with English Versions at Sentential Level (CCDA), mainly containing texts from press, science and finance. This is a sub-branch of Pool of Bilingual Parallel Corpora of Chinese Classics, a large on-line parallel corpus established by Shaoxing University. Since "*youyu*" is rarely used in oral Chinese [6: 629], CCDA is stylistically suitable for this research.

2.2 Research Methods

Firstly, researchers conducted a search in CCDA with "*youyu*" and obtained its English equivalents, such as "due to", "because", "because of", "result from" and "as a result of". The software AntConc3.3.4w was used to retrieve the node word "*youyu*" in TorCH2014 and its English equivalents in CLOB. After that, by using online searching function of CCDA, "*youyu*" was also searched in the parallel corpora. All concordances consisting these node words were downloaded and numbered.

If the concordances of a certain node word were less than 200, all of them were studied. If the total number was greater than 200, the concordances were then randomly selected by a software RandomMaker. These data, after manual screening, were used to demonstrate the frequency, ratio, and semantic prosody of "*youyu*" and its equivalents in these corpora.

2.3 Research Questions

This study seeks to answer the following questions: (1) Is the semantic prosody of "*youyu*" in TorCH2014 positive, neutral or negative? (2) Are semantic prosodies of the English equivalents of "*youyu*" in CLOB positive, neutral or negative? (3) Is the semantic prosody of "*youyu*" shown in CCDA positive, neutral or negative?

3   Results and Analysis

3.1   The Colligation and Semantic Prosody of "*Youyu*"

The lexical unit generally consists of five parts: core, collocation, colligation, semantic preference, and semantic prosody. The core and semantic prosody are essential components while the other three parts are optional [7: 37]. Apart from semantic prosody, the colligation, closely associated with word class, was also adopted in this research, for "*youyu*" is a conjunction as well as a preposition.

3.1.1   The Colligation of "*Youyu*"

The concordances of "*youyu*" extracted from TorCH2014 are 307, and 200 of them were randomly selected by RandomMaker, a software, for analysis. It can be seen that the colligation of "*youyu*" falls into 2 types:

1. *youyu* + NP

Eg. 1 **由于**   教育   资源   的   限制...

youyu   *jiaoyu   ziyuan   de   xianzhi*

Due to the limitations of educational resources...

2. *youyu* + clause

Eg. 2 **由于**   经常   吃 豆制品 ，   所以   不一定   要   喝   豆浆。

youyu *jingchang chi douzhipin   suoyi   buyiding yao   he   doujiang*

Because I usually eat bean products, I don't have to drink soybean milk.

The proportions of the colligation are shown in chart 1. The colligation "*youyu* + clause*" accounts for 79.5% of the total while "*youyu* + NP*" is 20.5%. This finding corresponds to the word class of "*youyu*", as it is a conjunction in "*youyu* + clause*" and a preposition in "*youyu* + NP*". The proportion of colligation in the pie chart also indicates that "*youyu*", a multi-class word, mainly serves as a conjunction, connecting clauses in modern Chinese.

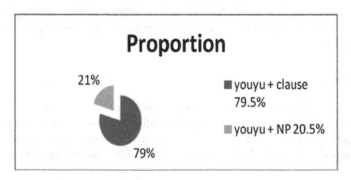

Chart. 1. The Colligation of "*Youyu*"

3.1.2   The Semantic Prosody of *"Youyu"*

In the study of collocation, the usual span for node words is four to five words, but it may be too short in the analysis of conjunctions. It is more appropriate to extend the span to the whole clause or even the sentence, as the primary function of semantic prosody is to express attitude or evaluation of the speaker/writer. Generally speaking, the semantic prosody of a node word can be judged by the words co-occurred with *"youyu"*.

Eg. 3 由于　　我们　　心心相印，　　　所以　我们　　结婚　　了。
youyu women xinxinxiangyin suoyi women jiehun le
Because we have mutual affinity, we are married.

Eg. 4 由于　　早期　　的 人类　　社会　　还　不发达
youyu zaoqi de renlei shehui hai bufada
Because the early human society was still underdeveloped

Eg. 5 由于　颜色　引发　的 情绪　与 人们　的　生活　经验　有关...
youyu yanse yinfa de qingxu yu renmen de shenghuo jingyan youguan
Because the emotions caused by color related to people's life experience...

In Eg. 3, the key word *"xinxinxiangyin"* (心心相印 'have mutual affinity') means "a strong mutual feeling between a couple who love and understand each other", thus its semantic prosody is definitely positive. Similarly, *"bufada"* (不发达 'underdeveloped') refers to "a country or an area that is poor and insufficient in modern industry", thus the semantic prosody of Eg. 4 is negative. In Eg. 5, the clause just states the fact and the semantic prosody is neutral.

In certain contexts, the key word of predict shows no preference and the semantic prosody appears to be neutral. However, when context is taken into consideration, the semantic prosody may be positive or negative, as shown in the following example:

Eg.6 由于　　互联网　　技术　的　介入...
youyu hulianwan jishu de jieru
Because of the intervention of internet technology...

In Eg. 6, "the intervention of internet technology" indicates a fact. When "new possibilities can be derived" from such "intervention", the semantic prosody of the *"youyu"* in this example should be judged as positive. To examine the hypothesized principles mentioned above, 200 selected concordances are numbered, classified, counted and labeled. The statistics of the semantic prosody of *"youyu"* were listed in the table below.

Table 1. The Semantic Prosody of *"Youyu"* in TorCH2014

	Positive		Neutral		Negative		Total	
	Hits	Ratio	Hits	Ratio	Hits	Ratio	Hits	Ratio
youyu + clause	32	20.13%	21	13.21%	106	66.66%	159	100%
youyu + NP	4	9.76%	6	14.63%	31	75.61%	41	100%
youyu	36	18.00%	27	13.50%	137	68.50%	200	100%

The statistics of "*youyu*" in TorCH2014 were listed in Table 1, which demonstrates the hits and ratio of positive, neutral and negative concordances. It exhibits that among 159 hits of "*youyu* + clause", the number of positive, neutral and negative hits is 32, 21 and 106, accounting for 20.13%, 13.21% and 66.66% respectively. With regard to "*youyu* + NP", the proportion of positive, neutral and negative is 9.76%, 14.63% and 75.61%.

It clearly shows that if "*youyu*" appears in a clause or a noun phrase, it is mostly used in a "derogatory" way, meaning that "*youyu*" mainly attracts clauses or phrases with negative meanings. Therefore, the semantic prosody of "*youyu*" is negative.

3.2   The English Equivalents of "*Youyu*" in CLOB and CCAD

Evidence can also be found in parallel corpus, a database composed of texts written in different languages. It displays observable corresponding entities and the explicit relationship between two language units, revealing the relationship of the bilingual communication activities. Key corresponding information of the equivalents can be observed clearly and directly in a parallel corpus.

In CCDA, 95 concordances containing "*youyu*" were found and downloaded. In general, "*youyu*" is converted to "due to", "because of", "because", "result from" and "as a result of", which will be regarded as prima facie equivalents. After that, AntConc3.3.4w was used to retrieve concordances with these node words from CLOB. A total of 200 concordances of each node words were selected randomly for the study (If the concordances are less than 200, all of them were analyzed). The semantic prosodies of these node words were also evaluated by the accompanying key words.

Eg. 7…the government liked the work **because** it reduced unemployment
Eg. 8…new bacteria always **result from** the division of a single parental cell.
Eg. 9…**due to** missing data on items, the number of observations in the analyses is lower than the full sample.

When the accompanied words of the predict and contexts are considered, Eg. 7–9 were deemed as positive, neutral, and negative respectively. In the light of above criteria, all downloaded concordances were evaluated and listed in the following table.

Table 2.   The Semantic Prosodies of English Equivalents of "*Youyu*" in CLOB

	Positive		Neutral		Negative		Total	
	Hits	Ratio	Hits	Ratio	Hits	Ratio	Hits	Ratio
Due to	18	16.98%	13	12.26%	75	70.76%	106	100%
Because	73	36.50%	53	26.50%	74	37.00%	200	100%
Because of	32	21.05%	36	23.69%	84	55.26%	152	100%
Result from	5	26.32%	7	36.84%	7	36.84%	19	100%
As a result of	11	28.21%	8	20.51%	20	51.28%	39	100%

Table 2 demonstrates the attitudinal preference of these node words, namely, "due to", "because", "because of", "result from" and "as a result of"". According to the table, the positive ratio of these English equivalents are 16.98%, 36.50%, 21.05%, 26.32% and 28.21% respectively; and the neutral proportion are 12.26%, 26.50%, 23.69%, 36.84%, and 20.51%; while the negative ones are 70.76%, 37.00%, 55.26%, 36.84%, and 51.28%. It seems that "due to", "because of", "as a result of" and "result from" have negative semantic prosodies, while the semantic prosody of "because" tends to be neutral.

To further investigate the attitudinal preference of these node words, this research introduces Prosody Strength (PS) [7], as shown in the formula below:

$$PS = (Pneg/F1) : (Ppos/F1). \tag{1}$$

PS is a relative ratio. In the above formula, F1 refers to the total frequency of the node words in the context, Pneg is the frequency of negative attitude expressed by node words in a specific context, and Ppos is the frequency of positive ones. In this formula, if the PS value is greater than 1, the node word has a negative connotation. If the PS value is less than 1, the node word demonstrates a positive connotation.

Based on the statistics in Table 2, the PS value of these node words can be calculated. The PS value and the frequency of node words are listed in the following table:

Table 3. The PS and Frequency of English Equivalents of "youyu" in CCDA

	Due to	Because of	Result from	As a result of	Because
PS	4.17	2.63	1.4	1.82	1.01
Hits	35	12	5	5	10
Ratio	36.84%	15.79%	5.26%	5.26%	10.53%

In CCDA, 95 hits of "youyu" were located, in which 28 of them were translated into other forms, namely, being omitted or paraphrased, and the rest were converted to equivalents in Table 3. Among which, 55 of them were converted into node words with negative semantic prosody, namely, "due to", "because of", "result from" and "as a result of", and 10 of them were translated into "because", a neutral word. It is interesting to note that "youyu", the node word with the highest PS value, appeared 35 times, accounting for 36% of the total hits. The PS value is positively correlated with the strength of its negative meaning and the probability of the node word expressing negative attitudes in the context. Therefore, the evidence from CLOB and CCDA also proves that "youyu" tends to have a negative semantic prosody.

4 Conclusion

By adopting AntConc3.3.4w and the integrated online search function, we retrieved and extracted concordances from two comparable corpora (TorCH2014 and CLOB) and one online parallel corpus (CCDA) to investigate the semantic prosody of "youyu". The downloaded concordances in the TorCH2014 shows that multi-category word "youyu" has two colligations, namely "youyu + clause" and "youyu + NP" and both of them have strong negative semantic prosodies. By retrieving CCDA with *"youyu"*, the English equivalents were located, and the identified equivalents such as "due to", "because of", "because", "result from" and "as a result of", were used as node words to search and extract concordances from CLOB. The extracted concordances and statistics demonstrate that "due to", a phrase with strong negative semantic prosody, appeared 35 times, and its frequency is much higher than the other node words with neutral semantic prosody. The research also shows that the combination of comparable corpora and parallel corpus provides a novel and effective approach for language research.

Acknowledgments. This research was financially supported by Hubei Provincial Department of Education (2017300, A Research on Interactive Translation Teaching Model - Take *Changjiang Weekly* Parallel Corpus as an Example).

References

1. Sinclair, J.: Reading Concordances: An Introduction. Longman, London (2003)
2. Stubbs, M.: Text and Corpus Analysis: Computer-Assisted Studies of Language and Institutions. Blackwell, Oxford (1996)
3. Dictionary Editing Room of Institute of Linguistics, Chinese Academy of Social Sciences: Modern Chinese Dictionary. Commercial Press, Beijing (2005)
4. Bin, Z.: Modern Chinese Dictionary of Function Words. The Commercial Press, Beijing (2001)
5. Shaobing, Q.: The semantic preference of the Youyu sentence. Stud. Chin. Lang. 1, 22–24 (2002)
6. Shuxiang, L.: Eight Hundreds Words of Modern Chinese. The Commercial Press, Beijing (2010)
7. Naixing, W.: Phraseology in Contrast: Evidence from English-Chinese Corpora. Foreign Language Teaching and Research Press, Beijing (2014)
8. Yinghua, Z.: New progress in the development of Chinese glossary in TCFL. J. Jianghan Univ. (Humanit. Sci. Ed.) (2011)

Corpus-Based Statistical Analysis of Polysemous Words in Legislative Chinese and General Chinese

Shan Wang[⊠] and Jiuhan Yin

Department of Chinese Language and Literature, University of Macau,
Taipa, Macau
shanwang@um.edu.mo

Abstract. Legislative language is an effective carrier of legal and judicial justice. It has many characteristics that are different from general language. However, currently the study of the language of legislation, especially legislative Chinese, is still relatively weak. This paper extracts high-frequency words from a legislative Chinese corpus and annotates their word meaning in this corpus. By taking them as target words, this paper then randomly extracts sentences from a large-scale general Chinese corpus (the CCL corpus or the corpus of National Language Committee) for word sense annotation. By comparing word meanings in legislative Chinese and general Chinese, this study finds that there are significant differences between them in terms of the total number of meanings and the frequency of meanings. The reasons of the differences are closely related to the accuracy, written style and contextual features of legislative Chinese in comparison with general Chinese. The comparative study between the two types of languages is helpful for exploring the characteristics of polysemous words in legislative Chinese, deepening the teaching and research of legislative Chinese, and providing references for lexical research in legislative Chinese.

Keywords: Legislative Chinese · General Chinese · Polysemous words · Corpora

1 Introduction

Polysemous words refer to words with two or more meanings. Chinese has a lot of polysemous words, which conforms to the economy principle in language, reflecting infinite things with limited words. But it also causes difficulties for us to understand and use them. In terms of general Chinese, researchers have carried out much research on polysemous words and have achieved fruitful results, such as the classification of these words, the causes of semantic evolution, and the relation between word meanings. The in-depth theoretical study is the premise of the research on teaching polysemous words, such as how to design teaching materials and improve teaching methods. The richness of Chinese polysemous words adds up to the difficulty of learning them for learners of

J.-F. Hong et al. (Eds.): CLSW 2019, LNAI 11831, pp. 661–673, 2020.
https://doi.org/10.1007/978-3-030-38189-9_67

Chinese as a second language. Compared to the fruitful research in general Chinese, the research on Chinese for specific purposes, especially legislative Chinese, is rare.

This paper extracts high-frequency polysemous words from a legislative Chinese corpus to annotates their meanings in this corpus. Then we use them as target words and randomly extract sample sentences with them from a general Chinese corpus, either the CCL corpus [1] or the corpus of National Language Committee (CNLC hereafter) [2] for word sense annotation. By comparing the usage of word meanings in legislative Chinese and general Chinese, this paper further examines the reasons that leads to the differences between legislative Chinese and general Chinese. This study not only deepens the study of the comparison between legislative Chinese and general Chinese, but also provides a useful reference for legislative Chinese teaching.

2   Related Research

Currently the existing research on Chinese polysemous words focuses on general Chinese. There is very little research on teaching Chinese for specific purposes, especially legislative Chinese.

2.1   Research on Polysemous Words in General Chinese

The studies of polysemous words in general Chinese can be classified into theoretical research and teaching research.

The theoretical research on polysemous words includes the distinction between polysemous words and homonyms, the classification of polysemous words' meanings, the analysis of the causes of polysemy from different perspectives, etc. Sun [3] pointed out that the connection between the meanings of polysemous words is the essential feature to distinguish them from homonyms. Fu [4] proposed that the meaning connection between polysemous words and homonyms should be distinguished by "etymology" and the "realistic language sense"; on this basis, Zhang [5] put forward a "synchronic – diachronic" double-defined principle of word meaning distinction and summarized the methods and principles to distinguish polysemous words and homonyms. Zhou [6] divided word meanings of polysemous words into the basic meaning and the transferred meaning. The transferred meaning is further divided into the metaphorical meaning and the extended meaning. Li [7] made a distinction between the metaphorical meaning and the extended meaning and pointed out that the relationship between a metaphorical meaning and an original meaning is whether they are similar, while the relationship between an extended meaning and an original meaning is whether they are related.

There are also many studies that analyze polysemous words based on different linguistic theories. For example, Li [8] analyzed Chinese polysemous words in the framework of cognitive linguistics, including their nature, characteristics, internal mechanism, and function realization. Wang [9] analyzed the causes of polysemy from the perspective of frame semantics and suggested that the prominence of different concepts that each word expresses resulted in the emergence of different word meanings. Xiong [10] indicated that the prototype theory focuses on the cognitive mechanism of polysemy, while metaphor and metonymy are the basic ways leading to word meaning change.

Research in language teaching focuses on case studies of specific words as well as different teaching methods. For example, scholars studied the acquisition order and semantic characteristics of 打 *dǎ* 'hit' [11], 算 *suàn* 'calculate' [12], 开 *kāi* 'open' [13], and 还是 *háishì* 'or' [14], and put forward some teaching suggestions. Chang and Wuliya [15] examined word meanings of polysemous words through the metaphor theory of cognitive linguistics and pointed out the significance of cognitive strategies for the teaching of polysemous words. Moreover, using contexts in teaching is an effective method of teaching polysemous words. Xie [16] pointed out that the teaching of polysemous words cannot be separated from contexts. Teachers use specific contexts to eliminate the ambiguity of polysemous words and enable students to use them in context. Research in this aspect is mainly studies on the teaching of general Chinese and there is very rarely research on the teaching of legislative Chinese.

2.2 Research on Polysemous Words in Legislative Chinese

Legal language is the type of language as used in the legal profession, including the language of legislation, court debates, and so on. The study of legal vocabulary in forensic linguistics mainly involves definitions, word meaning evolution, etc., among which there are very few studies on polysemous words. Legal vocabulary is usually univocal [17]; that is, one legal word always corresponds to one meaning. Even if there are two meanings of one word, when the word is used in different legal activities of legal departments, the meaning in use still remains univocal. Studying the application of commonly used words in legal Chinese can help to reduce ambiguity in use and better understand legal Chinese.

Legislative Chinese, as a part of the language of legal Chinese, have the style of written language, which requires its word meanings to be more accurate. There is very little research of polysemous words in legislative Chinese. Zhou [18] analyzed the mistranslation of polysemous words in legal texts. Luo and Wang [19] made a preliminary exploration of the word meaning of polysemous words in legislative Chinese through the self-constructed legislative Chinese corpus. They also investigated the different usages between 但 *dàn* and 但是 *dànshì* based on this corpus [20]. Based on this study, this paper makes a more specific classification and in-depth exploration.

This study extracts commonly used words from the legislative Chinese corpus [19, 21] and annotates their meaning in this corpus. 100 sentences with each of these words are randomly chosen from one general Chinese corpus (the CCL corpus or CNLC). this study then compares each word's meanings in the two different types of corpora and explains the reasons leading to the differences. This study fills the gaps in the study of polysemous words in legislative Chinese in comparison with general Chinese.

3   Comparison Between Polysemous Words in Legislative Chinese and General Chinese

Some commonly used words are extracted from the legislative Chinese corpus and their meaning in each sentence is manually annotated based on *The Contemporary Chinese Dictionary* [22]. In addition, we also randomly extract 100 sentences of each word from the CCL corpus or CNLC for word sense annotation to show their usage in general Chinese. The choice of one of the two corpora for a particular word is based on the coverage of the downloaded data: if the data is obviously from the same genre in one corpus, then the other corpus is selected. Based on the number of meanings used, this study divides the differences in the use of meanings into two categories depending whether the number of meanings in legal Chinese and that in general Chinese is different or the same. Details are given below.

3.1   Differences in the Number of Meanings Between Legislative Chinese and General Chinese

The number of word meanings in legislative Chinese and general Chinese is inconsistent, which means that some meanings of polysemous words are not used in legislative Chinese. Compared with the meanings of polysemous words in general Chinese, the meanings used in legislative Chinese are much reduced. They are divided into monosemy and polysemy and elaborated in the following sections.

3.1.1   Words that Are Polysemous in General Chinese but Monosemous in Legislative Chinese

Some polysemous words in general Chinese only has one meaning in legislative Chinese, as shown in Table 1. The meaning of 意见 *yìjiàn* in legislative Chinese is 'idea; view; opinion; suggestion'. In general Chinese, it also has a meaning 'objection; differing opinion; complaint'. 一般 *yībān* uses the meaning 'general; ordinary; common' in legislative Chinese, but it also means 'sort; kind' and 'same as; just like' in the general Chinese.

The general feature of this type of words is that a relatively small number of word meanings are in use. Most of the words have two meanings in general Chinese. One of the meanings that is not used in legal Chinese is also not frequently used in general Chinese, mostly between the percentages of 2% and 20%. The missing meaning in legislative Chinese usually has a colloquial characteristic, such as the meaning 'objection; differing opinion; complaint' of 意见 *yìjiàn* and the meaning 'sleeping' of 休息 *xiūxi*. Such kinds of vague meanings are not suitable for legislative Chinese, because it requires the language has a formal and written style.

Table 1. A comparison of word meanings between legislative Chinese and general Chinese

Word	Part of speech	No. of sentences in legislative Chinese	Percentage of meaning frequency in the legislative Chinese corpus	Percentage of meaning frequency in CNLC
意见 yìjiàn				
① idea; view; opinion; suggestion	Noun	93	100.00%	93.00%
② objection; differing opinion; complaint	Noun		0.00%	7.00%
一般 yìbān				
① same as; just like	Adjective		0.00%	7.00%
② sort; kind	Numeral		0.00%	0.00%
③ general; ordinary; common	Adjective	62	100.00%	93.00%
休息 xiūxi				
① take a breather; have (or take) a rest; rest	Verb	11	100.00%	88.00%
② sleep	Verb		0.00%	12.00%
建筑 jiànzhù				
① build (house, road, bridge, etc.)	Verb	55	0.00%	20.00%
② building; structure	Noun		100.00%	80.00%
原则 yuánzé				
① principle; tenet; criterion by which one speaks or acts	Noun	153	100.00%	98.00%
② in principle; in general	Noun		0.00%	2.00%

3.1.2 Words in Legislative Chinese Has Fewer Word Meanings Than in General Chinese

There are some words that are polysemous in both legislative Chinese and general Chinese, but they have fewer word meanings in legislative Chinese than in general Chinese, as shown in Table 2.

Such kind of words have many meanings in use, but general Chinese always uses more meanings than legislative Chinese. For example, according to the annotation result, the word 依据 yījù has three meanings in general Chinese, but only two in legislative Chinese; the word 吸收 xīshōu has four meanings in general Chinese, but only two in legislative Chinese; and the word 一定 yídìng has five meanings in general Chinese, but only three in legislative Chinese.

Table 2. A comparison of the number of word meanings of polysemous words in legislative Chinese and general Chinese

Word	Part of speech	No. of sentences in legislative Chinese	Percentage of meaning frequency in legislative Chinese corpus	Meaning frequency in CCL
依据 yījù				
① basis; foundation	Noun	55	63.64%	59.00%
② take sth. as a basis or foundation	Verb		0.00%	2.00%
③ according to; in the light of; on the basis of; judging by	Preposition		36.36%	39.00%
吸收 xīshōu				
① absorb; imbibe; draw; suck up; take sth. in through pores or interstices, such as a sponge adsorbing water; and charcoal absorbing gases	Verb	12	0.00%	0.00%
② assimilate; transform (food) into living tissue by the process of anabolism, such as the intestinal mucous membrane assimilating nutrition, and the root of a plant drawing water and inorganic salt, metabolize constructively	Verb		0.00%	4.00%
③absorb; take in; moderate the impact of certain phenomena or the role of sth., such as a spring absorbing shock and sound-insulating paper taking in sound	Verb		0.00%	2.00%
④ recruit; admit; enroll	Verb		8.33%	5.00%
⑤ absorb	Verb		91.67%	89.00%
一定 yīdìng				
① fixed; specified; certain; regular	Adjective	35	68.57%	14.00%
② surely; necessarily	Adjective		0.00%	5.00%
③ must; certainly	Adverb		0.00%	10.00%
④ given; particular; certain	Adjective		5.71%	36.00%
⑤ proper; fair; due	Adjective		25.71%	35.00%

It is obvious that some missing meanings in the legal language domain are specialized meanings in professional fields like biology and physics. A case in point is 吸收 xīshōu. Even if some meanings are used both in lexical legislative Chinese and general Chinese, their common meanings are sometimes inconsistent. For example, the common meaning of 一定 yīdìng is 'given; particular; certain' in general Chinese, but it is 'fixed; specified; certain; regular' in legislative Chinese. Moreover, the parts of speech distribution are broad in general Chinese compared to legislative Chinese. For example, 依据 *yījù* can be used as a verb, noun, and preposition in general Chinese, but the part of speech as a verb is not used in legislative Chinese.

3.2   Same Number of Word Meanings in Legislative Chinese and General Chinese

This section discusses the case when legislative Chinese and general Chinese have exactly the same number of meanings. In this situation, they often differ in word meaning frequency and the frequency order of different meanings. They are elaborated below.

3.2.1   Difference in Frequency Order of Word Meanings

When exactly the same number of meanings are used in legislative Chinese and general Chinese, there are often differences in the order of frequency, as shown in Table 3.

Table 3. Different frequency order of word meanings

Word	Part of speech	No. of sentences in legislative Chinese	Percentage of meaning frequency in legislative Chinese corpus	Percentage of meaning frequency in CCL
共同*gòngtóng*				
① belonging to; or shared by all	Adjective	143	52.00%	45.45%
② work together	Adverb		48.00%	53.15%
区别*qūbié*				
① distinguish; differentiate; discriminate; make a distinction between	Verb	8	75.00%	21.00%
② difference; distinction	Noun		25.00%	79.00%
教育*jiàoyù*				
① the work of training people according to certain requirements, mainly refers to the work of training people in schools	Noun	51	64.17%	88.00%
② cultivate according to certain requirements	Verb		19.61%	5.00%
③ expound; educate; guide; reasoned argument used to persuade sb. to follow rules, instructions or requests, etc.	Verb		15.69%	6.00%

Table 3 clearly shows the difference in the frequency of polysemous words' meanings in legislative Chinese and general Chinese. For example, though the most common meaning of 教育 *jiàoyù* in both legislative Chinese and general Chinese is the same (①), but the secondly ranked meaning is different. The comparison of frequency percentage is shown in Table 4. Knowing this difference is of great significance for the study of Chinese for specific purposes in the legal domain.

Table 4. A comparison of the frequency ranking of the meanings of polysemous words

Polysemous words	Legislative Chinese	General Chinese
共同 gòngtóng	①(52%) > ②(48%)	②(53.15%) > ①(45.45)
区别 qūbié	①(75%) > ②(25%)	②(79%) > ①(21%)
教育 jiàoyù	①(64.17%) > ②(19.61%) > ③(15.69)	①(88%) > ③(6%) > ②(5%)

3.2.2 Meanings with Same Frequency Order but a Large Frequency Difference

Although the meanings of some polysemous words have the same frequency order in both legislative Chinese and general Chinese, the frequency of the meanings can vary greatly. This paper takes the frequency difference of 30% as a boundary. The polysemous words with frequency difference greater than 30% are classified as the group with a large frequency difference, while the group with frequency difference less than 30% is classified as the group with a small frequency difference.

Table 5 shows the cases with a large frequency difference in word meanings. The most frequent meaning of 研究 *yánjiū* is 'study; research; look into' in both legislative Chinese and general Chinese, as high as 93.94% and 60.00% respectively. In contrast, the second meaning has a significantly lower frequency, as low as 6.06% and 20.00% respectively.

Table 5. The frequency of word meanings with a large frequency difference

Word	Part of speech	No. of sentences in legislative Chinese	Percentage of meaning frequency in legislative Chinese corpus	Percentage of meaning frequency in CNLC
研究yánjiū				
①study; research; look into	Verb	33	93.94%	60.00%
②consider; discuss; deliberate (opinions and issues)	Verb		6.06%	20.00%

3.2.3   Meanings with Same Frequency Order but a Small Frequency Difference

There are some words whose meanings not only have the same frequency order, but also a small frequency difference (less than 30%) between legislative Chinese and general Chinese. Table 6 shows such examples. The most common meaning of 及时 *jíshí* is 'promptly; without delay' in both legislative Chinese and general Chinese, and usage in legislative Chinese is just slightly higher. The most common meaning of 健康 *jiànkāng* is '(of the human body) good health; strong physique; be in good health; have a good condition in mental health and social adaptability' in both corpora and the use of this meaning in legislative Chinese is also just slightly higher.

Table 6. Word meanings with a small frequency difference between legislative Chinese and general Chinese

Word	Part of speech	No. of sentences in legislative Chinese	Percentage of meaning frequency in legislative Chinese corpus	Percentage of meaning frequency in CCL
及时 *jíshí*				
① timely; in time	Adjective	147	2.04%	11.00%
② promptly; without delay	Adverb		97.96%	89.00%
健康 *jiànkāng*				
① (of the human body) good health; strong physique; be in good health; have a good condition in mental health and social adaptability	Adjective	61	88.52%	78.00%
② healthy; sound; in normal condition; perfect	Adjective		11.29%	22.00%

4   Reasons for the Differences in the Use of Meanings in Legislative Chinese and General Chinese

The comparative analysis of frequency shows that the use of polysemous words have great differences in legislative Chinese and general Chinese, especially in the number of word meanings and the frequency order. The reasons for these differences are mainly reflected in the following aspects.

(1) Accuracy is the most important feature of legislative language. When talking about the style of legislative language, Pan [23] mentioned that accuracy is the soul and life of legislative language and "appropriate words" are the guarantee of the accuracy of legislative language. "Appropriate words" not only refer to the application of legal terminology, but also require the use of meanings, part of speech, and the usage scope of common words to conforms to the legislative context.

The accuracy characteristic of legislative Chinese leads to the difference in the number of word meanings used in legislative Chinese and general Chinese. Firstly, such accuracy limits the use of rhetoric in legislative language. The metaphorical meaning of certain polysemous words or the words used for metaphors rarely appear in legislative Chinese. For example, in the general Chinese corpus, the meaning 'same as; just like' of 一般 *yībān* has sentences shown in (a) and (b). These rhetorical usages do not appear in legislative language.

(a) 数不清的砖头、石块、玻璃瓶，冰雹一般落下来。

> Shǔbùqīng de zhuāntóu、shíkuài、bōli píng, bīngbáo yībān luò xiàlái.
> Numerous_DE_brick_stone_glass bottle, hailstone_like_fall_down
> 'Numerous bricks, stones, and glass bottles fall down, just like hailstones.'

(b) 上面按比例布置了房子，电线杆，树木，……犹如真街道一般。

> Shàngmian àn bǐlì bùzhì le fángzi, diànxiāngān, shùmù,……yóurú zhēn jiēdào yībān.
> on_according to_proportion_arrange_ASP_house, telegraph pole, tree,……just as_real_street_same as
> 'The houses, telegraph poles, trees… are arranged proportionally, just like the real street.'

Secondly, the accuracy of legislative language requires that legal documents avoid using omissions and substitutions in the text. For example, they use fewer elliptical words such as 等 *děng* "and so on", 之类 *zhīlèi* 'and the like' [24]. 原则 *yuánzé* only uses the meaning 'principle; tenet; criterion by which one speaks or acts' in legislative Chinese. The other meaning 'in principle; in general' is used in general Chinese shown in (c), which is not accurate enough to express the meaning.

(c) 讨论并原则通过了《三个暂行规定（草案）》。

> tǎolùn bìng yuánzé tōngguò le 《sān gè zànxíng guīdìng (cǎo'àn)》 .
> discuss_and_in principle_pass_ASP_three_CL_provisional_regulation_draft
> '…discussed and have in principle passed the Three Interim Regulations (Draft).'

(2) Different contextual requirements of legislative Chinese and general Chinese lead to the differences in the use of meanings. For example, the meaning 'absorb; take in; moderate the impact of certain phenomena or the role of sth., such as a spring absorbing shock and sound-insulating paper taking in sound' of 吸收 *xīshōu* belongs to the field of natural sciences. This kind of word meanings is not used in legislative Chinese because they do not belong to the legal domain.

(3) Legislative language has a written style, and thus it does not use colloquial meanings. The corpus of legislative Chinese we used in this study is derived from legislative texts of the Legal Work Committee of the National People's Congress, which is not for spoken language communication. Therefore, it is impossible to use the colloquial meanings of polysemous words. For example, in (d), the meaning of 意见 *yìjiàn* is 'objection; differing opinion; complaint'. This kind of meanings does not appear in legislative Chinese because of its colloquial feature.

(d) 朋友对这个问题提法却有意见。

  Péngyǒu duì zhège wèntí tífǎ què yǒu yìjiàn.
  friend_to_this_issue_formulation_but_has_differing opinion
  'The friend has a differing opinion on this issue.'

5 The Implications of the Comparative Study of Polysemous Words in Legislative Chinese and General Chinese

The comparative study of polysemous words in legislative Chinese and general Chinese is of great significance theoretically and practically. (1) It plays an important role in the study of legislative Chinese in teaching Chinese for specific purposes, the editing of teaching materials and the selection of teaching priorities. Knowing the difference between the meanings of polysemous words in legislative Chinese and general Chinese can make teaching more targeted: (i) Legislative Chinese sometimes uses different common meanings compared to general Chinese. For example, in legislative Chinese, the most frequent meaning of 共同 gòngtóng and 区别 qūbié is 'belonging to; or shared by all' and 'distinguish; differentiate; discriminate; make a distinction between' respectively, which is different from general Chinese. Therefore, when teaching such words, it is necessary to emphasize such common meanings. (ii) The difference in the number of meanings between legislative Chinese and general Chinese provides a reference for the choice of the scope and content of vocabulary teaching. Words like 意见 yìjiàn, 一般 yībān, 休息 xiūxi, and原则yuánzé are actually monosemantic in legislative Chinese. Their meaning in use in legislative Chinese should be emphasized during teaching. Other words such as 依据 yījù, 吸收 xīshōu, and 一定yīdìng in legislative Chinese lack certain meanings used in general Chinese, which enlightens us not to spend much time on such meanings. (2) As an important factor affecting reading, vocabulary plays an important role in language learning. Due to the diversity of word meanings, polysemous words should be the focus of vocabulary teaching. Based on the analysis of the differences in the meanings between legislative Chinese and general Chinese, we can pay more attention to the common meanings and unique meanings of the words in legislative Chinese, emphasize the particularity of legal style, and provide learners of legislative Chinese with more relevant teaching materials. (3) This study provides support for the research of Chinese legislative and forensic linguistics. The difference of word meanings between legislative Chinese and general Chinese not only complement the study of legislative words, but also provides evidence for the unique style and characteristics of legal language. For example, in legislative Chinese, 原则 yuánzé avoids using the fuzzy meaning 'in principle; in general', 一般 yībān avoids using the metaphorical meaning 'same as; just like', and 健康 jiànkāng rarely uses the metaphorical meaning 'healthy; sound; in normal condition; perfect'. These phenomena exactly reflect the characteristics of the accuracy of legal language.

6   Conclusions

In view of the lack of research in polysemous words in legislative Chinese, this study explores the differences in the usage of meanings between legislative Chinese and general Chinese and summarizes the types of differences through comparison. This study finds that the differences in the use of polysemous words in legislative Chinese and general Chinese can be divided into two large categories: the same number of word meanings and the different number of word meanings. Under the two categories, the frequency order and frequency ratio of polysemous words in legislative Chinese and general Chinese are quite different. This leads to different common meanings of some polysemous words in legislative Chinese and general Chinese. The reasons for these differences are inseparable from the accuracy and professionalism requirement of legislative language, the contextual characteristics of legislative Chinese, and the written style of legal documents. The results of this research not only enrich the study of Chinese legal vocabulary, but also provide an important reference for the teaching of legislative Chinese vocabulary in the choice of common meaning and the order of meanings. In future research, we will conduct a wider range of vocabulary comparison and explore the similarities and differences between legislative Chinese and general Chinese in a more in-depth manner.

Acknowledgement. This research is supported by the Start-up Research Grant of University of Macau (SRG2018-00126-FAH).

References

1. Zhan, W., Guo, R., Chen, Y.: The CCL Corpus of Chinese Texts: 700 million Chinese Characters, the 11th Century B.C. - present (2003)
2. Jin, G., Xiao, H., Fu, L.: The construction and deep processing of modern Chinese corpus (Xiàndài hànyǔ yǔliàokù jiànshè jí shēnjiāgōng). Appl. Linguist. **2**, 111–120 (2005)
3. Sun, C.: Chinese Vocabulary (Hànyǔ cíhuì). Jilin People's Press, Changchun (1957)
4. Fu, H.: Modern Chinese Vocabulary (Xiàndài hànyǔ cíhuì). Peking University Press, Beijing (1985)
5. Zhang, B.: The principles and methods for distinguishing perfect homonyms from polysemes in modern Chinese (Xiàndài hànyǔ tóngxíng tóngyīncí yǔ duōyìcí de qūfēn yuánzé hé fāngfǎ). Lang. Teach. Linguist. Stud. **4**, 36–45 (2004)
6. Zhou, Z.: Speeches on Chinese Vocabulary (Hànyǔ cíhuì jiǎnghuà). People's Education Press, Beijing (1962)
7. Li, L.: On the difference between the extended meaning and the metaphorical meaning of polysemous words (Shìtán duōyìcí de yǐnshēnyì hé bǐyùyì de qūbié). J. Huaiyin Teach. Coll. (Soc. Sci. Ed.) **1**, 57–60 (1983)
8. Li, Y.: The Chinese Polysemant Research in Cognitive Linguistics (Rènzhī yǔyánxué shìyù xià de hànyǔ duōyìcí yánjiū). Jilin University (2013)
9. Wang, R.: A study of polysemy in Chinese from the perspective of frame semantics (Kuàngjià yǔyì shìjiǎo xià hànyǔ yīcíduōyì de yánjiū). J. Heilongjiang Vocat. Inst. Ecol. Eng. **27**, 145–146 (2014)

10. Xiong, S.: A cognitive linguistic perspective on polysemy (Cóng rènzhī yǔyánxué jiǎodù kàn yīcíduōyì xiànxiàng). J. Anhui Vocat. Coll. Electron. Inf. Technol. 17, 58–61 (2018)
11. Dong, W.: The Research of the Polysemic Verb "da" in Modern Chinese Language in Teaching Chinese as a Foreign Language (Xiàndài hànyǔ duōyì dòngcí 'dǎ' de duìwàihànyǔ jiàoxué yánjiū). Hebei Normal University (2015)
12. Xu, Y.: A Study of the Foreign Students' Acquisition Error of Chinese "suàn" and the Related Words (Liúxuéshēng duōyì dòngcí "suàn" jíqí xiāngguān cíhuì de xídé piānwù fēnxī jí jiàoxué cèlüè). Shanghai International Studies University (2014)
13. Xu, H.: The Study of Acquisition Order in Polysemy "kāi" Semantic Item of Thailand Student and Teaching Advice (Tàiguó xuésheng hànyǔ duōyìcí "kāi" de gè yìxiàng xídé shùnxù yánjiū jí jiàoxué jiànyì). GuangXi University for Nationalities (2018)
14. Guo, X.: The Error Analysis of the Polysemous word "háishì" (Duōyìcí "háishì" de piānwù fēnxī). Yangzhou University (2018)
15. Chang, G., Wuliya, M.: Cognitive strategy applied in chinese polysemy teaching (Rènzhī cèlüè zài hànyǔ duōyìcí jiàoxué zhōng de yùnyòng). J. Beijing Univ. Posts Telecommun. (Soc. Sci. Ed.) 15, 99–104 (2013)
16. Xie, J.: The Application of the Contextual Teaching Method in the Advanced Stage of Polysemy Teaching in the Teaching Chinese to Speakers of Other Languages Students—A Case Study of the Polysemous Words in the Chinese Extensive Reading Course (Yǔjìng jiàoxuéfǎ zài duìwài hànyǔ zhōng gāojí jiēduàn duōyìcí jiàoxué zhōngde yùnyòng—yǐ zhōngwén guǎngjiǎo gāojí hànyǔ fàndú jiàochéng zhōngde duōyìcí wéilì). Yunnan Normal University (2017)
17. Yang, K.: Study on the Meaning Characteristics Extraction and Paraphrase Comparison of Legal Words (Fǎlǜ cíhuì de cíyì tèdiǎn tíqǔ jí shìyì duìbǐ yánjiū). Hebei Normal University (2016)
18. Zhou, D.: On the role of context theory in Chinese translation of polysemous words in WTO legal texts (Lùn yǔjìng lǐlùn zài WTO fǎlǜ wénběn zhōng duōyìcí hànyì de zuòyòng). Crazy Engl. Pro 2, 205–206 (2018)
19. Luo, H., Wang, S.: The construction and application of the legal corpus of Mainland China (Zhōngguó dàlù fǎlǜ yǔliàokù jíqí yīngyòng yánjiū). In: The 18th Chinese Lexical Semantics Workshop (CLSW2017). Leshan Normal University, Leshan, Sichuan, China (2017)
20. Luo, H., Wang, S.: The Transitional Sentences in Legislative Language in the Cases of the dan and danshi sentences (Lìfǎ yǔyán zhōng de dànshū hé fēidànshū yánjiū). Contemporary Rhetoric (Dāngdài xiūcí xué) 6, 77–89 (2018). [Reproduced by China Social Science Excellence Réndà fùyìn bàokān zīliào, 2, 130–139, 2019]
21. Luo, H., Wang, S.: The construction and application of the legal corpus. In: Wu, Y., Hong, J.-F., Su, Q. (eds.) Chinese Lexical Semantics, pp. 448–467. Springer, Cham (2018)
22. Dictionary Editing Room of Institute of Linguistics of China Academy of Social Sciences: The Contemporary Chinese Dictionary (Xiàndài hànyǔ cídiǎn). The Commercial Press, Beijing (2016)
23. Pan, Q.: Forensic Linguistics (Fǎlǜ yǔyánxué). China University of Political Science and Law Press, Beijing (2017)
24. Pan, Q.: On the general characteristics of Chinese legal style (Lùn hànyǔ fǎlǜ yǔtǐ de yībān tèzhēng). J. Shanghai Univ. (Soc. Sci. Ed.) 1, 8–14 (1985)

Corpus-Based Textual Research
on the Meanings of the Chinese Word "*Xífu(r)*"

Jingmin Wang(✉)

Luoyang Campus of the Information Engineering University, Luoyang, China
wangjingmin1972@163.com

Abstract. There are only two entries under the Chinese character "媳(*xí*)" in the 5th edition of *Modern Chinese Dictionary*, namely *xífù* and *xífur*. They are annotated as two words with completely different meanings, each with two meanings. By searching the Ancient Chinese Corpus and examining the use cases one by one, it is proved that *xífur* appeared later than *xífù*, but they are identical in lexical semantics. *Xífù* and *xífur* are actually a word, which should be classified as an entry. There are only three meanings in the word: son's wife, wife and young married woman in general.

Keywords: *Xífu(r)* · Term of meaning · Corpus · Textual research

1 Preface

At the beginning of any language's vocabulary, the signified corresponds to the signifier one by one, and the meaning of a word is single. With the continuous development of society, new things and new phenomena are emerging, and people's understanding of the world is also deepening. However, it is impossible for everything and phenomenon to be endowed with a new word. As a result, people give new meanings to the existing words, which results in the phenomenon of polysemy of words. Vocabulary generally evolves through ENT models such as expansion, narrowing and transfer. In the development of history, different meanings of a word either have kept pace with each other, or one meaning disappeared and the other grew [1]. After a long time of accumulation, they formed the meanings of modern words and have been included in dictionaries. The evolution of lexical meaning is an important part of lexical semantics, and the diachronic analysis of vocabulary is a common means to study the law of meaning evolution. In recent years, with the establishment of corpus, the diachronic quantitative analysis of lexical meanings has made it more scientific to explore the evolution of lexical meanings and to reveal and test the essence of the evolution of lexical meanings.

The Chinese character "媳(*xí*)" in the fifth edition of *Modern Chinese Dictionary*, compiled by the Dictionary Editorial Office of the Institute of Language Studies of the Chinese Academy of Social Sciences and published by the Commercial Press, is annotated as "daughter-in-law". There are only two entries below:

【媳妇】 *xífù* noun ①son's wife; daughter-in-law ②wife of a relative of the younger generation (with the younger generation appellation added in front)

【媳妇儿】 *xífur* <dialect> noun ①wife ②young married women in general

J.-F. Hong et al. (Eds.): CLSW 2019, LNAI 11831, pp. 674–680, 2020.
https://doi.org/10.1007/978-3-030-38189-9_68

The *Modern Chinese Study Dictionary* and the *Chinese Dictionary for Learners and Teachers*, published by the Commercial Press in 2010 and 2011, respectively, also treat the words *xifù* and *xifur* as two independent words under the Chinese character "媳(*xí*)", with identical annotations. As we all know, the Rhotic accent does have the function of distinguishing the meaning of words. For example, "*yìdiǎn*(一点, one o'clock)" and "*yìdiǎnr*(一点儿, a small amount, a little)", "*xìn*(信, letter)" and "*xìnr* (信儿, news)" and so on. But whether the *xifù* and *xifur* are two words with different meanings, and whether the four meanings correspond clearly to two words without crossing, are worth studying and discussing. Here, we use the ancient Chinese corpus developed by the Research Institute of Language and Character Application of the Ministry of Education to examine the meanings of *xifù* and *xifur* to see if their meanings overlap or not.

2 The Source of "*Xifù*"

The Chinese character "妇(*fù*)" refers to both women and wives during the pre-Qin period. For example, "神农之世，男耕而食，妇织而衣(*Shénnóng zhī shì, nán gēng ér shí, fù zhī ér yī*. In Shennong's time the men farmed in the fields to obtain food and the women were responsible for textile to make clothes). [*Shang Jun Shu* (*Book of Lord Shang Yang. Huace*)] "故昏姻之礼废，则夫妇之道苦，而淫辟之罪多矣(*Gù hūnyīn zhī lǐ fèi, zé fūfù zhī dào kǔ, ér yínpì zhī zuì duō yǐ*. Therefore, if the marriage etiquette is abandoned, the couple's morality will decline and they will live painfully together, and the crime of adultery and evil will increase a lot.) [*Li Ji* (*Book of Rites*) *Jingjie* (name of a chapter in the Book of Rites)].

According to the existing literature, the Chinese character "妇(*fù*)" first appeared in the late Tang Dynasty and Five Dynasties, and the word "*xifù*(媳妇, daughter-in-law)" began to appear [2]. In the Pre-Qin Dynasty, the term "*zǐfù*(子妇)" was generally used to denote the son's wife, that is, daughter-in-law. Huang Sheng, a scholar in the late Ming and early Qing Dynasties, said in his book *Yifu*: "In ancient times, the son was called *xī*(息), which means birth and life." So the son's wife was called "*xīfù*(息妇)". That is to say, the original writing form of *xifù* was "*xīfù*(息妇)". Later, the Chinese character "媳(*xí*)" was created by adding the female radical component "女 (*nǚ*, women)" next to the left of the Chinese character "*xī*(息)" and categorizing it [3]. Mei Yingzuo, a scholar of the Ming Dynasty, in his book *Vocabulary and Women's Department* explained: "Son's wife is commonly called *xī*(媳)" In other words, the word *xifù* of the son's wife belonged to the common name at that time.

However, Huang Sheng pointed out in his book *Yifu* that "Since the Han Dynasty, the son's wife has been called *xīnfù*(新妇, new woman)". Why was the son's wife called *xīnfù*? This is related to the meaning of the morpheme "*xīn*(新, new)". Huang Sheng said in *Yifu*: "Probably the woman who just married was called *xīnfù* at that time. Having been used to it for a long time, this call wasn't changed any more." Affected by this, the word *xīnfù* in the late Eastern Han Dynasty began to be used as a general term for married women. Until the Eastern Jin Dynasty, *xīnfù* had a new meaning, which could be expressed as "younger brother's wife". Guo Pu, a well-known scholar in the Eastern Jin Dynasty, commented on the statement in *Erya ShiQin* that "the elder brother's wife is called *sǎo*(嫂, sister-in-law), and the younger brother's wife is called *fù*(归)." He commented below: "As we say today, *xīnfù*." This meaning continued to Tang and Song Dynasties. So far, *xīnfù* has several meanings such as "bride", "son's wife", "married woman", "wife" and "younger brother's wife".

Because *xīnfù* has too many meanings, and it was very common to use "*xī*(息)" to refer to a son in Tang Dynasty and before. In addition, the pronunciation of "*xīn*(新)" and "*xī*(息)" is similar. So *xīnfù* was gradually transformed into "*xīfù*(息归)" in the spoken language and retained other meanings besides "bride". In the late Tang and Five Dynasties, the term "息归(*xīfù*)" was written as "媳归(*xífù*)" through radical categorization, and it was widely used in the Song Dynasty. The word *xīnfù* gradually declined and its meanings narrowed, and it is influenced by the morpheme meaning of "*xīn*(新, new)" and more means "bride". By the Ming and Qing Dynasties, *xífù* flourished in the northern mandarin, while *xīnfù* had no trace in the famous works such as *Jin Ping Mei*, *Water Margin*, *A Dream of Red Mansions*, and so on. So far, as a "language fossil", it has only been retained in some southern dialects [4].

3   Diachronic Quantitative Analysis of "*Xífù*(*Xífur*)"

In the 5th edition of *Modern Chinese Dictionary*, there are only two entries *xífù* and *xífur* under the Chinese character "媳(*xí*)". *Xífù* has two meanings, which refer to the son's wife and the wife of a relative of the younger generation. However, *xífur* has been marked "dialect" because of its Rhotic accent. It also has two meanings: wife and young married women in general. Judging from the dictionary notes, *xífù* and *xífur* are two words with completely different meanings, corresponding to four meanings respectively. So are these four meanings clearly divided into the two words? Do they overlap each other or not? If they overlap each other, it means that the two words are actually one word. These meanings should be interpreted as different meanings of the same word. Therefore, we need to make a diachronic linguistic study of *xífù* and *xífur*, and make a diachronic quantitative analysis of their meanings.

3.1   Textual Research on the Meanings of *Xífù*

Based on the Ancient Chinese Corpus (www.cncorpus.org) developed by the Institute of Applied Linguistics of Ministry of Education, we obtained 1640 cases of *xífù* in ancient Chinese of Song, Yuan, Ming and Qing Dynasties. After excluding the word *xífur*, a total of 1498 cases were obtained. After looking at the original contexts of these cases and analyzing them one by one, the distribution data of the meanings of the word *xífù* in ancient Chinese are as follows (Table 1):

Table 1. Distribution data of the meanings of *xifù* in ancient Chinese

媳妇 (*xifù*)	Song Dynasty	Yuan and Ming Dynasties	Qing Dynasty
1. son's wife	11	299	128
2. wife of a relative of the younger generation	1	15	23
3. wife	1	543	135
4. young married women in general	3	181	158
grand total: 1498	16	1038	444

From the statistical data above, it can be found that *xifù* did not appear before the Song Dynasty. It was seldom used in the Song Dynasty, yet it was widely used in the Yuan and Ming Dynasties. Its meaning "wife" has 543 cases in total, which has far exceeded 299 cases as its basic meaning "son's wife". In the Qing Dynasty, both the meanings "wife" and "young married women in general" had gone beyond the basic meaning of *xifù* as "son's wife", which showed that the main meanings of *xifù* were concentrated in these three aspects, even the meanings "wife" and "young married women" were more than that "son's wife". Another noticeable phenomenon is that in any dynasty, the four meanings of *xifù* as "son's wife", "wife of a relative of the younger generation", "wife" and "young married women in general" appeared at the same time, but not at a certain stage of language development. This shows that from the beginning, *xifù* has assumed the "important task" of expressing these four meanings.

When *xifù* expresses the meaning "wife", it can be used not only as a general term for "wife", but also as a back-appellation for husband to wife. When *xifù* refers to young married women, it is used not only in general terms, but also in address form or back-appellation, or even in self-styled address. It should be pointed out that *xifù* can also be called *xifùzi* when it refers to the meanings of wife and young married women. This usage is still in use today.

3.2 Textual Research on the Meanings of *Xifur*

There are 142 cases of *xifur* in the Ancient Chinese Corpus, which are far less than those of *xifù*. This is because the expression of "r"-retroflexed words is different in written and spoken language. The "r"-retroflexed words in spoken language are often expressed in writing as "sometimes with r-ending retroflexion, sometimes without r-ending retroflexion" or "generally without r-ending retroflexion in writing" [5]. In which form, people tend to be very casual. "It is not uncommon to use two forms of writing in a book or even an article [5]." Nevertheless, 142 cases of *xifur* have been acquired from the Ancient Chinese Corpus. After looking at the original contexts of these cases and analyzing them one by one, the distribution data of the meanings of the word *xifur* in ancient Chinese are as follows (Table 2):

Table 2. Distribution data of the meanings of *xifur* in ancient Chinese

媳妇儿(*xifur*)	Song Dynasty	Yuan and Ming Dynasties	Qing Dynasty
1. son's wife	0	38	13
2. wife of a relative of the younger generation	0	4	1
3. wife	0	62	9
4. young married women in general	0	4	11
grand total: 142	0	108	34

From the table above, we can see that the case of *xifur* was zero in the Song Dynasty. It shows that the retroflex suffixation of *xifù* did not appeared at that time. In the Yuan, Ming and Qing Dynasties, the political and cultural center of the country moved to Beijing. Influenced by the Rhotic accent in Beijing dialect, the "r"-retroflexed words had emerged in a large number in Mandarin, so appeared the retroflex suffixation of *xifù* — *xifur*. In the dramas of Yuan Dynasty and the novels of Ming Dynasty, *xifur* was mainly used for the meanings "wife" and "son's wife", with 62 and 38 cases respectively, while the meanings "wife of a relative of the younger generation" and "young married woman" were very few, with only 4 cases each. It was only in the Qing Dynasty that the use cases of *xifur* expressing "son's wife", "wife" and "young married woman" basically converged. Same as *xifù*, the four meanings of *xifur* above appeared at the same time. This shows that *xifur* is only the retroflex form of *xifù*, and the two words are identical in semantics, and there is no substantive difference between them.

The "r"-retroflexed word *xifur* was very common in the dramas of Yuan Dynasty and the novels of Ming and Qing Dynasties, which means "son's wife" and is used not only in general terms, but also in address form or back-appellation.

3.3 Use Case and Analysis of "Wife of a Relative of the Younger Generation"

It is rare for the word *xifù* to be used directly to refer to "wife of a relative of the younger generation". There is only one example in ancient Chinese. In other examples, when it is used to express the wife of a relative of the younger generation, the morphemes such as "younger brother, nephew, grandson" and so on are added before *xifù* to clarify the identity of the woman. In this case, *xifur* is exactly the same as *xifù*.

According to the componential analysis, there are some morphemes such as "younger brothers, nephews, grandson" and so on, which indicate the relatives of the younger generation, it is evident that *dìxifur*, *zhíxifur*, *sūnxifur* are the wives of the younger brother, nephew and grandson, not their daughters-in-law. Therefore, *xifù* actually means "wife" here, so the second item "wife of a relative of the younger generation" in the two statistical tables mentioned above can be merged into the third item "wife".

The best example of this is *érxífù*, because *xífù* itself can express the meaning "son's wife", but sometimes in order to emphasize or confirm, the term *érxífù* is often used, this usage is still in use today. Here *érxífù* and *xífù* obviously mean the same thing, but because of the morpheme *ér*(儿,son), *xífù* means "wife", which from one side proves the evolution of the meaning of *xífù* from "son's wife" to "wife", thus proving that *xífù* can be used to express both "son's wife" and "wife".

3.4 Summary

From the foregoing examples and statistics, we can see that *xífù* and *xífur* are identical in terms of meaning distribution. Whether with r-ending retroflexion or not, they are actually a word, and there is no distinct difference in meanings. This is because the Rhotic accent is widely used in spoken language, and sometimes specially showed in written writing, such as in literary works, and sometimes not.

4 Dispose of Other Dictionaries

There is only one word *xífù* under the Chinese character "媳(*xí*)" in *Wang Li's Ancient Chinese Dictionary*, which was edited by Professor Wang Li. It has two meanings: ①son's wife ②wife; young married women in general. He summed up the meanings "wife" and "young married women" as one. There is also only one entry *xífù* under the Chinese character "媳(*xí*)" in the 2010 edition of *Ciyuan* (*Etymology*, revision) of the Commercial Press, and there are also only two meanings: ①son's wife ②wife; married young women and old women. There is only one entry *xífù* under the Chinese character "媳(*xí*)" in the sixth edition of *Cihai* of Shanghai Lexicographic Press in 2009, and there are only two meanings as well: ①son's wife; also wife of a relative of the younger generation ②wife in northern dialect; also a general term for married women. For that, we can find that the two dictionaries *Ciyuan* and *Cihai* are consistent with *Wang Li's Ancient Chinese Dictionary*.

There is also only one entry *xífù* under the Chinese character "媳(*xí*)" in the *Ancient and Modern Chinese Dictionary* published by the Commercial Press in 2000, but it has four senses: ①son's wife ②<dialect> wife ③<dialect> young married women in general ④elderly woman servant. In fact, the fourth sense "elderly woman servant" still refers to married woman, but her social status is low. It is a historical word, which is no longer applicable today and can be classified into the third sense. It shows that the Commercial Press recognizes that the word *xífù* has the meanings of "son's wife", "wife" and "married woman". This is inconsistent with the standard of vocabulary annotation in the 5th edition of *Modern Chinese Dictionary*, which indirectly overthrows the strict distinction between *xífù* and *xífur*.

However, the 6th edition in 2012 and the 7th edition in 2016 of the *Modern Chinese Dictionary* of the Commercial Press still insist on setting up two entries and four senses. According to the Chinese character "妇(*fù*, woman)" with or without tone, the words *xífù* and *xífur* are annotated as two different words. Some sayings included in the *Folk Language Dictionary* (revised edition) and the *Chinese Folk Language Dictionary* (revised and enlarged edition) published by the Commercial Press in 2011, such as "娶了媳妇忘了娘(Qǔle xífu wàngle niáng. The man forgot his mother after getting married.)", "丑媳妇总得见公婆(Chǒu xífu zǒng děi jiàn gōngpó. Ugly daughter-in-law has to see her parents-in-law.)", "巧媳妇难为无米炊(Qiǎo xífu nán wéi wú mǐ chuī. The cleverest housewife cannot cook without rice)", whether they mean "son's wife" or "wife" or "married woman", all of them are written only as "媳妇" and annotated with the pronunciation xífu. Here the Chinese character "妇(fù, woman)" is read in neutral tone.

"媳妇" with frequency serial number 3352 and "媳妇儿" with frequency serial number 2514 are annotated with the pronunciation xífu and xífur respectively on page 571 of the *Lexicon of Common Words in Contemporary Chinese*, which was officially released by China's National Linguistics Work Committee and the Department of Language Information Management of the Ministry of Education in June 2008 and published by the Commercial Press. The pronunciation of xífu and xífur in the *Lexicon* reflects the opinions of state authorities, top-ranking linguists and experts in China, and can be regarded as the standard pronunciation of xífu. That means that the Chinese character "妇(fù, woman)" in word xífu should be read in neutral tone.

5 Conclusion

Xifu and *xífur* are both aimed at the same object of reference. The difference in meanings only reflects the different angles of address. For a married woman, she is the son's wife to her parents-in-law, the wife to her husband, and the married woman to other person. Its specific meanings can be judged from the speaker's point of view in the context, that is to say, it is not difficult to identify them and it does not cause confusion to the listener.

Therefore, *xífu* and *xífur* should be regarded as a word. It is suggested that only one entry *xífu* should be arranged under the Chinese character "媳(xí)" in the *Modern Chinese Dictionary*. It should be read in neutral tone and can add r-ending retroflexion, while it has only three senses: 【媳妇】 *xífu* (～儿) noun ①son's wife ②wife ③young married women in general.

References

1. Zhang, Z., Zhang, Q.: Lexical Semantics, 3rd edn. Commercial Press, Beijing (2012)
2. Shi, W.: On the formation of *Xífu*. Linguist. Sci. **2**, 212–219 (2008)
3. Luo, X.: Textual research of *Xífu*. Lexicograph. Stud. **1**, 128–131 (2001)
4. Yu, Y.: On the words *Xīnfu* and *Xífu*. Chin. Charact. Cult. **3**, 40–43 (2003)
5. Guo, Z.: On writing form of "X儿". Lang. Teach. Res. **4**, 67–70 (2000)

A Research into Third-Person Pronouns in *Lun Heng*(论衡)

Huiping Wang[1,2] and Zhiying Liu[1,2(✉)]

[1] Institute of Chinese Information Processing, Beijing Normal University,
Beijing, China
18013752306@163.com, liuzhy@bnu.edu.cn
[2] Ultra Power-BNU Joint Laboratory for Artificial Intelligence, Beijing, China

Abstract. *Lun Heng* was written in the East Han Dynasty. Because the writing style is of both Ancient Chinese and Middle Chinese, *Lun Heng* is worth researching. The third-person pronouns in *Lun Heng* are mainly "之(zhi)", "其(qi)", "彼(bi)", "厥(jue)", "若(ruo)", "夫(fu)", "此(ci)" and "是(shi)". Generally speaking, these third-person pronouns in *Lun Heng* inherit the existing usage in Ancient Chinese. However, they also have developments and changes, which involve the alteration of syntax functions and the appearance of new third-person pronouns.

Keywords: *Lun Heng* · The third-person pronouns · Metrology research

1 Introduction

From ancient times to mediaeval times, Chinese changed a lot. *Lun Heng* was written in the third year of Yuan He of Emperor Zhang of the Eastern Han (a method of counting the years in the name of the emperor in old China). This period is the key period when Ancient Chinese transfers to Middle Chinese. Therefore, the language style of *Lun Heng* is distinctive. On the one hand, it inherits a large number of grammatical forms already existed in the earlier period. On the other hand, it produces different grammatical phenomena.

This paper comprehensively investigates the syntactic and semantic characteristics of the third-person pronouns in *Lun Heng* through synchronic analysis and diachronic comparison. This paper mainly deals with the following questions: First, whether there are third-person pronouns in *Lun Heng*? Second, what are the characteristics of the syntactic and semantic functions of the third-person pronouns in *Lun Heng*? Third, what is the status of third-person pronouns in *Lun Heng* in the history of Chinese third-person pronouns system.

2 The Third-Person Pronouns System in *Lun Heng*

Are there third-person pronouns in *Lun Heng*? This problem can be traced back to whether there are third-person pronouns in Ancient Chinese and what third-person pronouns are. Over the past one hundred years, there has been a controversy on this

J.-F. Hong et al. (Eds.): CLSW 2019, LNAI 11831, pp. 681–689, 2020.
https://doi.org/10.1007/978-3-030-38189-9_69

issue in the community. The controversy is closely related to the multi-oriented pronouns (words that are both demonstrative pronouns and personal pronouns) in this period [1]. We think the correct approach to solve the controversy is to make a quantitative analysis of the usage of the third-person pronouns in the perspective of grammatical function and pragmatic frequency.

In the pronouns system of Ancient Chinese, "厥(jue)", "彼(bi)", "其(qi)", "之(zhi)" are the main multi-oriented pronouns [2]. Therefore, we chose several representative classical literatures, then made statistics on the usage of third-person pronouns (P) and demonstrative pronouns (D) of these four words respectively (Table 1).

Table 1. The pragmatic frequency of third-person pronouns and demonstrative pronouns in ancient representational documents

Document title	厥(jue)		彼(bi)		其(qi)		之(zhi)	
	P	D	P	D	P	D	P	D
Shang Shu	128	156	1	3	33	50	53	20
Shi Jing	36	24	28	281	373	116	486	73
Lun Yu	0	0	2	1	168	38	315	12
Meng Zi	1	0	24	13	439	105	689	91
Zuo Zhuan	8	0	46	5	1645	318	2225	1112
Lv Shi Chun Qiu	0	0	16	10	1719	79	1710	16
Xiao Jing	1	0	0	0	38	3	33	2
Gu Liang Zhuan	0	0	2	0	649	101	629	6

Lu and Guo believe that there are no real third-person pronouns in Ancient Chinese, because the words "其(qi)", "彼(bi)", "之(zhi)" are all transformed from demonstrative pronouns, none of them are fully developed [3, 4]. However, by examining the pragmatic frequency of the third-person pronouns and demonstrative pronouns, we can see that the words "其(qi)", "彼(bi)", "之(zhi)" have both demonstrative pronouns and third-person pronouns in ancient times. Moreover, with the development of history, the third-person pronouns gradually dominated, then completely transformed from demonstrative pronouns to independent third-person pronouns. Therefore, we believe that there is a third-person pronouns system in Ancient Chinese, and "彼(bi)", "其(qi)", "之(zhi)", "厥(jue)" are the main third-person pronouns.

There are eight third-person pronouns in Lun Heng, namely "之(zhi)", "其(qi)", "厥(jue)", "彼(bi)", "若(ruo)", "夫(fu)", "此(ci)" and "是(shi)", they eventually constitute the third-person pronouns system in Lun Heng (Table 2).

Table 2. Common third-person pronouns in *Lun Heng*

	之(zhi)	其(qi)	厥(jue)	彼(bi)	若(ruo)	夫(fu)	此(ci)	是(shi)
Subject	1	368	1	28	0	3	1	2
Object	2457	6	0	2	0	0	2	2
Attribute	5	1004	5	2	5	2	0	0
Concurrence	32	15	1	0	0	0	0	0
Total	2495	1393	7	32	5	5	3	4

3 The Usage of Third-Person Pronouns in *Lun Heng*

3.1 The Syntactic Characteristics of Third-Person Pronouns

3.1.1 '之(Zhi)'

"之(zhi)" is the third-person pronoun with the highest frequency in *Lun Heng*. It mainly acts as an object and a concurrence (a double-function element). The distribution of the syntactic functions is following (Table 3):

Table 3. The syntactic functions distribution of "之(zhi)" as a third-person pronoun.

Syntactic position	Subject	Object	Attribute	Concurrence	Total
Number	1	2457	5	32	2495
Proportion	0.05%	98.47%	0.20%	1.28%	100%

In *Lun Heng*, "之(zhi)" acts as the object mainly in the following three scenarios: The first is that "之(zhi)" is used as the object in the predicate-object structure. The predicate is mainly a verb, sometimes it can also be an adjective or a noun. Such as:

(1)夫鹤鸣云中,人闻声仰而视之,目见其形。（《艺增篇》）　　（verb+之）
(2)孔子抚其目而正之,因与俱下。（《书虚篇》）　　（adjective+之）
(3)击壤者无知，官之如何？（《艺增篇》）　　（noun+之）

The second is that "之(zhi)" is used as the object in the prepositional-object structure. Such as:

(4)杞梁氏之妻向城而哭,城为之崩。（《感虚篇》）

The third is that "之(zhi)" is used as the indirect object in the double-object structure. Such as:

(5)姓，因其所生赐之姓也。（《洁术篇》）

3.1.2 '其(qi)'

"其(qi)" is also a common third-person pronoun in *Lun Heng*. There are 1393 cases of "其(qi)" as a third-person pronoun in *Lun Heng*. The distribution of its syntactic function is following (Table 4):

Table 4. The syntactic functions distribution of "其(qi)" as a third-person pronoun.

Syntactic position	Subject	Object	Attribute	Concurrence	Total
Number	368	6	1004	15	1393
Proportion	26.41%	0.43%	72.07%	1.07%	100%

The third-person pronoun "其(qi)" in *Lun Heng* inherits the previous usage and mostly acts as an attributive [5]. Such as:

(6)至宜阳，为其主人入山作炭。（《吉验篇》）
(7)举世为佞者，皆以祸众。不能养其身，安能养其名?（《答佞篇》)

3.1.3 '厥(jue)' and '彼(bi)'

According to Zhang, the pronoun "厥(jue)" appeared in the Western Zhou Dynasty, mainly in the use of third-person pronouns [6]. In *Lun Heng*, "厥(jue)" is seldom used as a third-person pronoun. There are 5 examples in which "厥(jue)" is mainly used as an attributive. For example:

(8)珍贤圣之文，厥辜深重，嗣之及孙。（《佚文篇》)

There is one case in which "厥(jue)" is used as a subject and a concurrent respectively.

(9)《梓材》曰："强人有王开贤，厥率化民。"（《效力篇》） (subject)
(10) 文又曰："女於时，观厥刑於二女。"（《正说篇》） (concurrent)

There are 32 examples of "彼(bi)" acting as a third-person pronoun in *Lun Heng*, where 28 times as a subject, 2 times as an object and 2 times as an attributive.

(11)沈同曰：'燕可伐与?' 吾应之曰：'可。' 彼然而伐之。（《刺孟篇》） (subject)
(12)内有以相知，视彼犹我，取之不疑。（《初禀篇》） (object)
(13)孟子知言者也……见彼之问，则知其措辞所欲之矣。（《正说篇》） (attributive)

3.1.4 '若(ruo)' and '夫(fu)'

The frequency of "若(ruo)" as a third-person pronoun is not high. There are 5 times in *Lun Heng*, all of which are attributive. Such as:

(14)孔子生不知其父，若母匿之，吹律自知殷宋大夫子氏之世也。（《实知篇》）

The frequency of "夫" as a third-person pronoun is also not high. There are 5 cases in *Lun Heng*, including 3 cases as a subject and 2 cases as an attributive.

(15)夫得其术，虽不受命，犹自益饶富。（《率性篇》）   (subject)
(16)夫名在颜渊之上，当时所为，非子贡求胜之也。（《问孔篇》）   (attributive)

3.1.5 '此(ci)' and '是(shi)'

There are 3 cases in *Lun Heng* where "此(ci)" acts as a third-person pronoun, one as a subject and two as an object.

(17)壤於涂者，其志亦欲求食乎?此尚童子，未有志也。（《刺孟篇》）   (subject)
(18)邪人反道而受恩宠，与此同科，故合其名谓之《佞幸》。（《幸偶篇》）   (object)
There are 4 cases of "是(shi)", 2 times as a subject and two times as an object.
(19)行操之士亦怪毁之曰："是必乏於才知。"（《命禄篇》）   (subject)
(20)予原过竹二节，莫通，曰："为我以是遗赵无恤。"（《实知篇》）   (object)

3.2 The Semantic Characteristics of Third-Person Pronouns

The targets referred to by the third-person pronouns in *Lun Heng* can be roughly divided into the following categories: men and gods, animals, nation, concrete things and abstract things. Men and gods include human, heaven and earth, god, etc. Ancient people respected ghosts and gods, and often regarded them as life beings equal to human beings. The nation includes the states, cities and dynasties. The concrete things include rivers, clouds, rain, utensils, rice, gourds and other inanimate things. The abstract things include time, space, consciousness, existence, law, temperament, etc. The specific semantic features of each pronoun are as follows (Table 5):

Table 5. Semantic characteristics of common third-person pronouns in *Lun Heng*

	之	其	厥	彼	若	夫	此	是
Men and gods	+	+		+	+	+	+	+
Animals	+	+			+			
Nation	+	+	+					
Concrete things	+	+						+
Abstract things	+	+						

As third-person pronouns, why cannot the two most commonly used words "其 (qi)" and "之(zhi)" replace "厥(jue)" and "彼(bi)"? This accounts for their difference in semantics. "厥(jue)" is modest and respective title in Ancient Chinese. "厥(jue)"is commonly used in imperial edicts and petitions. [7] "厥(jue)" and "彼(bi)" are highly specialized. "厥(jue)" refers to almost all the names of emperors or dynasties, and "彼 (bi)" refers to almost men or gods. Such as:

(21)内有以相知，视彼犹我，取之不疑。(《初禀》)
(22)文又曰: "女於时，观厥刑於二女。" (《正说篇》)

Wang thinks there is no obvious boundary between singular and plural in personal pronouns in ancient times [8]. How are the singular and plural in personal pronouns in *Lun Heng*? We found that there are no distinctions of singular and plural in personal pronouns, third-person pronouns in *Lun Heng* can both represent singular or plural. "之 (zhi)", "其(qi)", "厥(jue)", "若(ruo)", "夫(fu)" mainly represent singular, for example:

(23)朔死，其妻有遗腹子。(《吉验篇》)
(24)夫言吴王杀子胥投之於江，实也。(《书虚篇》)
(25)文又曰: "女於时，观厥刑於二女。" (《正说篇》)
(26)夫得其术，虽不受命，犹自益饶富。(《率性篇》)
(27)孔子生不知其父，若母匿之，吹律自知殷宋大夫子氏之世也。(《实知篇》)
While "彼" mainly represents plural。For example:
(28)何则?欲专良善之名，恶彼之胜已也。(《累害篇》)

4 The Status of Third-Person Pronouns in *Lun Heng* in the History of the Third-Person Pronouns

Lun Heng was written in the East Han Dynasty. This period is in the transitional period from Ancient Chinese to Middle Chinese. It is necessary to make a diachronic comparative study of *Lun Heng* to determine the status of third-person pronouns in *Lun Heng* in the History of the third-person pronouns.

4.1 Third-Person Pronouns in *Lun Heng* Versus Third-Person Pronouns in Former Works

We chose *Zuo Zhuan*, the representative work in the Pre-Qin period, as the comparative target. The third-person pronouns in *Zuo Zhuan* mainly include "其(qi)", "之(zhi)", "彼 (bi)", "厥(jue)" and "夫(fu)" [9]. The use of third-person pronouns in *Zuo Zhuan* is as follows (Table 6):

Table 6. Common Third-Person Pronouns in *Zuo Zhuan*

	之(zhi)	其(qi)	厥(jue)	彼(bi)	夫(ruo)
Subject	0	232	8	40	16
Object	3668	10	0	5	0
Attribute	23	1521	10	2	1
Concurrence	19	0	0	0	1
Total	1775	1775	18	47	18

Generally speaking, the third-person pronouns in *Lun Heng* inherit the basic features of the grammatical system of the third-person pronouns in Ancient Chinese. Three pieces of evidence support this opinion: First, in the expression of the third-person pronouns, the common third-person pronouns in *Lun Heng* are almost the same as those in *Zuo Zhuan*. They are "其(qi)", "之(zhi)", "彼(bi)", "厥(jue)" and "夫(fu)". Second, in terms of the frequency of the use of third-person pronouns, *Lun Heng* and *Zuo Zhuan* are basically the same. The main third-person pronouns are "之(zhi)" and "其(qi)", which absolutely dominate. Third, on the grammatical function of the third-person pronouns, the third-person pronouns in *Lun Heng* inherit some functions of the previous period. For example, "彼(bi)" in *Lun Heng* and *Zuo Zhuan* both acts as the subject, object, attributive and concurrent.

However, the third-person pronouns in *Lun Heng* also have developments and changes, which involve the alteration of syntax function and the appearance of new third-person pronouns. First, some third-person pronouns in *Lun Heng* have more plentiful grammatic function. For example, there are 15 cases in *Lun Heng* where "其(qi)" acts as a double-function element, while there is 0 case in *Zuo Zhuan*. Second, new third-person pronouns appear in *Lun Heng*, they are "若(ruo)", "此(ci)" and "是(shi)". "之(zhi)", "其(qi)", "厥(jue)", "彼(bi)", "若(ruo)", "夫(fu)", "此(ci)" and "是(shi)" eventually constitute the third-person pronouns system in *Lun Heng*.

4.2 Third-Person Pronouns in *Lun Heng* vs Third-Person Pronouns in Later Works

We chose the representative work of Wei and Jin Dynasties, *San Guo Zhi*, as the comparative target. The third-person pronouns in *San Guo Zhi* are mainly "其(qi)", "之(zhi)", "厥(jue)", "彼(bi)" and "渠(qu)" [10]. The use of third-person pronouns in *San Guo Zhi* is as follows (Table 7):

Table 7. Common third-person pronouns in *San Guo Zhi*

	之(zhi)	其(qi)	厥(jue)	彼(bi)	渠(qu)
Subject	0	106	0	38	1
Object	989	147	0	14	0
Attribute	13	1518	36	8	0
Concurrence	12	62	0	0	0
Total	1014	1833	36	60	1

We can see that the biggest difference is that there is no third-person pronoun "渠 (qu)" in *Lun Heng*. Wang Li pointed out that new third-person pronouns such as "伊 (yi)", "渠(qu)" and "他(ta)" appeared in Middle Chinese, the following table supports this [11] (Table 8).

Table 8. The frequency of "渠(qu)" as a third-person pronoun

	San Guo Zhi	Shi Shuo Xin Yu	You Xian Ku	Han Shan Si	Zu Tang Ji
渠(qu)	1	0	18	13	29

Moreover, "其" has more and more grammatical functions. The use of "其" as an attributive is as common as *Lun Heng* in San Guo Zhi, while the use of "其" as the concurrent and object in San Guo Zhi is more than that in *Lun Heng*.

5 Conclusion

Through synchronic analysis and diachronic comparison, this paper comprehensively investigates the third-person pronouns in Lun Heng. Conclusively, on the one hand, these third-person pronouns in Lun Heng inherit the existing usage in Ancient Chinese. On the other hand, they also have developments and changes, which involve the alteration of grammatical functions and the appearance of new words.

The common third-person pronouns in *Lun Heng* mainly include "其(qi)", "之 (zhi)", "厥(jue)", "若(ruo)", "彼(bi)", "夫(fu)", "是(shi)" and "此(ci)". Among them, "之(zhi)" and "其(qi)" have the highest frequency, accounting for 63.26% and 35.32%.

In terms of grammatical functions, the third-person pronouns in *Lun Heng* act separately. "其(qi)" is mainly used as an attributive, followed by a subject, an object and a concurrent. "之(zhi)" is often used as an object. "厥(jue)" is mainly used as an attributive. "彼(bi)" is mainly used as a subject. "若(ruo)" is mainly used as an attributive. "夫(fu)" is mainly used as a subject, followed by an attributive. "是(shi)" and "此(ci)" are mostly used as a subject and an object.

In terms of semantic features, each third-person pronoun in *Lun Heng* represents both singular and plural. Among them, "其(qi)", "之(zhi)", "厥(jue)", "若(ruo)", "夫 (fu)", "是(shi)" and "此(ci)" mainly represent singulars, while "彼(bi)" is mainly plural. "其(qi)" and "之(zhi)" refer to a wide range of objects, while "彼(bi)" and "厥 (jue)" mainly refer to men, gods and nation.

In the ancient times, "其(qi)", "之(zhi)", "厥(jue)", "彼(bi)", "夫(fu)", etc.have developed into third-person pronouns. The third-person pronouns in *Lun Heng* inherit the existing usage in Ancient Chinese. In the middle times, on the one hand, third-person pronouns have more plentiful grammatical functions. On the other hand, a large number of new words are added to third-person pronouns system. In *Lun Heng*, the new third-person pronouns are "若(ruo)", "此(ci)" and "是(shi)". "之(zhi)", "其(qi)", "彼(bi)", "厥(jue)", "若(ruo)", "夫(fu)", "此(ci)" and "是(shi)" eventually constitute the third-person pronoun system in *Lun Heng*.

Acknowledgments. This work is supported by the Fundamental Research Funds for the Central Universities, National Language Committee Research Program of China (No. ZDI135-42) and National Social Science Fund of China (No. 18CYY029).

References

1. Sun, S.: Study on the third-personal pronouns in Ancient Chinese. J. Hubei Correspondence Univ. **24**(10), 96–97 (2011)
2. Jia, L.: An Investigation of Third-person Pronouns in Ancient Chinese. Shaanxi Normal University (2006)
3. Lu, S.: Chinese Grammar Summary, p. 154. Commercial Press (2010)
4. Guo, X.: Analects of the History of Chinese Language, p. 71. Commercial Press (1997)
5. Li, G.: Study on the personal pronoun "其"(qi). J. Guangxi Univ. (Philos. Soc. Sci. Edn.) 73–76 (1986)
6. Zhang, Y.: The discussion about the character of pronoun "厥(jue)" of the West Zhou language. J. Ancient Works Res. 57–62 (2005)
7. Zhang, Y.: The discussion about the original and flexible use of the character of pronoun "厥(jue)" of the West Zhou language. J. S. Chin. Norm. Univ. 88–96 (2006)
8. Wang, L.: Chinese Language History Draft, p. 258. Chinese Publishing house (2015)
9. Jia, S.: An Analysis of Pronoun Grammar in Zuozhuan. Northwest University (2014)
10. Zhao, X.: Research on Personal Pronouns in San Guo Zhi. Huaibei Normal University (2013)
11. Wang, L.: Chinese Language History Draft, p. 78. Chinese Publishing house (2015)

Analysis of the Collocation of "AA-Type Adjectives" Based on MLC Corpus

Junping Zhang[1(⊠)], Rui Song[1], Ting Zhu[2], Caihong Cao[1], and Mao Yuan[1]

[1] College of Chinese Studies, Beijing Language and Culture University, Beijing, China
jpzhang0315@126.com, songrui1990@126.com,
caocaihong@blcu.edu.cn, yuanmao5708@163.com
[2] Foreign Languages Department, Harbin Engineering University, Harbin, China
jessie0414@ruc.edu.cn

Abstract. Based on the data of *"Modern Chinese Dictionary"* and Media Language Corpus of Communication University of China, and according to the position of "AA-type adjectives" in the whole collocation, the 104 "AA-type adjectives" studied in this paper can be divided into three categories. They are post-positioning, pre-positioning, and the unlocated respectively. This paper focuses on the syntactic features of "AA-type adjectives and their collocations", the five types of their expressions, and the comparative analysis of the internal structural relationships of the five collocation types, with a view to provide ideas and clues to the semantic study of "AA-type adjectives and their collocations" and second language teaching.

Keywords: Collocation type · Internal structure relationship · Syntactic features

1 Introduction

In the author's teaching practice of Chinese as a foreign language and HSK corpus [1], it is found that foreign students tend to write such sentences, such as:

(1) *Dàn niánqīng rén yīnggāi zhīdào shàng yībèi rén yě céng yǒu guò niánqīng de shíqí, nà shíhòu, tāmen yě**xuèqì bóbó**, duì liàn'ài, jiùyè, péngyǒu huò liúxíng de shìwù hěn gǎn xìngqù.
(The error is "**xuèqì bóbó**".)

*但年轻人应该知道上一辈人也曾有过年轻的时期，那时候，他们也**血气勃勃**，对恋爱、就业、朋友或流行的事物很感兴趣。

(2) *Zài shèhuì yǐngxiǎng xià, zhōngxuéshēng kěyǐ cóng zázhì yǔ guāngdié dǒngdé liàn'ài zhīshì ér gèng **chǔnchǔn yīshì**.

J.-F. Hong et al. (Eds.): CLSW 2019, LNAI 11831, pp. 690–700, 2020.
https://doi.org/10.1007/978-3-030-38189-9_70

(The error is **"chǔnchǔn yī shì"**.)

*在社会影响下，中学生可以从杂志与光碟懂得恋爱之事而更**蠢蠢一试**。

(3) *……Cái zhīdào dōngjīng de nǔzǐ zhōngxué huò nánzǐ zhōngxué bìng bù yīdìng shì xiàng wǒ jiāxiāng de nǔzǐ zhōngxué nàyàng **chòumíng dǐngdǐng,** qíshí yǒu hěn duō suǒ míngpái xuéxiào.
(The error is "**chòumíng dǐngdǐng**".)

*……才知道东京的女子中学或男子中学并不一定是像我家乡的女子中学那样**臭名鼎鼎**，其实有很多所名牌学校。

(4) *Suīrán shuǐyuán lí tāmen nà hěn yuǎn, dànshì **xìngzhì bóbó de liǎng gèrén** táizhe shuǐtǒng dào shān dǐxia tiāo shuǐ hē.
(The error is **the order of "xìngzhì bóbó de" and "liǎng gèrén"**.)

*虽然水源离他们那很远，但是**兴致勃勃地两个人**抬着水桶到山底下挑水喝。

(5) *Yībān dōu shì rì chū ér zuò rìluò ér xī, yījiā rén zǎowǎn dōu pèngmiàn, xū hán wèn nuǎn, **róng róngqià qià** de.
(The error is "**róng róngqià qià**".)

*一般都是日出而作，日落而息，一家人早晚都碰面，嘘寒问暖，**融融洽洽的**。

(6) *Wǒmen yǐjīng bù shì háizi le, shì **tángtáng** yīgè chéngnián rén.
(The error is "**tángtáng**".)

*我们已经不是孩子了，是**堂堂**一个成年人。

(7) *Xiàtiān shí tiānqì **rè yányán**, nóngmín shēnshang mǎn shì hànshuǐ.
(The error is "**rè yányán**".)

*夏天时天气**热炎炎**，农民身上满是汗水。

It can be seen from the above errors that foreign students have not really grasped such "AA-type adjectives". The errors are reflected in the aspects of collocation, syntax, semantics, etc. and should be paid more attention to. Most of the current research about them is only mentioned in the discussion of ABB-type, AABB-type and other types, for example in [2] and [3]. There is a lack of special research on them. The questions to be discussed in this paper are as follows:

a How many "AA-type adjectives" are there in the "*Modern Chinese Dictionary*" [4]? How often are they used in large-scale standardized native speakers' corpus? Which are the more frequently used" AA-type adjectives"?

b What are the common collocations of these high-frequency "AA-type adjectives"? Do these words often appear on the left or right side of the "AA-type adjectives"? Which "AA-type adjectives" have the same collocation words on the left and right?

c What are the syntactic features of "AA-type adjectives and their collocations". What is the internal structure relationship of different collocation types?

Firstly, we searched 231 "AA-type adjectives" in "*Modern Chinese Dictionary*" one by one in the Media Language Corpus (MLC) of Communication University of China [5], and found that about 91.3% of "AA-type adjectives" in MLC can't be used alone or are used alone less[1]. By utilizing the function of "left sort" and "right sort" in MLC, the author collected statistics on the collocation of these 104 "AA-type adjectives", see Table 1. In this paper, the "104 AA-type adjectives and their collocations" in Table 1 is taken as the final research object, from the external collocation form to the internal structure relation, and then to the syntactic features to investigate and analyze.

In Table 1, on the definition of "collocation", Wei [6] has summed up the idea of Firth, the first British semantic scholar who has put forward the concept of collocation, as follows: (1) Collocation refers to the use of words in conjunction with words; (2) Collocation is a means of meaning; (3) Every partner of habitual word collocation expects and foresees each other; (4) Class joins are abstractions higher than word collocations. The third point of "mutual expectation and mutual foresight" is the problem of semantic compatibility, and the fourth point of "class join" is the problem of grammatical collocation.

The definition of collocation by Halliday, the representative of the New Firth school, seems to require recognition only of the linear co-occurrence of a term in some significant vicinity, either within a given region or within a truncated point. Wei [6] believes that this combination is the so-called collocation. Collocation is the chaperone behavior of words and words, which has a certain rule, at the same time, it has a certain probability, that is, it shows a certain probability characteristics (Shu [7]).

The author tends to combine the three points mentioned above, and she holds the view that collocation should be a companionship behavior in the proximity range and it is a habitual juxtaposition between the partners of mutual expectations and foresight, at the same time to conform to the grammar rules, and semantic compatibility. This kind

---

[1] Removing the 87 none results and irrelevant results in MLC (name of person or different pronunciation) is 144; these 144 AA adjectives have examples, but some appear fewer, such as: changchang 怅怅 (2), huanghuang 皇皇 (12), etc., so such AA-type adjectives with less frequency need to be excluded.

Table 1. Statistical examples of the frequency of right-left collocation of AA-type adjectives

Serial No.	total freque ncy	examples of left collocation and frequency	"AA-type adjectives"	examples of right collocation and frequency
1	107	白雪 47 / 白 10	+皑皑+	白雪 32
2	25	雄赳赳, 气 25	+昂昂+	×[a]
3	212	锈迹 78/血迹 51/劣迹 24	+斑斑+	血迹 19
4	106	文质 61	+彬彬+	有礼 45
5	892	兴致 343/生机 154/ 雄心 138/ 野心 20/ 生气 8	+勃勃+	生机 186
6	136	金 106/ 金光 16/ 黄 14	+灿灿+	×
7	165	白发 141/天 9	+苍苍+	×
8	96	死气 22/ 昏昏 18/ 阴 12/ 暮气 7/ 昏 5	+沉沉+	×
9	101	兴 65/怒气 21/气 10/急 5	+冲冲+	×
10	293	忧心 293	+忡忡+	×
11	1668	困难 344/疑点 83/顾虑 51/ 迷雾 47/矛盾 47/隐患 31/ 压力 26/问题 26/危机 25/ 心事 24/疑虑 24/危险 22/ 阻力 22	+重重+	困难 206/阻力 34/压力 32/障碍 28/包围 21/ 挑战 19
12	153	×	+蠢蠢+	欲动 152
13	840	急 116/行色 60/来去 26/ 脚步 14/步履 14	+匆匆+	
......				

[a] ""√"" means that the AA-type adjective has a matching component on the left/right side, and "×" indicates that no matching component is found on the left/right side. The same as follows.

of companionship behavior has a certain rule, this law has a certain probability, showing certain characteristics of probability. Therefore, the left-right collocations in Table 1 are used in close proximity to the AA-type adjectives.

2 Classification of "AA-Type Adjectives"

According to the characteristics of AA-type adjectives collocation in the left and right position and the location of AA-type adjectives in the whole collocation form, 104 AA-type adjectives are divided into three types, see Table 2.

Table 2. Distribution of 104 AA-type adjectives collocated at left-right side in MLC

Category	left [a] collocation	AA-type adjectives and the number of them		right collocation
class I post-positioning	√	昂昂、灿灿、苍苍、沉沉、冲冲、忡忡、匆匆、葱葱、眈眈、旦旦、霍霍、睽睽、淋淋、凛凛、便便、茸茸、洒洒、腾腾、汪汪、泓泓、痒痒、奕奕、翼翼、盈盈、凿凿	25	×
class II pre-positioning	×	蠢蠢、喋喋、泛泛、愤愤、乖乖、岌岌、津津、涓涓、侃侃、历历、寥寥、碌碌、缕缕、落落、冥冥、窃窃、区区、拳拳、冉冉、瑟瑟、姗姗、上上、莘莘、亭亭、团团、娓娓、栩栩、遥遥、殷殷、隐隐、郁郁、芸芸、谆谆、孜孜	34	√
class III the unlocated	√	Different at the left and right side	彬彬、楚楚、纷纷、耿耿、惶惶、恢恢、济济、炯炯、翩翩、平平、闪闪、堂堂、惺惺、奄奄、泱泱、扬扬、依依、熠熠、悠悠	19	√
		The left and right can be the same	皑皑、斑斑、勃勃、重重、鼎鼎、霏霏、滚滚、赫赫、款款、朗朗、累累2、累累3、粼粼、漫漫、茫茫、蒙蒙、绵绵、脉脉、袅袅、融融、滔滔、迢迢、习习、熊熊、炎炎、洋洋	26	
Total			104	

[a] Collocation components appeared at the left side is called "left collocation"; collocation components appear at the right side is referred to as "right collocation".

Class I– post-positioning, 25, AA can only appear at the right side, such as: jīnguāng/huáng + càncàn".

Class II–Pre-positioning, 34, AA can appear at the left side, such as: "liáoliáo+wújǐ/ shùyǔ/ kěshǔ/ jǐbǐ".

Class III–the unlocated, 45, AA can appear at either left or right side, which can be divided into two categories: (A) The left and right side are different; (B) The left and right side can be the same.

"Left and right side are different" refers to the AA-type adjectives have matching components on both sides, but the ingredients used in the left and right are different, such as: "wénzhì + bīnbīn" and "bīnbīn + yǒulǐ". "Left and right side can be the same" refers to the AA-type adjectives on the left and right side can not only appear collocation ingredients, but also are the same collocation elements, such as: "kùn-nànchóngchóng" and" chóngchóngkùnnán", "dàmíngdǐngdǐng" and" dǐngdǐngdàm-íng", "qiānlǐtiáotiáo" and" tiáotiáoqiānlǐ", "shēngjī + bóbó" and "bóbó + shēngjī" and so on.

AA-type adjectives in Class I and Class II are rather fixed and can't be changed freely when they are paired with other components. The position of these two types of words in the collocation can be strengthened separately during teaching.

Class III AA-type adjectives can be placed in front or back of the collocation structure, and the position is more random and less restrictive, so the difficulty of learning is increased to a certain extent.

In Class III (B), although AA-type adjectives are exactly the same, students are not allowed to memorize too many words, but their use tendency, semantic orientation, prosodic characteristics and so on will cause some difficulties to foreign students. This particular case will be discussed in particular.

3 Analysis of Five Types of "AA-Type Adjectives and Their Collocations" and Internal Structure Relation

There are five types of "AA-type adjectives and their collocations": AA + BC, BC + AA, B + AA, AA + B and $AA_1 + AA_2$, as shown in Table 3. Its characteristic is that the four-character form predominates and the BAA type is more than the AAB type.

It is found that in MLC, "AA-type adjectives and their collocations" appear most of the time in four-character form, accounting for 84.6% of the total number of collocation forms (370), among which AA + BC is the most and BC + AA is the second.

The appearance of more frequent appeared four-character patterns may be related to the rhythm and rhyme of ancient literary works or Chinese. Lv [8] has pointed out that the combination of double-tone and double-tone is a major rhythmic tendency in modern Chinese. BAA form is superior to AAB form in quantity and has its theoretical basis from the perspective of prosody. The form of BAA conforms to the three-syllable step rule of Feng [9] and [10], which is, "The Rules of Stress Constraint of Single and Double Branches and the Principle of Single and Light". The latter two syllables are the concrete description of the first syllable, and the two syllables are more important than one syllable in rhythm. In other words, the BAA form is rhythmic integrity and self-sufficient in rhythm, while the AAB is a broken step. This partly explains why BAA is more than AAB form.

The internal structure of five collocation types can be divided into three categories, as shown in Table 4.

(1) AA preposition: AA + BC and AA + B, both of which have the same structural relationship. When BC/ B is a noun, both of them are attribute and headword structures. When BC/B is a verb, it is a adverbial and headword structure.

(2) AA postposition: BC + AA and B + AA are similar in structure. When BC/B is a noun, both are subject-predicate structures. BCs in 129 BC + AA are nouns. AA is a predicate to indicate how or what the subject BC is.

Table 3. The five types of AA-type adjectives and their collocations

Type	number of collocation	for example
AA+BC	181	皑皑白雪、涓涓细流 [a]、彬彬有礼、洋洋得意……
BC+AA	129	白雪皑皑、困难重重、脚步匆匆、得意洋洋 [b]……
B+AA	42	白皑皑、天苍苍、圆滚滚、暖融融、慢腾腾 [c]……
AA+B	4	蒙蒙亮、上上签、上上策、团团转
AA$_1$+AA$_2$	3	纷纷扬扬、洋洋洒洒、郁郁葱葱
Other	11	惶惶不可终日、天网恢恢疏而不漏、郁郁不得志、拳拳赤子心……
aggregate	370	

[a] Nouns such as "Baixue", "Xiliu" and other BC forms account for 92; other predicate components of BC form take up 89.

[b] BC in BC + AA is almost all nouns, and only a few are non-nouns, such as pre-AA words in "láiqùcōngcōng, hǔshìdāndān, módāohuòhuò, hánqíngmòmò, xiǎoxīnyìyì, déyìyángyáng".

[c] 30 state adjectives are listed in "*Modern Chinese Dictionary*", such as: bái'áiái, jīncàncàn, huángcàncàn and so on. Another 12 collocations are not included in the Dictionary, which involves 9 AA-style adjectives. For example: tiāncāngcāng, hūnchénchén, jíchōngchōng, yǔfēnfēn, xuěfēnfēn, lùmànmàn, yúméngméng, huàngyōuyōu, wūyāngyāng, yáyǎngyang, shǒuyǎngyang, xīnyǎngyang.

Table 4. Internal relationship of 5 collocation types

Category	Structural relation			
	Type	BC/B as a noun	BC/B as a verb	BC/B as an adjective
AA preposition	AA + BC / AA + B	The relationship between attribute and headword	The relationship between the adverbial and headword	–
AA postposition	BC + AA / B + AA	Subject-predicate relation	–	– / The relationship between headword and complement
Special form	AA$_1$ + AA$_2$	**Cascade parallel bond: AA$_1$|AA$_2$ → AA$_1$ + AA$_2$**		

The difference is that there are 24 B adjectives in 42 B + AA. When B is an adjective, B + AA is a middle complement structure. The AA adjective is a further supplement to the nature and degree of B. Only 2 BC of 129 BC + AA are adjectives, namely "deyiyangyang" and "xiaoxinyiyi(be careful)".

(3) Special form: AA_1 + AA_2 is called special form. This kind of collocation is composed of two different AA adjectives. There are three words: "fēnfēn +yánngyáng", "yángyáng +sǎsǎ" and "yùyù +cōngcōng".

The Chinese Idiom Dictionary [11] holds that "fēnfēnyángyáng, yángyángsǎsǎ, yùyùcōngcōng" and "cōngcōng máng mang" are overlapping structures, which is inappropriate in the author's opinion. AA_1 + AA_2 is similar to the traditional AABB form, but superposition and duplication are two different grammatical means. The basic form of AABB is a double-syllable word, and the superposition is formed by the overlap and bond of two monosyllabic words respectively. According to the definition of broad superposition, the author holds that " "fēnfēnyángyáng, yángyángsǎsǎ, yùyùcōngcōng" belong to the superposition, not to the duplication. For example, "tēnfēnyángyáng" is overlapped by two monosyllabic syllabuses which include "fen" and "yang", and then they are combined to form "fēnfēnyángyáng". There is no such basic form as" fēnyáng", which means that" fēnyáng "is not a double-syllable word in Chinese. Therefore, the author calls it the overlapping-junction bond, its composition is: $AA_1|\ AA_2 \rightarrow AA_1 + AA_2$.

4 Syntactic Characteristics of "AA-Type Adjectives and Their Collocations"

There are three syntactic features in "AA-type adjectives and their collocations":

(1) No adverbs of degree

Most "AA adjective" have a degree of reinforcement in themselves. For example, "exuberant" describes the appearance of a strong spirit or desire, which indicates the degree of spirit or desire. Interpretation of "very" in "800 words in modern Chinese" [12]: Used in front of adjectives with a high degree of expression. In terms of the former's lexical meaning and the functional meaning of "very", both of them are of high degree, but the two methods of "high degree of expression" can't be used simultaneously in modern Chinese, and AA adjectives can't be modified with adverbs of degree.

The AA-type adjective indicates the meaning of degree, so the "AA adjectives and their collocations" inherits the meaning of degree as a whole, so it is not appropriate to add the degree adverb before them.

(2) Some syntactic components can be used separately, mainly on BC + AA and AA + BC.

As discussed above, BC + AA is mainly subject-predicate structure, which is syntactic function equivalent to predicative phrase. Therefore, BC + AA is often used as predicate in separate sentences, mostly at the end of sentences. E.g.:

(a) Xiànnián 25 suì de Màn níng duǎnfǎ, dài fù yǎnjìng, wénzhìbīnbīn.
(a)

   现年 25 岁的曼宁短发，   戴副眼镜， 文质彬彬。

(a) Mr Manning with gentlemanlike behaviors, 25, has short hair and glasses.
(b) Yī tào dǐngcéng fùshì dānwèi túrán rán qǐ dàhuǒ, nóng yān gǔngǔn.
(b)

   一套顶层复式单位突然燃起大火，   浓烟滚滚。

(b) a set of top-level duplex units burst into flames and smoke billowed.

  The structure of AA + BC can be divided into two types according to the nature of BC: the attribute and headword structure and the adverbial and headword structure, so AA + BC can be used separately in different syntactic components. When BC is a predicate component, AA + BC is an adverbial and headword structure, which is often used at the end of sentences. AA + BC is often used as predicate. For example:

(c) Bìmùshì xiànchǎng, dàjiā gòng xù yǒuyì, yīyī xībié.
(c)

   闭幕式现场，   大家共叙友谊， 依依惜别。

(c) At the closing ceremony, we share our friendship and leave.

  When BC is a noun, AA + BC is an attribute and headword structure, often at the beginning of a sentence, and can be used as a subject. For example:

(d) Lěilěi shuòguǒ, wèi yóukè píngtiānle xǔduō cǎizhāi de lèqù.
(d)

   累累硕果， 为游客平添了许多采摘的乐趣。

(d) The fruits are fruitful and add a lot of fun to the pickings.

  But sometimes there is no obvious syntactic component, more like a kind of a description with emotional component. Because AA-type adjectives have a certain degree of description and image color, in the use of them, it embodys vivid rhetoric function. AA + BC collocation also has the characteristics of this description.

(e) Mángmáng dàhǎi, guānbīng dà bùfèn shíjiān shì zài dāndiào zhōng dùguò.
(e)

   茫茫大海，官兵大部分时间是在单调中度过。

(e) In the vast sea, soldiers spend most of their time in boredom.
(f) Yányán xià rì, xiǎoshǔ de bīngqílín dà shòu huānyíng.

(f)

炎炎夏日，　消暑的冰淇淋大受欢迎。

(f) Ice cream is very popular in a blistering summer day.

(3) inheritance of four-sentence patterns

The form of expression can be directly composed of two or more four-character sentences, or a long sentence contains two or more four-character sentences, so as to enhance descriptive features, more abundant description of the state of things, and can express subjective evaluation and personal feelings, enhance momentum, and it is easy to express feelings.

The four-word sentence pattern is one of the most commonly used sentence patterns in classical Chinese, and "*Book of Songs*" is the peak of the application of this sentence pattern, which establishes its position in Chinese. Chinese idioms and allusions, slang and so on, are mostly shaped into four-word forms, and even consist of two four-word sentences, such as "The ruler is short, the inch is long" and "The other car depends on each other; when the lips are lost, the teeth feels cold". Idioms and proverbs are popular and frequently used, which have great influence on Chinese.

After replacing classical Chinese with vernacular Chinese, the Chinese sentence pattern tends to be more flexible. The four-word language style is still widely used in modern Chinese because of its neat, sharp rhythm, easy to read, easy to remember and so on, such as Mao Zedong's writing often uses the four-sentence pattern, "ziligengsheng, jiankufendou(Self-reliance, hard work)", "dijinwotui, dizhuworao(When the enemy proceeds, we retreats; when the enemy settles, we disturb" and so on. This is an inheritance to the ancient four-character sentence pattern.

(g) Wǔ huán shénshèng, shènghuǒ xióngxióng, hépíng fāzhǎn, gòngtóng jìnbù.
(g)

五环神圣，　　圣火熊熊 ，　　和平发展，　共同进步。

(g) Five rings are holy and the Olympic flame is burning; to pursue peaceful development and achieve common progress.

5 Conclusion

Structure relation and syntactic characteristic are two aspects of the comparison of tradition and foundation in grammar research: Looking inward to see what structure relation it is such as attributive or subject-predicate enbales us to know the inner structure relation between AA adjective and its collocation more deeply; Look outward to see what features and functions they embody in sentences, which can help us produce sentences that conform to grammatical rules and syntactic characteristics.

In the next step we plan to further study AA adjectives and their collocations' semantic characteristics, stylistic characteristics, how to divide them into different layers and stages in the second language teaching. Focus will be paid especially to the "AA adjective"" stone features, prosodic features, semantic points and error characteristics which can be matched before or after the same words. This is one of the difficulties for foreign students.

Acknowledgments. This research project is supported by Science Foundation of Beijing Language and Culture University (supported by "the Fundamental Research Funds for the Central Universities") (Approval number 19 YJ010110). The work is also supported by Major Program of National Social Science Foundation of China (18ZDA295). During the process of writing this paper, Zhimin Wang, Huizhou Zhao and Chunhong Liu has provided me with much support and encouragement. My thank also goes to Qingqing Lv and Yuan Qi, two post-graduates who provided assistance during the translation of this paper. The author is in charge of any improper matters in the article.

References

1. HSK. http://hsk.blcu.edu.cn/. Accessed 27 Mar 2019
2. Aimin, Z.: The comparison between adjective reduplicative adverbial and other elements. Lang. Teaching Linguist. Stud. **2**, 67–78 (1996)
3. Jingsong, Z.: The grammatical meaning of duplication adjectives. Linguis. Res. **3**, 9–17 (2003)
4. Dictionary Editing Office, Institute of Linguistics, Chinese Academy of Social Sciences (ed.): Modern Chinese Dictionary, 7th edn. The Commercial Press, Beijing (2016)
5. MLC: http://ling.cuc.edu.cn/RawPub/. Accessed 31 Dec 2018
6. Wei, N.: Corpus-based and corpus-driven approaches to the study of collocation. Contemp. Linguist. **2**, 101–114 (2002)
7. Shu, D. (ed.): Methods in Cognitive Linguistics. Shanghai Foreign Language Education Press, Shanghai (2013)
8. Shuxiang, L.: First Exploration of single and double words in Mandarin. Stud. Chin. Lang. **1**, 10–23 (1963)
9. Shengli, F.: On the "prosodic word" of Chinese. Soc. Sci. Chin. **1**, 161–176 (1996)
10. Shengli, F.: On the properties and pedagogy of written Chinese. Chin. Teach. World **4**, 98–106 (2006)
11. Xingguo, W. (ed.): Chinese Idiom Dictionary, 1st edn. Sinolingua, Beijing (2010)
12. Shuxiang, L. (ed.): 800 words in modern Chinese. The Commercial Press, Beijing (1999). Updated version

TG Network: A Model that More Effectively Identifies the Use of the Auxiliary Word "DE"

Chuang Liu[(⊠)], Hongying Zan, Xuemin Duan, Kunli Zhang,
and Yingjie Han

School of Information Engineering of Zhengzhou University, Zhengzhou, China
214674227@qq.com,
{iehyzan, ieklzhang, ieyjhan}@zzu.edu.cn,
xueminduan@stu.zzu.edu.cn

Abstract. In the knowledge base of function word usage of "trinity", the auxiliary word "DE" has the characteristics of high frequency and flexible usage. In this paper, a neural network model (TG network) is proposed to automatically recognize the usage of "DE". In this network, the self-attention mechanism is firstly adopted as the first-layer feature encoder and GRU (gated recurrent unit) as the second-layer semantic extractor, and the recognition accuracy rate reaches 82.8%. Experiments show that the recognition effect of TG network is better than that of previous methods. In further experiments, the larger the window, the better the effect of the model is proved by setting different windows. At the same time, the fine-grained analysis of each usage category is carried out. In the future, it is expected that this model will automatically recognize more function words and the recognition results can be applied to other natural language processing tasks.

Keywords: The knowledge base of function word · TG network · Self-attention mechanism · GRU · Natural language processing

1 Introduction

Grammar has long been one of the difficulties in Natural Language Processing (NLP). Since functional words are responsible for describing the semantic relationship between grammatical features and notional words in sentences, the study of functional words in sentences has always been an important part of grammar research. Among them, auxiliary words in function words have the characteristics of Chinese typology and are more closely related to various kinds of notional words. In modern Chinese, the auxiliary "DE" is used frequently and flexibly. The researches about it have a significant reference for Chinese grammar research, teaching Chinese as a foreign language and natural language processing.

Literature [1, 2] gives a review on the research of "DE" over the years, including the discussion of its usage. Yu et al. [3] put forward the idea of building a "trinity" (dictionary of function word usage, rule base of function word usage and a corpus of Functional Words) knowledge base of generalized function word usage. Since then, Zan et al. [2, 5, 6] and Zhang et al. [7] have constructed Chinese function word usage

© Springer Nature Switzerland AG 2020
J.-F. Hong et al. (Eds.): CLSW 2019, LNAI 11831, pp. 701–709, 2020.
https://doi.org/10.1007/978-3-030-38189-9_71

knowledge base (CFKB). It includes a dictionary of the usage of functional words of the auxiliary "DE", a rule base and a corpus of usage annotation. Han et al. [8] discussed rule-based automatic tagging of auxiliary usage in the construction of auxiliary usage knowledge base. Due to the complexity of the use of "DE", literature [7, 8] points out that the recognition effect of the rule-based method is very unsatisfactory, and the auxiliary "DE" is the function word with the highest frequency, so the automatic recognition of its use has become the focus of research. Then Liu [9] and others used rule-based, CRF-based and GRU-based in deep learning to automatically identify "DE". Chang et al. [10, 11] and Zhang et al. [12] applied functional word usage knowledge to machine translation tasks respectively.

According to the usage rules of "DE" in CFKB, there are 39 categories. Experiments were carried out on the corpus of People's Daily on April 2000 because of the high frequency of "DE". 64449 sequences were selected. The proposed TG network firstly uses the coding block in Transformer model [11] as the first layer feature coder of the sequence, automatically extracts the internal semantic relations of the sequence, and then extracts the semantics again with GRU model. Experiments show that the TG network can get better recognition effect.

2   Relative Works

2.1   Description of the Usage of the Auxiliary Verb "DE"

According to Lv's "Modern Chinese 800 Words", "Modern Chinese Dictionary" (5th Edition) [14] and Zhang's "Modern Chinese Functional Words Dictionary" [15], combined with the corpus of People's Daily on April 2000, an auxiliary usage dictionary is constructed, in which "DE" contains 11 kinds of usages and 39 kinds of subdivided usages. Through the attributes of interpretation, usage, example sentence, collocation and combination, the usage of the auxiliary "DE" is described in an all-round way. Table 1 lists only some of the attributes of usage rules. Detailed descriptions of these attributes are given in reference [6].

2.2   Research on Neural Network Model

In 2014, Yoon Kim tried to classify texts using a Convolutional Neural Network (CNN). Hinton proposed dropout method to prevent over-fitting in training. Because of the natural seriality of language, the cyclic neural network can effectively solve the seriality problem, so the application of cyclic neural network in natural language processing has also been widely used. However, the problem of gradient explosion or gradient disappearance exists in the cyclic neural network itself. To solve this problem, Long Short-Term Memory (LSTM) neural network [18] and other models are proposed. As an extension of the LSTM model, GRU model has fewer parameters, simpler structure and better results on small data sets [19]. In 2017, Vaswani et al. [11] put forward the Transformer model, which is a feature extractor based on the attention mechanism. It can solve the shortcomings of the circular neural network such as not

Table 1. The usage description of the auxiliary "DE"

ID	Interpretation	Direction	Example
u_de5_t2_1a	Form phrases containing "DE" to modify nouns. <b>	Noun + ~ +noun.<b> used between nouns and their modifiers, modifiers are nouns. <x><z>	He has a lot of luggage. I ~ Very little luggage <b><z>| before ~ unreasonable burden <z>
u_de5_t2_1b	Form phrases containing "DE" to modify nouns. <b>	Verb + ~ +noun.< b> used between nouns and their modifiers, modifiers are verbs. <x><z>	The original arrangement ~ It's Wednesday, now it's Friday. <b><z>| sing ~ voice is too low <b><z>
u_de5_t2_1ba	Form phrases containing "DE" to modify nouns. <b>	Verb + noun + ~ +noun. <b> used between nouns and their modifiers, modifiers are verb-object phrases. <x><z>	Telegraph ~ Expenses <b>| China will always maintain world peace and stability ~ Important power. <r>
u_de5_t2_1bb	Form phrases containinre3g "DE" to modify nouns. <b>	Verb + adjective + ~ +noun. <b> used between nouns and their modifiers, modifiers are verb complement structure. <x><z>	Be unable to enter ~ People continued to wait at the elevator door. <r>| Ding Yuzhen flushes well ~ The photo was given to Kong Ling. <r>
u_de5_t2_1bc	Form phrases containing "DE" to modify nouns. <b>	Used between nouns and their modifiers, modifiers are phrases with adverbial middle structures. Adverbials consist of adjectives, time words, adverbs, nouns, NOUN phrases, prepositional phrases, auxiliary verbs and localizer phrases. The central word is a verb or verb phrase. <x><z>	Work with the Common Center and Work Together ~ Democratic parties. <r>

parallel, and can effectively obtain the global information of the sequence. Different neural network models have their characteristics and shortcomings. For different tasks, using the characteristics of their models, designing different models is the mainstream method nowadays. The TG network proposed in this paper is the combination of the transformer model and GRU model to achieve better results.

3 TG Network Structure

3.1 Model Overview

The main structure of the model is shown in Fig. 1. The input sequence includes word information and part-of-speech information. Firstly, the sequence is encoded by the Transformer [11] coding block to find the internal structure of the model. Then, the encoded sequence is sent to the GRU model for further semantic extraction. Finally, it is output by the multi-layer perceptron. In Sect. 3.2, the Transformer coding layer is briefly described. In Sect. 3.3, the core part of the coding layer, namely the multi-head attention mechanism, will be emphasized. In the last 3.4 section, we introduce how to stitch the output of the coding layer, and then as the input of the semantic extraction layer.

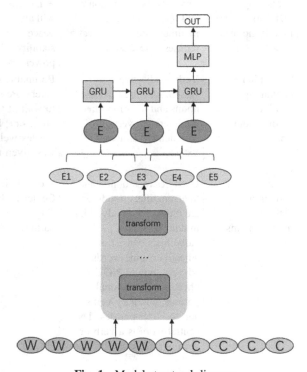

Fig. 1. Model structural diagram

3.2 Coding Layer

The coding layer uses the coding blocks in the Transformer model. N represents several identical modules. In this experiment, N is 1. Each coding layer contains two sublayers. The first is the multi-head self-attention mechanism, and the second is the fully connected feedforward neural network layer. The model uses a residual connection [20] around each of the two sub-layers, and then normalizes the layers [21], so the output of each sub-layer is LayerNorm (x + sublayer (x). And sublayer (x) is the output of each sub-layer.

3.3 Attention

Attention mechanism can effectively extract important parts of information. For example, in a text, through attention calculation, different words in the text will get different weights. The larger the weight, the higher the importance of the word in the text. In this paper, the attention equation can be described as mapping a pair of Q and a set of K and V pairs to the output, where Q, K, and V are vector representations. The output is the weighted sum of these values. When it is necessary to calculate the internal correlation of the sequence itself, Q = K = V, and all three inputs are input sequences. The advantage of this mechanism is that the information of each position in the sequence is correlated with the information of other locations, and the global information in the sequence can be obtained effectively.

Point Product Attention Equation. There are different ways to calculate attention. The attention equation used in the Transformer model is called "Scaled Dot-Product Attention" [22]. The input consists of Q, K, and V, where Q and K have the same vector dimension dk. Firstly, the dot product of Q and K is calculated and then divided by the square of dk. After a softmax function, the weight of V is obtained.

$$Attention(Q, K, V) = softmax\left(\frac{QK^T}{\sqrt{d_k}}\right)V$$

Multi-head Attention. Multi-head attention projects Q, K and V through h different linear transformations. Finally, different attention results are joined together. The size of H is set to 8, which means that the dimension of the word vector of the input sequence is divided into eight parts, each of which calculates its attention, and then splices the calculated results.

$$MultiHead(Q, K, V) = Concat(head1, \ldots, headh)Wo$$

Among them, Wo is a trainable parameter, $head_i$ = Attention(QW_i^Q, KW_i^K, VW_i^V).

Semantic Extraction Layer. After the first coding layer, we get the sequence that has acquired the internal semantic information, set a window of size 3, splice every three elements in the sequence, get a spliced sequence again, and then put the new sequence as input into the GRU model [9]. Finally, the output of the last hidden layer of GRU is classified into multi-layer perceptions.

4 Experimental Results and Analysis

The experimental data were collected from the functional word usage tagging corpus of People's Daily on April 2000. The training set and the test set are randomly divided into 9:1 scales. The usage of "DE" falls into 39 categories. Since there may be more than one "DE" in a sentence, the first step in the experiment is to set window size and select the words before and after "DE". For example, a window of size 3 means choosing the first three words and the last three words of "DE". The experiments were carried out under different window sizes (3, 5, 7).

In the experiment, the batch size is set to 64, the vector dimension d_k is set to 128, the layer number of transformer block is set to 1, and the head number is set to 8. In the experiment, Adam was chosen as the optimizer, and the learning rate was set randomly. The loss function uses a cross-entropy function. Table 2 compares the experimental results. The first three lines are the experimental results of Liu et al. [9] applying rule-based, CRF-based and GRU-based methods to the recognition of the auxiliary word "DE". The fourth line is the result of the recognition of the auxiliary word "DE" based on the Transformer encoder model alone. The last line is the result of the TG network proposed in this paper when the window size is 7. The experimental results. Precision, Recall, and F-Measure were selected as the criteria. Experiments show that the proposed method can effectively improve the effectiveness of usage recognition.

Table 2. Experimental comparison

Model	P%	R%	F%
Based on Rule	70.9	22.7	34.4
Based on CRF Model	76.8	78.3	77.5
Based on GRU Model	81.2	81.5	81.3
Based on transformer model	72.6	72.6	72.6
Based on TG Model (window size is 7)	**82.8**	**82.8**	**82.8**

From the experimental results, it is easy to find that the effect of the latter two models based on deep learning is better than that based on rules and CRF model. Compared with traditional methods, deep learning model can automatically extract features and reduce the workload of manual feature selection. After the sequence is encoded by the Transformer model, through its internal attention mechanism, the sequence automatically acquires the internal features, obtains the internal relations of the sequence itself, and then uses GRU network to extract the semantic information to get more accurate important information in the sequence, so as to get more accurate classification results.

At the same time, experiments show that different window sizes also have different effects on the recognition effect. Window settings have longer sequences than the general assembly. This also means that more information will be obtained when the model extracts features. The experimental results also confirm that longer sequences will have better recognition effect. As shown in Table 3.

Table 3. Experimental comparison of different window sizes based on TG model

Window size	F%
Based on TG model (Window size is 3)	77.8
Based on TG model (Window size is 5)	81.6
Based on TG model (Window size is 7)	**82.8**

Due to the different frequencies of different usages in corpus, in order to further understand the recognition effect of each usage category, when the window is 7, this paper chooses the category whose usage frequency is at least 1 time in the test set and calculates its P, R and F values respectively, as shown in Table 4. The experimental results show that the frequency distribution of the "DE" usage is uneven, and the use of more training sets is easier to obtain better accuracy, which shows that the effect of the model is also affected by the size of the data set.

Table 4. Fine-grained recognition results for the use of the auxiliary "DE" (window size 7)

ID	P%	R%	F%	Frequency of training sets	Frequency of test set occurrence
u_de5_t2_1c	95.0	95.0	95.0	5374	536
u_de5_t2_1a	89.7	92.5	91.1	24171	2781
u_de5_t2_1bc	74.7	66.9	70.6	4630	509
u_de5_t2_1e	81.4	75.9	78.5	2418	272
u_de5_t2_1bd	53.5	53.7	53.6	1761	239
u_de5_t2_1ca	86.8	92.2	89.4	1800	175
u_de5_t2_1cb	79.3	92.0	85.2	735	81
u_de5_t2_1g	74.2	73.5	73.8	6758	720
u_de5_t2_5a	72.1	75.0	73.5	988	134
u_de5_t2_1ba	66.7	69.0	67.8	4052	436
u_de5_t2_1b	81.4	81.1	81.3	2356	237
u_de5_t2_1bb	75.7	54.9	63.6	503	57
u_de5_t2_2d	39.1	51.4	44.4	350	28
u_de5_t2_2cd	30.0	33.3	31.6	119	16
u_de5_t2_2cc	46.3	42.2	44.2	351	40
u_de5_t2_2ca	66.7	41.7	51.3	230	33
u_de5_t2_2b	75.0	57.7	65.2	178	14
u_de5_t2_1be	89.8	91.7	90.7	429	53
u_de5_t2_2a	62.5	50.0	55.6	99	9
u_de5_t2_2ce	75.0	75.0	75.0	44	5
u_de5_t2_1d	100.0	60.0	75.0	114	13
u_de5_t2_2cb	100.0	12.5	22.2	33	6
u_de5_t2_3c	28.6	22.2	25.0	72	5
u_de5_t2_9	0.0	0.0	0.0	4	1
u_de5_t2_2c	0.0	0.0	0.0	173	15
u_de5_t2_3b	0.0	0.0	0.0	14	1
u_de5_t2_2ba	76.5	86.7	81.3	156	16
u_de5_t2_5b	0.0	0.0	0.0	17	4
u_de5_t2_4b	0.0	0.0	0.0	41	3

5 Summary and Outlook

This paper redesigns an end-to-end neural network to recognize the use of the auxiliary word "DE", and achieves better results than previous methods. This provides another way of thinking for function word usage recognition algorithm and also proves that the more complex network can better extract sequence information, to recognize more effectively. It lays a foundation for the application of functional word usage knowledge base.

The next step is:

1. To improve the network design and use better network as feature extractor;
2. To carry out usage recognition experiments for functional words with flexible usage and high frequency;
3. In natural language processing applications, deep learning has its unique advantages, but it lacks explainability. There are shortcomings of generalization ability in the same task. Introducing external knowledge has become one of the important ways to solve this problem. How to better integrate functional word usage knowledge into natural language processing tasks will also be one of the following key tasks. It can be predicted that the integration of functional word usage knowledge can better improve the understanding and reasoning ability of natural language, and complement existing deep learning algorithms.

Acknowledgments. We thank the anonymous reviewers for their constructive comments, and gratefully acknowledge the support of the National Key Basic Research and Development Program under Grant No. 2014CB340504; The National Social Science Fund of China under Grant No. 18ZDA315; the Key Scientific Research Program of Higher Education of Henan under Grant No. 20A520038; the science and technology project of Science and Technology Department of Henan Province under Grant No. 192102210260 and the international cooperation project of Science and Technology Department of Henan Province No. 172102410065.

References

1. Xu, Y.: Functional Word "DE" and Related Issues. China Social Science Press, Beijing (2006)
2. Lu, S.: 800 Words in Modern Chinese. Commercial Press, Beijing (1980)
3. Yu, S., Zhu, X., Liu, Y.: Construction of the knowledge base of generalized functional words in modern Chinese. J. Chin. Lang. Comput. 13(1), 89–98 (2003)
4. Ying, X., Zhu, X.: Research on chinese functional words for natural language processing and construction of generalized functional words knowledge base. Contemp. Linguis. 11(2), 124–135 (2009)
5. Ying, X., Zhang, K., Chai, Y., et al.: Research on knowledge base of functional words in modern Chinese. J. Chin. Inf. Sci. 21(5), 107–111 (2007)
6. Zan, H., Zhang, K., Zhu, X., et al.: Research on the Chinese function word usage knowledge base. Int. J. Asian Lang. Process. 21(4), 185–198 (2011)
7. Zhang, K., Ying, X., Yumei, C., et al.: A summary of the construction of knowledge base on the usage of functional words in modern Chinese. J. Chin. Inf. Sci. 29(3), 1–8 (2015)

8. Han, Y., Ying, X., Zhang, K., et al.: Rule-based automatic identification of common auxiliary usage in modern Chinese. Comput. Appl. **31**(12), 3271–3274 (2011)

9. Liu, Q.: Research on automatic recognition of the usage of auxiliary words "DE". J. Peking Univ. (Nat. Sci. Edn.) **54**(3) (2018)

10. Chang, D., Jurafsky, P.-C., Dan, M., Disambiguating, C.D.: "DE" for Chinese-English machine translation. In: The Workshop on Statistical Machine Translation, pp. 215–223. Association for Computational Linguistics (2009)

11. Vaswani, A., et al.: Attention is all you need. CoRR, abs/1706.03762 (2017a)

12. Zhang, K., Xu, H., Xiong, D., Liu, Q., Zan, H.: Improving Chinese-English neural machine translation with detected usages of function words. In: Huang, X., Jiang, J., Zhao, D., Feng, Y., Hong, Yu. (eds.) NLPCC 2017. LNCS (LNAI), vol. 10619, pp. 741–749. Springer, Cham (2018). https://doi.org/10.1007/978-3-319-73618-1_64

13. Lu, S.: 800 Words in Modern Chinese. Commercial Press, Beijing (2006)

14. Lu, S.: Modern Chinese Dictionary. Commercial Press, Beijing (2007)

15. Zhang, B.: Dictionary of Functional Words in Modern Chinese. Commercial Press, Beijing (2006)

16. Kim, Y.: Convolutional neural networks for sentence classification. arXiv preprint arXiv: 1408.5882 (2014)

17. Srivastava, N., Hinton, G., Krizhevsky, A., et al.: Dropout: a simple way to prevent neural networks from overfitting. J. Mach. Learn. Res. **15**(1), 1929–1958 (2014)

18. Hochreiter, S., Schmidhuber, J.: Long short-term memory. Neural Comput. **9**(8), 1735–1780 (1997)

19. Chung, J., Gulcehre, C., Cho, K.H., et al.: Empirical evaluation of gated recurrent neural networks on sequence modeling. arXiv preprint arXiv:1412.3555 26 (2014)

20. He, K., Zhang, X., Ren, S., Sun, J.: Deep residual learning for image recognition. In: Proceedings of the IEEE Conference on Computer Vision and Pattern Recognition, pp. 770–778 (2016)

21. Ba, J.L., Kiros, J.R., Hinton, G.E.: Layer normalization. arXiv preprint arXiv:1607.06450 (2016)

22. Bahdanau, D., Cho, K., Bengio, Y.: Neural machine translation by jointly learning to align and translate. CoRR, abs/1409.0473 (2014)

A Comparative Study of the Collocations in Legislative Chinese and General Chinese

Shan Wang[✉] and Jiuhan Yin

Department of Chinese Language and Literature,
University of Macau, Taipa, Macau
shanwang@um.edu.mo

Abstract. Remarkable achievements have been made in the study of lexis in general Chinese, such as lexical meanings, word structures, lexical systems, and semantic evolution. However, these studies can hardly reflect the unique characteristics of Chinese for special purposes, such as legal Chinese, travel Chinese, and business Chinese. Taking the commonly used word 管理 *guǎnlǐ* 'manage; management' as an example, this paper explores the characteristics of legislative Chinese in terms of semantic categories and saliency of collocations by comparing them in a legislative Chinese corpus and the BCC corpus. This study finds that 管理 *guǎnlǐ* 'manage; management' is mainly used as a modifier and a modified term in legislative Chinese. The collocated words cover less semantic categories compared to general Chinese. Most of the collocated words are nouns, whose semantic categories mainly come from the political, social and economic fields. The study of the usage of commonly used words in legislative Chinese can not only help to explore the characteristics of legislative Chinese itself and its differences compared with general Chinese, but also provide references for law revision, legal lexicography, and legal Chinese teaching.

Keywords: Collocations · Semantic categories · Saliency · Legislative Chinese · General Chinese

1 Introduction

Legislative language is a unique language style formed in the process of formulating and implementing laws and regulations. It is considered as a technical language different from natural languages [1]. It has a strict requirement in the choice of vocabulary concerning word meanings, parts of speech, emotional coloring, and the applicable scope. It sometimes even sacrifices certain readability for accuracy, but as law adjusts the relationship and interests of all members of society, it should be understood as much as possible for the majority of members of society [2, 3].

The characteristics of collocations in legislative Chinese are closely related to the contents, linguistic features and the style of legislative Chinese texts. Understanding their characteristics is of great significance for further study of legislative Chinese. This paper takes 管理 *guǎnlǐ* 'manage; management' as a target word, extracts the corpus data from legislative Chinese and general Chinese, and explores its characteristics by comparing its collocated words' semantic categories and saliency of common

© Springer Nature Switzerland AG 2020
J.-F. Hong et al. (Eds.): CLSW 2019, LNAI 11831, pp. 710–724, 2020.
https://doi.org/10.1007/978-3-030-38189-9_72

collocations. This study can not only deepen the understanding of the semantic tendency of legislative Chinese, but also contribute to exploring the differences of vocabulary between legislative Chinese and general Chinese.

2 Related Research

Significant achievements have been made in the study of lexis in general Chinese, such as lexical meanings, word structures, lexical systems, and semantic evolution. However, the research on the lexis of legal Chinese is limited. The related research is reviewed below.

2.1 Legal Chinese in the Study of Chinese for Specific Purposes

There are only a few studies on legal Chinese for specific purposes and they are focused on the study of teaching materials. Two legal Chinese textbooks that have been published are *Specialized Chinese Course of Chinese Law* [4] and *Legal Chinese-Commercial Chapter* [5]. Some scholars have conducted research on two published legal Chinese textbooks [6, 7]. Liu [6] compared the guiding ideology, targeted users, material sources, the structure, and components of the textbooks. Wang [7] pointed out that the former is a transitional advanced Chinese textbook and does not involve systematic legal knowledge, while the latter is concentrated on the civil and commercial laws.

Other textbooks containing legal information are only a few business Chinese textbooks with some commercial law contents [8]. For example, *Intermediate Business Chinese Practical Conversation* [9] involves legal words like 索赔 *suǒpéi* 'claims' and 仲裁 *zhòngcái* 'arbitration'; *International Business Chinese Course* [10] involves legal language points related to international trade activities, such as 资信调查 *zīxìn diàochá* 'credit investigation' and 报关与清关 *bàoguān yǔ qīngguān* 'customs declaration and customs clearance'.

In addition, Yang [11] summarized the legal vocabulary in *The Graded Chinese Syllables, Characters and Words for Teaching Chinese as a Foreign Language* [12] and analyzed their meanings and example sentences. However, this syllabus is not designed for legal Chinese. Therefore, it is difficult to reflect the characteristics of legal Chinese.

2.2 Research on Legal Chinese Vocabulary

The research on legal Chinese vocabulary mainly focuses on the classification of legal vocabulary, legal vocabulary in ancient classics, and the semantic evolution of legal vocabulary.

There are different opinions on the classification of legal vocabulary. Pan [13] divided the vocabulary in the legal domain into legal vocabulary and general vocabulary. He also divided the vocabulary in legal documents into legal terms, judicial idioms, classical Chinese words, and general words. Liu [14] divided the legal vocabulary into legal terms and the basic terms of laws according to whether it has legal meaning. Wang and Wang [15] further classified legal terms, dividing them into specific lexical items and expressive function words. The classification of legal terms provides a clear research direction for further study of legal vocabulary.

The study of legal vocabulary in ancient classics includes *Bamboo Strips from Tombs of Qin Dynasty at Shuihudi* [16], *San Yan Er Pai* [17], *Comments on Law of Tang Dynasty* [18], *Records of the Grand Historians* [19], *Collected Cases of the Judgement by Famous Good Officials* [20], etc. In addition to analyzing word meanings from the literature itself, these studies paid attention to legal culture changes reflected in the literature; some studies included the comparison with modern Chinese vocabulary. They attempted to reveal the causes and characteristics of the semantic evolution of legal Chinese through synchronic and diachronic comparison.

Some scholars also analyze the causes and types of semantic evolution of legal vocabulary. For example, Wu [21] suggested that historical factors and foreign language influences are the main factors that cause the semantic evolution of legal vocabulary. The main way of semantic evolution of legal vocabulary includes semantic expansion, semantic reduction, semantic transfer, metaphorical evolution, and referential evolution. Liu [22] supplemented the factors of legal vocabulary semantic evolution with language factors, psychological factors, and the emerging needs of new concepts or things. On the basis of this study, Zhang and Zuo [23] investigated the cognitive mechanism of semantic evolution of legal vocabulary, pointing out that metaphor and metonymy play an important role in semantic evolution.

In general, the studies of legal Chinese vocabulary focuses on the field of forensic linguistics: they classified legal Chinese vocabulary from a synchronic perspective, analyzed the legal vocabulary in ancient classics, and explored the semantic evolution of legal Chinese in a diachronic way. However, the lack of comparative studies between modern legal Chinese and general Chinese make them have limited reference value for the analysis on the characteristics of modern legal Chinese vocabulary.

2.3   Comparative Studies of Modern Legal Vocabulary

In modern Chinese, there are only a few comparative studies on vocabulary between legal Chinese and general Chinese. Tang [24] explored the differences of the word 适用 *shìyòng* 'applicable' between general language and legal language. He indicated that in general language, 适用 *shìyòng* 'applicable' is used as 'suitable for use' and can be an adjective or a verb. It can be a predicate, object or attributive in sentences. In legal language, 适用 *shìyòng* 'applicable' is used as 'apply'. It can be a verb and act as a predicate in sentences. Yang [25] investigated the differences of 可以 *kěyǐ* 'can' in legal language and daily language. She pointed out that in legal documents, the meaning of 可以 *kěyǐ* 'can' should be clear. But legislators are often consciously or unconsciously influenced by daily language style and interprets it as having discretionary power, which causes contradictions with the accuracy of legal language.

Legislative Chinese is a kind of legal Chinese. That is, the language used in making laws. Recently, Luo and Wang [26, 27] made a preliminary exploration of the differences in the use of legislative Chinese vocabulary and general Chinese vocabulary using a self-built legislative Chinese corpus and a general Chinese corpus. They further collected all 但 *dàn* and 但是 *dànshì* sentences in the self-built legislative Chinese corpus and classified them into two types as proviso and non-proviso clauses to investigate how to use 但 *dàn* and 但是 *dànshì* sentences appropriately in legislative Chinese [28]. Based on this work, the current study conducts a further analysis through looking at the collocations of a commonly used word 管理 *guǎnlǐ* 'manage; management'.

3   Research Methods

3.1   Corpus Selection

The legal Chinese corpus used in this paper is the legislative corpus of mainland China [27]. The corpus selected 35 legislative laws covering the seven legal categories of the Sub-Committee of Legislative Affairs of the Standing Committee of the National People's Congress, including the constitution, the administrative laws, the criminal laws, the civil and commercial laws, the economic laws, the social laws, and the procedural laws. In this paper, we select the commonly used word 管理 *guǎnlǐ* 'manage; management' from this corpus. We also extract all the sentences with it from this corpus.

The general Chinese corpus used in this paper is the BCC multi-domain corpus of Beijing Language and Culture University [29]. The contents of the corpus include newspapers, literature, microblogs, science and technology, ancient Chinese, etc., and the proportion of each source is relatively balanced. In this paper, 10000 sentences with the target word 管理 *guǎnlǐ* 'manage; management' are randomly downloaded to form a general Chinese corpus.

3.2   Sketch Engine

Sketch Engine [30] was originally used in the production of the Macmillan English Dictionary. It is a corpus analysis tool that can automatically analyze lexical collocation characteristics in the corpus. Originally it can only be used in English, but after the engine was developed in 2002, it is gradually applied to most language systems, such as Chinese [31]. It can help us to get the grammatical relations, the collocational words, and the saliency score of each collocation. This paper uses it to get such information of 管理 *guǎnlǐ* 'manage; management' from the data of both legislative Chinese and general Chinese.

3.3   A Thesaurus of Modern Chinese

A Thesaurus of Modern Chinese [32] is a dictionary that is arranged according to the meaning of words. The materials used to acquire the vocabulary is new and complete, which makes its vocabulary coverage wide. The classification of semantic categories is hierarchical and logical; the arrangement of semantic categories is orderly and cultural. This dictionary plays an important guiding role in the research of lexical system, core words and comparative study of vocabulary [33].

This paper uses the nine first-level semantic categories of *A Thesaurus of Modern Chinese* to classify the semantic categories of the collocated words of 管理 *guǎnlǐ* 'manage; management'. They are: Living Things, Concrete Objects, Abstract Things,

Time and Space, Biological Activities, Social Activities, Movement and Change, Nature and State, and Auxiliary Words. In addition, this paper has also analyzed the second-level semantic categories of Abstract things, which take up a large proportion out of all collocated words. There are 10 second-level semantic categories of Abstract Things: Things, Attributes, Consciousness, Society, Politics, Military, Economy, Science and Education, Culture, Sports and Health, and Quantity Units.

4 A Comparison of the Collocations of Common Grammatical Relations

In the legislative Chinese corpus, there are 78 high-frequency words with a frequency of more than 500. 管理 guǎnlǐ 'manage; management' has a frequency of 880, which is a commonly used word in legislative Chinese. Therefore, this paper selects it to do a case study to show the comparison between legislative Chinese and general Chinese.

This paper first uses Sketch Engine to get the grammatical relations of 管理 guǎnlǐ 'manage; management' in legislative Chinese and general Chinese. It then manually annotates the semantic categories of its collocated words to compare their semantic preference. In addition, their common collocated words of the common grammatical relations in legislative Chinese and general Chinese are compared according to the saliency score [30].

4.1 A Comparison of Common Grammatical Relations Between General Chinese and Legislative Chinese

Sketch Engine is used to get the grammatical relations of 管理 guǎnlǐ 'manage; management'. The comparison results show that the common grammatical relations in general Chinese and legislative Chinese include *Modifies*, *N_Modifier*, *Object_of*, and *and/or*, which respectively mean "*guǎnlǐ* XX" (*guǎnlǐ* as a modifier), "XX *guǎnlǐ*" (*guǎnlǐ* as a modified term), "XX *guǎnlǐ*" (*guǎnlǐ* as an object), and "XX *hn/huò guu*" "X" (a coordinate relation). Because there are only a few examples of the coordinate relation in legislative Chinese, this paper only discusses the first three grammar relationships. The quantity of collocated words for the three grammatical relations is shown in Table 1.

Some of the automatically extracted collocated words are wrong; they are screened out only because of the high co-occurrence frequency with 管理 guǎnlǐ 'manage; management'. For example, 来源 láiyuán 'sources' in *Modifies* is not a collocated word of 管理来源 guǎnlǐ láiyuán 'manage sources', but appears at the end of each sentence as "source of the data". It is extracted as a collocated word because of its high frequency. After manually deleting such words, the number of collocated words is shown in Table 1.

By comparison, out of the three common grammatical relations, the most common one in general Chinese is *N_Modifier*; that is, 管理 guǎnlǐ 'manage; management' is used as a modified term. The number of collocated words is 231, which is almost five times that of *Object_of*. The number of *Modifies* is the least. In legislative Chinese, the most common grammatical relation is *Modifies*; that is, 管理 guǎnlǐ 'manage; management' is used as a modifier.

Table 1. Number of collocated words of different grammar relations in legislative Chinese and general Chinese

Grammatical relations	Modifies		N_Modifier		Object_of	
	Original data	Selected data	Original data	Selected data	Original data	Selected data
Number of collocated words in general Chinese	45	38	237	231	81	49
Number of collocated words in legislative Chinese	23	23	18	18	5	5

4.1.1 Semantic Analysis of the Collocated Words in the *Modifies* Relation

The grammatical relations of *Modifies* refers to "*guǎnlǐ* XX" (*guǎnlǐ* as a modifier). The collocated words of 管理 *guǎnlǐ* are mostly nouns, forming a modifier-head relation, such as 管理人员 *guǎnlǐ rényuán* 'management personnel'. There is one verb phrase, which is 管理网络 *guǎnlǐ wǎngluò* 'manage network'. General Chinese is also dominated by modifier-noun phrases, such as 管理人 *guǎnlǐ rén* 'manager'. There are only four verb phrases, such as 管理财产 *guǎnlǐ cáichǎn* 'manage property'.

We manually annotated the semantic categories of each collocated words based on *A Thesaurus of Modern Chinese*. The distribution of the first-level semantic categories of the collocated words in general Chinese and legislative Chinese is shown in Table 2.

The semantic categories of the collocated words in the *Modifies* relation of general Chinese are mainly Abstract Things, accounting for 78.95%, followed by Living Things, accounting for 7.89%. There is only one word 规定 *guīdìng* 'regulations' that belongs to Social Activities, accounting for 2.63%. Out of the 23 collocated words in legislative Chinese, 21 refer to Abstract Things, accounting for 91.3%; the remaining two words referring to Living Things are 人 *rén* 'people' and 人员 *rényuán* 'personnel'. Compared with legislative Chinese, general Chinese has more semantic categories, in addition to Abstract Things, there are Concrete Objects, Movement and Change, and Social Activities.

Since the main semantic category is Abstract Things, the second-level semantic categories under Abstract Things are further analyzed. The statistical results are shown in Table 2. Overall, the collocated words of general Chinese have more second-level semantic categories than legislative Chinese. In general Chinese, they are mainly focused on the three semantic categories of Attributes, Things, and Politics, taking up 30%, 20%, and 20% respectively; there are fewer collocated words in Science and Education, Consciousness, and Society, accounting for 13.33%, 10%, and 6.67% respectively. In comparison, the proportion of Politics in legislative Chinese is relatively large, accounting for 52.38%; there are fewer collocated words in Things and Society, both accounting for 19.05% respectively. In addition, legislative Chinese also has collocated words in Economy, accounting for 9.52%, which is unique compared with general Chinese.

In the *Modifies* relation, there are nine common collocated words between legislative Chinese and general Chinese, which are 机构 *jīgòu* 'institution', 部门 *bùmén* 'department', 人员 *rényuán* 'personnel', 职责 *zhízé* 'responsibility', 公司 *gōngsī* 'company', 工作 *gōngzuò* 'job', 制度 *zhìdù* 'system', 办法 *bànfǎ* 'method', and 信息 *xìnxī* 'information'. Eight of them belong to Abstract Things and the remaining one (人员 *rényuán* 'personnel') belongs to Living Things. Among the eight words belonging to Abstract Things, their second-level semantic categories are: four represent Politics, accounting for 50%; two represent Things and another two represent Society.

In summary, in the *Modifies* relation of the word 管理 *guǎnlǐ* 'manage; management', Abstract Things are the main semantic categories of its collocated words. The semantic categories of general Chinese are more widely distributed than legislative Chinese.

Table 2. Semantic category distribution of the collocated words in the *Modifies* relation in general Chinese and legislative Chinese

Semantic level	Semantic category	General Chinese		Legislative Chinese	
		Quantity	Percent	Quantity	Percent
First-level semantic category	Abstract Things	30	78.95%	21	91.3%
	Living Things	3	7.89%	2	8.7%
	Movement and Change	2	5.26%	0	0
	Concrete Objects	2	5.26%	0	0
	Social Activities	1	2.63%	0	0
	Biological Activities	0	0	0	0
	Nature and State	0	0	0	0
	Total	38	100%	23	100%
Second-level semantic category of Abstract Things	Attributes	9	30%	0	0
	Things	6	20%	4	19.05%
	Politics	6	20%	11	52.38%
	Science and Education	4	13.33%	0	0
	Consciousness	3	10%	0	0
	Society	2	6.67%	4	19.05%
	Economy	0	0	2	9.52%
	Total	30	100%	21	100%

4.1.2   Semantic Analysis of the Collocated Words of in the *N_Modifier* Relation

The *N_Modifier* relation refers to cases like "XX *guǎnlǐ*" (*guǎnlǐ* is a modified word). The collocated words XX are mostly nouns.

Among the collocated words of general Chinese, seven of them are not in *A Thesaurus of Modern Chinese*, namely, 出入境 *chūrùjìng* 'entry and exit', 供应链 *gōngyìng liàn* 'supply chain', 高效能 *gāoxiàonéng* 'high efficacy', 专户 *zhuānhù* 'special account', 药事 *yàoshì* 'pharmacy', 专账 *zhuānzhuàng* 'separate account', and 目标化 *mùbiāohuà* 'target quest'. These words are classified by referring to their synonyms. The comparison of the first-level semantic categories of the collocated words in general Chinese and legislative Chinese in *N_Modifier* is shown in Table 3.

Table 3. Semantic category distribution of the collocated words in the *N_Modifier* relation in general Chinese and legislative Chinese

Semantic level	Semantic category	General Chinese		Legislative Chinese	
		Quantity	Percent	Quantity	Percent
First-level semantic category	Abstract Things	135	58.44%	12	70.59%
	Social Activities	25	10.82%	1	5.88%
	Concrete Objects	19	8.23%	1	5.88%
	Movement and Change	15	6.49%	0	0
	Time and Space	10	4.33%	2	11.76%
	Living Things	10	4.33%	0	0
	Biological Activities	9	3.90%	0	0
	Nature and State	8	3.46%	1	5.88%
	Total	231	100%	17	100%
Second-level semantic category of abstract things	Attributes	26	19.26%	2	16.67%
	Things	26	19.26%	3	25%
	Society	25	18.52%	3	25%
	Economy	17	12.59%	4	33.33%
	Science and Education	15	11.11%	0	0
	Politics	11	8.15%	0	0
	Consciousness	6	4.44%	0	0
	Quantity Units	5	3.71%	0	0
	Culture, Sports and Health	2	1.48%	0	0
	Military	2	1.48%	0	0
	Total	135	100%	12	100%

Abstract Things account for the largest proportion in both general Chinese and legislative Chinese, taking up 58.44% and 70.59% respectively, which are lower than the *Modifies* relation. The collocated words in general Chinese has the widest semantic categories in this type of grammatical relation. Eight out of the nine first-level semantic categories are involved. The semantic category distribution of legislative Chinese is less than that of general Chinese. It only covers four of them, but it involves more semantic categories than the *Modifies* relation.

Since Abstract Things is the largest category, we further examine its second-level semantic categories. The results are also shown in Table 3. In the second-level semantic categories of Abstract Things, general Chinese covers all the secondary categories. Attributes and Things are most common, both accounting for 19.26% respectively; *Military* and *Culture, Sports and Health* are least common, both only accounting for 1.48% respectively. In comparison, legislative Chinese involves less semantic categories than general Chinese and the largest category is Economy, followed by Society and Things.

There are 12 common words in the collocated words between legislative Chinese and general Chinese, namely 土地 *tǔdì* 'land', 行政 *xíngzhèng* 'administration', 治安 *zhì'ān* 'public security', 资产 *zīchǎn* 'assets', 风险 *fēngxiǎn* 'risk', 质量 *zhìliàng* 'quality', 内部 *nèibù* 'internal', 基金 *jījīn* 'fund', 集体 *jítǐ* 'collective', 登记 *dēngjì* 'registration', 金融 *jīnróng* 'finance', and 社会 *shèhuì* 'society'. Among them, 9 belong to Abstract things, accounting for 75%; one Space and Time (内部 *nèibù* 'internal'), one Concrete Object (土地 *tǔdì* 'land'), and one Social Activity (登记 *dēngjì* 'registration'). Among the 9 belonging to Abstract Things, their second-level semantic categories are: three belong to Society (社会 *shèhuì* 'social', 金融 *jīnróng* 'financial', 集体 *jítǐ* 'collective'), three Things (行政 *xíngzhèng* 'administration', 治安 *zhì'ān* 'public security', 风险 *fēngxiǎn* 'risk'), two Economy (资产 *zīchǎn* 'assets', 基金 *jījīn* 'funds') and one Attributes (质量 *zhìliàng* 'quality').

In summary, in the *N_Modifier* relation, general Chinese has wider semantic categories than legislative Chinese. The category of Abstract Things is also the largest semantic category in both general Chinese and legislative Chinese. Regarding the second-level category of Abstract Things, General Chinese takes up a larger proportion in the categories of Attributes and Things, while legislative Chinese takes up a larger proportion in the category of Economy.

4.1.3 Semantic Analysis of the Collocated Words of in the *Object_of* Relation

The *Object_of* relation refers to "XX *guǎnlǐ*" (*guǎnlǐ* as an object) and XX is verb. These collocated words are also classified into different semantic categories according to *A Thesaurus of Modern Chinese*. 重在 *zhòngzài* 'to emphasize' does not exist in this dictionary. It is classified into the category Movement and Change according to the classification of 在 *zài* 'in, at, on' and 在于 *zàiyú* 'lie in' in this dictionary.

The comparison of the number of semantic categories of the collocated words in general Chinese and legislative Chinese is shown in Table 4. In the grammatical relation of *Object_of*, the main semantic category of 管理 *guǎnlǐ* 'manage; management' in general Chinese is Social Activities, accounting for 36.74%. Movement and Change, Biological Activities also take up a large proportion, accounting for 28.57% and 24.49% respectively. In legislative Chinese, the dominant semantic category is also Social Activities, accounting for 57.14%, which is more prominent than in general Chinese; the collocated words also appear in the categories of *Movement and Change* and *Biological Activities*.

Table 4. First-level semantic category distribution of the collocated words in the *Object_of* relation in general Chinese and legislative Chinese

First level semantic category	General Chinese		Legislative Chinese	
	Number	Percent	Number	Percent
Social Activities	18	36.74%	2	40%
Movement and Change	14	28.57%	2	40%
Biological Activities	12	24.49%	1	20%
Auxiliary Words	3	6.12%	0	0
Nature and State	2	4.08%	0	0
Total	49	100%	5	100%

In the *Object_of* relation, there are two common collocated words in both general Chinese and legislative Chinese, which are 加强 *jiāqiáng* 'strengthen, enhance' and 实施 *shíshī* 'implement'. Among them, 加强 *jiāqiáng* 'strengthen, enhance' belongs to Movement and Changes, 实施 *shíshī* 'implement' belongs to Biological Activities.

Overall, in the *Object_of* relation, the main semantic category of general Chinse is Social Activities; the main semantic categories of legislative Chinese are Social Activities and Movement and Changes. In addition, the semantic categories in general Chinese is more widely distributed than in legislative Chinese.

Table 5 summaries the semantic categories of the collocated words of 管理 *guǎnlǐ* 'manage; management' in general Chinese and legislative Chinese under three grammatical relations *Modifies*, *N_Modifier*, and *Object_of*. The first-level semantic categories in *A Thesaurus of Modern Chinese* is labeled in order as follows: ① Living Things, ② Concrete Objects, ③ Abstract Things, ④ Time and Space, ⑤ Living Activities, ⑥ Social Activities, ⑦ Movement and Change, ⑧ Nature and State, ⑨ Auxiliary Words. The second-level semantic categories of Abstract Things are numbered in sequence as follows: (1) Things, (2) Attributes, (3) Consciousness, (4) Society, (5) Politics, (6) Military, (7) Economy, (8) Science and Education, (9) Culture, Sports and Health, and (10) Quantity Units. The underlined numbers in Table 5 show the shared semantic categories of general Chinese and legislative Chinese. The italic (7) *Economy* is the unique semantic category of legislative Chinese. Overall, the semantic categories of general Chinese are wider than legislative Chinese.

Table 5. The distribution of semantic categories of the collocated words of 管理 *guǎnlǐ* 'manage; management'

	Modifies		N_Modifier		Object_of	
	General Chinese	Legislative Chinese	General Chinese	Legislative Chinese	General Chinese	Legislative Chinese
First-level semantic categories	③, ⑤, ②, ⑦, ⑥	③, ⑤	③, ④, ②, ⑥, ①, ⑤, ⑦, ⑧	③, ④, ②, ⑥	⑤, ⑥, ⑦, ⑧, ⑨	⑤, ⑥, ⑦
Second-level semantic categories under Abstract Things	(1), (4), (5), (2), (3), (8)	(1), (4), (5), (7)	(1), (2), (4), (7), (10), (3), (5), (6), (8), (9),	(1), (2), (4), (7), (10)	/	/

4.2   A Comparison of the Saliency of the Common Collocations

Among the three grammatical relations of *Modifies*, *N_Modifier* and *Object_of*, the number of common collocations in legislative Chinese and general Chinese is 9, 12 and 2, respectively. In the following, by using the saliency [30] of legislative Chinese as the ranking basis, these common collocations of legislative Chinese and general Chinese are compared.

The saliency comparison of the common collocations in the three grammatical relations are shown in Table 6. The absolute value of saliency difference shows the difference between saliency in general Chinese and legislative Chinese. The larger the value, the greater the difference is, which is an indicator of the collocation preference of general Chinese and legislative Chinese.

In the *Modifies* relation, the collocations with an obvious saliency difference between legislative Chinese and general Chinese (saliency difference > 2) are 管理机构 *guǎnlǐ jīgòu* 'management organization', 管理部门 *guǎnlǐ bùmén* 'management department', and 管理职责 *guǎnlǐ zhízé* 'management responsibility'. In these collocations, 管理 *guǎnlǐ* 'management' collocates with nouns to form modifier-noun phrases. The collocated words 机构 *jīgòu* 'organization', 部门 *bùmén* 'department', and 职责 *zhízé* 'responsibility' all belong to the second-level semantic category Politics under the first-level semantic category Abstract Things.

In the *N_Modifier* relation, the collocations with significant saliency difference (saliency difference > 3) are 土地管理 *tǔdì guǎnlǐ* 'land manage', 集体管理 *jítǐ guǎnlǐ* 'collective manage', and 治安管理 *zhì'ān guǎnlǐ* 'public security manage', which are modifier-head phrases. 土地 *tǔdì* 'land' belongs to the second-level semantic category Natural Objects under the first-level semantic category Concrete objects. 集体 *jítǐ* 'collective' and 治安 *zhì'ān* 'public security' belong to the second-level semantic category Society and Things under the first-level semantic category Abstract Things.

In the *Object_of* relation, there are two common collocations, which are 加强管理 *jiāqiáng guǎnlǐ* 'strengthen management' and 实施管理 *shíshī guǎnlǐ* 'implement manage'. They are verb-object structures, the semantic category of 加强 *jiāqiáng* 'strengthen' is Movement and Changes and 实施 *shíshī* 'implement' is Biological Activities.

In sum, among the common collocations, in the *Modifies* relation, 管理 *guǎnlǐ* 'management' as a modifier tends to collocate with words of Politics; in the *N_Modifier* relation, 管理 *guǎnlǐ* 'manage' as a modified word often collocates with words of Society, Things, and Natural Objects. In the *Object_of* relation, 管理 *guǎnlǐ* 'management', as the object of a verb, often collocates with words of *Biological Activities* and *Movement and Changes*.

Table 6. The saliency comparison of the common collocations between general and legislative Chinese

Grammar relation	Common collocated words	Legislative Chinese		General Chinese		Saliency difference	
		Frequency	Saliency	Frequency	Saliency	Saliency of legislative Chinese minus general Chinese	Absolute value of saliency difference
Modifies (guǎnlǐ XX)	机构 jīgòu 'institution'	278	12.7	18	8.49	4.21	4.21
	部门 bùmén 'department'	98	11.56	25	8.91	2.65	2.65
	人员 rényuán 'personnel'	86	10.71	23	8.82	1.89	1.89
	职责 zhízé 'responsibility'	18	9.63	6	7.03	2.6	2.6
	公司 gōngsī 'company'	14	8.73	14	8.07	0.66	0.66
	工作 gōngzuò 'job'	12	8.66	42	9.43	-0.77	0.77
	制度 zhìdù 'system'	8	8.37	31	9.17	-0.8	0.8
	办法 bànfǎ 'method'	7	8.32	19	8.65	-0.33	0.33
	信息 xìnxī 'information'	5	7.72	8	7.34	0.38	0.38
N_Modifier (XX guǎnlǐ)	土地 tǔdì 'land'	32	10.82	28	6.79	4.03	4.03
	行政 xíngzhèng 'administration'	87	10.72	121	8.87	1.85	1.85
	治安 zhì'ān 'public security'	12	10.54	39	7.27	3.27	3.27
	资产 zīchǎn 'asset'	11	10.21	60	7.93	2.28	2.28
	风险 fēngxiǎn 'risk'	6	9.49	54	7.73	1.76	1.76
	质量 zhìliàng 'quality'	5	9.06	53	7.69	1.37	1.37
	内部 nèibù 'internal'	3	8.57	67	8.04	0.53	0.53
	基金 jījīn 'fund'	3	8.5	13	5.69	2.81	2.81
	集体 jítǐ 'collective'	3	8.16	6	4.58	3.58	3.58
	登记 dēngjì 'registration'	3	8.08	13	5.69	2.39	2.39
	金融 jīnróng 'finance'	3	7.97	12	5.56	2.41	2.41
	社会 shèhuì 'society'	3	7.86	31	6.91	0.95	0.95
Object_of (XX guǎnlǐ)	加强 jiāqiáng 'strengthen, enhance'	6	10.66	473	12.08	-1.42	1.42
	实施 shíshī 'implement'	4	8.75	41	8.87	-0.12	0.12

5 Conclusion

The research of lexis in general Chinese is fruitful. However, it cannot reflect the characteristics of legislative Chinese. This paper selects a commonly used word 管理 *guǎnlǐ* 'manage; management' and extracts the grammatical relations and collocations using Sketch Engine based on the data of the legislative Chinese corpus and the multi-domain BCC corpus. It then compares their grammatical relations in legislative Chinese and general Chinese and explores three of their common grammatical relations *Modifies*, *N_Modifier*, and *Object_of*. This study further compares the collocated words from the perspectives of their semantic categories and saliency, which reveal the collocational preference.

The analysis of the use of 管理 *guǎnlǐ* 'manage; management' in legislative Chinese and general Chinese provides a case study for further exploration of vocabulary differences. Meanwhile, the in-depth analysis of the grammatical relations, semantic categories of collocated words, and saliency differences provides references for textbook compilation, dictionary compilation, and language teaching in legislative Chinese. In the future, more commonly used words will be examined to deepen the comparative study of legislative Chinese and general Chinese.

Acknowledgement. This research is supported by MYRG2019-00013-FAH, University of Macau.

References

1. Pan, Q.: Law language is a technical one different from national language (Fǎlǜ yǔyán shì yìzhǒng yǒubiéyú zìrán yǔyán de jìshù yǔyán). J. Jianghan Univ. (Humanit. Sci.) **2**, 23 (2004)
2. Hu, F.: The logical starting point of the study of humorous language, lies, legal languange, rhetoric of the images of organizations, empirical rhetoric, etc.: thinking through "new speech act analysis" (Yōumò yǔyán, huǎngyán, fǎlǜ yǔyán, jīgòu xíngxiàng xiūcí, shíyàn xiūcíxué yánjiū de luójí qǐdiǎn—jīyú "xīnyányǔ xíngwéi fēnxī" de sīkǎo). Humanit. Soc. Sci. **47**, 1–9+164 (2015)
3. Hu, F.: A study of legal language based on the theory of "speech act" (Jīyú "yányǔ xíngwéi fēnxī" de fǎlǜ yǔyán yánjiū). J. East Chin. Norm. Univ. (Philos. Soc. Sci.) **37**, 87–93+125 (2005)
4. Wang, R.: Specialized Chinese Course of Chinese Law (Zhōngguó fǎlǜ zhuānyè hànyǔ jiàochéng). Peking University Press, Beijing (2007)
5. Zhang, T.: Legal Chinese—Commercial Chapter (Fǎlǜ hànyǔ—shāngshì piān). Peking University Press, Beijing (2008)
6. Liu, Z.: Comparative Case of Legal Chinese Textbooks—Take Specialized Chinese Course of Chinese Law and Legal Chinese-Commercial Chapter as the Case. School of Chinese as a second language (Fǎlǜ hànyǔ jiàocái gè'àn bǐjiào—yǐ 'Zhōngguó fǎlǜ zhuānyè hànyǔ jiàochéng' hé 'fǎlǜ hànyǔ—shāngshì piān' wéilì). Beijing Postgraduate Forum on Teaching Chinese as a Foreign Language, vol. 10 (2013)
7. Wang, H.: Research on Legal Chinese Textbooks Based on the ESP Theory (Jīyú ESP lǐlùn de fǎlǜ hànyǔ jiàocái yánjiū). Liaoning Normal University (2013)

8. Song, C.: A study on the training of legal talents and the development of legal Chinese textbooks ("Yīdài yīlù" fǎlù réncái péiyǎng yǔ fǎlù hànyǔ jiàocái yánfā de ruògān wèntí). Chin. Legal Educ. Res. **2**, 16–31 (2018)
9. Huang, W.: Practical Conversation of Intermediate Business Chinese (Zhōngjí shāngwù hànyǔ shíyòng huìhuà). Beijing Language and Culture University Press, Beijing (2008)
10. Zhang, T.: International Business Chinese Course (Guójì shāngwù hànyǔ jiàochéng). Peking University Press, Beijing (2008)
11. Yang, D.: Research on Legal Vocabulary in Teaching Chinese as a Foreign Language (Duìwài hànyǔ jiàoxué zhōngde fǎlù cíhuì yánjiū). Heilongjiang University (2012)
12. The project team: The Graded Chinese Syllables, Characters, and Words for the Application of Teaching Chinese to the Speakers of Other Languages (Hànyǔ guójì jiàoyù yòng yīnjié hànzì cíhuì děngjí huàfēn). Beijing Language and Culture University Press, Beijing (2010)
13. Pan, Q.: Judgment of Chinese Forensic Language (Zhōngguó fǎlù yǔyán jiànhéng). Truth & Wisdom Press, Shanghai (2004)
14. Liu, H.: Forensic Linguistics (Fǎlù yǔyánxué). Peking University Press, Beijing (2007)
15. Wang, D., Wang, L.: The classification of legal terms in the context of the legislative standardization and the scientific pursuit of legal language (Lìfǎ guīfànhuà, kēxuéhuà shìjiǎo xià de fǎlù shùyǔ fēnlèi yánjiū). Appl. Linguist. **3**, 108–114 (2010)
16. Li, M.: The Study of Laws in Bamboo Slips Excavated from Ancient Tombs of the Qin Dynasty in Shuihudi ("Shuì hǔ dì qín mù zhújiǎn" fǎlù yòngyǔ yánjiū). Southwest Universit (2003)
17. Cao, G.: A Study of Legal words in "San Yan Er Pai" ("Sān yán èr pāi" fǎlù cíyǔ yánjiū). Xiamen University (2007)
18. Wang, D., Wang, L.: An analysis of the semantic features of the legal meaning of comments on law of tang dynasty ("Táng lǜ shū yì" fǎlù cíyì de yǔyì tèzhēng fēnxī). Ludong Univ. J. (Philos. Soc. Sci. Edn.) **4**, 87–92 (2007)
19. Xiang, J.: A Study of the Legal Words in Records of the Grand Historian ("Shǐjì" fǎlù cíyǔ yánjiū). General South University (2011)
20. Xu, H.: Study on Vocabulary of The Collected Case of the Judgement by Famous Good Officials ("Míng gōng shū pàn qīng míng jí" cíhuì yánjiū). Shandong University (2011)
21. Wu, Y.: The Semantic Changes of Chinese Legal Words (Zhōngguó fǎlù cíhuì de yǔyì yǎnjìn). Southwest University of Political Science & Law (2012)
22. Liu, Q.: On the causes of the semantic evolution of chinese legal vocabulary (Zhōngguó fǎlù cíhuì yǔyì yǎnbiàn de yòuyīn tàntǎo). Soc. Sci. Rev. **29**, 70–71 (2014)
23. Zhang, S., Zuo, Y.: Wa(Fǎlù cíhuì yǔyì yǎnbiàn de fāngshì yǔ rènzhī jīzhì). Foreign Lang. Lit. (Bimonthly) **31**, 84–89 (2015)
24. Tang, J.: The difference of "apply" in common language and legal language ("Shìyòng" zài pǔtōng yǔyán hé fǎlù yǔyán zhōngde qūbié). J. Mianyang Norm. Univ. **22**, 71–74 (2003)
25. Yang, S.: Semantic analysis of Chinese "kěyǐ" in legal and daily context ("Kěyǐ" yīcí zài fǎlù yǔtǐ yǔ rìcháng yǔtǐ zhōngde yǔyì chāyì fēnxī). J. Lüliang Univ. **3**, 15–18 (2013)
26. Luo, H., Wang, S.: The construction and application of the legal corpus of Mainland China (Zhōngguó dàlù fǎlù yǔliàokù jiànshè jíqí yìngyòng yánjiū). In: 18th Chinese Lexical Semantics Workshop (CLSW2017). Leshan Normal University (2017)
27. Luo, H., Wang, S.: The construction and application of the legal corpus. Chinese Lexical Semantics. LNCS (LNAI), vol. 10709, pp. 448–467. Springer, Cham (2018). https://doi.org/10.1007/978-3-319-73573-3_41
28. Luo, H., Wang, S.: The Transitional Sentences in Legislative Language in the Cases of the dan and danshi sentences (Lìfǎ yǔyán zhōng de dànshū hé fēidànshū yánjiū). Contemporary Rhetoric (Dāngdài xiūcí xué) 6, 77–89 (2018). [Reproduced by China Social Science Excellence Réndà fùyìn bàokān zīliào, 2, 130–139, 2019]

29. Xun, E., Rao, G., Xiao, X., Zang, J.: The construction of the BCC Corpus in the age of big data (Dà shùjù bèijǐng xià BCC yǔliàokù de yánzhì). In: Corpus Linguistics, vol. 3 (2016)
30. Kilgarriff, A., et al.: The sketch engine: ten years on. Lexicography 1, 7–36 (2014)
31. Huang, C.-R., et al.: Chinese Sketch engine and the extraction of grammatical collocations. In: The Fourth SIGHAN Workshop on Chinese Language Processing (2005)
32. Su, X.: A Thesaurus of Modern Chinese (Xiàndài hànyǔ fēnlèi cídiǎn). The Commercial Press, Beijing (2013)
33. Wang, S.: The concept system construction: a case study on the "economy and trade" category (Cíhuì xìtǒng jiàngòu de xīn shíjiàn—yǐ "jīngjì màoyì" lèi wéilì). In: The 19th Annual Conference of the International Association of Chinese Linguistics (IACL-19), Nankai University, Tianjin (2011)

From Modern to Ancient Chinese:
A Corpus Approach to Beneficiary Structure

Yu-Yin Hsu[1(✉)] and Tao Wang[2]

[1] Hong Kong Polytechnic University, Kowloon, Hong Kong SAR, China
yyhsu@polyu.edu.hk
[2] Sichuan University, Chengdu, Sichuan, China
caichcd@Outlook.com

Abstract. This study reports the results of two sets of corpus studies on the use of beneficiary structures (*wèi-dòng shì*), one in modern and the other in ancient Chinese. First, we analyzed the semantic associations of the word *wèile* 'do something for something/someone' in modern Chinese, using two corpora and the word-embedding model. The results were in line with semantic analyses proposed in the Semantic-Map Model. Second, based on an examination of all the sentences expressing beneficiary meanings in *Zuo's Commentary* and *Mencius*, we established that the beneficiary structure in those works involves a light-verb structure that should be syntactically distinguished from other such structures that introduce causative and intentional events. As well as providing some new evidence regarding the semantic content of the *wèi-dòng shì* in modern Chinese, we present structural evidence of its source, which can be dated to the pre-Qin period, as shown by the examples in the two target ancient-Chinese texts.

Keywords: *Chinese GigaWord Corpus 2.0* · Beneficiary structure · Word-embedding · Modern Chinese · *Zuo's Commentary* · *Mencius*

1 Introduction

In linguistic typology, Modern Chinese is generally regarded as an isolated language. Recently, some Chinese sentences have appeared with uncommon forms that are unlike the typical analytical structures of modern Chinese, and more similar to structures in synthetic languages. These include examples like *bèi xiāngqīn* 'to be match-made', in which the verb *xiāngqīn* 'blind-date' usually does not allow passivization, and *děng-huǐ'er duǎn wǒ* 'Later, P(rivate)-M(ail) me', the meaning of which was formerly expressed using the ditransitive verb *fā* 'send' (as in *děng-huǐ'er fā duǎnxìn gěi wǒ* 'Later, send a P-M to me'). In the recent linguistic literature, light-verb theory is often used to account for these special sentences in modern Chinese [1, 2]. However, in light of another special type of sentence in modern Chinese that will be discussed in Sect. 2, below, we will argue that in addition to the light-verb structures proposed in the existing literature, there is another type, called *wèi-dòng* beneficiary structure, whose existence is related to linguistic typology's semantic-map theory [3] (see Sect. 3), and its syntactic source can be found based on data from the pre-Qin dynasty era around the time in 221 BC. Our results for modern Chinese are obtained via inspecting the dynamic meanings of the *wèi-dòng* structure as used in the *Chinese GigaWord Corpus 2.0* [4], and via applying the word-embedding model [5] to more than 240,000 sentences containing *wèi-le* 'do something for something/someone' in modern Chinese (see Sect. 4). With the semantic results obtained from Sects. 3 and 4, we then analyze the sentences conveying *wèi-dòng* meanings in the ancient historical texts, *Zuo's Commentary* and *Mencius*, and use the results to provide a syntactic analysis of the *wèi-dòng* structure in Sect. 5. Lastly, we briefly sum up our findings and their implications in Sect. 6.

2 Meanings of *Wèi-Dòng* and the Light-Verb Theory

Recently, researchers have called attention to some special Chinese sentences that are atypical of an analytic language [1, 2]. For example, the meaning of (1a) has traditionally been said in the form (1b), and the sentence in (2a) has until recently been more commonly said as (2b).

(1) a. Qǐng   diàn   fúwù.chù.
       please call   service.center
       'Please call the service center.'
    b. Qǐng   dǎ.diànhuà   gěi   fúwù.chù.
       please make.phone.call   to   service.center
       'Please make a phone call to the service center.'

(2) a. Lǎo Wáng   bèi   tuìxiū   le.
       Lao Wang   passive   retire   perfective
       'Lao Wang got retired.'
    b. Lǎo Wáng   bèi   qiǎngpò tuìxiū   le.
       Lao Wang   passive   force   retire   perfective
       'Lao Wang has been forced to retire.'

Tsai [2] has pointed out how light-verb theory can explain such special patterns in modern Chinese. According to that theory, the three most common light verbs are *shǐ-dòng* (causative), *chéng-shì* (become), and *yì-dòng* (intentional) [6–8]. A classic analysis of light verbs is that they are denominal verbs [7]. As shown in (3), in the lexical-syntax derivation (L-Syntax), the word *shelf* is moved up from its original position inside a post-verbal prepositional phrase, and merges first with the light verb

BECOME, and then moves and merges with the light verb CAUSE, so that *shelf* obtains verbal functions and can serve as a verb taking a direct object, *the book*. Another similar example is 'bench the player' meaning 'order the player to sit on the bench rather than continue playing baseball', where the noun 'bench' serves as a verb taking a direct object.

(3) 'shelf the book' ([5], citing [7])

       ⋯⋯ [$_{vP}$ CAUSE   [$_{VP}$ the book [$_{V'}$ BECOME [$_{PP}$ ON the shelf]]]

       → [[[shelf + ON] + BECOME] + CAUSE]   t$_V$ t$_P$ t$_N$

Light-verb theory has also been used to analyze *shǐ-dòng* and *yì-dòng* expressions found in ancient Chinese [9, 10]: e.g., in *xiǎo tiānxià* in (4), *xiǎo* 'small' is not an adjective, but expresses a verbal meaning 'to regard something as small'. This structure is illustrated in (4b) [9, 10].

(4) a. Dēng Tài.shān   ér   xiǎo   tiānxià.

      Climb Mount-Tai   and   small   world

       'Climb Mount Tai and deem the world to be small.'

      b.

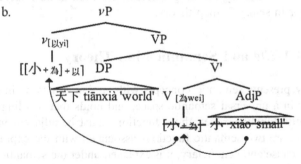

However, we would like to point out that some expressions in modern Chinese cannot be explained semantically by the three common types of light verbs mentioned above. For example, the (a) examples in (5) and (6) would typically be uttered as the (b) examples, and their meanings are not related to causation, becoming, and/or the intentional 'taking/being' as specified in the theoretical literature on light verbs. However, all of them contain a sense of 'benefiting someone or something' or 'benefiting the public good' (i.e., 'beneficiary'). Other examples involved with similar structures include *bèi zhàn* 'prepare for war', *jiù huǒ* 'fight fires', *yǎng bìng* 'recuperate from illness', and *máng kǎoshì* 'be busy preparing for exams'.

(5) a. dǎsǎo   wèishēng
       clean   hygiene
       'to clean up for the sake of hygiene/health'

   b. wèile   wèishēng   ér   dǎsǎo
      for/consider   hygiene   then   clean
      'to clean up for the sake of hygiene/health'

(6) a. xiàn   shēn   zǔguó
       sacrifice   body   nation
       'to sacrifice one's life for one's country'

   b. wèile   zǔguó   ér   fèngxiàn zìjǐ.de   shēngmìng
      for/consider   nation   then   sacrifice one's   life
      'to sacrifice one's life for one's country'

We also note that the meanings of *wèile* in (5) and (6) show some differences; that is, in (5) it is related to the aim or goal that "causes" the events to happen, whereas in (6) *wèile* shows a beneficiary relationship. In semantic theory, these different meanings are identified as cause and beneficiary meanings. Below, we look at how these two concepts are accounted for in semantic-map theory.

3   The Meanings of *Wèile* and Semantic-Map Theory

The semantic-map theory presents semantic concepts with similar functions in close proximity to one another in a universal semantic space, and holds that if a linguistic form assumes two semantic functions, then those functions must be adjacent in that space [3]. Given that there are two semantic functions associated with the expression *wèile* in modern Chinese, cause and beneficiary, it is expected under the semantic-map model that these two functions should be adjacent in the universal semantic map, and this is indeed the case: in the base graph of the "tools" class (7), the beneficiary function ('for someone') is adjacent to the cause function ('for this reason') [3].

(7)   The semantic base of the "tools" class ([3])

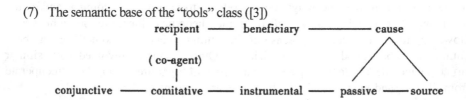

Based on an examination of the semantic base of the "tools" class of the universal semantic map [3] in light of materials from modern Chinese dialects, Zhang [11]

constructed a semantic map (8) corresponding to the semantic relations of *passivity*, *causation*, *disposal*, *tool*, and *beneficiary* in Chinese.

(8) Base map of "disposal" and its related functions ([11])

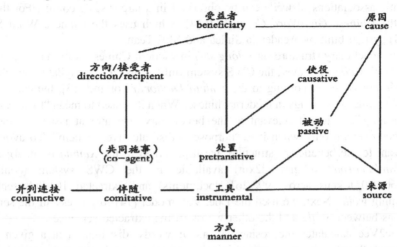

The adjacency of the beneficiary function and the cause function in (8) provides important theoretical support for the dual meanings associated with the form *wèile*. Based on this result, in Sect. 4, we use a corpus to investigate the correlation between *wèile*'s meaning and the meanings of some other terms related to the cause and beneficiary functions in modern Chinese. Then, in Sect. 5, we go on to investigate how such meanings are expressed as well as the related argument types in the corpora of *Zuo's Commentary* and *Mencius*.

4 *Wèile* in Modern Chinese and Word Embedding

4.1 *Wèile* in the Center for Chinese Linguistics Corpus

From the Center for Chinese Linguistics (CCL) corpus of modern Chinese [12], we extracted the entries containing *wèile*. After eliminating redundant components, this yielded more than 130,000 effective linguistic data points. Since the CCL corpus has not been annotated with parts of speech, we performed our semantic analysis based on a 1:100 scale distributed sampling of 1,300 records.

Overall, purpose and beneficiary meanings were *wèile*'s major senses in the dataset. However, many more *wèile* sentences expressed purpose than beneficiary readings (939 vs. 361). These two types of meanings appeared in various linguistic units such as words, phrases, and clauses. It was because of these observations that we decided to perform a more robust semantic analysis of the use of *wèile* in modern Chinese.

4.2   A Word-Embedding Analysis of *Wèile*

Given that modern Chinese allows some synthetic structures like (5a) and (6a), and the fact that the use of the term *wèile* in modern Chinese can introduce either a purpose or beneficiary meaning, we used the word-embedding model [5] to examine whether such semantic associations of *wèile* can be observed in a larger-scale corpus. For this, we used the *Chinese GigaWord Corpus 2.0* [4], which uses the Chinese Word Sketch (CWS) system built by Academia Sinica and SkE Team.

The beneficiary structure (*wèi-dòng shì*) in modern Chinese is primarily expressed by the verb *wèile*. However, the CWS system only allows a monosyllabic search of 为 *wèi/wéi*, not *wèile*. According to the *Xinhua Dictionary* online [13], the character 为 has two primary meanings in modern Chinese. When it is used to mean "to do, execute, and become", it is not relevant to the beneficiary structure at issue. Its meaning becomes relevant only when it is "purpose, substitute/give, or help". To avoid uses irrelevant to the beneficiary structure, we first turned to the *Xinhua News Agency in Mainland corpus* (Gigaword2xin) available in the CWS system (containing 311,660,000 tokens across 992,261 documents) and extracted 101,868 sentences containing *wèile*. Next, we used the Doc2Vec model [14] to process all the semantic relations between *wèile* and the other words in the extracted sentences.

Doc2Vec calculates the cosine vectors of words' distribution in a given set of documents through an unsupervised learning algorithm, and returns the probability of the occurrence of each word while retaining information on word order within sentences and the semantic impacts of words in their respective contexts. Possibly because the limit of textual content, our results showed that among the 50 words most closely related to *wèile*, only five were adjectives or verbs, with all others being nouns. Moreover, the word with the highest degree of similarity to *wèile* was only 0.5254 similar. Next, we compared four terms that are often discussed as semantic features of the beneficiary structure, i.e., *yuányīn* 'reason/cause', *mùdì* 'purpose', *hǎochù* 'advantage', and *shòuhuì* 'benefit'. In (9), we present the degree of similarity of these words to *wèile* as calculated by the Doc2Vec model. The fact that the results are influenced by the contexts of words in a given text is evidenced by the differences among the results for each word pair.

(9) Degrees of similarity of *wèile* to words associated with the beneficiary structure (Gigaword2xin)

　　a. *wèile* and *yuányīn* 'reason/cause': 0.2450   c. *wèile* and *hǎochù* 'advantage':   0.2401

　　b. *wèile* and *mùdì* 'purpose': 0.1443   d. *wèile* and *shòuhuì* 'benefit':   0.2197

The Doc2Vec model also allows us to calculate the semantic associations formed between more than one word and their relations to other words in a given text. Our comparison of the semantic intersection of *wèile* with reason (*wèile+reason/cause*) and *wèile* with benefit (*wèile+benefit*) yielded a very high similarity, 0.895. This suggests that *wèile* expressing 'cause for reasons' and 'for benefit' are closely correlated.

The Doc2Vec model was also used to identify the 50 words in the *wèile* dataset we extracted from Gigaword2xin that had the highest degree of semantic association with *wèile+reason/cause* (i.e., meaning 'for the sake of the reason') and with *wèile+benefit* (i.e., meaning 'for the sake of the good'). Interestingly, we observed a difference in how these two semantic combinations of *wèile* affected other words. As shown in (10), both these subsets had the same number of adjectives and of prepositions/conjunctions. However, the 'for the sake of the reason' subset included many more nouns than verbs, while the 'for the sake of the good' had a similar number of nouns and verbs. Although further research will be needed to explain the correlation between word-class selection and the associated semantic set, differences related to word-class distributions may indicate some qualitative differences between semantic and structural interactions.

(10) Counts (percentages) of the parts of speech represented by the 50 words most similar to *wèile*

	Nouns	Verbs	Adjectives	Prepositions/ Conjunctions
wèile + reason/cause	30 (60%)	5 (10%)	10 (20%)	5 (10%)
wèile + benefit	17 (34%)	18 (36%)	10 (20%)	5 (10%)

In addition, we wanted to establish if there were differences between the use of causative structure (*shǐ-dòng*) and intentional structure (*yì-dòng*) in our chosen corpora. Again using the Doc2Vec model, we took the words *zàochéng* 'cause' and *dǎozhì* 'result in' as representative of the causative structure and compared them with *wèile*. Likewise, we compared *wèile* against the words *rènwéi* 'consider' and *yǐwéi* 'think', taken as representative of intentional structure. The results show that the semantic similarity between these concepts and *wèile* in the corpus texts tended to be low, as shown in (11a-b) for *shǐ-dòng* structure, and in (11c-d) for *yì-dòng* structure.

(11) Degrees of similarity of *wèile* to *shǐ-dòng* and *yì-dòng* in Gigaword2xin

    a. *wèile* and *zàochéng* 'to cause': 0.1048   c. *wèile* and *rènwéi* 'to consider':   0.1285

    b. *wèile* and *dǎozhì* 'to result in': 0.1507   d. *wèile* and *yǐwéi* 'to think':   0.1506

We then extracted further 147,613 sentences containing *wèile* from another, larger corpus in the CWS system, the *Central News Agency in Taiwan corpus* (Gigaword2cna), which contains 501,456,000 tokens across 1,769,953 documents. We used the same Doc2Vec model to identify the semantic relationships within this dataset, and as shown in (12), the results were basically similar to those we derived from the other corpus, but with lower levels of correlation between *wèile* and the class of reason/purpose. Nonetheless, when we compared the semantic intersections *wèile+reason/cause* and *wèile+benefit*, we again found very similar results (0.9017). In other words, in Gigaword2cna as in Giga-word2xin, *wèile*'s semantic association with related words are similar.

(12) Degrees of similarity of *wèile* to words associated with the beneficiary structure (Gigaword2cna)

    a. *wèile* and *yuányīn* 'reason/cause': 0.1449 c. *wèile* and *hǎochù* 'advantage':   0.2297

    b. *wèile* and *mùdì* 'purpose': 0.1355   d. *wèile* and *shòuhuì* 'benefit':   0.219

In summary, supplementing our findings arrived at via semantic map theory, our corpus studies indicate that the beneficiary structure as expressed by *wèile* in modern Chinese differs from both *shǐ-dòng* (causative) and *yì-dòng* (intentional). In the next section, we further test this analysis through a syntactic examination of the beneficiary structure in two ancient Chinese documents, *Zuo's Commentary* and *Mencius*.

5 The Beneficiary (*Wèi-Dòng*) Structure in Ancient Chinese

The foregoing study of modern Chinese has shown that the similarity between "cause for reasons" and "cause for benefit" is high: i.e., 0.895 and 0.902 cosine similarities. Our preliminary examination of ancient Chinese texts suggested that the meaning of phrases in the class 'benefit' was associated not only with people but also with events. Likewise, it indicated that the associations of phrases in the class of 'purpose' may not be limited to a single event, and can also be persons. These findings imply that there could be two sub-classes of the beneficiary structure, since the object selections of these two meanings are not complementary, and either can take either a human or an event as its argument.

Zuo's Commentary (375-351 BC) is an ancient Chinese narrative history covering a period from 722 to 468 BC. *Mencius* (ca. late 4th century BC) is a collection of anecdotes about and conversations of the philosopher Mengzi and other Confucian thinkers, on topics in moral and political philosophy. In *Zuo's Commentary* [15], we identified 110 occurrences of the *wèi-dòng* structure, of which 22 were "benefited" and 88 showed a purpose relationship. We found that the purpose structure could be used to describe a cause-purpose event related to a person or to an event or process. For example, *qǐ zhī* in (13a) means 'to open the door for him,' while *qín wáng* in (13b) means 'to help and support actions related to the best interests of the emperor/the royal family.'

(13) a. 大叔   完 聚⋯⋯   將   襲 鄭,   夫人   將   啓   之。
       Tàishū   wán jù⋯⋯   jiāng   xí Zhèng,   fūrén   jiāng   qǐ   zhī.
       Taishu   finish gathering   about.to   attack state.Zheng   lady   plan   open.for him
       'Taishu finished gathering troops and was about to attack the state of Zheng. The lady was planning to open the gate for him to enter.'

     b. 求   諸侯,   莫   如   勤   王。
       Qiú   zhūhóu, mò   rú   qín   wáng
       request   duke not   like   support.for   emperor
       'It is better to facilitate actions related to the best interests of the Emperor than to make a request to the Dukes.'

Similarly, the "benefit" function in *Zuo's Commentary*, while most commonly associated with a person, is not limited to persons as objects. For example, *bǔ qī Jìngzhòng* in (14) means 'to perform divination about marrying his daughter to Jingzhong'. Thus, in this beneficiary structure, the verb phrase *qī Jìngzhòng* 'to marry [his daughter to] Jingzhong' functions as the object.

(14) 懿氏   卜   妻   敬仲。
    Yishi   bǔ   qī   Jìngzhòng
    Yishi   tell.fortune   marry   Jingzhong
    'Yishi performed divination about marrying his daughter to Jingzhong.'

In *Mencius*, we identified 26 occurrences of the *wèi-dòng* structure, of which only two were of the 'for benefit' class. One of these two examples, *sǐ qí-zhǎng* in (15a), means 'to sacrifice their lives for the leader', and *mò zhī sǐ* in (15b) interpreted as *mò sǐ zhī* meaning 'not to sacrifice their lives for the officers.'

(15) a.

斯民	親	其	上,	死	其	長	矣。
Sīmín	qīn	qí	shàng,	sǐ	qí	zhǎng	yǐ
people	close	their	leader	die	their	leader	then

'People worship their leader and are willing to die for him.'

b.

吾	有	司	死	者	三十三.人,	而	民	莫	之	死	也。
Wú	yǒu	sī	sǐ	zhě	sānshísān.rén, ér		mín	mò	zhī	sǐ	yě.
I	have	officer	die	one	33.person	and	people	not	them	die.for	particle

'I have 33 officers died, but no one is willing to die for them.'

The other 24 examples all had cause-purpose meanings. For example, *qǐng shìshī* in (16) means 'to request for the position of being a judge.'

(16)

子	之	辭	靈丘	而	請	士師
Zǐ	zhī	cí	Língqiū	ér	qǐng	shìshī
you	possessive	leave	Lingqiu	then	ask.for	judge

'Your leaving Lingqiu to request for the position of being a judge.'

We also found one example of a prepositional phrase being used to modify the beneficiary structure. In (17), *wèi wǒ* 'for me' modifies the purpose structure (i.e., *yán zhī* 'to say good things for it') to introduce the person (*wǒ* 'I') who receives the benefit.

(17)

子	盍	為	我	言	之?
Zǐ	hé	wèi	wǒ	yán	zhī
you	why	for	me	say	it

'Why don't you put in a good word about it for me?'

Based on the semantic distinction between the beneficiary structure under discussion and other typical light-verb structures, along with prior light-verb analyses of ancient Chinese [9, 10], we propose that two light verbs that are unique to the beneficiary sense should be identified in syntax. They are 緣由*yuányóu* (for cause-purpose) and 受惠*shòuhuì* 'benefit'. For example, the phrase *qín wáng* 'to help and support actions related to the emperor/the royal family' in (13b) can be derived through a structure like (18), in which the verb *qín* 'to support' moves to pick up the active feature (for direct object) and then moves to the light verb *yuányóu* to express cause-purpose regarding the object noun phrase, *wáng*. The verb *qín* then moves to *v* to license the subject of the whole event.

(18) 勤王 *qín wáng* 'to help and support actions related to the emperor/the royal family'

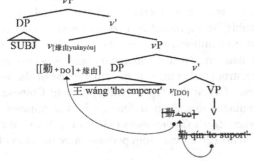

When the sentence is about doing something for benefit, another light verb *shòuhuì* 'benefit' is involved. For example, *sǐ qí-zhǎng* 'to die for one's leader' in (15) can be derived through a structure like (19), in which the verb *sǐ* 'to die for' moves to pick up the active feature (for direct object) and then raises to the light verb *shòuhuì* to express beneficial effects on the object noun phrase *qí-zhǎng*. The verb *sǐ* then moves to *v* to license the subject of the whole event.

(19) 死其長 *sǐ qí-zhǎng* 'to die for one's leader'

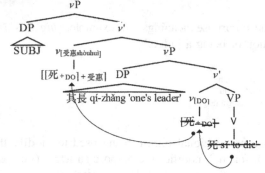

6 Conclusion

Important theoretical support for the semantics of different uses of *wèile*, in keeping with semantic-map theory, has been provided by this paper's investigation of the semantic correlations between *wèile* in modern Chinese, on the one hand, and on the other, some key words related to typical light-verb structures in two corpora of modern Chinese and those sentences expressing the beneficiary structure in two ancient Chinese texts.

For our corpus study of modern Chinese, we used a basic word-embedding model to analyze all those sentences containing *wèile* in two large datasets within the *Chinese GigaWord Corpus 2.0*. Our results, though preliminary, show that analyses of word embedding can indeed be influenced by which texts are used, and yet, that the beneficiary structure at issue is semantically different from the three typical light-verb structures that we also investigated. Then, our corpus study of ancient Chinese usage of the beneficiary structure, based on *Zuo's Commentary* and *Mencius*, found that such structure differed semantically from other light-verb structures, and that the subtypes of its beneficiary associations could probably be expressed by two light verbs, one being *yuányóu* (used to express cause-purpose meanings), and the other, *shòuhuì* 'benefit'.

The existence of the beneficiary data in *Zuo's Commentary* and *Mencius* also provides some evidence regarding the structural source of the emergence of the new beneficiary structure in modern Chinese. Given that typologically, ancient Chinese is generally considered a synthetic language, the sharp difference in the number of beneficiary-structure occurrences between *Zuo's Commentary* (110 cases), which dates from the early 4th century BC, and *Mencius* (26 cases), from perhaps three quarters of a century later, may imply a transformation of the Chinese language from a synthetic to an analytic one, partially within that period.

References

1. Huang, C.T.J., Liu, N.: The syntax and semantics of the new non-canonical bei(被) XX construction. Linguist. Sci. **13**(03), 225–241 (2014). (In Chinese)
2. Tsai, W.-T.: On the distribution and interpretation of inner and outer light verbs in Chinese. Linguist. Sci. **15**(04), 362–376 (2016). (In Chinese)
3. Haspelmath, M.: The geometry of grammatical meaning: semantic maps and cross-linguistic comparison. In: Tomasello, M. (ed.) New Psychol. Lang., vol. 2, pp. 211–243. Erlbaum, New York (2003)
4. Huang, C.R.: Tagged Chinese Gigaword version 2.0 (2009). https://catalog.ldc.upenn.edu/LDC2009T14
5. Mikolov, T., Yih, W.-T., Zweig, G.: Linguistic regularities in continuous pace word representations **1**, 746–751 (2013)
6. Lin, T.-H.J.: Light verb syntax and the theory of phrase structure. Ph.D. dissertation, University of California, Irvine (2001)
7. Hale, K., Keyser, S.J.: Prolegomenon to a Theory of Argument Structure. The MIT Press, Cambridge (2002)
8. Tang, S.-W.: Formal Chinese Syntax. Shanghai Educational Publishing House, Shanghai (2010). (In Chinese)
9. Feng, S.-L.: Light verb movement in modern and classical Chinese. Linguist. Sci. **1**, 3–16 (2005). (In Chinese)
10. Feng, S.-L.: Light verb syntax between English and classical Chinese. In: Li, Y.-H.A., Simpson, A., Tsai, W.-T. (eds.) Chinese Syntax in a Cross-linguistic Perspective, pp. 229–250. Oxford University Press, New York (2015)
11. Zhang, M.: On the 'constant' and 'variables' towards spatial maps and semantic maps: take the relation of passive, causative, disposal, tools, beneficiaries in Chinese. Lecture at the Institute of languages of the Chinese Academy of Social Science (2008). (In Chinese)
12. Zhan, W.-D., Guo R., Chen, Y.-R.: The CCL corpus of Chinese texts: 700 million Chinese characters, the 11th century B.C. – present, Peking University (2003). http://ccl.pku.edu.cn:8080/ccl_corpus
13. XinHua Dictionary Online (Xianshang XingHua Zidian). https://zidian.911cha.com/zi4e3a.html
14. Le, Q., Mikolov, T.: Distributed representations of sentences and documents. In: The 31th International Conference on Machine Learning, Beijing, vol. 32, pp. 1188–1196 (2014)
15. Yang, B.-J.: Annotation of Zuo's Commentary (Revised Edition). Zhonghua Book Company, Beijing (1990). (In Chinese)

Research on Gender Tendency of Foreign Student's Basic Chinese Vocabulary

Zhimin Wang[1](✉) and Huizhou Zhao[2]

[1] Research Institute of International Chinese Language Education, Taipei, China
wangzm000@qq.com
[2] School of Information Science, Beijing Language and Culture University,
Beijing 100083, China
zhaohuizhou@blcu.edu.cn

Abstract. The paper designs a basic vocabulary sequencing model and explores the differences in basic vocabulary output between male and female. It is found that male and female have obvious preferences in choosing needed vocabulary. Males are sensitive to numbers and concerned about the world issues, while females show dependence on living and learning environment and evaluation of their mood. Moreover, men tend to output abstract vocabulary, thinking about issues relatively macro. On the contrary, women tend to output specific vocabulary, consider things relatively micro. At the same time, there are also obvious differences between male and female in kinship terms and personal pronouns. Male and female give priority to the output of kinship terms with the same sex, but male give priority to the output of personal pronouns before female. The study reveals the differences in Chinese vocabulary output among learners of different genders.

Keywords: Sequencing model · Vocabulary output · Basic vocabulary · Gender tendency

1 Introduction

Language and gender studies have always been one of the hot topics in sociolinguistics. Lakoff [1] described the main features of "female language" in American culture from the perspective of discourse, which makes language research and gender differences an important part of sociolinguistics.

Linguists have found that there are many stylistic differences in language between male and female. For example, men interrupt women's words more and control the turn longer [2]. Female discourse is more euphemistic and polite than male discourse [3]. Fishman [4] suggested that women provide more positive feedback during conversation.

Several studies have shown that there are some differences in Chinese between male and female. For example, women are more euphemistic and polite than men in requests, compliments and responses [5, 6]. As far as the frequency of interruptions is concerned, men can also control topics more successfully after interrupting female

© Springer Nature Switzerland AG 2020
J.-F. Hong et al. (Eds.): CLSW 2019, LNAI 11831, pp. 736–745, 2020.
https://doi.org/10.1007/978-3-030-38189-9_74

discourse [7]. In addition, there are gender differences in politeness, social appellation, movie and television catchwords and Internet language [8].

The previous research shows that there are many research had been made on gender differences in the first language, but little attention has been paid to the second language. Both male and female participate together in the process of learning second language but what are the similarities and the differences of their vocabulary output? Can they write out basic vocabulary that they have in mind within a time given? Problems above deserve further study.

Therefore, the paper investigates the basic vocabulary data of foreign students, and summarizes and analyses their gender differences. The rest of the paper is organized as follows. In Sect. 2, we propose a basic vocabulary survey. In Sect. 3, we present a sequencing model of the basic vocabulary. In Sect. 4, we explore basic vocabulary differences made by male and female. Finally, we conclude with a summary and an outline of further research.

2   A Survey of Basic Vocabulary Based on Psychological Word List Hypothesis

In learning Chinese, foreign students have their own understanding of which vocabulary they should master first. With the increasing of learning Chinese experience, their understanding is also deepening. We propose a psychological wordlist hypothesis that no matter how the environment changes, everyone has a vague set of basic vocabulary in his mind. Therefore, the basic vocabulary survey is based on this psychological hypothesis. The questionnaire is designed as follows.

Dear classmates:

With your experience in learning Chinese, what do you think are the 300 most commonly used words for international students? These words may relate to learning, survival, daily communication, daily activities and so on. Please fill in your ideas on the form.

The purpose of this survey is to acquire the psychological experience of foreign students by actively outputting their own basic vocabularies without any references. There are 808 valid questionnaires and 42 invalid questionnaires. The sample of male students is 297, accounting for 37%, and female students is 511, accounting for 63%. The number of basic vocabulary is more than 240,000.

Students' Chinese proficiency determines their vocabulary output ability. Principally, the higher the grade, the stronger the ability of vocabulary output. Considering the output ability of students, second year or above undergraduate foreign students have been selected to participate in the survey, as shown in Table 1.

Table 1. Grade Distribution of Male and Female Samples

Grade	Number	Proportion	Proportion of grade
2nd Year (Senior)	40	4.95%	4.95%
3rd Year (Junior)	348	43.07%	72.77%
3rd Year (Senior)	240	29.70%	
4th Year (Junior)	126	15.59%	22.27%
4th Year (Senior)	54	6.68%	

Table 1 shows that most foreign students participated in this survey are concentrated in third year, with 348 students and 240 students respectively, accounting for the total of 72.77%. The number of fourth year students are 126 and 54 respectively, accounting for 22.27%. The number of students in second year are the least, accounting for 4.95%.

Word output ability is closely related to language learning time. The longer they spend time in learning Chinese, the easier they can output the basic vocabulary. Therefore, the survey mainly focused on second year and above students. The length of Chinese learning time for 808 foreign students is shown in Fig. 1.

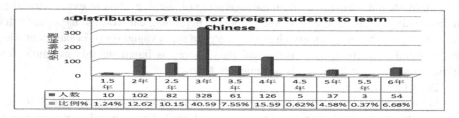

Fig. 1. Distribution of time for foreign students to learn Chinese

According to data above, only 10 students have studied Chinese for 1.5 years. The remaining 798 students have studied Chinese for more than two years. Among them, 328 students, 61 students, 126 students and 5 students studied for 3, 3.5, 4 and 4.5 years, accounting for 40.59%, 7.55%, 15.59% and 0.62% respectively; 102 students, 82 students, accounting for 12.62% and 10.15% in 2 and 2.5 years respectively. The remaining students are those who have studied for 5 years, 5.5 years or more. Senior students have been chosen to participate in this survey mainly due to the fact that the higher the grades are, the deeper the understanding of language learning will be. Moreover, they have their own consideration when outputting vocabularies.

3   Analysis of Basic Vocabulary Characteristics of Male and Female

There are some differences between male and female basic vocabulary output, and also similarities between them. We extracted vocabularies for male and female over 200 times and find that *mobile phone, teacher, study* are the most frequent vocabularies output by foreign students. Mobile phone ranks the top third, reflecting its importance in people's lives. It is also found that there were differences in the ranking of *mobile phone* between male and female samples. For male samples, *mobile phone* ranked first but it ranked second for female samples.

Frequency can be regarded as an important factor of word ranking, but it is not the only criterion. Lexical position is also considered as a factor affecting word ranking. For any vocabularies, frequency can only reflect on how many learners can think of and output in the given time, and the position of the output reflects the importance of the vocabularies in learners' minds. If a vocabulary is effectively output by learners, but part of the output is in the front, part of the output is in the back, and opinions are not uniform, then the vocabulary ranking needs to take into account the multiple factors such as frequency and position.

Vocabulary ranking is the basis of comparing male and female sample sets. We need to find an effective way to measure the two sample sets. Standard deviation is a measure of the degree of data dispersion. The standard deviation of *mobile phones, teachers and study* is shown in the table below.

Table 2. Samples of basic vocabularies

Male	Freq	Standard deviation of position	Female	Freq	Standard deviation of position
Mobile phone(手机)	234	77.6197	Teacher (老师)	456	56.9232
Teacher (老师)	232	72.7802	Mobile phone(手机)	447	71.7920
Study (学习)	232	47.875	Study (学习)	396	46.5278

The data in Table 2 reveals that although the frequency of *study* ranks third, the standard deviation of position is much lower than *teacher, mobile phone*. Apparently, the position of *study* in the minds of students is relatively stable and consistent. Therefore, frequency and location information are taken as important parameters to measure basic vocabularies in this paper.

In order to compare two samples, we design a sequencing model that can measure the sample set of male and female. It considers not only the frequency of vocabulary output, but also the position of vocabulary output. The sequencing model is designed as follows.

$$SM = \frac{F^2}{Stdev(K)} \tag{1}$$

In formula (1), *SM* is the sequencing model of basic vocabulary. F is the average frequency of words, *stdev*(k) is the word standard deviation of position. In formula (2) "*n*" refers to the number of average frequency $\bar{k}$.

$$Stdev(K) = \sqrt{\frac{\sum (k - \bar{k})^2}{n - 1}} \tag{2}$$

According to the sequencing model, we calculate the sample sets of male and female separately and get the sample set of male and female. As shown in Table 3.

Table 3. Sequential Set for Male and Female Samples

Male				Female			
Order	Vocabulary	Freq	SM	Order	Vocabulary	Freq	SM
1	study (学习)	232	194393.0879	1	teacher (老师)	456	652101.7932
2	mobile phone (手机)	234	145111.0974	2	study (学习)	396	562901.8239
3	teacher (老师)	232	135388.6632	3	mobile phone (手机)	447	545141.8249
4	time (时间)	199	89333.7044	4	friend (朋友)	388	370304.2953
5	I (我)	185	87547.7364	5	汉语 Chinese (汉语)	357	344318.1921
6	Chinese (汉语)	191	84230.6350	6	time (时间)	381	316800.9848
7	friend (朋友)	191	81504.5489	7	I (我)	334	306879.6512
8	student (学生)	168	77803.7615	8	mother (妈妈)	350	303681.4961
9	China (中国)	184	75647.8290	9	school (学校)	324	292412.4176
10	father (爸爸)	176	75545.9341	10	student (学生)	312	291542.0654
11	school (学校)	160	72740.1709	11	hello (你好)	253	279407.8510
12	computer (电脑)	180	72589.5477	12	China (中国)	342	275049.8792
13	mother (妈妈)	175	71070.9047	13	father (爸爸)	325	266784.1274

According to the sequencing model, all vocabularies in the sample set of male and female are sorted separately. Although male and female have common understanding of basic vocabulary, such as giving priority to *learning, teacher, Chinese, mobile phone, mother, father* output, these vocabularies are essential in modern language life and can be used as basic vocabulary for teaching. The study found that the output order of male and female vocabularies are totally different. If the output order of male and female is hugely different, it can reflect the linguistic differences in gender to a certain extent.

4   Analysis of Common Vocabulary Between Male and Female

In this paper, the first 300 words of two samples are extracted. Statistics show that there are 249 common vocabularies and 51 unique vocabularies in each set. This shows that male and female have the same understanding of the most basic vocabularies, and that common vocabularies will be the core of future vocabulary learning.

Although both male and female have their own choices of common vocabulary, the sequence is totally different. Those vocabulary order differences are an effective means to observe the vocabulary characteristics. Therefore, the paper labels 300 vocabularies and define the order difference = female [order n]- male [order n]. For example, the order difference of *mobile phone* is marked as female [3] - male [2] = 1, and the order difference of *mother* is marked as female [8] - male [13] = −5. Four forms of [positive value] [negative value] [0] [special vocabulary] are obtained by sequential difference labeling.

[Positive value] represents male output before female output; [negative value] refers to female output before male output. [0] stands for male and female with the same serial number, only the *younger brother* is [0]. [special vocabulary] correspond to the unique vocabulary of male and female in the first 300 words, with 51 words in each set.

The paper makes a further study of whether words tend to be male-output-first or female-output-first. As shown in Table 4.

Table 4. Interval Statistics of Order Differences

	Male Priority Output	Female Priority Output
Order difference interval	1 until 165	-1 until -225
Between 0 and 50	72	72
Between 50 and 100	43	31
More than 100	14	16
Total number of order differences	129	119

The data presented in Table 4 identified that 129 words are given priority to output for male, 119 words take precedence over output for female, and males priority output are slightly higher than females. However, from the order differences interval point of view, female order differences interval "−1 to −225" is much higher than male "1 to 165". Generally, the highest order differences can reflect the difference between male and female priority output. In this paper, the top 24 basic vocabularies with the highest order differences are shown in Table 5.

Table 5. Order Differences between Male and Female

Order Difference of Male				Order Difference of Female			
Male	SD	Male	SD	Female	SD	Female	SD
Need 需要	165	Two 二	100	Dormitory 宿舍	-225	Super market 超市	-108
Idea 想法	161	Economy 经济	100	Canteen 食堂	-197	Sorry 对不起	-108
Nine 九	149	One 一	99	Dictionary 词典	-194	Cold 冷	-105
Seven 七	137	West 西	97	Understand 明白	-178	Expensive 贵	-105
Five 五	134	Four 四	97	Tea 茶	-175	Tell 告诉	-96
Six 六	127	Three 三	95	Pretty 漂亮	-146	English 英语	-89
Is 是	121	Can 会	94	Colour 颜色	-145	Doctor 医生	-89
Eight 八	114	World 世界	90	Happy 开心	-138	Winter 冬天	-89
North 北	107	Relation 关系	88	Spoken language 口语	-135	Room 房间	-88
Korea 韩国	106	America 美国	85	Waitress 服务员	-133	Happy 快乐	-87
Ten 十	105	Already 已经	82	Summer 夏天	-133	Younger Sister 妹妹	-78
Do 做	105	Cooperation 合作	81	Birthday 生日	-130	Travel 旅游	-75

Table 5 shows that male vocabulary preferences are numbers and vocabularies that related to world issues. Specifically, males are very sensitive towards numbers, and numbers are very important in their minds, so their output order of numbers is significantly higher than women. The order difference of ten numbers is positive, all of which are given priority by men, and the difference is almost 100 and over. To some extent, this explains why men have stronger abstract thinking ability than women, especially in the calculation of numbers. In addition, vocabularies such as *need, idea, Korea, economy, world, relation, America, already, cooperation* which men give priority to output are mostly abstract words, which shows the problems they think about are relatively macroscopic.

The 24 vocabularies with the highest order difference output shows female's concern about living, learning environment and their mood preferences. Specifically, compared to male, their vocabularies output are related to their study and life, such as *dormitory, canteen, supermarket, room* and learning daily necessities, such as *dictionary, spoken language, English, tea, younger sister, doctor*. Female also pay attention to the expression of mood, such as *happy, color, beautiful, cold, expensive, travel*, season and special dates, such as: *birthday, summer, winter*.

Most of the above-mentioned vocabularies are specifically related to their living environment, and also the evaluation of their mood. Female vocabulary output is relatively specific, and their problems thinking are relatively microscopic. What is the difference between male and female in the output location of kinship? The paper also makes an investigation between male and female in the output of kinship terms in Table 6.

Table 6. Output Differences of Kinship

Kinship Terms	Location of Female Sample	Location of Male Sample	Order difference
brother (哥哥)	57	48	9
father (爸爸)	13	10	3
Younger brother (弟弟)	65	65	0
mother (妈妈)	8	13	-5
grandparent (爷爷)	228	241	-13
older sister (姐姐)	28	59	-21
grandmother (奶奶)	195	257	-62
Younger sister (妹妹)	64	143	-78

For the important appellation such as *father* and *mother*, male and female are given priority to output, but the order of output is different. The output location of *mother* in female samples is prior to *father,* while for male samples is just the opposite. By calculating the order difference, we can see that male and female pay more attention to gender identity. Men give priority to their same-sex kinship terms such as *brothers, fathers, brothers*, while female give priority to their same-sex kinship terms such as *mother, older sister, grandmother, younger sister,* and also give priority to their *grandparent*, which shows that female tend to have a deeper relationship with family. In addition to kinship terms, there is also a group of personal pronouns referring to people or things, which have the functions of nouns and adjectives. The paper also examines the order in which male and female output personal pronouns, and finds interesting phenomena. As shown in Table 7.

Table 7. Output Differences of Personal Pronouns

Personal Pronouns	Female Sample Position	Male Sample Position	Order difference
I (我)	7	5	2
you (你)	23	21	2
he/she (他)	46	30	16
we (我们)	148	89	59
they (他们)	241	175	66
you (你们)	Beyond 300	258	----

As shown in Table 7, personal pronouns are output in the order of *I, you, he, we, they and you,* and the order of male and female is in consensus. However, in terms of order difference, all personal pronouns are produced by male before female. Personal pronouns have more abstract meanings than specific kinship terms, which show that male is better at abstract thinking. Personal pronouns are also outputted by male before female, which again reflects that male prefer to use abstract vocabulary output. In addition to 249 common vocabularies, there are 51 unique vocabularies in male and female samples. These words are also important data for finding gender differences between male and female. In this paper, we extract the top 10 unique vocabularies in each sample, as shown in Table 8 below.

Table 8. Top 10 Unique Vocabularies in Male and Female Samples.

Unique Vocabularies for males				Unique Vocabularies for females			
order	word	order	word	order	word	order	word
105	bread (面包)	167	flower (花)	101	football (足球)	199	survey (调查)
135	family (家人)	173	raining (下雨)	121	of (的)	200	experience (经验)
146	office (办公室)	184	autumn (秋天)	127	think (认为)	203	but (可是)
156	library (图书馆)	185	milk (牛奶)	176	such as (比如)	204	white (白)
161	cold (感冒)	186	spring (春天)	192	hit (打)	208	survival (生存)

Table 8 shows male and female have obvious preferences in vocabulary output. For example, male output *football* in 101 order and female output *bread* in 105 order. This reflects how important *football* is to male, while female show a preference for *bread* and *food*. Additionally, male in the first 200 positions have begun to output function words and non-substantive verbs with relatively abstract part-of-speech meanings, such as, auxiliary word *de*, conjunction word *but*, verbs *think, investigate* and so on. Female tends to output specific life words, which include nouns and verbs. For example, nouns *office, library, flower, autumn, spring, milk* and verbs *cold, raining.* This shows women are concern of their surroundings.

5 Conclusion

The paper designs a vocabulary sequencing model based on the questionnaire of basic vocabulary of foreign students. By analyzing the vocabulary characteristics of male and female students, we find that there are obvious differences in the basic vocabulary usage between male and female students. Males are sensitive to numbers and pay attention to world issues. Females, however, show a dependence on the living and learning environment as well as description of their mood. Males tend to output abstract vocabularies and think macroscopically while females tend to output specific vocabularies and think microcosmically. At the same time, obvious differences between male and female in terms of kinship terms and personal pronouns are discovered. The exploration provides helpful reference for further research on the characteristics of vocabulary output of Chinese learners of different genders.

Acknowledgments. The work was supported by Major Program of National Social Science Foundation of China (18ZDA295); Funding Project of Education Ministry for Development of Liberal Arts and Social Sciences (16YJA740036); Top-ranking Discipline Team Support Program of Beijing Language and Culture University(JC201902); the Fundamental Research Funds for the Central Universities (18YBT03).

References

1. Lakoff, R.: Language and Woman's Place. Harper and Row, New York (1975)
2. Zimmerman, D., West, C.: Sex roles, interruptions and silences in conversation. In: Thorne, B., Henley, N. (eds.) Language and Sex: Difference and Dominance. Newbury House, Rowley (1975)
3. Holmes, J.: Hedging your bets and sitting on the fence: some evidence for hedges as support structures. Te Reo **27**, 47–62 (1984)
4. Fishman, P.: Interaction: the work women do. In: Thorne, B., Kramarae, C., Henley, N. (eds.) Language, Gender and Society, pp. 89–101. Newbury House, Rowley (1983)
5. Ding, F.: Gender Differences in Speech Acts of Chinese Request. Journal of Xi'an Foreign Languages University. **10**(1), 46–50 (2002). (in Chinese)
6. Quan, L.-H.: Sex-based differences in the realization patterns of compliments and compliment responses in the Chinese context. Mod. Foreign Lang. (Q.) **27**(1), 62–69 (2004). (In Chinese)
7. Zhang, Q.: A study of gender difference in interruptions and overlaps by young people in conversations. Lang. Teach. Linguist. Stud. **4**, 74–80 (2005). (in Chinese)
8. Xie, X.-H., Zhou, H.: A study on the gender difference of college students' language application. J. ShangRao Norm. Univ. **30**(5), 75–77 (2010). (in Chinese)
9. Ma, J.-H.: Theoretical shifts of language and gender studies in the west. Foreign Lang. Lit. (Bimonthly) **28**(2), 53–56 (2012). (In Chinese)
10. Wei, Y.-Z.: A study of gender differences in compliments. J. Xi'an Foreign Lang. Univ. **9**(1), 1–5 (2001). (in Chinese)

The Restrictions on the Genitive Relative Clauses Triggered by Relational Nouns

Xin Kou(⌗)

School of Literature, Shandong University, Jinan, China
snjdkx@163.com

Abstract. Genitives can be relativized in Mandarin Chinese. This article focuses on the genitive relative clauses triggered by relational nouns. The constructions are restricted to some conditions grammatically and semantically. First, the relational nouns in the relative clauses must serve as subjects. And the predicates of the relative clauses are usually intransitive verbs. In addition, the kinship terms cannot construct the qualified genitive relative clauses. Finally, the research explains these restrictions in terms of the theory of prominence condition.

Keywords: Relational nouns · Genitive relative clauses · Prominence condition

1 Introduction

This study focuses on a single class of constructions - attribute clauses in Mandarin Chinese, which a noun or a noun phrase is modified by an adnominal subordinate clause to form a complex noun phrase. In typology, attribute clauses can be classified according to the gaps in the clauses. If a clause has a gap which co-indexes with an argument, then this clause is a relative one, like (1a). While if there is no gap in the clause, then it is noun-complement clause, like (1b).

(1) a. the girl who I met yesterday
     b. the fact that I met a girl yesterday

In typological studies of relative clauses, [1] has put forward the famous principle of relativization: *noun phrase accessibility hierarchy*, which reveals that in the sequence of *subject, direct object, indirect object, oblique, genitive and object of comparison*, the possibility of relativization of the noun phrases reduces progressively. Simultaneously, the grammatical strategy of relativization turns complicated gradually. Likewise, this principle works to the relativization patterns in Mandarin Chinese, except that the genitives in Chinese can be relativized [2, 3].

There are different kinds of genitive relative clauses in Mandarin Chinese, the majority of which are triggered by relational nouns. In these constructions, the subjects in the relative clauses are always relational nouns, such as attribute nouns and part nouns, which have correlations with other nouns [4], such as:

© Springer Nature Switzerland AG 2020
J.-F. Hong et al. (Eds.): CLSW 2019, LNAI 11831, pp. 746–752, 2020.
https://doi.org/10.1007/978-3-030-38189-9_75

(2) a. 面积 等于　　　 45平方厘米 的　三角形
    area be equal to   45cm^3 DE triangle
    the triangle whose area is equal to 45cm^3

    b. 三　个　边　相等　的　三角形
    three CL edge equal DE triangle
    the triangle whose edges are equal

This paper concentrates on this type of genitive relative clauses. The following sections will analysis the semantic and syntactic features of this construction, and furthermore, find the interpretation of the special performance of the genitive relative clauses in Mandarin Chinese.

2 The Restrictions on Genitive Relative Clauses

Though genitive can be relativized in Chinese, its relative clause is apparently different from subject or object relative clause semantically and syntactically. The difference can be reduced to three restrictions as follows.

2.1 Subject-Object Asymmetry

In the genitive relative clause triggered by relational nouns, the relational nouns usually appear as subjects rather than objects or other constituents. If the relational nouns serve as objects, the constructions will be incorrect.

(3) a. ? 我　算　　不　出　面积　的　三角形
    1sg calculate NEG out area DE triangle
    a triangle whose area I cannot calculate

    b. ? 你　要　测量　　三　个　边　的　三角形
    2sg. need measure three CL edge DE triangle
    a triangle whose edges you need to measure

Actually many previous studies have found that subjects are more likely to be extracted than objects in Mandarin Chinese [5, 6]. The asymmetry between subjects and objects can also be found in *island effect*:

(4) a. 这 位　作家，　写　的　书　非常　有意思。
    this CL author write DE book very interesting
    The books written by this author are very interesting.

    b. 这 家　出版社，　出版　的　图书　都　很　精美。
    this CL publisher publish DE book all very delicate
    The books published by this publisher are all very delicate.

The *island effect* describes a syntactic phenomenon that nouns embedded in relative clauses cannot be extracted out of the relative constructions. However, (4a-b) demonstrate that if the relative construction appears as the subject of the main clause, then the subject of the relative clause can be extracted out of the construction to be the topic in the main clause. However, this exception of island effect could not work on objects of relative clauses or main clauses. Therefore, we can see that the subject-object asymmetry influences the Chinese syntactic constructions systematically.

2.2   Asymmetry Between Transitive Verb and Intransitive Verb

Besides, the predicates in genitive relative clauses are usually intransitive verbs. According to Corpus of Center for Chinese Linguistics, in 93 genitive relative clauses triggered by relational nouns, 91.40% (85/93) constructions have intransitive verbs as their predicates. Moreover, most transitive verbs cannot serve as predicates in genitive relative clauses:

(5) a. ? 速度　提到　　了　80迈　的　车
   speed　increase　LE　80mph　DE　car
   the car whose speed has increased to 80mph

  b. ? 腿　碰　　了　凳子　的　人
   leg　bump　LE　stool　DE　person
   the person whose leg has bumped against a stool

In (5a-b), the subjects of the relative clauses are relational nouns, such as 车 (*speed*) and 腿 (*leg*). But both of these constructions are unacceptable because of the transitive verbs serving as predicates in the clauses.

2.3   Limitation of Kinship Terms

Typical relational nouns contain attribute nouns, part nouns and kinship terms. Though part nouns and attribute nouns can construct well-formed structures, kinship terms are limited in this construction. Most kinship terms cannot form genitive relative clauses, even if they serve as subjects of the clauses:

(6) a. *妈妈　很　爱　干净　的　孩子
   mum　really　like　clean　DE　child
   the children whose mums are really clean

  b. *妹妹　漂亮　的　男孩儿
   sister　pretty　DE　boy
   the boys whose sisters are pretty

  c. *儿子　不　听话　的　父母
   son　NEG　obedient　DE　parents
   the parents whose sons are not obedient

Based on the analysis above, the genitive relative clauses triggered by relational nouns are constrained by certain syntactic and semantic features: first, the relational nouns must serve as subjects of the relative clauses; second, the predicates of the relative clauses tend to be intransitive verbs; and third, kinship terms, which are also relational nouns, cannot trigger the genitive relative clauses.

3   The Relativization of Genitive and Prominence of Nouns

It has been pointed out that the relativization of nouns is psychologically real, which reflects the diverse difficulty of syntactic processing on various sentence constituents [1]. Influenced by this view, the new calculating method of filler-gap domain has been put forward. In this method, the different difficulties involved in the relativization of various nominal constituents can be explained reasonably [7]. According to this study, either the genitives of subject or object is hard to be relativized. Therefore, genitive relative clauses are not very easy to form, which generates a question for us: why Mandarin Chinese has genitive relative clauses? This section is devoted to solving this question and also explaining why there are certain constraints on the genitive relative clauses.

3.1   Prominence of Nouns

As existing researches point out, licence of a certain syntactic constituent is based on local syntactic structure as well as semantic and pragmatic conditions which could be called prominence [8, 9]. So both locality and prominence can affect the formation of constructions. Locality condition means that in syntactic system, the most local constituent is processed first, while prominence condition demands that the most prominent constituent should be processed preferentially. In Mandarin Chinese, the prominence condition is prior to locality condition. Because of this, there are lots of mismatch and replacement between agents and patients in Chinese. Turning to this

Table 1. Aspects on prominence of nouns.

Aspects	Level
Animacy	Human > inhuman (animal) > inanimate things > abstraction
Reference	Personal pronoun > proper name > definiteness > indefiniteness > non-specific reference
Person	First person, second person > third person

research, we believe that prominence condition can help to explain the formation of genitive relative clauses.

The prominence of nouns can be defined in many different perspectives, such as animacy, reference and person [9] (Table 1).

Moreover, the prominence of nouns is influenced by options for topic in a context [10]. When there is only one noun as the option to be topic, then the entity which the noun refers to has the highest prominence. Nevertheless, if there are several nouns waiting to be chosen as the topic of the context, the prominence of a certain noun will decrease.

In summary, the facts that impact on the prominence of nouns can be reduced as follows: (a) the syntactic and semantic features of nouns, and (b) the amount of nouns in a clause. Hereby, the restrictions on the genitive relative clauses triggered by relational nouns can be interpreted uniformly by means of prominence condition. That is, all the restrictions are the means for increasing prominence of the relational nouns in relative clauses, because only when the relational nouns have the highest prominence, can the constructions be well-formed.

3.2 The Interpretation of Genitive Relative Clauses from Perspective of Prominence Condition

The relational nouns serve as subjects in genitive relative clauses, and the predicates of the clauses are intransitive verbs. Both of the features can be explained by prominence condition. Because when the relational nouns work as subjects of intransitive verbs, there will be no other nouns in the clauses competing to get more prominence. However, if the relational noun appears as object or the predicate of the clause is a transitive verb, then, there must be another noun competing with the relational noun. Based on the information structure of Chinese, subjects are always definite and code old information. At the same time, objects tend to be indefinite and code new information. Moreover, most relational nouns refer to abstractions, which must be less prominent than the objects of transitive verbs. And then, in this situation the prominence of relational nouns will decrease. Though, if there is a subject that is inanimate and indefinite, while the relational noun object is definite, then the relational noun will be more prominent in the clause, which leads to a more acceptable genitive relative clause, such as:

(7) 谁　　都　算　　不　出　得数　的　方程
whoever all calculate NEG out result DE formula
a formula whose result no one can calculate

The relational noun 得数 (result) serves as the object of the relative clause in (7). And this construction seems well-formed. Because the subject 谁 (whoever/no one) is indefinite, which is less prominent than the object relational noun, so the relational noun still gets high prominence which may guarantee the eligibility of the genitive relative clause.

When the predicates of the clauses are transitive verbs, the relational noun subjects are always less prominent. So this kind of clauses tend to be incorrect. However, if the objects of the intransitive verbs are indefinite or non-specific, the construction can be accepted:

(8) a. 成绩　没　达到　标准　　的　学生
grade NEG get standard DE student
the students whose grades did not reach the standard
　　b. 双手　撑　杠　的　体操　　运动员
hands prop bar DE gymnastics athlete
the gymnast whose hands are propping against the bar

In (8a-b), though the predicates of the relative clauses are transitive verbs, the objects are less prominent, so the whole constructions are still well-formed. Due to the requirement that the relational nouns should have the highest prominence in the constructions, the genitive relative clauses are used to describe the attribute of entities. Furthermore, from the perspective of pragmatic function, relative clauses are informative relative clauses which are used to describe the head nouns and add new background information.

3.3   The Particularity of Kinship Terms

Kinship terms are usually definite in context and most of the time they refer to human beings, so kinship terms should have high prominence. But this kind of relational nouns is so restricted in genitive relative clauses. Coincidentally the restriction happens in topic construction as well.

It has been discovered that in Mandarin Chinese the kinship terms are hard to be topicalized [11]:

(9) a. * 张三，   我 认为 妹妹 很   漂亮。
       Zhangsan 1sg think sister very pretty
       Speak of Zhangsan, I think his sister is very pretty.
   b. *小张   我 看见 了 爸爸。
       Xiaozhang I   see   LE dad
       Speak of Xiaozhang, I saw his dad.

The study gave a plausible explanation of this phenomenon: the kinship terms are very special compared to other relational nouns in their self-center meaning. This meaning makes the kinship terms prone to be interpreted from the perspective of speakers, i.e. the kinship terms are prior to be understood as the relatives of speakers. Because of this, kinship terms are usually hard to be topicalized. Similarly, the kinship terms in genitive relative clauses tend to be interpreted as the speaker's rather than the head noun's, therefore the whole construction is misunderstood.

4   Conclusion

In Mandarin Chinese, the possessors of subjects can be relativized to form genitive relative clauses. This construction is triggered by the relational noun. Typical relational nouns contain attribute nouns, part nouns and kinship terms, but the kinship terms are restricted in genitive relative clauses. Besides, this construction has two other features: first, the relational nouns always serve as subjects in relative clauses; and second, the predicates of the clauses tend to be intransitive verbs. This research focuses on the explanation on the formation and restrictions of genitive relative clauses in terms of prominence condition. The prominence of relational nouns exerts profound influence on genitive relative clauses. That is, all the syntactic and semantic restrictions are used to guarantee that the relational nouns in the relative clauses have the highest prominence, which is the basic prerequisite for the formation of genitive relative clauses.

Acknowledgements. I am grateful to the anonymous reviewers of CLSW 2019 for helpful suggestions and comments. This research was supported by the Ministry of Education (MOE) Humanities and Social Sciences Fund of China under grant No. 18YJC740058 (Metadiscourse Function of Mandarin Emphatic Markers) to Lu Ying. All errors remain my own.

References

1. Keenan, E., Comrie, B.: Noun phrase accessibility and universal grammar. Linguist. Inq. **8**(1), 63–99 (1977)
2. Xu, Y.: Noun phrase accessibility and relativization in Chinese and English: a contrastive study from the perspective of linguistic typology. Foreign Lang. Teach. Res. **5**, 643–657 (2012). (in Chinese)
3. Ning, C.: The overt syntax of relativization and topicalization in Chinese, Ph.D. dissertation, University of California, Irvine (1993)
4. Yuan, Y.: A cognitive study on one-valence nouns. Stud. Chin. Lang. **4**, 241–253 (1994). (In Chinese)
5. Huang, C.-T.J.: Logical relations in Chinese and the theory of grammar, Ph.D. dissertation. MIT, Cambridge (1982)
6. Huang, C.-T.J.: On the distribution and reference of empty pronouns. Linguist. Inq. **15**(4), 531–547 (1984)
7. Hawkins, J.: Cross-linguistic Variation and Efficiency. Oxford University Press, Oxford (2014)
8. Hu, J.: Prominence and locality in grammar: the syntax and semantics of wh-question and reflexives. Ph.D. dissertation, The City University of Hong Kong, Hong Kong (2002)
9. Hu, J.: The distribution and selection of arguments: prominence and locality in grammar. Stud. Chin. Lang. **1**, 3–20 (2010). (in Chinese)
10. Givón, T.: Topics, pronoun and grammatical agreement. In: Li, C.N. (ed.) Subject and Topic, pp. 149–188. Academic Press, New York (1976)
11. Lu, S., Pan, H.: The licensing conditions of Chinese possessive topicalization. Contemp. Linguist. **1**, 15–30 (2014). (In Chinese)

The Construction of Interactive Environment for Sentence Pattern Structure Based Treebank Annotation

Shiyu Guan, Weiming Peng[(⊠)], Jihua Song[(⊠)], and Zhiping Xu

Beijing Normal University, No. 19, XinJieKouWai St., Haidian District,
Beijing 100875, People's Republic of China
{pengweiming, songjh}@bnu.edu.cn

Abstract. In the construction of treebanks, manual annotation is inefficient, and unable to ensure the consistency of results. Based on the existing graphical syntax annotation platform for the sentence pattern structure, this paper adds the automatic syntax analysis function and constructs a human-computer interactive annotation environment. It shows that, compared with manual mode, human-computer interactive mode can greatly improve the efficiency.

Keywords: Treebank · Sentence pattern structure · Automatic parsing · Human-computer interaction

1 Introduction

In the research of Theoretical Linguistics and Computational Linguistics, treebanks are important resources. In Theoretical Linguistics, treebanks provide linguists with a large amount of data based on real language materials, which lays a solid foundation for verifying and improving traditional linguistic theories. In Computational Linguistics, automatic parsing and various upper-level applications rely on the size and quality of treebanks.

At present, according to different grammar systems, treebanks can be divided into two categories: phrase structure treebanks and dependency structure treebanks. After years of research, the accuracy of automatic parsing has reached about 90%. So, the construction of treebanks usually adopts the mode of computer automatic parsing and manual correction to improve the efficiency, especially dependency structure treebanks [1]. Because the dependency structure directly annotates the syntactic relationships in the word sequence, and it is relatively convenient to modify the errors of automatic parsing.

Since the introduction of the sentence pattern structure system [2], a treebank of 100,000 sentences has been established using the graphical syntax annotation platform. The platform can complete the graphical annotation of sentence pattern structure through simple sentence component segmentation operations. However, at present, the mode of annotation is purely manual, and the efficiency still needs to be improved.

J.-F. Hong et al. (Eds.): CLSW 2019, LNAI 11831, pp. 753–763, 2020.
https://doi.org/10.1007/978-3-030-38189-9_76

Like phrase structure, sentence pattern structure also highlights the hierarchy of sentences. Taking this feature into account, this paper will build an interactive environment for sentence pattern structure treebank annotation, and improve the efficiency of treebank construction by using the human-centered and computer-assisted annotation mode.

2   Overview of Sentence Pattern Structure

2.1   Grammar System

Sentence pattern structure is a formal structure system based on Mr. Li Jinxi's sentence-based grammar. Figure 1 shows diagrams of the basic sentence pattern, the extended sentence pattern and the complex sentence pattern. These sentence patterns are the basis of the sentence pattern structure.

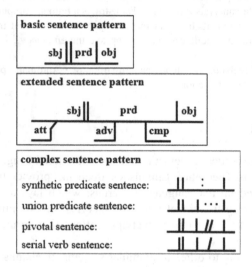

Fig. 1. Sentence pattern classification in the sentence pattern structure system

For the convenience of the following article, Table 1 lists element marker set used in the sentence pattern structure system.

Table 1. Element marker set in the sentence pattern structure system

Tag	Component	Tag	POS
ju	Sentence	**n**	Noun
xj	Clause	**t**	Time word
sbj	Subject	**f**	Localizer
prd	Predicate	**m**	Numeral
obj	Object	**q**	Measure word
att	Attribute	**r**	Pronoun
adv	Adverbial	**v**	Verb
cmp	Complement	**a**	Adjective
ind	Independent element	**d**	Adverbial
pp	Position of preposition	**p**	Preposition
cc	Position of conjunction	**c**	Conjunction
uu	Position of auxiliary word	**u**	Auxiliary word
ff	Position of localizer	**e**	Exclamation
		o	Onomatopoeia
		w	Punctuation
		x	Default

2.2 Graphical Annotation Platform

With the evolution of the grammar system, the graphical annotation platform has undergone several versions. Yang constructed the first version, which shows the hierarchical structure of sentences from trunk to branches and leaves in a form close to human understanding [3]. Zhao redesigned the platform to make the annotation process more efficient in view of the shortcomings of the first version in annotation mode and system design [4]. The latest annotation platform is mainly Zhao's version.

Figure 2 illustrates in detail how purely manual mode accomplishes the sentence pattern structure annotation of a sentence. As can be seen in the figure, in order to divide a component from a sentence, the annotator just needs to place the cursor at the boundary between the component and other sentence components, and then clicks the corresponding component button. This method is convenient, but the components need to be divided one by one, so the annotator needs to constantly locate the cursor and divide a component.

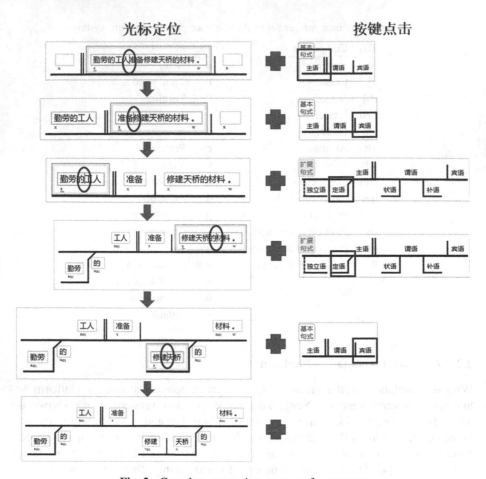

Fig. 2. Complete annotation process of a sentence

Sentence pattern structure emphasizes the concept of hierarchy, which can be seen in the diagram result: a long horizontal line corresponds to a sentence level, and components at the same level are linearly arranged; the vertical distance between long horizontal lines corresponds to the depth of the sentence level [5].

The results of graphical annotation can be stored in the form of sentence structure expression [6]. Sentence structure expression marks sentence components by inserting separators in the text: attributive, adverbial and complement are surrounded by parentheses, middle parentheses and angle parentheses respectively; the separators of trunk components and function words are basically consistent with the graphical form. In this paper, the results of automatic sentence pattern structure analysis will be given in the form of sentence structure expression. Figure 3 shows the annotation result of the sentence "您对国家不允许名人以患者的身份出现在广告中的规定怎么看?".

The sentence structure expression corresponding to each level is as follows:

(1) The 0th level: 您||看?
(2) The 1th level: 您||[对∧规定][怎么]看?
(3) The 2th level: 您||[对∧(国家||允许|名人//出现▷的)规定][怎么]看?
(4) The 3th level: 您||[对∧(国家||[不]允许|名人//[以∧身份]出现<在∧广告□中>▷的)规定][怎么]看?
(5) The 4th level: 您||[对∧(国家||[不]允许|名人//[以∧(患者▷的)身份]出现<在∧广告□中>▷的)规定][怎么]看?

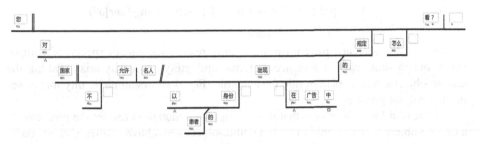

Fig. 3. Graphical annotation result of a multi-level sentence

3 Sentence Pattern Structure Analysis Based on Word and Part of Speech

3.1 Analysis Principle

Sentence pattern structure analysis is aimed to find out all kinds of sentence components in a sentence under the constraints of sentence structure. The general sentence pattern structure of Chinese sentences is "attributive + subject + adverbial + predicate + complement + attributive + object". According to a large amount of annotation experience, each component has a part of speech tendency, such as:

(1) Subject and object: Noun or pronoun.
(2) Predicate: Verb or adjective. When the predicate is an adjective, the sentence has no object.
(3) Attributive: Adjective, quantifier, noun or demonstrative pronoun.
(4) Adverbial: Adverb, time word or prepositional phrase.
(5) Complement: The structure guided by the word "得".

According to the above understanding, a regular expression can be used to describe each sentence component. A complete regular expression is formed by combining the regular expressions of each component in order. It is stored as a rule in the rule base for automatic analysis of sentence pattern structure. The following is a sentence pattern rule for double-object sentences:

- $(mq|[anr]u?) * [nr] v [nr](mq|[anr]u?) * [nr]$

When analyzing sentence pattern structure, the part of speech sequence of word segmentation is used as input for regular matching. In order to record the component information of each part and its matching subsequence, the actual rules use "Named Group" [7]. When the regular matching succeeds, the result sentence structure expression can be obtained by adding the component separator to the corresponding text according to the component information of the captured content. The actual rule is as follow:

- $(? <att> mq|[anr]u?)$
  $*(? <sbj> [nr])(? <prd> v)(? <obj> [nr])(? <att> mq|[anr]u?)$
  $*(? <obj> [nr])$

Most of the sentence pattern rules can only restrict the part of speech of components, but in some special sentence patterns, ambiguity can easily arise. Taking the double-object sentence pattern rule for example, Fig. 4 shows an ambiguity that arises in the analysis process.

The reason for ambiguity is that it is not rigorous enough to restrict the predicate of a double-object sentence only to verb. Double-object verbs have relatively clear scope, usually with granting or teaching meanings. Therefore, it is necessary to restrict words for specific components in a regular expression. This is especially obvious in the rules of complex sentence patterns:

(1) Synthetic predicate sentence: The predicate is "auxiliary verb/copula verb + verb".
(2) Pivotal sentence: The predicate verbs are mainly divided into two categories: the imperative meaning words and the affirmative meaning words, such as "请", "让" and "称".
(3) Union predicate sentence: Some conjunctions between two verbal phrases, such as "而", "越……越……".
(4) Serial verb sentence: The first verb is a trend verb such as "来" and "去".

Words in specific components are extracted from the annotated large-scale corpus and form vocabulary sets after manual screening (e.g. double-object verb set "vooprd"). In complex sentence pattern rules, vocabulary sets are used to replace part of speech in those specific components.

However, the traditional regular matching mechanism will no longer be applicable to the word and part of speech mixed rules. One is that the input is no longer a part of speech sequence, but a "word/part of speech" sequence; the other is that the traditional regular expression cannot express words.

3.2 System Implementation

In order to match the word and part of speech mixed rules with the "word/part of speech" sequence, the system extends the traditional regular expression. The new regular expression retains the basic function of string matching and achieves the matching of multi-attribute atom sequence.

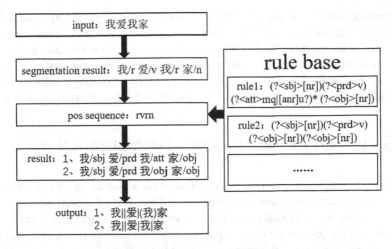

Fig. 4. Sentence pattern structure analysis based on part of speech

In the new regular expression, a word is defined as a "Word Object" for the convenience of attribute expansion. It contains two attributes: word and part of speech. A sentence will be transformed into "Word Object" sequence after word segmentation as shown in Figs. 5 and 6.

There are two main differences between the new regular expression and the traditional regular expression:

(1) Traditional regular matching is to determine whether the characters in the string match the symbols in the corresponding position of the regular expression. In the new regular expression, as long as an arbitrary attribute of the "Word Object" matches the symbol in the corresponding position, it is considered that the current "Word Object" matches the current symbol successfully. Table 2 illustrates this difference by taking "subject" matching for example.

Table 2. Differences between character matching and "word object" matching

	traditional	new	new
regular expression	(?<sbj>[nr])	(?<sbj>[nr])	(?<sbj>[你我他])
input	r	word: 我 pos: r	word: 我 pos: r
result	success	success	success

(2) The traditional regular expression is based on characters, which cannot reflect the concept of word. For example, the regular expression "可以" represents two characters "可" and "以". If it is used to match "Word Object" sequence, the

sequence should consist of a "Word Object" with word of "可" and a "Word Object" with word of "以". But the real requirement is to match only one "Word Object" with word of "可以". Therefore, the new regular expression defines a new form to express a word by wrapping a word of arbitrary length with "[#" and "]". It can also be used to represent a vocabulary set by separating words with "|". For example, the regular expression of the double-object verb set is as follows:

- (?<prd>[#告诉|授予|给予|赋予|赔偿|请教|通知|赠|还|送|给|欠|找|借|叫])

With such a regular expression that can match the sequence of "Word Object", we can analyze the sentence pattern structure by word and part of speech mixed rules. We still take the double-object sentence pattern for example to illustrate the form of rules in the system and the process of sentence pattern structure analysis. For a specific double-object sentence pattern, the following rule is added to the rule base:

- att + sbj + adv + vooprd + obj + att + obj

Inside this rule:

(1) att: (?<att>[rm]m?q?|[antrv]的?)*
(2) sbj: (?<sbj>[nr])
(3) adv: (?<adv>[vadt]地?|p[ntr]f?)*
(4) vooprd: (?<prd>[#告诉|授予|给予|赋予|赔偿|请教|通知|赠|还|送|给|欠|找|借|叫])
(5) obj: (?<obj>[nr])

When the input sentence is "我爱我家", the word segmentation result is shown in Fig. 5. During the matching process, "vooprd" cannot find the matching item, so the matching fails.

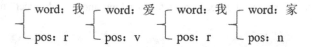

Fig. 5. The word segmentation result of "我爱我家"

When the input sentence is "他给我钱", the word segmentation result is shown in Fig. 6. During the matching process: The part of speech of the first word "r" matches "sbj"; The word of the second word "给" is in the double-object verbs set, so it matches "vooprd"; The part of speech of the third word "r" matches "obj"; The part of speech of the fourth word "n" matches "obj"; The matches of "att" and "adv" may not exist. So, the sentence matches the rule successfully and outputs the sentence structure expression "他||给|我|钱" as the analysis result.

word：他 word：给 word：我 word：钱
pos：r   pos：v   pos：r   pos：n

Fig. 6. The word segmentation result of "他给我钱"

In the system, basic sentence pattern rule base, extended sentence pattern rule base and complex sentence pattern rule base are established respectively. For attributive, adverbial, complement and other components, phrase pattern rule bases are also established. The rule bases of the system can analyze not only complete sentences of the initial input, but also the phrases generated in the annotation process.

4 Interactive Environment

For dependency structure, when the automatic parsing achieves high accuracy, the annotation mode of computer automatic parsing and manual correction can greatly improve the annotation efficiency. However, for phrase structure and sentence pattern structure, the same mode is no longer suitable considering the feature of hierarchical nesting. Only the mode of analysis and correction layer-by-layer can improve the efficiency practically.

Therefore, based on the existing platform, we construct an interactive annotation environment, which is human- centered and computer-assisted. It has the following three features:

(1) Only the single-layer structure of sentences will be analyzed.
(2) It can analyze not only complete sentences but also phrases of various components.
(3) For an input text, the system can provide zero or one or more analysis results.

The above (1) and (2) ensure that in the annotation environment, sentences are analyzed and corrected layer by layer. For a complete sentence, the system first analyzes its single-layer structure. After correction, the annotator takes the modifiers as input to analyze its structure. (3) explains that the system is only responsible for providing the reference analysis results. In the case of limited size of rule base, the system may not be able to analyze the current text. At that time, the annotator still needs to annotate manually. When the system can provide the reference analysis results, the annotator needs to decide whether to adopt the results. If a result is adopted, one-key annotation can be achieved by choosing it. If none is adopted, manual annotation is still needed. So, in this annotation environment, annotator's judgment still occupies a dominant position. Figure 7 illustrates the annotation process in this interactive environment.

This is the same sentence as the example in Fig. 2. A thorough analysis of sentence pattern structure goes through two human-computer interaction processes. It only needs two focus selection operations and two result selection operations to complete the annotation. A lot of time was saved compared with the manual mode in Fig. 2.

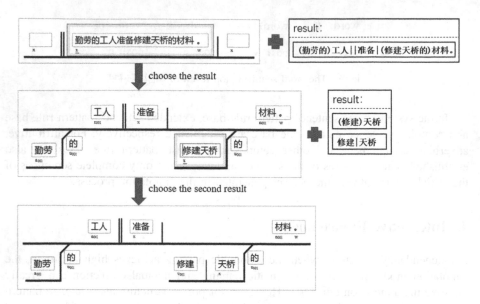

Fig. 7. Interactive annotation process based on sentence pattern structure

5   Conclusion

This paper constructs an interactive environment for sentence pattern structure based treebank annotation, which replaces manual annotation mode with a human-centered and computer-assisted mode. In this environment, the computer uses word and part of speech mixed rules to analyze the single-level structure of sentences and phrases. When the analysis result is correct, the annotator can select it to realize the one-key annotation of single-layer structure. At present, the annotation efficiency can reach 10224 words (345 sentences)/day (7 h). Compared with the purely manual annotation mode of 4128 words (238 sentences)/day (7 h), the efficiency has been greatly improved.

However, current rules can only analyze some simple sentences. On the one hand, it is impossible to determine the component only based on word and part of speech information. On the other hand, current rules do not fully consider the case of hierarchical nesting. So, the existing rules can just analyze limited sentences and are prone to ambiguity. In future research, we will add semantic information to rules to improve this problem.

References

1. Qiu, L.: Theoretical study and practice on multi-view Chinese treebank building. Postdoctoral Research Report of Peking University (2012)
2. Peng, W., Song, J., Sui, Z.: Formal schema of diagrammatic Chinese syntactic analysis. In: Workshop on Chinese Lexical Semantics, pp. 701–710. Springer, Cham (2015)

3. Tianxin, Y., Weiming, P., Jihua, S.: High efficiency syntax tagging system based on the sentence pattern structure. J. Chin. Inf. Process. **28**(4), 43–49 (2014)
4. Min, Z., Weiming, P., Jihua, S.: Development and optimization of syntax tagging tool on diagrammatic treebank. J. Chin. Inf. Process. **28**(6), 26–33 (2014)
5. He, J., Peng, W., Song, J.: Digitalization of Chinese sentence structure patterns: improvement of sentence-based grammar and diagrammatic syntactic analysis. J. Beijing Norm. Univ. (Nat. Sci.) **52**(4), 413–419 (2016)
6. Shuqin, Z., Weiming, P., Jihua, S.: The extraction of Chinese sentence pattern instance based on diagrammatic treebank. J. Chin. Inf. Process. **31**(5), 32–39 (2017)
7. Friedl, J.E.F.: Mastering Regular Expressions, 2nd edn. Oreilly Media (2002)

Semantic Representations of Terms in Traditional Chinese Medicine

Qinan Hu[1,2], Ling Zhu[3(✉)], Feng Yang[4], Jinghua Li[3], Qi Yu[3], Ye Tian[3], Tong Yu[3], and Yueguo Gu[1,2]

[1] Institute of Linguistics & Research Center for Lexicology and Lexicography, Chinese Academy of Social Sciences, Beijing, China
[2] China Multilingual and Multimodal Corpora and Big Data Research Centre, Beijing, China
[3] Institute of Information on Traditional Chinese Medicine, China Academy of Chinese Medical Sciences, Beijing, China
[4] Institute of Acupuncture and Moxibustion, China Academy of Chinese Medical Sciences, Beijing, China
qinan.hu@qq.com, jjzhuling@163.com, yangfengzhuling@163.com, zingarlee@hotmail.com, yuqi0948@sina.com, weierty@163.com, yutongoracle@hotmail.com, gyg@beiwaionline.com

Abstract. Word embeddings have been widely used in lexical semantics and neural networks in Natural Language Processing. This article investigates the semantic representations using word embedding technologies by verifying them on a human constructed domain ontology. The domain of Traditional Chinese Medicine (TCM) is used as a workbench in this study, because this domain is knowledge-rich and has a large-scale domain ontology with well-defined entity types and relation types. This article releases a dataset, named "TCMSem", to capture TCM domain experts' intuitions of semantic relatedness. This data set is designed to cover the medical entities and relations with as many semantic types as possible so as to initiate a diverse and comprehensive evaluation on word embeddings. Experimental results show that word embeddings have demonstrated higher proficiencies in the detection of synonyms and collocations than other types of semantic relations. Furthermore, the semantic relatedness of thousands of terms of major categories in TCM is visualized using the taxonomy defined in the ontology.

Keywords: Semantic representation · Word embeddings · Evaluation · TCMSem

1 Introduction

Nowadays, word embeddings take a predominant role as the inputs to Natural Language Processing (NLP) tasks, such as text classification [1], question answering [2], machine translation [3], and the construction of knowledge graphs [4].

This study investigates the semantics represented with word embeddings by verifying them on a human constructed domain ontology. The main components

© Springer Nature Switzerland AG 2020
J.-F. Hong et al. (Eds.): CLSW 2019, LNAI 11831, pp. 764–775, 2020.
https://doi.org/10.1007/978-3-030-38189-9_77

of an ontology are entities and relations. The entity type is a semantic classification of entities and concepts. Semantic relations [6,7] are the relationships between entities, which can be used to form a semantic network to represent the domain knowledge.

Being a knowledge-rich domain, Traditional Chinese Medicine (TCM) has a large-scale domain ontology with well-defined entity types and relation types. Researchers have defined as many as 58 relation types and 51 entity types in Traditional Chinese Medicine Language System (TCMLS) [14,15]. TCMLS provides a workbench to observe word embeddings' capabilities to capture the semantics of the entities and relations of various types.

Synonyms are ubiquitous in TCM literatures. Take the term 失眠 (insomnia) as an example. A doctor of TCM may use many different terms such as 不寐 (sleeplessness), 难于安卧 (difficult to repose), 寐差 (poor sleep quality) 辗转难眠 (toss about in bed), 失寐 (unable to sleep), 眠少 (little sleep) to express this symptom. It's observed there are even more than 1,500 terms to express the symptom of insomnia in TCM.

There have been several studies on TCM from the lexical semantic perspective [8–12]. They generally use literal string matching to measure semantic similarities among terms. However, semantic relatedness in TCM has gone far beyond literal similarities. It originates not only from synonyms, but also from antonyms, hypernyms, hyponyms, collocations and analogical words, among many others [13].

On the other hand, medical literatures have played an important role in the inheritance of medical knowledge from generation to generation in TCM. Physicians in TCM record the symptoms, syndromes, formula and many other useful information in terms of written texts in their researches or daily clinical practices. These literatures include valuable semantic information.

2 Related Studies

Word embeddings represent a word as a low-dimensional dense vector with continuous values, making them more suitable for computation in neural networks. The hypothesis underlying is that words occurring in the same contexts tend to have similar meanings [16]. In other words, word embeddings capture semantic relatedness based on the contexts of words, rather than their literal forms.

The methods of generating word embeddings have emerged one after another in recent years [17–21]. These methods stand in two lines, i.e. "predictive" [17,19–21] and "count-based" [18].

Predictive models learn word vectors by minimizing the loss of predicting the target words from the context words (or vice versa) given the vectors learned. Word2Vec [17] is predictive and the most widely used word embedding model. Although Word2Vec has a straightforward architecture, it has beaten many computationally heavy models. FastText [21] takes advantage of the morphology of words. It learns embeddings for character n-grams, and uses the sum of n-gram embeddings to represent a word. It accordingly allows to compute word embeddings for out-of-vocabulary words.

Count-based models essentially conduct dimensionality reduction on a co-occurrence matrix, which is build upon a large corpus. A row in the matrix corresponds to a word in the vocabulary; a column, a context; and a value, the frequency of co-occurrences. This matrix is then decomposed to yield a lower-dimensional matrix. Count-based models learn word vectors by minimizing the loss of decomposition to get the lower-dimensional representations which can explain most of the variances in the high-dimensional representations. GloVe is a popular "count-based" model [18].

There have been several studies on word embeddings for Chinese. Li et al. [5] build a dataset conveying the morphological and semantic relations in Chinese. They explore the influences of vector representations, context features, and corpora on word embeddings using analogical reasonings. The reasonings are defined on the basis of one-to-one semantic relations, such as the relationships between countries and capitals [29]. However, it is difficult to apply analogical reasoning tasks to the TCM domain. Because there are massive one-to-many semantic relations in TCM domain, the answers to analogical reasoning tasks are typically non-unique. For example, the same symptom may be caused by several different syndromes, giving rise to multiple reasoning answers.

Word embeddings have also shed a light on TCM-related studies. They are utilized as knowledge-enriched inputs to down-stream TCM applications. Zhu et al. [22] adopted a deep learning approach to damp-heat syndrome differentiation using medical records represented in terms of word embeddings and TF-IDF. Yao et al. [23] studied medical record classification using document embedding technologies enriched with domain knowledge [24, 25].

3   TCMSem

An evaluation on word embeddings [26–28] is either extrinsic or intrinsic. In an extrinsic evaluation, word embeddings are used as the inputs to a downstream task. They are evaluated indirectly using changes of the performance of that specific task. Consequently, the extrinsic evaluation results are task-specific.

This article adopts an intrinsic approach. In an intrinsic evaluation, word embeddings are used to measure the semantic relatedness between words in a

Table 1. TCMSem sample entries

Entity 1	Entity 2	Score	Relation Type	Entity Type 1	Entity Type 2
神倦 (lassitude of the spirit)	精神不振 (low spirits)	8.33	Near-synonym	Symptom	Symptom
人参 (Ginseng)	安神 (soothe the nerves)	6.67	Property of	Property of	Function of Chinese medicinal
牡蛎 (Oyster)	龙骨 (fossil fragments)	6.00	Co-occur with	Chinese medicinal	Chinese medicinal
睛明穴 (Jingming point)	腹泻 (diarrhea)	0.00	N.A.	Acupuncture point	Symptom

Table 2. Correlation strengths of different relation types

Relation type	Correlation
同义词 synonym	9.64
现象表达 manifestation of	6.29
治疗 treat	5.82
归经 meridian entry	5.5
有... 特性 property of	5.31
父子关系 is a	5.08
较大概率地同时存在 cooccur with	4.59

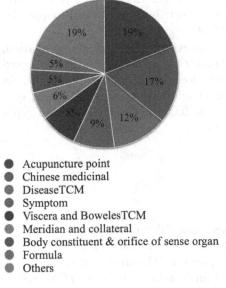

- Acupuncture point
- Chinese medicinal
- DiseaseTCM
- Symptom
- Viscera and BowelesTCM
- Meridian and collateral
- Body constituent & orifice of sense organ
- Formula
- Others

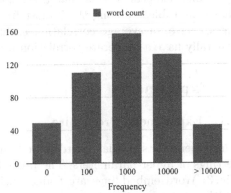

Fig. 1. Distributions of major entity types

Fig. 2. Distributions of word frequencies

predefined word list [29]. In this study, the relatedness will be compared against a man-made ground-truth using correlation coefficients.

In previous studies, researchers have proposed several data sets [29,30] for intrinsic evaluations. These data sets differed in word types, relation types, and scales, ranging from dozens to thousands of words. Schnabel et al. pointed out that these data sets are often dominated by specific types of words, and poorly calibrated to corpus statistics, giving rise to biased evaluation results [27]. They emphasized on the diversity in data sets, and proposed to increase diversity by selecting words of various frequencies, parts of speech, and abstractness vs. concreteness.

To capture TCM domain experts' intuitions to semantic relatedness, a man-made ground-truth named "TCMSem" is released. It is composed of 500 triples (*entity1, entity2, score*) representing the averaged *score* of semantic relatedness between word pairs *entity1* and *entity2* assigned by TCM experts, as illustrated in Table 1.

Following [27], the data set is designed to cover the semantic types of medical entities and relations defined in TCMLS [14,15] as many as possible to inspire a diverse and comprehensive evaluation. TCMSem includes 37 relation types and 51 entity types in total. The distributions of the entity types in TCMSem are shown in Fig. 1.

When domain experts further construct pairs of words complying with the selected relation types and entity types, they also take the diversity of word frequencies into consideration. Figure 2 shows the distributions of word frequencies in the corpus *700 ancient literatures on TCM*.

Correlation scores are rated on a 0~10 scale, wherein 0 stands for no relation, and 10, the most highly related. A close observation on the correlations assigned by human experts reveals that these correlations are of various strengths. As shown in Table 2, among the 7 most frequent relation types in TCMSem, synonyms have an average correlation as high as 9.64; while other relation types generally have a moderate correlation around 5.

4   Experiments

4.1   Experiment Procedure

The experiments evaluate word embeddings using different corpora and different settings of hyper-parameters. The procedure of the experiments is described in Fig. 3. Word embeddings are trained using two TCM corpora of varying specialties and sizes. One is composed of 700 ancient literatures on TCM, which is publicly available on the web (in Baidu wenku). The other is the Formula database from CINTCM (http://cowork.cintcm.com/engine/windex.jsp), which has a specialty in symptoms and formulas. The former contains 90 million Chinese characters; and the latter, 13 million Chinese characters.

The tokenizer in Language Technology Platform (https://github.com/HIT-SCIR/ltp) is used to tokenize the corpora. This tokenizer is equipped with a glossary of more than 170,000 terms in TCM. In addition, because fastText is capable to generate embeddings based on character n-grams, the corpora are also tokenized into characters for fastText.

This article investigates the semantic representations using three word embedding technologies, i.e. Word2Vec, GloVe, and fastText. It computes semantic relatedness between a pair of words defined in TCMSem using cosine similarity. The similarities are further compared against the scores assigned by domain experts in TCMSem, using Spearman's rank correlation coefficient.

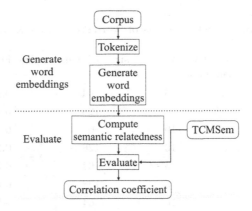

Fig. 3. Experiment procedure

4.2   Experimental Results

In general, the correlation coefficients shown in Table 3 indicate that the consistency between word embeddings and TCMSem are in the following order: GloVe < Word2Vec < fastText (word) < fastText (character). Specifically, the semantic relatedness captured by the character-level skip-gram model in fastText shows the highest consistency with experts' intuitions to semantic relatedness.

The corpora have different coverages. 86.98% of word pairs defined in TCM-Sem are covered in the word embeddings generated by Word2Vec and GloVe using the corpus of 700 ancient literatures; and only 55.03%, using the formula corpus. Word2Vec and GloVe have shown a significant preference to the larger corpus, because it gives rise to less out-of-vocabulary words. On the other hand, due to the compositionality of the word embeddings generated using fastText, most words in TCMSem can have a reasonable embedding no matter whether it appears in the training corpus or not. Accordingly, the preference to larger corpus is not salient in word embeddings generated by fastText.

The word embeddings are evaluated generated using word2vec on different entity types, as shown in Fig. 4. Among the 8 most frequent entity types in TCMSem, the semantic relatedness of "acupuncture points" and "meridians and collaterals" are captured more accurately than other entity types. It is observed that these two entity types generally have fixed contexts, since they are semantically related to fewer types of entities than other entity types.

The word embeddings are further measured on different relation types. As shown in Fig. 5, among the 7 most frequent relation types in TCMSem, word embeddings have demonstrated higher proficiencies in the detection of synonyms and collocations than other types of semantic relations. An explanation is that similar contexts exist for synonyms and collocations, while the contexts for other semantic relation types are more diverse.

Table 3. Evaluation results of Word2Vec, GloVe & fastText

No.	Model	Negative sampling	Window size	Dimension size	Formula	700 Ancient Books
1	word2vec	5	5	64	0.16	0.21
2	word2vec	5	5	128	0.17	0.25
3	word2vec	5	10	64	0.16	0.26
4	word2vec	5	10	128	0.21	0.29
5	GloVe		5	64	0.00	−0.07
6	GloVe		5	128	0.01	−0.03
7	GloVe		10	64	−0.02	−0.03
8	GloVe		10	128	0.01	−0.02
9	fastText (word)	5	5	64	0.25	0.22
10	fastText (word)	5	5	128	0.24	0.20
11	fastText (word)	5	10	64	0.29	0.23
12	fastText (word)	5	10	128	0.25	0.21
13	fastText (word)	10	5	64	0.26	0.21
14	fastText (word)	10	5	128	0.24	0.20
15	fastText (word)	10	10	64	0.28	0.23
16	fastText (word)	10	10	128	0.24	0.20
17	fastText (char)	5	5	64	0.29	0.28
18	fastText (char)	5	5	128	0.28	0.24
19	fastText (char)	5	10	64	0.31	0.30
20	fastText (char)	5	10	128	0.30	0.28
21	fastText (char)	10	5	64	0.29	0.29
22	fastText (char)	10	5	128	0.29	0.29
23	fastText (char)	10	10	64	**0.32**	0.30
24	fastText (char)	10	10	128	0.31	0.29

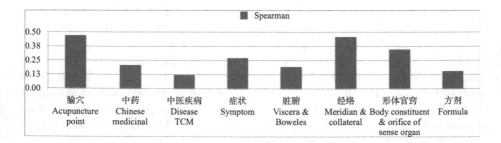

Fig. 4. Word2Vec coefficients on different entity types

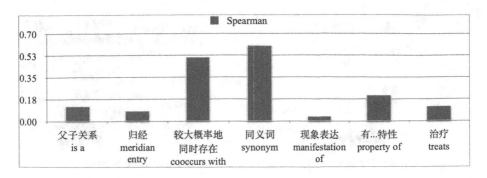

Fig. 5. Word2Vec coefficients of different relation types

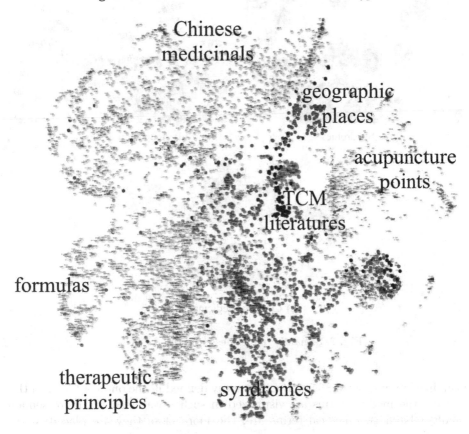

Fig. 6. Overview of semantic relatedness of terms in TCM

4.3 Visualizations and Intuitions

To intuitively understand the semantic relatedness captured by word embeddings, the word vectors of thousands of medical terms of major categories are

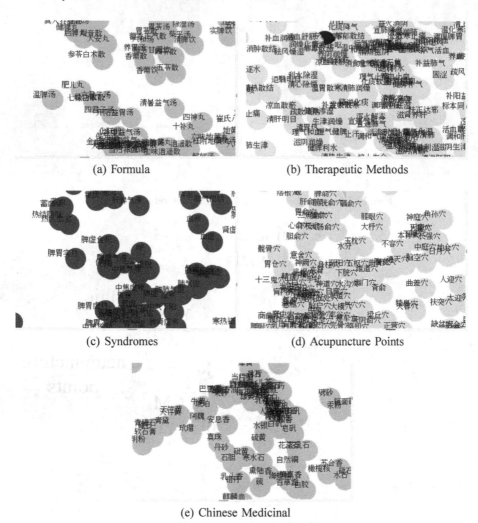

(a) Formula

(b) Therapeutic Methods

(c) Syndromes

(d) Acupuncture Points

(e) Chinese Medicinal

Fig. 7. Semantic relatedness of terms in TCM

visualized in one picture using the taxonomy defined in the ontology [31]. In this picture, the medical terms are visualized in such a way that the more semantically related two medical terms are, the more close they are placed in the picture.

Over 6,000 terms in Traditional Chinese Medical Subject Headings (TCMesh) appear in the 700 ancient literatures. TCMesh consists of 13,905 medical terms in 15 categories. This study chooses the terms in TCMesh to visualize because the categorical information in TCMesh can be used to label the terms using different colors for an easy perception and verification of the semantic relatedness. As shown in Fig. 6, it is observed that the entities of the same categories are generally close to each other to form clusters.

The figures in Fig. 7 below shows enlarged snippets of formulas, therapeutic principles, syndromes, acupuncture points and Chinese medicinals in details. Figure 7(a) shows a snippet of formulas. It is observed that the clusters of formulas are generally consistent with their therapeutic principles. For instance, the formulas for invigorating spleen, such as the shenling baizhu powder (参灵白术散), jianpi pills (健脾丸), and qipi pills (启脾丸), are placed at the upper-left corner; while those for dispersing stagnated liver qi for relieving qi stagnation, such as xiaoyao powders (逍遥散) and modified xiaoyao powders (加味逍遥散) are at the lower-right corner.

Figure 7(b) indicates a snippet of therapeutic principles. At the lower-middle part, perfusing qi movement (宣通气机) is close to perfusing lung qi (宣通肺气); while both of them are distant from astringing (固涩).

Figure 7(c) illustrates a snippet of syndromes. Stomach qi deficiency (胃气虚), spleen-stomach qi deficiency (脾胃气虚), sunken middle qi (中气下陷), and sunken spleen qi (脾气下陷) are semantically close, taking a lower-middle position in the figure; while blood cold (血寒) is close to blood deficiency (血虚), taking a upper-right position.

5 Conclusions

This study uses the domain of TCM as a workbench for an investigation on semantic representations. The semantics are originally expressed implicitly in large volumes of TCM literatures. In order to build a comprehensive evaluation system, the ground truth dataset TCMSem is released, covering as many kinds of medical entities and relations in TCMLS as possible. Then several word embedding methods are experimented on domain-specific corpora of varying specialty and size. When evaluated on TCMSem, it is discovered that word embeddings are proficient to detect synonyms and collocations.

Acknowledgments. (1) The 13th Five-Year Plan for National Key R&D Program of China (2018YFC1705401) *Literature mining and evidence-based research on ulcerative colitis*; (2) State Natural Science Fund Project 81873390 *Study on pedigree construction of ancient knowledge of acupuncture and moxibustion based on text vector.*

References

1. Joulin, A., Grave, E., Bojanowski, P., Mikolov, T.: Bag of tricks for efficient text classification, arXiv preprint arXiv:1607.01759
2. Li, J., Monroe, W., Ritter, A., Galley, M., Gao, J., Jurafsky, D.: Deep reinforcement learning for dialogue generation, arXiv preprint arXiv:1606.01541
3. Klein, G., Kim, Y., Deng, Y., Senellart, J., Rush, A.M., OpenNMT: open-source toolkit for neural machine translation, arXiv preprint arXiv:1701.02810
4. Lin, Y., Liu, Z., Sun, M., Liu, Y., Zhu, X.: Learning entity and relation embeddings for knowledge graph completion, In: AAAI, vol. 15, pp. 2181–2187 (2015)

5. Li, S., Zhao, Z., Hu, R., Li, W., Liu, T., Du, X.: Analogical reasoning on chinese morphological and semantic relations. In: ACL 2018 (2018)
6. Bodenreider, O.: The unified medical language system (UMLS): integrating biomedical terminology. Nucl. Acids Res. 32(suppl_1), D267–D270 (2004)
7. McCray, A.T.: The UMLS semantic network. In: Proceedings, Symposium on Computer Applications in Medical Care, American Medical Informatics Association, pp. 503–507 (1989)
8. Chen, H., et al.: Towards a semantic web of relational databases: a practical semantic toolkit and an in-use case from traditional Chinese medicine. In: Cruz, I., et al. (eds.) ISWC 2006. LNCS, vol. 4273, pp. 750–763. Springer, Heidelberg (2006). https://doi.org/10.1007/11926078_54
9. Wang, Y., Yu, Z., Jiang, Y., Xu, K., Chen, X.: Automatic symptom name normalization in clinical records of traditional Chinese medicine. BMC Bioinform. 11(1), 40 (2010)
10. Wu, Z., Chen, H., Jiang X.: Modern Computational Approaches to Traditional Chinese Medicine. Elsevier (2012)
11. Gu, P., Chen, H., Yu, T.: Ontology-oriented diagnostic system for traditional Chinese medicine based on relation refinement. In: Computational and Mathematical Methods in Medicine (2013)
12. Ruan, T., et al.: An automatic approach for constructing a knowledge base of symptoms in Chinese. J. Biomed. Semant. 8(1), 33 (2017)
13. Ballatore, A., Bertolotto, M., Wilson, D.C.: An evaluative baseline for geo-semantic relatedness and similarity. GeoInformatica 18(4), 747–767 (2014)
14. Cui, M., et al.: Current status of traditional Chinese medicine language system. In: Li, Shaozi, Jin, Qun, Jiang, Xiaohong, Park, James J.(Jong Hyuk) (eds.) Frontier and Future Development of Information Technology in Medicine and Education. LNEE, vol. 269, pp. 2287–2292. Springer, Dordrecht (2014). https://doi.org/10. 1007/978-94-007-7618-0_280
15. Yu, T., Cui, M., Li, H.-Y., et al.: Semantic network framework of traditional Chinese medicine language system: an upper-level ontology for traditional Chinese medicine. China Digit. Med. 9, 44–47 (2014)
16. Harris, Z.: Distributional structure. Word 10(23), 146–162 (1954)
17. Mikolov, T., Sutskever, I., Chen, K., Corrado, G.S., Dean, J.: Distributed representations of words and phrases and their compositionality. In: Advances in Neural Information Processing Systems, pp. 3111–3119 (2013)
18. Pennington, J., Socher, R., Manning, C.: Glove: global vectors for word representation. In: Proceedings of the 2014 Conference on Empirical Methods in Natural Language Processing (EMNLP), pp. 1532–1543 (2014)
19. Bojanowski, P., Grave, E., Joulin, A., Mikolov, T.: Enriching word vectors with subword information, arXiv preprint arXiv:1607.04606
20. Peters, M.E., et al.: Deep contextualized word representations, arXiv preprint arXiv:1802.05365
21. Jacob, D., et al.: Bert: pre-training of deep bidirectional transformers for language understanding. arXiv preprint arXiv:1810.04805 (2018)
22. Zhu, W., Zhang, W., Li, G.-Z., He, C., Zhang, L., et al.: A study of damp-heat syndrome classification using word2vec and TF-IDF. In: 2016 IEEE International Conference on Bioinformatics and Biomedicine (BIBM). IEEE, pp. 1415–1420 (2016)
23. Yao, L., Zhang, Y., Wei, B., Li, Z., Huang, X.: Traditional Chinese medicine clinical records classification using knowledge-powered document embedding. In: 2016 IEEE International Conference on Bioinformatics and Biomedicine (BIBM). IEEE, pp. 1926–1928 (2016)

24. Zhou, X., Peng, Y., Liu, B.: Text mining for traditional Chinese medical knowledge discovery: a survey. J. Biomed. Inform. **43**, 650–660 (2010)

25. Zhou, Xuezhong, Liu, Baoyan, Zhang, Xiaoping, Xie, Qi, Zhang, Runshun, Wang, Yinghui, Peng, Yonghong: Data mining in real-world traditional chinese medicine clinical data warehouse. In: Poon, Josiah, Poon, Simon K. (eds.) Data Analytics for Traditional Chinese Medicine Research, pp. 189–213. Springer, Cham (2014). https://doi.org/10.1007/978-3-319-03801-8_11

26. Baroni, M., Dinu, G., Kruszewski, G.: Don't count, predict! a systematic comparison of context-counting vs. context-predicting semantic vectors, In: Proceedings of the 52nd Annual Meeting of the Association for Computational Linguistics (Volume 1: Long Papers), vol. 1, pp. 238–247 (2014)

27. Schnabel, T., Labutov, I., Mimno, D., Joachims, T.: Evaluation methods for unsupervised word embeddings. In: Proceedings of the 2015 Conference on Empirical Methods in Natural Language Processing, pp. 298–307 (2015)

28. Nayak, N., Angeli, G., Manning, C.D.: Evaluating word embeddings using a representative suite of practical tasks. In: Proceedings of the 1st Workshop on Evaluating Vector-Space Representations for NLP, pp. 19–23 (2016)

29. Mikolov, T., Chen, K., Corrado, G., Dean, J.: Efficient estimation of word representations in vector space. arXiv preprint arXiv:1301.3781

30. Finkelstein, L., et al.: Placing search in context: the concept revisited. In: Proceedings of the 10th International Conference on World Wide Web. ACM, pp. 406–414 (2001)

31. van der Maaten, L., Hinton, G.: Visualizing data using t-SNE. J. Mach. Learn. Res. **9**(Nov), 2579–2605 (2008)

32. Wen, X., et al.: Terms selecting study of revision of traditional chinese medical subject headings thesaurus, based on terms frequency statistics of subject headings indexes and keywords.". Chin. J. Inf. Tradit. Chin. Med. **10**, 16–18 (2013)

The *Hé*-Structure in the Subject Position Revisited

Fanjun Meng[1,2(✉)]

[1] School of English Language, Literature and Culture,
Beijing International Studies University,
1 Dingfuzhuangnali, Chaoyang District, Beijing 100024, China
`louismeng2010@gmail.com`
[2] Department of Foreign Languages and Literatures, Tsinghua University,
1 Qinghuayuan, Haidian District, Beijing 100084, China

Abstract. The syntactic status of *hé* '和' in Chinese [DP₁ *hé* DP₂] structure is ambiguous when occurring in the subject position. It could be a preposition or a conjunctor. Three views have been proposed to account for this ambivalence, namely, the context-deterministic account, the multi-categorizer account, and the preposition-taking-all account. This paper argues against the preposition-taking-all account from both theoretical and empirical perspectives, and proposes that *hé* in the subject position can be and should be coordinative. On the basis of this, this paper presents several advantages of reinstalling the preposition-conjunction dichotomy analysis of *hé*.

Keywords: *Hé* · Syntactic category · Semantics · Preposition · Conjunctor

1 Introduction

The identification of the syntactic status of *hé* in the subject position in Mandarin is a controversial issue. It falls into three basic types, which are referred to as the context-deterministic account, the multi-categorizer account, and the preposition-taking-all account. Firstly, it could be a preposition or a conjunctor and the specific identification of it is context-dependent [1–3]. The second typical view takes this syntactic object as a special hybrid category. To be specific, *hé* is taken to be a prepositional conjunction [1] or a conjunctive preposition [4, 5].[1] Recently, there is a third view which takes all occurrences of *hé* in the subject position to be prepositional [6, 7]. Specifically, there is no conjunctive *hé* in the subject position in Chinese, and there is a subject-object asymmetry given that the specific instantiation of *hé* in the object position is exclusively conjunctive. In other words, the categorial identification of *hé* is a function of the syntactic position it occurs, namely when it occurs in the subject position it is a preposition, and it is a conjunctor when occurring in the object position. Therefore, there is not any corresponding counterpart of the English-type conjunctor *and* in

---

[1] [4] and [5] argue that the primary function of *hé* is prepositional but is with the function of a conjunctor. [1] instead argues that it is a conjunctor but with a prepositional function. Notice that this special hybrid category can also be taken as alternating between a conjunctor and a preposition.

© Springer Nature Switzerland AG 2020
J.-F. Hong et al. (Eds.): CLSW 2019, LNAI 11831, pp. 776–787, 2020.
https://doi.org/10.1007/978-3-030-38189-9_78

Mandarin. Given this, [7] proposes that *hé*-DP$_2$ is a secondary predicate (SP) with *hé* as the relator, and that *hé* projects a maximal projection SP with PRO as its specifier according to [8]. Furthermore, [7] argues that SP is a *zhuijia chenshu*[2] to the first conjunct DP$_1$. Syntactically, it is a secondary predicate to DP$_1$; semantically, SP yields a conjunctive reading between PRO and DP$_2$ which is originally thought to be the result of DP$_1$-*hé*-DP$_2$, and this is done by co-indexing PRO and DP$_1$.

This article revisits the syntactic status of *hé* from both the theoretical and the empirical perspectives, and argues that the conjunctive *hé* is indeed present in Mandarin on the basis of a bunch of syntactic and semantic evidence. Furthermore, the evidence provided in [7] in arguing against the conjunctive status of *hé* is not as valid as they appear to be. Thus, the prepositional re-identification of the conjunctive *hé* in the subject position needs to be reexamined. Then we point out some of the advantages of assuming a preposition-conjunctor dichotomy.

2   A Preposition or a Conjunctor: Is This a Question?

The nominal *hé*-structure can appear in argument positions. When it occurs in the object position, the syntactic status of *hé* is a conjunctor. However, when it occurs in the subject position, its syntactic status is controversial. As introduced above, the identification of it is either context-dependent or multi-categorial. Take (1a) as an example. As indicated in (1b-c), it is ambiguous between two different readings: it refers to either the case where I watched a movie with her, or the case where both she and I each watched a movie. The two readings of (1a) can be disambiguated from each other by some syntactic or semantic diagnostics. For instance, the item *hé* in (1b) is a conjunctor which indicates the case where there are two different movies watched. On the contrary, the insertion of "zuowan (yesterday)" before "*hé* ta (with her)" in (1c)

(1) a. Wo hé ta kan   le   yichang dianying.

    I   hé her watch ASP one-CL movie

    'I watched a movie with her. / Both she and I watched a movie.'

  b. Wo hé ta dou   kan   le   yichang dianying.

    I   hé she DOU watch ASP one-CL movie

    'Both she and I each watched a movie last night.'

  c. Wo zuowan hé ta kan   le   yichang dianying.

    I   last.night hé her watch ASP one-CL movie

    'I watched a movie with her last night.'

  d. Wo kanjian le   Zhangsan hé Lisi.

    I   saw   ASP Zhangsan hé Lisi

    'I saw Zhangsan and Lisi.'

---

[2] The term *Zhuijia chenshu* is not defined in [7]. According to the context it appears in [7], the essential meaning of it is to add something to a preceding element.

suggests that the syntactic status of *hé* is prepositional, as indicated by the English translation. However, when the DP_1-*hé*-DP_2 structure occurs in the object position, *hé* is unanimously identified to be a conjunctor [1, 4].

Instead of keeping a conjunctor-preposition dichotomy, [7] argues that *hé* is not grammaticalized into a conjunctor yet, but at the same time departs from its verbal origin. Thus, they argue the only possible reading for *hé* is a preposition. Given the conjunctive reading of *hé* in the object position maintains, [7] claims there is a subject-object asymmetry. One of the benefits of this idea lies in its ability to formulate a specific condition to distinguish the conjunctive *hé* from the prepositional one; to be specific, it takes the category identification of *hé* as a function of its syntactic distribution: when it appears in the object position, it is a conjunctor, whereas the occurrence of *hé* in the subject position yields a prepositional reading only. This is called the preposition-taking-all account. In what follows, we will review the main claims and argumentations in [7].

[7] compares the differences between the subject *hé* construction with the object *hé* construction from the following four perspectives, namely, the tolerance of CSC (Coordinate Structure Constraint, CSC, [9]) violation in topicalization, focalization only of the first conjunct, the negation and A-not-A reduplication of the conjunctor. We have demonstrated that all these exceptions can be accounted for by mechanisms that are motivated independently, so that there is no need to assume the prepositional status of *hé*. They observe that the subject *hé* construction tolerates all these four violations. For the sake of space, this paper only focuses on the topicalization of the first conjunct, as indicated in (3), with (2) as the baseline example, which are all cited from [7].

(2) Zhangsan hé$_{conj}$ Lisi zai butongde xuexiao xuexi.

    Zhangsan hé   Lisi at different school study

    'Zhangsan and Lisi study at different schools.'

(3) a. Zhangsan, wo renwei *t* hé Lisi zai butongde xuexiao xuexi.

      Zhangsan , I think   hé Lisi at different school study

      'lit. Lisi, I think Zhangsan and *t* studies at different schools.'

   b. *Lisi, wo renwei Zhangsan hé *t* zai butongde xuexiao xuexi.

      Lisi I think Zhangsan hé at different school study

      'lit. Lisi, I think Zhangsan and *t* studies at different schools.'

   c. *Zhangsan, wo renwei baba xihuan *t* hé Lisi.

      Zhangsan I think father like   hé Lisi

      'lit. Zhangsan, I think my father likes *t* and Lisi.'

   d. *Lisi, wo renwei baba xihuan Zhangsan hé *t*.

      Lisi I think father like   Zhangsan hé

      'lit. Lisi, I think my father likes Zhangsan and *t*.'

(3a) shows that the topicalization of the first conjunct is immune from the CSC, whereas (3b) demonstrates that the same cannot be applied to the second conjunct and the object *hé*-constructions. [7] argues that given CSC is a universal constraint that could not be violated; therefore, the grammaticality of (3a) suggests *hé* in the subject position is not a conjunctor, but a preposition.

In this paper, we will show that the CSC violation exceptions can be accounted for by independently motivated mechanisms without assuming the prepositional status of *hé* and the preposition-conjunctor dichotomy should be maintained.

3   The Potential Problems of the Prepositional Account of *Hé*

This section examines the potential problems of the prepositional account from both the theoretical and empirical perspectives, even though the preposition-taking-all account as proposed in [7] could capture the asymmetries in (3).

3.1   Theoretical Problems

Firstly, syntactic operations like passivization have changed the category of the subject and the object due to the different categorial identifications of *hé*. If *hé* in the subject position is prepositional, but a coordinator in the object position, then we predict that the demotion of the subject and the promotion of the object during the process of passivization as shown in (4a) to (4b) should not alter the syntactic category.[3] However, this is not as predicted as the account claimed in [7] given that the prepositional *hé* originally contained in the subject position is altered to a conjunctive instance; similarly, the conjunctive *hé* generated originally in the object position is changed to a prepositional one. Evidently, this goes against the structure-preserving principle.

(4) a. Zhangsan hé$_{prep}$ Chenliu kanjian le   Lisi hé$_{conj}$ Wangwu.

      Zhangsan hé   Chenliu see   ASP Lisi hé   Wangwu

      'Zhangsan and Chenliu saw Lisi and Wangwu.'

   b. Lisi hé$_{prep}$ Wangwu bei   Zhangsan hé$_{conj}$ Chenliu kanjian le.

      Lisi hé   Wangwu passive-marker Zhangsan hé   Chenliu see   ASP

      'Lisi and Wangwu were seen by Zhangsan and Chenliu.'

Secondly, *hé*-DP$_2$ as a SP is a *zhuijia chenshu* to the first conjunct DP$_1$, thus it is non-*at-issue* [10], which resembles the comitative structure, not coordination. Put differently, the prepositional account seems to target a comitative construction, not a coordinative structure.

---

[3] Leave aside the debate on the labeling of *hé*-structure as an &P or a DP.

(5) a. Zhangsan hé$_{conj}$ Lisi zai chifan.

  Zhangsan hé   Lisi at   eat.rice

  'Zhangsan and Lisi are eating.'

 b. Zhangsan hé$_{prep}$ Lisi zai chifan.

  Zhangsan hé   Lisi at   eat.rice

  'Zhangsan is eating with Lisi.'

In (5a), the agent of eating is *Zhangsan* and *Lisi*; in (5b), it could only be *Zhangsan*, and *Lisi* is just a participant accompanying the eater. We may wonder whether the prepositional account could capture this difference given that they are both comitative in nature according to [7]. The entity denoted by DP$_2$ does not appear to be the agent of the eventive predicate since SP and the predicate are not in a syntactic predication relation. The *zhuijia chenshu* function of SP constitutes another piece of collaborating evidence. The conjunctive reading forces SP to be an argument of the eventive predicate via predication; if so, the assumed null element PRO is also taken to be one of the participants of the eating event. However, this line of reasoning has three problems: Firstly, SP can only be an argument when it is identified to be nominal, but in fact it is a maximal projection headed by a prepositional phrase in the analysis of [7]. That is to say, it is difficult for the prepositional SP to form a predication relation in terms of syntax with the eventive predicate. Secondly, the eventive predicate is predicated of the same entity as its argument in figuring out the right semantics of (5b): one is PRO and the other is its controller—the first conjunct DP$_1$. If SP is the source of plurality, then it seems possible to equally take all these three elements, namely the first conjunct DP$_1$, PRO and the second conjunct DP$_2$, to be the eventive agents. Clearly, it is not so. Thirdly, *Zhangsan* and *Lisi* in the conjunctive construction share the same theta role. According to Uniformity of Theta Assignment Hypothesis (UTAH) as proposed in [11], *Zhangsan* and *Lisi* should occupy the same syntactic position, namely the subject position. However, according to the SP analysis, *Zhangsan* and *Lisi* are assigned divergent theta roles while both holding the subject position. If UTAH is equally applied in Chinese, then the phenomenon that two DPs with different theta roles are assigned to the same syntactic position turns out to be mysterious.

Thirdly, the distribution of PRO awaits further explanation. Following [8], [7] assumes the subject of the secondary predicate turns out to be PRO. That is to say, they end up claiming that PRO could be the null subject of a prepositional phrase. The cross-linguistic research on the distribution of PRO demonstrates that it distributes almost exclusively in non-finite clauses. Along with [7], we are led to take the prepositional phrase to be non-finite. However, this inference is unreasonable. Moreover, the topicalization of the first conjunct indicates that the DP$_1$-*hé*-DP$_2$ structure is contained within a finite clause since the domain that A'-movement crosses over can only be finite in nature [12]. Therefore, the observation of topicalization of the first conjunct in the first place invalidates the existence of PRO.

Fourthly, the interpretation of PRO is mysterious. Even if the SP analysis is reasonable, whether PRO is obligatorily controlled or not is still an issue. For this, we need to evaluate the satisfaction of the OC Signature [13], namely locality and PRO as a variable. As indicated in (6a), PRO as a variable is bound by the closest antecedent *Zhangsan*, whereas (6b) shows that PRO as a variable could also be controlled by *Zhangsan*, a further DP crossing over *Lisi*. Thus, we are now in a dilemma. On the one hand, the OC Signature is met, on the other, it is not. Furthermore, [6] takes "Zhangsan, Lisi" in (6b) as a real conjunction in Chinese, which is expected to control PRO as a single unit, yet this is not possible since *Zhangsan* could control PRO independently. What's worse, Chinese is a pro-drop language, thus the subject can be silent. In this case, the real subject according to the SP analysis is the element before SP, namely *Zhangsan* in (6a) or "*Zhangsan, Lisi*" in (6b). When the subject is dropped, PRO ends up being controlled by a silent element *per se*, which itself needs to be identified by an entity in the discourse to begin with. This further confirms the non-obligatoriness of PRO construal in (6b) which reaches a conflicting conclusion as compared with (6a). Besides, the topicalization of the first conjunct leaves PRO to be bound by a lambda operator, which is rarely seen in the study of control.

(6) a. [(Zhangsan)$_i$[$_{SP}$ PRO$_i$ hé Lisi]] dou lai   le.

     Zhangsan   hé Lisi DOU come SFP

     'Both Zhangsan and Lisi came.'

   b. [(Zhangsan$_i$, Lisi$_j$) [$_{SP}$ PRO$_{i/*j}$ hé tade$_i$ mama] dou lai   le.

     Zhangsan Lisi   hé his   mother DOU come SFP

     'Zhangsan, Lisi and his mother came.'

Fifthly, the semantic type of *hé* is dubious. Along with the prepositional account of *hé* and the Likeness Condition, when the first conjunct is of type *e*, the second conjunct should automatically be of the same type as well. As shown in (7), *e*-typed DP$_1$ requires the semantic type of DP$_2$ to be *e*, and SP to be <e, e> in order to yield an *e*-typed DP that is subsequently plugged into VP. The type of PRO is *e*, thus the semantic type of *hé*-DP2 is calculated to be <e, e, e>. Finally, the semantic type of *hé* is calculated out, namely <e, e, e, e>. However, this type is not the semantic type for a typical transitive preposition, which should be <e, e>. In other words, the semantic type assigned to the prepositional *hé* in this structure is a new type (What's worse, we haven't considered the case where DP is considered to be a generalized quantifier, and it will yield another different semantic type for *hé*). Put differently, we end up with two types of prepositional *hé* at least in terms of semantic type, though the conjunctive *hé* in the subject position and the prepositional one is reduced to a single syntactic category. Whether the reduction of syntactic category is preferred over the complication of semantic type requires further investigation.

(7)

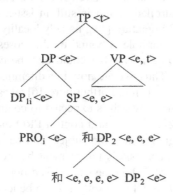

Sixthly, the case assignment of *hé* poses a problem for the Chinese language. If *hé* in the subject position can be reanalyzed as a preposition, then it will assign an accusative case to its internal argument, and a nominative case and (together with the internal argument) an agent theta role to its subject, namely PRO. However, PRO must be ungoverned, thus it could not be case-marked, at least not a nominative case. Recently, cross-linguistic research observes that PRO could carry a morphological case in languages like Icelandic. If the distribution and the case-marking of PRO are indeed independent from each other as argued in [13],[4] then PRO could be assigned a nominative case. If so, could we claim that the overt realization of PRO in Chinese could also be assigned a case? It seems reasonable to make such a claim. However, whether the DPs in Chinese could be case-marked is a controversial issue and is contradictory to Hu himself [14]. Therefore, the assumption of PRO brings in a controversial issue in Chinese which awaits further research.

Lastly, the labeling of the maximal projection SP is problematic. To project the unit that is composed of the subject and SP is another issue for [7]. For instance, when the *hé*-construcion is in the object position, it is a self-contained projection;[5] however, passivization leaves the label of it unspecified. This is undesirable since a syntactic operation should not change the category of a syntactic object. The prepositional analysis in [7] clearly goes against this generalization.

To sum up, the assumption of the prepositional status of *hé* faces a range of theoretical problems.

3.2   The Empirical Problems and the CSC Violation Revisited

For the sake of space, we only focus on the topicalization of the first conjunct.[6] We will demonstrate that the subject-object asymmetry of the CSC violation in topicalization

---

[4] We leave aside details of this controversial issue in this paper.

[5] Set aside the specific label for it in this paper, be it &P or DP.

[6] For more detailed responses to the remaining issues, please refer to [17].

could be accounted for without assuming the prepositional status of *hé* in the subject position.

Cross-linguistically, the violation of CSC is not unique to Chinese, nor is it unique to topicalization. Evidently, the prepositional analysis of *hé* faces empirical problems. For instance, [15] observes that the first conjunct in Serbo-Croatian as indicated in (8) is topicalized and is a clear violation of CSC, yet the remaining sentence still remains grammatical. The reason lies in the cliticization of the conjunctor *i* to the second conjunct *filmove*, leaving the trace of the conjunctor *-marked and deleted at PF. According to [15], this phenomenon could be explained by the following observation, that is, the nullification of the head deprives the islandhood of any given island. Put differently, the violation of CSC in the topicalization of the first conjunct could be saved by the head movement of the conjunctor and the deletion of the *-marked structure at PF; thus, the fact that the remaining sentence as shown in (8) is grammatical is captured.

(8) ?Knjige$_i$ je Marko [$_{ConjP}$ t$_i$ i   filmove] kupio

     book   is Marko   conj film   buy

     'Marko bought a book and a film ticket.'

Another important correlation is that CSC islandhood voidance occurs only in article-less languages. This implies that the scenario observed in Chinese might be treated on a par since Chinese is a language that lacks articles.

The next step is to argue that the Chinese conjunctor *hé* behaves like a clitic. Note that we are not trying to argue that *hé* is a clitic, but simply demonstrating that *hé* displays the properties of a clitic.

The theoretical approach adopted here is the framework of canonical typology. The canonical clitic theory as proposed in [16] claims that clitics stands between functional words and affixes. In other words, [16] argues that clitics exhibit a dual feature of these two categories: its formal properties are those of the canonical affix, but its distributional properties are those of the canonical function word. The first criterion (9i-a) is modified by [18] to accommodate the tonal Chinese languages. To be specific, the clitic should not be the same as the full-form functional item with regard to tonal patterns. Furthermore, this clitic should not bear stress or focus. According to the first distributional criterion, the clitic should be attached to its host, which is a phrase as well as a word, and it remains in the same position as its full-formed counterpart. The second distributional criterion holds that the clitic should take wide scope over conjuncts. We will go through them one by one. [7]

---

[7] It should be pointed out that the Canonical Clitic Theory does not require the item to meet each individual criterion. As for whether *hé* meets the requirement in (9) awaits more solid evidence from experimental phonetics.

(9) (i) Canonical form properties for clitics

    a. Toneless or tonally unspecified; [18]

    b. Lacking prosodic prominence and hence is prosodically dependent on its host. [16]

  (ii) Canonical distributional properties for clitics

    c. A clitic is placed with respect to the syntactic phrase bearing the functional property expressed by the clitic (a.k.a. the phrasal placement criterion); [16]

    d. A clitic canonically takes wide scope over a coordinated phrase with which aitis in construction (also known as wide scope criterion). [16]

The linker *hé* is not stressed and is prosodically dependent on its host. [19] claims that functional elements are "generally stressless...even phonologically null." As indicated in (10a), the cliticization of *hé* to the second conjunct is evidenced by the intolerance of the insertion of a plural suffix *-lia*, and the item could only be placed after the lexical string *hé Lisi*. Furthermore, since *-lia* could only be attached to a plurality element, which lends more support to the conjunctive reading of *hé*. Furthermore, (10a) and (10b) show that the host that *hé* is attached to is phrasal, not lexical. Finally, the scope of *hé* is over *meiyige nanhai* and *meiyige nühai*. As indicated in (10b), the meaning is "Every boy came and every girl came", which could only be possible when *hé* takes the wide scope over the two conjuncts.

(10) a. Zhangsan hé (*lia) Lisi (lia) dou   canhui   le.

    Zhangsan hé   two Lisi two DOU attend.meeting SFP

    'Both Zhangsan and Lisi attended the meeting.'

    b. Meiyige nanhai hé meiyige nühai dou   lai   le.

    Every   boy   hé every   girl   DOU come SFP

    'Every boy and Every girl came.'

On the basis of the above analysis, the conjunctor *hé* could be considered as a clitic-like element. If so, the conjunctor-as-clitic movement account that is originally proposed for (8) could be extended to (3a-b), repeated here as (11a-b). This analysis could not only account for the grammaticality of (11a) but also the ungrammaticality of (11b). To be specific, the head movement of *hé* in the way of cliticization could invalidate the islandhood of *Zhangsan hé Lisi* in the subject position, and thus the extraction of *Zhangsan* is permitted and consequences of the CSC violation is repaired.

(11) a. Zhangsan, wo renwei *t* hé Lisi zai butongde xuexiao xuexi.

     Zhangsan, I   think   hé Lisi at different   school   study

     'lit. Lisi, I think Zhangsan and *t* studies at different schools.'

   b. *Lisi, wo renwei Zhangsan hé *t* zai butongde xuexiao xuexi.

     Lisi I   think   Zhangsan hé   at   different   school   study

     'lit. Lisi, I think Zhangsan and *t* studies at different schools.'

In contrast, the ungrammaticality of (11b) could be captured in the following way: the item *hé* behaves like a clitic, thus it could not be stranded from its host — the second conjunct. In this way, the obligatory extraction of the host of *hé* from the coordination island is strongly prohibited. This stranding prohibition account could be extended to (3d) as well, repeated here as (12a). That is, the topicalization of *Lisi* to the left periphery leaves *hé* being stranded.

(12) a. *Lisi, wo renwei baba   xihuan Zhangsan hé *t*.

     Lisi I   think   father like   Zhangsan hé

     'lit. Lisi, I think my father likes Zhangsan and *t*.'

   b. *Zhangsan, wo renwei baba   xihuan *t* hé Lisi.

     Zhangsan I   think   father like   hé Lisi

     'lit. Zhangsan, I think my father likes *t* and Lisi.'

The next question is how to explain the subject-object asymmetry regarding the tolerance of CSC violation in the subject *hé*-conjunction. That is, how to capture the grammaticality contrast between (11a) and (12b). We argue that the contrast could be attributed to the topic prominence of the Chinese languages. In other words, we could assume that there is an interpretable Topic feature on the Chinese subject. To put it more formally in syntactic terms, we assume that there is a [uToP] in the left periphery of the matrix clause that needs to be checked off, and the closest possible candidate for the valuation of this very feature is the subject that carries an [iTop] feature. It cannot be the object since it does not have a Topic feature; otherwise the locality condition of Agree is violated [20]. To be specific, if *Zhangsan hé Lisi* in the subject position as indicated in (11a) and (12b) is conjunctive in nature, then according to the feature percolation mechanism through spec-head agreement as implemented in [21], the [iTop] feature carried by the whole structure could be passed onto the first conjunct which occupies the spec position of this conjunctive structure, rather than the object occupying the complement position. Therefore, the [iTop] feature is transferred to *Zhangsan* only, not *Lisi*. Thus, the contrast between (11a) and (12b) is accounted for.

    To sum up, the subject-object asymmetry with regard to the violation of CSC in topicalization of the first conjunct is captured without reanalyzing the conjunctive *hé* as a preposition. Furthermore, this paper has aligned Chinese with other article-less languages which at the same time permit the CSC violation. In this way, the apparent CSC-violation is proved to be not unique to Chinese.

4 The Merits of a Preposition-Conjunctor Dichotomy

This section points out two pieces of merits of holding a preposition-conjunctor dichotomy.

Firstly, the dichotomy avoids the problems that are brought in by reducing the two instances of *hé* in the subject position to a single prepositional one.

One reviewer asks whether and what kind of consequences we will have to face in Chinese language teaching or in the computational parsing of Chinese if we identify the syntactic status of *hé* in a wrong way. The answer is definitely yes. Let's take the construction of Combinatory Categorial Grammar Bank as an example, the incorrect syntactic categorization will wrongly label the conjunctive *hé* as the prepositional one, and this will mislead us to choose and apply the wrong combinatory rules, consequently, a wrong syntactic parsing is yielded.

Secondly, holding a preposition-conjunctor dichotomy could account for the apparent CSC violation from a cross-linguistic perspective; meanwhile, it will also put Chinese into a cluster of languages that lack articles yet at the same time tolerate CSC violation. In this way, the CSC violation no longer stays as a rigid generalization and is violated as more empirical facts unfold.

5 Conclusion

This paper reexamines the syntactic status of *hé* in the subject position from both the theoretical and empirical perspectives. We argue that the reduction of the two instances of *hé* to a single syntactic category, namely preposition, could not cover all the linguistic facts that the conjunctive *hé* and prepositional *hé* pose, and might incur a range of potential problems as illustrated above. In this paper, we emphasize to return to the traditional preposition-conjunctor dichotomy. In this sense, the Occam's Razor could not be applied freely and should be keyed to the empirical facts; otherwise, oversimplification arises which will invite more problems than solutions.

Acknowledgments. The author thanks Professor Jinglian Li, Professor Honghua He, Professor Xiaolu Yang, Dr. Ge Guo and Dr. Changsong Wang for their encouragements and discussion. Special thanks extend to Dr. Yang Chen for her understanding and support. My gratitude also goes to the anonymous reviewers and the conference organizers for their insightful comments and suggestions. Finally, I want to thank the financial support from the Ministry of Education Fund (Grant No. 18YJA740022).

References

1. Chao, Y.: A Grammar of Spoken Chinese. University of California Press, Berkeley (1968)
2. Zhu, D.: Lectures on Grammar. The Commercial Press, Beijing (1982). (in Chinese)
3. Lü, Sh. (ed.): Modern Chinese 800 Words. The Commercial Press, Beijing (1984). See also Ji, X. (ed.): Lü Shuxiang's Anthology. Northeast Normal University Press, Changchun (2002). (In Chinese)
4. Jiang, L.: Sources, paths, and types of the grammaticalization of Chinese conjunction-prepositions. Stud. Chin. Lang. **349**, 291–308 (2012). (In Chinese)
5. Jiang, L.: The origins of spatial functions and the heterogeneity of Chinese conjunction-prepositions. Stud. Chin. Lang. **363**, 483–497 (2014). (In Chinese)
6. Yang, M.: The syntax of the functional word *hé*. Master Dissertation of Chinese Academy of Social Sciences, Beijing (2014). (in Chinese)
7. Yang, M., Hu, J.: The syntax of *hé*. Lang. Teach. Linguist. Stud. **191**, 58–70 (2018). (In Chinese)
8. Napoli, J.: Predication Theory: A Case Study for Indexing Theory. Cambridge University Press, Cambridge, UK (1989)
9. Ross, J.: Constraints on variables in syntax. Ph.D. dissertation of MIT, Cambridge (1967)
10. Haspelmath, M.: Coordination. In: T. Shopen, (ed.) Language Typology and Linguistic Description, vol. 2, Complex Constructions, pp. 1–51. Cambridge University Press, Cambridge (1997)
11. Baker, M.: Incorporation: A Theory of Grammatical Function Changing. University of Chicago Press, Chicago (1988)
12. Chomsky, N.: Lectures on Government and Binding: The Pisa Lectures. Foris, Dordrecht (1981)
13. Landau, I.: Control in Generative Grammar: A Research Companion. Cambridge University Press, Cambridge (2013)
14. Hu, J.: On the distribution and selection of arguments: the prominence and locality in grammar. Stud. Chin. Lang. **334**, 3–20 (2010). (in Chinese)
15. Bošković, Ž.: On the Coordinate Structure Constraint, islandhood, phases, and rescue by PF deletion. Ms. UConn (2017)
16. Spencer, A., Luís, A.: The canonical clitic. In: Brown, D., Chumakina, M., Corbet, G. (eds.) Canonical Morphology and Syntax, pp. 123–150. Oxford University Press, Oxford (2012)
17. Meng, F.: The syntax and semantics of Chinese *hé*-conjunction. Ph.D. dissertation of Tsinghua University, Beijing (2019)
18. Ye, K., Pan, H.: The syntactic nature of *gei* in *ba* construction. Foreign Lang. Teach. Res. **5**, 656–665 (2014). (In Chinese)
19. Abney, S.P.: The English noun phrase in its sentential aspect. Ph.D. dissertation of MIT, Cambridge (1967)
20. Citko, B.: Phase Theory: An Introduction. Cambridge University Press, Cambridge (2014)
21. Zhang, N.: The Syntax of Coordination. Cambridge University Press, Cambridge (2009)

On the Semantics of Suffix *-men* and NP-*men* in Mandarin Chinese

Yan Li[(✉)]

School of Foreign Languages, Nanjing Normal University, Nanjing, China
lucie6995@163.com

Abstract. The paper deals with the semantics of the nominal suffix *-men* and NP-*men* in Mandarin Chinese. In sinologist linguistics, *-men* is often considered as a "marker of collectivity". However, the term *collectivity* (*jíhé* in Chinese) is confusing. In semantics, the concept of collectivity refers to two things, collective interpretation of a NP and collective nouns. This distinction can help to clarify the usage of *-men*. The author argues that the suffix *-men* is a marker of plurality, with [plural] feature due to its denotation. It indicates the number of a NP. In this paper, *-men* will be analyzed with respect to the collective and distributive interpretations of NP-*men*, especially when it co-occurs with different types of predicates.

Keywords: Nominal suffix *-men* · NP-*men* · Collectivity · Formal semantics · Mandarin Chinese

1 Introduction

This paper discusses the nominal suffix *-men* in Mandarin Chinese. *-Men* is placed after a noun, which only refers to a human, as in (1).

(1)

a. *Xiǎopéngyǒu-men*	b.*Shítóu-men*	c. *Shīzǐ-men*
child-*men*	stone-*men*	lion-*men*
'Children'	intended: 'stones'	intended: 'lions'

In literature, the suffix *-men* has been broadly discussed. Some linguists analyze diachronically the origin of *-men* and its evolution, such as [2, 3], etc. Some explain why *-men* is not compatible with classifiers, like [4]. Synchronically, some linguists concentrate on what *-men* marks: (i) *-men* is a collective marker [5]; (ii) *-men* is a plural marker [6–8]; (iii) it is a both collective (*jíhé* in Chinese) and plural marker [9, 10] etc.

In the next section, the author will show the two main points of view about *-men*, which are markers of collectivity and plurality.[1]

---

[1] Abbreviations: Acc: Accomplished aspect; Cl: individual classifier; $Cl_{collective}$: collective classifier; $Cl_{container}$: container classifier; Cl_{group}: group classifier; DE: relater *de* (的); Pl: plural.

J.-F. Hong et al. (Eds.): CLSW 2019, LNAI 11831, pp. 788–800, 2020.
https://doi.org/10.1007/978-3-030-38189-9_79

1.1   *Men* as a Collective Marker

[5] indicates that -*men* marks collectivity. His arguments are shown below. (i) NP-*men* cannot give rise to a generic interpretation; (ii) NP-*men* always refers to the definite, which blocks its presence in an existential sentence, as (2) illustrates.

(2)

a. *Yŏu rén-men	b. *Yŏu rén*
have man-*men*	have man
	'There is a person.'

(iii)   NP-*men* cannot co-occur with a numeral (Num)-classifier (Cl) phrase, as in (3).

(3)

> **Sān   gè   xuéshēng-men*
> three Cl   student-*men*
> intended: 'Three students'

(iv)   NP-*men* can refer to a group of people, as in (4) (in which (4b) is taken from [11], cited in [5]), the group is defined by the speaker relative to himself. NP-*men* refers to a situationally anchored and defined group. When a noun is suffixed by -*men*, there is generally a subject-locator in the context, relative to whom the group is viewed.

(4)

> a. *Sīmăguāng hé   xiăopéngyŏu-men*
>   Simaguang and   kid-*men*
>   'Simaguang and kids'
> b. *Xiăoqiáng-men* ([11], p.342 cited in [5])
>   Xiaoqiang-*men*
>   'Xiăoqiáng's group'

In his view, NP-*men* refers to a temporary group, explicitly presented as subjective. -*Men* marks a subjective location, which shows the subjectivity of the speaker because the speaker sees several individuals as a group and the number of individuals is indeterminate. For him, the *collectivity* means a **group** of people.

1.2   -*Men* as a Plural Marker

[6] disagrees with the analysis of [5] that -*men* marks collectivity. Firstly, she points out that besides collectivity, NP-*men* can denote different notions of plurality. For instance, *Xiaoqiang-men* can be interpreted as (i) several persons that are called Xiaoqiang; (ii) multiple persons who share an identical property with Xiaoqiang. Secondly, according to [6], in some cases, -*men* can co-occur with numeral expressions, as illustrate (5) which are acceptable for the author.

(5)

    a. *Wǒ qǐng   tāmen sāngè   (hái zǐ) chīfàn.*
      I invite them three-Cl (child) eat
      'I treated them a meal, those three kids.'
    b. *Wǒ qǐng   Xiǎoqiáng-men sāngè   (rén)   chīfàn.*
      I invite Xiaoqiang-*men* three-Cl person eat
      'I invited Xiaoqiang and two others (in the group) for a meal.'

Moreover, NP-*men* can co-occur with the adverb *dōu*, as in (6).

(6)

        Xuéshēng-men dōu   líkāi   le.
        student-*men*   all   leave Acc
        'Students all left.'

[6] argues that since *dōu* is an operator of distributivity, as in (6), it can hardly treat -*men* as a collective marker. Therefore, [6] analyzes -*men* as a plural morpheme, which can be attached to noun with human feature or personal pronoun only. This does differentiate from the plural morpheme in Indo-European languages, like -*s* in English. Syntactically, the suffix -*men*, a plural morpheme occupies a position in D (Determiner) position, higher than numeral phrase and Cl phrase, as (7) shows while -*s* in English positions at N (Noun) level, as (8) illustrates.

(7)

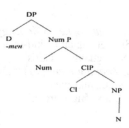

(8)

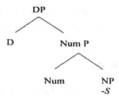

[6] thinks due to -*men* at D position, NP-*men* in DP is marked as definite in Mandarin Chinese.

Comments. [6]'s syntactic analysis can account for the co-occurrence of -*men* with personal pronouns, common nouns and proper nouns since -*men* marks the plurality of those types of nouns. [5] studies the semantics of -*men* and its usage. However, the author founds that in [5] and other past studies, the notion of *collectivity* is not clearly defined, neither is what collectivity refers to. This subject deserves to be elaborated in detail.

1.3 Research Questions

In view of the above two analyses of -*men*, it is needed to clarify what *collectivity* refers to and understand what the suffix -*men* denotes in respect to plurality. Therefore, this paper is aimed at answering the following questions:

  (i)   What does *collectivity* mean?
 (ii)   What does -*men* mark?
(iii)   What interpretation does -*men* give rise to? Collective or plural interpretation?

In this paper, the suffix -*men* will be analyzed from the formal semantic perspective. Firstly, the concept of collectivity will be presented, and this will help to explore the function of -*men*. Secondly, the basic semantic properties of the suffix -*men* will be established. Thirdly, the co-occurrence of the suffix -*men* with different types of predicates will be studied in order to demonstrate the denotation of NP-*men*.

2 Collectivity and Collective Nouns

Collectivity can be used in at least two ways in formal semantics. One is to see collectivity as a property of form-meaning pairs: collective nouns. The other refers to a type of meaning, which is often analyzed in semantics: a certain property is attributed to a set of individuals, viewed as a whole.

Collective nouns have a singular form but plural reference. [12] lists the types of collective nouns in English: (a) General "collection nouns": *collection, set, group, multitude*; (b) Nouns denoting multitudes of humans or animals: *crowd, herd, swarm*; (c) Nouns denoting particular spatial configurations of multiple objects: *stack, pile, heap, bunch*; (d) Nouns denoting institutions or groups of humans formed for some official purpose: *committee, council, team, army* (See [13–15]). These nouns refer to the notion of *group* and they are compatible with some predicate like *gather* which applies to a set of individuals but viewed as a single individual, as in (9) and they are not compatible with predicates such as *have bleu eyes* which applies to individuals in the set (See [15]: 117–118, (5) and (6a)).

(9) The committee gathered/$^?$has blue eyes.

Collectivity with respect to application of predicates over its NP subject refers to **collective interpretation**. [15] indicates that the plural NP, *the boys* in (10) can give rise to two interpretations, distributive and collective. [15] thinks that the collective interpretation is an interpretation of group and the group is defined as (11). In (11), 'a' denotes a sum, then *a* can be seen as a group by operation ↑(a). The operator '↑'

indicates the group formed by a sum. The group is considered as a complex individual, a single entity without composed individuals.

(10)   **The boys** carried the piano upstairs.
(11) $\uparrow(a) := a$ as a group

These two notions of collectivity can help to show the inference of the suffix *-men* when it co-occurs with a NP.

In the next section, the author will begin with collective nouns.

3   Nature of NP-*Men*

In this section, the author proposes to compare so-called collective nouns in sinologist linguistics with collective nouns used in formal semantics. Then, the author will show the properties of NP-*men*.

3.1   NP-*Men* Does not Refer to a Traditional Collective Noun

According to [16], collective nouns in Chinese include the three types of nouns below. (i) Nouns listed in (12), which is labeled as collective nouns of conjunctive form.

(12)

a. *Fù   mǔ*
   Father mother
   'parents'

b. *Shī   shēng*
   teacher student
   'teachers and students'

c. *qīn   yǒu*
   relative friend
   'relatives and friends'

d. *jūn   huǒ*
   army weapon
   'army and weapon'

(ii)   Compound nouns consist of a noun and a classifier, as (13) illustrates. The author names them as noun-classifier collective nouns.

(13)

a. *zhǐ   zhāng*
   paper Cl
   'paper'

b. *rén   qún*
   people Cl$_{group}$
   'cluster of people'

(iii)   NPs suffixed with *-men*, like in (14).

(14)

	a. *zhànshì-men*		b. *lǎoshī-men*
	soldier-*men*		teacher-*men*
	'soldiers'		'teachers'

In addition, in formal semantics, the nouns which refer to a set of entities, which are viewed as a group and singular, are defined as **collective nouns**, as (15) shows.

(15)

	a. *wěiyuánhuì*	b. *jūnduì*	c. *tuántǐ*
	'committee'	'army'	'group'

Now, the author will show the compatibility of the four types of nouns mentioned above with classifiers and the animacy property. Nominal classifiers involve different types of classifiers, including individual classifiers such as *gè*, kind classifiers such as *zhǒng*, collective classifiers such as *qún*, *duī*, standard measure words such as *gōngjīn*, container measure words such as *píng*, *xiāng*, partitive classifiers such as *piàn* (cf. [9, 17–19]). Only individual classifiers, collective classifiers and container measure words will be used in order to test their compatibility with collective nouns.

Firstly, the class of nouns that the author named as collective nouns of conjunctive form is not compatible with individual classifiers, as illustrates (16), but it is compatible with collective classifiers and *xiē*, another element denoting plurality, as shows in (17).

(16)

	a.* *yī*	*gè*	*fùmǔ*	b.* *wǔ*	*gè*	*shīshēng*
	a	Cl	parents	five	Cl	teacher and student

(17)

	a. *yī*	*qún*	*fùmǔ*	b. *yīxiē*	*qīnyǒu*
	a	Cl$_{group}$	parents	some	relative and friend
		'a cluster of parents'		'some relatives and friends'

Secondly, noun-classifier nouns cannot co-occur with individual classifiers (18), but it can co-occur with container measure words and collective classifiers (19). (see [20]).

(18)

	a.* *liǎng zhāng*	*zhǐzhāng*	b.* *sān*	*gè*	*rénqún*
	two Cl	paper-Cl	three	Cl	people-Cl$_{group}$

(19)

a. *liǎng   xiāng   zhǐzhāng*   b. *yī   pái   chēliàng*
   two   Cl_{container} paper-Cl   one Cl_{collective} car-Cl
   'two boxes of papers'   'a line of cars'

However, NP-*men* cannot co-occur with individual classifiers nor collective classifiers, as (20) shows.

(20)

a. **sān   bǎi   gè háizǐ-men*   b. **liǎng   qún   lǎoshī-men*
   three hundred Cl child-*men*   two   Cl_{group} teacher-*men*

Contrarily, *committee, army* are compatible with individual classifiers, not collective classifiers, as (21) illustrates.

(21)

a. *liǎng gè   wěiyuánhuì*   b. **liǎng   qún   jūnduì*
   two   Cl   commettees   two   Cl_{group} army
   'two committees'

Therefore, the above tests show that NP-*men* is different from the three other noun types with respect to their compatibility with classifiers. NP-*men* cannot combine with any type of classifiers while collective nouns of conjunctive form, noun-classifier nouns and collective nouns are able to co-occur with classifiers, especially individual ones, collective ones or container measure words.

Next, the author will examine the animacy of these types of nouns. Nouns with conjunctive form such as *fùmǔ* 'parents', *shīshēng* 'teachers and students' are animate but *jūnhuǒ* 'army and weapon' are inanimate. Likewise, noun-classifier nouns can be animate or inanimate, for instance, *rénqún* 'cluster of people' is animate while *chēliàng* 'car' is inanimate. Collective nouns such as *wěiyuánhuì* 'committee', *tuántǐ* 'group' or *jūnduì* 'army' refer to animate entities. NP-*men* must refer to plural entities with animate feature, which is a plurality of animate individuals because -*men* attaches to human nouns or personal pronouns only.

In our view, NP-*men* differs from collective nouns which include those of conjunctive form (12) and noun-classifier ones (13) in respect to syntax and semantics. It is not appropriate to categorize them in the same class and name them as collective nouns. However, NP-*men* shares some syntactic and semantic properties with collective nouns in the view of formal semantics like *wěiyuánhuì* 'committee' and *jūnduì* 'army'. Can NP-*men* be identified as collective nouns of formal semantics? Is -*men* a marker of collectivity? In the next section, NP-*men* will be compared with collectivity nouns.

3.2   NP-*Men* Is not Collective in Semantics

In formal semantics, collective nouns such as *wěiyuánhuì* 'committee', *tuántǐ* 'group' or *jūnduì* 'army' which denote a set of entities are viewed as a single individual. This means individuals in the set are not distinct. Collective NPs cannot give rise to a distributive interpretation when they co-occur with distributive predicates, which are discussed in [21], as shows in (22).

(22)

> **Wěiyuánhuì /jūnduì   hěn   gāoxìng.*
> committee   army   very   happy

Collective nouns can become non collective as long as the individuals in the set are focused and explicated with the form *wěiyuánhuì de chéngyuán* 'members in the committee' and *jūnduì de jūnrén* 'soldiers in the army', as in (23). These NPs do infer a distributive interpretation and they are compatible with distributive predicates.

(23)

> *Wěiyuánhuì   de   chéngyuán /jūnduì de   jūnrén hěn   gāoxìng.*
> Committee DE   member   army DE   soldier very   happy
> 'The members in the committee / the soldiers in the army were happy.'

Furthermore, collective nouns cannot co-occur with *dōu*, the adverb of quantification which is an operator of distributivity, as (24) illustrates.

(24)

> **Zhè gè wěiyuánhuì /zhè   gè   jūnduì dōu   mǎi   le   yī   běn   shū.*
> this Cl committee   this   Cl   army   all   buy   Acc   a   Cl   book

Also, collective nouns are not compatible with the predicate, *bùtóng* 'be different', as in (25) because it has to select a NP subject which denotes a set of individuals in order to operate over individuals. The collective noun *wěiyuánhuì* 'committee' denotes a set viewed as single only.

(25)

> a.*Wěiyuánhuì zǒujìn   le   bútóng   de   jiāoshì.*
>   committee   walk into Acc   different   DE   classroom
> b.*Wěiyuánhuì zhùzài bútóng   de   chéngshì.*
>   committee   live in   different   DE   city

However, NP-*men* can co-occur with distributive predicates, as in (26). This shows NP-*men* denotes distributive interpretation, *i.e.* the property of *be happy/cry* applies to each member of the plural subject individually.

(26)

>Xuéshēng-men   hěn   gāoxìng /kū   le.
>student-*men*   very   happy   cry   Acc
>'Students are happy/cried.'

Different from collective nouns, NP-*men* can co-occur with *dōu*, as (27) illustrates.

(27)

>Xuéshēng-men   dōu chī   le   píngguǒ.
>Student-*men*   all   eat   Acc   apple
>'Students all ate apples.'

In (27), each individual in the set denoted by NP-*men* ate apples due to the operation of *dōu*. Therefore, it can be seen that distributive predicates and the adverb *dōu* can give rise to a distributive interpretation of NP-*men*.

Besides, NP-*men* can co-occur with the predicate *bùtóng* 'be different', as (28) illustrates.

(28)

>a. *Xuéshēng-men zǒujìn   le   bùtóng   de   jiāoshì.*
>   Student-*men*   walk into Acc   different   DE   classroom
>   'Students walked into different classrooms.'
>b. *Xuéshēng-men zhùzài bùtóng   de   chéngshì.*
>   Student-*men*   live in   different   DE   city
>   'Students live in different cities.'

bùtóng 'be different' must select a NP subject which denotes a set of individuals, viewed as individuals so that the comparison between any two individuals in the set is allowed. The above tests indicate clearly that NP-*men* denotes a set of individuals viewed as individuals, *i.e.* distributive interpretation. This differs from collective nouns like *wěiyuánhuì* 'committee', *jūnduì* 'army' which denote a complex atom (a set of atoms viewed as an individual), and refer to the notion of group. Therefore, it is shown that NP-*men* is not correspondent to the notion of collectivity, *i.e. group* defined in formal semantics.

In sum, the author has shown that NP-*men* is not collective noun traditionally defined nor collective nouns used in formal semantics.

In the next section, the author will analyze the semantics of NP-*men*.

4   Semantics of -*Men*

In this section, the author will show the interpretation of N-*men* and describe the semantics of -*men*. Linguists like Landman (cf. [15]), Link, Winter (cf. [21, 22]) among others discuss that according to the types of predicates, its NP subject selected denotes either distributive or collective interpretation. Some types of predicates will be used in this section. They are mixed predicates, collective predicates and symmetric predicates which will help to distinguish the interpretation of NP-*men*.

Mixed predicates apply to NP denoting distributive or collective interpretations (cf. [21]). The mixed predicate *jiāo* 'hand in' is attributed to the NP NP-*men* (29) which denotes a set of individuals.

(29)

> Xuéshēng-men   jiāo   le   bàogào.
> Student-*men*   hand in   Acc   report

Collective reading: 'Students handed in their **report**.'
Distributive reading: 'Students handed in their **reports**.'

When the predicate applies to individuals in the set, NP-*men* is interpreted as distributive. When the predicate operates over the set of individuals, it is interpreted as collective. (29) is read as collective at first. Then the distributive NP is the secondary interpretation. According to [21], such distributive interpretation is due to D-operation over the mixed predicate, attributed to individuals in the set denoted by NP subject.

Collective predicates must apply to a set of individuals. For example, the verb like *xuǎnchū* 'elect' in (30) operates over the NP-*men* which is yielded to a collective interpretation only.

(30)

> Xuéshēng-men   xuǎnchū   le   dàibiǎo.
> Student-*men*   elect   Acc   representative
> 'Students elected the representative.'

The above sentence means: the group of students as a unit elected a representative. The NP-*men* denotes a set of individuals, viewed as a group, or a unit.

However, NP-*men* cannot co-occur with other collective predicates, like *shì yī-gè hěn hǎo-de tuánduì* 'be a good team', *hěnduō* 'be numerous' (cf. [22]), as (31) illustrates.

(31)

> a. *Xuéshēng-men   shì   yī gè   hěnhǎo   de   tuánduì.
>   Student-*men*   be   a Cl   very good   DE   team
> b. *Xuéshēng-men   hěnduō.
>   Student-*men*   be numerous

NP-*men* denotes a set of individuals but it does not behave the same as the NP consisting of a demonstrative and the suffix *xiē* like *zhè-xiē* and *nà-xiē* and denotes also a set of individuals, as (32) illustrates.

(32)

    a. *Zhèxiē   xuéshēng   shì yī gè hěnhǎo de tuánduì.*
       these   student   be a Cl very good DE team
       'These students are a very good team.'
    b. *Nàxiē   xuéshēng   hěnduō.*
       Those   student   be numerous
       'Those students are numerous.'

A type of collective predicates including *jùjí, jíhé* 'gather' have to apply to a set of individuals, viewed as group, as (33) illustrates.

(33)

            Xuéshēng-men zài   dàtīng jíhé.
            Student-*men* in   hall   gather
            'Students gathered in the hall.'

Symmetric predicates like *be a couple, get married* must apply to a set of individuals which is divided into sub-sets of two individuals. These predicates such as *shì fūqī* 'be husband and wife', *shì péngyǒu* 'be friends', as in (34) cannot be attributed to the set denoted by NP-*men*.

(34)

    a. **Xuéshēng-men shì fūqī.*
       Student-*men*   be husband and wife
    b. **Xuéshēng-men shì péngyǒu.*
       Student-*men*   be friend

Such predicates require two individuals with certain identity. The contrast between (33) and (34) shows NP-*men* denotes a set of individuals in which sub-sets can involve two or more than two individuals without discrimination of identity. In other words, the individuals denoted by NP-*men* are identical.

Therefore, the above demonstrations indicate that the interpretation of NP-*men* denoting a set of individuals either distributive or collective depends on the types of predicates that co-occur with. Consequently, -*men* should not be labelled as a marker of collectivity but a marker of plurality over individuals. In addition, the plural denotation of NP-*men* is semantically differentiated from *zhè-xiē/nà-xiē*-NP.

In the next section, the author will analyze the connotation of the suffix -*men*.

5 *Men* with [Plural] Feature

This paper has shown the plural denotation of NP-*men* which is a set of individuals. In this section, the author will adopt the idea of [1] about numbers to explain the plurality of -*men*.

[1] shows the notion of "general number", which refers to NPs without singular-plural distinction. He points out that in English, there is not notion of general number because the distinction between singular and plural exists, as his figure, (35) illustrates.

(35)   ([1]: 19, Fig. 2.4)

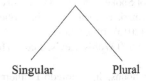

Singular   Plural

In English, *a*/*the professor* is the singular form and refers to singular entities while the NP suffixed by -*s*, (*the*) *professors* indicates plural entities. However, different from English, Mandarin Chinese uses the general number. For instance, in a sentence like: *zhuō shàng yǒu xīguā* 'there is/are (a) watermelon(s).', the bare noun *xīguā* 'water-melon' can refer to one entity or several entities, and it is possible to refer this bare noun to one slice or slices of watermelon according to its context (cf. [23]).

The author claims that the suffix -*men* marks a NP as plural because it makes distinction between singular and plural denotation. NP-*men* expresses the number of the entities denoted by NP is more than two, without exact number indication. For example, *zhànshì-men* 'warriors' denotes a plurality of warriors, without exact quantity. Thus, the suffix -*men* can be regarded as a number marker, and a marker of plurality. The suffix itself is featured with [plural].

6   Conclusion

This paper has shown that the suffix -*men* marks the plurality over NPs. This suffix conveys the number on the one hand, and it encodes NP as plural so as to help NP to denote a plurality of entities on the other hand. Since NP-*men* is plural denoting, it can give rise to two interpretations, collective and distributive according to the types of predicates that apply to the NP subject. As the paper clarified, the term, collectivity concerns two aspects of meaning in fact, collective nouns and collective interpretation of a NP.

References

1. Corbett, G.G.: Number. Cambridge University Press, Cambridge (2000)
2. Zhu, Q.Z.: On the probable relationship between the creation of the noun and personal pronoun's plural mark in Chinese and the translation of Indian Buddhist texts in ancient times. J. Chin. Linguist. **16**, 10–43 (2014). (in Chinese)
3. Jiang, L.S.: Wu as the origin of men. Stud. Chin. Lang. **3**, 259–273 (2018). (in Chinese)
4. Li, A.Y.H., Shi, Y.Z.: The story of -*men*. Contemp. Linguist. (1), 27–36 (2000)
5. Iljic, R.: Quantification in Mandarin Chinese: two markers of plurality. Linguistics **32–1** (329), 91–116 (1994)

6. Li, A.Y.H.: Plurality in a classifier language. J. East Asian Linguist. **8**, 75–99 (1999)
7. Huang, J.C.T., Li, A.Y.H., Li, Y.F.: The Syntax of Chinese. Cambridge University Press, Cambridge (2009)
8. Jiang, L.J.: Mandarin associative plural -men and NPs with -men. Int. J. Chin. Linguist. **4**(2), 191–256 (2017)
9. Chao, Y.R.: A Grammar of Spoken Chinese. University of California Press, Berkeley (1968)
10. Yang, Y.H.: The plural marker "men" and the collective marker "men". Lang. Teach. Linguist. Stud. **6**, 78–88 (2015). (in Chinese)
11. Lü, S.X.: Modern Chinese: 800 Words, Revised edn. The Commercial Press, Beijing (1999). (in Chinese)
12. De Vries, H.: Collective nouns. In: Cabredo-Hofherr, P., Doetjes, J. (eds.) The Oxford Handbook of Grammatical Number. OUP (2018)
13. Scha, R.: Distributive, collective and cumulative quantification. In: Groenendijk, J., Janssen, T., Stokhof, M. (eds.) Formal Methods in the Study of Language, pp. 483–512. Mathematical Center Tracts, Amsterdam (1981)
14. Landman, F.: Groups I, II. Linguist. Philos. (12), 559–605, 723–744 (1989)
15. Landman, F.: Events and Plurality. The Jerusalem Lectures. Kluwer, Dordrecht (2000)
16. Zhu, D.X.: Lectures on Grammar. The Commercial Press, Beijing (1982). (in Chinese)
17. Li, C., Thompson, S.: Mandarin Chinese: a functional reference grammar. University of California Press, Berkeley (1981)
18. Paris, M.C.: Problèmes de syntaxe et de sémantique en linguistique chinoise. Institut des Hautes Etudes Chinoises. Collège de France, Paris (1981)
19. Zhang, N.N.: Classifier Structures in Mandarin Chinese. De Gruyter Mouton (2013)
20. Vinet, M.T., Liu, X.Y.: Plurality in Chinese with a restricted class of noun-classifier words. In: Clarke, S., Manami, H., Kim, K., Sue, E. (eds.) Proceedings of the International Conference on East Asia Linguistics, Coll, Toronto WorkingPapers in Linguistics, no. 28, Toronto, University of Toronto, pp. 357–373 (2008)
21. Link, G.: The logical analysis of plural and mass terms: a lattice theoretic approach. meaning, use and interpretation of language. In: Bäuerle, R., Schwarze, C., Von Stechow, A. (eds.) Mouton de Gruyter, Berlin (1983)
22. Winter, Y.: Atoms and sets: a characterization of semantic number. Linguist. Inq. **33**(3), 493–505 (2002)
23. Zhang, N.N.: Expressing number productively in Mandarin Chinese. Linguistics **52**(1), 1–34 (2014)

Analysis of the Foot Types and Structures of Chinese Four-Syllable Abbreviations

Wei Ying[1] and Jianfei Luo[2(✉)]

[1] School of Humanities and International Education,
Zhejiang University of Science and Technology, Hangzhou, China
grace_ying0192@163.com
[2] College of Advanced Chinese Training,
Beijing Language and Culture University, Beijing, China
jluo@blcu.edu.cn

Abstract. This paper investigates the foot types and structural features of Chinese four-syllable abbreviations. Results show that: first, whether for the need of complexity and richness of ideogram or for avoiding ambiguity, these abbreviated forms are an essential and meaningful part of modern Chinese abbreviations; second, most of the existing forms tend to present 2 + 2 balanced prosodic structure, nominal attribute in part of speech. In addition, there also exists the possibility of abbreviation further to less than four syllables under some specific conditions.

Keywords: Four-syllable · Abbreviation · Foot type · Syntactic structure

1 Introduction

Abbreviations are a short form of words or phrases produced by abbreviating, aimed at making the complicated and lengthy expressions present much more efficiently, concisely, clearly and conveniently. Compared with the original forms, the number of syllables in abbreviations has been reduced obviously, with their original meanings unchanged completely, which meets the psychological requirements of people seeking convenience and the economic principle of language expression.

The previous studies of Chinese abbreviations mainly focus on the following three aspects. The first one is concerned about the definition of abbreviations. Lu & Zhu refer to abbreviations as 简称 [jiancheng] (shorter form) [1]. Xiao thinks of abbreviations as 词语减缩 [ciyu jiansuo] (word reduction) [2]. Li holds the view that the abbreviated forms mean the achievement of transformation successfully from a complicated and lengthy expression to a shorter structural form [3]. The second one is related to discussing the principles of abbreviating words or phrases. For instance, Yuan points out that abbreviations have to conform to the structural characteristics of words in modern Chinese language [4]. Niu holds that the most important feature of modern Chinese abbreviations is conciseness, thus the abbreviations should be the simplest form without affecting their conceptual expression [5]. The last one examines the varied number of abbreviated forms' syllables. For example, Liu believes that abbreviations mostly use two-syllable combination. According to his research, two-syllable

© Springer Nature Switzerland AG 2020
J.-F. Hong et al. (Eds.): CLSW 2019, LNAI 11831, pp. 801–809, 2020.
https://doi.org/10.1007/978-3-030-38189-9_80

abbreviations represent about 84.2% of the total quantity of authentic language materials investigated by him, while the figure for others using more than two syllables is just around 15.8% [6].

As discussed above, it is evident that many previous studies usually explore Chinese abbreviations in definition, principle and quantity of syllables, taking the general abbreviated forms as their research object. Most arguments revolve around disyllables, which are widely regarded as the best choice among Chinese abbreviations. However, there still exist three syllables, four syllables, or even more than four syllables in Chinese abbreviations. Actually, all of them play an important role in our daily conversations, and also deserve to be investigated and researched fully. Therefore, this paper attempts to concentrate on examining the foot types and syntactic structure features of Chinese four-syllable abbreviations, including exploring the necessity of their existence and the possibility of abbreviation further.

2 Types of Four-Syllable Abbreviations

This paper exhaustively searches all four-syllable abbreviations in *Modern Chinese Abbreviation Dictionary* (2002 edn.), totaling 866. According to their features of the prosody and internal syntactic structure, they can be divided into two categories: balanced type and unbalanced type. The number of the former is 794, accounting for 91.6% of the total four-syllable abbreviations; the latter is only 72, representing 8.3% merely.

Further observation demonstrates that the balanced category can be classified into five types: "[1 + 1] + [1 + 1]", "[1 + 1] + 2", "1 + 1 + 1 + 1", "2 + [1 + 1]" and "2 + 2", while the unbalanced category only includes three types of "3 + 1", "1 + 3" and other types. The specific quantitative statistics of each type or situation are shown in Table 1.

Table 1. Distribution of different types of four-syllable abbreviations

Type	Balanced type (794)					Unbalanced type (72)			Total
	[1 + 1] + [1 + 1]	[1 + 1] + 2	1 + 1 + 1 + 1	2 + [1 + 1]	2 + 2	3 + 1	1 + 3	Others	
Number	114	290	46	206	138	26	43	3	866
Proportion	13.2%	33.5%	5.3%	23.8%	15.9%	3.0%	5.0%	0.3%	100%

3 Balanced Four-Syllable Abbreviations

3.1 [1 + 1] + [1 + 1] Type

There are 114 abbreviations of this "[1 + 1] + [1 + 1]" type in the dictionary, representing 13.2% of the total balanced type, for example:

(1) a. [V+N] + [V+N]

减	租	减	息
jian	zu	jian	xi
reduce	rent	lower	interest

reduce the rent and lower the interest rate

b. [M+N] + [M+N]

两	弹	一	星
liang	dan	yi	xing
two	bombs	one	satellite

atomic and hydrogen bombs and man-made satellite

c. [M+V] + [M+V]

三	清	一	打
san	qing	yi	da
three	clean	one	reduce

clean up the book market, the audio and video market, printing enterprises and fight against lawbreakers who engage in producing pornographic and illegal publications.

The number of the above three subtypes is shown in the following Table 2.

Table 2. "[1 + 1] + [1 + 1]" internal category distribution

Type	[V + N] + [V + N]	[M + N] + [M + N]	[M + V] + [M + V]	Others	Total
Number	50	32	13	19	114
Proportion	44%	28%	11%	17%	100%

3.2   [1 + 1] + 2 Type

The number of this "[1 + 1] + 2" type is 290, accounting for 36.5% of the total balanced abbreviations, which is the largest of the above five sorts of the balanced, as shown in Table 1. Further analysis shows that the position of "2" in this type is mostly an existing disyllabic noun. In addition, a small number of existing disyllabic verbs are used, and very few adjectives and adverbs appear, as shown in Table 3.

Table 3. "[1 + 1] + 2" internal category distribution

Type	Noun	Verb	Adjective	Total
Number	261	27	1	190
Proportion	90.0%	9.3%	0.3%	100%

3.2.1   Type of Head Word [N]

The Table 3 demonstrates that there are 261 abbreviations which place a disyllabic noun in the core position of 2, accounting for 90% of the total number of "[1 + 1] + 2" type. Further observation shows that the grammatical structure combination of the part "1 + 1" is not single, but presents a variety of forms. According to whether the two parts of speech of "1" are identical, this paper divides them into two basic categories, and then subdivides them further, considering the different grammatical properties of morphemes.

One basic category is that "[1 + 1]" consists of two monosyllabic morphemes with the same part of speech, and the morphemes are juxtaposed to modify the subsequent noun headwords. The number of these abbreviated forms is 152, accounting for 58.2% of the total "[1 + 1] + 2" abbreviations with the headword N. For example:

(2) a. [N+N] + N

客　　　　　货　　　　　列车
ke　　　　　huo　　　　　lieche
passenger　freight　　　train
passenger trains and freight trains

b. [NR+NR] + N

马　　　　　恩　　　　　著作
Ma　　　　　En　　　　　zhuzuo
Marx　　　　Engels　　　work
works of Marx and Engels

c. [V+V] + N

采　　　　　编　　　　　人员
cai　　　　　bian　　　　renyuan
interview　　edit　　　　staff
interviewers and editors

The other is that the part of "[1 + 1]" is composed of two monosyllabic morphemes with different part of speech. The quantity of this sort is 110, representing 42.1% of the "[1 + 1] + 2" forms with the headword N. For example:

(3) a. [M+A] + N

四	大	名绣
si	da	mingxiu
four	great	famous embroidery

Suzhou embroidery, Hunan embroidery, Sichuan embroidery and Guangdong embroidery

b. [M+Q] + N

三	项	制度
san	xiang	zhidu
three	item	system

the personnel system, the salary system and social security system

c. [N+V] + N

船	保	协会
chuan	bao	xiehui
ship	assure	association

China Shipowners Mutual Assurance Association

3.2.2 The Type of the Head Word [V]

"2" in the type of "[1 + 1] + 2" can also be a disyllabic verb, which occupies the core position of the abbreviations. The amount of this type is 27, accounting for only 9.3% of the total number of [1 + 1] + 2. The subdivisions of this type can be illustrated as the following aspects.

Firstly, there are 10 [1 + 1] parts composed of two parallel monosyllabic nominal morphemes, such as 陆空联运 [lu kong lianyun] (the united transportation by railway and road and by air), 体脑倒挂 [ti nao daogua] (the social phenomenon of physical labor earning more than mental labor).

Secondly, an independent monosyllabic numeral morpheme combined with an independent monosyllabic quantifier or noun or adjective or verb morpheme occupies the position of [1 + 1]. The number of this sort abbreviations in the dictionary is 9, for example, 三网养鱼 [san wang yangyu] (the three ways to breed fish), 三大改造 [san da gaizao] (Three Great Remolding), 三防训练 [san fang xunlian] (the training aimed to prevent the army being damaged by nuclear, chemical and biological weapon).

In addition, there is a monosyllabic noun morpheme combined with a monosyllabic verb morpheme to form the part of [1 + 1], i.e. [N + V] + V type, such as自筹投资 [zi chou touzi] (self-raised investment), 机助制图 [ji zhu zhitu] (computer-aided mapping).

3.3 "1 + 1 + 1 + 1" Type

The abbreviations of "1 + 1+1 + 1" type are mainly composed of four independent monosyllabic morphemes with identical parts of speech. The morphemes are juxtaposed, and semantically express four different aspects of the same category. There are 46 items of such type in the dictionary, of which 40 items are nominal morphemes, accounting for 87% of the total "1 + 1 + 1 + 1" type and only 6 items are verbal morphemes, accounting for 13%.

3.3.1   Composition of Nouns

The abbreviations of "1 + 1 + 1 + 1" type, which consist of four nominal monosyllabic morphemes, for example, 煤电油运 [mei dian you yun] (coal industry, electric power industry, petroleum industry and transportation), 理工农医 [li gong nong yi] (science, engineering, agriculture and medicine).

3.3.2   Composition of Verbs

Although most of the abbreviations of "1 + 1 + 1 + 1" are composed of nominal morphemes, there also exist some examples of verbal morphemes (V + V + V + V). There are six items of this category in the dictionary, such as 采分编流 [cai fen bian liu] (purchase, classify, catalog and circulate), 产供销贸 [chan gong xiao mao] (produce, supply, sell and trade), 关停并转 [guan ting bing zhuan] (shut down, production halts, merge with other enterprises, switch to other products).

3.4   "2 + [1 + 1]" Type

The total number of abbreviations of "2 + [1 + 1]" in the dictionary reaches 206, accounting for 25.9% of the total balanced type, ranking the second place among the five types. In this type, the position of "2" is mostly a disyllabic noun, such as 长江三峡 [Changjiang san xia] (the Three Gorges of Yangtze River), and a few are disyllabic adjectives or verbs, for instance, 民主妇联 [minzhu fu lian] (Democratic women's federation), 运输工联 [yunshu gong lian] (Transport workers' Federation).

In the part of "[1 + 1]", the former "1" is mostly monosyllabic nominal morphemes, such as 中国足协 [Zhongguo zu xie] (Chinese Football Association). There also exist a few forms that are monosyllabic verbal or monosyllabic numerals. For example, 外贸扩权 [waimao kuo quan] (Expand autonomy in foreign trade), 台湾两岸 [Taiwan liang' an] (the both sides of the straits of Taiwan).

3.5   "2 + 2" Type

The abbreviation of "2 + 2" is mainly composed of two existing disyllabic words. In the dictionary, 138 abbreviations are of this type, accounting for 17.4% of the total balanced abbreviations.

Further observation shows that the latter "2" of such type is mainly composed of disyllabic nouns or disyllabic verbs.

3.5.1   Ending with Nouns

The number of "2 + 2" abbreviations ending with nouns in the dictionary is as high as 124, accounting for 89.9% of the total number of these abbreviations. According to the grammatical differences of morphemes in the first "2" position, it can be summarized into three categories:

The first category is "N + N" type, in which "2" is a disyllabic noun, such as 荧光光谱 [yingguang guangpu] (fluorescence spectrum), 液体导弹 [yeti daodan] (liquid missile) and so on. The first "2" can also be a proper noun, such as 麦道公司 [maidao gongsi] (Mc Donnell Douglas), 巴黎公约 [Bali gongyue] (Paris Convention).

The second category is "V + N" type. The first "2" in the structure is the existing disyllabic verb, such as 浮动价格 [fudong jiage] (floating prices), 引进项目 [yinjin xiangmu] (introduced project).

The third category is "A + N" type. The first "2" in the structure is the existing disyllabic adjective, such as 临时价格 [linshi jiage] (temporary price) and 民用燃具 [minyong ranju] (the burner for civil use).

3.5.2   Ending with Verbs

There are only 14 abbreviations of "2 + 2" ending with verbs in the dictionary, accounting for only 10.1% of the total. Further observation shows that most of them are "N + V" type, such as 工业审计 [gongye shenji] (an audit of industrial enterprise), 人口老化 [renkou laohua] (aging population). The remaining item is "V + V" type, for instance, 留用察看 [liuyong chakan] (be kept in office on a probationary basis).

4   Unbalanced Four-Syllable Abbreviations

As shown in Table 1 above, there are only 73 unbalanced four-syllable abbreviations, accounting for 8.4% of the total number of four-syllable abbreviations in the dictionary. Further, according to its internal prosodic, syntactic, semantic features and the degree of cohesion between morphemes, it can be divided into two basic types, namely "3 + 1" type and "1 + 3" type. In addition, the prosodic structure of the remaining three abbreviations is different from the two basic types, such as 教职员工 [jiaozhi yuangong] (faculty and staff) (1 + 2 + 1), 文教科技 [wen jiao keji] (culture, education, science and technology) (1 + 1 + 2) and so on.

4.1   "3 + 1" Type

There are 26 abbreviations of "3 + 1" type, accounting for 36.1% of the total unbalanced abbreviations, for example, 大中小学 [da zhong xiao xue] (university, middle school and primary school), 高中低档 [gao zhong di dang] (high-end, medium-end and low-end) and so on.

4.2   "1 + 3" Type

There are 43 abbreviations of "1 + 3" type, accounting for 59.1% of the total unbalanced abbreviations, which is much higher than that of "3 + 1" type. They can be divided into three sub-categories of "1 + [1 + 1]," "1 + [1 + 2]" and "1 + 3".

(4) 1+ [1+1+1] type

港	康	体	局
Gang	kang	ti	ju
Hong Kong	health	sports	board

Hong Kong sports development board

(5) 1+ [1+2] type

上	投	公司
shang	tou	gongsi
Shanghai	invest	corporation

Shanghai Investment Corporation

(6) 1+3 type

核	反应堆
he	fanyingdui
nuclear	reactor

a nuclear reactor

5 Conclusion

This paper mainly concentrates on investigating and analyzing Chinese four-syllable abbreviations from different perspectives. The results are in the following aspects:

First, these four-syllable abbreviations have strong function of description in semantics, by which people can express the diversity and complexity of events efficiently. For example, as in (1), each abbreviation contains rich semantic content, including all aspects of the event. Further reduction of the number of these abbreviated forms' syllables at random possibly results in failing to express the original meaning completely. Take two universities' name as another example: 华中师大 [Huazhong shida] (Central China Normal University), 华东师大 [Huadong shida] (East China Normal University). They can be abbreviated to "Huashi" by their students or local residents. But if people communicate with individuals with no related information, they cannot accurately understand which university "Huashi" refers to. Thus, the four-syllable abbreviations obviously have the necessity and value of existence in modern Chinese abbreviations.

Second, the four-syllable abbreviations mainly present the form of 2 + 2 balanced type. From Table 1, it can be found that the proportion of balanced and unbalanced abbreviations is basically 10:1. Some unbalanced types are usually linked with the expressions of representing three kinds of things or phenomena once, such as 大中小学 [da zhong xiao xue] (university, middle school and primary school), 海陆空军 [hai lu kong jun] (navy, army and air force). Furthermore, most of the four-syllable abbreviations are noun-based, which means that noun (or nominal morpheme) is the core of word formation, such as (2) and (3).

The last one, the four-syllable abbreviations can continue to be abbreviated further under some specific conditions, such as 二次大战 [er ci dazhan] (World War II) and 二战 [er zhan] (World War II), 甲型肝炎 [jiaxing ganyan] (Hepatitis A) and 甲肝 [jia gan] (Hepatitis A).

References

1. Lu, S., Zhu, D.: Speech on Grammar and Rhetoric. The Commercial Press Beijing (1951). (in Chinese)
2. Xiao, W.: On the reduction of modern Chinese words. Stud. Chin. Lang. (3), 125–128 (1959). (in Chinese)
3. Li, X.: On the contraction of Chinese words. Chin. Mod. (1), 88 (1983). (in Chinese)
4. Yuan, H., Ruan, X.: Modern Chinese Abbreviation Dictionary. Language and Culture Press, Beijing (2002). (in Chinese)
5. Niu, X.: Research and standardization of modern Chinese abbreviations. Master thesis, Hebei Normal University, Shijiazhuang (2004). (in Chinese)
6. Liu, J.: An analysis of modern Chinese abbreviations. Master thesis, Anhui University, Wuhu (2004). (in Chinese)

Similarities and Differences Between Chinese and English in Sluicing and Their Theoretical Explanation

Yewei Qin[1] and Jie Xu[1,2(✉)]

[1] School of Chinese Language and Literature,
Wuhan University, Wuhan 430072, China
jiexu@um.edu.mo
[2] Department of Chinese Language and Literature,
University of Macau, Taipa, Macao, China

Abstract. Sluicing refers to a certain type of compound sentence in which one clause is a *wh* question where all sentential elements, but the *wh*-phrase itself alone, are omitted. In semantic interpretation, a sluicing sentence is comparable to a full *wh* interrogative. The study of sluicing sentence involves several important aspects of syntactic theory. Zhang and Xu in their article provide a unified account of sluicing in Chinese and English from the perspective of predicative Empty Category [1]. It is demonstrated in this article that one important issue still remains to be resolved regarding the similarities and differences between Chinese and English in sluicing: What remains after deletion in English is the *wh*-phrase alone, but there must be a copular verb going with the retained *wh*-phrase in Chinese. As the major new viewpoint articulated in this article, the above cross-linguistic contrast is illustrated to be more principally explainable by appealing to the theory of focus rather than by using ad hoc stipulations.

Keywords: Sluicing · Focus · Chinese · English · Similarity · Difference

1 Introduction

In its classical sense, the so-called "sluicing" refers to the type of construction as illustrated by the English sentences below.

(1) John saw Mary somewhere, but I don't know where.
(2) Jack resigned, but I don't know why.
(3) John opened the locked door but I don't know how.
(4) John went out with somebody, guess who.
(5) Abby was reading something, but I don't know what.
(6) I know he bought something, but I don't know what.

The second parts of the above sentences have the interpretation equivalent to "but I don't know where John saw Mary" for (1), and "but I don't know why Jack resigned" for (2) respectively. Sluicing has been taken as an important construction to show what is not seen visibly or heard phonetically can still be syntactically active. This kind of

© Springer Nature Switzerland AG 2020
J.-F. Hong et al. (Eds.): CLSW 2019, LNAI 11831, pp. 810–820, 2020.
https://doi.org/10.1007/978-3-030-38189-9_81

phenomenon has attracted attentions from a number of linguists including the authors of [2–6], and [7] among others. The clause is presumably derived by moving a *wh*-phrase to the clause periphery position, exactly in the way a *wh*-question is formed in languages like English. The clause following the moved *wh*-phrase, an Inflectional Phrase (IP) using a generative grammatical term, is simply left unpronounced, or deleted at the level of Phonetic Form.

2 Similarities and Differences Between Chinese and English in Sluicing

The above movement and deletion approach to English sluicing phenomenon leads to a natural question of what happens in *wh-in-situ* languages such as Mandarin and other Chinese dialects. Two approaches to this problem are remarkable. One is to argue that sluicing also exists in Chinese and the similar movement and deletion operations are at work in the Chinese language as well as in English. The other is to deny that Chinese has a counterpart of English in sluicing. Instead, Chinese has a base-generated pseudo-sluicing structure containing a *pro* subject, a copular verb "be", and *wh*-phrase in the form of "*pro* + *be* + *wh*", with "be" being optional in some cases, obligatory in others.

Superficially, Chinese has the exact counterpart of the English sluicing examples [8–10].

(7) Zhangsan zai mouge difang kandao Lisi, dan wo bu zhidao shenme difang.
 Zhangasn at some place see Lisi,   but I not know what place.
 "Zhangsan saw Lisi at some place, but I don't know where."
(8) Zhangsan cizhi le, dan wo bu zhidao weishenme.
 Zhangsan resign ASP but I not know why
 "Zhangsan resigned, but I don't know why."
(9) Zhangsan zou le yi-duan lu, dan wo bu zhidao duoyuan-de lu.
 Zhangsan walk ASP one section road but I not know how far
 "Zhangsan walked for a certain distance, but I don't know how far."

Having acceptable counterparts like those in the sentences above might suggest that Chinese also features sluicing derived by movement and deletion. The two languages might just differ in the motivation that triggers the movement of the *wh*-phrase or in which component of the language mechanism the movement operation takes place. The *wh*-phrase is motivated to move by its own *wh*-feature as is generally assumed in English while the phrase is motivated to move by its focus feature as argued in [8] and [11]. Zhang and Xu offers a unified account of sluicing in Chinese and English from the perspective of predicative Empty Category, arguing that Chinese sluicing, just like its English counterpart, is also a grammatical form containing a predicative Empty Category, and that the contrast between two the languages mainly lies in the difference as where and how the derivational operations take place ([1]). In particular, the movement takes place in syntax in English while that happens at the level of Logical Form in Chinese. In either way, a unified account of Chinese and English sluicing and maximal level of universality have both been achieved.

However, it is noted that the two constructions in Chinese and English are not exactly the same. In additional to some other less significant differences, one important difference between the two languages is that a copular verb (i.e., *shi* "be" in Mandarin Chinese) is often used before the remaining *wh*-phrase in contrast to a sole *wh*-phrase in English ([7–10]).

> (10) Zhangsan renshi mouge ren, dan wo bu zhidao shi shei.
>   Zhangsan know some person but I not know be who
>   "Zhangsan knows someone, but I don't know who."
> (11) Zhangsan dadao le dongxi, dan wo bu zhidao shi shenme.
>   Zhangsan hit ASP thing   but I not know be what
>   "Zhangsan hit something, but I don't know what."

Even more notably, although the copular verb *shi* "be" is required to be used before the remaining *wh*-phrase obligatorily in (10) and (11) above when the remnant *wh* phrases are arguments, such a copular verb does not have to be used when the remnant *wh* phrases are adjuncts. The copular verb *shi* is used before remnant wh adjuncts largely when the adjuncts *wh*-phrases have been put in emphasis (as put in bold below).

> (12) Zhangsan zai mouge difang kandao Lisi,dan wo bu zhidao
>   Zhangasn at some place see Lisi,   but I not know
>   (shi) SHENME DIFANG.
>   be what place.
>   "Zhangsan saw Lisi at some place, but I don't know WHERE."
> (13) Zhangsan cizhi le, dan wo bu zhidao (shi) WEISHENME.
>   Zhangsan resign ASP but not know be why
>   "Zhangsan resigned, but I don't know WHY."
> (14) Zhangsan zou le yi-duan lu, dan wo bu zhidao (shi) DUOYUAN-DE LU.
>   Zhangsan walk ASP one section road but I not know how far
>   "Zhangsan walked for a certain distance, but I don't know HOW FAR."

However, what is most crucial here for points discussed in this article is not whether the copular verb is optional or obligatory, but rather is whether it may or may not be used before a remnant *wh*-phrase. It has been made clear at least that the copular verb may be used in (pseudo-) sluicing construction in Chinese but not in English. This is the most important cross-linguistic contrast that demands a reasonable explanation. It will be argued below that this crucial cross-linguistic difference can be recast as part of an overall picture of a language typology of focus, which has much more general effects than those observed in sluicing constructions cross-linguistically.

3   A Language Typology of Focus Phenomena

As is discussed quietly extensively in some early generative literature, one aspect of semantic interpretation of a sentence is a division of its reading into FOCUS and PRESUPPOSITION. As a working definition, Jackendoff proposes that FOCUS is the information conveyed in a sentence that is assumed by the speaker not to be shared by him/her and the hearer, and that PRESUPPOSITION is the information in a sentence that is assumed by the speaker to be shared by him/her and the hearer. For a normal sentence, the question is whether the FOCUS is reflected syntactically or not, rather than whether it has a FOCUS at all [12]. The unshared information is assumed by the speaker to be known to the speaker himself in a declarative sentence, whereas it is known to the hearer, but not to the speaker in an interrogative sentence. Note that the above definition of Focus is formulated more semantically or pragmatically than formal-syntactically. It is argued in some works including [13, 14] and [15] that Focus can also be characterized as a purely formal syntactic feature "[+F]", which gets assigned to constituents at an appropriate level of syntactic representation through a syntactic operation of "Focus-Assignment." Although every sentence, by definition, has at least one focused constituent, degree of focalization on different constituents may vary from one sentence to another. There are obviously two types of Focus that are formal-linguistically relevant: "Strong Focus" and "Weak Focus".

3.1 Focus Marking

The [+F] marking resulting from Focus-Assignment may trigger phonological and/or syntactic processing. The phonological processing of [+F], such as primary stress and higher pitch, is well discussed in the literature (e.g., [12] and [13], among others). As for syntactic processing, the most conceivable way is simply to insert an overt Focus Marker in a sentence, whatever the marker may be in a particular language. This possibility is attested in Chinese (readers are referred to works of [15, 16] and [17] among others for details). For example:

> (15) Shi wo mingtian cheng huoche qu Guangzhou.
>   be I tomorrow ride train   go Guangzhou
>   "It is I who will go to Guangzhou by train tomorrow."
> (16) Wo shi mingtian cheng huoche qu Guangzhou.
>   I be tomorrow ride train   go Guangzhou
>   "It is tomorrow when I will go to Guangzhou by train."

As seen in the above examples, *shi* "be" is used to mark a focused constituent in Chinese. Of course, this is not the only usage of *shi* in the language. *Shi*, just like its English counterpart *be*, may also be a regular copular verb.

In *wh*-question, only the *wh*-phrase, not any others, can be the strongly focused constituent. This is because *wh*-phrases have already been assigned the Strong Focus feature [+F] in the lexicon of the language and carry the feature into syntax when they themselves are composed into a phrase marker. Such a lexical marking interacts with the syntactic marking in an interesting way. For example:

(17) Shui[+F] mai-le neiben zidian?
Who buy-ASP that dictionary
"Who bought that dictionary?"
(18) Ni shenme shihou[+F] nian-de daxue?
you what   time   attend-ASPcollege
"When did you attend college?"

If necessary, the Focus Marker *shi* may also be inserted to intensify focus feature of the constituent concerned, giving rise to sentences like the following.

(19) Shi shui[+F] mai-le neiben zidian?
be who buy-ASP that dictionary
"WHO bought that dictionary?"

(20) Ni shi shenme shihou[+F] nian-de daxue?
you be what   time   attend-ASP college
"WHEN did you attend college?"

3.2 Focus Fronting

The insertion of a Focus Marker just represents one way of syntactically processing a strong focus. An alternative device is to move the focused constituent to a more prominent position, as is observed in such languages as Archaic Chinese, Hungarian, and English as illustrated in the work of [15].

Focus-Fronting in Archaic Chinese
The basic word order of Archaic Chinese, just like that of modern Chinese, is of Subject-Verb-Object pattern. A focused object Noun Phrase (NP), however, may be fronted to a pre-verbal position.

(21) Wu shui qi? qi tian hu? (Lun Yu. Zihan)
I who cheat cheat God Particle
"Who do I cheat? Do (I) cheat the God?"
(22) Yu wei li shi shi. (Zuozhuan. Chenggong Year 15)
I only interest that care
"I care about INTEREST only."

Focus-Fronting in Hungarian
Similar phenomena are also observed in Hungarian, as reported in [14] among others. In terms of the basic word order, Hungarian is also a Subject-Verb-Object language. But if an object NP is a *wh*-phrase or a focused constituent, it will not remain in a post-verbal position. Rather, it has to move to a pre-verbal position, otherwise the sentence will be ungrammatical. The movement exhibited in the sentences is also triggered by the strong focus feature [+F] in Hungarian in the same way as in Archaic Chinese to satisfy a syntactic condition. It is important to note that the formal syntax of Hungarian

is not sensitive to whether a focused constituent is a *wh*-phrase or not either. It will be fronted as long as it has the feature [+ Fs].

> (23) Attila a foldrengestol$_i$ felt t$_i$.
> Attila the earthquake-from feared
> "It was the earthquake that Attila was afraid of."
> (24) Mari mit$_i$ telt az asztalra t$_i$?
> Mary what-Acc put the table-onto
> "What did Mary put on the table?"

Focus-Fronting (+Focus Marking) in English

As demonstrated above, cross-linguistically there are two types of syntactic processing of [+F]-marked constituents. One is "Focus Marking," which inserts a Focus Marker (e.g., the copular verb *shi*) before the focused constituent, as attested in modern Chinese, and the other is "Focus-Fronting," which moves the focused constituent to a pre-verbal position, as observed in Archaic Chinese and modern Hungarian. The question now is whether these two devices can be used jointly in a single construction in a single language. It is illustrated in [15] that the formation of the English cleft-sentence is a perfect example of this logical possibility's instantiation. The Focus Marker in English is also a copular verb *be*. These two devices are both triggered by the same feature mark [+F]. And the insertion of pronoun *it* before the copular verb *be* in English is due to a completely different requirement of the English grammar which requires a subject position to be filled with a lexical NP. Archaic Chinese and modern Hungarian pattern with each other, differing from modern English minimally in the landing sites of focused constituents.

> (25) It is the new house$_i$ that John will buy t$_i$ for his mother tomorrow.
> (26) It is tomorrow$_i$ when John will buy the new house for his mother t$_i$.

The movement of focused constituent is triggered by the strong focus mark [+F] for both *Wh*-phrases and non-*Wh*-phrases. *Wh*-phrases and non-*Wh*-phrases differ minimally in how and where the mark [+F] is assigned, but not in whether the mark is assigned in the first place or in whether the [+F]-marked constituents will move. [+F] is assigned to *Wh*-phrase in the lexicon and is carried along with *Wh*-phrase into the syntax, whereas the feature is assigned to non-*Wh*-phrase at the level of deep structure. However, there appears to be an obvious problem when one takes a second look at the English phenomenon. As demonstrated above, the formation of English cleft-sentence involves the joint application of two focus devices: The insertion of the Focus Marker *be*, and the fronting of the focused constituent. The formation of *Wh*-question, however, seems to involve the fronting of the *Wh*-phrase only, but no insertion of a Focus Marker. That is presumably because a *Wh*-phrase itself represents a sort of overt Focus Marking which is comparable to the lexical item *be* in English at an abstract level.

> (27) What$_i$ will you buy t$_i$?
> (28) Why did you leave New York City t$_i$?

A language typology of focus device is established by which there are two devices: the insertion of the Focus Marker before focused constituent, and the fronting of focused constituent. English represents the third type of language, which makes use of both devices in case of normal focused constituent and only the device focus fronting in case of *wh* focused constituent [15].

4 Deriving the Cross-Linguistic Contrasts in Sluicing

It can be seen obviously that the similarities and differences between Chinese and English in sluicing pattern perfectly with that between the two languages in focus device, we thus may make a generalization that superficial properties of sluicing in the above two languages have no independent theoretical status. They are various types of instantiation of focus devices in various languages, or a special subtype of focus construction. What makes it "special" is that there is only operation of focus device in typical focus construction whereas there are operations of focus device plus deletion of unfocused constituents in sluicing sentence in Chinese, English, and perhaps in all other human languages.

4.1 Sluicing in Chinese: Focus Marking + Deletion of Unfocused Elements

As is discussed above, the focus device in Chinese is the insertion of focus marker (i.e., copular verb *shi*), and that a *wh*-phrase is usually a focused constituent, whose focus nature could even be intensified though the use of the focus marker.

(29) Shi wo mingtian cheng huoche qu Guangzhou.
　　　be I   tomorrow ride train   go Guangzhou
　　　"It is I who will go to Guangzhou by train tomorrow."
(30) Shui mai-le neiben zidian?
　　　who buy-ASP that dictionary
　　　"Who bought that dictionary?"
(31) Shi shui mai-le neiben zidian?
　　　be who   buy-Asp that   dictionary
　　　"WHO bought that dictionary?"

Sluicing sentence patterns with focus contraction in Chinese in that the focus marker *shi* is used before the focused constituent. The two constructions differ minimally in that the unfocused constituents are still there, and perhaps just weakened phonetically in typical focus construction whereas unfocused constituents are all deleted after the insertion of a focus marker in a sluicing sentence.

(32) Youren ganggang likai le – caicai shi shui ~~ganggang likai le~~.
    Somebody just left   guess be who just left.
    "Somebody just left – guess who."

(33) *Youren ganggang likai le – caicai shui ~~ganggang likai le~~.
    Somebody just left – guess who just left.
    Intended: "Somebody just left – guess who."

(34) Lisi jueding cizhi, keshi wo bu zhidao shi weishenme ~~Lisi jueding cizhi~~.
    Lisi decide resign but I not know be why Lisi decide resign.
    "Lisi decided to resign, but I don't know why.'

(35) Lisi jueding cizhi, keshi wo buzhidao weishenme ~~Lisi jueding cizhi~~.
    Lisi decide resign but I not know why Lisi decide resign.
    "Lisi decided to resign, but I don't know why.'

Furthermore, as illustrated in the above examples, the focus marker *shi* can but does not have to be used in typical focus construction and in those sluicing sentences with adjunct *wh*-phrases whereas the focus marker has to be used there obligatorily in sluicing sentences with argument *wh*-phrases. In other words, there is a clear argument-adjunct asymmetry of *shi* support. First of all, it is not an issue as why a focus marker could be optional in focus construction since we all know that focus is a relative conception, the degree of focalization varies in different contexts and the focus maker is used only when the degree of focalization is relatively high enough. The very same logic applies to cases of sluicing with adjunct *wh*-phrases. The issue simply is why the focus marker has to be used all the time in all sluicing sentences with argument *wh*-phrases. Although this could be a very interesting and very important question, its full scale exploration will go beyond the scope of this short article. We would rather leave the question open, but simply point out that among various attempts in tackling this problem, the Case requirement approach is the most promising one. By the Case requirement approach to the issue, *shi* as copular verb has to be placed before a *wh*-argument since the argument cannot get Case assignment elsewhere other than from the copular verb itself whereas support of *shi* before a *wh*-adjunct is not necessary because those adjuncts don't require a Case assignment in the first place [11]. Additional evidence in support of this Case requirement approach comes from the following example in which the copular verb *shi* does not have to be used even before a *wh*-argument in case the argument may get Case-marked by such a preposition as *gen* "with" [11].

(36) Laowu gen ren qu wan, keshi wo bu zhidao gen shui.
    Laowu with person go play but I not know with who
    "Laowu went playing with someone, but I don't know with who."

What is even more remarkable is that Chinese sluicing sentence, very much like regular focus construction, does not have to have a *wh* –phrase in the focus position, rather, it may just have a normal noun phrase there, and the focus marker *shi* needs to be placed obligatorily before the non-*wh* noun phrase as well.

(37) Youren ganggang likai, wo zhidao shi Zhang Xiansheng.
　　Somebody just left – I know be Zhang Mr.
　　"Somebody just left – I know it is Mr. Zhang."
(38) Lisi cizhi le, wo zhidao shi zhege yuanyin
　　Lisi resign ASP, I know be this reason
　　"Lisi resigned, I know it is because of this reason."
(39) *Youren ganggang likai – wo zhidao Zhang Xiansheng.
　　Somebody just left – I know Zhang Mr.
　　Intended: "Somebody just left – I know it is Mr. Zhang."
(40) *Lisi cizhi le, wo zhidao zhege yuanyin
　　Lisi resign ASP, I know this reason
　　Intended: "Lisi resigned, I know it is because of this reason."

4.2　Sluicing in English: Focus-Fronting (+Focus Marking) + Deletion of Unfocused Elements

The focus device in the English language is the obligatory use of Focus – Fronting and optional use of Focus-Marking. When it operates in a declarative sentence it will be joint application of Focus-Fronting and Focus-Marking, when it operates in an interrogative sentence, it will be Focus-Fronting only.

(41)　It is the new house$_i$ that John will buy t$_i$ for his mother tomorrow.
(42)　What$_i$ will you buy t$_i$?

　　That is exactly what happens in comparable sluicing sentence in the language. Once again, the only difference between typical focus construction and sluicing sentence in English, just like that in Chinese, is that there is a deletion of unfocused constituents following the operation of focus device.

(43) Somebody just left— guess who ~~just left~~.
(44) Abby was reading, but I don't know what ~~Abby was reading~~.

　　Furthermore, in case the focused constituent is a non-*wh* noun phrase, rather than a *wh*-phrase, the copular verb *be*, together with a dummy subject *it*, will need to be used obligatorily.

(45)　Somebody just left, and I know it is John ~~who just left~~.
(46)　*Somebody just left – I know John ~~who just left~~.
(47)　Abby was reading, and I know it was a short novel ~~Abby was reading~~
(48)　*Abby was reading, and I know a short novel ~~Abby was reading~~

4.3 A Universal of Sluicing Form: Focus Device + Deletion of Unfocused Elements

Given the forgoing discussion about relevant phenomena observed in Chinese and English, we now have come up with a generalization with regard to a language universal of sluicing form. That is, all sluicing phenomena in human languages can be derived by joint application of various forms of focus device in various language and the deletion of unfocused constituents. This generalization can be diagrammed as follows.

Focus Device (i.e., Focus Fronting OR Focus Marking)

+

Deletion of Unfocused Constituents

v

The So-Called "(Pseudo-) Sluicing" Phenomena in Chinese, English, and…

5 Conclusion

Starting with a brief discussion of the advantages and remaining problems of previous proposals regarding similarities and differences in sluicing between Chinese and English, this article argues that some crucial cross-linguistic contrasts can be explained properly by appealing to the theory of focus. In particular, the so-called "sluicing sentence" does not have an independent status in the theory of grammar, but all the relevant phenomena previously assumed to be of sluicing can be recast as various forms of instantiation of various focus devices in various languages. In this regard, a language universal regarding sluicing can be generalized so that all "(pseudo-) sluicing" phenomena in all human languages can be derived from the joint application of focus device and deletion of unfocused constituent, with the focus device being either focus marking, focus fronting or both.

References

1. Zhang, Y., Xu, J.: Predicative empty category in sluicing sentence and its theoretical significance. Paper Presented at the International Symposium on Syntactic Deletion in Chinese. Central China Normal University, Wuhan, 12–14 October (2018)
2. Ross, J.R.: Guess who? In: Binnick, R., et al. (eds.) Proceedings of the Fifth Regional Meeting of the Chicago Linguistic Society. The University of Chicago Press, Chicago (1969)
3. Takahashi, D.: Sluicing in Japanese. J. East Asian Linguist. 3(3), 265–300 (1994)
4. Chung, S., Ladusaw, W.A., McClosky, J.: Sluicing and logical form. Nat. Lang. Semant. 3, 239–292 (1995)
5. Kizu, M.: Sluicing in Wh-in-situ languages. In: Proceedings of the Thirty-Third Regional Meeting of the Chicago Linguistic Society (The Main Section), pp. 231–244 (1998)

6. Merchant, J.: The Syntax of Silence: Sluicing, Islands, and the Theory of Ellipsis. Oxford University Press, Oxford (2001)
7. Li, Y.-H.A., Wei, T.-C.: Sluicing, sprouting and missing objects. Stud. Chin. Linguist. **38**(2), 63–92 (2017)
8. Wang, C.: On sluicing in Mandarin Chinese. M.A. thesis. Taiwan Tsing Hua University, Taiwan (2002)
9. Adams, P.W.: The structure of sluicing in Mandarin Chinese. Proc. Pa. Linguist. Colloq. **27**, 1–16 (2003)
10. Wei, T.: Predication and sluicing in Mandarin Chinese. Ph.D. dissertation. National Kaohsiung Normal University, Taiwan (2004)
11. Wang, C.A., Wu, H.I.: Sluicing and focus movement in *Wh*-in-situ languages. In: Eilam, A., Scheffler, T., Tauberer, J. (eds.) University of Pennsylvania Working Papers in Linguistics, vol. 12, no. 1, pp. 375–387 (2006)
12. Jackendoff, R.: Semantic Interpretation in Generative Grammar. MIT Press, Cambridge (1972)
13. Culicover, P.W. Rochemont, M.S.: Stress and focus in English. Language **59**(1) (1983)
14. Horvath, J.: FOCUS in the Theory of Grammar and the Syntax of Hungarian. Foris Publications, Dordrecht (1986)
15. Xu, J.: Formal Aspects of Chinese Grammar. World Scientific, London (2016)
16. Xu, J., Li, Y.: Focus and the two non-linear grammatical categories: "Neg" and "Wh". Zhongguo Yuwen **2**, 81–92 (1993)
17. Xu, J.: Grammatical Principles and Grammatical Phenomena. Peking University Press, Beijing (2001)

An Investigation of Heterogeneity and Overlap in Semantic Roles

Long Chen$^{(\boxtimes)}$ and Weidong Zhan$^{(\boxtimes)}$

Peking University, Peking, China
{chenlong,zwd}@pku.edu.cn

Abstract. An inventory of semantic roles is needed in semantic role labelling. However, no matter what are semantic roles defined as, there is always heterogeneity and overlap in semantic roles. The semantic properties of members in the same semantic role can be different, while those of members in different semantic roles can be similar. It is a widespread phenomenon in semantic role types. The paper analyzes the cause of heterogeneity and overlap, and points out the close connection between them. The severity of heterogeneity and overlap described in the article is assessed using data from the survey of Chinese PropBank, a publicly available semantic resource consisting of two parts: the Frameset and the corpus.

Keywords: Semantic roles · Heterogeneity · Overlap · Chinese PropBank

1 Introduction

Semantic role labeling is an essential task in natural language processing. It helps computers understand the meanings of natural language, and provides necessary semantic information for tasks such as machine translation, question and answering, and information extraction [1]. Based on the theory of semantic roles in linguistics, semantic role labeling requires an inventory of semantic roles.

Linguists set up the concept of semantic roles, and classify the semantic roles of verbs, in order to summarize the meanings of verbs, find out the commonality of participants of events described by some verbs, and distinguish them from other participants with different semantic properties. However, in the practice of semantic role labeling, it is often discovered that there are members of the same semantic role with different semantic properties, and members of different semantic roles with similar semantic properties.

(1a) 小明刚才哭了，现在正在擦脸。(Xiaoming cried, and is wiping his face now.)

(1b) 小明刚才哭了，现在正在擦眼泪。(Xiaoming cried, and is wiping out his tears.)

(2a) 张三想办法吵醒了李四。(Zhangsan managed to awaken Lisi.)

© Springer Nature Switzerland AG 2020
J.-F. Hong et al. (Eds.): CLSW 2019, LNAI 11831, pp. 821–833, 2020.
https://doi.org/10.1007/978-3-030-38189-9_82

(2b) 闹钟吵醒了李四。(The alarm clock awakened Lisi.)
(2c) 张三吵醒了李四。(Zhangsan awakened Lisi.)

In sentence (1a) and (1b), "脸" (face) and "眼泪" (tears) should both be labeled as patients according to PKU SemBank Annotation Guidelines [2] and the classification of Peking University Netbank [3]. However, they are semantically heterogeneous. The former exists after the action happens, and the latter disappears.

According to the semantic role systems listed above, agents are defined as volitional actors of actions. "张三" (Zhangsan) in (2a) matches the definition of agents, and "闹钟" (alarm clock) in (2b) does not. In (2c) it is hard to tell whether "张三" (Zhangsan) is volitional, as he might have awakened "李四" (Lisi) intentionally, or accidentally. Therefore it is hard to classify the semantic role of "张三" (Zhangsan). It can be concluded that there are overlaps between agents and other proto-agent roles like experiencers.

The heterogeneity and overlap of semantic roles cause trouble to natural language understanding. Take the task of question and answering as an example, if the computer is asked "什么被小明擦了？(What's wiped out by Xiaoming?)", the answer should be "眼泪" (tears), no matter which sentence in (1) is the input sentence. However, if the question and answering system is based on the result of semantic role labeling, it may yield a wrong answer, since "脸" (face) in (1a) and "眼泪" (tears) in (1b) are both labeled as patients. If we try to annotate them as different roles, the difficulty of semantic role labeling will increase, and when the object is in the intersection of the different roles, its annotation is still tricky.

In this paper, the heterogeneity and overlap of semantic roles in Chinese and the way Chinese PropBank reflects and deals with the problem are examined.

2   The Heterogeneity of Semantic Roles

If some participants in the events expressed by a verb have different semantic properties, but we have to label these participants as the same semantic role, this semantic role is considered heterogeneous in this paper. Heterogeneity occurs in many types of semantic roles.

2.1   The Heterogeneity of Agents

Agents are usually defined as the volitional actor of an action. In sentence (3a), "小明" (Xiaoming) fits with the definition of agents. However, "风" (wind) in (3b) has no life or volition, and cannot be viewed as a typical agent. Lin [4] suggested that natural forces can be treated as agents, and Fillmore [5] regarded winds as instruments. In fact, no matter what semantic role is labeled on "风" (wind), it is an atypical member of that semantic role, which has different semantic properties from other typical members. Some linguists noticed the different semantic properties of natural forces from those of typical agents and proposed to set up a role called "Actor" or "Force" to label natural forces [6].

(3a) 小明把门关上了。(Xiaoming closed the door.)

(3b) 风把门关上了。(The Wind closed the door.)

Generally, as for any semantic roles, the members that do not fit with the definitions well are usually heterogeneous with other members. They have different semantic properties with the typical members. In semantic role labeling, for the consistency in annotation, we can classify the atypical members through analogy. In shallow semantic parsing, the differences between the typical members and atypical ones of a semantic role can be omitted sometimes, but further semantic analyses are needed to uncover the real accurate meanings of the atypical members and the differences between them and the typical members.

Besides the atypical agents, some members that match the definition of agents well may also have different meanings with other members.

(4a) 小赵去修自行车了。(Xiaozhao is going to have his bike repaired.)

(4b) 修车师傅在给小赵修自行车。(A repairer is repairing the bike for Xiaozhao.)

(5) 小赵去洗澡了。(Xiaozhao is going to take a bath.)

In sentence (4a), "小赵" (Xiaozhao) indeed volitionally caused the event "修自行车" (repairing the bike) to happen, so "小赵" (Xiaozhao) matches the definition of agents. According to the theory of proto-roles [7], "小赵" (Xiaozhao) has volition and causality, which are two properties of proto-agents. Compared with sentence (4b), "修车师傅" (bike repairer) seems to be the "real agent", and "小赵" (Xiaozhao) in (4b) is more likely to be labeled as a beneficiary. This phenomenon is not rare in Chinese. It also appears with verbs like "上课" (take/have a lesson), "看病" (see a doctor/patient), "理发" (cut one's hair). A type of the subjects of these verbs has the semantic feature of being benefited, while another type has the semantic feature of "doing by oneself". If we label the "benefited subjects" as beneficiaries, then it will be hard to classify "小赵" (Xiaozhao) in sentence (5), because on one hand, "小赵" (Xiaozhao) is both being benefited from the action and is doing it by himself, and we cannot label two semantic roles on it; on the other hand, we can imagine an occasion when "小赵" (Xiaozhao) cannot take a bath on his own for some reasons and needs to take a bath with the assist of others. Above all, these sentences reflect the heterogeneity of agents. Typical agents have the semantic feature of volition, but there are also agents who do not have the semantic feature of volition, and agents who have the semantic feature of being benefited.

2.2 The Heterogeneity of Patients

The role patient is also a heterogeneous semantic role. Patients are usually defined as the participants affected by the events or actions. Yuan [8] pointed out that patients have the semantic feature of being affected, being changed, and existing. They are general and abstract semantic properties of patients. More meticulously, patients of different verbs change in different ways. Patients of "吃" (eat) and "销毁" (demolish) change in existence, patients of "洗" (wash) and "污染" (pollute) change in cleanness, and patients of "扔" (throw) and "放" (put) change in positions. Hownet [9] classifies verbs according to the types of change of their patients. These indicate that patients are heterogeneous.

If heterogeneous patients appear on complementary occasions, or heterogeneity does not happen on patients of the same verb, the heterogeneity of patients will not affect natural language understanding, since the types of change can be inferred from the types of verbs. The classification of Hownet also acquiesces that patients of the same verb are not heterogeneous. However, sentences (6a) and (6b) show that the same verb can take patients of different semantic properties.

(6a) 小王正在扫地。(Xiaowang is sweeping the floor.)
(6b) 小王正在扫雪。(Xiaowang is sweeping the snow.)

"地" (floor) in (6a) and "雪" (snow) in (6b) are both patients of the verb "扫" (sweep), but the former changes in cleanness, and the latter changes in existence. The label "patient" cannot reflect the differences between their meanings.

2.3   The Heterogeneity of Other Semantic Roles

Besides core semantic roles like agents and patients, other non-core semantic roles such as places are heterogeneous. Zhu [10] and Zhan [11] pointed out that there are two types of places by constructional transformations. Some places are entity places, like "纸" (paper) in sentence (7a) is the place of the entity "论文" (essay), and some places are event places, like "教室" (classroom) in (7b) is the place of the event "写论文" (write an essay).

(7a) 张三在纸上写论文。(Zhangsan is writing an essay on a paper.)
(7b) 张三在教室里写论文。(Zhangsan is writing an essay in the classroom.)

Instruments, materials, and some other non-core semantic roles are also heterogeneous. Their heterogeneity is not discussed due to the limitation of the length of the article.

3   The Overlap Between Semantic Roles

If it is hard to classify a participant of an event, and it fits with the definitions of different semantic roles, these semantic roles overlap. Like the heterogeneity, the overlap is also common among semantic roles.

3.1   The Overlap Among Proto-Agents

In the Chinese semantic role systems, the division of proto-agents varies from one another. An agreement is only reached on the definition of agents. For example, PKU SemBank Annotation Guidelines [2] divides proto-agents into agents and experiencers; Yuan [8] divides proto-agents into agents, experiencers, causers and themes; Lu [12] divides proto-agents into agents, experiencers, and possessors. However, no matter how proto-agents are divided, agents always overlap the rest part of the proto-agents. Agents also overlap non-core semantic roles like causes and instruments.

(8a) 马克思证明了共产主义制度的优越性。(Marx proved the superiority of communism.)

(8b) 事实证明了共产主义制度的优越性。(Facts proved the superiority of communism.)

"马克思" (Marx) in (8a) meets the definition of agents, but "事实" (facts) in (8b) are causative, but not volitional. In the sentences (2b) and (2c) listed above, "闹钟" (alarm clock) is not volitional, and it is hard to determine whether "张三" (Zhangsan) is volitional. According to PKU SemBank Annotation Guidelines [2], "马克思" (Marx) and "张三" (Zhangsan) in the sentences above are agents, "事实" (facts) is an instrument, and "闹钟" (alarm clock) is a cause; according to Yuan's definitions [3], "马克思" (Marx) is an agent, and the other three should be labeled as causers. In Lin's semantic role system [4], they are all agents. It can be concluded that it is hard to distinguish these semantic roles. According to Chen's calculation [13], among the verbs whose proto-agents are agents in the semantic role system of Hownet, agents of 85 verbs overlap experiencers. These verbs have the semantic feature of causation, but do not have the feature of volition.

3.2   The Overlap Among Proto-Patients

Proto-patients are usually divided into semantic roles such as patients, results, contents, and objects. These roles also overlap. Take patients, results, and objects as examples, according to PKU SemBank Annotation Guidelines [2] and Yuan [3], the difference between patients and results is the semantic feature of existence; the former exist before the actions or events happen and the latter come into being after the actions; the difference between patients and results is the semantic feature of being affected; patients are affected by the actions or events and results are not. However, sentence groups (9) and (10) indicate that it is sometimes hard to tell whether a proto-patient exists before the action and whether it is affected by the action.

(9a) 这位教练培养了许多运动员。(This coach cultivated many athletes.)

(9b) 这位教练培养出了许多人才。(This coach cultivated many talents.)

(9c) 这位教练培养出了姚明。(This coach cultivated Yao Ming.)

(10a) 我正在修改的论文超过一万字。(The essay I am revising is more than 10 thousand words.)

(10b) 我修改的论文不到一万字。(The essay I revised is less than 10 thousand words.)

In (9a), "运动员" (athletes) exist before the event, and should be annotated as patients or objects; in (9b), "人才" (talents) usually do not exist before being cultivated, and the preposition "出" (out) also indicates they are results. However, in (9c), "姚明" (Yao Ming) obviously exists before cultivation, but there is also preposition "出" (out) in the sentence, indicating that "姚明" (Yao Ming) in the sentence may not be the same one before cultivation, so it is hard to determine which role should it be labeled as.

In (10a), "论文" (essay) refers to the essay before revision finishes, so it should be labeled as an object; in (10b), "论文" (essay) refers to the essay after revision, which can be seen as a new essay, so it can be labeled as a result. Many other verbs can also cause the overlap, such as "升级" (upgrade), "优化" (improve), "翻译" (translate). The meanings of these verbs are all concerned with changing the state of existence of their proto-patients.

3.3 The Overlap Between Proto-Agents and Proto-Patients

Overlapping not only exists among the roles of proto-agents and proto-patients. Proto-agents and proto-patients overlap as well. Dowty [7] pointed out that semantic roles are not discrete classes, so it is hard to define them precisely, and therefore, he proposed to use proto-roles to describe the semantic roles. Semantic properties like volition, sentience, and movement are relevant to proto-agents, and semantic properties like the change of state and being affected are related to proto-patients. According to the proto-role theory, there is, of course, no clear boundary between proto-agents and proto-patients. Dowty mentioned that movement was also a change of state.

Semantic roles of ergative verbs are also in the overlapping areas of proto-agents and proto-patients. For example, in (11a), "中国" (China) is a proto-agent and "经济" (economy) is a proto-patient; in (11c), "中国" (China) is a proto-agent; in (11b), it is hard to tell whether "中国" (China) and "经济" (economy) are proto-agents or proto-patients.

    (11a) 中国正在发展经济。(China is developing the economy.)
    (11b) 中国经济正在发展。(Chinese economy is developing.)
    (11c) 中国正在发展。(China is developing.)

Other semantic roles also overlap. For example, instruments and materials overlap each other, and instruments and places also overlap each other.

4   The Causes for Heterogeneity and Overlap of Semantic Roles and Their Correlations

4.1 The Causes for Heterogeneity and Overlap of Semantic Roles

From the introduction of heterogeneity and overlap, it can be inferred that there is a similar cause for the phenomenon. Semantic roles are usually defined according to the semantic properties or semantic features of a participant of the event. However, the semantic features of a semantic role of a verb do not always match the definition perfectly. For example, as is discussed above, some of the proto-agents of the verb "吵醒" (awaken) are volitional, and some are not. It leads to the overlap of agents and experiencers.

More basically, the heterogeneity and overlap of semantic roles are caused by two properties of people's cognition of things, fuzziness and generality.

There are often no clear boundaries between different semantic features. In the case of the heterogeneity of agents, the boundary of semantic features of volition and non-volition is fuzzy. It is often hard to determine whether the person's behavior that awakened another one is volitional or not.

Moreover, because of the generality in our understanding of events, when someone woke up because of another person, Chinese use the verb "吵醒" (awaken) to describe the event, regardless of the volition of the person. Thus, the proto-agent of the verb "吵醒" (awaken) can be either volitional or non-volitional, resulting in the heterogeneity or the overlap of relevant semantic roles.

Therefore, the heterogeneity and overlap cannot be eliminated in semantic role labeling. In order to overcome the trouble the phenomena bring to natural language understanding, more in-depth semantic analysis is needed on relevant verbs.

4.2 The Heterogeneity and Overlap Are Closely Related and Cannot Be Eliminated

From the perspective of the fineness of semantic role systems, heterogeneity and overlap of semantic roles are closely related to each other. Yuan [3] pointed out that there can be three levels of fineness of semantic role systems, which are macro levels, medium levels and, micro levels. Most semantic role systems in linguistic theories belong to medium-level systems; the proto-role theory proposed by Dowty belongs to the macro-level systems. In language resources, the semantic role classification of PropBank is a macro-level system; the semantic role system of VerbNet is a micro-level system; the rest are medium-level semantic role systems.

If several semantic roles overlap each other in a medium-level semantic role system, heterogeneity will happen in a large semantic role in a macro-level system, and vice versa. For example, in medium-level semantic role systems, patients, results, and objects overlap each other, and in a macro-level system, they are all proto-patients. The overlap of the three roles turns into the heterogeneity of proto-patients. Therefore changing the fineness cannot eliminate the heterogeneity and overlap.

Even in the same level of fineness, because of the difference in the definitions of semantic roles, the heterogeneity of a semantic role in one semantic role system may appear as the overlap of some semantic roles in another semantic role system. In example (2b), some semantic role systems classify "风" (wind) as an agent, so it is a heterogeneous agent; some classify it as an instrument, so in these systems agents and instruments overlap. In a word, adjusting the definitions of semantic roles on the same fineness level can only transform heterogeneity into overlap, or transform overlap into heterogeneity; it cannot eliminate the phenomena.

Above all, heterogeneity and overlap are closely related. They are twin phenomena caused by the fuzziness and generality of language. This problem cannot be solved by adjusting the definitions of semantic roles or changing the fineness of semantic role systems.

5   The Heterogeneity and Overlap of Semantic Roles in Chinese PropBank

Based on the proto-role theory, PropBank [14] uses "ArgN" to represent semantic roles of verbs, where N is an integer from 0 to 6. It also annotates the meanings of each semantic role of verbs. Chinese PropBank [15] is a Chinese semantic role labeling corpus based on Chinese Treebank. It contains over 80000 tokens of 11000 verbs. The annotation of Chinese PropBank is the same as the one of English PropBank. It is also a macro-level semantic role system. Apart from the corpus, there is also a Frameset in PropBank, which contains the meanings of semantic roles of verbs. For example, the Frameset of the verb "培养" (cultivate) contains two arguments, named as "Arg0" and "Arg1". Their meanings are "agent" and "entity cultivated" respectively. The meanings in the Frameset can be seen as micro-level or medium-level semantic roles. In other words, there are semantic roles on different fineness levels in the Frameset.

According to the analysis above, there is always heterogeneity and overlap in any semantic role systems. In this section, the phenomena in Chinese PropBank are examined.

5.1   The Heterogeneity and Overlap in the Frameset

The heterogeneity and overlap of some semantic roles can be discovered by only examining the Frameset. If an ArgN of a verb has more than one meaning label in the Frameset, the ArgN of the verb is heterogeneous; if a meaning label is used to describe Arg0s of some verbs, and Arg1s of other verbs, the label reveals the overlap of Arg0 and Arg1. In this paper, the heterogeneity and overlap of Arg0s and Arg1s is examined.

The Heterogeneity of Arg0s and Arg1s. In the Frameset of Chinese PropBank, the meanings of semantic roles can reveal the heterogeneity of semantic roles. After traversing the Arg0s and Arg1s of all verbs in the Frameset of Chinese PropBank, many instances of heterogeneity are discovered. The meaning of the Arg0 or Arg1 of a verb may be described as "Label1/Label2", where there are two meanings separated by a "/" or ",". This implies the Arg0 or Arg1 is heterogeneous. For example, the meaning of the Arg0 of the verb "破坏" (destroy) is described as "agent/cause", indicating that the Arg0 of the verb is sometimes volitional, and sometimes causative.

There are 1345 heterogeneous Arg0s in the Frameset, accounting for 5.07% of all 26534 Arg0s in the Frameset. Some heterogeneous Arg0s with high frequency are listed in Table 1.

Table 1. Examples of heterogeneity of Arg0s

Heterogeneous Arg0s	Frequency	Examples
agent/cause	1070	破坏(destroy), 振兴(revitalize)
agent/entity	20	残害(cruelly injure), 吓唬(threaten)
agent/causer	16	放大(magnify), 打破(break)
theme/entity	8	下坠(fall), 上火(inflame)
agent/organization	7	编撰(compile), 转载(reprint)
agent/experiencer	5	中意(like), 侧目(glance)

Among all the heterogeneous Arg0s, 1186 of them are tagged as agents and other semantic roles. As is shown in Table 1, 1070 Arg0s are tagged as agents and causes. There are also some Arg0s tagged as agents and entities, causers, organizations or fine-grained semantic labels like drivers, guardians, helpers. These data show that the main reason for the heterogeneity of Arg0s is the overlap of agents and other roles like causes and entities. Causes and entities are usually causative but not volitional.

Similar phenomena occur in other languages. In English PropBank, there are 67 Arg0s tagged as "agent/cause", and 125 Arg0s tagged as "agent/causer". Thus, the heterogeneity in Arg0s may be a universal problem across the languages.

In fact, there should be more heterogeneous Arg0s. First, in addition to "," and "/", "or" is also used in heterogeneous Arg0s. However, in the Frameset, some homogeneous arguments may also contain "or", like the Arg1 of the verb "逗" (amuse) is tagged as "person Arg0 amuses or teases". Thus, in this paper, heterogeneous arguments represented by "or" are not calculated. Second, not all heterogeneous Arg0s are tagged correctly with both meanings. According to Chen [13], among the heterogeneous Arg0s of the 85 verbs, some are only tagged as "agent", and fine-grained meaning labels conceal the heterogeneity of the Arg0s of some verbs. For instance, the Arg0 of "违反" (violate) is tagged as "violator", and the Arg0 of "促使" (prod) is tagged as "entity prodding".

There are 659 heterogeneous Arg1s in the Frameset, accounting for 4.72% of all 13968 Arg1s in the Frameset. Some high-frequency heterogeneous Arg1s are listed in Table 2.

Table 2. Examples of heterogeneity of Arg1s

Heterogeneous Arg1s	Frequency	Examples
theme/entity	29	推开(push away), 孵化(hatch)
time/place	12	进入(enter), 建于(build on)
locative/destination	11	折返(turn back), 流落(wander)
theme/locative	3	留驻(be stationed), 穴居(live in caves)

Some of the heterogeneity of Arg1s is caused by non-core roles like time and place, because if a verb has only one core semantic role, its non-core role may become its Arg1 in the semantic role system of PropBank. It revealed the heterogeneity of place, since a "locative" or "place" is the static location of an entity, and a "destination" is the location into which an entity moves.

In the heterogeneity caused by core roles, there are 57 heterogeneous Arg1s concerning themes, and 29 of them overlap entities.

The heterogeneity of Arg1s is not fully revealed in the Frameset as well. For example, according to the observation above, the Arg1s of "培养" (cultivate) and "修改" (revise) are heterogeneous, but in the Frameset of "培养" (cultivate), its Arg1 is tagged as "entity cultivated", which is close to a patient; there are two Framesets for "修改" (revise), its Arg1s are both tagged as "thing revised", which is close to a patient, and one of its Frameset contains an Arg2 tagged as "thing Arg1 becomes after revision", which is close to a result. Neither of the two representations reveals the heterogeneity of their Arg1s.

The Overlap of Arg0 and Arg1. The Frameset also reveals the overlap of Arg0 and Arg1. One hundred and fifty meaning labels are used to tag both Arg0s and Arg1s. The most frequent one is the theme, which is used 505 times in Arg0s and 3291 times in Arg1s. Some high-frequent medium-level semantic roles in the overlapping areas of Arg0 and Arg1 are listed in Table 3.

Table 3. Semantic roles in the intersection of Arg0 and Arg1 and their frequencies

Semantic roles	Frequency in Arg0	Examples	Frequency in Arg1	Examples
theme	505	翻滚(roll)	3291	划分(divide)
agent	13624	目击(collect)	20	减缓(relieve)
experiencer	124	担心(worry)	12	煎熬(suffer)
cause	117	惊吓(frighten)	43	得意(benefit)
source	6	产生(produce)	23	来自(come from)
beneficiary	3	获益(benefit from)	31	撑腰(support)
recipient	10	启蒙(enlighten)	159	拍马(flatter)

These data show heterogeneity and overlap is common among semantic role types, but verbs whose arguments are heterogeneous account for about 5%. Therefore, the current semantic role labeling task can express the meanings of most verbs accurately.

5.2   The Heterogeneity of Semantic Roles in the Corpus

We can also observe the heterogeneity of semantic roles from the corpus of the Chinese PropBank, and examine whether the annotation of the Chinese PropBank reflected the heterogeneity.

According to the examination above, the Frameset may not reflect the heterogeneity of Arg1s. This will influence the semantic role labeling of relevant verbs. The study examined the sentences containing verbs "培养" (cultivate) and "修改" (revise), and discovered that their heterogeneous Arg1s are not distinguished in the annotation.

In the corpus of Chinese PropBank, 39 sentences contain the verb "培养". In 31 of them, the Arg1s of "培养" do not have the semantic feature of existence. These Arg1s include "人才" (talent), "能力" (ability), "兴趣" (interest). In three sentences, the Arg1s of "培养" exist before cultivation. The Arg1s are "小孩" (child), "我们" (we), and "病毒" (virus). There are also four sentences where it is hard to tell whether the Arg1s of the verb exist before cultivation. The Arg1s are "警犬" (police dog), "律师" (lawyer), "接班人" (successor), and "将士" (soldiers). Taking "警犬" (police dog) as an example, before cultivation, it may be a police dog, and it may also be an ordinary dog.

Besides, there is a sentence in the corpus "把他们培养成为美国服务的精英" (cultivate them into elites serving the USA), where the existent proto-patient "他们" (them) co-occurs with the non-existent proto-patient "为美国服务的精英" (elites serving the USA). According to the Frameset, "他们" (them) corresponds to the meaning of the Arg1 of the verb, and "为美国服务的精英" (elites serving the USA) cannot be correctly annotated. In Chinese PropBank, this sentence is not annotated.

Forty-five sentences contain the verb "修改" (revise). In 34 of them, the Arg1s exist before revision. They correspond to the meaning of the Arg1 in the Frameset. There are only two sentences where the proto-patients of the verb do not have the semantic feature of existence. Although the proto-patients in these two sentences are closer to the meaning of the Arg2 in the Frameset, they are still labeled as Arg1s. There are two sentences where the existent proto-patients co-occur with the non-existent proto-patients. The existent proto-patients are labeled as Arg1s, and the non-existent proto-patients are labeled as Arg2s. In the rest seven sentences, the Arg1s of the verb do not appear.

According to our examination, the Frameset of "培养" and "修改" cannot help represent the heterogeneity of their proto-patients in the annotation of the corpus. If the meanings of their proto-patients are described as the meaning of the Arg1 of "培养", which is closer to a patient, it can only represent the existent proto-patients accurately, and the ones that do not have the semantic roles of existence cannot be expressed precisely; If the meanings of their proto-patients are described closer to a result, the existent proto-patients cannot be represented accurately; If the meanings of their proto-patients are described in the form of "patient/result", it enabled further semantic analysis on relevant heterogeneous proto-patients, but this type of representation fails when the existent and non-existent proto-patients co-occur in one sentence. If the existent proto-patient and the non-existent proto-patient are labeled as Arg1 and Arg2 respectively, like the Frameset of "修改", it can cope with the problems mentioned above, but it will be challenging to annotate a sentence when the existence of the proto-patient is hard to determine. Therefore in the task of semantic role labeling, there is no perfect solution for the heterogeneity and overlap of semantic roles.

Above all, according to the examination of sentences containing "培养" and "修改", heterogeneity of Arg1s exists in the corpus of Chinese PropBank. The Frameset of Chinese PropBank fails to represent the meanings of the Arg1s accurately and fails to reflect the heterogeneity.

6   Conclusive Remarks

The phenomena of heterogeneity and overlap in semantic roles occur among all semantic roles types. From the perspective of the fineness of semantic roles, the study discovers that heterogeneity and overlap are twin problems caused by the generality and fuzziness of language. The problems cannot be solved by adjusting the definitions or the fineness of the semantic roles.

The study examines relevant language resources, and discovers the heterogeneity and overlap in Chinese PropBank. In the Frameset of Chinese PropBank, 1345 Arg0s and 659 Arg1s exhibit heterogeneity, each accounting for 5% of all Arg0s and Arg1s. Meanwhile, some heterogeneous arguments are not accurately labeled in the Frameset, like the Frameset of "培养" (cultivate) and "修改"(revise). The semantic role labeling of these verbs in the corpus cannot reflect the precise meanings of their proto-patients. The heterogeneity of these proto-patients needs to be discovered by deeper semantic analysis.

There are other problems in semantic role labeling, such as the difficulties in representing the meanings of syntactic constructions and irregular compositions. These problems need further investigation.

Acknowledgment. This paper was supported by Major Project of Humanities and Social Sciences of Ministry of Education, P.R.China (Project NO.15JJD740002).

References

1. Palmer, M., Gildea, D., Xue, N.: Semantic Role Labeling. Morgan & Claypool Publishers, San Rafael (2010)
2. PKU SemBank Annotation Guidelines (2015). http://klcl.pku.edu.cn/xwdt/231664.htm
3. Yuan, Y.: The fineness hierarchy of semantic roles and its application in NLP. J. Inf. Proc. **21**(4), 10–20 (2007)
4. Lin, X.: Lexical Semantics and Computational Linguistics. Language & Culture Press, Beijing (1999)
5. Fillmore, C.J.: The case for case. In: Bach, E., Harms, R.T. (eds.) Universals in Linguistic Theory. Holt, Rinehart & Winston, New York (1968)
6. Saeed, J.I.: Semantics. Wiley-Blackwell, Hoboken (2009)
7. Dowty, D.: Thematic proto-roles and argument selection. Language **67**(3), 547–619 (1991)
8. Yuan, Y.: On the hierarchical relation and semantic features of the thematic roles in Chinese. Chin. Teach. World **61**(3), 10–22 (2002)
9. Dong, Z., Dong, Q.: Construction of a knowledge system and its impact on Chinese research. Contemp. Linguist. **3**(1), 33–44 (2001)

10. Zhu, D.: "Zai heiban shang xiezi" and relevant constructions. Lang. Teach. Linguist. Stud. **3**, 4–18 (1978)
11. Zhan, W.: Argument structure and constructional transformation. Stud. Chin. Lang. **300**, 209–221 (2004)
12. Lu, C.: Semotactic network of Chinese. Appl. Linguist. **26**, 82–88 (1998)
13. Chen, L., Zhan, W.: An investigation of inconsistency in semantic role labeling: a case study of agent. J. Inf. Proc. **33**(1), 1–10 (2019)
14. Palmer, M., Gildea, D., Kingsbury, P.: The proposition bank: an annotated corpus of semantic roles. Comput. Linguist. **31**(1), 71–106 (2005)
15. Xue, N., Palmer, M.: Adding semantic roles to Chinese Treebank. Nat. Lang. Eng. **15**(1), 143–172 (2009)

Cognitive Semantics and Its Application on Lexicography: A Case Study of Idioms with *xīn* in Modern Chinese Dictionary

Qian Li[✉]

Guangdong University of Foreign Studies, Guangzhou, China
lqchristina@gdufs.edu.cn

Abstract. This paper attempts to explore the role of cognitive semantics in Chinese lexicography. Specifically, the paper focuses on the multiple senses of the headword 心 *xīn* and the related idioms in *Modern Chinese Dictionary* (现代汉语词典 *xiàn dài hàn yǔ cí diǎn*). It first provides a cognitive analysis of the related idioms of 心 *xīn* based on its different underlying conceptual metaphors and metonymies. From the cognitive perspective, 心 *xīn* and 'heart' in both Chinese and English can be further classified into different categories. As a result, different idioms with 心 *xīn* may be interpreted in various ways to match their conceptual categories. In present *Modern Chinese Dictionary*, all the idioms are arranged according to their *pinyin* alphabetical sequence. This arrangement mixes up with the headword's conceptual mechanisms, and may bother some dictionary users, especially the Chinese language learners. This paper suggests that in Chinese dictionary compilation, the arrangement of those idioms may take the underlying conceptual mechanisms into account.

Keywords: Cognitive semantics · Lexicography · Idioms

1 Introduction

Lexicology and lexical semantics can be the theoretical counterparts in lexicography, when the latter is considered to belong to the domain of applied linguistics [1]. Atkins and Bouillon [2] also note that before the compilation of a dictionary, a solid theoretical basis for the analysis of word behavior is fundamental. Without theoretical foundation, the collection of facts will be patchy and inconsistent, and it cannot be sure that no important aspect of the word's behavior has been overlooked. Similarly, during synthesis of headwords, it is hard to ensure that their approach to these tasks is consistent from A to Z of the dictionary.

The present paper explores the advantages of employing cognitive semantics as a theoretical foundation for lexicographical theory and practice.

J.-F. Hong et al. (Eds.): CLSW 2019, LNAI 11831, pp. 834–839, 2020.
https://doi.org/10.1007/978-3-030-38189-9_83

2 Senses in the Dictionary

For most headwords in a dictionary, several different senses can be identified. For instance, the word 安 *ān* can be used as an adjective and a verb in modern Chinese, and it can be used as a pronoun for questioning in ancient Chinese. In addition, there are different senses when 安 *ān* is used as an adjective.

How to arrange and sequence these senses and the related idioms is a major issue in lexicography. There were different lexicographical traditions approaching this issue. For example, in ancient Chinese dictionaries, senses were arranged in a way to indicate the evolving of new meanings in time within the arbitrary alphabetical ordering. The sense sequence started from literal meanings and moved on to evolving figurative meanings. Consequently, the sense arrangement follows the developing conceptual structure clearly. In contrast, some modern Chinese dictionaries prefer to organize senses and sub-entries according to their frequency of use. In this way, figurative meanings often appear before literal meanings. These dictionaries may ignore the conceptual sequence and present senses in a mixed-up way.

Some recent cognitive linguistic research [3] argued that the meaning structure of polysemous words can be interpreted and presented in a conceptually systematic way. Thus, meanings and sub-entries within a polysemous entry may be organized with the help of conceptual mechanisms such as metonymy and metaphor.

3 Metaphor and Metonymy for 心 *xīn* in Chinese Idioms

Conceptual Metaphor Theory, the approach introduced by Lakoff, includes two basic ideas: firstly, the view that metaphor is a cognitive phenomenon rather than a purely lexical one; secondly, the view that metaphor should be analyzed as a mapping between two domains, i.e. the source and the target domain [4, 5]. To take the conceptual metaphor 'ARGUMENT IS WAR' as an example. The source domain here is WAR, and the target domain is ARGUMENT; the concept of ARGUMENT is comprehended via the concept of WAR.

Compared to metaphor, metonymy is probably even more basic in language and cognition. Metonymy consists of a mapping within the same experiential domain or conceptual structure [6, 7]. For instance, 'Westminster', a borough of London in the United Kingdom, could be used as a metonym for the country's government.

Metaphor and metonymy may interact with each other [8]. Researchers [9] argued that a language user's understanding of an expression depends upon his/her awareness of the subtle conceptual links between the source and the target domain. In the following, examples of 心 *xīn* in Chinese culture will be taken into account. Specifically, its metaphorical or metonymical underpinning will be analyzed so as to classify the conceptual structure of 心 *xīn* into different categories. All the examples are four-word Chinese idioms or phrases which are extracted from *Modern Chinese Dictionary*. In addition, to give a comparison between Chinese and other language cultures, English examples of 'HEART' will be adopted as the counterpart. Most of the English examples are from *Roget's Thesaurus*.

heart stands for the person

On the most specific level of meaning, 心 *xīn* stands metonymically for the whole person. It is the most salient body part in the understanding model of emotions. In English culture, the most prototypical emotion connected with HEART is love. In Chinese culture, this emotion may be broadened into all the precious feelings among people.

Ex 1:   *set one's heart on somebody*
Ex 2:   *great heart*
Ex 3:   心 心 相 印 *(xīn xīn xiāng yìn) (to be in love with each other so that the two hearts are matching)*
Ex 4:   心 有 灵 犀 *(xīn yǒu líng xī) (hearts which beat in unison are linked so that two people understand each other well and think of the same idea spontaneously).*

In the English examples, HEART is conceptualized both as A MOVABLE OBJECT or CHANGEABLE IN SIZE. In Ex 1, the movement of heart refers to the movement of one's attention or love. In Ex 2, HEART is supposed to be changed in size and to house a variety of good feelings. But in the Chinese examples, instead of representing A MOVABLE OBJECT, 心 *xīn* is conceptualized as AUTONOMOUS ENTITY. In Ex 3 and 4, the two persons in love can still own their hearts but in a special way. The two hearts may echo and match even in a long distance. This way of love expression is so different from that of Western cultures. In Chinese culture, love is hidden deeply at the bottom of a heart, rather than expressed openly. The difference in cultures results in the different metaphorical interpretations of HEART in the two languages.

HEART AS AN OBJECT

HEART can be the symbol of other feelings either positive or negative.

Ex 5:   *soft heart*
Ex 6:   *heart of iron, heart of stone*
Ex 7:   心慈手软 *(xīn cí shǒu ruǎn) (to describe a person with soft heart and soft action)*
Ex 8:   铁石心肠 *(tiě shí xīn cháng) (to describe a person who has a heart of iron)*
Ex 9:   刻骨铭心 *(kè gǔ míng xīn) (to describe something unforgettable which is carved in the heart).*

The above examples show the sub-folk model of HEART AS AN OBJECT. This object may be made of either a soft or a hard material. In both English and Chinese cultures, a soft material is metaphorically related with the feeling of tenderness. Therefore, HEART (in Ex 5 and 7) stands for the benevolence, sympathy or compassion. In contrast to soft materials, heavy and hard materials are metaphorically connected with an unyielding attitude and hard feelings. Thus, HEART (in Ex 6 and 8) in this sense indicates the coldness or stubborn attitude. In addition, Ex 9 in Chinese culture is different from the above conceptualization. Hard materials (such as stone, iron, steel) in Chinese culture are metaphorically endowed with a firm and strong willingness.

In Ex 8, the hardness of the material can be reinterpreted as two senses—the coldness or the firmness. The two types of hardness are metaphorically mapped onto the domain of heart.

HEART AS A LIVING ORGANISM

When another sub-folk model is concerned, HEART may be counted as a living organism.

Ex 10:   *heart-burning*
Ex 11:   *an aching heart*
Ex 12:   心急如焚 *(xīn jí rú fén) (similar to heart-burning)*
Ex 13:   心病难医 *(xīn bìng nán yī) (mental worries cannot be cured by medicine).*

Research shows that the metonymic effects on perceivers depend on our sensory experience in both source and target domains [6]. In the above metaphors, HEART exists independently with its own feeling. One's heart can be hurt or destroyed by various means; one's heart may break, ache, bleed or be pieced or burned. All the above metaphors are based on a prior metonymic understanding. A person undergoes a specific type of physiological pain when hurt or disappointed, or suffering a loss. This pain is experienced and expressed differently among various persons [9]. Chinese and English metaphors above share the same metonymic base, because the basic feelings resulting from aching, being pierced and burned are the same among different cultures. When HEART is conceptualized as a LIVING ORGANISM, the metonymies involved are less basic or even appear as metaphors. Nevertheless, the underlying metonymic basis is kept intact in both Chinese and English examples.

HEART AS A CONTAINER

CONTAINER is one of the most pervasive and common metaphors in everyday language usage. It can also be applied to the human body and its major parts, such as the head, the heart, and the chest.

Ex 14:   *to open one's heart*
Ex 15:   *a heart overflowing (with gratitude)*
Ex 16:   *find somebody in one's heart*
Ex 17:   心胸狭窄 *(xīn xiōng xiá zhǎi) (a narrow/small-sized heart; to describe a person who is not willing to accept, or even be hostile to any opposite suggestion or criticism; the opposite of 'generous' to some extent).*

Here HEART is viewed as a container. Some containers are with lids (in Ex 14); some can be filled up (in Ex 15); some can be measured (in Ex 17); some can be further regarded as a storehouse (in Ex 16). In different cultures, HEART can be treated as different types of containers with different facets, such as an INTERNAL CONTAINER, and a STOREHOUSE CONTAINER.

Compared with the culture-specific nature of HEART, the container schema seems to be a more universal schema and applies to different cultural contexts.

4 Suggestions for Idiom Presentation in Chinese-English Learners' Dictionaries

The above analysis reveals new ways of dealing with multiple senses of lexical items. In the literature, Zhanj and Barnden [10] and Morrissey and Schalley [11] argued for more attention to the motivational link between core senses and figurative sub-senses. Such motivational links could involve conceptual metaphors in the Lakovian sense, or even image schemas. In lexicography, this suggestion has been developed by Adamska-Sałaciak [12]. In the actual practice of lexicography, the *Macmillan English Dictionary for Advanced Learners* incorporates 'metaphor boxes', showing the conceptual metaphors behind common expressions.

In *Modern Chinese Dictionary*, under the entry of 心 *xīn* there are altogether 5 senses, and among them three are the common uses of the headword. Based on the analysis in the above section, it is found that those senses in MCD correspond to different metaphors or metonymies. In other words, there are different conceptual structures when 心 *xīn* is used in various idioms. However, there is a problem for the grouping and sequencing of the 35 idioms with 心 *xīn* as sub-entries included. All the idioms are presented in the alphabetical order of *pinyin*. In this way, the metaphor- and metonymy-based senses were mixed up, and it is a challenge for language users to recognize the different senses among those idioms.

The proposed way to present 心 *xīn* idioms in Chinese-English dictionary goes as follows. The basic principle is that all idioms, as shown in Table 1, are categorized into different groups according to their different cognitive underpinnings, rather than according to the *pinyin* alphabetical order.

Table 1. Idiom Box in Chinese-English Dictionary.

Metaphors & Metonymies	Idioms
HEART stands for a person	心心相印 *xīn xīn xiāng yìn* 心有灵犀 *xīn yǒu líng xī*
HEART AS AN OBJECT	心慈手软 *xīn cí shǒu ruǎn* 铁石心肠 *tiě shí xīn cháng*
HEART AS A LIVING ORGANISM	心急如焚 *xīn jí rú fén* 心病难医 *xīn bìng nán yī*
HEART AS A CONTAINER	心宽体胖 *xīn kuān tǐ pán* 心腹之患 *xīn fù zhī huàn*

With the help of such idiom box, it will be easier for dictionary users, especially Chinese language learners to fully understand the meaning differences among the idioms.

5   Conclusion

Considering lexicography, lexical semantics and cognitive linguistics can play their own roles in dictionary research and practice. This paper explores a new way of arranging idioms in Chinese-English dictionary sub-entries according to different conceptual metaphors and metonymies. It is expected that such arrangement may provide a clear conceptual clue for dictionary users, especially for Chinese language learners. Further research is necessary to provide systematic conceptual analysis of polysemous headwords for the application of cognitive semantics on lexicography.

Acknowledgements. This study was supported by the Research Funding for Social Science Original Program in Guangdong Universities (2016WTSCX027).

References

1. Geeraerts, D.: Lexicography and theories of lexical semantics. In: Durkin, P. (ed.) The Oxford Handbook of Lexicography, pp. 425–438. Oxford University Press (2015)
2. Atkins, B.T.S., Bouillon, P.: Relevance in dictionary making: sense indicators in the bilingual entry. In: Bowker, L. (ed.) Lexicography, Terminology, and Translation: Text-Based Studies in Honor of Ingrid Meyer, pp. 25–43. University of Ottawa Press (2006)
3. Wishart, R.A.: Hierarchical and distributional lexical field theory: a critical and empirical development of Louw and Nida's semantic domain model. Int. J. Lexicogr. **31**(2), 394–419 (2018)
4. Lakoff, G., Johnson, M.: Metaphors We Live By. University of Chicago Press, Chicago (1980)
5. Lakoff, G., Johnson, M.: Metaphors We Live By, 2nd edn. University of Chicago Press, Chicago (2003)
6. Barcelona, A.: On the plausibility of claiming a metonymic motivation for conceptual metaphor. In: Barceloca, A. (ed.). Metaphor and Metonymy at the Crossroads: A Cognitive Perspective, pp. 327–358. Mouton de Gruyter (2000)
7. Kovecses, Z., Radden, G.: Metonymy: developing a cognitive linguistic view. Cogn. Linguist. **9**(1), 37–77 (1998)
8. Radden, G., Kovecses, Z.: Towards a theory of metonymy. In: Panther, J., Radden, G. (eds.) Metonymy in Language and Thought, pp. 635–648. John Benjamins Publishing Company, Amsterdam/Philadelphia (1999)
9. Niemeier, D.: Straight from the heart—metonymic and metaphorical explorations. In: Barceloca, A. (ed.) Metaphor and Metonymy at the Crossroads: A Cognitive Perspective, pp. 105–139. Mouton de Gruyter (2000)
10. Zhanj, L., Barnden, J.: Affect sensing using linguistic, semantic and cognitive cues in multi-threaded improvisational dialogue. Cogn. Comput. **4**(4), 368–429 (2012)
11. Morrissey, L., Schalley, A.C.: A lexical semantics for refugee, asylum seeker and boat people in Australian English. Aust. J. Linguist. **37**(4), 489–513 (2017)
12. Adamska-Sałaciak, A.: Translation of dictionary examples notoriously unreliable? Atti del XII Congresso Internazionale di Lessicografia, vol. 1, pp. 493–501 (2006)

Research on Quantifier Phrases Based on the Corpus of International Chinese Textbooks

Dongdong Guo, Jihua Song[✉], Weiming Peng, and Yinbing Zhang

College of Information Science and Technology,
Beijing Normal University, Beijing, China
{dongdongguo, zhangyinbing}@mail.bnu.edu.cn,
{songjh, pengweiming}@bnu.edu.cn

Abstract. Chinese is rich in quantifiers, and studying various types and structures of quantifier phrases is a necessary way to master quantifiers. Using Chinese information processing technology to study quantifier phrases in international Chinese teaching is conducive to promoting deep integration in the two fields. This paper first constructs a knowledge base of quantifier phrase structural modes by tagging the quantifier phrases in the corpus of a certain scale of international Chinese textbooks. Then the characteristics of quantifier phrases in the field of international Chinese teaching are analyzed through the constructed structural mode knowledge base. Finally, on the basis of the structural mode knowledge base, automatic recognition of quantifier phrases in the corpus of international Chinese textbooks is studied.

Keywords: Quantifier phrase · Structural mode · International Chinese textbooks · Automatic recognition · Chinese information processing

1 Introduction

Compared with the Indo-European language family, Chinese has a rich category of quantifiers. In modern Chinese, the use of quantifiers is very common, but there is no clear division of quantifiers in English, which has different forms of measurement from Chinese.

Quantifier usage is a very common language phenomenon in international Chinese teaching. It is also the key content of vocabulary and grammar teaching in international Chinese teaching. Quantifier content occupies a large proportion in the Hanyu Shuiping Kaoshi (HSK) syllabus, where the vocabulary syllabus (5000 vocabulary items) contains 82 quantifier vocabulary items, and the language point syllabus also has many explanations on quantifiers [1–6]. Quantifiers generally do not act as syntactic components alone. They mainly attach to numerals, demonstrative pronouns, interrogative pronouns and other words to form quantifier phrases, and then act as various components of sentences [7]. Therefore, it is necessary to study the types and structures of quantifier phrases to master the use of quantifiers. The linguistic circle and the field of international Chinese teaching have made a lot of research on quantifier phrases, and

J.-F. Hong et al. (Eds.): CLSW 2019, LNAI 11831, pp. 840–852, 2020.
https://doi.org/10.1007/978-3-030-38189-9_84

have made many achievements [8, 9]. However, there is a lack of related research on quantifier phrases by using Chinese information processing technology in international Chinese teaching. Using Chinese information processing technology [10, 11] to study quantifier phrases in international Chinese teaching is conducive to promoting deep integration in the two fields.

Quantifier phrases can be divided into numeral-quantifier phrases (phrases consisting of numerals or numeral phrases combined with quantifiers, such as "一个[yi ge]"), demonstrative-quantifier phrases (phrases consisting of demonstrative pronouns or the words such as "上[shang]", "下[xia]", "前[qian]" or "后[hou]" combined with quantifiers or numeral-quantifier phrases, such as "这幅[zhe fu]", "那五位[na wu wei]" and "上回[shang hui]"), interrogative-quantifier phrases (phrases consisting of interrogative pronouns combined with quantifiers or numeral-quantifier phrases, such as "哪个[na ge]" and "多少斤[duoshao jin]") and other quantifier phrases (phrases consisting of other types of words combined with quantifiers, such as "满瓶[man ping]" and "全套[quan tao]") according to the different components combined with quantifiers. This paper first constructs a knowledge base of quantifier phrase structural modes by tagging quantifier phrases in the corpus of a certain scale of international Chinese textbooks. Then the characteristics of quantifier phrases in the field of international Chinese teaching are analyzed through the constructed structural mode knowledge base. Finally, on the basis of the structural mode knowledge base, automatic recognition of quantifier phrases in the corpus of international Chinese textbooks is studied.

2 Construction of the Quantifier Phrase Structural Mode Knowledge Base

2.1 Representation of Structural Modes

In the field of Chinese information processing, a quantifier phrase should be treated as a whole, that is, as a dynamic word [12–15]. Therefore, when describing structural modes of quantifier phrases, we first determine a whole part of speech for each quantifier phrase. For example, the whole part of speech of a numeral-quantifier phrase is represented by the numeral "m", the whole part of speech of a demonstrative-quantifier phrase or interrogative-quantifier phrase is represented by the pronoun "r", and the whole parts of speech of other quantifier phrases are determined by their structures and meaning characteristics, which are generally classified into the numeral "m" or the pronoun "r". In order to effectively reflect and describe the structural features of quantifier phrases from the two levels of the whole and the internal structure, the following four types of information are used to express knowledge of their structural modes: the whole part of speech of a quantifier phrase, the part of speech of each internal component, the syllable number of each internal component and structural relationships between the internal components. The structural relationships between the internal components are shown in Table 1.

Table 1. Symbols of structural relationships

Relationship	Symbol	Sample	Note
数量 (Quantity)	-	一-个 这-种 下-一-次 一-整-条	Represents the numeral-quantifier relationship or the combination relationship of pronouns/localizers/adjectives with quantifiers or numeral-quantifier phrases (affix/auxiliary word structures are also connected with "-", such as "圈-儿")
概数 (Approximation)	_	几_百万-吨 一千_多-年 五-天_左右	Represents the combination relationship of numerals with approximate numbers or front and back additives
重叠 (Reduplication)	·	一-个·个	Represents the reduplication relationship
并列 (Coordination)	...	六...七-里	Represents the coordination relationship
动宾 (Verb-object)	|	上|万-次	Represents the verb-object relationship
状中 (Adverbial-headword)	→	很→多-家	Represents the adverbial-headword relationship

The knowledge of the quantifier phrase structural mode is expressed as follows:

- <structural mode> :: = <the whole part of speech of the quantifier phrase>: <the part of speech of the internal component><the syllable number of the internal component>[<structural relationship symbol><the part of speech of the internal component><the syllable number of the internal component>]+
- <the whole part of speech of the quantifier phrase> :: = m | r
- <the part of speech of the internal component> :: = n | t | f | m | q | r | v | a | d | p | c | u | e | o | Ug
- <the syllable number of the internal component> :: = <NULL> | 2 | 3 | 4 | ...
- <structural relationship symbol> :: = - | _ | · | ... | | | →

The "n", "t", "f", "m", "q", "r", "v", "a", "d", "p", "c", "u", "e", "o" and "Ug" denote nouns, time words, localizers, numerals, quantifiers, pronouns, verbs, adjectives, adverbs, prepositions, conjunctions, auxiliary words, interjections, onomatopoetic words and affix morphemes (such as the category of the suffix "儿[er]" in "玩儿 [wan er]") respectively. Punctuation symbols are denoted by the "w". The syllable number is the default value 1 when it is NULL. In addition, the numeral is separated directly from the punctuation mark "-" or "∼" which denotes continuity by a space.

Examples of quantifier phrase structural modes are shown in Table 2. The structural mode of "五公斤[wu gongjin]" is "m: m-q2". In "m: m-q2", the "m" before the colon indicates the whole part of speech is a numeral; the "m" after the colon indicates the part of speech of the internal component "五[wu]" is a numeral and its syllable number is 1 (The syllable number is the default value 1 when it is NULL); the "q2" indicates the part of speech of the internal component "公斤[gongjin]" is a quantifier and its syllable number is 2; the "-" means the relationship between the internal component "五[wu]" and the "公斤[gongjin]" is the quantity relationship.

Table 2. Examples of quantifier phrase structural modes

Quantifier Phrase	Mode	Quantifier Phrase	Mode
一个	m: m-q	五公斤	m: m-q2
这种	r: r-q	这两个	r: r-m-q
下次	r: f-q	下一个	r: f-m-q
整条	m: a-q	一小块	m: m-a-q
一件件	m: m-q·q	一天一天	m: m-q·m-q
十几个	m: m_m-q	三十多岁	m: m2_m-q

2.2 Annotation of Corpus

The vocabularies included in the Modern Chinese Dictionary (Sixth Edition) are relatively stable [16]. In order to get the information about quantifier phrase structural modes in the corpus of international Chinese textbooks, we use the vocabularies and parts of speech in the Modern Chinese Dictionary as the basis to annotate the internal components of quantifier phrases. To analyze structural modes of quantifier phrases, we need to separate their internal components first. The criterion of segmentation is based on the unity of structure and meaning, until the corresponding sense and part of speech information of each internal component can be found in the Modern Chinese Dictionary. For example, for the quantifier phrase "十几个[shi ji ge]", as the Modern Chinese Dictionary contains "十[shi]", "几[ji]" and "个[ge]", so the correct segmentation results should be "十[shi]_几[ji]-个[ge]". For any real number and ordinal number, even if they are not included in the Modern Chinese Dictionary, they are no longer segmented, and the parts of speech are marked as the numeral "m".

Undergraduates and postgraduates in the background of linguistics are invited to annotate the quantifier phrase structural mode in the corpus of international Chinese textbooks (include *New Practical Chinese Textbook, Happy Chinese, Great Wall*

Chinese, Chinese with Me, Mandarin Teaching Toolbox, Contemporary Chinese, Chinese Paradise and other international Chinese textbooks) manually, and the contents of annotating include the corresponding senses of the internal components of quantifier phrases in the Modern Chinese Dictionary and the structural mode information of quantifier phrases. Each sense of a vocabulary in the Modern Chinese Dictionary is uniquely identified by the sense code (three digits).

In order to ensure the accuracy and consistency of the annotating results, the text of the same paragraph is annotated by two students at least, and the annotating result is audited by experts. The annotated data which is consistent and approved will be regarded as valid data. If the annotating results are inconsistent or not approved, the annotators and the auditor will be required to discuss and decide the final results.

2.3   Construction of the Knowledge Base

A total of 29465 sentences (498965 words) of corpus data of international Chinese textbooks annotated are obtained in this paper. Use regular expressions to match and extract quantifier phrases and their structural mode information in the annotated corpus. A regular expression is a formula to match a class of strings in a certain pattern, consisting of a number of ordinary characters and special characters (meta characters). Ordinary characters include small and medium letters, numbers, and Chinese characters, and meta characters refer to special characters with special meanings. The rule of quantifier phrases and their structural mode information in the annotated corpus is clear. Using the regular expression, all the information to be extracted can be accurately matched. According to the statistical analysis of the extracted information, a quantifier phrase structural mode knowledge base with 75 structural modes is established. The structure of the quantifier phrase structural mode knowledge base is shown in Table 3, and the 75 structural modes are shown in Table 4 by the order of the frequency of quantifier phrases corresponding to the structural mode in the corpus from high to low.

Table 3. The structure of the structural mode knowledge base

Field	Illustration
Id	The serial number of the quantifier phrase structural mode
Mode	The quantifier phrase structural mode
POS	The whole part of speech of the quantifier phrase
Syllable	The syllable number of the quantifier phrase
Rule	The regular expression rule for the quantifier phrase
Frequency	The frequency of quantifier phrases corresponding to the structural mode in the corpus
Class	The number of the types of quantifier phrases corresponding to the structural mode in the corpus
Detail	Every quantifier phrase corresponding to the structural mode, its internal component sense code and its frequency

Table 4. 75 quantifier phrase structural modes

ID	Mode/Example	Fre-quency	ID	Mode/Example	Fre-quency	ID	Mode/Example	Fre-quency
1	m: m-q 一个	4664	26	r: r-q…r-q 各种各样	14	51	m: m4_m_m-q 1000多万吨	2
2	r: r-q 这种	1993	27	r: r2-q 若干张	14	52	m: m7-q 五千七百八十五块	2
3	m: m2-q 十四岁	734	28	m: m3_m-q 500多名	11	53	r: r-q-Ug 这点儿	2
4	r: r-m-q 这一回	417	29	m: m2 w m2-q 55-60岁	10	54	m: a-q·a-q 大片大片	1
5	m: m3-q 二十二块	177	30	m: m3-q-m 三十八度七	9	55	m: a-q…a-q 大包小包	1
6	r: f-q 下个	142	31	m: m_m-q2 数千英里	9	56	m: d→a-q 很多家	1
7	m: m-q-m 两块五	98	32	m: m-q-m2 一米六三	7	57	m: m-a-q·m-a-q 一大批一大批	1
8	m: m2_m-q 一千多年	93	33	m: m2…m-q 十八九岁	7	58	m: m-f-m-q 零下十度	1
9	m: m4-q 两万八千册	86	34	m: m4_m-q2 五千五百多公里	7	59	m: m-f-m2-q 零上二十度	1
10	m: m_m-q 几千条	69	35	m: m6-q 三千四百一十元	7	60	m: m-q-m2_m 一米七十多	1
11	m: m-q2 两公分	56	36	m: m…m-q2 六七公里	6	61	m: m-q3 六平方米	1
12	m: m2-q2 15公里	53	37	m: m-q-Ug 一圈儿	5	62	m: m2-q_f2 一千块左右	1
13	m: m…m-q 两三天	51	38	m: m2_m-q2 五千多公里	5	63	m: m2-q_v2 二十岁出头	1
14	m: a-q 多种	43	39	m: m4-q2 1000公斤	5	64	m: m2_m-q3 10多平方米	1
15	r: f-m-q 下一个	40	40	m: m3_m_m-q 800多万辆	4	65	m: m2_m_m-q 60多万对	1
16	m: m-q·q 一张张	33	41	m: m5_m-q 10000多只	4	66	m: m4 w m4-q 6000~8000双	1
17	m: m3-q2 五十八公斤	19	42	m: m_m2-q 几百万吨	4	67	m: m4-q_f2 1000只左右	1
18	m: m4_m-q 一万三千多里	17	43	r: d-m-q 再一次	4	68	m: m5-q-m 一百八十六块四	1
19	r: a-m-q 同一个	17	44	m: m-q…m-q 一天两天	3	69	m: m5-q2 2000万美元	1
20	m: m-a-q 一小块	16	45	m: m w m-q 1~5分	2	70	m: m_u-q 十来个	1
21	m: m-q·m-q 一次一次	16	46	m: m w m2-q 6-10粒	2	71	m: m_u-q2 十来分钟	1
22	m: m2-q-m 十二块五	15	47	m: m-q_f2 五天左右	2	72	m: v l m-q 上万亩	1
23	m: m-q·d·m-q 一遍又一遍	14	48	m: m2-q3 五十平方米	2	73	m: v l m..v l m-q 成百上千只	1
24	m: m5-q 一百六十五块	14	49	m: m3 w m3-q 15万-18万辆	2	74	r: a-m2-q 近30年	1
25	m: m…m2-q 一二百年	14	50	m: m3_m-q2 180多公斤	2	75	r: r-m2-q 每90秒	1

The specific content of the structural mode "m: m-a-q" in the knowledge base is shown in Table 5. The "[单音节数词集合][大小整][单音节量词集合]" in the "rule" field is a regular expression rule. The "单音节数词集合" and "单音节量词集合" refer to the collection of monosyllabic numerals and the collection of monosyllabic quantifiers in the Modern Chinese Dictionary respectively. The regular expression rule of each structural mode is mainly determined by the characteristics of the mode itself, the annotated data and the various vocabulary sets in the Modern Chinese Dictionary. The "rule" field defines the information of syllable number, the part of speech and syllable number of each internal component and the fixed component of quantifier phrases corresponding to the specific structural mode. Thus, the "rule" field is the key information to recognize quantifier phrases and their structural modes.

Table 5. The structural mode "m: m-a-q" in the knowledge base

Id	Mode	POS	Syllable	Rule	Frequency	Class	Detail	
20	m: m-a-q	m	3	[单音节数词集合][大小整][单音节量词集合]	16	11	【一[001]小[001]块[002]】	3
							【一[001]大[001]堆[005]】	2
							【一[001]小[001]撮[007]】	2
							【一[001]整[001]天[004]】	2
							【一[001]大[001]片[106]】	1
							【一[001]大[001]批[203]】	1
							【两[001]大[001]杯[001]】	1
							【一[001]大[001]排[005]】	1
							【一[001]大[001]包[005]】	1
							【三[001]小[001]瓶[001]】	1
							【一[001]大[001]束[002]】	1

3   Analysis of Quantifier Phrases

The total number of quantifier phrases corresponding to 75 structural modes in the knowledge base is 9066, and the total number of classes of quantifier phrases reaches 1873. The 15 quantifier phrases with the highest frequency in the corpus are "一个[yi ge]" (1312), "每天[mei tian]" (244), "一种[yi zhong]" (206), "这种[zhe zhong]" (191), "两个[liang ge]" (179), "每个[mei ge]" (121), "一张[yi zhang]" (112), "一家[yi jia]" (110), "一件[yi jian]" (109), "一位[yi wei]" (106), "一次[yi ci]" (94), "这件[zhe jian]" (89), "这次[zhe ci]" (75), "几点[ji dian]" (70) and "这位[zhe wei]" (62) respectively, and the sum of their frequencies reaches 3080, accounting for 33.97% of the total number of quantifier phrases. All quantifier phrases contain 200 kinds of quantifiers. Among them, the 15 quantifiers with the highest frequency are "个[ge]", "种[zhong]", "天[tian]", "次[ci]", "年[nian]", "位[wei]", "块[kuai]", "件[jian]", "岁[sui]", "张[zhang]", "家[jia]", "点[dian]", "条[tiao]", "本[ben]" and "只[zhi]" respectively. Their specific frequencies are shown in Fig. 1.

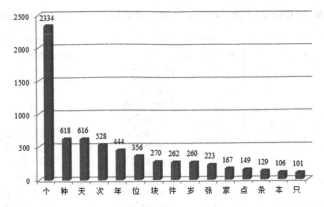

Fig. 1. The 15 quantifiers with the highest frequency in quantifier phrases

Table 6. The structural modes whose frequencies are in the top ten

Id	Mode	Frequency	Frequency/total	Class	Class/total
1	m: m-q	4664	51.44%	468	24.99%
2	r: r-q	1993	21.98%	195	10.41%
3	m: m2-q	734	8.10%	315	16.82%
4	r: r-m-q	417	4.60%	148	7.90%
5	m: m3-q	177	1.95%	138	7.37%
6	r: f-q	142	1.57%	8	0.43%
7	m: m-q-m	98	1.08%	45	2.40%
8	m: m2_m-q	93	1.03%	62	3.31%
9	m: m4-q	86	0.95%	71	3.79%
10	m: m_m-q	69	0.76%	29	1.55%

Ten structural modes with the highest frequency of corresponding quantifier phrases in the knowledge base are shown in Table 6. The number of quantifier phrases corresponding to the ten structural modes accounts for 93.46% of the total number of quantifier phrases, and the number of the classes of quantifier phrases accounts for 78.96% of the total number of all classes. As shown in Table 6, in the international Chinese textbooks, quantifier phrases composed of monosyllabic, disyllabic, trisyllabic and quadrisyllabic numerals and monosyllabic quantifiers have a higher frequency, ranking first, third, fifth and ninth respectively; demonstrative-quantifier phrases or interrogative-quantifier phrases composed of monosyllabic pronouns and monosyllabic quantifiers, monosyllabic pronouns and numeral-quantifier phrases (monosyllabic numerals and monosyllabic quantifiers), and monosyllabic localizers and monosyllabic quantifiers are also very common, ranking second, fourth and sixth in frequency respectively.

Table 7. Quantifier phrase types and their corresponding structural modes

ID	Type	Frequency	Structural Mode	
1	数词+量词	5829	m: m-q m: m3-q m: m5-q m: m7-q m: m2-q2 m: m4-q2 m: m-q3 m: m-q-Ug	m: m2-q m: m4-q m: m6-q m: m-q2 m: m3-q2 m: m5-q2 m: m2-q3 m: m-q…m-q
2	数词+量词+数词	130	m: m-q-m m: m2-q-m m: m5-q-m	m: m-q-m2 m: m3-q-m
3	相邻系数词或位数词连用+量词	78	m: m…m-q m: m2…m-q	m: m…m2-q m: m…m-q2
4	含概数词或附加成分的概数+量词	239	m: m_m-q m: m2_m-q m: m4_m-q m: m_m-q2 m: m3_m-q2 m: m2_m-q3 m: m3_m_m-q m: m_u-q m: m-q-m2_m m: m2-q_f2 m: m2-q_v2 m: v ǀ m…vǀ m-q	m: m_m2-q m: m3_m-q m: m5_m-q m: m_m-q2 m: m4_m-q2 m: m2_m_m-q m: m4_m_m-q m: m_u-q2 m: m-q_f2 m: m4-q_f2 m: v ǀ m-q
5	数值范围+量词	17	m: m w m-q m: m2 w m2-q m: m4 w m4-q	m: m w m2-q m: m3 w m3-q
6	量词短语重叠	65	m: m-q·q m: m-q·d·m-q m: m-a-q·m-a-q	m: m-q·m-q m: a-q·a-q
7	代词/方位词+量词/数量短语	2623	r: r-q r: r-q…r-q r: r-m-q r: f-q	r: r-q-Ug r: r2-q r: r-m2-q r: f-m-q
8	（数词+）形容词+量词	60	m: m-a-q m: a-q…a-q	m: a-q
9	其他量词短语	25	r: a-m-q r: d-m-q m: m-f-m-q	r: a-m2-q m: d→a-q m: m-f-m2-q

By sorting out and analyzing 75 structural modes in the knowledge base, quantifier phrases can be further divided into nine types: (1) 数词+量词 (numerals and quantifiers), (2) 数词+量词+数词 (numerals, quantifiers and numerals), (3) 相邻系数词或位 数词连用+量词 (adjacent numerals and quantifiers), (4) 含 "几, 数, 多少, 若干" 等概数词或前后附加 "成, 上, 数", "来, 多, 上下, 左右, 开外" 等成分的概数+量词 (approximate numbers that include "几[ji]", "数[shu]", "多少[duoshao]", "若干[ruogan]" or the components such as "成[cheng]", "上[shang]", "数[shu]", "来[lai]", "多[duo]", "上下[shangxia]", "左右[zuoyou]", "开外[kaiwai]" and quantifiers), (5) 数值范围+量词 (numerical ranges and quantifiers), (6) 量词短语重叠 (reduplication of quantifier phrases), (7) 代词/方位词+量词/数量短语 (pronouns/localizers and quantifiers/numeral-quantifier phrases), (8) (数词+)形容词 +量词 ((numerals) adjectives and quantifiers), (9) 其他量词短语 (other quantifier phrases). As shown in Table 7, the number of quantifier phrases corresponding to the nine types is 5829 (64.30%), 130 (1.43%), 78 (0.86%), 239 (2.64%), 17 (0.19%), 65 (0.72%), 2623 (28.93%), 60 (0.66%) and 25 (0.28%) respectively.

Based on the quantifier phrase structural mode knowledge base, combined with the corpus of international Chinese textbooks, this chapter makes a preliminary analysis of the high-frequency quantifiers and quantifier phrases in international Chinese teaching, the distribution of quantifier phrase structural modes and the main types of quantifier phrases. The knowledge base contains a lot of valuable information. By further mining, it is certain to have a more comprehensive and deeper understanding of quantifier phrases in international Chinese teaching.

4 Automatic Recognition of Quantifier Phrases

The realization of automatic recognition of quantifier phrases is of great importance to the research and teaching of quantifiers and quantifier phrases in international Chinese teaching, and it is also an important part of realizing automatic word segmentation, part-of-speech tagging and automatic parsing for international Chinese teaching. In addition, rule-based recognition of quantifier phrases can be an important complement to statistical-based methods of word segmentation and part-of-speech tagging. It plays an important role in dealing with ambiguities, correcting errors and improving the accuracy of word segmentation and part-of-speech tagging. On the basis of the previous work, this chapter explores automatic recognition of quantifier phrases in the raw corpus of international Chinese textbooks based on a rule method. In the process of recognition, the information of the "rule" field and the "frequency" field in the knowledge base are mainly used. In order to objectively reflect the automatic recognition effect of quantifier phrases, 3 groups of raw corpus of international Chinese textbooks (tens of thousands to hundreds of thousands of Chinese characters in each group) were selected randomly for the experiment.

4.1 Recognition Process

The automatic recognition process of quantifier phrases in the raw corpus of international Chinese textbooks is as follows. In the process of automatic recognition, the sequence of quantifier phrases in different structural modes is different.

1. Sort quantifier phrase structural modes according to the number of internal components from more to less, the number of syllables of internal components from large to small, and the frequency of corresponding quantifier phrases from high to low. For example, the mode "m: m-a-q·m-a-q" has the largest number of internal components, so it ranks first.
2. Match and recognize quantifier phrases corresponding to the current quantifier phrase structural mode in order. The specific method is to match the combinations of the raw corpus satisfying the regular expression in the "rule" field, and the result is the set of quantifier phrases corresponding to the current structural mode. Once identified as quantifier phrases, the contents are no longer involved in the following recognition operations.
3. Compare automatic recognition results of quantifier phrases with the results of manual labeling and auditing, and calculate the accuracy rate, recall rate and F-measure of automatic recognition of quantifier phrases.

4.2 Experimental Result

According to the above recognition process, the automatic recognition experiment is carried out on 3 groups of randomly selected raw corpus, and the recognition results of 9 types of quantifier phrases are shown in Table 8 (the accuracy rate and recall rate are presented as average).

According to the experimental result, it can be concluded that the quantifier phrase structural mode knowledge base has a significant effect on automatic recognition of quantifier phrases. The accuracy and recall rate of automatic recognition are both high, and the recognition effect is relatively ideal. Through analysis of the experimental data, it is found that the reasons for reducing the accuracy rate and recall rate of automatic recognition are mainly the following aspects:

1. Due to the limited size of the corpus of international Chinese textbooks, the quantifier phrase structural modes can't cover all the classes of quantifier phrases actually appearing in the texts of international Chinese textbooks, which limits the recall rate of recognition. For the conversion of nouns into quantifiers, the recognition method has not made a response strategy. In the later stage, it is necessary to further improve the structural mode knowledge base and make a thorough study on the conversion of nouns into quantifiers.
2. It's difficult to avoid some ambiguities in the process of recognition. The recognition error occurs during the specific recognition process. For example, the "一路 [yilu]" in the sentence "我跟他一路来的[wo gen ta yilu lai de]" should be used as an adverb, but there is a quantifier "路[lu]" in the Modern Chinese Dictionary, so it is easy to be classified into the set of quantifier phrases with the structural mode "m: m-q". The ambiguities need to be further eliminated by improving the recognition process.

3. Regular expression rules can only reflect the partial external characteristics of quantifier phrases, and can't fully reflect the characteristics of quantifier phrases. It is necessary to dig out more effective and quantifiable feature information with the in-depth study of quantifier phrases.

Table 8. Recognition results of quantifier phrases

ID	Type	Accuracy Rate	Recall Rate	F-measure
1	数词+量词	86.85%	90.56%	88.67%
2	数词+量词+数词	91.53%	83.08%	87.10%
3	相邻系数词或位数词连用+量词	91.36%	94.87%	93.08%
4	含概数词或附加成分的概数+量词	83.03%	96.23%	89.15%
5	数值范围+量词	85.00%	100%	91.89%
6	量词短语重叠	92.65%	96.92%	94.74%
7	代词/方位词+量词/数量短语	84.46%	97.83%	90.66%
8	（数词+）形容词+量词	86.15%	93.33%	89.60%
9	其他量词短语	82.14%	92.00%	86.79%

5 Conclusion

In this paper, we use the method of knowledge engineering to analyze the quantifier phrases in the corpus of international Chinese textbooks, and preliminarily build a knowledge base of quantifier phrase structural modes for international Chinese teaching. Then based on the structural mode knowledge base and the corpus of international Chinese textbooks, the characteristics of quantifier phrases in the field of international Chinese teaching are analyzed. Finally, on the basis of the structural mode knowledge base, quantifier phrases in the corpus of international Chinese textbooks are automatically recognized. This research method can be further extended to the study of other composite structures in the corpus of international Chinese textbooks, so as to better serve international Chinese teaching and information processing for international Chinese teaching.

Acknowledgments. This study was supported by the National Natural Science Foundation of China (61877004) and the Natural Science Foundation for the Higher Education Institutions of Anhui Province of China (KJ2019A0592).

References

1. Confucius Institute Headquarters (Hanban): HSK Test Syllabus Level 1. People's Education Press, Beijing (2015)
2. Confucius Institute Headquarters (Hanban): HSK Test Syllabus Level 2. People's Education Press, Beijing (2015)
3. Confucius Institute Headquarters (Hanban): HSK Test Syllabus Level 3. People's Education Press, Beijing (2015)
4. Confucius Institute Headquarters (Hanban): HSK Test Syllabus Level 4. People's Education Press, Beijing (2015)
5. Confucius Institute Headquarters (Hanban): HSK Test Syllabus Level 5. People's Education Press, Beijing (2015)
6. Confucius Institute Headquarters (Hanban): HSK Test Syllabus Level 6. People's Education Press, Beijing (2015)
7. Zhang, B.: The Modern Chinese Grammar. Commercial Press, Beijing (2010)
8. Guo, R.: A Study on Parts of Speech in Modern Chinese. Commercial Press, Beijing (2002)
9. Lu, J.: An investigation on the insertion of adjectives in quantifiers. Lang. Teach. Linguist. Stud. **04**, 53–72 (1987)
10. Yu, S.: An Introduction to Computational Linguistics. Commercial Press, Beijing (2003)
11. Song, J., Yang, E., Wang, Q.: Chinese Information Processing Tutorial. Higher Education Press, Beijing (2011)
12. Peng, W., Song, J., Sui, Z., Guo, D.: Formal schema of diagrammatic Chinese syntactic analysis. In: Lu, Q., Gao, H. (eds.) CLSW 2015. LNCS (LNAI), vol. 9332, pp. 701–710. Springer, Cham (2015)
13. Guo, D., Zhu, S., Peng, W., Song, J., Zhang, Y.: Construction of the dynamic word structural mode knowledge base for the international Chinese teaching. In: Dong, M., Lin, J., Tang, X. (eds.) CLSW 2016. LNCS (LNAI), vol. 10085, pp. 251–260. Springer, Cham (2016)
14. Guo, D., Song, J., Peng, W.: Research on dynamic words and their automatic recognition in Chinese information processing. In: Wu, Y., Hong, J.F., Su, Q. (eds.) CLSW 2017. LNCS (LNAI), vol. 10709, pp. 479–488. Springer, Cham (2018)
15. Guo, D.: Analyzing on Dynamic Words and Their Structural Modes in Building the Sentence-based Treebank. Beijing Normal University, Beijing (2016)
16. The Dictionary Editing Room in the Linguistics Institute of Chinese Academy of Social Sciences: Modern Chinese Dictionary. Commercial Press, Beijing (2012)

Two-Fold Linguistic Evidences on the Identification of Chinese Translation of Buddhist Sutras: Taking *Buddhacarita* as a Case

Bing Qiu[✉]

Department of Chinese Language and Literature, Tsinghua University,
Beijing, China
qiubing@mail.tsinghua.edu.cn

Abstract. As a product of language contact, the Chinese translation of Buddhist sutras has significant linguistic values. However, it is hard to accurately identify the translators and years of translation for some early translated sutras. Considering that the Chinese translation was mingled with elements loaned from foreign languages, mostly Sanskrit, in order to conduct identification from the linguistic perspective, it is necessary to adopt both external and internal linguistic evidences. The external approach is to analyze the characteristics of translation embodied in the corpus using the proofreading method with parallel texts of the Sanskrit original and the Chinese translation, and the internal approach is to examine the language style and language use habits of the translator according to the identification criteria of the Chinese corpora. Taking *Buddhacarita* as a case, this paper attempts to identify its translator with both the two-fold evidences, which provides a new way for the identification of such corpora.

Keywords: Chinese translation of Buddhist sutras · Language contact ·
Proofreading of Buddhist sutras with parallel Sanskrit-Chinese texts ·
Buddhacarita

1 Introduction

The Chinese translation of Buddhist sutras forms an important part of historical corpora. It not only records numerous emerging linguistic phenomena in the Chinese language of the time but also reflects the influence of external languages on Chinese, and therefore, has significant linguistic values for the study of the history of the Chinese language. Owing to their age-old history, Chinese translated sutras might have been subjected to scribal errors or changes during the process of copying. The translators and the years of translations for some sutras, especially the early translated ones, are difficult to accurately identify, which to some extent undermines the effectiveness of tapping into this corpus and bringing its values into full play. In practice, the identification of early Chinese translations of Buddhist sutras is usually done through examination of documentary records or with linguistic methods. Since language is in a

© Springer Nature Switzerland AG 2020
J.-F. Hong et al. (Eds.): CLSW 2019, LNAI 11831, pp. 853–858, 2020.
https://doi.org/10.1007/978-3-030-38189-9_85

perpetual state of evolving, a particular language will display distinctive features in grammar and lexis across different time periods; even within the same period, authors vary from each other in their language use [1]. With illustrations, linguistic identification can be used to determine the authenticity of ancient literature and its years of composition [2].

Yet it should be noted that the nature of the Chinese translation of Buddhist sutras is unique. Different from indigenous Chinese literature, it represents a non-natural language variant in Chinese historical literature: As a typical product of the non-natural language contact induced by literature translation, the language in Chinese translated sutras comprises components loaned from foreign languages, most Sanskrit. Therefore, the identification of such Chinese translated sutras requires both external and internal linguistic evidences. The external evidences are obtained from language contact, using the proofreading method to examine the parallel texts of the Sanskrit original and the Chinese translation in a word-by-word and sentence-by-sentence manner, in order to find out the translation characteristics of various translators and across different time periods. The internal evidences are based on the linguistic features of the Chinese language proper, adopting identification criteria characterized by popularity, regularity, interrelativity, and chronological peculiarity, in an attempt to examine the language style and language use habits of each translator. Only by taking such two-fold linguistic evidences, i.e. external and internal, can it be possible to make a comprehensive analysis and survey on the translators and years of translation of the Chinese translated Buddhist sutras.

Buddhacarita was the first epic poem (Kāvya) in the history of Indian Sanskrit literature. It was translated and introduced into China in the Middle Ages and thereafter exerted a great influence on Chinese literature. However, records about its translator were divergent and controversial. This paper, therefore, intends to take *Buddhacarita* as a case and probe into the identification of its translator, by proofreading parallel Sanskrit-Chinese texts in the external respect and examining its lexical identifiers in the internal respect. In fact, this paper provides a new approach of adopting two-fold evidences, i.e. external and internal, for the identification of Chinese translated Buddhist sutras in general.

2 Controversy over the Translator of *Buddhacarita*

The Sanskrit original *Buddhacarita* was composed by Aśvaghoṣa, a Buddhist philosopher, poet and dramatist from ancient India in the first century CE. It was translated and introduced into China in the Southern and Northern dynasties. *Fo suoxing zan* (佛所行讚) (abbr. *FSXZ*), the Chinese translation of *Buddhacarita* is included in *Taisho Tripitaka* (大正藏) Volume 4, Original Conditions Division Number 192, with a preface reading "Composed by Bodhisattva Aśvaghoṣa,. Translated by Tan Wuchen (昙无谶, Dharmakṣema), a Tripiṭaka Master from India in the Northern Liang dynasty". This statement has been widely acknowledged by many scholars [3, 4], etc., who deemed the translator of *Buddhacarita* to be Tan Wuchen from the Northern Liang dynasty.

In fact, claims of the translator and translation years of *F FSXZ* are inconsistent in scholars' records across history. Some examined its translator by virtue of historical documents, mostly with reference to the record by Sengyou from the Liang period of the Southern dynasties in his *Chu Sanzang Jiji*, which is the compilation of notes on the translation of the Tripitaka. Zhou Yiliang believed that Sengyou's record of the translator of *FSXZ* being Baoyun was more reliable, as Sengyou and Baoyun "were contemporaries living not afar. Baoyun died in the 26th year during the rule of Yuanjia in the Liu Song dynasty (449) and Sengyou died in the 17th year during the rule of Tianjian in the Liang dynasty (518), close to the death of Baoyun. Baoyun translated sutras in Liuhe, within the borders of the Southern dynasties. Therefore, Sengyou was supposed to be clear about Baoyun's work" [5]. Besides Zhou, scholars Tokiwa Daijo, Kanakura Ensho and Charles Willemen also held the same opinion [6–8].

In scholars' records across history, the translation of *FSXZ* was majorly attributed to Baoyun or Tan Wuchen. Therefore, in order to identify its translator from the linguistic perspective, it is necessary to conduct a linguistic comparison between *FSXZ* and other Chinese translated sutras by Baoyun and Tan Wuchen, respectively. The existing translated sutras by Baoyun are quite limited in number, which, according to documentary records, comprise of two pieces of work only: The controversial *Fo Benxing Jing* (佛本行经, *Buddhacarita-sūtra*, abbr. *FBXJ*), which was confused with *FSXZ* for the names of the two sutras are similar, and the relatively more reliable *Fo Shuo Si Tianwang Jing* (佛说四天王经, *The Buddha Speaks of the Four Heavenly Kings Sutra*), which was a joint effort by Baoyun and Zhiyan and thus cannot represent the language style and translated characteristics of Baoyun in a complete picture. In contrast, Tan Wuchen left a richer collection of translated sutras. In this paper, we choose *Jin Guangming Jing* (金光明经, *Golden Light Sutra*, abbr. *JGMJ*), rendered by Tan Wuchen, to be in comparison with *FSXZ*.

3 External Evidences for the Identification of Chinese Translated Buddhist Sutras

From the external perspective of language contact, we can take a proofreading approach to study the parallel Sanskrit-Chinese texts in order to examine how each translator completed such rendering.

When rendering Buddhist sutras, translators usually took either transliteration or semantic translation approach to convert the Sanskrit words into Chinese. Concepts and items new to the target language, such as Buddhist terms and proper nouns, were mostly transliterated by being mapped to Chinese words of the same or similar pronunciation. Such transliterated words, as a result, carried a distinctive exotic style. Words from semantic translation, in contrast, were new words coined with Chinese word-building material in compliance with Chinese word-forming rules, regardless of the phonetic form of the original words in the source language. Such semantically translated words are hardly different from aboriginal words in form, and their foreign roots are difficult to identify if they are not examined in parallel texts of the original and the translated works. Different translators, when rendering the same Sanskrit word, tended to adopt different approaches, transliteration or semantic translation. Even if they ended up adopting the same approach, they might have adopted different Chinese equivalents. Therefore, by proofreading the Sanskrit-Chinese parallel texts, we can find out critical external evidences for the identification of Chinese translated sutras.

FSXZ and *FBXJ*, both depictions of the life of the Buddha, contained a great number of identical concepts and proper nouns such as the names of people and places. However, when these two classics were rendered into Chinese as *FSXZ* and *FBXJ* respectively, the vast majority of those identical Sanskrit words are found to have been put into different Chinese words, as is shown in Table 1. The few exceptions are like "Moluo" (摩罗), "Mojie Yu" (摩竭鱼), "Shelifu" (舍利弗), "Xumi" (须弥), which ended up the same.

Table 1. Comparison of the translation style of *FSXZ* and *FBXJ*

Sanskrit word	Chinese translation in *FSXZ*	Chinese translation in *FBXJ*
śuddhodana	Jingfan (净饭)	Baijingwang (白净王)
māyā	Moye Hou (摩耶后)	Miao Hou (妙后)
asura	Axiuluo (阿修罗)	Axulun (阿须伦)
rāhula	Luohouluo (罗睺罗)	Luoyun (罗云)
asita	Asituo (阿私陀)	Ayi (阿夷)
licchavi	Liche (离车)	Lijian (离犍)
nirvāpayati	Niepan (涅盘)	Damie (大灭)

FSXZ and *FBXJ* display different translation styles for the same concepts and proper nouns. Take "māyā" (name of the Buddha's mother) as an instance. It was transliterated into "Moye Hou" (摩耶后) in *FSXZ* yet semantically translated into "Miao Hou" (妙后) in *FBXJ*. Even when with the same translation method, the two translated sutras adopted different Chinese characters. For example, "asita" (one of the eight demi-gods and semi-devils) was transliterated into "Axiuluo" (阿修罗) in *FSXZ* and "Axulun" (阿须伦) in *FBXJ*. To sum up, considering their distinctive translation styles, *FSXZ* and *FBXJ* were supposedly rendered by different translators.

Moreover, *FSXZ* and *JGMJ* should be attributed to different translators in light of translation style. The details are omitted here due to space constraints.

4   Internal Evidences for the Identification of Chinese Translated Buddhist Sutras

Cao Guangshun, Yu Xiaorong established some identification criteria for a given corpus: "Such criteria must enjoy high popularity and strong regularity, which will guarantee their wide use and lead to precise and reliable conclusions." [1] Fang further put forward four principles, namely popularity, regularity, interrelativity, and chronological peculiarity. Popularity, a principle in quantity, refers to a certain frequency of use. Regularity, a principle in analogizability, means that the lexical identifiers should be rule-framed and deducible. Interrelativity emphasizes the interrelation and comparison between words. Chronological peculiarity, a principle of change and alternation, manifests the linguistic evolution by telling which words were not in use at a given time and then taking a further step to point out the words in use back in the time period under study [9].

Adverbs of generalization had a relatively high frequency of use in Chinese translated Buddhist sutras, and moreover, their use varied significantly across different time periods and among different translators. Therefore, these words, which demonstrated better popularity, regularity and chronological peculiarity compared to the general vocabulary, can function as lexical identifiers in order to identify the translators and years of translation for Chinese translated sutras. This paper, based on existing studies, intends to compare and analyze the use of adverbs of generalization in the three Chinese translated sutras, in an attempt to investigate on the translator of *FSXZ* in light of the internal characteristics of the Chinese language.

In *FSXZ*, "xi" (悉), with a frequency of 190, was the most frequently used adverb of generalization, while "jie" (皆) (52), "ju" (俱) (33), "xian" (咸) (14), "ju" (具) (6) ranked after it and "dou" (都), with a frequency of merely 1, was the least frequently used. In FBXJ, "jie" (皆), with a frequency of 144, was the most frequently used adverb of generalization, while "ju" (俱) (over 50), "xi" (悉) (22), "ju" (具) (11), and "xian" (咸) (1) followed. It is worth noting that the occurrences of "dou" (都) totaled 16. In JGMJ, the most frequently used adverbs of generalization were "xi" (悉) (89) and "jie" (皆) (35), and the less frequently used were "ju" (俱) (6), "dou" (都) (2) and "xia" (咸) (1).

Adverbs of generalization in *FSXZ* and *JGMJ*, apart from being used on their own, were also frequently used in conjunction, while in *FBXJ*, however, such conjunct use cannot be found. The frequency of use for adverbs of generalization in *FSXZ*, *FBXJ*, and *JGMJ* is shown as in Table 2.

Table 2. The comparison of the use of adverbs of generalization in *FSXZ*, *FBXJ*, and *JGMJ*

Adverbs of generalization	In *FSXZ* frequency/percentage	In *FBXJ* frequency/percentage	In *JGMJ* frequency/percentage
xi (悉)	190/59.6%	22/8.7%	89/52.7%
jie (皆)	52/16.3%	144/56.9%	35/20.7%
ju (俱)	33/10.3%	59/23.3%	6/3.6%
ju (具)	6/1.9%	11/4.4%	3/1.8%
xian (咸)	14/4.4%	1/0.4%	4/2.4%
dou (都)	1/0.3%	16/6.3%	2/1.1%
xijie (悉皆)	19/5.9%	0/0	16/9.4%
jiexi (皆悉)	4/1.3%	0/0	14/8.3%
Total	**319/100%**	**253/100%**	**169/100%**

It can be summarized from the table above that in *FSXZ* and *JGMJ*, the most frequently used adverb of generalization was "xi" (悉), accounting for 60% and 53% of all occurrences of adverbs of its kind in each sutra, while the most frequently used adverb in *FBXJ* was "jie" (皆), accounting for 57%. In addition, "dou" (都) was seldom used in *FSXZ* and *JGMJ*, accounting for merely 0.3% and 11.%, respectively, while in *FBXJ*, it accounted for 6.3%. It demonstrates, to a certain degree, that *FBXJ* differs from *FSXZ* and *JGMJ* in its use of adverb of generalization.

5 Conclusion

This paper takes *Buddhacarita* as a case and conducts an investigation on its Chinese translation *FSXZ* from both external and internal perspectives. Externally, based on the Sanskrit-Chinese parallel texts, it is found that *FSXZ* had a translation style different from both *FBXJ* and *JGMJ*, the latter of which was translated by Tan Wuchen. Such external evidences prove that the translator of *FSXZ* was neither that of *FBXJ* nor that of *JGMJ*. Internally, according to the statistical survey on the use of the adverbs of generalization in *FSXZ*, *FBXJ*, and *JGMJ*, it is observed that there was major deviation in the language characteristics among the three Chinese translated sutras, which reflected the specific language use habits of each translator.

As is discussed above, records about the translator of *FSXZ* were divergent in documents across history. Taking the two-fold linguistic evidences, both external and internal, into consideration, we find that the record by Sengyou from the Liang period of the Southern dynasties shows no sign of contradiction.

Acknowledgments. The work was supported by the Beijing Social Science Foundation (Grant No. 17YYC019). The author would also like to thank the national youth talent support plan in the Ten Thousand Talent Program.

References

1. Yu, X., Cao, G.: Also on the translator's problem of Liu Du Ji Jing and Jiu Za Pi Yu Jing from the perspective of language [也从语言上看《六度集经》与《旧杂譬喻经》的译者问题]. Ancient Chin. Stud. [古汉语研究] **2**, 4–7 (1998)
2. Hu, C.: A summary of the identification of middle ancient chinese corpus [中古汉语语料鉴别述要]. J. Chin. Hist. [汉语史学报] **1**, 270–279 (2005)
3. Hu, S.: History of Vernacular Literature [白话文学史]. Shanghai Press [上海书局], Shanghai (1929)
4. Rao, Z.: Collection of Sanskrit Knowledges [梵学集]. Shanghai Ancient Books Publishing House, Shanghai (1993)
5. Zhou, Y.: The History of the Wei, Jin, Southern and Northern Dynasties [魏晋南北朝史论集]. Peking University Press, Beijing China (1997)
6. Daijo, T.: Record of the translation of Buddhist scriptures from the Eastern Han dynasty to Southern Qi dynasty [后汉至宋齐译经总录]. Institute of Oriental Culture, Tokyo Institute [东方文化学院东京研究所], Tokyo, Japan (1938)
7. Kanakura, E.: Works of Aśvaghoṣa [马鸣の著作]. Religious research [宗教研究], vol. 153 (1957)
8. Willemen, C.: Chinese buddhacarita. In: Proceedings of the World Buddhist Forum (2006)
9. Fang, Y.: Extraction and principle of common identification words: focusing on early Chinese translation of buddhist scriptures [普通鉴别词的提取及原则—以早期汉译佛经鉴别为中心]. Lang. Res. [语文研究] **2**, 8–16 (2009)

Author Index

Printed in the United States
By Bookmasters